DESIGN AND INTEGRATION OF CHEMICAL PROCESS

Design and Integration of Chemical Process

Editor:

Asad Rahman

Delve Publishing

www.delvepublishing.com

Design and Integration of Chemical Process

Editor: **Asad Rahman**

708 3rd Avenue, 6th Floor
New York, NY 10017
United States of America
www.delvepublishing.com

ISBN: 978-1-68095-894-2
Library of Congress Control Number: 2016951755

About the Editor

Asad Rahman

Asad Rahman obtained MSc Chemical Engineering from University of Nottingham in 2012. Currently He is working towards PhD in Energetic Material Synthesis and its Performance Analysis in Micro Thrusters. His research interests are in Ultrasound based particle treatment, Energetic material synthesis and Micro thruster applications.

Contents

List of Contributors

Kew Hong Chew
Process Systems Engineering Centre (PROSPECT), Faculty of Chemical Engineering, Universiti Teknologi Malaysia, 81310 UTM Johor Bahru, Johor, Malaysia

Jiří Jaromír Klemeš
Centre for Process Integration and Intensification—CPI^2, Research Institute of Chemical and Process Engineering—MŰKKI, Faculty of Information Technology, University of Pannonia, Egyetem u. 10, Veszprém H-8200, Hungary

Sharifah Rafidah Wan Alwi
Process Systems Engineering Centre (PROSPECT), Faculty of Chemical Engineering, Universiti Teknologi Malaysia, 81310 UTM Johor Bahru, Johor, Malaysia

Zainuddin Abdul Manan
Process Systems Engineering Centre (PROSPECT), Faculty of Chemical Engineering, Universiti Teknologi Malaysia, 81310 UTM Johor Bahru, Johor, Malaysia

Andrea Pietro Reverberi
Department of Chemistry and Industrial Chemistry (DCCI), University of Genova, Via Dodecaneso 31, Genova 16146, Italy

H. Machguth
Department of Geography, University of Zurich, Zurich, Switzerland

R. S. Purves
Department of Geography, University of Zurich, Zurich, Switzerland

J. Oerlemans
Institute for Marine and Atmospheric Research, Utrecht University, Utrecht, The Netherlands

M. Hoelzle
Department of Geography, University of Zurich, Zurich, Switzerland

F. Paul
Department of Geography, University of Zurich, Zurich, Switzerland

Ewa Brychcy
Department of Animal Products Technology and Quality Management, Wroclaw University of Environmental and Life Sciences, Chelmonskiego 37/41, 51-630 Wroclaw, Poland

Dominika Kulig
Department of Animal Products Technology and Quality Management, Wroclaw University of Environmental and Life Sciences, Chelmonskiego 37/41, 51-630 Wroclaw, Poland

Anna Zimoch-Korzycka,
Department of Animal Products Technology and Quality Management, Wroclaw University of Environmental and Life Sciences, Chelmonskiego 37/41, 51-630 Wroclaw, Poland

Krzysztof Marycz
Electron Microscope Laboratory, Department of Animal Hygiene and Ichthyology, Wroclaw University of Environmental and Life Sciences, Kożuchowska 5B, 51-630 Wroclaw, Poland

Andrzej Jarmoluk
Department of Animal Products Technology and Quality Management, Wroclaw University of Environmental and Life Sciences, Chelmonskiego 37/41, 51-630 Wroclaw, Poland

Ana L. Gonçalves
LEPABE, Departamento de Engenharia Química, Faculdade de Engenharia, Universidade do Porto, Rua Dr. Roberto Frias, Porto 4200-465, Portugal

Maria C. M. Alvim-Ferraz
LEPABE, Departamento de Engenharia Química, Faculdade de Engenharia, Universidade do Porto, Rua Dr. Roberto Frias, Porto 4200-465, Portugal

Fernando G. Martins
LEPABE, Departamento de Engenharia Química, Faculdade de Engenharia, Universidade do Porto, Rua Dr. Roberto Frias, Porto 4200-465, Portugal

Manuel Simões
LEPABE, Departamento de Engenharia Química, Faculdade de Engenharia, Universidade do Porto, Rua Dr. Roberto Frias, Porto 4200-465, Portugal

José C. M. Pires
LEPABE, Departamento de Engenharia Química, Faculdade de Engenharia, Universidade do Porto, Rua Dr. Roberto Frias, Porto 4200-465, Portugal

Vincenzo Spallina
Chemical Process Intensification, Chemical Engineering and Chemistry, Eindhoven University of Technology, 5600MB Eindhoven, The Netherlands

Tommaso Melchiori
Chemical Process Intensification, Chemical Engineering and Chemistry, Eindhoven University of Technology, 5600MB Eindhoven, The Netherlands

Fausto Gallucci
Chemical Process Intensification, Chemical Engineering and Chemistry, Eindhoven University of Technology, 5600MB Eindhoven, The Netherlands

Martin van Sint Annaland
Chemical Process Intensification, Chemical Engineering and Chemistry, Eindhoven University of Technology, 5600MB Eindhoven, The Netherlands

João G. R. Cardoso
The Novo Nordisk Foundation Center of Biosustainability, Technical University of Denmark, Hørsholm, Denmark

Mikael Rørdam Andersen
Department of Systems Biology, Technical University of Denmark, Lyngby, Denmark

Markus J. Herrgård
The Novo Nordisk Foundation Center of Biosustainability, Technical University of Denmark, Hørsholm, Denmark

Nikolaus Sonnenschein
The Novo Nordisk Foundation Center of Biosustainability, Technical University of Denmark, Hørsholm, Denmark

Borbála Erdei
Lund University, Department of Chemical Engineering, SE-221 00 Lund, Sweden.

Balázs Frankó
Budapest University of Technology and Economics, Department of Applied Biotechnology and Food Science, Szt Gellért tér 4, 1111 Budapest, Hungary

Mats Galbe
Lund University, Department of Chemical Engineering, SE-221 00 Lund, Sweden.

Guido Zacchi
Lund University, Department of Chemical Engineering, SE-221 00 Lund, Sweden.

Chih-Cheng Lu
Institute of Mechtronic Engineering, National Taipei University of Technology, Taipei 106, Taiwan

Yong-Sheng Huang
Institute of Mechtronic Engineering, National Taipei University of Technology, Taipei 106, Taiwan

Jun-Wei Huang
Institute of Mechtronic Engineering, National Taipei University of Technology, Taipei 106, Taiwan

Chien-Kuo Chang
Graduate Institute of Mechanical and Electrical Engineering, National Taipei University of Technology, Taipei 106, Taiwan

Sheng-Po Wu
Institute of Mechtronic Engineering, National Taipei University of Technology, Taipei 106, Taiwan

Qinglei Jiang
Department of Chemical and Biological Engineering, Institute of Chemical Machinery and process equipment, Zhejiang University, Hangzhou 310027, China
Engineering Research Center of High Pressure Process Equipment and Safety, Ministry of Education, Hangzhou 310027, China

Lulu Zhai
Department of Chemical and Biological Engineering, Institute of Chemical Machinery and process equipment, Zhejiang University, Hangzhou 310027, China
Engineering Research Center of High Pressure Process Equipment and Safety, Ministry of Education, Hangzhou 310027, China

Leqin Wang
Department of Chemical and Biological Engineering, Institute of Chemical Machinery and process equipment, Zhejiang University, Hangzhou 310027, China
Engineering Research Center of High Pressure Process Equipment and Safety, Ministry of Education, Hangzhou 310027, China

Dazhuan Wu
Department of Chemical and Biological Engineering, Institute of Chemical Machinery and process equipment, Zhejiang University, Hangzhou 310027, China
Engineering Research Center of High Pressure Process Equipment and Safety, Ministry of Education, Hangzhou 310027, China

Giuliano Dall'O'
Architecture, Building Environment and Construction Engineering (A.B.C.) Department, Polytechnic of Milan, Via E. Bonardi 9, Milano 10133, Italy

Annalisa Galante
Architecture, Building Environment and Construction Engineering (A.B.C.) Department, Polytechnic of Milan, Via E. Bonardi 9, Milano 10133, Italy

Nicola Sanna
Architecture, Building Environment and Construction Engineering (A.B.C.) Department, Polytechnic of Milan, Via E. Bonardi 9, Milano 10133, Italy

Karin Miller
YR&G Sustainability Consulting, Education, and Analysis, Chicago, IL 60661, USA

Charles H. Williams
Division of Cardiovascular Medicine, Department of Medicine, Vanderbilt University School of Medicine, Nashville, TN 37232, USA

Charles C. Hong
Division of Cardiovascular Medicine, Department of Medicine, Vanderbilt University School of Medicine, Nashville, TN 37232, USA
Research Medicine, Veterans Affairs TVHS, Nashville, TN 37212, USA

Mumtaj Shah
Department of Chemical Engineering, SOET, ITM University, Gwalior 475001, India

S. K. Garg
Department of Chemical Engineering, HBTI, Kanpur 208002, India

Shu-zhi Gao
College of Information and Engineering, Shenyang University of Chemical Technology, Shenyang 110142, China

Jing Yang
Multi-Functional Design and Research Academy, Zhengzhou University, Zhengzhou 450001, China

Jie-sheng Wang
School of Electronic and Information Engineering, University of Science and Technology Liaoning, Anshan 114044, China

N. El Boulifi
Department of Chemical Engineering, Faculty of Chemistry, Complutense University, 28040 Madrid, Spain

A. Bouaid, M. Martinez
Department of Chemical Engineering, Faculty of Chemistry, Complutense University, 28040 Madrid, Spain

J. Aracil
Department of Chemical Engineering, Faculty of Chemistry, Complutense University, 28040 Madrid, Spain

Tai-Yin Chiu
Department of Physics, National Taiwan University, Taipei, Taiwan,
Graduate Institute of Electronics Engineering, National Taiwan University, Taipei, Taiwan,

Hui-Ju K. Chiang
Graduate Institute of Electronics Engineering, National Taiwan University, Taipei, Taiwan,
EPI Lifeware, Inria Paris-Rocquencourt, Rocquencourt, France,

Ruei-Yang Huang
Graduate Institute of Electronics Engineering, National Taiwan University, Taipei, Taiwan,

Jie-Hong R. Jiang
Graduate Institute of Electronics Engineering, National Taiwan University, Taipei, Taiwan,
Department of Electrical Engineering, National Taiwan University, Taipei, Taiwan

François Fages
EPI Lifeware, Inria Paris-Rocquencourt, Rocquencourt, France,

Elena Zemtsova
Saint Petersburg State University, Universitetskii Pr. 26, Saint Petersburg 198504, Russia

Denis Yurchuk
Saint Petersburg State University, Universitetskii Pr. 26, Saint Petersburg 198504, Russia

Vladimir Smirnov
Saint Petersburg State University, Universitetskii Pr. 26, Saint Petersburg 198504, Russia

Bhaskar Garg
Department of Chemistry, National Tsing Hua University, Hsinchu 30013, Taiwan

Tanuja Bisht
Department of Chemistry, Government Degree College, Champawat 262523, Uttarakhand, India

Yong-Chien Ling
Department of Chemistry, National Tsing Hua University, Hsinchu 30013, Taiwan

Roberto Agromayor
Área de Máquinas y Motores Térmicos, Escola de Enxeñería Industrial, Universidade de Vigo, Vigo E-36310, Spain

David Cabaleiro
Departamento de Física Aplicada, Facultade de Ciencias, Universidade de Vigo, Vigo E-36310, Spain

Angel A. Pardinas
Área de Máquinas y Motores Térmicos, Escola de Enxeñería Industrial, Universidade de Vigo, Vigo E-36310, Spain

Javier P. Vallejo
Departamento de Física Aplicada, Facultade de Ciencias, Universidade de Vigo, Vigo E-36310, Spain

Jose Fernandez-Seara
Área de Máquinas y Motores Térmicos, Escola de Enxeñería Industrial, Universidade de Vigo, Vigo E-36310, Spain

Luis Lugo
Departamento de Física Aplicada, Facultade de Ciencias, Universidade de Vigo, Vigo E-36310, Spain

Brahim Aïssa
Qatar Environment and Energy Research Institute (QEERI), Qatar Foundation, Doha, Qatar
Department of Smart Materials and Sensors for Space Missions, MPB Technologies Inc., Montreal, QC, Canada

Nasir K. Memon
Qatar Environment and Energy Research Institute (QEERI), Qatar Foundation, Doha, Qatar

Adnan Ali
Qatar Environment and Energy Research Institute (QEERI), Qatar Foundation, Doha, Qatar

Marwan K. Khraisheh
Qatar Environment and Energy Research Institute (QEERI), Qatar Foundation, Doha, Qatar

Iftikhar Ahmad
Center of Excellence for Research in Engineering Materials, Advanced Manufacturing Institute, King Saud University, Riyadh 11421, Saudi Arabia

Bahareh Yazdani
College of Engineering, Mathematics and Physical Sciences, University of Exeter, Exeter EX4 4QF, UK

Yanqiu Zhu
College of Engineering, Mathematics and Physical Sciences, University of Exeter, Exeter EX4 4QF, UK

Iftikhar Ahmad
Center of Excellence for Research in Engineering Materials, Advanced Manufacturing Institute, King Saud University, Riyadh 11421, Saudi Arabia

Bahareh Yazdani
College of Engineering, Mathematics and Physical Sciences, University of Exeter, Exeter EX4 4QF, UK

Yanqiu Zhu
College of Engineering, Mathematics and Physical Sciences, University of Exeter, Exeter EX4 4QF, UK

Caitlin Fisher
Plasma Nanoscience Centre Australia (PNCA), CSIRO Materials Science and Engineering, Lindfield, NSW 2070, Australia
Plasma Nanoscience at Complex Systems, School of Physics, The University of Sydney, Sydney, NSW 2006, Australia

Amanda E. Rider
Plasma Nanoscience Centre Australia (PNCA), CSIRO Materials Science and Engineering, Lindfield, NSW 2070, Australia
Plasma Nanoscience at Complex Systems, School of Physics, The University of Sydney, Sydney, NSW 2006, Australia

Zhao Jun Han
Plasma Nanoscience Centre Australia (PNCA), CSIRO Materials Science and Engineering, Lindfield, NSW 2070, Australia
Plasma Nanoscience at Complex Systems, School of Physics, The University of Sydney, Sydney, NSW 2006, Australia

Shailesh Kumar
Plasma Nanoscience Centre Australia (PNCA), CSIRO Materials Science and Engineering, Lindfield, NSW 2070, Australia
Plasma Nanoscience at Complex Systems, School of Physics, The University of Sydney, Sydney, NSW 2006, Australia

Igor Levchenko
Plasma Nanoscience Centre Australia (PNCA), CSIRO Materials Science and Engineering, Lindfield, NSW 2070, Australia
Plasma Nanoscience at Complex Systems, School of Physics, The University of Sydney, Sydney, NSW 2006, Australia

Kostya (Ken) Ostrikov
Plasma Nanoscience Centre Australia (PNCA), CSIRO Materials Science and Engineering, Lindfield, NSW 2070, Australia
Plasma Nanoscience at Complex Systems, School of Physics, The University of Sydney, Sydney, NSW 2006, Australia

Preface

Design and integration of Chemical Process is the step required to convert laboratory process into industrial scale. It calculates the requirement of mass of raw materials and products, energy, fabrication materials and auxiliaries such as pumps, motors, electricity, heaters, boilers etc. On the other hand, in laboratory environment, a chemical reaction is developed based only on reaction success in glass beakers and final yield without much consideration to its cost of production. Therefore, when that reaction is transferred to large industrial scale for bulk production, several other aspects are also considered. Among these aspects or parameters, the first and the most important parameter is mass balance and process type i.e. batch or continuous production. It tells us about the requirement of each raw material required for the process, its cost, and supply source. Then, the total energy requirement is calculated, it includes determination of energy released or absorbed during each step of the reaction, and how it can be provided. Sometimes, energy from an exothermic reaction step is extracted by heat exchangers and used in another energy deficient reaction step. Once mass and energy requirements are calculated, then process design considerations are determined. It includes resident time of reactants and products in each step, its physical condition, and type of material required for fabrication of reaction vessels and pipes. This entire process is iterative in nature therefore, performed several times to obtain the most optimized process design parameters. Furthermore, each of the above-mentioned steps are specialization in itself and requires utmost attention to details. Therefore, in this book several researches related to different approaches of design and integration of chemical process are included.

Editor
Asad Rahman

CHAPTER 1 TOTAL SITE HEAT INTEGRATION CONSIDERING PRESSURE DROPS

Kew Hong Chew[1], Jiří Jaromír Klemeš [2], Sharifah Rafidah Wan Alwi [1], Zainuddin Abdul Manan [1], and Andrea Pietro Reverberi [3]

[1]Process Systems Engineering Centre (PROSPECT), Faculty of Chemical Engineering, Universiti Teknologi Malaysia, 81310 UTM Johor Bahru, Johor, Malaysia

[2]Centre for Process Integration and Intensification—CPI[2], Research Institute of Chemical and Process Engineering—MŰKKI, Faculty of Information Technology, University of Pannonia, Egyetem u. 10, Veszprém H-8200, Hungary

[3]Department of Chemistry and Industrial Chemistry (DCCI), University of Genova, Via Dodecaneso 31, Genova 16146, Italy

ABSTRACT

Pressure drop is an important consideration in Total Site Heat Integration (TSHI). This is due to the typically large distances between the different plants and the flow across plant elevations and equipment, including heat exchangers. Failure to consider pressure drop during utility targeting and heat exchanger network (HEN) synthesis may, at best, lead to optimistic energy targets, and at worst, an inoperable system if the pumps or compressors cannot overcome the actual pressure drop. Most studies have addressed the pressure drop factor in terms of pumping cost, forbidden matches or allowable pressure drop constraints in the optimisation of HEN. This study looks at the implication of pressure drop in the context of a Total Site. The graphical Pinch-based TSHI methodology is extended to consider the pressure drop factor during the minimum energy requirement (MER) targeting stage. The improved methodology provides a more realistic estimation of the MER targets and valuable insights for the implementation of the TSHI design. In the case study, when pressure drop in the steam distribution networks is considered, the heating and cooling duties increase by 14.5% and 4.5%.

INTRODUCTION

Pressure drop is an important factor to consider during a Heat Integration (HI) system design [1]. It is especially so with Total Site Heat Integration (TSHI) when distances between the different plants are large and the heat exchangers are often installed at different elevations within a plant. Pressure drop is mainly due to frictional losses as the fluids flow through pipes and fittings as well as pressure losses across the heat exchangers. When the fluids are liquid phase, there is additional pressure loss due to elevation changes. Failure to include the pressure drop factor in the early stages of design can lead to serious problems at the later stages. Exclusion of pressure drops when targeting minimum energy requirement (MER) may lead to too optimistic energy targets resulting in undersizing of central utilities systems. Neglecting pressure drops at the heat exchanger network (HEN) synthesis stage may render a proposed design infeasible if the actual pressure drops are higher than that what is allowable by the pumps or compressors. The need to replace the pumps or compressors may outweigh the savings from Heat Integration.

Most studies on pressure drop issues are associated with the retrofitting or synthesis of HEN for a single process.

Citation: Kew Hong Chew, Jiří Jaromír Klemeš, Sharifah Rafidah Wan Alwi, Zainuddin Abdul Manan, and Andrea Pietro Reverberi, Total Site Heat Integration Considering Pressure Drops, Energies 2015, 8, 1114-1137; doi:10.3390/en8021114

Mathematical Programming (MP)-based methodologies were mostly used to address the impact of pressure drop in the optimisation of HEN. Polley *et al.* [2] introduced the concept of pressure drop targeting in HEN retrofits where the pressure drop is correlated to the heat exchange area and heat transfer coefficient. The allowable pressure drop is used as an objective to optimise the heat exchange area. Ciric and Floudas [3] addressed the pressure drop issue based on the distances between heat exchangers and used a piping cost factor to minimise HEN modification costs. Ahmad and Hui [4] considered the pressure drop issue, in terms of distance between processes, by grouping the processes into "areas of integrity" and incorporated the impact in the methodology in the form of forbidden matches. Sorsak and Kranvanja [5] extended the Mixed Integer Non-linear Programming (MINLP) model of Yee and Grossman [6] to optimise the pressure drop and heat transfer coefficient. The pressure drop across the heat exchangers, both tube and shell sides, were estimated and considered in terms of pumping costs. Nie and Zhu [7] considered pressure drops in HEN retrofits by first estimating the pressure drop limits and then tackled the pressure drop constraints by optimising the area allocation, shell arrangement and use of heat transfer enhancement option. Panjeshahi and Tahouni [8] proposed a procedure whereby the pressure drop is considered together with the possibility of pump/compressor replacement when optimising area and utility costs. Soltani and Shafiei [9] introduced a new procedure which uses a genetic algorithm along with linear programming to retrofit HEN, including pressure drops. Stream pressure drop is correlated to area and heat transfer coefficient and the allowable pressure drops are introduced as constraints in the network optimisation.

Few studies have addressed pressure drops in the MER targeting stage. Zhu and Nie [10] considered the pressure drop aspect simultaneously with area and utility costs during the targeting and design stages. The pressure drop estimated for the heat exchanger is used to determine the optimum minimum approach temperature (ΔT_{min}) along with area and utility cost in the targeting stage. Inclusion of pressure drop (for heat exchangers only) in the proposed MP model led to different network structures and costs. Chew *et al.* [11] highlighted the significance of considering distribution piping pressure drop on steam generation from a Site Source. In the case study, the amount of steam recovered from the Site Source is significantly reduced when steam has to be generated at a higher pressure level to overcome the pressure drop in the pipes. Without considering pressure drops, the estimated utility targets maybe too optimistic and would result in undersizing of central steam generation systems. Liew *et al.* [12] extended the numerical algorithms, Total Site Problem Table Algorithm and Total Site Utility Distribution, to consider pressure drops and heat losses in steam pipes. The utility targets are based on a steam level which is at higher pressure (*i.e.*, to overcome the pressure drop) and superheated (*i.e.*, at a sufficient degree of superheat such that after heat loss the steam will reach the user at saturated conditions).

The studies so far have addressed the pressure drop factor in the optimisation of HEN in terms of pumping costs (based on distance or heat exchanger pressure drops), allowable pressure drop as constraints or objectives, or forbidden matches. The consideration of pressure drops in MER targeting has been at the heat exchanger (ΔT_{min}) or due to distance (steam pipes). None had looked at the pressure drop implications in a Total Site (TS) context which would encompass distance, equipment and utility distribution systems. Moreover, the MP-based methods provide few design insights required by designers [1]. In this paper, the graphical pinch-based TSHI methodology is extended to consider the pressure drop factor during the MER targeting stage. The methodology provides a more realistic estimation of MER targets and better understanding of the TSHI design for implementation later.

PRESSURE DROP FACTOR IN TSHI

In the established TSHI methodology, the utility targeting are based on temperatures and heat loads. The overall heat surplus (Source) and deficit (Sink) of the processes in a TS are represented by the Total Site Profile (TSP). The potential utility generation from the source and heating requirement of the Sink are shown by the Site Utility Composite Curves (SUCC) which are then used to set the targets for site heating and cooling utilities requirements [13]. The steam utilities are generated (from Site Source) and utilised (at Site Sink) at the same temperatures, seeFigure 1a,b.

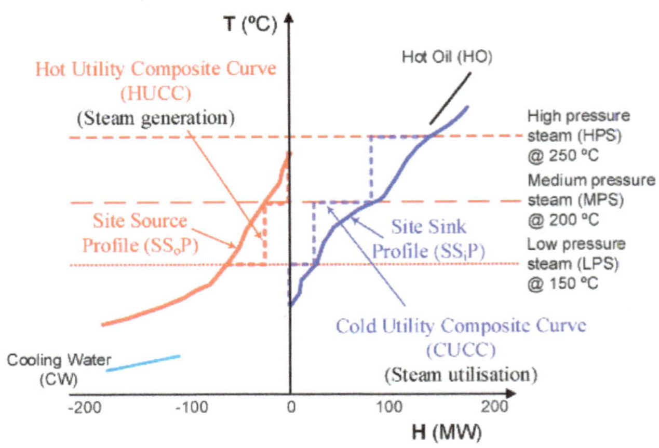

(**a**) TSP

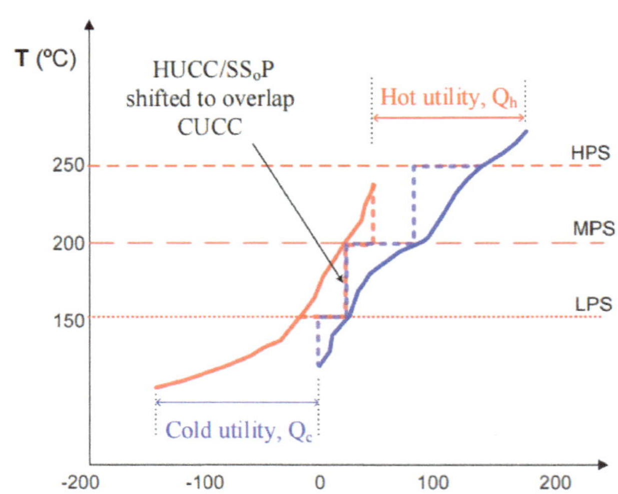

(**b**) Utilities targeting

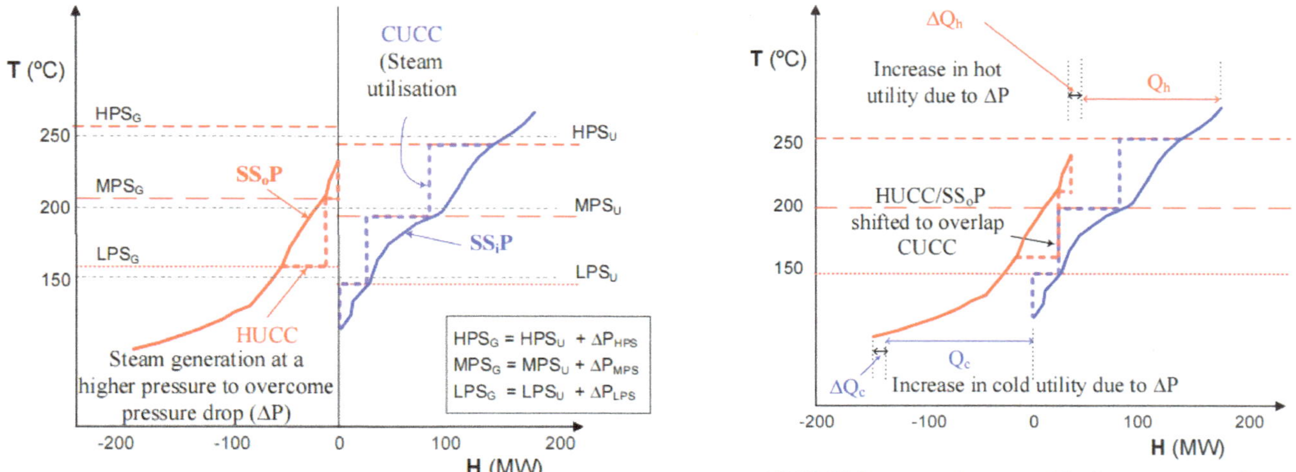

(**c**) TSHI considering pressure drop

(**d**) Utilities targeting considering pressure drop

Figure 1: TSHI utilities targeting (adapted from Klemeš *et al.* [13]).

In a TS, the utilities are distributed by an array of headers, sub-headers and pipes. Figure 2 gives a flow schematic of a TS comprising four plants with hot oil (HO), high pressure steam (HPS), medium pressure steam (MPS), low pressure steam (LPS) and cooling water (CW) utilities.

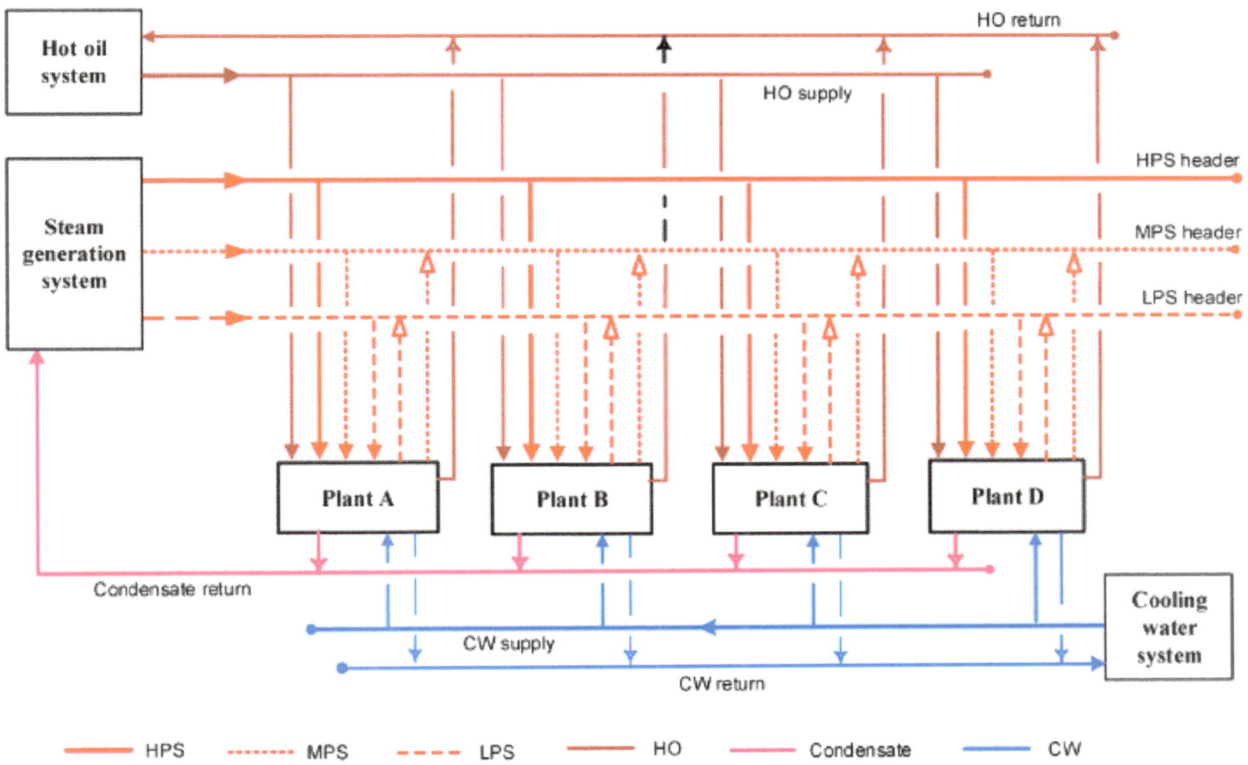

Figure 2: Schematic of a typical utilities distribution system at a TS with HO, HPS/MPS/LPS and cooling water.

Steams: HPS, MPS, LPS

The main headers take supply from the boilers and various steam generators which recover heat from the Site Source. Steam is then distributed to the various plants via sub-headers and distribution pipes. The main header operates at a sufficiently high pressure to supply steam to the furthest steam users. Figure 3 gives the process flow diagram of a typical steam generation and distribution system. Because of pressure drops in the headers, pipes and equipment, steam will be generated at a higher pressure and used at a lower pressure. The pressures and pressure drops of the steam distribution system are summarised in Figure 4. As saturated steam temperature is a function of its pressure, the difference in pressures between generation and usage can be represented by the difference in temperatures for generation and usage as shown in Figure 1c,d. As shown, consideration of the pressure drop factor will increase the heating (ΔQ_h) and cooling (ΔQ_c) utilities. In addition, the discharge head of the boiler feed water (BFW) pumps will have to be specified accordingly and the information used as input in the cost optimisation exercise.

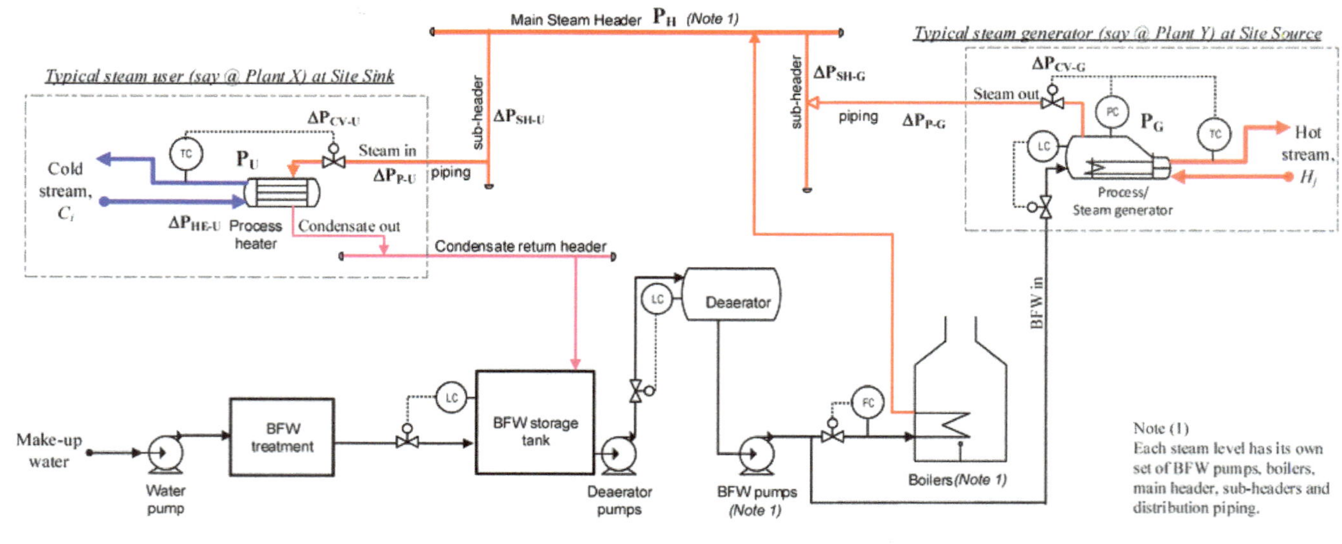

Figure 3: Process flow diagram—a typical steam generation, distribution and utilisation system at a TS.

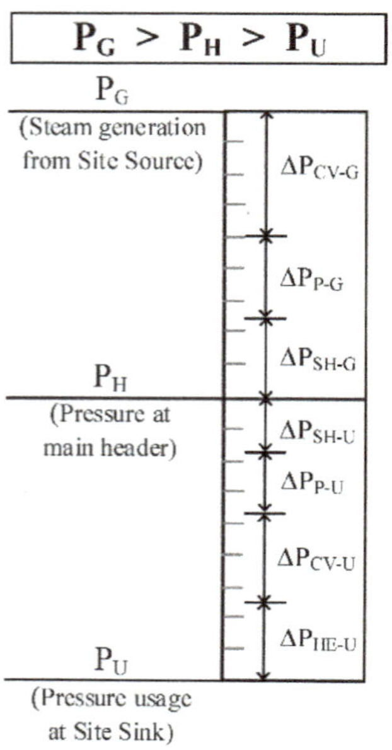

Figure 4: Pressures and pressure drops at a steam distribution system.

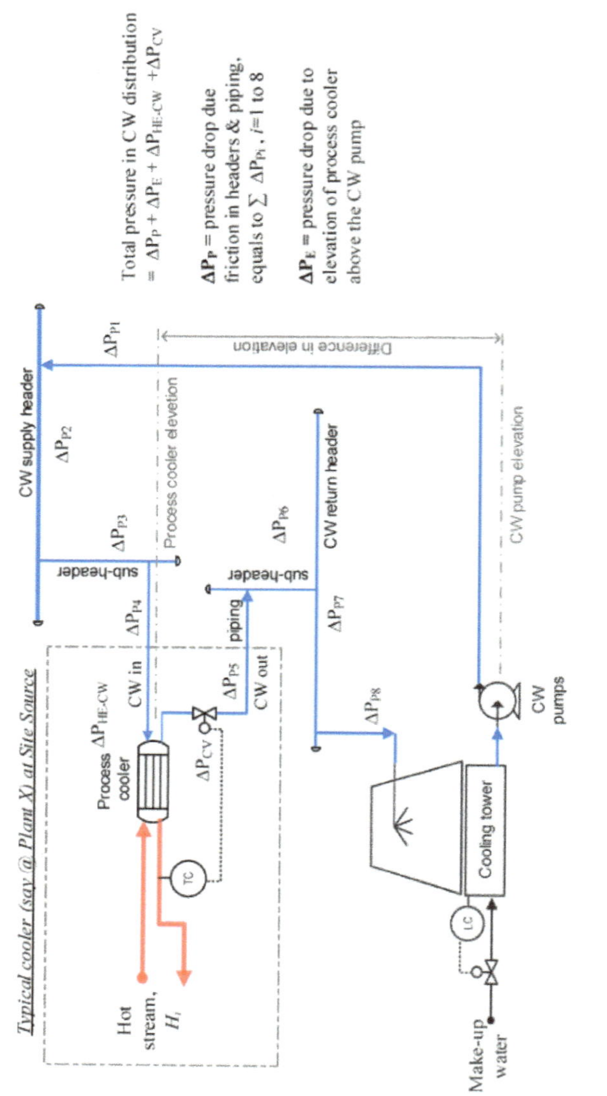

Figure 5: Process flow diagram—cooling water distribution system.

Cooling Water

Figure 5 is the flow diagram of a typical CW distribution system. The CW pumps deliver CW to the various users, *i.e.*, the process coolers, on the TS via the supply header, sub-headers and distribution piping. Warm CW exiting the coolers is routed to the cooling towers via the return header. For liquids like water, the temperature is not affected by its flow pressure. As long as there is no phase change, the pressure drop does not affect the MER targeting. However, the CW pumps have to be specified for a sufficient discharge head so as to overcome the pressure drop in the distribution system to ensure adequate volumes of the utility are delivered to the users as required. For liquid utilities such as CW, the elevation pressure drop due to liquid column static head (above the pump) is important and has to be considered. The required pump discharge head can then be used as an input parameter in the cost optimisation exercise.

Hot Oil

Like CW, pressure drops do not affect MER targeting. Pressure drops due to elevation have to be included when estimating the discharge head for the HO circulation pumps and used as an input to the cost optimisation exercise. As with the liquid utilities, the impact of pressure drop on the process streams are seen in the penalty of additional pumping or compression costs so long as there is no phase changes, *i.e.*, liquid remains as liquid and gas stays as gas in the pipes. The impact of pressure drop on TSHI is summarised in Table 1.

Table 1: Impact of pressure drop on TS.

Fluid	MER Targeting	Cost Optimisation
Steam e.g., HPS, MPS, LPS	Increase ΔQ_h and ΔQ_c	Higher BFW pump capital and pumping costs
Liquid utilities (e.g., CW, HO, *etc.*)	No impact	Higher utility circulation pump capital and pumping costs
Process—liquids [(a)]	No impact	Higher pump capital and pumping costs
Process—gas [(b)]	No impact	Higher compressor capital and compressing costs

[(a)] Assume no phase change, liquid remains as liquid in the pipes; [(b)] assume no phase change, gas stays as gas in the pipes

PRESSURE DROP ESTIMATES

Steam Distribution System

Figure 3 is a process flow diagram of a steam generation and distribution system. The main steam header takes supply from the boilers and process/steam generators which recover heat from the Site Source. At the process/steam generator, a pressure control valve regulates the pressure at the heat exchanger ensuring that steam is generated at a sufficient pressure for delivery to the main header via the sub-header. The pressure drops between the process/steam generator and the main header, $\Delta P_{G\text{-}S}$:

$$\Delta P_{G\text{–}S} = \Delta P_{CV} + \Delta P_{P} + \Delta P_{SH} + \Delta P_{H} \quad (1)$$

The pressure drops between the process/steam user and the main header, $\Delta P_{S\text{-}U}$:

$$\Delta P_{S\text{–}U} = \Delta P_{H} + \Delta P_{SH} + \Delta P_{P} + \Delta P_{CV} + \Delta P_{HE} \quad (2)$$

where ΔP_{CV} is the pressure drop across the control valve, ΔP_{HE} is the pressure drop across the heat exchanger and ΔP_{P}, ΔP_{SH}, ΔP_{H} are the frictional pressure drops in the distribution pipe, sub-header and header.

The steam is assumed to be saturated and dry throughout the distribution network. Any condensate dropouts due to heat losses from the insulated pipe to the ambient and/or due to the Joule-Thompson effects of pressure drops are removed by steam traps located at strategic locations [14]. Heat loss from a steam distribution system occurs in several ways. In addition to the heat loss from the insulated pipes to the ambient a majority of the heat loss is through leaks in steam pipes, condensate return lines as well as steam traps. It is more appropriate to account for steam losses (which have to be made up by extra steam generation) as a percentage of steam consumption than to use a degree of superheating in the steam temperature as proposed by [12] to account for heat losses.

The frictional pressure drop in steam lines can be calculated using the Babcock equation [15]:

$$\Delta P_f = 2489 \left\{\frac{d + 3.6}{d^6}\right\} \frac{W^2 L}{\rho} \quad (3)$$

where W is the mass flow (kg/h), L is the pipe length (m), ρ is the single phase density (kg/m^3) and d is the pipe internal diameter (mm).

Alternatively, a steam line sizing nomograph, see Appendix 1, can be used for quick estimate of steam line pressure drops [14]. Commercial software such as Pipe module, which estimate pressure drop and heat loss in pipes, in the Aspen-HYSYS process simulator can also be used [16].

Cooling Water Distribution System

In Figure 5, pressure drop ΔP_{CW} at the CW distribution system, for a process/CW cooler, can be described as:

$$\Delta P_{CW} = \Delta P_{P} + \Delta P_{E} + \Delta P_{HE} + \Delta P_{CV} \quad (4)$$

where ΔP_{P} is the frictional pressure drop, ΔP_{E} is the elevation pressure drop, ΔP_{HE} and ΔP_{CV} are as described before. Equation (4) can generally be used for other liquid phase utilities such as HO, *etc.*

Frictional Pressure Drop in Liquid and Gas Lines, ΔP_P

Fluid flow always results in energy losses due to friction. The frictional losses will be have to be overcome by additional head required on the pump. The pressure drop due to friction can be estimated by the well-known Darcy-Weisbach equation [17]:

$$\Delta P_f = 0.5\, \rho\, f_m\, L\, V^2/d \quad (5)$$

where ρ is the density (kg/m^3), L is the length (m), V is the velocity (m/s) and d is the internal diameter of pipe (mm). f_m is the Moody friction factor, which depends on the Reynolds number (Re) and ε, the absolute roughness of the pipe for turbulent flow, typical of fluids flow in plant. Appendix 2 gives the values of ε and f_m for different pipe materials. These values are the iterative solution of the Colebrook correlation [17]:

$$\frac{1}{\sqrt{f_m}} = -2 \log_{10} \left\{\frac{\varepsilon}{3.7\, d} + \frac{2.51}{Re\, \sqrt{f_m}}\right\} \quad (6)$$

Equation (5) can be directly applied for liquid lines.

To estimate pressure drop in gas lines within plant or battery limits, the Darcy-Weisbach formula can be written in a simple form, assuming that the pressure drop through the line is less than 10% of the line pressure [17]. Pressure drop per 100 m of equivalent pipe length can be written as:

$$\Delta P_{100} = \frac{W^2}{\rho} \left\{\frac{62\,530\,(10^2) f}{d^5}\right\} \quad (7)$$

where W is the mass flow (kg/h), ρ is the single phase density (kg/m^3), f is the friction factor and d is the pipe internal diameter (mm).

Elevation Pressure Drop for Liquid Lines, ΔP_E

For liquid lines, the pressure drop due to static head of liquid column above the utility circulation pump need to be included. The elevation pressure drop has to be calculated separately using the following equation which is based on Bernoulli's Theorem:

$$\Delta P_E = 0.00981\, \rho_l\, Z_E \quad (8)$$

where ρ_l is the liquid density (kg/m^3) and Z_E is the elevation of the heat exchanger above the utility circulation pump centre line (m).

Pressure Drop across Heat Exchanger, ΔP_{HE}

During conceptual design, the type of heat exchanger or detailed geometry of the heat exchanger are often not available. Typical values of pressure drop based on company's guidelines or designer's experience can be used. Alternatively, the heat exchangers pressure drop can be estimated using established equations with some explicit assumptions on the heat exchanger geometries, for e.g., shell and tube heat exchangers: number of passes, tube diameter, tube length, tube pitch, tube configurations, baffle cuts, *etc.* [18].

Pressure Drop across Control Valve, ΔP_{CV}

The pressure drop across a control valve can be estimated if the characteristics of the control valve, C_v, is known. A larger pressure drop will increase pumping costs while a smaller pressure drop will increase valve costs. During the conceptual stage, when the details of the valves are not known, the usual rule of thumb is to use an allowable pressure drop of 10%–15% of total pressure drop, or 70 kPa, whichever is greater [19].

METHODOLOGY

The proposed methodology is presented in Figure 6 and described as follows.

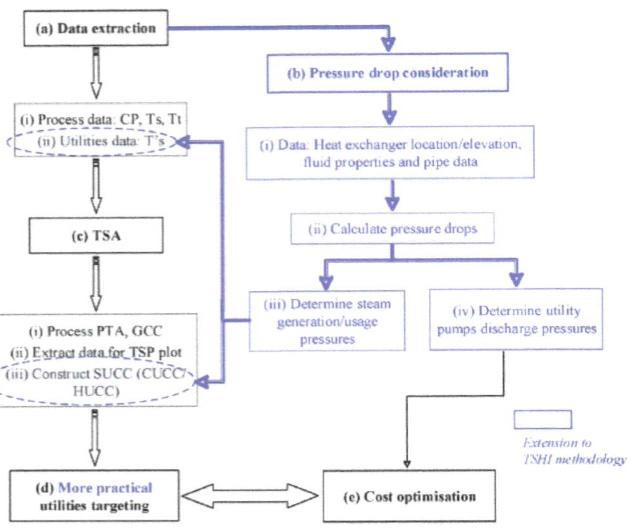

Figure 6: Algorithm to consider pressure drops in TSHI.

- Data extraction—extract stream and utilities data, *i.e.*, heat capacities (CP) and temperatures.
- To consider the pressure drop factor in TSHI:
- Information on the location of the heat exchangers, fluid properties and pipe data are required in order to estimate the pressure drops. Location and elevation of the heat exchangers can be obtained from the site plot plan, individual plant layout and elevation drawings. Fluid properties such as mass flow and density can be extracted from the heat and mass balances. Pipe data required are the internal diameter and roughness factor. For each plant on site, determine the header, sub-header and pipe lengths based on the process/utility heat exchangers located furthest from the reference point and the process/utility heat exchanger at the highest elevation.
- The pressure drops can be estimated using the equations given in Section 3. Alternatively, pressure drops can be based on the typical ΔP per unit length for pipes, control valves and heat exchangers available from company guidelines or designer's experiences.
- Determine steam generation/usage pressure and corresponding steam saturation temperatures. Referring to Figure 3 again, the steam usage pressure, P_U, is the steam pressure at the steam/process heater, furthest from the utility reference point:

$$P_U = \Delta P_{S-U} + P_H \quad (9)$$

where, P_H is the header pressure.

The steam generation pressure, P_G, is the steam pressure at the process/steam generator furthest from the utility reference point

$$P_G = P_H + \Delta P_{G-S} \quad (10)$$

The steam saturation temperatures at P_U and P_G can be obtained from the steam tables.

I. DETERMINE UTILITY PUMPS DISCHARGE PRESSURE.

Referring to Figure 5, the CW pump discharge pressure reads as:

$$P_{PUMP} = P_{DES} + \Delta P_{CW} \quad (11)$$

where subscript DES denotes destination, at the process/CW cooler furthest from the CW pumps.

A. CARRY OUT TS ANALYSIS:

i. Prepare the TSP from individual process PTA and GCC [13]. The utility usage and generation are directly interpolated on the TS-PTA at the respective utilities temperatures [20]. An example of the TS-PTA is given in Table 3.

ii. A graphical representation of the SUCC can be obtained from the TS-PTA, see Figure 8.

b. Utilities targeting—Steam is generated and used at different temperatures due to the pressure drops in the steam distribution network. The TS energy targets are determined using the pinch-based graphical and algebraic method [20].

c. Pressure drops determined for liquid utility systems can be used as an input to the cost optimisation in terms of higher pumping cost and the constraints in allowable ΔP.

ILLUSTRATIVE EXAMPLES

The TS consists of four plants A, B, C and D with hot oil (HO), HPS, MPS, LPS and cooling water (CW) utility systems as depicted in Figure 2. A simplified plot plan and elevation drawing is given in Figure 7.

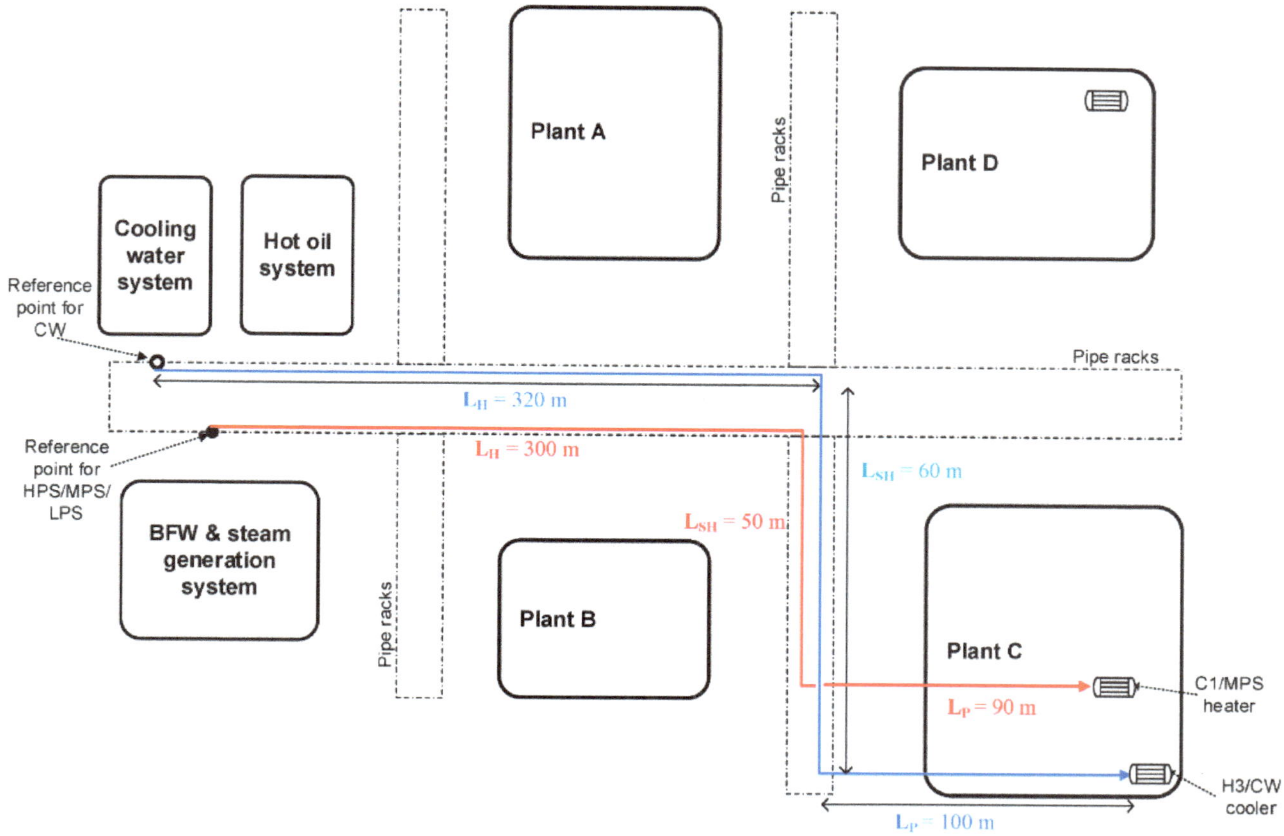

(**a**) TS plot plan

(**a**) TS plot plan

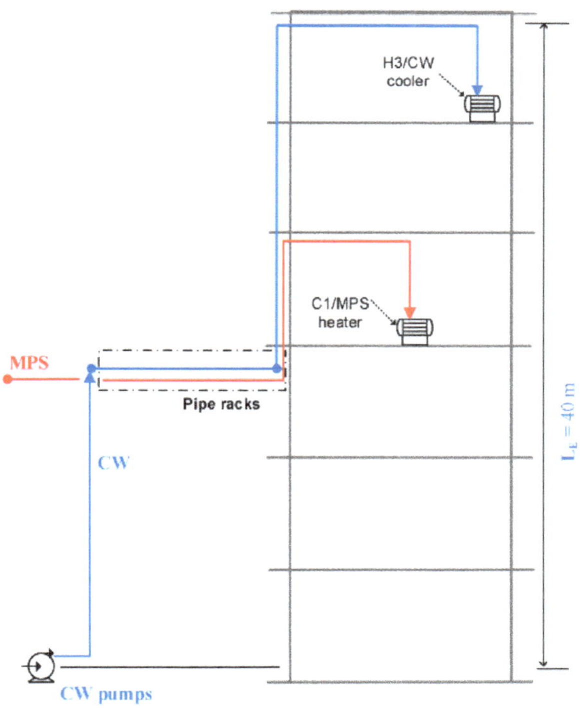

(**b**) Plant C elevation view

Figure 7: Simplified plot plan and elevation drawing for the TS.

For LPS, $P_{U\text{-}LPS}$ is governed by stream C1/LPS heater at Plant D, located furthest from the main LPS header, while $P_{G\text{-}LPS}$ is governed by stream H1/LPS steam generator at the same Plant D. For MPS, $P_{U\text{-}MPS}$ is governed by stream C1/MPS heater at Plant C located furthest from the main MPS header while $P_{G\text{-}MPS}$ is governed by stream H1/MPS steam generator at the same Plant C. For HPS, $P_{U\text{-}HPS}$ is governed by stream C1/HPS heater at Plant C, located furthest from the main HPS header. There is no HPS steam generation on site.

A summary of the stream data, layout and elevation information for the estimation of pressure drops is given in Table 2. Figure 8 shows the TSP and SUCC of the TS. The results of the pressure drops estimation and the corresponding steam generation and usage temperatures for the steams and CW distribution networks are summarised in Table 3. Table 4 gives the modified TS-PTA by which the utilities usage and generation are interpolated from the Site Sink and Site Source PTA. The revised SUCC, with consideration for pressure drops, are superimposed on Figure 8.

Table 2: Summary of input data for TS analysis

Process	Stream	CP (MW/°C)	T_s (°C)	T_t (°C)	Length, *L*/Elevation, E [(1)]				
					L_H (m)	L_{SH} (m)	L_P (m)	E (m)	Heat exch. Furthest from Utility Reference Point
A	H1	0.35	260	225					
	H2	1.15	260	195					
	H3	0.50	195	130					
	C1	1.25	240	255					
	C2	0.65	175	260					
	C3	0.20	155	205					
B	H1	0.36	260	175					
	H2	0.60	260	115					
	H3	0.75	175	95					
	C1	1.10	175	255					
	C2	0.20	110	175					
	C3	0.89	95	155					
C	H1	0.62	225	155	300	50	90		H1/MPS
	H2	0.32	195	95					
	H3	1.00	130	85	320	60	100	40	H3/CW
	C1	0.60	110	240	300	50	85		C1/HPS, C1/MPS
	C2	0.40	155	240					
	C3	0.70	110	175					
D	H1	0.41	130	85	300	50	90		H1/LPS
	H2	0.10	110	80					
	H3	0.15	95	70					
	C1	0.20	90	140	300	50	80		C1/LPS
	C2	0.50	60	110					
	C3	0.40	50	100					
Utility	**T_s (°C)**		**T_r (°C)**						
HO	300		260				$\Delta T_{min\text{-}pp}$ is	20	°C
HPS	250		-				$\Delta T_{min\text{-}pu}$ is	15	°C
MPS	200		-						
LPS	150		-						
CW	25		45						

[(1)] Piping lengths are only extracted for those heat exchangers furthest from the utilities reference point, *i.e.*, which govern the steam generation and usage levels and utility circulation pump sizing. Only steam and CW are considered.

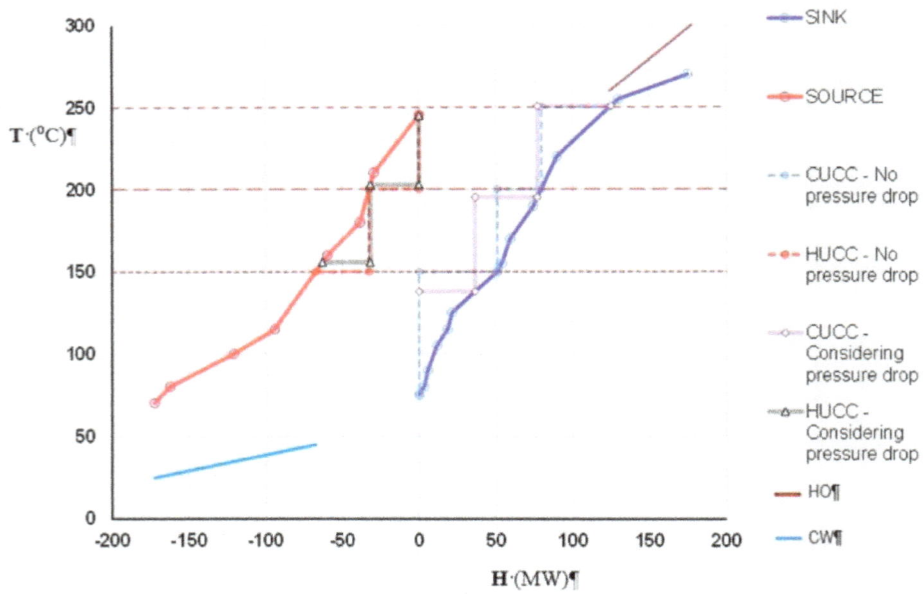

Figure 8: Simplified plot plan and elevation drawing for the TS.

Due to pressure drops in the headers, sub-headers, piping and across control valves and heat exchangers, steams have to be generated at a higher value than their usage. A comparison of utility targeting with and without consideration of pressure drop is given in Table 5. The impact of pressure drop is more notable for steam at low pressure, due to its higher volumetric flow. From Table 3, the differences in steam usage and generation temperatures are 18.1 °C and 7.7 °C for LPS and MPS. From Table 5, the overall heating utilities increases by 14.7 MW (14.5%) and the cooling utility increases by 4.7 MW or (4.5%) when pressure drop is taken into consideration. The HPS requirement increases by 3.4 MW, the MPS usage increases by 11.9 MW while LPS usage reduces by 14.4 MW. Excluding pressure drop could lead to over estimation of the amount of steam that can be raised at the Site Source for HPS and MPS leading to the undersizing of central HPS and MPS generation capacities.

Table 3: Summary of pressure drops estimation for steam and CW distribution networks

Operating parameter			LPS Usage	LPS Generation	MPS Usage	MPS Generation	HPS Usage	CW
Header pressure		kPag	375	375	1453	1453	3831	-
Header temp.		°C	150	150	200	200	250	-
Pressure drops [1]:	ΔP_H	kPa	33	33	33	33	33	-
	ΔP_{SH}	kPa	5.5	5.5	5.5	5.5	5.5	-
	ΔP_P	kPa	10.8	9.6	9	9.6	9	-
	ΔP_{CV}	kPa	35	35	50	50	50	70
	ΔP_{HE}	kPa	50	-	50	-	50	100
	ΔP_f	kPa	-	-	-	-	-	218 [2]
	ΔP_E	kPa	-	-	-	-	-	392.4
Total pressure drops		kPa	134	83.1	148	98.1	148	780.4
Pressure @ user	P_U	kPag	240.7	-	1305.5	-	3683.5	
Temperature @ user	T_U	°C	137.8	-	195.1	-	247	
Pressure @ generation	P_G	kPag	-	458.1	-	1551	-	
Temperature @ generation	T_G	°C	-	155.9	-	202.8	-	
ΔT between usage and generation		°C	18.1		7.7			
Pressure @ CW pump		kPag						980.4 [3]

[1] A frictional pressure drop of 0.11 kPa/m has been assumed for the headers and sub-headers and 0.12 kPa/m for the piping; [2] Total frictional pressure drops at supply/return headers, sub-headers and piping; [3] Based a destination pressure, *i.e.*, pressure at the H3/CW cooler within Process C, of 200 kPag.

Table 4: TS-PTA for Site Source and Site Sink with utilities usage and generation

(a) Site Sink PTA

T** (°C)	ΔT (°C)	Process CP A (MW/°C)	B (MW/°C)	C (MW/°C)	D (MW/°C)	Σ CP (MW/°C)	ΔH (MW)	Cascade H (MW)	H [1] (MW)	Utility Usage H (MW)
75						0	0	0		
80	5				0.65	0.65	3.3	3.3		
90	10				0.24	0.24	2.4	5.7		
105	15				0.39	0.39	5.9	11.5		
115	10				0.69	0.69	6.9	18.4		
125	10				0.29	0.29	2.9	21.3		
137.8									36.4	LPS = 36.4
150	25			0.98	0.20	1.18	29.5	50.8		
155	5			0.36	0.20	0.56	2.8	53.9		
170	15			0.36		0.36	5.4	59.0		
190	20			0.76		0.76	15.2	74.2		
195.1									76.9	MPS = (76.9 − 36.4) = 40.5
220	30		0.14	0.38		0.52	15.6	89.8		
250.7									124.8	HPS = (124.8 − 76.9) = 47.9
255	35		0.14	1.00		1.14	39.9	129.7		
270	15	1.90	1.10			3.00	45.0	174.7		
275	5	0.65				0.65	3.3	178.0		

[1] Interpolate at the steam temperatures.

(b) Site Source PTA

T^{**} (°C)	ΔT (°C)	Process CP A (MW/°C)	B (MW/°C)	C (MW/°C)	D (MW/°C)	Σ CP (MW/°C)	ΔH (MW)	Cascade H (MW)	H ⁽¹⁾ (MW)	Utility Usage H (MW)
245						0	0	0		
210	35	0.85				0.85	29.8	29.8		
202.8									31.9	MPS = 31.9
180	30	0.30				0.30	9.0	38.8		
160	20	0.30	0.76			1.06	21.2	60.0		
155.9									63.1	LPS = (63.1 − 31.9) =31.2
115	45	0.50	0.26			0.76	34.2	94.2		
100	15		0.46	1.32		1.78	26.7	120.9		
80	20		0.75	1.32		2.07	41.4	162.3		
70	10			1.00		1.00	10.0	172.3		

T^{**} Double shifted temperature for TSP plot and TS-PTA, °C; (1) Interpolate at the steam temperatures.

Table 5: Impact of pressure drop on TS

Utilities	Base Case (No Pressure Drops) Usage (MW)	Generation (MW)	Nett (MW)	Case 1 (with Pressure Drops) Usage (MW)	Generation (MW)	Nett (MW)	(Case 1)—(Base Case) Usage (MW)	Generation (MW)	Nett (MW)
HO	54.0	-	54.0	53.2	-	53.2	−0.8	-	−0.8
HPS	44.6	-	44.6	48.0	-	48.0	+3.4	-	+3.4
MPS	28.6	32.8	−4.2 (1)	40.5	31.9	8.6	+11.9	−0.9	+8.6
LPS	50.8	34.8	11.8 (1)	36.4	31.2	5.2	−14.4	−3.8	−6.8
Total heating			100.4			115.0			+14.6
CW	-	-	104.7			109.2			+4.7

(1) Excess MPS generated is used for LPS heating. Nett LPS heating is (50.8 − 34.8 − 4.2) = 11.8 MW.

The pressure drop in liquid utilities does not affect TSHI MER targeting, however it should be considered and used as an input parameter when evaluating the TSHI options for economic evaluation. Exclusion of pressure drops will lead to undersizing of pumps or compressors leading to infeasible design solutions, and expensive re-design at the detailed design stage.

CONCLUSIONS

A systematic methodology that considers pressure drops in TSHI utility targeting has been developed. The case study proved that ignoring pressure drops in TSHI design led to optimistic MER targets and resulted in undersizing of external steam generation capacity. While pressure drops of liquid utilities such as water do not affect MER targeting, the pressure drop information should be incorporated in the economic evaluation of TSHI options. Pressure drops due to pipe friction, elevation changes and pressure drops across control valve and heat exchangers all need to be accounted for. Incorporation of pressure drops leads to closer to real life MER targeting and design. The proposed methodology can benefit from the visualisation advantages of the graphical method and from the precision of the numerical method and should be of the benefit to both industry and academia [21].

ACKNOWLEDGMENTS

The authors gratefully acknowledge the financial supports from the Universiti Teknologi Malaysia (UTM) Research University Grant under Vote No. Q.J130000.2509.07H35 and the EC FP7 project ENER/FP7/296003/EFENIS "Efficient Energy Integrated Solutions for Manufacturing Industries—EFENIS". The support from the Hungarian project Társadalmi Megújulás Operatív Program "TÁMOP-4.2.2.A-11/1/KONV-2012-0072—Design and optimisation of modernisation and efficient operation of energy supply and utilisation systems using renewable energy sources and ICTs" significantly contributed to the completion of this analysis.

AUTHOR CONTRIBUTIONS

Jiří Jaromír Klemeš, Sharifah Rafidah Wan Alwi, Zainuddin Abdul Manan and Andrea Pietro Reverberi conceived the idea of research, provide guidance and supervision; Kew Hong Chew implemented the research, performed the analysis and wrote the paper. All authors have contributed significantly to this work.

APPENDIX

Appendix 1: Steam Line Sizing Chart—Pressure Drop (Spirax Sarco, 2014).

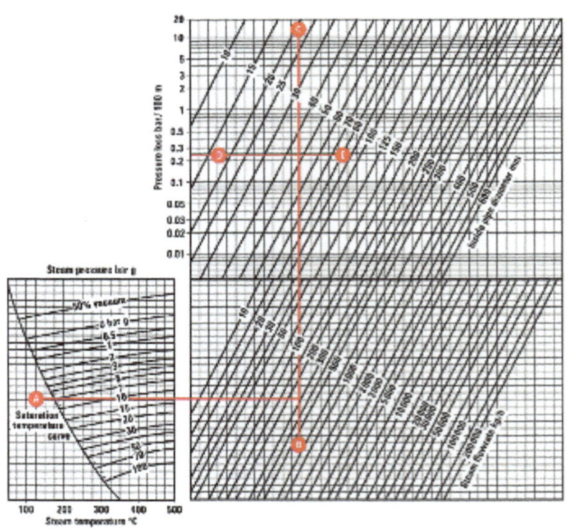

- Select the steam pressure at the saturated steam line (7 barg), A;
- From A, draw a horizontal line to the steam flowrate (286 kg/h) and mark B;
- From B, draw a vertical line to the top of nomograph, C;
- Draw a horizontal line from 0.24 bar/100 m (allowable DP) on the pressure loss scale (DE);
- Point which BC crosses DE will indicate the pipe size required.

Appendix 2: Relative Roughness of Pipe Materials and Friction Factors for Complete Turbulence (GPSA, 1998).

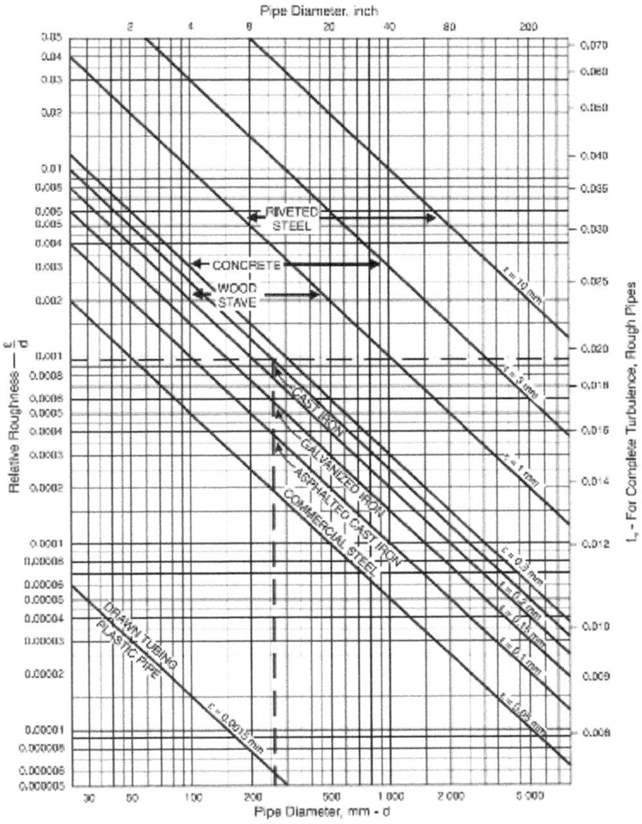

REFERENCES

1. Klemeš, J.J.; Varbanov, P.S.; Kravanja, Z. Recent developments in process integration. *Chem. Eng. Res. Des.* 2013,*91*, 2037–2053.
2. Polley, G.T.; Panjehshahi, M.H.; Jegede, F.O. Pressure drop considerations in the retrofit of heat exchanger networks. *Chem. Eng. Res. Des.* 1990, *68*, 211–220.
3. Ciric, A.M.; Floudas, C.A. A mixed integer nonlinear programming model for retrofitting heat-exchanger networks. *Ind. Eng. Chem. Res.* 1990, *29*, 239–251.
4. Hui, C.W.; Ahmad, S. Minimum cost heat recovery between separate plant regions. *Comput. Chem. Eng.* 1994,*18*, 711–728.
5. Sorsak, A.; Kranvanja, Z. Simultaneous MINLP synthesis of heat and power integrated heat exchanger network. *Comput. Chem. Eng. Suppl.* 1999, S143–S147.
6. Yee, T.F.; Grossman, I.E. Simultaneous optimisation models for heat integration—II. Heat exchanger network synthesis. *Comput. Chem. Eng.* 1990, *14*, 1165–1184.
7. Nie, X.R.; Zhu, X.X. Heat exchanger network retrofit considering pressure drop and heat transfer enhancement. *AIChE J.* 1999, *45*, 1239–1254.
8. Panjaeshahi, M.H.; Tahouni, N. Pressure optimisation in debottlenecking of heat exchanger networks. *Energy*2008, *33*, 942–951.
9. Soltani, H.; Shafiei, S. Heat exchanger networks retrofit with considering pressure drop by coupling genetic algorithm with LP (linear programming) and ILP (integer linear programming) methods. *Energy* 2011, *36*, 2381–2391.
10. Zhu, X.X.; Nie, X.R. Pressure drop considerations for heat exchanger network grassroots design. *Comput. Chem. Eng.* 2002, *26*, 1661–1676.
11. Chew, K.H.; Klemeš, J.J.; Wan Alwi, S.R.; Abdul Manan, Z. Industrial implementation issues of Total Site Heat Integration. *Appl. Therm. Eng.* 2013, *61*, 17–25.
12. Liew, P.Y.; Wan Alwi, S.R.; Klemeš, J.J. Total Site Heat Integration targeting algorithm incorporating plant layout issues. *Comput. Aided Chem. Eng.* 2014, *33*, 1801–1806.
13. Klemeš, J.J.; Dhole, V.R.; Raissi, K.; Perry, S.J.; Puigjaner, L. Targeting and design methodology for reduction of fuel, power and CO_2 on total sites. *Appl. Therm. Eng.* 1997, *17*, 993–1003.
14. Spirax Sarco. Available online: spiraxsarco.com/resources/steam-engineering-tutorials/steam-distribution/pipes-and-pipe-sizing.asp (accessed on 23 September 2014).
15. *Flow of Fluids through Valves, Fittings and Pipe*; Technical Paper No. 410; Crane Co.: Chicago, IL, USA, 1988.
16. *Aspen-HYSYS (Version 2006) (computer software)*; AspenTech: Burlington, MA, USA, 2014.
17. Gas Processing Supplier Association (GPSA). Fluid flow and piping. In *GPSA Engineering Data Book*, 11th ed.; GPSA: Tulsa, OK, USA, 1998; Volume 17, pp. 1–28.
18. Smith, R. *Chemical Process—Design and Integration*; McGraw-Hill: New York, NY, USA, 2005.
19. Choudhury, A.A.S.; Nwaoha, C.; Vishwasrao, S.V. *Process Plant Equipment: Operation, Control and Reliability*; Holloway, M.D., Nwaoha, C.N., Onyewuenyi, O.A., Eds.; Wiley: New York, NY, USA, 2012; pp. 9–15.
20. Chew, K.H.; Klemeš, J.J.; Wan Alwi, S.R.; Manan, Z.A. Process Modifications to Maximise Energy Savings in Total Site Heat Integration. *Appl. Therm. Eng.* 2014.
21. Klemeš, J.J. Industry-academia partnership. *Clean Technol. Environ. Policy* 2013, *15*, 861–862.

CHAPTER 2

EXPLORING UNCERTAINTY IN GLACIER MASS BALANCE MODELLING WITH MONTE CARLO SIMULATION

H. Machguth[1], R. S. Purves[1], J. Oerlemans[2], M. Hoelzle[1], and F. Paul[1]

[1]Department of Geography, University of Zurich, Zurich, Switzerland

[2]Institute for Marine and Atmospheric Research, Utrecht University, Utrecht, The Netherlands

ABSTRACT

By means of Monte Carlo simulations we calculated uncertainty in modelled cumulative mass balance over 400 days at one particular point on the tongue of Morteratsch Glacier, Switzerland, using a glacier energy balance model of intermediate complexity. Before uncertainty assessment, the model was tuned to observed mass balance for the investigated time period and its robustness was tested by comparing observed and modelled mass balance over 11 years, yielding very small deviations. Both systematic and random uncertainties are assigned to twelve input parameters and their respective values estimated from the literature or from available meteorological data sets. The calculated overall uncertainty in the model output is dominated by systematic errors and amounts to 0.7 m w.e. or approximately 10% of total melt over the investigated time span. In order to provide a first order estimate on variability in uncertainty depending on the quality of input data, we conducted a further experiment, calculating overall uncertainty for different levels of uncertainty in measured global radiation and air temperature. Our results show that the output of a well calibrated model is subject to considerable uncertainties, in particular when applied for extrapolation in time and space where systematic errors are likely to be an important issue.

INTRODUCTION

A wide range of approaches to the modelling of mass balance exist, ranging from simple temperature index correlations (e.g., Braithwaite, 1981; Reeh, 1989) through to complex physical models of energy balance and associated melt (e.g., Brock et al., 2000; Arnold et al., 2006). Typically, models are developed for the general case of modelling mass balance, but are calibrated and validated at a few point locations and therefore for a particular set of topographic and climatic conditions. Thus, for example, temperature index approaches require very little input data (positive degree days and a degree days factor) and could be applied in regions with sparse measurements (e.g., Reeh, 1989). However, they require calibration for each area in order to consider local charateristics (Braithwaite, 1995). By contrast, more complex physical models are assumed to require less tuning, and thus to be more suitable for extrapolating mass-balance in both space and time but at the expense of a higher demand for input data. Several studies exist where models of various complexities have been extrapolated over, for example individual glaciers or mountain ranges to produce seasonal values for mass balance (e.g. Arnold et al., 1996), and a key

Citation: Machguth, H., Purves, R. S., Oerlemans, J., Hoelzle, M., and Paul, F.: Exploring uncertainty in glacier mass balance modelling with Monte Carlo simulation, The Cryosphere, 2, 191-204, doi:10.5194/tc-2-191-2008, 2008.

question in the development of such methods is the uncertainty associated with them.

Any approach to quantifying uncertainty must firstly consider potential sources and techniques for the quantification of uncertainty. Mass balance models typically require both meteorological inputs and snow or ice parameterisations representing the point(s) at which the model is being run. Although many models have been developed using data measured at the same point as mass balance measurements, such an approach is not viable for extrapolating mass balance in space and time, where typical inputs to such a model have to be interpolated from point measurements or the projections of, for instance, general circulation models (GCM) or regional climate models (RCM). Equally, if we wish to explore mass balance in the past, point data are normally not available from the glacier of interest, and meteorological measurements from some long term data series, assumed to be correlated with the glacier location are generally used. In both cases, the meteorological inputs to the mass balance model must be extrapolated or interpolated in space, and perhaps interpolated in time, to derive values appropriate to the modelled mass balance.

Uncertainties in input parameters to mass balance models can thus be characterised as stemming from, in the case of measured data, errors and uncertainties in measured values, and in the case of modelled values, differences between modelled and measured values (which in turn stem from differences between the spatial extent over which these values can be considered to be representative). The methods used to extrapolate/interpolate these data from the measurement location to the model location are other potential sources of uncertainty. Further uncertainties arise as a result of the abstraction of processes themselves within the mass balance model, and resulting generalisation of the real world system – for example, few mass balance models represent changes in snowpack form during melt (e.g. formation of ablation hollows or sun-cups during the ablation season) and the resulting increase in roughness length and change in turbulent energy fluxes. In considering uncertainties only in measured values of meteorological data there are two important sources of uncertainty to consider: random and systematic errors. Random errors are usually assumed to be related to the device making the measurement and its notional precision and are either temporally uncorrelated or only temporally correlated over short periods, whilst systematic errors are some constant offset or trend in measured values over long periods of time. Systematic errors are a well-known problem in long term measurement series and can arise from, for example, incorrect calibration of an instrument or changes in a measurement site (e.g. Bohm et al. ¨ , 2001).

Despite our understanding of the likely sources of uncertainty in mass balance modelling, most papers in the literature limit their exploration of uncertainty to sensitivity studies which explore modelled responses to variation in individual parameters. While such approaches provide useful information and may be adequate for models calibrated and run at the same point in space, they are insufficient to understand the uncertainties in modelled mass balance extrapolated in time and space. For example, Van der Veen (2002) argued that sensitivity studies were inadequate in modelling mass balance for polar ice sheets because they do not provide a probability for a certain result, but rather only the range of possible results for variation in a given input parameter. Furthermore, sensitivity studies do not allow a full exploration of the parameter space and resulting non-linear effects as a result of combined uncertainties in multiple parameters. Parametric uncertainty analysis, in contrast to sensitivity tests, aims to evaluate the multi-dimensional response of a model to combined uncertainty in input parameters with a probability density function as an output (Tatang et al., 1997; Van der Veen, 2002).

In this paper we set out to estimate the uncertainty in mass balance calculations made with a glacier surface energy balance model of intermediate complexity (Klok and Oerlemans, 2002) applied to the Morteratsch valley glacier in the Swiss Alps. The paper has three key aims:

- to estimate uncertainty for individual input parameters used in Klok and Oerlemans's 2002 model;
- to calculate a probability density function (PDF) for mass balance as a function of the uncertainties in input parameters at a point on the Morteratsch glacier; and
- to assess the modelled uncertainties for studies extrapolating glacier mass balance forward in time and space.

In the first part of the paper we introduce the data and test area for the model, before describing the mass balance model. The basic model is compared to measurements over 11 years to validate that it can reproduce measured values. We then explore the size and form of random and systematic errors in the model's input parameters, before running Monte Carlo simulations to derive the overall uncertainty in modelled mass balance. In order to explore the likely implications of our uncertainty study for climate change projections, we then calculate PDFs for two key input parameters which are also outputs of most typical climate models: air temperature and global radiation.

TEST SITE AND TIME FRAME

Morteratsch is a mid-sized valley glacier in the south-eastern Swiss Alps, extending from approximately 4000 m a.s.l. down to 2020 m a.s.l. and covering an area of about 16 km2 (Fig. 1). Mass balance measurements on Morteratsch are embedded in a relatively long term study of the glacier surface energy balance which was initiated by the IMAU, Utrecht, in 1995 (e.g. Oerlemans, 2000; Oerlemans and Klok, 2002). Since then an automatic weather station (AWS), a sonic ranger for continuous surface height measurements and three stakes have been operated on the tongue of the glacier at approximately 2100 m a.s.l. Mass balance measurements were initiated in 1999 at two other sites and in the following year at a fourth site. In this paper we make use of sonic ranger data and stake measurements from all four sites.

The present study focuses on the mass balance at the AWS over 400 days, starting from 18 October 1998 and ending on 20 November 1999. In the following, this time period is referred to as the "calculation period".

Data from four meteorological stations, operated by MeteoSwiss, are used in this study as input data for the model or for the assessment of uncertainties: Corvatsch (3315 m a.s.l., located on a summit, 8 km west of the point AWS on

Morteratsch Glacier), Hospizio Bernina (2307 m a.s.l., located at a pass, 7 km east), Samedan (1705 m a.s.l., located on a wide and flat valley floor, 12 km north) and Weissfluhjoch (2690 m a.s.l., located on a summit, 45 km north) (Fig. 1). Hospizio Bernina is a manual weather station and only daily means are available – we acquired daily mean 2 m air temperatures (T_a). The other stations are automatic and we obtained from all three stations hourly means of T_a, relative humidity (e_a) and air pressure (p). In addition, global radiation ($S_{\text{in}_{\text{meas}}}$) and precipitation (P) were acquired from Corvatsch.

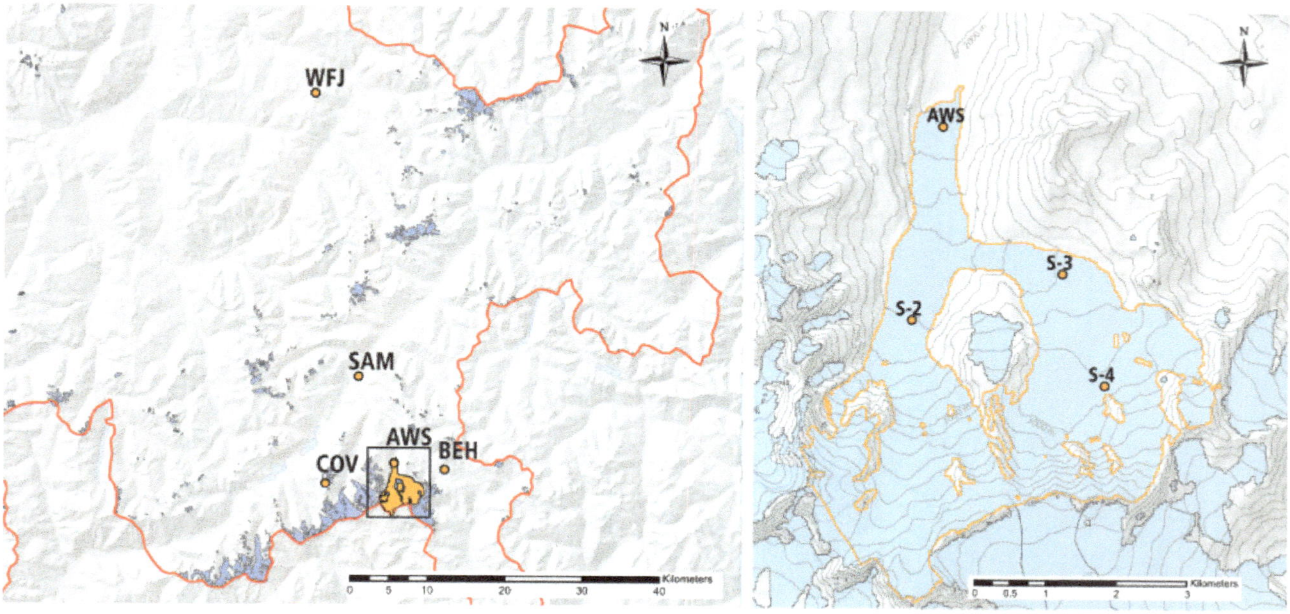

Figure 1: Left: Location of Morteratsch Glacier (orange) with the Automatic Weather Station (AWS) and the weather stations operated by MeteoSwiss that were used in this study: Corvatsch (COV), Hospizio Bernina (BEH), Samedan (SAM) and Weissfluhjoch (WFJ). The Border of Switzerland is marked in red and glaciers within Switzerland are depicted in blue. The area of the detail map of Morteratsch glacier is indicated with a rectangle. To the Right: Map of Morteratsch Glacier (orange outlines) and the test sites used in this study. Contour lines every 100 m. Glacier outlines are from Paul et al. (2004). The digital elevation model (DHM25 level2) is reproduced by permission of swisstopo (BA081413).

THE MASS BALANCE MODEL

Description of the model

In this study we investigate the numerical mass balance model developed by Klok and Oerlemans (2002). We selected this model because of the detailed and clear description of parameterizations, model output and validation procedure. Furthermore, the original model or parts of it have already been used in other studies e.g. (e.g., Klok and Oerlemans, 2004; Arnold et al., 2006). According to Klok and Oerlemans (2002), the model is based upon the following equation describing the specific mass balance, M(kg m^{-2}):

$$M = \int \left(\frac{Q_m}{L_m} + P + \frac{Q_l}{L_s} \right) dt \tag{1}$$

Q_m is the melt energy involved in melting and Q_l the energy involved in sublimation or riming. L_m (3.34×10^5 J kg^{-1}) is the latent heat of melting, and L_s (2.83×10^6 J kg^{-1}) is the latent heat of sublimation. P represents the accumulation due to snowfall. The surface energy heat flux (F) supplies energy for melting (Q_m) or for the glacier heat flux (G), which implies the warming or the cooling of the snow or ice pack.

$$F = Q_m + G = S_{\text{in}} - S_{\text{out}} + L_{\text{in}} - L_{\text{out}} + Q_h + Q_l \tag{2}$$

S_{in} and S_{out} are incoming and reflected solar radiation; L_{in} and L_{out} are incoming and emitted longwave radiation. Sensible and latent heat fluxes are represented by Q_h and Q_l . Melting can occur only when the surface temperature is at 0 °C and F is positive. If the latter is the case but the surface temperature is below zero, then F=G and the snow pack or ice is heated. The model is driven by data from synoptic weather stations located outside of the glacier boundary layer. Required input from meteorological measurements are: T_a, $S_{\text{in}_{\text{meas}}}$, P, e_a and p. Energy fluxes at the glacier surface are parameterized according to Klok and Oerlemans (2002). In their model parameterizations from Oerlemans and Grisogono (2002) are applied to calculate turbulent fluxes. Katabatic flow is parametrized therein and thus, measured wind speed is not required for input in the model. While writing our program code we closely followed the explanations given by Klok and Oerlemans (2002). Some of the original parameterisations have been modified and are described here.

We calculated potential clear sky global radiation according to Corripio (2003) and Iqbal (1983). The snow albedo (α_s) is calculated according to Klok and Oerlemans (2002) and modified formulas from ECMWF, using snowageing functions for melting (exponential) and non-melting conditions (linear).

Furthermore, in contrast to Klok and Oerlemans (2002), the new value of α_s after a snow fall event is not only a function of total snow depth and the underlying ice albedo (α_i), but also a function of α_s, the albedo of the old snow surface before the snow fall event.

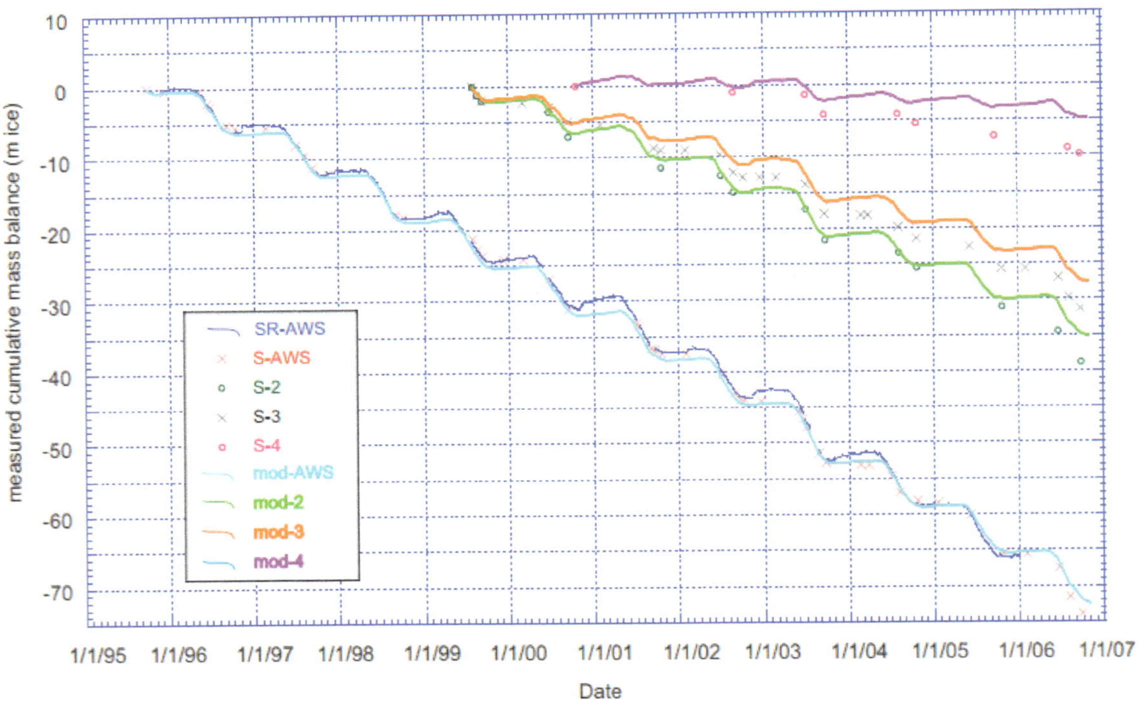

Figure 2: Comparison of modelled mass balance and measured mass balance at four points on Morteratsch-Glacier. "SR" stands for sonic ranger, "S" for stake measurements and "mod" for modelled values. The elevation of the four points are: AWS: 2100 m a.s.l., 2: 2500 m a.s.l., 3: 2700 m a.s.l. and 4: 2950 m a.s.l. All data is given in meters ice except for the snow accumulation measured by the sonic ranger which is depicted in meters snow height.

Klok and Oerlemans (2002) calculate precipitation (P) from measurements at Samedan and two manual weather stations (Pontresina and Bernina-Curtinatsch) in combination with a multiplication factor. Here we simply use measured P from Corvatsch station in combination with a tuning factor ($_{Pcorr)}$. While in the original model a single threshold air temperature (T_{snow}) of 1.5°C is used to distinguish snowfall and rain, we apply a gradual transition between 1°C and 2°C.

Testing of the mass balance model

The original model is reported as having delivered results in good agreement with measurements on Morteratsch for both a two and a five year model run (Klok and Oerlemans, 2002, 2004). In order to calibrate our modified model, we first conducted a model run for the 400 days calculation period (Sect. 2). We adjusted P_{corr} over this period to achieve good agreement between modelled and observed date of the disappearence of the winter snow cover at the AWS in spring 1999. Tuning was performed only at the AWS but the value we found for P_{corr} was applied uniformly at all points where mass balance was calculated. Except for the new calculation of P_{corr}, we did not apply any other new tuning to the model.

The modelled cumulative mass balance before tuning (P_{corr}=1) was −6.63 m w.e. and after tuning (P_{corr}=2.1) a value of −5.97 m w.e. resulted. According to Klok and Oerlemans (2002), measured melt at the AWS during the melt season of 1999 was −5.9 m w.e., which approximately equals to total observed mass balance for the full 400 days – according to the sonic ranger measurements, the 1998 melt period ended on 18 October and little accumulation (approx. 0.05 m w.e.) occurred between the end of the melt period in 1999 and 20 November 1999. Hence, the cumulative mass balance calculated for the tuning period agrees well with measurements. According to Klok and Oerlemans (2002) the snow cover at the AWS disapeared on 18 May 1999. After tuning of P_{corr} modelled melt out occurs on the same day.

Finally, to test the robustness of the calibrated model, an eleven year model run for four points on Morteratsch Glacier was conducted and the results compared to measurements (Figs. 2 and 3). For the point AWS, modelled cumulative mass balance is −72.5 m ice (m ice) which agrees very well with a measured value of −74 m ice (Fig. 2). The two curves are very similar, deviations during winter time are only apparent since the model results are shown in m ice and the sonic ranger measures surface elevation which corresponds to snow height in winter. The three other points with measurements available were also included in the comparison. The agreement for S-2 and S-3 is also good, however, the model systematically overestimates mass balance at S-4. The comparison of measured and modelled mass balance in a scatter plot (Fig. 3) shows that for individual stake readings deviations occur. There is one clear outlier where measured melt at S-2 is strongly underestimated. The disagreement is likely due to an uncertain stake reading.

UNCERTAINTY ASSESSMENT

The uncertainty model

Different approaches exist to determine the PDF of the output of a model. Analytical solutions are often desirable, however, they become complex or impossible if the set of uncertain parameters is large and nonlinear effects are present. Consequently, methods such as probabilistic collocation (e.g., Tatang et al., 1997) or Monte Carlo simulations are commonly applied to approximate uncertainty. Although computationally expensive, Monte Carlo simulations are popular because they are relatively simple to apply even when working with complex models and large numbers of uncertain parameters. For instance, Van der Veen (2002) used Monte Carlo simulations to assess uncertainty in the mass balance of the Greenland ice sheet.

In a Monte Carlo simulation a certain calculation (in our case a model run) is repeated many times and uncertain input parameters are varied within their uncertainty ranges. Model outputs are stored and a histogram is constructed to obtain the PDF for the desired output variable.

In the context of this paper the calculation being repeated is the modelling of the cumulative mass balance at a point AWS on Morteratsch Glacier over the time span between 18 October 1998 and 20 November 1999, the period over which we tuned the parameter P_{corr}.

For every uncertain parameter we estimated random and systematic uncertainty based on the literature and measured values of these parameters. An explanation of the uncertainties of individual parameters is given in Sect. 4.2 and the values chosen are listed in Table 1.

These uncertainties were then multiplied by normally distributed random numbers with a mean at 0 and a standard deviation of 1, resulting in systematic (ε_s) and random uncertainties (ε_r). The random numbers for the two classes of uncertainties as well as for the different parameters are independent of each other, implying the simplifying assumption that input parameter uncertainty is not correlated. Calculated ε_s and ε_r are then added to the measured values of the corresponding input parameters. The ε_s are calculated at the beginning of every run and remain constant throughout the run. The ε_r are treated as either fully random or temporally auto correlated. In the first case, every second numerical time step (every hour) a new ε_r is calculated. In the case of temporal autocorrelation ε_r at time step t is correlated to ε_r at time steps t+1, t+2, . . . , t+(t_{decor}−1) and only at time step t+t_{decor} is the uncertainty decorrelated from the original uncertainty at t. Where temporal autocorrelation was applied, a typical mean time span t_{decor} and a related standard deviation were defined. Based on the latter two values and another array of normal distributed random numbers, we calculated a series of consecutive time spans, of variable length with a mean length of t_{decor}. At the start of every time span a value for ε_r is calculated and finally linear interpolation was applied between two successive ε_r. For some parameters the addition of ε_s and ε_r can result in impossible values (e.g. relative humidity of more than 100%). Physically defined limits were set where necessary (Table 1) and whenever such a limit was violated, the related parameter was set to its limiting value.

Parameters and related uncertainties

Twelve input parameters were assigned uncertainties: All five directly measured values (T_a, Sin_{meas}, P, e_a and p) as well as the parameters used in spatial interpolation, the lapse rate (Γ_{T_a}) and precipitation gradient (Γ_P). Furthermore, from the parameters selected by Klok and Oerlemans (2002) for sensitivity testing, we included ice albedo (α_i), T_{snow}, the thickness of the surface layer (z_1), cloudiness (n) and the turbulent exchange coefficient (C_b). No uncertainties were assigned to the temperature of the lowermost layer in the three layer subsurface model since sensitivity testing of this parameter has previously shown it to have a neglible influence on mass balance (Klok and Oerlemans, 2002).

Table 1: The selected uncertainties (ε_s, ε_r) and thresholds (min, max) and their respective units.

parameter	symbol	ε_s	ε_r	unit	min	max	unit
measured air temperature	T_a	0.3	0.3	°C			
measured global radiation	Sin_{meas}	7.5	7.5	%	0		$W m^{-2}$
measured precipitation	P	25	25	%	0		m
relative humidity	e_a	5	5	%	0	100	%
measured air pressure	p	100	100	pa			
temperature lapse rate	Γ_{T_a}	0.0002	0.001	$°C m^{-1}$			
precipitation gradient	Γ_P	0.0001	0.0004	$m m^{-1} a^{-1}$			
fixed albedo value for ice	α_i	0.05	–				
threshold temperature of snowfall	T_{snow}	0.3	0.5	°C			
background turbulent exchange coefficient	C_b	0.0005	–				
cloudiness	n	0.03	0.2		0	1	
thickness of the surface layer	z_1	0.055	–	m			

We aimed to give independent estimates of random and systematic uncertainties for each parameter. However, where both types of uncertainties are of similar magnitude or available data and literature did not allow for individual estimates of uncertainty, we made the assumption that ε_s=ε_r .

The present uncertainty analysis aims to illustrate an approach to estimating uncertainty in mass balance modeling. However, not all potential sources of uncertainty could be considered. In particular, no uncertainty was assigned to the parameterization of the snow albedo and potential clear sky global radiation. Furthermore, uncertainties in precipitation are treated in a simplified manner.

Measured air temperature

Strasser et al. (2004) use a measurement error of 0.3°C for unventilated thermometers of an AWS in their study. The ventilated thermometers from MeteoSwiss are believed to measure T_a very accurately. Nevertheless, we used 0.3°C here for both stations because the microclimate at the respective stations (e.g. the Corvatsch Station is located on the roof of a house) may not be fully representative for nearby locations at the same altitude. Additionally, a systematic error of the same magnitude was introduced. The uncertainty assigned to T_a refers only to the uncertainty of the measurement at the synoptic weather stations. Total uncertainty in T_a at the point AWS is larger because uncertainty in the lapse rate (cf. Sect. 4.2.6) is multiplied by the difference in altitude and combined with the uncertainty of the measurement to obtain total uncertainty in air temperature.

Measured global radiation

A detailed study on measurement errors in $S_{in_{meas}}$ published by Meteo Swiss concluded that after data corrections the remaining uncertainties are of the order of 5 to 10% (Moesch and Zelenka, 2004). We therefore assigned both systematic and random uncertainties of 7.5%.

Measured precipitation

Measuring precipitation, in particular snowfall, is related to large uncertainties reaching 50% or more (Sevruk, 1985, 1989). In the present model, P is tuned by means of P_{corr}. Hence, we cannot directly apply values on uncertainties in precipitation measurements from the literature. Tuning measured P to observed accumulation on the glacier introduces systematic errors since the observations on the glacier are also related to considerable uncertainties: only accumulation can be observed which is not identical to precipitation because the snow cover is subject to snow drifting, melt, riming and sublimation. Furthermore, spatial variability of snow accumulation is large (e.g., Machguth et al., 2006) and there are difficulties in determining the spatial and temporal variability of snow density. We therefore assigned systematic and random uncertainties of 25%. The latter is treated as fully random because precipitation intensity is spatially and temporally highly variable.

Measured relative humidity

It was difficult to find information on uncertainty in e_a that goes beyond technical specifications from typical measurement devices. A systematic and a random uncertainty of both 5% was assigned to e_a. The chosen values are a rough estimate to take into account uncertainties in the measurement itself, in the assumption that the values are representative and that they can be interpolated linearly inbetween the two stations. Consequently, the uncertainties at the two stations are assumed to be not correlated.

Measured air pressure

Air pressure has a very small influence on the model outcome and thus uncertainty in this parameter is not discussed in detail here. A systematic and a random uncertainty of 100 pa was assigned to p, uncertainties at the two stations are considered to be uncorrelated.

Lapse rate

To assess the variability of Γ_{T_a} we calculated mean lapse rates from measured T_a at the four selected meteorological stations in the vicinity of Morteratsch. The mean hourly Γ_{T_a} over the time span from 1995–2006 (approx. 100 000 records for each station) is 0.0049°C m^{-1} for CorvatschSamedan (13 km apart, 1607 m difference in altitude) and 0.0056°C m^{-1} for Corvatsch-Weissfluhjoch (46 km apart, 625 m difference in altitude). The corresponding standard deviations are 0.0037°C m^{-1} and 0.0022°C m^{-1}, respectively. For Corvatsch-Hospizio Bernina (15 km apart, 1007 m difference in altitude), the mean daily lapse rate for the same time period amounts to 0.0062°C m^{-1} with a standard deviation of 0.0022°C m^{-1} . Taking into consideration the implication that a greater lapse rate will result in correspondingly higher values of T_a at the point AWS, it is at first glance surprising that running the model for the 400 days with data from Corvatsch and Weissfluhjoch results in a less negative mass balance (−5.19 m w.e.) than for Corvatsch and Samedan (−5.97 m w.e., see Sect. 2). Replacing Samedan with Weissfluhjoch also results in modifications of e_a and p. However, on closer inspection it becomes apparent that the main reason for reduced melt is the difference in summer lapse rates: from 1 May to 30 September, mean hourly Γ_{T_a} (1995–2006) amounts to 0.0061°C m^{-1} for CorvatschSamedan, 0.0058°C m^{-1} for Corvatsch-Weissfluhjoch and mean daily Γ_{T_a} is 0.0074°C m^{-1} for Corvatsch-Hospizio Bernina.

Although closer to the glacier than Weissfluhjoch, it is questionable as to whether Samedan better represents meteorological conditions at Morteratsch. Samedan is located on a wide valley floor with large diurnal and annual temperature fluctuations. Furthermore, linear regression of hourly T_a yields R^2=0.61 for Corvatsch-Samedan and R^2=0.96 for Corvatsch-Weissfluhjoch. On the other hand, Γ_{T_a} calculated from the latter pair of stations seems rather low, most probably because Weissfluhjoch is situated further to the north, in an area more open to colder air currents from the north and north-west, whereas the area around Morteratsch is more in-fluenced from the south and south-west. These comparisons show that the calculation of Γ_{T_a} is very sensitive to the selection of the synoptic weather stations and that there is considerable uncertainty in its value. However, the small number of available station pairs makes it difficult to determine the magnitude and type of the uncertainty. As a rough estimate we assumed a normally distributed systematic uncertainty of 0.0002°C m^{-1} .

Furthermore a temporally autocorrelated random uncertainty of 0.001°C m^{-1} is assigned to Γ_{T_a}. The assumption of temporal autocorrelation is essential here because otherwise strong hourly fluctuations and jumps in Γ_{T_a} would result. To determine the typical time span of temporal autocorrelation, we analyzed both twelve year time series of hourly Γ_{T_a} for their respective lag autocorrelation (Since Hospizio Bernina are not available on a hourly basis, this analysis could not be carried out for all three station pairs). Semivariograms of the temporal correlation of both time series are depicted for lags of between 1 and 24 000 h in Fig. 4. In the case of Corvatsch-Samedan clear daily and seasonal variations are present whereas the time series Corvatsch-Weissfluhjoch decorrelates rapidly with little daily or seasonal variation. From these figures we conclude, that Γ_{T_a} decorrelate within roughly 24 h. Once again, Fig. 4 shows the dilemma of the two pairs of stations: The valley station shows strong diurnal and seasonal fluctuations which are not present when comparing two summit stations. Although these strong fluctuations are large compared to the conditions at the point AWS (e.g., Oerlemans, 2001), in particular during the melt season, the mean values of Γ_{T_a} calculated with data from the valley station may still be more representative for the glacier.

Vertical precipitation gradient

According to Schwarb (2000), Γ_P is to a certain extent a virtual parameter because vertical and horizontal components of the precipitation distribution can never be fully distinguished. Klok and Oerlemans (2002) obtained Γ_P =0.0004 m m^{-1} a^{-1} from the same author who applied a comprehensive set of rain-gauge data to a complex interpolation scheme in order to derive spatially distributed Γ_P and P at approximately 2 km resolution. Based on the assumption that the methodology of Schwarb (2000) provides a reliable mean value for Γ_P in the Morteratsch area while temporal variability around that mean is large, we assigned a moderate systematic uncertainty (0.0001 m m^{-1} a^{-1}) and a large random uncertainty (0.0004 m m^{-1} a^{-1}). Random uncertainties are not temporally correlated and the occurrence of negative Γ_P is allowed.

Ice albedo

In the present model α_i was fixed to 0.34 in order to have a good representation for the snow free part of Morteratsch Glacier (Klok and Oerlemans, 2002). It is generally stated that the ice albedo is subject to significant small scale variability over short distances (e.g. Knap et al., 1999), thus a single mean value will result in either under- or overestimations for different parts of the glaciers ice surface (according to Klok and Oerlemans (2002), measured α_i at the AWS in summer 1999 was significantly lower than 0.34). In order to approximate the errors that result from assigning a fixed mean albedo to a glacier surface with an albedo distribution varying in both space and time (Klok and Oerlemans, 2004; Paul et al., 2005), we assigned a normal distributed systematic uncertainty of 0.05. No random uncertainty was assigned here because α_i is not subject to significant random changes at an hourly time scale and the determination of a typical time span for a temporally autocorrelated random error seems rather difficult and uncertain.

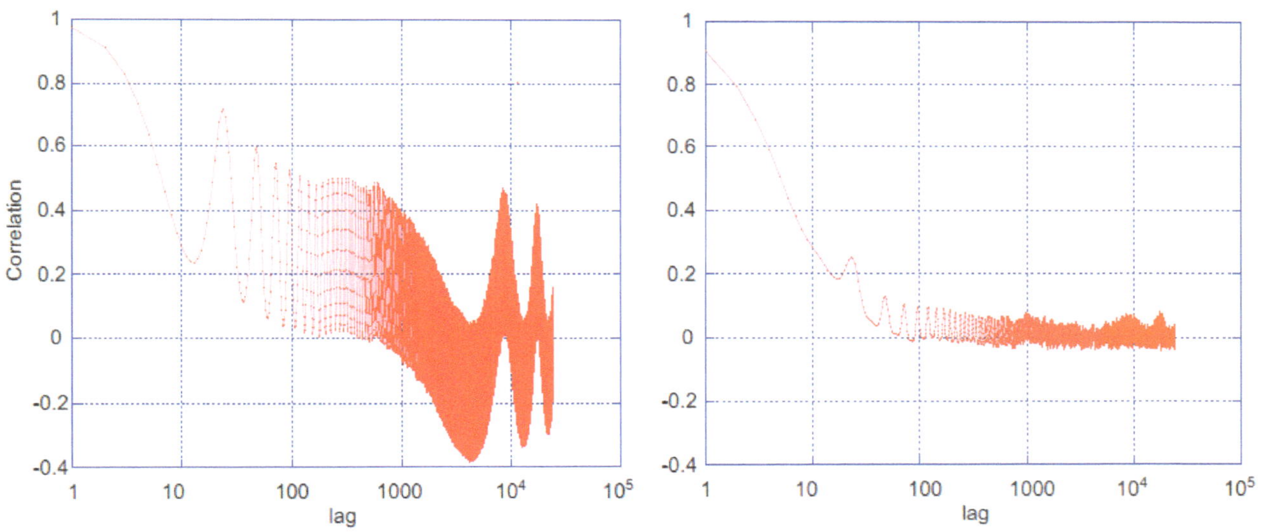

Figure 4: Lag autocorrelation for two time series (1995–2006) of hourly lapse rates. To the left the lag autocorrelation for the lapse rates obtained from air temperature measured at Corvatsch and Samedan is depicted and to the right for Corvatsch-Weissfluhjoch.

Threshold temperature snowfall

Long term observations of air temperature and snow – rain transitions, compiled by Rohrer (1989) for a meteorological station in Davos (1590 m a.s.l., 45 km north of Morteratsch) show that the average of the transition range from rain to mixed precipitation to snow is somewhere between 0.75°C and 1.5°C with a standard deviation of roughly 0.3–0.5°C. Furthermore, Rohrer (1989) shows that for the example of Davos, a change in both instrumentation and the measurement site resulted in significant change of the mean and the spread of the transition range. We applied a systematic error of 0.3°C and a random error of 0.5°C .

Thickness of the surface layer

The present mass balance model contains a three layer subsurface model to compute heat fluxes to and from the glacier. Since melt can only occur when the surface layer has reached the melting point, the chosen thickness of the surface layer (z_1) influences mass balance by controlling the time available for melt. Klok and Oerlemans (2002) varied z_1 by 0.11 m, here we assigned a normally distributed systematic uncertainty of 0.055, but since this parameter is initialised as a constant over an individual model run, it is not assigned a random uncertainty

Cloudiness

In the present model cloudiness (n) plays an important role. It is derived from the ratio:

$$T_{cl} = \frac{S_{\text{in}_{\text{meas}}}}{S_{\text{in}_{cs}}} \tag{3}$$

where $S_{\text{in}_{cs}}$ is potential global radiation under clear sky conditions. T_{cl} is used as a reduction factor to compute S_{in} and to derive n according to the following relationship, given by Greuell et al. (1997):

$$T_{cl} = 1.0 - 0.233n - 0.415n^2 \tag{4}$$

Consequently, errors in both $S_{\text{in}_{\text{meas}}}$ and $S_{\text{in}_{cs}}$ affect cloudiness. If, for example, $S_{\text{in}_{\text{meas}}}$ is above the real value, this will result in an overestimation of T_{cl}. However, n is derived from T_{cl} and will be lowered. Finally, since L_{in} depends on cloudiness, it will also be lowered. Cloudiness during night time is interpolated from n before sunset and after sunrise. Consequently additional S_{in} is present only during daytime while Lin is affected 24 h a day. Both effects, enhanced Sin and lowered Lin are of the same order of magnitude. An error in $S_{\text{in}_{\text{meas}}}$ will therefore shift the ratio of short- to longwave radiation balance but not have a large influence on their summed value.

In order to reduce this back-coupling effect and to account for uncertainties in the parameterisation of n, we first calculated T_{cl}, applied it as a reduction factor, computed n and only afterwards modified n by adding an uncertainty. According to Greuell et al. (1997) Eq. (3) performed very well in explaining the relationship between observed cloudiness (n_{obs}: eight classes, from 0 to 1 in steps of 1/8) and the mean observed T_{cl} per class of n_{obs}. On the other hand, the mean T_{cl} per class were computed from a larger set of individual values of T_{cl} which showed a large variance. Thus we assigned normally distributed uncertainties, consisting of small systematic (0.03) and a larger random uncertainties (0.2). Since the observed variance of T_{cl} for the individual classes of n might be due to differing effects of the various types of clouds, we introduced a temporal autocorrelation to n with a mean time span of 12 h because cloud types typically persist for more than one hour.

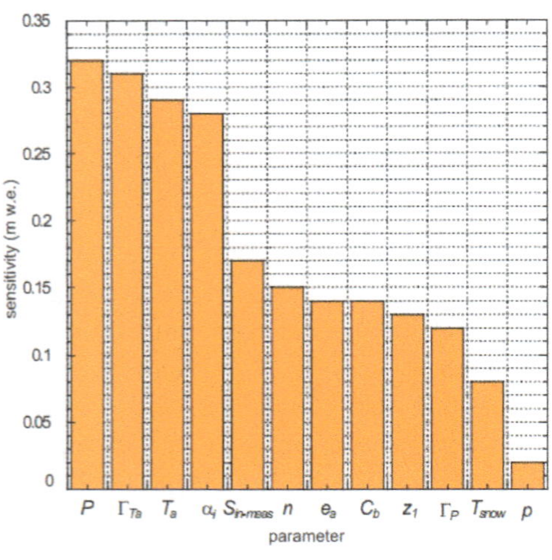

Figure 5: Sensitivities to combined random and systematic uncertainties (according to Table 1) for the individual parameters. All values are given in m w.e.

Background turbulent exchange coefficient

The value for C_b was found by matching measured and modelled melt at AWS for the year 1999 (Klok and Oerlemans, 2002). Since the measurements are not error free (e.g., Braithwaite et al., 1998), the tuning will result in a systematic over- or underestimation of C_b. Consequently, we have assigned a systematic uncertainty to this parameter and we selected the same value (0.0005) as used for the parameter sensitivity testing by Klok and Oerlemans (2002). As with surface layer thickness, this parameter is constant over an individual model run, and is thus not assigned a random uncertainty.

RESULTS

Sensitivity tests

Individual sensitivity studies were performed for every parameter listed in Table 1 to assess the combined influence of random and systematic uncertainties (Fig. 5). For reasons of simplicity, sensitivity studies were also conducted by means of Monte Carlo simulations. Modelled mass balance is most sensitive to the prescribed uncertainties in P and Γ_{T_a}, followed by T_a and α_i. Uncertainties in $S_{\text{in}_{\text{meas}}}$ have a much smaller influence while the impact of the remaining parameters' uncertainties are less than half the most sensitive parameter.

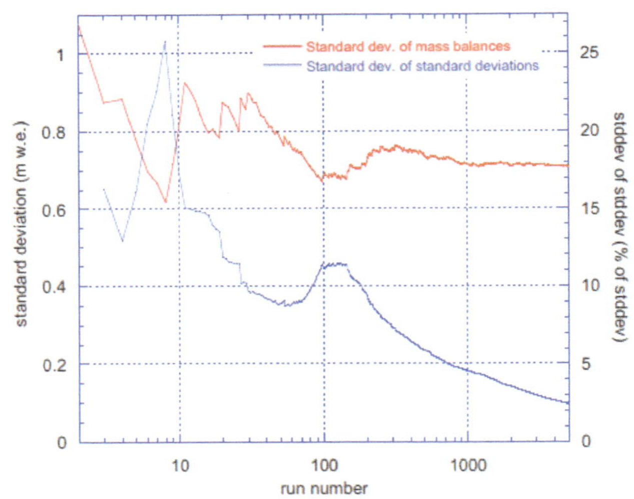

Figure 6: Evolution of the standard deviation with increasing number of runs and related standard deviation of the standard deviations. Note the logarithmic scale on the X-axis, starting with run number two.

Parametric uncertainty analysis

In order to assess the required number of runs for the Monte Carlo simulations, we plotted the evolution of the standard deviation, and standard deviation of standard deviation, of modelled mass balances over 5000 runs in a Monte Carlo simulation where all systematic and random uncertainties according to Table 1 were applied. Both parameters are depicted in Fig. 6, indicating that fluctuations in standard deviations become small after roughly 1000 runs, suggesting that the chosen number of runs (5000) is likely to deliver stable results.

Figure 7 shows the PDF of the model outcome, resulting from a Monte Carlo simulation applying the full set of systematic and random uncertainties. The mean modelled mass balance is −6.02 m w.e., maximum and minimum values are −3.72 m w.e. and −8.69 m w.e., respectively. The standard deviation is 0.71 m w.e. or roughly 10% of total cumulative melt. The PDF shows one distinctive peak at −5.7 m. However, this peak could not be reproduced in further experiments with different initiation values for the random numbers. From the output of the same Monte Carlo simulation, we depicted the temporal evolution of mean mass balance over the calculation period, growth of the related standard deviation and most and least negative of all 5000 runs in Fig. 8. In winter uncertainty grows with snow fall events, while during dry periods there is virtually no growth in uncertainty. With the onset of melt, uncertainty starts to grow continuously. Note that both the most and least negative of all runs cannot be considered as obviously unrealistic: They show an accumulation and an ablation period. Winter accumulation is a few centimetres water equivalent in the most negative and almost one meter of water equivalent in the least negative. These values roughly mark the bounds of natural variability as observed on Morteratsch (cf. Fig. 2). Furthermore, the dates of the disapearence of the snow cover were stored throughout the simulation and their PDF is shown in Fig. 9. The mean date of melt out is day number 138 (according to Klok and Oerlemans (2002) the melt out happened one day earlier). The probability for snow cover disappearance is clearly not normally distributed but shows distinct peaks and troughs. Finally, the Monte Carlo simulations were repeated to estimate the contribution of random and systematic errors separately. Figure 10 shows PDFs accounting for all systematic or all random uncertainties. Corresponding standard deviations are 0.69 m w.e. and 0.14 m w.e., respectively. Obviously, overall uncertainty is dominated by the systematic uncertainty.

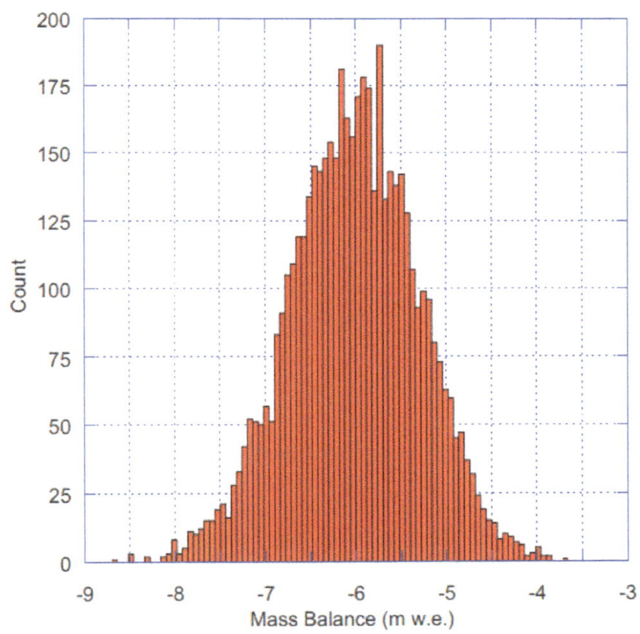

Figure 7: Probability density function of the modelled cumulative mass balance (18 October 1998–20 November 1999) at point AWS (2100 m a.s.l.) on Morteratsch Glacier.

Table 2: Change in overall uncertainty (m w.e.) for different uncertainties in T_a and $S_{in_{meas}}$. Note that the means of the resulting PDFs are shifting with growing uncertainty, from −5.97 m w.e. at 0 °C/0% to −6.47 m w.e. at 2°C/20%.

			$S_{in_{meas}}$		
unc T_a unc	0%	5%	10%	15%	20%
0°C	0.56	0.57	0.61	0.68	0.79
0.5°C	0.72	0.73	0.76	0.82	0.92
1°C	1.09	1.10	1.12	1.16	1.23
1.5°C	1.51	1.52	1.54	1.57	1.62
2°C	1.96	1.97	1.99	2.01	2.05

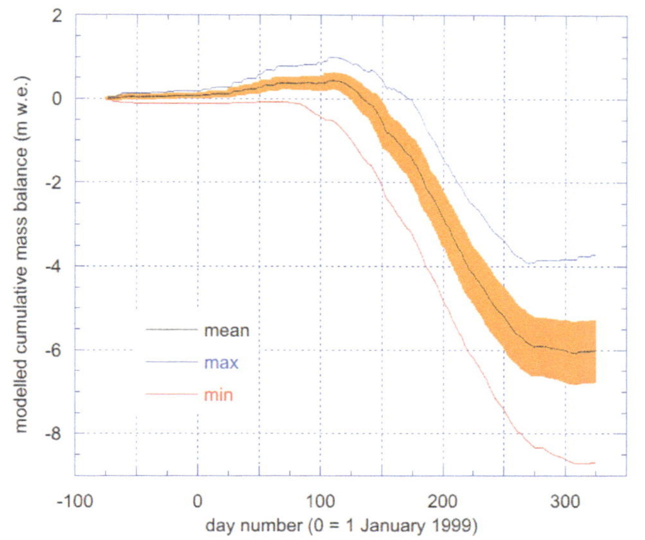

Figure 8: Temporal evolution of the modelled cumulative mass balance (18 October 1998–20 November 1999) at point AWS (2100 m a.s.l.) on Morteratsch Glacier. The mean value is the mean of all 5000 runs in the full Monte Carlo simulation. Calculated uncertainty (±1 standard deviation) is depicted with the orange shading. The most and least negative of all runs are also shown.

However, uncertainty will clearly vary according to the data set used for model input. In order to make this study applicable to a broader audience working with different data (e.g. data from climate models), a further experiment was conducted, evaluating model response to different levels of uncertainty in $S_{in_{meas}}$ and T_a. Here, we varied uncertainty in T_a from 0°C to 2°C in steps of 0.5°C and uncertainty in $S_{in_{meas}}$ from 0% to 20% in steps of 5%. The uncertainties in all other parameters were varied in the Monte Carlo simulation according to the values given in Table 1, except for uncertainties in Γ_{T_a} which were set to zero in order to have full control over the uncertainty in T_a. The number of runs was reduced to 1000 for every combination of uncertainties in the two parameters and resulting model uncertainties are listed in Table 2. Of course, it would be an interesting experiment to vary P similar to $S_{in_{meas}}$ and T_a since uncertainties in precipitation are particularly large. However, to our opinion a detailed assessment of the various sources of uncertainty in P is a prerequisite and should be addressed in future studies.

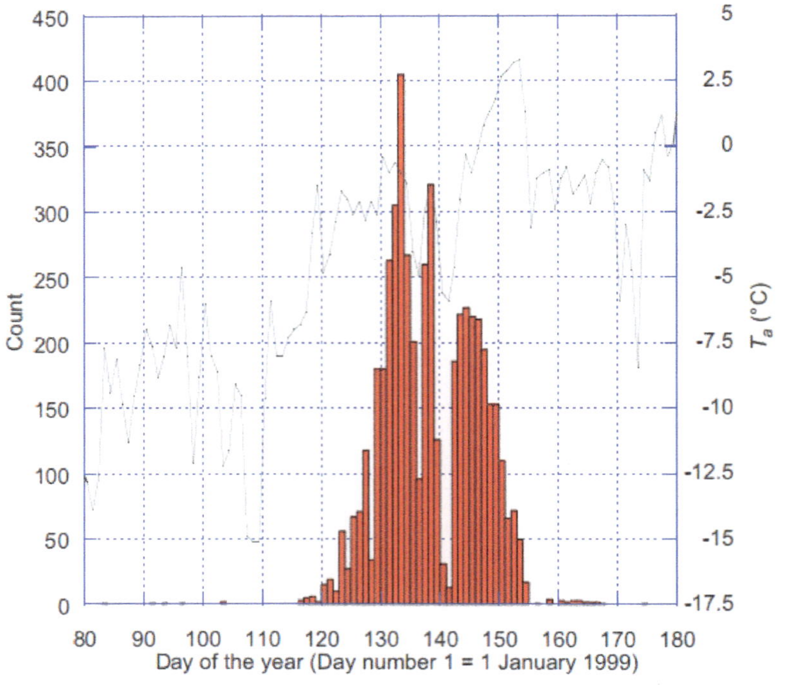

Figure 9: Probability density function of the modelled date of melt out at the point AWS (2100 m a.s.l.) on Morteratsch Glacier. Day no. 1 corresponds to 1 January 1999. Daily means of air temperature measured at Corvatsch are depicted with a grey line.

DISCUSSION

Estimating parameter uncertainties

The first step in any sensitivity test, or a parametric uncertainty analysis, is the selection of parameters for sensitivity testing and estimation of their associated ranges. Most previous research on energy and mass-balance modelling has focussed only on sensitivity testing, with parameter ranges based on a variety of sources. In this work, we have recognised the importance of not only random uncertainty, which can be considered akin to instrument precision and accuracy, but also systematic uncertainties which have generally been ignored. An important limitation is the difficulty in estimating values for both random and systematic uncertainties, and where appropriate, temporal autocorrelation of random uncertainties. However, we believe that an approach based on the selection of parameters through literature is an appropriate one, and that all of these sources of uncertainty must be modelled.

Sensitivity tests

Sensitivity testing was conducted to estimate the contribution of individual parameters to overall model uncertainty. The model appears to be most sensitive to the prescribed uncertainties in P and Γ_{T_a}. However, since air temperature at the AWS is calculated from T_a at Corvatsch, Γ_{T_a} and the difference in altitude from Corvatsch and AWS, the latter has an amplifying effect on the impact of uncertainties in Γ_{T_a}

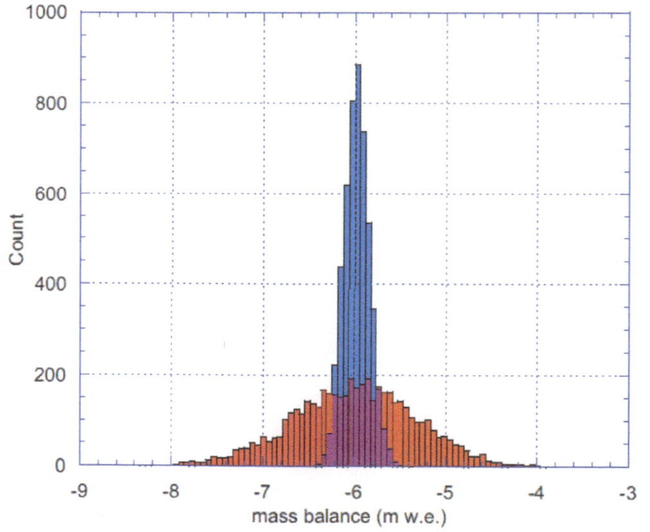

Figure 10: Probability density function of modelled cumulative mass balance (18 October 1998–20 November 1999) at point AWS (2100 m a.s.l.) on Morteratsch Glacier, calculated separately with only systematic (red bars) and only random uncertainties (blue bars).

Nevertheless, the discussion in Sect. 4.2.6 indicates that the lapse rate can vary significantly depending on which station pair is chosen. Air temperature is generally considered a well-known parameter, however the sensitivity to the prescribed uncertainties in T_a, which might be considered conservative, is among the largest. Consequently, the common assumption of neglible uncertainties in air temperature must be regarded as questionable because air temperature has a large influence on the glacier mass balance (Ohmura, 2001). Sensitivity to uncertainties in $S_{in_{meas}}$ are quite small although realistic levels of uncertainty have been applied. However, the impact of errors in $S_{in_{meas}}$ is partly compensated due to the coupling of shortwave and longwave radiation balance through the parameterisation of cloudiness.

Furthermore, it should be noted that the impact of the chosen uncertainties in the two investigated internal model parameters (C_b and z_l) is smaller than the influence of most of the uncertainties related to the meteorological data or ice albedo.

Parametric uncertainty analysis

Several striking results are apparent from the parametric uncertainty analysis. Firstly, the standard deviation in cumulative melt of 0.71 m w.e. seems at first glance to be contradicted by the very good agreement of modelled and observed melt shown in Fig. 2. However, this good result was obtained by a model tuned to a particular location, over 400 days in the same location as to which it is applied. The uncertainty shown in Fig. 7 indicates the impact that typical uncertainties in measurement could have on such a point measurement – and which are in this case accounted for, at least to some degree, by the tuning of the model to the point. Furthermore, Fig. 2 also shows the corresponding drift of modelled melt from measured for sites at which the model was not tuned – these differences are much more similar to the uncertainty associated with our PDF, and also indicated by the increase in uncertainty over time as shown in Fig. 8.

Secondly, whilst the PDF of cumulative mass balance is normally distributed, the PDF of the day of melt out has a distinct bimodal appearance (Fig. 9). This is because uncertainty has a much smaller influence on the average melt out day, since meteorological events which occur over relatively short periods have a strong influence on the timing of melt out – for example, an influx of cold air around day 140 reduces the probability of melt out to almost zero, irrespective of uncertainty. To illustrate this in more detail, daily mean T_a at Corvatsch is shown in Fig. 9. Until day 110 air temperatures are low, and only in 6 runs has the snow cover already disappeared by then. Temperatures then rise sharply and reach a first peak at day 119 which also marks the onset of a rise in probability for melt out. However, since the snow cover is still quite thick in most runs and its temperature most likely below 0°C, the rise in probability for snow cover disappearance lags temperature. T_a stays high and the longer the period of warm temperatures last, the higher becomes the probability for melt out. A further rise in T_a results in a peak at day 133 where the snow cover disapears in more than 8% of all runs. The number of runs with melt out and T_a then show a distinct correlation with two sharp drops in probability of melt out, both caused by an influx of cold air. Around day 145 the probability for melt out starts to diminish regardless of the still increasing T_a. By this date temperatures have been favourable for snow melt for almost one month (since approx. day 119), the probability that snow cover persists is already low and thus T_a and the day of melt out decorrelate. This result points to the importance of choosing appropriate parameters for model validation – the day of melt out is much less sensitive to uncertainty than cumulative mass balance.

Thirdly, we observe that systematic uncertainties contribute much more to overall uncertainty than random uncertainties which tend to cancel one another out. Although this result is perhaps rather obvious, it indicates the importance of considering techniques for estimating and characterising systematic uncertainties, which are generally ignored despite their well-known importance in, for example, the homogenisation of long term temperature series (Bohm et al¨ , 2001). There is no reason to assume that measurements or projections made today are not subject to systematic uncertainties – for example, consistent under or over estimation of albedo through a melt season (Paul et al., 2005) – and these should be accounted for in uncertainty analysis.

We calculated uncertainty to be approximately 10% of total melt at one particular point on one glacier. However, for two main reasons, the results of this study should be transferred to other glaciers with caution. Firstly, as already discussed the uncertainties in the input parameters may differ strongly depending on the data sets used. Second, the uncertainty model contains simplifications which may not apply to other sites. The study focuses on a test site where the total melt is about one order of magnitude larger than accumulation. Consequently only a rough estimate of uncertainty in the latter was applied and thus, the uncertainty model cannot yet be applied to an entire glacier surface. Although it would be of particular interest to assess the uncertainty of the mass balance calculation for an entire glacier instead of a single point, such an uncertainty assessment would require a more profound analysis and description of the uncertainties in accumulation modelling, and also consideration of appropriate methods for spatially autocorrelated uncertainty analysis.

Impact of variation in individual parameter uncertainty

Many projections in the future are based on relatively simple models, where one or two parameters are varied to explore the response of a system (for example, increases in air temperature). However, these projections are in themselves subject to the uncertainty found in all input parameters. Thus, we have carried out Monte Carlo simulations on two sets of scenarios based on particularly important parameters – air temperature and measured global radiation. The results demonstrate that the growth in overall uncertainty is not a linear function of uncertainty in T_a or $S_{in_{meas}}$, but rather an exponential one. Uncertainty of more than 1 m w.e. is reached for 1°C of uncertainty in T_a which is still a rather conservative estimate, in particular where a mass balance model is applied to unmeasured glaciers or driven by data from climate models. Uncertainty in measured global radiation is of clearly lower impact for reasons discussed in Sect. 4.2.11. This approach allows the assessment of not only scenarios of future change, but the sensitivity of these scenarios to uncertainty to be estimated.

Implications

According to Anderson and Woessner (1992) an uncertainty analysis should be built into modelling strategies from the onset. Models are often validated by comparing their output to measurements. However, since both observations and model results may be uncertain, meaningful model validation requires not only the mean outputs but also their PDFs (Tatang et al., 1997). A parametric uncertainty analysis can contribute to process understanding by helping to identify modelled values, for example in Fig. 3, that show larger deviations from measurement than their level of uncertainty (obviously, uncertainties of the measurements have to be considered as well). For these values it is likely that the model failed due to conceptualisation problems and not due to errors in input parameters. Identifying such problems by explaining why the model has failed at particular points will lead to an improved mass balance model and potentially add to the understanding of process.

Monte Carlo simulations are also a valuable tool in assessing whether output of a mass balance model is more sensitive to the choice of tuning parameters or to uncertainties stemming from model input and parameterizations.

Finally, particularly when we wish to extrapolate in space or time, a parametric uncertainty analysis becomes essential. Since, uncertainties in data values are likely to increase as we extrapolate further from our observations in space and time, it is important to realise that uncertainties are also unlikely to be constant in space or time and modellers should take care not to over interpret results which are simply the mean of PDFs with potentially large standard deviations.

CONCLUSIONS

In this work we have calculated the uncertainty in a glacier mass balance model, estimated at a point on the Morteratsch Glacier in Switzerland over a period of 400 days, using uncertainty values for individual meteorological and model parameters. Despite good agreement of the tuned model with observed mass balance over a period of 11 years, we estimate uncertainty in cumulative mass balance of approximately 10%. The uncertainty is dominated by systematic uncertainties in parameters, which in most studies are not considered. Thus, we believe it is important in future work to consider methods of estimating and quantifying systematic uncertainties in typical parameters.

Our estimate of overall uncertainty is most likely a conservative one: Not all potential sources of uncertainty could be considered and precipitation was tuned without assigning uncertainty to the respective tuning parameter. However, to consider all the different aspects of uncertainty in precipitation or accumulation is not trivial. More detailed studies on this subject are required and would add to a more comprehensive understanding of uncertainties in mass balance modelling.

However, the main implication of this work is related to the extrapolation of model results in time and space. This paper shows, that for a well-tuned model with relatively low values for individual uncertainties, one can expect considerable uncertainty in modelled outputs. This in turn implies that, for a glacier where appropriate data are unavailable for tuning, and where the measured input data are less certain, one could expect increased uncertainties in cumulative mass balance. In future work we will explore how we can integrate parametric uncertainty analysis into models which extrapolate glacier mass balance in both space and time.

ACKNOWLEDGEMENTS

We would like to thank Javier Corripio for making his code for global radiation calculation available to us. MeteoSwiss provided the meteorological data and Swisstopo the digital elevation model. The careful and constructive reviews of two anonymous reviewers as well as the interactive comment by Ruzica Dadic helped to improve this paper. H. Machguth is funded by the grant 20-117761 of the Swiss National Science Foundation.

REFERENCES

1. Anderson, M. and Woessner, W.: The role of the postaudit in model validation, Adv. Water Resour., 15, 167–173, 1992.
2. Arnold, N., Willis, I., Sharp, M., Richards, K., and Lawson, W.: A distributed surface energy balance model for a small valley glacier: I. Development and testing for Haut Glacier d'Arolla, Valais, Switzerland, J. Glaciol., 42, 77–89, 1996.
3. Arnold, N., Rees, W., Hodson, A., and Kohler, J.: Topographic controls on the surface energy balance of a high arctic valley glacier, J. Geophys. Res., 111, F02011, doi:10.1029/2005JF000426, 2006.
4. Bohm, R., Auer, I., Brunetti, M., Maugeri, M., Nanni, T., and ¨ Schoner, W.: Regional temperature variability in the European ¨ Alps: 1760–1998 from homogenized instrumental time series, Int. J. Climatol., 21, 1779–1801, doi:10.1002/joc.689, 2001.
5. Braithwaite, R.: On glacier energy balance, ablation, and air temperature, J. Glaciol., 27, 381–391, 1981.
6. Braithwaite, R., Konzelmann, T., Marty, C., and Olesen, O.: Errors in daily ablation measurements in northern Greenland, 1993– 1994, and their implications for climate studies, J. Glaciol., 44, 583–588, 1998.
7. Braithwaite, R. J.: Positive degree-day factors for ablation on the Greenland ice sheet studied by energy-balance modelling, J. Glaciol., 41, 153–160, 1995.
8. Brock, B., Willis, I., Sharp, M., and Arnold, N.: Modelling seasonal and spatial variations in the surface energy balance of Haut Glacier d'Arolla, Switzerland, Ann. Glaciol., 31, 53–62, 2000.
9. Corripio, J.: Vectorial algebra algorithms for calculating terrain parameters from DEMs and solar radiation modelling in mountainous terrain, Int. J. Geogr. Inf. Sci., 17, 1–23, doi:10.1080/713811744, 2003.
10. Greuell, W., Knap, W., and Smeets, P.: Elevational changes in meteorological variables along a mid-latitude glacier during summer, J. Geophys. Res., 102, 25 941–25 954, 1997.
11. Iqbal, M.: An Introduction to Solar Radiation, Academic Press, Toronto, 1983.
12. Klok, E. J. and Oerlemans, J.: Model study of the spatial distribution of the energy and mass balance of Morteratschgletscher, Switzerland, J. Glaciol., 48, 505–518, 2002.
13. Klok, E. J. and Oerlemans, J.: Modelled climate sensitivity of the mass balance of Morteratschgletscher and its dependence on albedo parameterization, Int. J. Climatol., 24, 231–245, 2004.
14. Knap, W. H., Brock, B. W., Oerlemans, J., , and

Willis, I. C.: Comparison of Landsat TM-derived and ground-based albedos of Haut Glacier d'Arolla, Switzerland, Int. J. Remote Sens., 20, 3293–3310, doi:10.1080/014311699211345, 1999.

15. Machguth, H., Eisen, O., Paul, F., and Hoelzle, M.: Strong spatial variability of snow accumulation observed with helicopter-borne GPR on two adjacent Alpine glaciers, Geophys. Res. Lett., 33, L13503, doi:10.1029/2006GL026576, 2006.
16. Moesch, M. and Zelenka, A.: Globalstrahlungsmessung 1981–2000 im ANETZ, Arbeitsbericht 207, MeteoSchweiz, 2004.
17. Oerlemans, J.: Analysis of a 3 year meteorological record from the ablation zone of Morteratschgletscher, Switzerland: energy and mass balance, J. Glaciol., 46, 571–579, 2000.
18. Oerlemans, J.: Glaciers and Climate Change, A.A. Balkema Publishers, Lisse, 2001.
19. Oerlemans, J. and Grisogono, B.: Glacier winds and parameterisation of the related surface heat fluxes, Tellus, 54A, 440–452, 2002.
20. Oerlemans, J. and Klok, E.: Energy balance of a glacier surface: analysis of automatic weather station data from Morteratschgletscher, Switzerland, Arct., Antarct. Alp. Res., 34, 477– 485, 2002.
21. Ohmura, A.: Physical basis for the temperature-based melt-index method, J. Appl. Meteorol., 40, 753–761, 2001.
22. Paul, F., Ka¨ab, A., Maisch, M., Kellenberger, T. W., and ¨ Haeberli, W.: Rapid disintegration of Alpine glaciers observed with satellite data, Geophys. Res. Letters, 31, L21402, doi:10.1029/2004GL020816, 2004.
23. Paul, F., Machguth, H., and Ka¨ab, A.: On the impact of glacier ¨ albedo under conditions of extreme glacier melt: the summer of 2003 in the Alps, in: EARSeL eProceedings, 4, 139–149, 2005.
24. Reeh, N.: Parametrizations of melt rate and surface temperature on the Greenland ice sheet, Polarforschung, 59, 113–128, 1989.
25. Rohrer, M.: Determination of the transition air temperature from snow to rain and intensity of precipitation., in: WMO/IAHS/ETH International Workshop on Precipitation Measurements St. Moritz, 1989 Switzerland, ETH Zurich, Switzerland, 475–482, 1989.
26. Schwarb, M.: The alpine precipitation climate: evaluation of a high-resolution analysis scheme using comprehensive rain-gauge data, ETH Dissertation No. 13911, ETH Zurich, Institute for Climate Research, http://e-collection.ethbib.ethz.ch/show?type= diss&nr=13911, 2000.
27. Sevruk, B.: Correction of precipitation measurements: Summary report., in: Proceedings Workshop on the Correction of Precipitation Measurements 1985 Zurich, 13–23, 1985.
28. Sevruk, B.: Reliability of precipitation measurements, in: WMO IAHS ETH, International Workshop on Precipitation Measurement, St. Moritz, 1989 Switzerland, 13–19, 1989.
29. Strasser, U., Corripio, J., Pellicoti, F., Burlando, P., Brock, B., and Funk, M.: Spatial and temporal variability of meteorological variables at Haut Glacier d'Arolla (Switzerland) during the ablation season 2001: measurements and simulations, J. Geophys. Res., 109, D03103, doi:10.1029/2003JD003973, 2004.
30. Tatang, M., Pan, W., Prinn, R., and McRae, G.: An efficient method for parametric uncertainty analysis of numerical geophysical models, J. Geophys. Res., 102(D18), 21 925–21 932, 1997.
31. van der Veen, C. J.: Polar ice sheets and global sea level: how well can we predict the future?, Global Planet. Change, 32, 165–194, 2002.

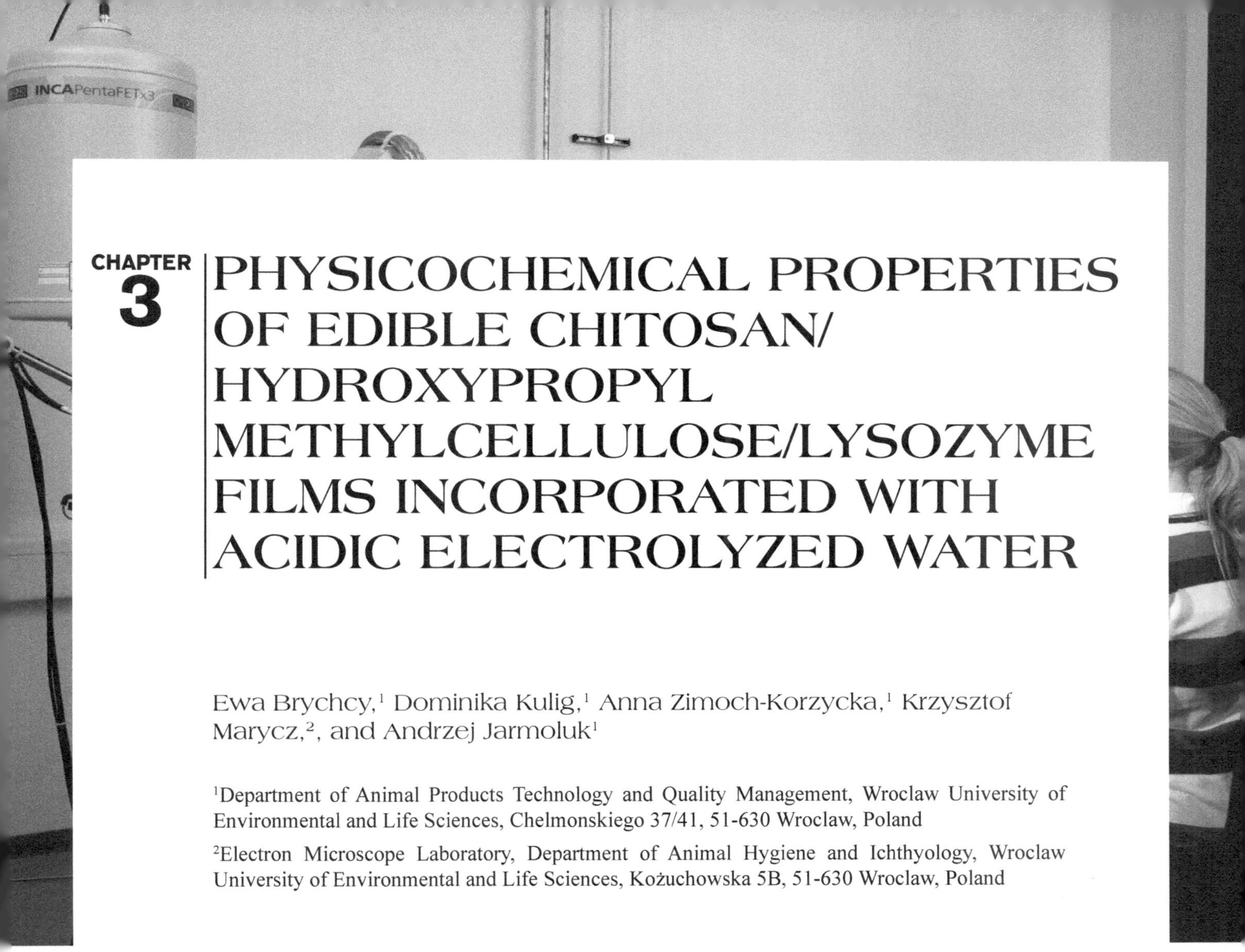

CHAPTER 3 PHYSICOCHEMICAL PROPERTIES OF EDIBLE CHITOSAN/ HYDROXYPROPYL METHYLCELLULOSE/LYSOZYME FILMS INCORPORATED WITH ACIDIC ELECTROLYZED WATER

Ewa Brychcy,[1] Dominika Kulig,[1] Anna Zimoch-Korzycka,[1] Krzysztof Marycz,[2], and Andrzej Jarmoluk[1]

[1]Department of Animal Products Technology and Quality Management, Wroclaw University of Environmental and Life Sciences, Chelmonskiego 37/41, 51-630 Wroclaw, Poland

[2]Electron Microscope Laboratory, Department of Animal Hygiene and Ichthyology, Wroclaw University of Environmental and Life Sciences, Kożuchowska 5B, 51-630 Wroclaw, Poland

ABSTRACT

The treatment with acidic electrolyzed water (AEW) is a promising disinfection method due to its effectiveness in reducing microbial population. The aim of the study was to evaluate physicochemical properties of chitosan/HPMC films incorporated with lysozyme and acidic electrolyzed water. In the composite films, decreasing film solubility and increasing concentration of sodium chloride solution and prolongation of electrolysis time were observed. Electrolysis process with sodium chloride induces spongy network of film structure. The use of AEW has not changed chemical composition of films which was proved by ^{1}H NMR, MALDI-TOF, and FT-IR spectroscopy. The research confirmed that electrolysis significantly improved thermomechanical properties of the examined films. The contact angle values of the films were quite similar and ranged between 56° and 73°. The increase of salt concentration used in the electrolysis process had an impact on increasing flexibility of samples. Application of electrolyzed water in commonly used food processing systems is possible. Fusion of AEW and biopolymers may provide better integration with coated food product and multidirectional protecting effect.

INTRODUCTION

Packaging is used to maintain appropriate sensory characteristics and nutritional value of food products. It also has protective function against the adverse effect of environment during transportation and storage; additionally, it has an impact on appearance of the final product. In recent years, the application of edible films and coatings based on biopolymers like proteins, polysaccharides, and lipids has raised attention of many researchers [1, 2] mainly because of the fact that biodegradability of protective coatings containing natural polymers reduces the application of synthetic polymers difficult for utilisation [3]. Requirements for the packaging from natural materials depend on the characteristics of food products and their changes during production and storage [4]. Many authors focus on modification of hydrocolloids designed to produce packaging materials with the most preferable properties and multifunctional applications [5, 6].

Citation: Ewa Brychcy, Dominika Kulig, Anna Zimoch-Korzycka, Krzysztof Marycz, and Andrzej Jarmoluk, "Physicochemical Properties of Edible Chitosan/Hydroxypropyl Methylcellulose/Lysozyme Films Incorporated with Acidic Electrolyzed Water," International Journal of Polymer Science, vol. 2015, Article ID 604759, 10 pages, 2015. doi:10.1155/2015/604759

Chitosan and cellulose derivatives as hydroxypropyl methylcellulose (HPMC) are promising materials for creating edible coatings or films. HPMC is a water soluble cellulose ether hydrocolloid with good film forming characteristics. It is used as a film former, tablet binder, and coating, stabilizing, suspending, and viscosity increasing agent [7, 8]. Cellulose derivative films are poor water vapour barriers, which are the result of the inherent hydrophilic nature of polysaccharides and cellulose derivative films' poor mechanical properties. Improving the moisture barrier would be by incorporation of hydrophobic compounds, such as fatty acids, into the cellulose ether matrix to develop a composite film [9].

Chitosan is a derivative of chitin, a polysaccharide with linear chain of linked 2-acetamido-2-deoxy-D-glucopyranose units. This material is biodegradable, biocompatible, and nontoxic. Due to its film forming properties and multiple uses in applications of coatings production, drug delivery, nutrients, and controlled release of food ingredients and separation techniques, it is a valuable polymer [10]. It has a high modulus along with low elongation at break owing to the high glass transition temperature (Tg) and crystallinity. Therefore, mixing or copolymerizing chitosan with different polymers can influence its morphology and plasticity [11].

Edible films are not good barriers against water vapour; however, they could be used as a carrier of active compounds, antimicrobial agents, or preservatives, which protect food quality [12]. Such active ingredients of edible films can be lysozyme or acidic electrolyzed water.

Lysozyme is muramidase (N-acetyl-muramyl-hydrolase) which decomposes β-1,4-glycosidic linkage between N-acetylmuramic acid and N-acetylglucosamine in the cell wall of polysaccharides in bacteria. It possesses lytic properties and can be applied as an antiseptic substance [13].

Acidic electrolyzed water is an antimicrobial agent generated by membrane electrolysis of sodium chloride solution. It is characterized by low pH, high oxidation-reduction potential, and free chlorine which is a major germicidal component [14, 15]. The free chlorine in electrolyzed acidic water is associated with a combination of hypochlorous acid and chlorine gas [15]. AEW has the advantages of nonirritating response of mucous membranes and skin tissue [16]. Acidic electrolyzed water can be also used as a component in biomolecular system, such as hydrogels. Polymer hydrogel composites have been synthesised and characterised for their application in electrochemically controlled drug release devices. Polymer hydrogels give the possibility of obtaining various advanced functional polymers. Electrosensitive hydrogels are commonly made of polyelectrolytes and an insoluble, swellable, polymer network with ionic groups [11].

To the best of our knowledge, there is no published study regarding characterization of edible films incorporated with acidic electrolyzed water. Interactions between hydrocolloids and acidic electrolyzed water could affect physicochemical properties of edible films.

Therefore the aim of the present study was to evaluate the physicochemical properties of chitosan/HPMC films incorporated with lysozyme and acidic electrolyzed water. Further investigations will be correlated with antibacterial properties of examined films and their effectiveness in food products.

MATERIALS AND METHODS

Apparatus

Electrolyzed salt solutions (acidic electrolyzed water, AEW) were generated using a water generator (own design batch type generator, equipped with two titanium electrodes coated with 0.6 μm layer of platinum) by membrane electrolysis of diluted salt solutions (0.001, 0.01, and 0.1%) in various time spans (0, 5, and 10 minutes). The diluted aqueous solutions of sodium chloride (analytical quality, POCH) were added to both the anode and the cathode chambers (2 L in each chamber) of the ionizer.

Material

A low molecular weight chitosan (CH) obtained from shrimp shells was supplied from Sigma-Aldrich with 20–200 cP viscosity and DD = 75–85%; hydroxypropyl methylcellulose (HPMC) in the METHOCEL SX product was obtained from Dow Chemical Company; lysozyme isolated from hen egg whites with the 2000 U/mg activity was purchased from Ovopol Company (Poland), in addition to lactic acid (PURAC FCC 80), glycerol (analytical quality, POCH), and 80% sodium chloride (analytical quality, POCH).

Preparation of Polymer Films

The film stock solutions were prepared by dissolving known amount of film material (chitosan, HPMC) in (non)electrolyzed sodium chloride solutions in the amount shown in Table 1 and 0.5% of lactic acid continuously stirred, using mechanic stirrer CAT R-250 (stirring rate 400 rpm), for 16 hours. The lysozyme solution was prepared by dissolving known amount of lysozyme in (non)electrolyzed sodium chloride solutions. Glycerol (as a plasticizer) was added to the homogenous solutions; then all the solutions were mixed together to obtain the final concentration of components presented in Table 1. The final solutions were degassed by centrifugation and poured into a glass Teflon coated plate of 80 mm × 200 mm and dried for 72 hours. Dehydration was carried out in Binder KBF-LOC 240 chamber at 4°C and relative humidity of 60%. Dried films were taken off and cut out for further tests.

Table 1: Experimental design

Variants	Variability factors		Constant factors				
	NaCl (%)	Electrolysis time (min)	Chitosan (%)	HPMC (%)	Lysozyme (%)	Glycerol (%)	Lactic acid (%)
N0.001E0	0.001	0	1	1	0.2	25 (of dry weight of used polymers)	0.5
N0.01E0	0.01						
N0.1E0	0.1						
N0.001E5	0.001	5					
N0.01E5	0.01						
N0.1E5	0.1						
N0.001E10	0.001	10					
N0.01E10	0.01						
N0.1E10	0.1						

Film Characterization

Film Solubility in Water

The analysis was performed according to the method described by Pinotti et al. [17]. The percentage of total soluble matter (% solubility) was calculated by the following formula:

$$\% \text{ solubility} = \frac{\text{Initial dry weight} - \text{Final dry weight}}{\text{Initial dry weight}} \times 100. \quad (1)$$

The samples were analysed at least in triplicate.

Scanning Electron Microscopy (SEM)

Cross section of the obtained films was performed using EVO LS 15 Zeiss scanning electron microscope. The samples were sputtered with gold for 150 s using Scancoat 6 type (Edwards, London, England) and examined using 20 kV voltage.

Fourier Transform Infrared Spectroscopy (FT-IR)

The ATR FT-IR method was used for each sample testing. The spectra were recorded at a resolution of 2 $cm^{-1}$by 64 scans between 450 and 4000 cm^{-1} in Infinity AR60 spectrometer (ATI Mattson).

^{1}H NMR Spectroscopy

The samples were then characterized by nuclear magnetic resonance (NMR) spectroscopy. The nuclear magnetic resonance measurements were performed on a Bruker Avance III. The solid-state CP MAS NMR experiments were performed using the technique of cross-polarization with magic-angle spinning. The spectra for ^{13}C and ^{1}H nuclei were obtained, respectively, at 100.61 MHz and 400.15 MHz frequency in a MAS BB DVT wide-band probe with 4 mm diameter of zirconium (ZrO_2) rotor. The isotopically labeled L-[1-13C]-tyrosine (Tyr) was applied in order to optimize the parameters for the CP MAS ^{13}C NMR and also to attain the first-order Hartmann-Hahn matching condition. Spectroscopic parameters are as follows: for ^{13}C CP MAS spectrum: measurement temperature: 298 K, rotation speed: 8 kHz, relaxation time: 3 s, pulse: 90° for ^{1}H 4 µs, contact time: 2 ms, spectral width SWH: 40 kHz, TD = 3.5 k, and SPINAL decoupling; for ^{1}H MAS spectrum: measurement temperature: 298 K, rotation speed: 8 kHz, relaxation time: 2 s, pulse: 90° for ^{1}H 4 µs, spectral width SWH: 40 kHz, and TD = 16 k.

MALDI-TOF Mass Spectrometry

The samples for the experiment were prepared by mixing 10 µL of examined solution with 10 µL of matrix solution (an aqueous solution of 2,5-dihydroxybenzoic acid (DHB) at a concentration of 10 mg/mL) in Eppendorf tube with a capacity of 0.5 mL. 2 µL of the sample was placed on a measuring plate and left to evaporate the solvent. The plate was then placed in an ion source of the mass spectrometer Voyager-Elite (PerSeptive Biosystems, Framingham, CT, USA). The ionization technique which was applied was matrix assisted laser desorption/ionization (MALDI). The wavelength of the laser radiation and accelerating voltage were at 337 nm and 20 kV, respectively. The positive ions were subjected to registration using ion time of flight (TOF) analyser with the reflection of ions (reflectron). The recorded mass range varied from 300 to 3500 (m/z). Mass spectrum is the sum of 200 spectra. The measurements were taken using an external mass calibration based on defined reference mixture spectrum of polyethylene glycols. Processing of spectra was carried out using Data Explorer v. 4 (Applied Biosystems, Foster City, CA, USA).

Dynamic Mechanical Thermal Analysis (DMTA)

The tests were performed using Rheometric Scientific DMTA Mk III. The loss modulus, storage modulus, and loss tangent ($\tan\delta$) were measured at the temperature range from −80°C to +50°C, with heating rate 2°C/min and frequency of 1 Hz.

Contact Angle Measurements

The measurements were taken with Contact Angle Analyzer (Surface Electro Optics Company) using the sessile drop method. The drop of distilled water (6 µL) was placed on the surface (2 cm^2) of the coating sample.

Statistical Analysis

The experiments were made in triplicate. The effects of two independent categorical variables such as time of electrolysis and concentration of sodium chloride were evaluated. The obtained data were analyzed using a 2-way factor analysis of variance (ANOVA) using Statistica 10 (StatSoft, Poland). Differences between means were established by Duncan test with 5% significance.

RESULTS AND DISCUSSION

Film Solubility in Water

It was found out that the use of the acidic electrolyzed water significantly influences solubility of edible protective films. The films containing 0.1% sodium chloride solution after 10 minutes of electrolysis (N0.1E10) present the lowest solubility of 6% (Figure 1). The highest solubility of 44% was observed in N0.01E0 film. The decreasing film solubility with increasing concentration of sodium chloride solution and prolongation of electrolysis time were observed in composite films (Figure 1). The solubility can be adjusted to enhance the possible applications by controlling the parameters of the solvent used in film formulation [17]. Water solubility is an indicator of the film's water affinity [18]. High solubility may be an advantage for some applications. In some cases, a water-insoluble film is preferred in order to provide water resistance and improve food integrity; in other cases, edible films with high water solubility may be required [1]. The required degree of solubility of the material may be changed depending on intended applications [19].

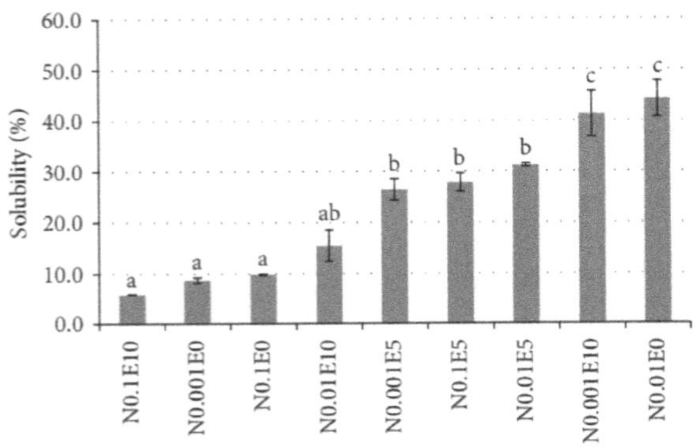

Figure 1: Film solubility in water. Different letters indicate significantly different groups determined by Duncan's test ($p<0.05$).

Scanning Electron Microscopy (SEM)

The scanning electron micrographs of films are shown in Figure 2. The SEM (a), (b), and (c) micrographs showed smooth, compact, and homogeneous surface, and no pores or cracks were detected. These results are consistent with those described by Yin et al. [20], Rotta et al. [21], and de Moura et al. [22], where micrographs of cross section confirmed the partial miscibility between chitosan and HPMC. The N0.001E0 and the N0.01E5 films revealed multilayer structures, which could be associated with limited chain mobility [17]. It is related to specificity of used polymers (HPMC, chitosan, and chitooligomers) and causes differences in formation of their chains and functional groups. Electrolysis process with NaCl induces spongy network of film structure, which is shown in Figure 2. Process of electrolysis with NaCl as electrolyte causes lack of creation of crystal structure of components in LCAEW. By subjecting the electrodes to direct current voltage, two kinds of electrolyzed water with different characteristics were produced. In anode chamber of ionizer are present only chloride compounds, such as Cl_2, HOCl, HCl, and chloride ions [15]. Sodium ions are transferred to the cathode as a result of membrane electrolysis and obtained water is not used in this study.

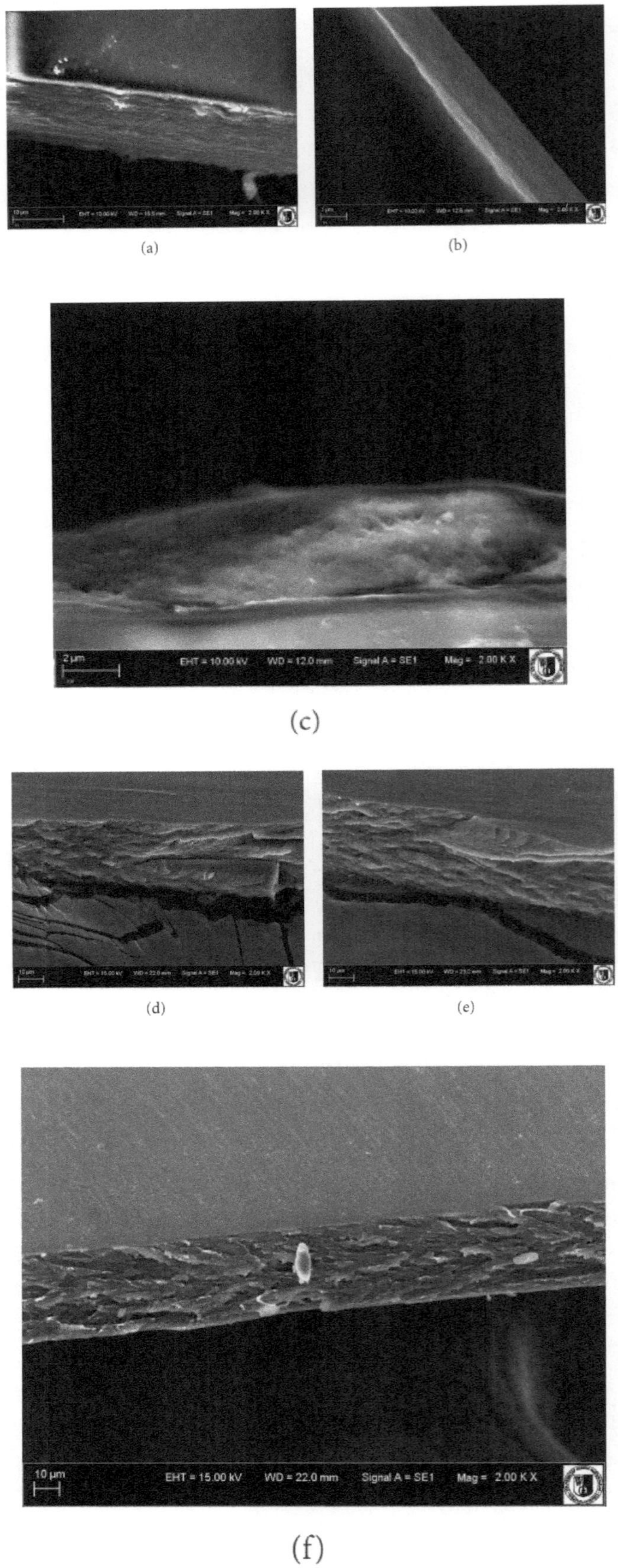

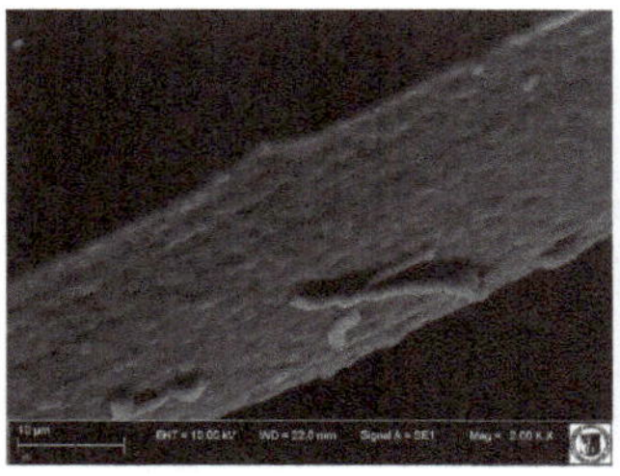

(g)

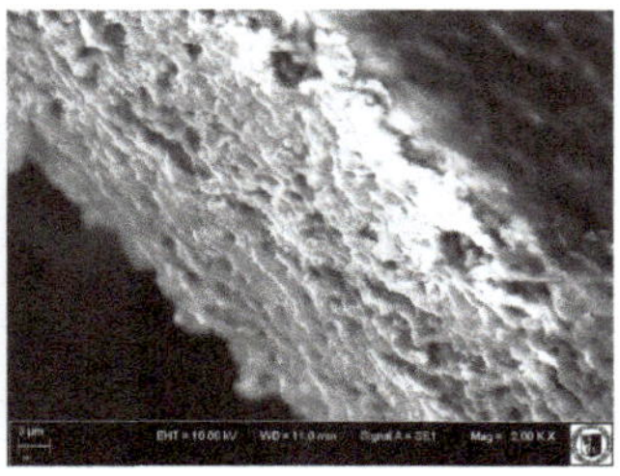

(h)

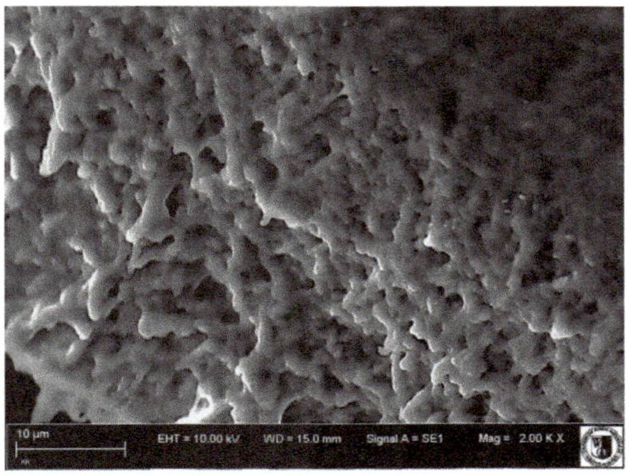

(i)

Figure 2: SEM micrographs of the cross section of (a) N0.001E0, (b) N0.001E5, (c) N0.001E10, (d) N0.01E0, (e) N0.01E5, (f) N0.01E10, (g) N0.1E0, (h) N0.1E5, and (i) N0.1E10 films.

Fourier Transform Infrared Spectroscopy (FT-IR)

The FT-IR spectra of examined films and their ingredients are shown in Figure 3. The spectrum of N0.2E30 sample shows the absorption peak at about 3232 cm^{-1}, which is assigned to stretching vibrations from overlapping O–H and N–H bonds. These findings are in accordance with the results obtained by Leceta et al. [23]. The vibrations indicate inter- and intramolecular hydrogen bonds interaction. The signals observed at 3426 cm^{-1}, 3377 cm^{-1}, 3303 cm^{-1}, and 2878 cm^{-1} were associated with O–H stretching vibration, NH_2asymmetric stretching, N–H stretching, and C–H stretching, respectively. Similar peak absorption was also reported in the work of Leceta et al. [23]. Characteristic absorption bands of NAG units were observed for C–O stretching of secondary amides at 1656 cm^{-1} and for N–H bending (primary amine) at 1597 cm^{-1}. The absorption bands near 1378 cm^{-1} are associated with the symmetrical skeletal deformation of CH_3 group and those at 1421 cm^{-1} with the O–H (primary alcohol). The absorption bands at 1032 cm^{-1} (stretching vibration of C–O–C ring) are characteristic for the glucopyranose ring. Praxedes et al. [24] indicated that the band peaks at 1409 cm^{-1} and 1316 cm^{-1} correspond to the vibration of –OH and –CH groups in the pyranose ring. Specific bands of (1,4)-glycosidic bridges were observed at 1153 and 897 cm^{-1} [25, 26]. Praxedes et al. [24] observed at the peak of antisymmetric stretching C–O–C glycosidic linkage at 1080 cm^{-1}. Many authors present different wave numbers at which there are adequate absorption bands [27]. The small band with low absorption at 1733 cm^{-1} refers to the acylated group –OH [28]. No significant differences in the positions and intensities of the bands of all the samples were observed. It means that polysaccharides were not degraded by lysozyme or acidity of electrolyzed water. Despite the fact that electrolyzed water has low pH, basically it is a weak acid [15].

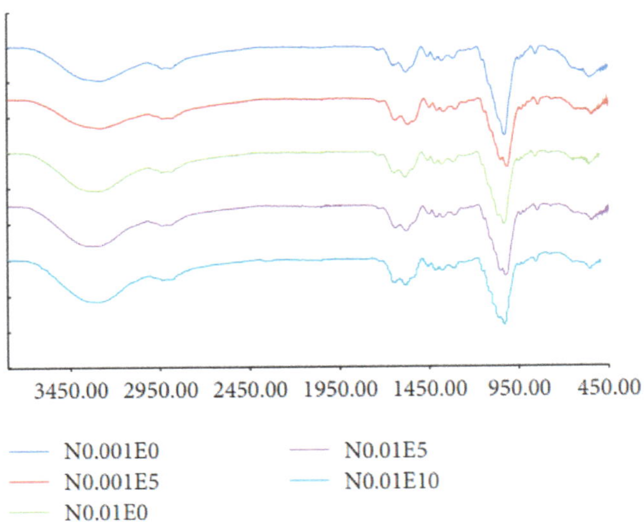

Figure 3: Spectra of Fourier Transform Infrared (FT-IR) for edible films. Reference films: N0.01E0 and N0.001E0; films incorporated with acidic electrolyzed water: N0.01E5 and N0.01E10.

^{1}H NMR and ^{13}C NMR Spectroscopy

No significant differences of intensities and widths of ^{13}C individual resonance signals for experimental films were observed (Figure 4). The stability of the fundamental structure of the films was proved by the lack of significant changes in ^{13}C CP MAS NMR spectra. The interactions between large amounts of protons as well as high value of their dipole coupling resulted in broadening of the proton signals in ^{1}H MAS NMR spectra. Some significant differences of ^{1}H MAS NMR spectra were noted between N0.001E0, N0.01E0, N0.001E5, N0.01E5, and N0.001E10 and N0.01E10, N0.1E0, N0.1E5, and N0.1E10 samples. The intense signal at about 4.4 ppm was observed in spectra of N0.001E0, N0.01E0, N0.001E5, N0.01E5, and N0.01E10 films. The peak originates predominantly from water protons. Three strong signals at 5, 3.5, and 1.5 ppm were noted. They originate from water protons or glycerol, CH group of chitosan and HPMC (or CH_2 of glycerol), and CH_3 of HPMC. Changes in intensity and width of these signals are probably correlated with different content and cross-linking structures of the films.

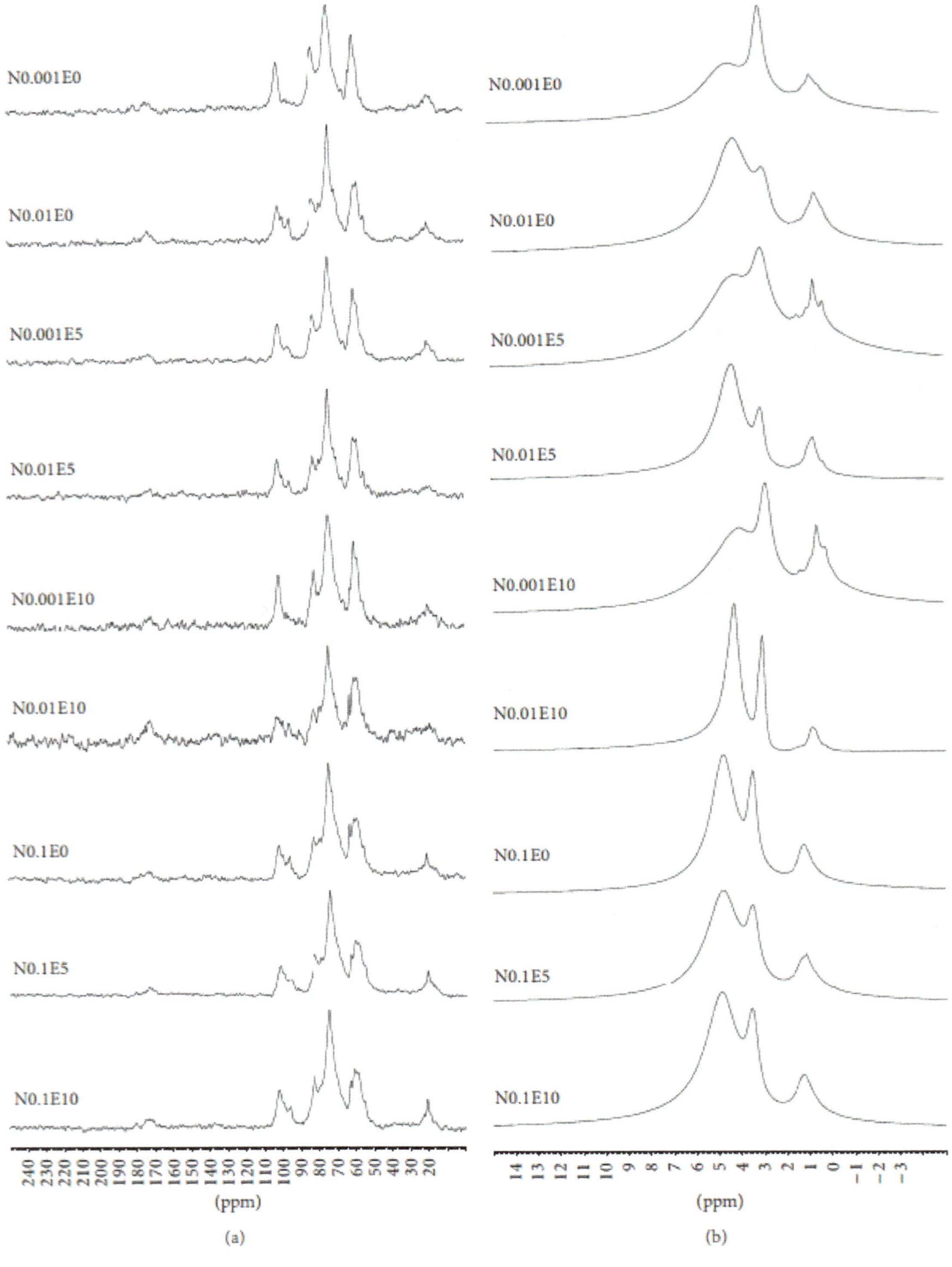

Figure 4: ^{13}C NMR (a) and ^{1}H NMR (b) spectra of examined edible films.

MALDI-TOF Mass Spectrometry

The analysis of the compounds consisting of sugar units conducted by MALDI enables the observation of the oligosaccharides in the molecular weight range from a few to several thousand daltons [29]. Although lysozyme was added to potentially increase antimicrobial activity, it also could cause polysaccharides hydrolysis, which was investigated by others [30]. Park et al. [31] and Zimoch-Korzycka and Jarmoluk [13] noted that lysozyme may cause polysaccharides hydrolysis and their products enhanced inhibition efficacy against many bacteria species. Therefore the research was carried out to identify potential oligomeric products of chitosan and HPMC hydrolysis. The intense peaks of matrix dominate in the mass range from 300 to 500 m/z; therefore the mass range over 500 m/z was taken into account (Figure 5). No peaks of other m/z values, which would create a series of constant mass differences between peaks, indicating the presence of oligomeric products with different composition, were reported. The optimum temperature required for the maximum activity of lysozyme proved to be 37°C [32]. Uncontrolled enzymatic degradation was inhibited by sample storage under cooling condition. Oligomeric products of chitosan hydrolysis were not found, which was confirmed by MALDI-TOF analysis. Chitosan degradation could lead to reduction of film homogeneity which was not desirable in this study.

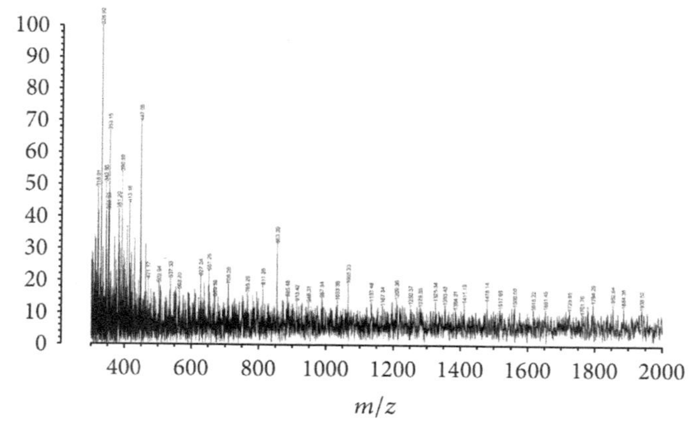

(a)

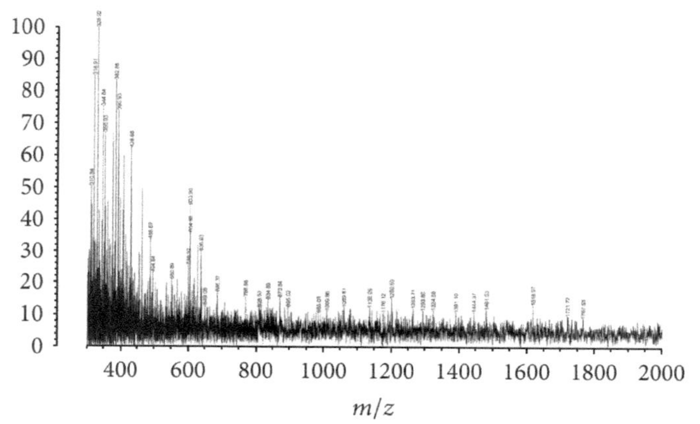

(b)

Figure 5: MALDI-TOF mass spectrometry analysis of edible films based on acidic electrolyzed water (N0.1E10) (a) and its reference (N0.1E0) (b).

Dynamic Mechanical Thermal Analysis (DMTA)

Figure 6 presents the dependence on dynamic mechanical behaviour for selected films (N0.01E0, N0.01E5, N0.01E10, N0.001E0, and N0.1E10) in order to show possible differences caused by AEW. A remarkable signal at 0°C for N0.001E0 and N0.01E0 was noted and probably is associated with the phase transition. The analysis of N0.001E0 and N0.01E5 films showed less intense and more significant peaks at 0°C and −30°C, respectively. Sample N0.01E10 (like N0.1E10 sample) comprises the second loss tangent peak at about 0°C. Rotta et al. [21] and Martínez-Camacho et al. [33] observed a Tg value of chitosan and hydroxypropyl methylcellulose at 203°C and 164°C, respectively, whereas Kittur et al. [34] and Neto et al. [35] have not found any evidence of Tg. The maximum in the $\tan\delta$ graph corresponds to transition temperature of the amorphous regions of the polymer (α relaxation) [36]. Tg is strongly dependent on both the film composition and the moisture content and can denote stability of a film. As water content of the amorphous material increases, Tg decreases. Double tangent peak was observed in N0.1E10 film, which suggests the presence of two phase transitions or weak homogeneity of the material. Two separate peaks on the graph of the loss tangent were probably associated with higher NaCl content. Aggregation of fillers would result in the heterogeneous collections of polymer matrix and filler. It was found that with increasing NaCl concentration in examined films α relaxation moves toward higher temperatures and its maximum is lower. The use of AEW caused a shift of the main relaxation to a higher temperature which probably indicates restricted molecular movement [37]. Close to 0° loss tangent values indicate high flexibility of the sample. Process of electrolysis significantly improved thermomechanical properties of polymeric coatings. It was observed that 5-minute electrolysis improved N0.01E5 elasticity by 27.28% and elongation of electrolysis time promotes reduction of $\tan\delta$ by 54.55% compared to the untreated sample N0.01E0 (0.55°). Electrolysis process created spongy network of films, which was confirmed by SEM analysis. When the porosity of material increases, the elasticity increases also [38].

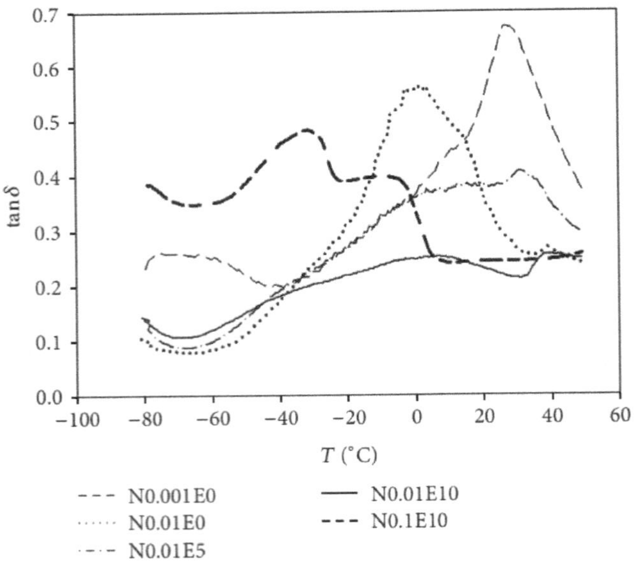

Figure 6: DMTA spectra of $\tan\delta$ for N0.001E0, N0.01E0, N0.01E5, N0.01E10, and N0.1E10 films.

Contact Angle Measurements

The photographs of wetting films are shown in Figure 7. The study revealed that there was no significant deformation between films which contained AEW. The destruction of films is caused by dissolution of the sample by its contact with water. This process is accompanied by the formation of stress, which results in deformation of the sample. Among the tested samples N0.01E0 film was subject to insignificant deformation. There are two types of terminal groups in the chemical structure of examined films: hydrophobic methyl group ($-CH_3$) and hydrophilic hydroxyl group (CH_3-OH). During the first phase of the analysis methyl groups were exposed to the outside, which prevented water penetration into the interior of the surface film. The tested surface of the coating was wetted with water, which probably caused its dynamic reorganization. Hydrophilic hydroxyl groups were gradually appearing on the material surface. The water drop was extended, but the penetration was insignificant. Chemical stability of N0.1E0, N0.001E5, N0.1E5, and N0.1E10 films was observed during the time of measurement. It was noted that the contact angle values of the films were quite similar and were in the range 56°–73°. This evidenced insignificant differences in chemical structure surface of the test samples. Contact angles greater than 90° indicate surface hydrophobicity and below 90° correspond to hydrophilicity of the surface [39]. Contact angles below 90 degrees can indicate that examined films are characterized by high wettability. Park et al. [31] observed that the addition of lysozyme causes increased hydrophobic side chains in film structure and may be responsible for the decrease in hydrophilicity of lysozyme-chitosan films.

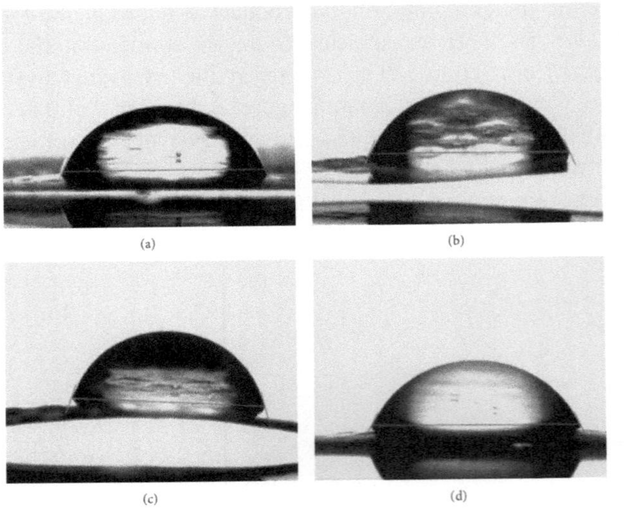

Figure 7: Images of contact angle for (a) N0.001E0, (b) N0.01E0, (c) N0.01E5, and (d) N0.01E10 films.

CONCLUSIONS

Control of AEW production parameters allows modifying properties of edible protective films, including their solubility and wettability, while electrolysis process improves elasticity of the polymeric films. The use of AEW has not caused undesirable changes in chemical composition of films which was proved by ^{1}H NMR, MALDI-TOF, and FT-IR analysis. The use of low salt concentration guarantees obtaining a homogeneous coating surface and desirable cohesion with all components. Application of acidic electrolyzed water generator is possible to be implemented into already existing food processing system. Fusion of AEW with biopolymers and lysozyme is easy to introduce and will provide better integration with coated food product.

ACKNOWLEDGMENTS

This research was supported by Wroclaw Centre of Biotechnology, programme The Leading National Research Centre (KNOW) for years 2014–2018. The authors are indebted to researchers from Centre of Molecular and Macromolecular Studies, Polish Academy of Sciences, in Lodz for their support.

REFERENCES

1. B. W. S. Souza, M. A. Cerqueira, J. A. Teixeira, and A. A. Vicente, "The use of electric fields for edible coatings and films development and production: a review," Food Engineering Reviews, vol. 2, no. 4, pp. 244–255, 2010.
2. V. Enujiugha, R. Ayodele, and K. Seidu, "Protein-lipid films: applications in food systems," International Journal of Agriculture and Food Security, vol. 4, no. 18, pp. 553–570, 2013.

3. K. Sztuka and I. Kołodziejska, "Jadalne folie oraz powłoki powierzchniowe z polimerów naturalnych stosowane do opakowań żywności," Polimery, vol. 9, pp. 625–706, 2008.

4. S. Guilbert, N. Gontard, and L. G. M. Gorris, "Prolongation of the shelf-life of perishable food products using biodegradable films and coatings," LWT—Food Science and Technology, vol. 29, no. 1-2, pp. 10–17, 1996.

5. M. Ye, H. Neetoo, and H. Chen, "Control of Listeria monocytogenes on ham steaks by antimicrobials incorporated into chitosan-coated plastic films," Food Microbiology, vol. 25, no. 2, pp. 260–268, 2008. · ·

6. M. D. R. Moreira, M. Pereda, N. E. Marcovich, and S. I. Roura, "Antimicrobial effectiveness of bioactive packaging materials from edible chitosan and casein polymers: assessment on carrot," Journal of Food Science, vol. 76, no. 1, pp. M54–M63, 2011.

7. F. A. Osorio, P. Molina, S. Matiacevich, J. Enrione, and O. Skurtys, "Characteristics of hydroxy propyl methyl cellulose (HPMC) based edible film developed for blueberry coatings," Procedia Food Science, vol. 1, pp. 287–293, 2011. ·

8. Siddaramaiah, P. Kumar, K. Divya, B. Mhemavathi, and D. Manjula, "Chitosan/HPMC polymer blends for developing transdermal drug delivery systems," Journal of Macromolecular Science, Part A: Pure and Applied Chemistry, vol. 43, no. 3, pp. 601–607, 2006.

9. O. Skurtys, C. Acevedo, F. Pedreschi, J. Enrione, F. Osorio, and J. Aguilera, "Food hydrocolloid edible films and coatings," Food Science and Technology, vol. 3, pp. 6–9, 2010.

10. F. Al-Sagheer and S. Muslim, "Thermal and mechanical properties of chitosan/SiO_2 hybrid composites,"Journal of Nanomaterials, vol. 2010, Article ID 490679, 7 pages, 2010.

11. S. Kaewpirom and S. Boonsang, "Electrical response characterisation of poly(ethylene glycol) macromer (PEGM)/chitosan hydrogels in NaCl solution," European Polymer Journal, vol. 42, no. 7, pp. 1609–1616, 2006. ·

12. R. Ahmadi, A. Kalbasi-Ashtari, A. Oromiehie, M.-S. Yarmand, and F. Jahandideh, "Development and characterization of a novel biodegradable edible film obtained from psyllium seed (Plantago ovataForsk)," Journal of Food Engineering, vol. 109, no. 4, pp. 745–751, 2012.

13. A. Zimoch-Korzycka and A. Jarmoluk, "The use of chitosan, lysozyme, and the nano-silver as antimicrobial ingredients of edible protective hydrosols applied into the surface of meat," Journal of Food Science and Technology, 2014.

14. Y. Zhao, H. Xin, D. Zhao et al., "Free chlorine loss during spraying of membraneless acidic electrolyzed water and its antimicrobial effect on airborne bacteria from poultry house," Annals of Agricultural and Environmental Medicine, vol. 21, no. 2, pp. 249–255, 2014.

15. S. Hati, S. Mandal, P. S. Minz et al., "Electrolyzed Oxidized Water (EOW): non-thermal approach for decontamination of food borne microorganisms in food industry," Food and Nutrition Sciences, vol. 03, no. 06, pp. 760–768, 2012. ·

16. C.-Y. Chuang, S. Yang, H.-C. Huang et al., "Applying the membrane-less electrolyzed water spraying for inactivating bioaerosols," Aerosol and Air Quality Research, vol. 13, no. 1, pp. 350–359, 2013.

17. A. Pinotti, M. A. García, M. N. Martino, and N. E. Zaritzky, "Study on microstructure and physical properties of composite films based on chitosan and methylcellulose," Food Hydrocolloids, vol. 21, no. 1, pp. 66–72, 2007.

18. A. I. Bourbon, A. C. Pinheiro, M. A. Cerqueira et al., "Physico-chemical characterization of chitosan-based edible films incorporating bioactive compounds of different molecular weight," Journal of Food Engineering, vol. 106, no. 2, pp. 111–118, 2011.

19. J. Rotta, R. Á. Ozório, A. M. Kehrwald, G. M. de Oliveira Barra, R. D. de Melo Castanho Amboni, and P. L. M. Barreto, "Parameters of color, transparency, water solubility, wettability and surface free energy of chitosan/hydroxypropylmethylcellulose (HPMC) films plasticized with sorbitol," Materials Science & Engineering C, vol. 29, no. 2, pp. 619–623, 2009.

20. J. Yin, K. Luo, X. Chen, and V. V. Khutoryanskiy, "Miscibility studies of the blends of chitosan with some cellulose ethers," Carbohydrate Polymers, vol. 63, no. 2, pp. 238–244, 2006.

21. J. Rotta, E. Minatti, and P. L. M. Barreto, "Determination of structural and mechanical properties, diffractometry, and thermal analysis of chitosan and hydroxypropylmethylcellulose (HPMC) films plasticized with sorbitol," Ciência e Tecnologia de Alimentos, vol. 31, no. 2, pp. 450–455, 2011.

22. M. R. de Moura, F. A. Aouada, R. J. Avena-Bustillos, T. H. McHugh, J. M. Krochta, and L. H. C. Mattoso, "Improved barrier and mechanical properties of novel hydroxypropyl methylcellulose edible films with chitosan/tripolyphosphate nanoparticles," Journal of Food Engineering, vol. 92, no. 4, pp. 448–453, 2009. · ·

23. I. Leceta, P. Guerrero, and K. de la Caba, "Functional properties of chitosan-based films," Carbohydrate Polymers, vol. 93, no. 1, pp. 339–346, 2013.

24. A. P. P. Praxedes, A. J. C. da Silva, R. C. da Silva et al., "Effects of UV irradiation on the wettability of chitosan films containing dansyl derivatives," Journal of Colloid and Interface Science, vol. 376, no. 1, pp. 255–261, 2012.

25. P. Gomes, C. A. R. Gomes, M. K. S. Batista, L. F. Pinto, and P. A. P. Silva, "Synthesis, structural characterization and properties of water-soluble N-(γ-propanoyl-amino acid)-chitosans," Carbohydrate Polymers, vol. 71, no. 1, pp. 54–65, 2008.

26. G. Socrates, Infrared and Raman Characteristic Group Frequencies Tables and Charts, John Wiley & Sons, New York, NY, USA, 3rd edition, 2001.

27. C. Le Tien, M. Lacroix, P. Ispas-Szabo, and M.-A. Mateescu, "N-acylated chitosan: hydrophobic matrices for controlled drug release," Journal of Controlled Release, vol. 93, no. 1, pp. 1–13, 2003. · ·

28. J. Xu, S. P. McCarthy, R. A. Gross, and D. L. Kaplan, "Chitosan film acylation and effects on biodegradability," Macromolecules, vol. 29, no. 10, pp. 3436–3440, 1996.

29. P. Mischnick, "Mass spectrometric characterization of oligo- and polysaccharides and their derivatives,"Advances in Polymer Science, vol. 248, pp. 105–174, 2012.

30. A. Zimoch-Korzycka, C. Gardrat, A. Castellan, V. Coma, and A. Jarmoluk, "The use of lysozyme to prepare biologically active chitooligomers," Polímeros, vol. 25, no. 1, pp. 35–41, 2015. ·

31. S.-I. Park, M. A. Daeschel, and Y. Zhao, "Functional properties of antimicrobial lysozyme-chitosan composite films," Journal of Food Science, vol. 69, no. 8, pp. M215–M221, 2004.

32. V. Sidhan and S. Gurnani, "Kinetic characterization of rat liver nuclear lysozyme," Journal of Biosciences, vol. 4, no. 2, pp. 191–195, 1982.

33. A. P. Martínez-Camacho, M. O. Cortez-Rocha, J. M. Ezquerra-Brauer et al., "Chitosan composite films: thermal, structural, mechanical and antifungal properties," Carbohydrate Polymers, vol. 82, no. 2, pp. 305–315, 2010.

34. F. S. Kittur, K. V. H. Prashanth, K. U. Sankar, and R. N. Tharanathan, "Characterization of chitin, chitosan and their carboxymethyl derivatives by differential scanning calorimetry," Carbohydrate Polymers, vol. 49, no. 2, pp. 185–193, 2002.

35. C. G. T. Neto, J. A. Giacomettib, A. E. Jobb, F. C. Ferreirab, J. L. C. Fonsecaa, and M. R. Pereiraa, "Thermal analysis of chitosan based networks," Carbohydrate Polymers, vol. 62, no. 2, pp. 97–103, 2005. ·

36. T. H. Kim, J. Ahn, H. Choi, Y. Choi, and C. Cho, "A novel mucoadhesive polymer film composed of carbopol, poloxamer and hydroxypropylmethylcellulose," Archives of Pharmacal Research, vol. 30, no. 3, pp. 381–386, 2007.

37. K. Piyada, S. Waranyou, and W. Thawien, "Mechanical, thermal and structural properties of rice starch films reinforced with rice starch nanocrystals," International Food Research Journal, vol. 20, no. 1, pp. 439–449, 2013. ·

38. A. Muscat, U. Prüße, and K.-D. Vorlop, "Stable suport materials for the immobilization of viable cells," inImmobilized Cells: Basics and Application, R. H. Wijffels, R. M. Buitelaar, C. Bucke, and J. Tramper, Eds., pp. 55–61, Elsevier Science, Amsterdam, The Netherlands, 1996.

39. Y. Yuan and T. Lee, "Contact angle and wetting properties," in Surface Science Techniques, G. Bracco and B. Holst, Eds., pp. 3–34, Springer, Berlin, Germany, 2013.

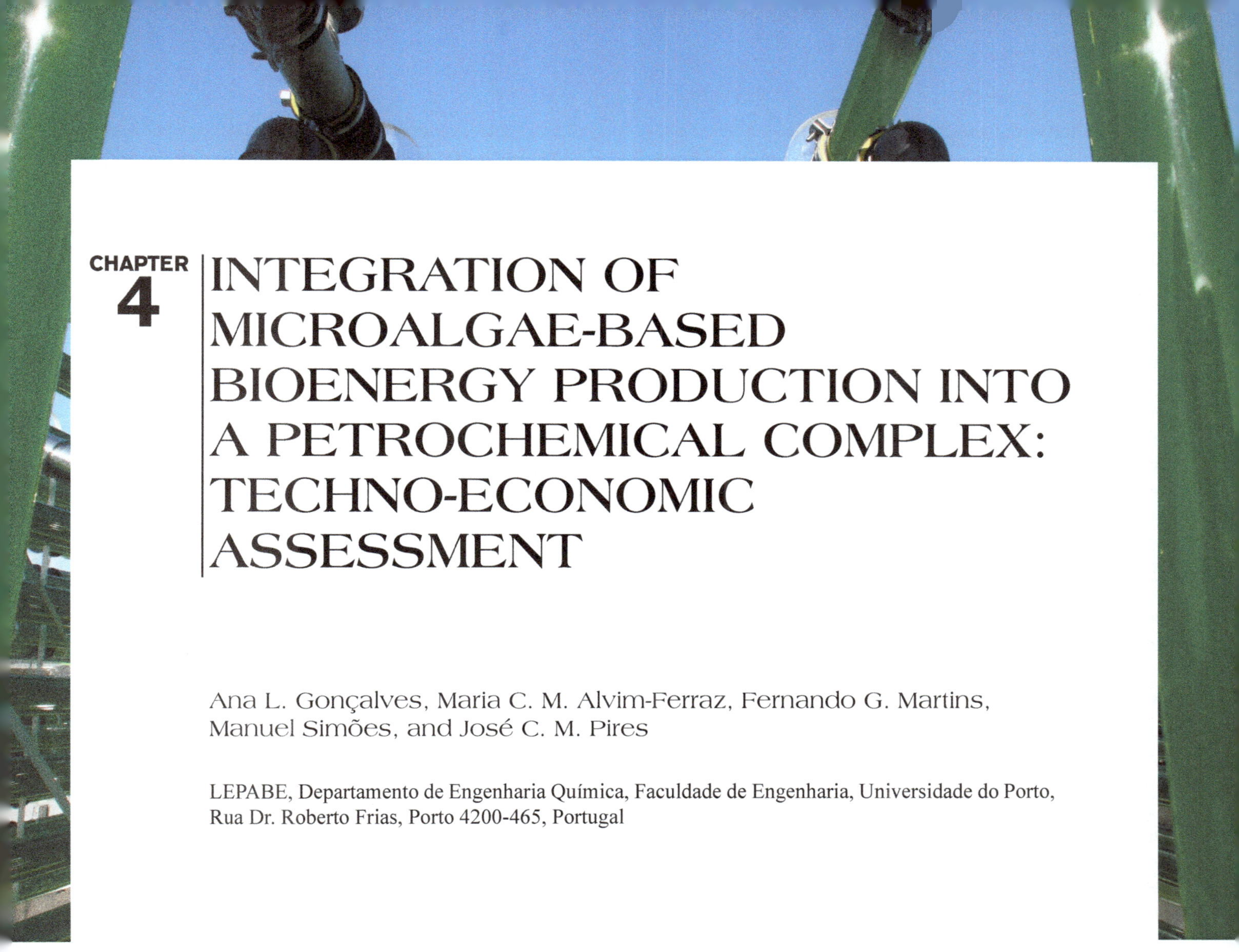

CHAPTER 4

INTEGRATION OF MICROALGAE-BASED BIOENERGY PRODUCTION INTO A PETROCHEMICAL COMPLEX: TECHNO-ECONOMIC ASSESSMENT

Ana L. Gonçalves, Maria C. M. Alvim-Ferraz, Fernando G. Martins, Manuel Simões, and José C. M. Pires

LEPABE, Departamento de Engenharia Química, Faculdade de Engenharia, Universidade do Porto, Rua Dr. Roberto Frias, Porto 4200-465, Portugal

ABSTRACT

The rapid development of modern society has resulted in an increased demand for energy, mainly from fossil fuels. The use of this source of energy has led to the accumulation of carbon dioxide (CO_2) in the atmosphere. In this context, microalgae culturing may be an effective solution to reduce the CO_2 concentration in the atmosphere, since these microorganisms can capture CO_2 and, simultaneously, produce bioenergy. This work consists of a techno-economic assessment of a microalgal production facility integrated in a petrochemical complex, in which established infrastructure allows efficient material and energy transport. Seven different scenarios were considered regarding photosynthetic, lipids extraction and anaerobic digestion efficiencies. This analysis has demonstrated six economically viable scenarios able to: (i) reduce CO_2 emissions from a thermoelectric power plant; (ii) treat domestic wastewaters (which were used as culture medium); and (iii) produce lipids and electrical and thermal energy. For a 100-ha facility, considering a photosynthetic efficiency of 3%, a lipids extraction efficiency of 75% and an anaerobic digestion efficiency of 45% (scenario 3), an economically viable process was obtained (net present value of 22.6 million euros), being effective in both CO_2 removal (accounting for 1.1×10^4 t per year) and energy production (annual energy produced was 1.6×10^7 kWh and annual lipids productivity was 1.9×10^3 m^3).

INTRODUCTION

In the last decades, greenhouse gas (GHG, mainly carbon dioxide, CO_2) emissions to the atmosphere have increased drastically, leading to atmospheric concentrations 40% higher than those observed in the pre-industrial period [1]. This concentration increase is associated with the planet's climate change, strongly concerning different entities worldwide, because the associated environmental changes tend to be uncontrolled and unacceptable [2,3,4]. Therefore, there is an urgent need to develop efficient CO_2 mitigation systems and also to use carbon neutral energy sources. Microalgae culturing is considered a promising alternative to efficiently remove CO_2 from atmosphere or from flue gas emissions; microalgae are photosynthetic microorganisms that convert CO_2 into organic compounds in the presence of light [5,6].

Citation: Ana L. Gonçalves, Maria C. M. Alvim-Ferraz, Fernando G. Martins, Manuel Simões, and José C. M. Pires, Integration of Microalgae-Based Bioenergy Production into a Petrochemical Complex: Techno-Economic Assessment, Energies 2016, 9(4), 224; doi:10.3390/en9040224

Although terrestrial plants are also capable of reducing CO_2 levels, microalgae present some notable advantages [7,8,9,10]: (i) higher growth and biomass production rates; (ii) shorter times to maturity; (iii) no need for arable land; and (iv) the capability to grow using waste as nutrients. To reduce the costs associated with nutrient supply and the environmental impact of using freshwater resources, microalgal culture can be performed using: (i) domestic [11,12]; (ii) leachate [13,14]; (iii) agricultural [15,16]; (iv) refinery [17]; and (v) industrial [18] wastewater as culture medium. This procedure simultaneously promotes nutrient removal from wastewaters (a costly treatment process) and biomass production [19,20,21]. In addition, microalgal biomass can play an important role in biofuel production [22,23,24]: (i) the fatty acids produced by microalgae can be extracted and converted into biodiesel; and (ii) residual biomass can be fermented to produce ethanol or methane. This study aims to develop a techno-economic analysis of microalgal cultivation near a petrochemical complex located in Sines (Portugal) for bioenergy production using CO_2 from local emissions and domestic wastewater as culture medium.

MICROALGAL PRODUCTION PLANT SITING

Under autotrophic conditions, microalgae reproduce through photosynthesis, requiring CO_2, water and inorganic salts to convert light energy into chemical energy in the form of organic compounds [25]. Growth medium requires the presence of the inorganic elements that constitute algal cells: nitrogen and phosphorus. Marine microalgae are commonly grown in sea water supplemented with nitrate and phosphate fertilizers, whereas other microalgae can be cultivated in wastewaters, thus reducing culturing costs and providing wastewater treatment [26]. In commercial scale systems, microalgal culturing represents up to 30% of the total oil production costs [27]. The utilization of wastewater as culture medium can significantly reduce the operational costs associated to the culturing step. In addition, CO_2 from the atmosphere or from flue gas emissions should be supplied continuously during light periods, thus enabling CO_2 mitigation. Therefore, the selection of the location for microalgal culturing should not focus exclusively on the weather conditions, but also on the availability of water, nutrients (carbon, nitrogen and phosphorus) and land [28,29]. In addition, the proximity to an established infrastructure will also be important to allow the efficient material transport (raw materials and process products) and utility needs. The reduction of transportation costs will enhance the economic viability of the process and improve its overall energy balance. In biofuel production using other feedstocks, the transportation costs may represent 12%–50% of the total production costs [28].

Proposed Location and Site Description

The present study proposes the construction of a high rate pond facility in Sines, Portugal. Sines is a municipality from the district of Setúbal, which is located in the Alentejo Litoral region. Total area of this municipality is approximately 203 km^2 and in 2011, its population density was 70 inhabitants km^{-2}, corresponding to a total of 14,210 inhabitants [30].

Microalgal culturing in open ponds strongly depends on environmental factors, such as temperature, solar light irradiation and evaporation rates. Therefore, selection of an adequate site for the installation of an algal facility should take into account these parameters. Daily and annual fluctuations in temperature can result in significant microalgal productivity losses. Temperatures ranging from 15 to 26 °C have been reported as optimal growth temperatures for some microalgal species [31]. The average annual temperature observed in the region of Sines is around 17.2 °C (monthly minimum, maximum and average temperatures are presented in Figure 1A). Horizontal light irradiation in Sines municipality (Figure 1B) presents an annual average of 5.21 $kWh{\cdot}m^{-2}{\cdot}day^{-1}$ [32]. Water evaporative losses in open ponds are very common. Evaporation rate depends on different factors, such as [33]: (i) water temperature in the air-water surface; (ii) air-water surface area; and (iii) air temperature. Evaporation rates in lagoons from the south of Portugal were determined by Rodrigues [33]. In this study, the author has demonstrated average evaporation rates of 0.075 $m{\cdot}month^{-1}$.

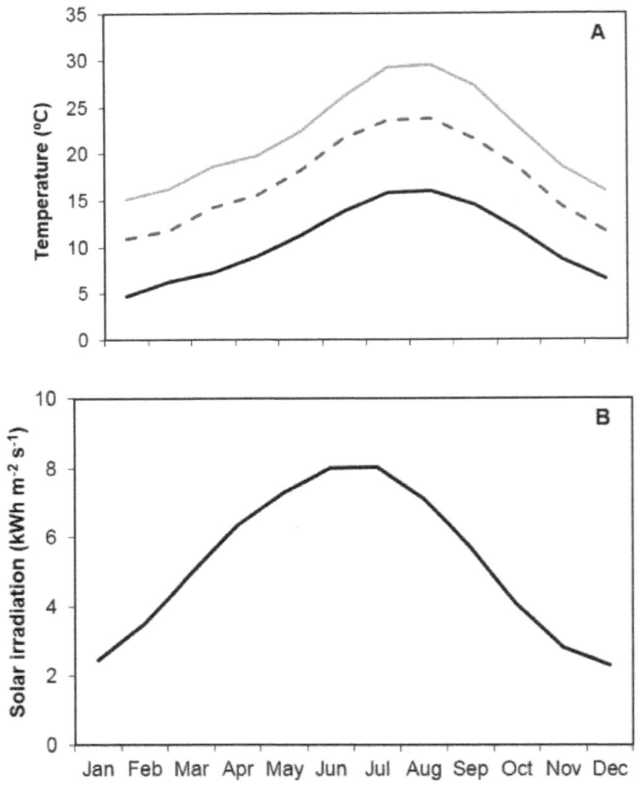

Figure 1: Minimum (black line), maximum (grey line) and average (dashed line) temperature (**A**) and horizontal solar irradiation (**B**) per month observed in Sines municipality, Portugal [32,34].

Selection of this local site was based on these environmental factors. However, other characteristics were considered: (i) the flat topography of this region, which avoids the need for land preparation before open pond construction; (ii) the presence of a thermoelectric power plant in this area, which can supply the facility with the required CO_2 and other utilities, such as steam; (iii) the presence of a biodiesel production plant able to generate energy from raw materials, such as oils and animal fats, with a production capacity of 27 $kt{\cdot}year^{-1}$; (iv) the closeness to the coast, so that seawater can be easily used if required; and (v) the availability of sufficient domestic wastewater to feed the algal facility (the wastewater load at a typical Portuguese wastewater treatment plant is about 1000

$m^3 \cdot h^{-1}$ for a population equivalent of 170,000 inhabitants and the number of inhabitants of Setúbal district is approximately 867,000 [30]).

PROCESS FLOWSHEET AND SCENARIOS DESCRIPTION

Process Flowsheet Description

Cultivation of microalgae can be carried out in closed or open bioreactors. Microalgal production in open systems is less expensive in terms of construction and operation and has larger production capacity [35,36,37]. However, biomass productivities achieved with these systems are lower than those achieved with closed systems, which is mainly due to insufficient mixing. Additionally, these systems are more susceptible to the diffusion of CO_2 to the atmosphere, evaporative losses of water, poor light utilization by cells, oscillations in the culture conditions and microbial contaminations [36,37,38,39].

As open pond systems are more commonly used for commercial scale applications [35,40,41] and lower investment and operational costs (important advantage in bioenergy production) are required, they are proposed in this study for *Chlorella vulgaris* growth (a fast-growing microalga widely applied in wastewater treatment processes and biofuels production, due to its high biomass productivities and lipid contents and to their high ability for nutrients removal and resistance to contaminations [7,42,43,44]) using domestic wastewater as culture medium (SWW). According to Cai *et al.* [45], nitrogen and phosphorus concentrations typically range between 15–90 and 5–20 $mg \cdot L^{-1}$, respectively. In this study, nitrogen and phosphorus concentrations in the domestic wastewater were assumed to be 50 and 10 $mg \cdot L^{-1}$, respectively. To enhance biomass productivities, CO_2 can be provided to cultures using flue gases from a thermoelectric power plant and some refinery processes (e.g., steam methane reforming). Microalgal biomass (S02—output stream of open ponds) is then harvested by flocculation in a clarifier (S03) followed by centrifugation (S04) to a concentration of about 20% (w/w). To avoid the use of chemicals, cell disruption procedure was the continuous pulsed electric field (PEF) tested for lipids extraction from *C. vulgaris* by Flisar *et al.* [46]. The extracted lipids (S05) will be sold to the biodiesel plant located nearby. The remaining biomass (S06) is then forwarded to the anaerobic digestion process. The output streams of this process are the following: (i) biogas that is burned in the combined heat and power (CHP) generation unit (S07); (ii) fertilizer, considered a product of the process (S08); and (iii) wastewater that is recycled to the open ponds (SAD). The process diagram of this plant is presented in Figure 2.

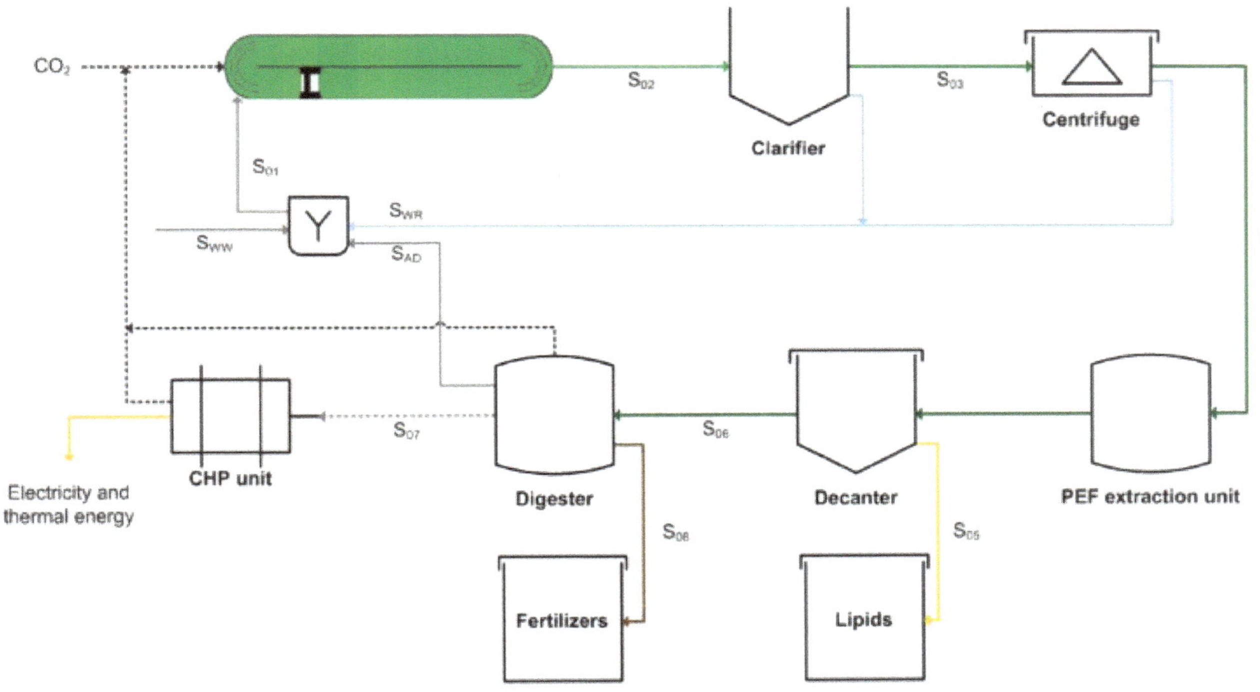

Figure 2: Process diagram. CHP—combined heat and power; PEF—pulsed electric field.

Scenarios Description

In this study, seven scenarios were considered, characterized by different efficiencies in some of the most important steps of the microalgal facility (see Table 1). In the scenarios 1 (base scenario), 2 and 3, the effect of photosynthetic efficiency (2%, 1% and 3%, respectively) was analysed. These values were already determined for open ponds. Photosynthetic efficiencies achieved in open systems lie between one tenth and one third of the "theoretical" value of 10%, due to several losses [40,47]: (i) inactive photon absorption; (ii) reflection; (iii) respiration; (iv) light saturation; and (v) photo-inhibition [48]. In the scenarios 1, 4 and 5, the efficiency of lipids extraction (75%, 60% and 90%, respectively) was assessed. PEF extraction efficiencies between 60% and 90% have already been reported in the literature [46,49,50]. Since lipids extraction efficiency is easier to control than cell lipid content, this value was assumed to be constant (25%). In the scenarios 1, 6 and 7, the efficiency of anaerobic digestion (45%, 30% and 60%, respectively) was compared, since different studies have reported anaerobic digestion efficiencies within this range of values [51,52].

Table 1: Characterization of the seven scenarios (Sc) evaluated in this study.

Scenarios' Assumptions	Sc1	Sc2	Sc3	Sc4	Sc5	Sc6	Sc7
Photosynthetic efficiency (%)	2	1	3	2	2	2	2
Biomass productivity ($g \cdot m^{-2} \cdot day^{-1}$)	16.4	8.2	24.6	16.4	16.4	16.4	16.4
Lipids extraction efficiency (%)	75	75	75	60	90	75	75
Anaerobic digestion efficiency (%)	45	45	45	45	45	30	60

TECHNO-ECONOMIC ASSESSMENT

Mass Balance

The overall process is schematically represented in Figure 2 and the mass balance to the different streams involved in the process is presented in Table 2. Briefly, the process comprises four different steps: (i) microalgal growth; (ii) microalgal harvesting; (iii) combined cell disruption and lipids extraction; and (iv) anaerobic digestion followed by electricity production. The following sections include a description of each step and all the considerations assumed to determine mass balances for each process unit.

Table 2: Mass balance to the flow streams involved in the process determined in each of the studied scenarios (Sc).

Streams	Sc1	Sc2	Sc3	Sc4	Sc5	Sc6	Sc7
S_{01}—water input in the cultivation step ($\times 10^4$ $m^3 \cdot day^{-1}$)	3.5	1.9	5.2	3.5	3.5	3.5	3.5
S_{02}—biomass flow rate after the cultivation step [a] ($\times 10^4$ $m^3 \cdot day^{-1}$)	3.3	1.6	4.9	3.3	3.3	3.3	3.3
S_{03}—biomass flow rate after the pre-concentration step ($\times 10^3$ $m^3 \cdot day^{-1}$)	8.2	4.1	12	8.2	8.2	8.2	8.2
S_{04}—biomass flow rate after the centrifugation step ($m^3 \cdot day^{-1}$)	78	39	117	78	78	78	78
S_{05}—extracted lipids flow rate ($m^3 \cdot day^{-1}$)	3.4	1.7	5.1	2.7	4.1	3.4	3.4
S_{06}—biomass flow rate after the lipids extraction step ($m^3 \cdot day^{-1}$)	75	37	112	75	74	75	75
S_{07}—biogas flow rate after the anaerobic digestion step ($t \cdot day^{-1}$)	6.6	3.3	10	6.9	6.3	4.4	8.9
S_{08}—flow rate of the residue produced in the anaerobic digestion step ($t \cdot day^{-1}$)	7.0	3.5	11	7.3	6.6	8.9	5.1
S_{EV}—water flow rate required to compensate evaporation losses ($\times 10^3$ $m^3 \cdot day^{-1}$)	2.5	2.5	2.5	2.5	2.5	2.5	2.5
S_{WW}—wastewater flow rate required to feed the culture ($\times 10^4$ $m^3 \cdot day^{-1}$)	2.4	0.62	5.2	2.4	2.4	2.4	2.4
S_{WR}—recycling water flow rate required to feed the culture ($\times 10^4$ $m^3 \cdot day^{-1}$)	1.1	1.3	0	1.1	1.1	1.1	1.1
S_{AD}—anaerobic digestion effluent flow rate required to feed the culture ($m^3 \cdot day^{-1}$)	62	31	93	62	62	62	62

[a] Biomass flow rate after the cultivation step was determined considering annual average biomass productivities and assuming a final biomass concentration of $0.5\ g \cdot L^{-1}$ [40,47].

Microalgal Growth

The proposed algal facility consists of 25 similar high rate ponds with 0.3 m height, performing a total pond area of 100 ha. These open ponds may operate during diurnal periods, since photosynthetic growth does not occur at night. Therefore, this period may be used for shut down for cleaning and maintenance. Average biomass productivities were determined taking into account the operation time of open ponds, average horizontal light irradiance observed in Sines (Figure 1B) and the assumed photosynthetic efficiencies (1% to 3%). Photosynthetic efficiencies ranging between approximately 1.5% and 4.5% were already reported for *C. vulgaris* [53]. Accordingly, annual average biomass productivity determined for the base scenario in this region is approximately 16 $g\ m^{-2} \cdot day^{-1}$, which is similar to the values determined by Doucha and Lívanský [54] for *Chlorella* sp. grown in open ponds (25 $g\ m^{-2} \cdot day^{-1}$). Considering annual average biomass productivities and the pond volume ($3.0 \times 10^5\ m^3$) and considering the average evaporation rate reported in Section 2.1 ($0.075\ m \cdot month^{-1}$), the input stream of water and nutrients (S01) required for microalgal growth on a daily basis corresponds to 3.5×10^4 $m^3 \cdot day^{-1}$(in the base scenario). This water input is obtained from domestic wastewater (SWW, 2.4×10^4 $m^3 \cdot day^{-1}$), from water recycling (SWR, 1.1×10^4 $m^3 \cdot day^{-1}$) and from water resulting from the anaerobic digestion step (SAD, 62 $m^3 \cdot day^{-1}$). Flow rates of the recycling water (SWR) can be regulated to avoid excessive dilution of the cultures in rainy days. Assuming that biomass concentration achieved during the cultivation step is $0.5\ g \cdot L^{-1}$, biomass flow rate after the cultivation step (S02) is, in the base scenario, 3.3×10^4 $m^3 \cdot day^{-1}$. Input and output streams of microalgal production in the studied scenarios are summarized in Table 2.

As referred before, domestic wastewater will be used as culture medium. Wastewater will provide nutrients, such as nitrogen and phosphorus, to microalgae. On the other hand, CO_2 resulting from the thermoelectric power plant and refinery processes will be supplied to the cultures. Taking into account the annual average biomass productivities and the typical molecular formula described for microalgae, $CO_{0.48}H_{1.83}N_{0.11}P_{0.01}$ [7], theoretical nitrogen, phosphorus and carbon removal rates were estimated. For these determinations, it was assumed that all nitrogen, phosphorus and carbon removed from the wastewater or from the flue gas were incorporated into microalgal biomass. Although the typical molecular formula of microalgal biomass was not determined for *C. vulgaris*, it has already been applied by several authors to determine C, N and P removal rates by microalgae from the genus *Chlorella* [55,56,57]. Considering the base scenario, nitrogen, phosphorus and carbon removal rates are 1.1, 0.22 and 8.43 g·m^{-2}·day^{-1}. With these values, minimum concentrations of these nutrients required in the feed stream were determined. Accordingly, for the same scenario, minimum nitrogen and phosphorus concentrations in the feed stream are 31 and 6.2 mg·L^{-1}, respectively, whereas CO_2 requirements correspond to 39 t·day^{-1}.Table 3 presents average removal rates determined for nitrogen, phosphorus and carbon in the studied scenarios, as well as minimum required concentrations of these nutrients.

Table 3: Average removal rates of nitrogen, phosphorus and carbon and minimum concentrations required for microalgal cultivation in each of the studied scenarios (Sc).

Nutrients Loads and Removal Rates	Sc1	Sc2	Sc3	Sc4	Sc5	Sc6	Sc7
r_N (g· m^{-2}· day^{-1})	1.1	0.54	1.6	1.1	1.1	1.1	1.1
$[N]$ (mg· L^{-1})	31	15	46	31	31	31	31
r_P (g· m^{-2}· day^{-1})	0.22	0.11	0.33	0.22	0.22	0.22	0.22
$[P]$ (mg· L^{-1})	6.2	3.1	9.2	6.2	6.2	6.2	6.2
r_C (g· m^{-2}· day^{-1})	8.4	4.2	13	8.4	8.4	8.4	8.4
$[C]$ (t· day^{-1})	39	19	58	39	39	39	39

Microalgal Harvesting

The proposed harvesting techniques for this study include a pre-concentration step through flocculation followed by centrifugation, where biomass concentration achieved was assumed to be 200 g·L^{-1} [48,58]. The use of a pre-concentration step aims the reduction of the flow rate to be processed in the centrifugation step, which may result in significant savings in terms of energy. In the pre-concentration step, flocculation may be induced by the addition of NaOH as flocculant. The amount of flocculant used was assumed to be 9 mg·g$^{-1}_{biomass}$, as reported by Vandamme [59]. In the last step, a harvesting efficiency of 95% was also assumed. Harvesting efficiencies higher than 94% were obtained in different studies, when applying centrifugation for the harvesting of microalgal biomass [60,61]. With this harvesting efficiency and the average biomass productivities, the flow rate of the output stream from the centrifugation step (S04) corresponds to 78 m^3·day^{-1} (in the base scenario).

Knowing the initial composition of the domestic wastewater, as well as elemental composition of microalgae and total biomass collected per harvesting, it is possible to determine effluent composition. Considering that all nitrogen and phosphorus removed from wastewater is incorporated into microalgal biomass, effluent composition in nitrogen and phosphorus (for the studied scenarios) range between 3.2–5.3 and 0.57–1.0 mg·L^{-1}, respectively. Nitrogen and phosphorus concentrations in the resulting effluent are lower than the limits established by EU legislation for the discharge of domestic effluents (15–20 and 1–2 mg·L^{-1} for nitrogen and phosphorus, respectively) [62,63], which means that the proposed process promotes the efficient treatment of domestic wastewaters.

Cell Disruption and Lipids Extraction

PEF technology is a non-thermal method usually applied in food processing applications for inactivation of microbes, helping to maintain the food quality for human consumption [64,65]. This technique uses short and high voltage pulses, which induce the non-thermal permeabilization of cell membranes and, in determined conditions, the complete disruption of cells into fragments. It is a rapid (treatment time is less than a second), flexible and energy-efficient method (heat is minimized) that avoids the use of organic solvents, usually toxic, thus not affecting the biochemical composition of microalgal biomass [46,64,66]. Regarding microalgal products, PEF is considered to have high potential for the extraction of compounds, due to the low energy consumption, easy scale-up and low operational costs. This extraction method does not use any toxic extraction solvent (not requiring a solvent recovery step) and is highly effective when directly applied to wet feedstocks [49,65,67]. PEF was already applied to extract lipids from *C. vulgaris* [46,68,69]. It was considered a clean, cheap and quick extraction process, being a promising method for the production of biodiesel and pharmaceutical and dietary products.

Considering the base scenario, presenting a PEF efficiency of 75% [49], total microalgal oil extracted, with a density of 0.86 kg·L^{-1} [70], corresponds to 3.4 m^3·day^{-1}. For the other studied PEF efficiencies, 60% (scenario 4) and 90% (scenario 5), total lipids extracted are 2.7 and 5.1 m^3·day^{-1}, respectively. For the base scenario, biodiesel production through transesterification of the extracted lipids results in a biodiesel productivity of 3.0 t·day^{-1} (3.47 m^3·day^{-1}).

Anaerobic Digestion

Biomass resulting from the oil extraction step (75 m^3·day^{-1} in the base scenario) is subjected to anaerobic digestion followed by electricity production. In this step different process efficiencies were evaluated (45%, 30% and 60%). Taking into account the elemental composition of microalgal biomass after lipids extraction and considering that fractions of CH_4 and CO_2 in the biogas are respectively 60% and 40% (v/v) (typical composition of the biogas consists of 55%–70% CH_4 and 30%–45% CO_2 [71]), resulting biogas stream in the base scenario has the following composition: 2.3 t·day^{-1} of CH_4 and 4.3 t·day^{-1}of CO_2. Additionally, a residue rich in nitrogen and phosphorus is produced (approximately 7.0 t·day^{-1} in the base scenario). Due to its high content in nitrogen, this residue can then be used as fertilizer [43].

Net CO_2 Balance

Based on a recent study performed for *C. vulgaris* [72], the use of a CO_2 concentration of about 5% (v/v) was proposed, which has shown to be optimal for microalgal growth. Taking into account the results obtained in this study, it is expected a slight decrease in the pH of the culture to about 6.5, which will not be harmful for microalgae. During this process different CO_2 streams are involved. In the cultivation step, CO_2 is fed into microalgal cultures at different rates (Table 3), depending on the studied scenarios. On the other hand, anaerobic digestion and electricity production in the CHP generation unit release CO_2. Since CO_2 will be mainly supplied from the flue gas of a thermoelectric power plant working with natural gas, it is expected that this flue gas presents residual sulphite and nitrite concentrations. Accordingly, there is no need of a purification step prior to addition in the ponds. Considering a CO_2 uptake efficiency of 80%, net CO_2 balances were determined for the studied scenarios (Table 4).

Table 4: Net CO_2 balance in each of the studied scenarios (Sc)

CO_2 Streams	Sc1	Sc2	Sc3	Sc4	Sc5	Sc6	Sc7
CO_2 required for microalgal growth ($t \cdot day^{-1}$)	39	19	58	39	39	39	39
CO_2 resulting from the anaerobic digestion ($t \cdot day^{-1}$)	4.3	2.1	6.4	4.5	4.1	2.9	5.7
CO_2 resulting from CHP generation ($t \cdot day^{-1}$)	6.4	3.2	9.7	6.7	6.1	4.3	8.6
Net CO_2 balance ($t \cdot day^{-1}$)	−20	−10	−30	−20	−21	−24	−17

CHP—combined heat and power.

These results have shown negative values in all the scenarios (net CO_2 balances range between −30 and −10 $t \cdot day^{-1}$), which means that the proposed process is a net zero emission process able to efficiently uptake CO_2 from the flue gases of a thermoelectric power plant. Comparing the studied scenarios, it is possible to conclude from Table 4 that the most effective in CO_2 uptake is the scenario 3, the one assuming the highest photosynthetic efficiency (3%). In this scenario, annual CO_2uptake corresponds to 1.1×10^4 t.

Energy Balance

In this process, energy is mainly required in the following steps: microalgal cultivation, microalgal harvesting and cell disruption and lipids extraction. Figure 3 shows the electrical requirements of each of these processes for the studied scenarios.

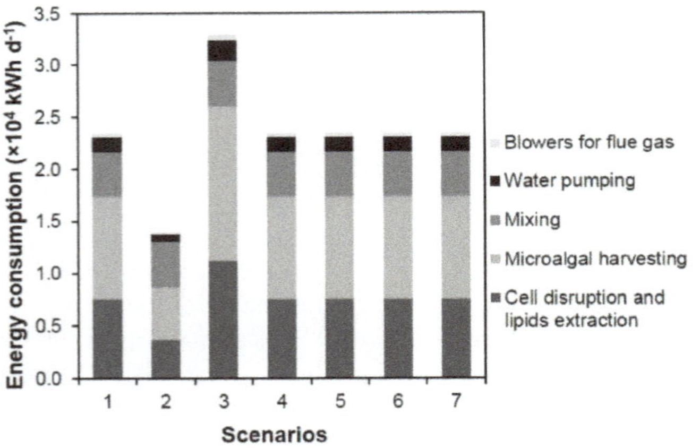

Figure 3: Electrical requirements determined for different steps of the microalgal facility in each of the studied scenarios.

In the cultivation step, energy is required in three different stages: (i) mixing; (ii) water pumping; and (iii) blowers for flue gas. Considering a mixing velocity of 0.23 $m \cdot s^{-1}$ and all the head losses occurring in the open ponds (corresponding to head losses around the bends, 0.010 m, through the sumps, 0.026 m, and down the straightaways, 0.11 m), it is possible to determine the power required to overcome all these head losses. Assuming 25 similar open ponds operating at the same time, energy required daily can be determined. Therefore, assuming a 12:12 light:dark period, the total energy required for mixing corresponds to 4.4×10^3 $kWh \cdot day^{-1}$. Energy needed in water pumping is calculated based on the flow rate of the input (S01). Pump and motor efficiencies of 88% and 83% were assumed [73]. Therefore, for the base scenario, a water input of 3.5×10^4 $m^3 \cdot day^{-1}$ corresponds to an energy input of 1.4×10^3 $kWh \cdot day^{-1}$. Energy required for CO_2 distribution was determined based on the CO_2 requirements for each studied scenario (Table 3), assuming a CO_2 concentration in the flue gas of 5% (v/v). Additionally, air blower efficiency was considered to be 77% [74]. Accordingly, the energy consumption associated to the air blowers was 4.4×10^2 $kWh \cdot day^{-1}$ for the base scenario.

Energy consumption in biomass harvesting corresponds to the energy required for centrifugation. Accordingly, this value was determined taking into account the flow rate resulting from the pre-concentration step (S03) and the specific energy consumption, 1.2 $kWh \cdot m^{-3}$, commonly reported for microalgal harvesting through centrifugation [75]. For the base scenario, the harvesting step corresponds to a power consumption of 9.8×10^3 $kWh \cdot day^{-1}$.

Regarding cell disruption and lipids extraction through the continuous PEF method, energy required was determined assuming the specific energy consumption reported by Flisar *et al.* [46]. According to the authors, the energy required to process 1 L of culture broth for 1 h is 14.4 kJ, which corresponds to 4.0 $kW \cdot m^{-3}$. Taking into account the flow rates to be processed in this unit (S04) for each scenario, energy consumption for this step was determined. For the base scenario, an energy input of 7.5×10^3 $kWh \cdot day^{-1}$ is required.

Although energy is required in several processes, in the CHP generation unit there is an energetic output composed by electrical (40% of the total energy) and thermal energy (45% of the total energy) [76]. Assuming that the biogas produced by anaerobic digestion of microalgae presents a chemical composition similar to the one obtained from household waste, inferior and superior calorific power correspond to 6.0 and 6.6 $kWh \cdot m^{-3}$, respectively [77]. With the flow rates resulting from the anaerobic digestion process (S07) for each scenario and considering an average value between inferior and superior calorific power, values for electrical and thermal energy produced were determined. Considering the base scenario, total electrical and thermal energy produced is 1.4×10^4 and 1.6×10^4 $kWh \cdot day^{-1}$, respectively.

The energy balance performed to the microalgal facility allowed the evaluation of the energetic performance for each of the studied scenarios (Table 5). Analysing the ratio between the energy produced by the microalgal facility (corresponding to the energy obtained from the extracted lipids and the one obtained in the CHP generation unit) and the total energy required, the energy returned on energy invested (EROEI) was determined. For all studied scenarios, EROEI was higher than one, which means that the studied scenarios are energetically efficient. With an EROEI of 3.0, the scenario 7, which assumes an anaerobic digestion efficiency of 60%, is the most efficient in terms of energy.

Table 5: Net energy balance in each of the studied scenarios (Sc).

Energetic Streams	Sc1	Sc2	Sc3	Sc4	Sc5	Sc6	Sc7
Energy required in microalgal cultivation ($\times 10^3$ kWh· day^{-1})	6.2	5.3	7.0	6.2	6.2	6.2	6.2
Energy required in microalgal harvesting ($\times 10^3$ kWh· day^{-1})	9.8	4.9	15	9.8	9.8	9.8	9.8
Energy required in cell disruption and lipids extraction ($\times 10^3$ kWh· day^{-1})	7.5	3.7	11	7.5	7.5	7.5	7.5
Energy obtained from the extracted lipids ($\times 10^4$ kWh· day^{-1})	3.0	1.5	4.6	2.4	3.6	3.0	3.0
Electrical energy produced in the CHP generation unit ($\times 10^4$ kWh· day^{-1})	1.4	0.69	2.1	1.4	1.3	0.92	1.8
Thermal energy produced in the CHP generation unit ($\times 10^4$ kWh· day^{-1})	1.6	0.77	2.3	1.6	1.5	1.0	2.1
EROEI	2.5	2.1	2.7	2.3	2.7	2.1	3.0

CHP—combined heat and power; EROEI—energy returned on energy invested.

Economic Assessment

The economic analysis of an industrial process should take into account the investment capital, also known as fixed capital, as well as annual production costs and annual revenues. With these parameters, it is possible to determine the economic viability of the project. Therefore, the next sections present a detailed economic analysis of the proposed system of wastewater treatment and energy production (electricity and biofuels) using microalgae.

Fixed Capital

Fixed capital is the total investment cost needed to create the facility. It includes the equipment acquisition and installation costs, piping and electrical costs and also the costs of buildings, yard improvements, service facilities and land. Acquisition costs were determined for almost all the equipment required in this process: high rate pond, air blowers, clarifier, centrifuge, decanter, digester and CHP generation unit (Table 6).

Table 6: Total purchase costs (in k€) of the major equipment in each of the studied scenarios (Sc)

Equipments	Sc1	Sc2	Sc3	Sc4	Sc5	Sc6	Sc7
High rate pond [a]	3479	3479	3479	3479	3479	3479	3479
Air blowers [b]	123	62	185	123	123	123	123
Clarifier [c]	325	163	488	325	325	325	325
Centrifuge [d]	14	14	14	14	14	14	14
Decanter [e]	44	44	89	44	44	44	44
Digester and CHP generation unit [f]	1399	1399	1399	1399	1399	1399	1399
Total	5753	5530	6023	5753	5753	5753	5753
Total purchase costs	5983	5734	6281	5983	5983	5983	5983

[a] 34,000 USD per ha (2009) [78], including the costs associated to paddlewheels and liners; [b] 2500 € per 200 m^3· h^{-1} (2012) [79]; [c] 948,000 USD per 23,200 m^3 (2010) [73]; [d] 4500 USD per ha (1996) [80]; [e] 45,000 € per 4 m^3· h^{-1} (2012) [79]; [f] 10,000 USD per ha (1996) [80]; CHP—combined heat and power.

These costs were defined according to values already reported in the literature for different years and were normalized to 2014 cost basis, using the Chemical Engineering Plant Cost Index, *I*, according to Equation (1):

$$\text{Cost} = \text{Base cost}\frac{I}{I_{base}} \quad (1)$$

For the studied scenarios, total acquisition costs determined ranged between 5.7 and 6.3 million euros. This value was obtained assuming that the estimated costs account for 90% of total purchase costs. The correct estimation of total capital investment should also include cost factors associated to direct and indirect costs (Table 7).

Table 7: Total capital (fixed capital) cost estimation (in k€) in each of the studied scenarios (Sc)

Costs	Factor [a]	Sc1	Sc2	Sc3	Sc4	Sc5	Sc6	Sc7
Directcosts								
Total purchase costs	1.00	5983	5734	6281	5983	5983	5983	5983
Purchased equipment installation	0.20	1197	1147	1256	1197	1197	1197	1197
Instrumentation and control	0.15	897	860	942	897	897	897	897
Piping	0.20	1197	1147	1256	1197	1197	1197	1197
Electrical	0.10	598	573	628	598	598	598	598
Buildings	0.15	897	860	942	897	897	897	897
Yard improvements	0.05	299	287	314	299	299	299	299
Service facilities	0.20	1197	1147	1256	1197	1197	1197	1197
Indirectcosts								
Engineering and supervision	0.3	1795	1720	1884	1795	1795	1795	1795
Construction expenses	0.05	299	287	314	299	299	299	299
Contractor's fee	0.03	179	172	188	179	179	179	179
Contingency	0.08	479	459	502	479	479	479	479
Total capital cost		15,017	14,391	15,765	15,017	15,017	15,017	15,017

[a] Fraction of the total purchase costs [79].

In the direct costs, typical factors that should be considered are: (i) installation costs; (ii) instrumentation and control; (iii) piping; (iv) electrical equipment and materials; (v) buildings; (vi) yard improvements; and (vii) service facilities. Indirect costs comprise engineering and supervision, construction expenses, contractor's fee and contingency [81]. These factors typically represent a fraction of the total purchase costs. Including these factors in the calculus of fixed capital results in a total capital investment between 14.4 and 15.8 million euros (Table 7). For the studied scenarios, this value is mainly influenced by the expenses associated to the acquisition costs, followed by those associated to engineering and supervision, equipment installation, piping and service facilities.

Annual Production Costs

To evaluate the economic viability of a project, it is also necessary to determine the annual production costs, which are presented in detail in Table 8. These costs include variable and fixed costs. Variable costs oscillate according to productivity rates. In this study, these costs comprise raw materials, miscellaneous materials, utilities, such as electricity and steam, costs associated to the pre-concentration step with NaOH and to PEF extraction and shipping and packaging. On the other hand, fixed costs are constant and do not oscillate with productivity rates. In these costs are included maintenance, operating labour, laboratory costs, supervision, plant overheads, insurance, local taxes and royalties. Normally, estimation of these costs is done by attributing a percentage of fixed capital or other variables to each of the referred parameters [82].

Table 8: Estimation of the annual production costs (in k€) in each of the studied scenarios (Sc)

Costs	Sc1	Sc2	Sc3	Sc4	Sc5	Sc6	Sc7
Variablecosts							
Raw materials	0	0	0	0	0	0	0
Miscellaneous materials [a]	75	74	77	75	75	75	75
Utilities	857	510	1205	857	857	857	857
Pre-concentration with NaOH	35	17	52	35	35	35	35
PEF extraction	11	5	16	9	13	11	11
Shipping and packaging [b]	0	0	0	0	0	0	0
Fixedcosts							
Maintenance [c]	751	738	766	751	751	751	751
Operating labour	120	120	120	120	120	120	120
Laboratory costs [d]	24	24	24	24	24	24	24
Supervision [d]	24	24	24	24	24	24	24
Plant overheads [e]	60	60	60	60	60	60	60
Insurance [f]	150	148	153	150	150	150	150
Local taxes [g]	300	295	306	300	300	300	300
Royalties [f]	150	148	153	150	179150	150	150
Annual production costs	2557	2164	2956	2555	2559	2557	2557

[a] 10% of the Maintenance costs; [b] Usually negligible; [c] 5% of the total fixed capital; [d] 20% of the Operating labour costs; [e] 50% of the Operating labour costs; [f] 1% of the total fixed capital; [g] 2% of the total fixed capital; PEF—pulsed electric field.

In this case-study, it was assumed that raw materials costs were negligible because all the process requirements can be found in the local site of the facility: nutrients are supplied in the domestic wastewater that is daily fed into the algal ponds and CO_2 is obtained from flue gas emissions from the thermoelectric power plant located in Sines and from the anaerobic digestion and combined heat and power generation processes. Regarding the utilities, electrical energy was considered the most important one. According to the energetic balance, total energy required oscillates between the studied scenarios, being 2.3×10^4 kWh·day^{-1} in the base scenario. Considering electricity costs of 0.10 € kWh^{-1} [81], total annual costs for utilities are, for the base scenario, 857 thousand euros. Assuming NaOH requirements of 9 mg·g$^{-1}_{biomass}$ and NaOH costs of 0.682 USD·kg^{-1} [83], the costs associated to the pre-concentration step were determined. Production costs for PEF extraction were determined assuming lipids extraction costs of 10 €·t^{-1} [84]. Operating labour costs were calculated assuming 10 operators with an average salary per month of 1000 €. The other parameters were determined basing on the percentages proposed by Sinnott and Towler [82] and showed in Table 8. Resulting annual production costs are, for the base scenario, approximately 3.0 million euros.

Annual Revenues

Although the major aims of this process are the production of lipids and energy (both electrical and thermal), credits from wastewater treatment, CO_2 capture and from the production of an N-rich residue that can be used as fertilizer should also be considered.

Therefore, considering lipids sales of 1 €·kg^{-1}, electricity and steam sales of 0.10 €·kWh^{-1} [81], a credit of 3.50 €·kg^{-1} of nitrogen removed and 2.40 €·kg^{-1} of phosphorus removed [76], a credit of 30 € t^{-1} of CO_2 captured [85] and an income from fertilizers sales of 0.40 €·kg^{-1} [79], total annual revenues for the studied scenarios oscillate between 2.5 and 7.4 million euros (Figure 4), being the best scenario, the one assuming a photosynthetic efficiency of 3% (scenario 3), and the worst, the one considering a photosynthetic efficiency of 1% (scenario 2). The other studied scenarios have shown similar annual revenues of about 5.0 million euros.

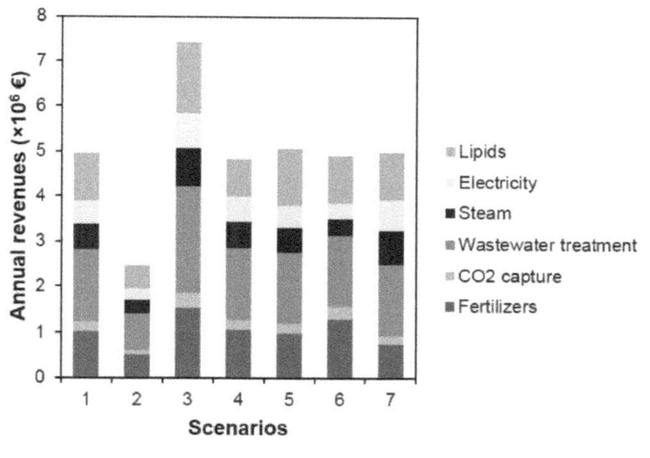

Figure 4: Annual revenues obtained in each of the studied scenarios.

Economic Viability

The viability of a project can be evaluated through the determination of net present value (NPV) and internal rate of return (IRR). NPV is the sum of present values of the individual cash-flows (revenues minus costs). When NPV is positive, the viability of the project is ensured because it implies that net income is higher than costs. On the other hand, IRR is the rate of return that makes NPV of all cash flows (both positive and negative) from a particular investment equals to zero. In other words, IRR is the minimum interest value for which there is no income, but there are no other costs. Therefore, a project is economically feasible when this value is higher than the interest rate, so that revenues are higher than costs [86].

NPV and IRR were determined for the process here described, assuming a 10% interest rate and a 30-year bond to fund the facility construction (Table 9). According to these values, it is possible to state that the project is economically viable for all the studied scenarios (except the scenario 2), as they present a positive NPV, ranging between 4.3 and 22.6 million euros. However, the best scenario is the third one (which assumes a photosynthetic efficiency of 3%), since the IRR determined for this scenario (26%) is much higher than the assumed interest rate. Additionally, a payback time of about 4 years was determined for this scenario, whereas for the other studied scenarios payback times determined range between 7 and 8 years. These results indicate that at the end of the 30-year bond term, this project would be fully amortized and debt-free for these scenarios.

Table 9: Economic viability of the proposed project in each of the studied scenarios

Economic Viability Parameters	Sc1	Sc2	Sc3	Sc4	Sc5	Sc6	Sc7
Interest rate (%)	10	10	10	10	10	10	10
Lifetime (years)	30	30	30	30	30	30	30
Net present value (NPV, k€)	5287	−12,124	22,609	4267	6307	4940	5634
Internal rate of return (IRR, %)	14	n.a.	26	13	15	14	14
Payback time (years)	8	n.a.	4	8	7	8	8

n.a.—not applicable; NPV—net present value; IRR—internal rate of return.

Taking into account the NPV obtained in each of the studied scenarios, a sensitivity analysis was performed to evaluate which are the crucial conditions for an economically viable process (Figure 5). Analysis of Figure 5 shows that photosynthetic efficiency is the most important factor influencing NPV: for photosynthetic efficiencies ranging between 1% and 3%, NPV oscillates between −12.1 and 22.6 million euros.

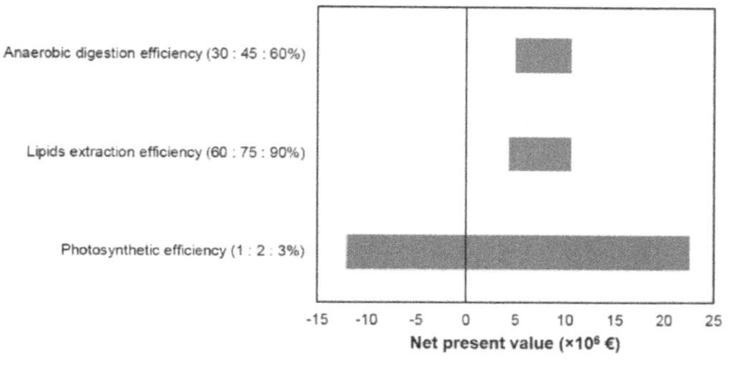

Figure 5: Sensitivity analysis of the process considering the different studied scenarios.

Regarding the other assumptions considered, lipids extraction and anaerobic digestion efficiencies, the evaluated ranges have not strongly influenced NPV, being these values positive for all studied scenarios. For lipids extraction efficiencies ranging between 60% and 90%, NPV obtained, ranged between 4.3 and 6.3 million euros. On the other hand, anaerobic digestion efficiencies between 30% and 60% have resulted in NPV of 4.9 and 5.6 million euros.

CONCLUSIONS

This study presents an economically viable process of microalgal production in Portugal concerning wastewater treatment, CO_2 emission saving and bioenergy production purposes. For this process, seven scenarios were considered, assuming different efficiencies in some of the most important steps of microalgal processing. From the considered scenarios, six were economically viable. From those, the one assuming a photosynthetic efficiency of 3%, a lipids extraction efficiency of 75% and an anaerobic digestion efficiency of 45% (scenario 3) was considered the most effective in terms of: (i) CO_2uptake (1.1×10^4 t per year); (ii) energy production (annual energy produced was 1.6×10^7 kWh and annual lipids productivity was 1.9×10^3 m^3); and (iii) economic viability (NPV of 22.6 million euros with an IRR of 26% and a payback time of 4 years). In addition, since this project assumes the use of domestic wastewater as culture medium, this scenario is also effective in nitrogen and phosphorus removal, processing 1.9×10^7 m^3 of wastewater per year.

ACKNOWLEDGMENTS

This work was financially supported by: Project UID/EQU/00511/2013-LEPABE (Laboratory for Process Engineering, Environment, Biotechnology and Energy—EQU/00511) by FEDER funds through Programa Operacional Competitividade e Internacionalização—COMPETE2020 and by national funds through FCT—Fundação para a Ciência e a Tecnologia; SFRH/BD/88799/2012 and FEUP-ON2-26-FM-Eng Sistemas.

AUTHOR CONTRIBUTIONS

Ana L. Gonçalves performed the techno-economic assessment and wrote the manuscript; Maria C. M. Alvim-Ferraz helped in manuscript revision; Fernando G. Martins helped in the economic assessment and manuscript revision; Manuel Simões helped in manuscript revision; José C. M. Pires conceived and designed this study and helped in the mass and energetic balances and in manuscript revision.

REFERENCES

1. Trends in Atmospheric Carbon Dioxide. Available online: http://www.esrl.noaa.gov/gmd/ccgg/trends/ (accessed on 19 February 2016).
2. Allen, M.R.; Frame, D.J.; Huntingford, C.; Jones, C.D.; Lowe, J.A.; Meinshausen, M.; Meinshausen, N. Warming caused by cumulative carbon emissions towards the trillionth tonne. *Nature* 2009, *458*, 1163–1166.
3. Moss, R.H.; Edmonds, J.A.; Hibbard, K.A.; Manning, M.R.; Rose, S.K.; van Vuuren, D.P.; Carter, T.R.; Emori, S.; Kainuma, M.; Kram, T. The next generation of scenarios for climate change research and assessment. *Nature* 2010, *463*, 747–756.
4. Rockström, J.; Steffen, W.; Noone, K.; Persson, Å.; Chapin, F.S.; Lambin, E.F.; Lenton, T.M.; Scheffer, M.; Folke, C.; Schellnhuber, H.J. A safe operating space for humanity. *Nature* 2009, *461*, 472–475.
5. Bilanovic, D.; Andargatchew, A.; Kroeger, T.; Shelef, G. Freshwater and marine microalgae sequestering of CO_2 at different C and N concentrations—Response surface methodology analysis. *Energy Conv. Manag.* 2009, *50*, 262–267.
6. Sayre, R. Microalgae: The potential for carbon capture. *Bioscience* 2010, *60*, 722–727.
7. Chisti, Y. Biodiesel from microalgae. *Biotechnol. Adv.* 2007, *25*, 294–306.
8. Clarens, A.F.; Resurreccion, E.P.; White, M.A.; Colosi, L.M. Environmental life cycle comparison of algae to other bioenergy feedstocks. *Environ. Sci. Technol.* 2010, *44*, 1813–1819.
9. Gonçalves, A.L.; Pires, J.C.; Simões, M. Green fuel production: Processes applied to microalgae. *Environ. Chem. Lett.*2013, *11*, 315–324.
10. Lee, A.K.; Lewis, D.M.; Ashman, P.J. Energy requirements and economic analysis of a full-scale microbial flocculation system for microalgal harvesting. *Chem. Eng. Res. Des.* 2010, *88*, 988–996.
11. Rawat, I.; Ranjith Kumar, R.; Mutanda, T.; Bux, F. Dual role of microalgae: Phycoremediation of domestic wastewater and biomass production for sustainable biofuels production. *Appl. Energy* 2011, *88*, 3411–3424.
12. Yang, J.; Li, X.; Hu, H.; Zhang, X.; Yu, Y.; Chen, Y. Growth and lipid accumulation properties of a freshwater microalga, *Chlorella ellipsoidea* YJ1, in domestic secondary effluents. *Appl. Energy* 2011, *88*, 3295–3299.
13. Lin, L.; Chan, G.; Jiang, B.; Lan, C. Use of ammoniacal nitrogen tolerant microalgae in landfill leachate treatment.*Waste Manag.* 2007, *27*, 1376–1382.
14. Mustafa, E.-M.; Phang, S.-M.; Chu, W.-L. Use of an algal consortium of five algae in the treatment of landfill leachate using the high-rate algal pond system. *J. Appl. Phycol.* 2012, *24*, 953–963.
15. González, L.E.; Cañizares, R.O.; Baena, S. Efficiency of ammonia and phosphorus removal from a Colombian agroindustrial wastewater by the microalgae *Chlorella vulgaris* and *Scenedesmus dimorphus*. *Bioresour. Technol.* 1997, *60*, 259–262.
16. Lefebvre, S.; Hussenot, J.; Brossard, N. Water treatment of land-based fish farm effluents by outdoor culture of marine diatoms. *J. Appl. Phycol.* 1996, *8*, 193–200.
17. Chojnacka, K.; Chojnacki, A.; Górecka, H. Trace element removal *Spirulina* sp. from copper smelter and refinery effluents. *Hydrometallurgy* 2004, *73*, 147–153.
18. Woertz, I.; Feffer, A.; Lundquist, T.; Nelson, Y. Algae grown on dairy and municipal wastewater for simultaneous nutrient removal and lipid production for biofuel feedstock. *J. Environ. Eng.* 2009, *135*, 1115–1122.
19. Boelee, N.C.; Temmink, H.; Janssen, M.; Buisman, C.J.N.; Wijffels, R.H. Nitrogen and phosphorus removal from municipal wastewater effluent using microalgal biofilms. *Water Res.* 2011, *45*, 5925–5933.
20. Park, J.B.K.; Craggs, R.J.; Shilton, A.N. Wastewater treatment high rate algal ponds for biofuel production. *Bioresour. Technol.* 2011, *102*, 35–42.
21. Silva-Benavides, A.; Torzillo, G. Nitrogen and phosphorus removal through laboratory batch cultures of microalga*Chlorella vulgaris* and cyanobacterium *Planktothrix isothrix* grown as monoalgal and as co-cultures. *J. Appl. Phycol.* 2012,*24*, 267–276.
22. Benemann, J.R. Hydrogen production by microalgae. *J. Appl. Phycol.* 2000, *12*, 291–300.
23. Singh, S.; Kate, B.; Banerjee, U. Bioactive compounds from cyanobacteria and microalgae: An overview. *Crit.*

Rev. Biotechnol. 2005, *25*, 73–95.

24. Spolaore, P.; Joannis-Cassan, C.; Duran, E.; Isambert, A. Commercial applications of microalgae. *J. Biosci. Bioeng.* 2006,*101*, 87–96.
25. Falkouski, P.G.; Raven, J.A. *Aquatic Photosynthesis*, 2nd ed.; Princeton University Press: Princeton, NJ, USA, 2007; pp. 5–21.
26. Pires, J.C.M.; Alvim-Ferraz, M.C.M.; Martins, F.G.; Simões, M. Wastewater treatment to enhance the economic viability of microalgae culture. *Environ. Sci. Pollut. Res.* 2013, *20*, 5096–5105.
27. Richardson, J.W.; Johnson, M.D.; Zhang, X.; Zemke, P.; Chen, W.; Hu, Q. A financial assessment of two alternative cultivation systems and their contributions to algae biofuel economic viability. *Algal Res.* 2014, *4*, 96–104.
28. Slegers, P.; Leduc, S.; Wijffels, R.; van Straten, G.; van Boxtel, A. Logistic analysis of algae cultivation. *Bioresour. Technol.* 2015, *179*, 314–322.
29. Venteris, E.R.; McBride, R.C.; Coleman, A.M.; Skaggs, R.L.; Wigmosta, M.S. Siting algae cultivation facilities for biofuel production in the United States: Trade-offs between growth rate, site constructability, water availability, and infrastructure. *Environ. Sci. Technol.* 2014, *48*, 3559–3566.
30. Densidade Populacional Segundo os Censos em Portugal. Available online: http://www.pordata.pt/Portugal/Densidade+populacional+segundo+os+Censos-412 (accessed on 5 March 2013). (In Portugese).
31. Kumar, A.; Ergas, S.; Yuan, X.; Sahu, A.; Zhang, Q.; Dewulf, J.; Malcata, F.X.; van Langenhove, H. Enhanced CO_2fixation and biofuel production via microalgae: Recent developments and future directions. *Trends Biotechnol.* 2010,*28*, 371–380.
32. Photovoltaic Geographical Information system—Incident Global Irradiation. Available online: http://re.jrc.ec.europa.eu/pvgis/apps4/pvest.php (accessed on 20 Jaunary 2013).
33. Rodrigues, C.M.M. Cálculo da evaporação de albufeiras de grande regularização do sul de Portugal. Ph.D. Thesis, Universidade de Évora, Évora, Portugal, October 2009. (In Portugese).
34. Chazarra, A.; Barceló, A.M.; Pires, V.; Cunha, S.; Mendes, M.; Neto, J. *Iberian Climate Atlas—Air Temperature and Precipitation (1971–2000)*; AEMET-IM—Agencia Estatal de Meteorología de España and Instituto de Meteorologia de Portugal: Madrid, Spain, 2001.
35. Borowitzka, M.A. Commercial production of microalgae: Ponds, tanks, and fermenters. *Prog. Ind. Microbiol.* 1999, *35*, 313–321.
36. Posten, C.; Schaub, G. Microalgae and terrestrial biomass as source for fuels—A process view. *J. Biotechnol.* 2009, *142*, 64–69.
37. Pulz, O. Photobioreactors: Production systems for phototrophic microorganisms. *Appl. Microbiol. Biotechnol.* 2001, *57*, 287–293.
38. Lee, Y.-K. Microalgal mass culture systems and methods: their limitation and potential. *J. Appl. Phycol.* 2001, *13*, 307–315.
39. Ugwu, C.; Aoyagi, H.; Uchiyama, H. Photobioreactors for mass cultivation of algae. *Bioresour. Technol.* 2008, *99*, 4021–4028.
40. Jiménez, C.; Cossío, B.R.; Labella, D.; Niell, F.X. The feasibility of industrial production of *Spirulina* (*Arthrospira*) in Southern Spain. *Aquaculture* 2003, *217*, 179–190.
41. Tredici, M.R.; Materassi, R. From open ponds to vertical alveolar panels: The Italian experience in the development of reactors for the mass cultivation of phototrophic microorganisms. *J. Appl. Phycol.* 1992, *4*, 221–231.
42. Lardon, L.; Hélias, A.; Sialve, B.; Steyer, J.-P.; Bernard, O. Life-cycle assessment of biodiesel production from microalgae. *Environ. Sci. Technol.* 2009, *43*, 6475–6481.
43. Mata, T.M.; Melo, A.C.; Simões, M.; Caetano, N.S. Parametric study of a brewery effluent treatment by microalgae*Scenedesmus obliquus*. *Bioresour. Technol.* 2012, *107*, 151–158.
44. Wang, L.; Min, M.; Li, Y.; Chen, P.; Chen, Y.; Liu, Y.; Wang, Y.; Ruan, R. Cultivation of green algae *Chlorella* sp. in different wastewaters from municipal wastewater treatment plant. *Appl. Biochem. Biotechnol.* 2010, *162*, 1174–1186.
45. Cai, T.; Park, S.Y.; Li, Y.B. Nutrient recovery from wastewater streams by microalgae: Status and prospects. *Renew. Sust. Energy Rev.* 2013, *19*, 360–369.
46. Flisar, K.; Meglic, S.H.; Morelj, J.; Golob, J.; Miklavcic, D. Testing a prototype pulse generator for a continuous flow system and its use for *E. coli* inactivation and microalgae lipid extraction. *Bioelectrochemistry* 2014, *100*, 44–51.
47. Norsker, N.-H.; Barbosa, M.J.; Vermuë, M.H.; Wijffels, R.H. Microalgal production—A close look at the economics.*Biotechnol. Adv.* 2011, *29*, 24–27.
48. Williams, P.J.L.B.; Laurens, L.M.L. Microalgae as biodiesel & biomass feedstocks: Review & analysis of the biochemistry, energetics & economics. *Energy Environ. Sci.* 2010, *3*, 554–590.
49. Joannes, C.; Sipaut, C.; Dayou, J.; Yasir, S.; Mansa, R. Review Paper on Cell Membrane Electroporation of Microalgae Using Electric Field Treatment Method for Microalgae Lipid Extraction. In Proceedings of the 9th CUTSE International Conference, Miri, Sarawak, Malaysia, 3–4 December 2014; IOP Publishing: Miri, Sarawak, Malaysia, 2015.
50. Show, K.-Y.; Lee, D.-J.; Tay, J.-H.; Lee, T.-M.; Chang, J.-S. Microalgal drying and cell disruption—Recent advances.*Bioresour. Technol.* 2015, *184*, 258–266.
51. Kinnunen, V.; Craggs, R.; Rintala, J. Influence of temperature and pretreatments on the anaerobic digestion of wastewater grown microalgae in a laboratory-scale accumulating-volume reactor. *Water Res.* 2014, *57*, 247–257.
52. Tran, K.C.; Mendoza Martin, J.L.; Heaven, S.; Banks, C.J.; Acien Fernandez, F.G.; Molina Grima, E.

Cultivation and anaerobic digestion of *Scenedesmus* spp. grown in a pilot-scale open raceway. *Algal Res.* 2014, *5*, 95–102.

53. Hirayama, S.; Ueda, R.; Sugata, K. Evaluation of active oxygen effect on photosynthesis of *Chlorella vulgaris*. *Free Rad. Res.* 1996, *25*, 247–254.

54. Doucha, J.; Lívanský, K. Novel outdoor thin-layer high density microalgal culture system: Productivity and operational parameters. *Algol. Stud.* 1995, *76*, 129–147.

55. Gonçalves, A.L.; Simões, M.; Pires, J.C.M. The effect of light supply on microalgal growth, CO_2 uptake and nutrient removal from wastewater. *Energy Conv. Manag.* 2014, *85*, 530–536.

56. Gouveia, L. *Microalgae as a Feedstock for Biofuels*; Springer: Heidelberg, Germany, 2011; pp. 30–32.

57. Wang, B.; Lan, C.Q. Biofixation of carbon dioxide (CO_2) by microorganisms. In *Developments and Innovation in Carbon Dioxide (CO_2) Capture and Storage Technology*, 1st ed.; Maroto-Valer, M.M., Ed.; Woodhead Publishing Limited: Cambridge, UK, 2010; Volume 2.

58. Davis, R.; Aden, A.; Pienkos, P.T. Techno-economic analysis of autotrophic microalgae for fuel production. *Appl. Energy* 2011, *88*, 3524–3531.

59. Vandamme, D. Flocculation Based Harvesting Processes for Microalgae Biomass Production. Ph.D. Thesis, KU Leuven, Belgium, May 2013.

60. Barros, A.I.; Gonçalves, A.L.; Simões, M.; Pires, J.C.M. Harvesting techniques applied to microalgae: A review. *Renew. Sustain. Energy Rev.* 2015, *41*, 1489–1500.

61. Molina Grima, E.; Belarbi, E.H.; Acién Fernández, F.G.; Robles Medina, A.; Chisti, Y. Recovery of microalgal biomass and metabolites: Process options and economics. *Biotechnol. Adv.* 2003, *20*, 491–515.

62. Directive 1991/271/EEC. Directive of the European Council of 21 May 1991 Concerning Urban Waste-Water Treatment.*Off. J. Eur. Union* 1991, *L 135*, 40–52.

63. Directive 1998/15/EC. Directive of the European Comission of 27 February 1998 amending Council Directive 91/271/EEC with respect to certain requirements established in Annex I thereof. *Off. J. Eur. Union* 1998, *L 67*, 29–30.

64. Guderjan, M.; Töpfl, S.; Angersbach, A.; Knorr, D. Impact of pulsed electric field treatment on the recovery and quality of plant oils. *J. Food Eng.* 2005, *67*, 281–287.

65. Taher, H.; Al-Zuhair, S.; Al-Marzouqui, A.H.; Haik, Y.; Farid, M.M. A review of enzymatic transesterification of microalgal oil-based biodiesel using supercritical technology. *Enzyme Res.* 2011.

66. Guderjan, M.; Elez-Martínez, P.; Knorr, D. Application of pulsed electric fields at oil yield and content of functional food ingredients at the production of rapeseed oil. *Innov. Food Sci. Emerg. Technol.* 2007, *8*, 55–62.

67. Halim, R.; Danquah, M.K.; Webley, P.A. Extraction of oil from microalgae for biodiesel production: A review. *Biotechnol. Adv.* 2012, *30*, 709–732.

68. Foltz, G. Algae Lysis with Pulsed Electric Fields. Master's Thesis, California State Polytechnic University, San Luis Obispo, CA, USA, May 2012.

69. Luengo, E.; Condón-Abanto, S.; Álvarez, I.; Raso, J. Effect of pulsed electric field treatments on permeabilization and extraction of pigments from *Chlorella vulgaris*. *J. Membr. Biol.* 2014, *247*, 1269–1277.

70. Schlagermann, P.; Göttlicher, G.; Dillschneider, R.; Rosello-Sastre, R.; Posten, C. Composition of algal oil and its potential as biofuel. *J. Comb.* 2012, *2012*, 1–14.

71. Sialve, B.; Bernet, N.; Bernard, O. Anaerobic digestion of microalgae as a necessary step to make microalgal biodiesel sustainable. *Biotechnol. Adv.* 2009, *27*, 409–416.

72. Gonçalves, A.L.; Rodrigues, C.M.; Pires, J.C.M.; Simões, M. The effect of increasing CO_2 concentrations on its capture, biomass production and wastewater bioremediation by microalgae and cyanobacteria. *Algal Res.* 2016, *14*, 127–136.

73. Lundquist, T.J.; Woertz, I.C.; Quinn, N.; Benemann, J.R. *A Realistic Technology and Engineering Assessment of Algae Biofuel Production*; Energy Biosciences Institute: Berkeley, CA, USA, 2010.

74. Fans—Efficiency and Power Consumption. Available online: http://www.engineeringtoolbox.com/fans-efficiency-power-consumption-d_197.html (accessed on 22 December 2015).

75. Tredici, M.R.; Bassi, N.; Prussi, M.; Biondi, N.; Rodolfi, L.; Chini Zittelli, G.; Sampietro, G. Energy balance of algal biomass production in a 1-ha "Green Wall Panel" plant: How to produce algal biomass in a closed reactor achieving a high Net Energy Ratio. *Appl. Energy* 2015, *154*, 1103–1111.

76. Zamalloa, C.; Vulsteke, E.; Albrecht, J.; Verstraete, W. The techno-economic potential of renewable energy through the anaerobic digestion of microalgae. *Bioresour. Technol.* 2011, *102*, 1149–1158.

77. Biogas renewable energy—The biogas composition. Available online: http://www.biogas-renewable-energy.info/biogas_composition.html (accessed on 22 December 2015).

78. Davis, R.; Fishman, D.; Frank, E.D.; Wigmosta, M.S. *Renewable Diesel from Algal Lipids: An Integrated Baseline for Cost, Emissions, and Resource Potential from a Harmonized Model*; U.S. Department of Energy Biomass Program: Oak Ridge, TN, USA, 2012.

79. Acién, F.; Fernández, J.; Magán, J.; Molina, E. Production cost of a real microalgae production plant and strategies to reduce it. *Biotechnol. Adv.* 2012, *30*, 1344–1353.

80. Benemann, J.R.; Oswald, W.J. *Systems and Economic Analysis of Microalgae Ponds for Conversion of CO_2 to Biomass—Final Report*; Department of Civil Engineering of the California University: Berkeley, CA, USA, 1996.

81. Peters, M.S.; Timmerhaus, K.D. *Plant. Design and Economics for Chemical Engineers*, 4th ed.; McGraw-Hill: New York, NY, USA, 1991.

82. Sinnott, R.K.; Towler, G. *Chemical Engineering Design*, 4th ed.; Elsevier Butterworth-Heinemann: Oxford, UK, 2009; pp. 243–268.

83. Chemical Market Insight and Foresight on a Single Page—Caustic Soda. Available online: http://www.orbichem.com/userfiles/CNF%20Samples/cas_13_11.pdf (accessed on 22 Feburay 2016).

84. FAQ about Pulsed Electric Field Processing. Available online: https://www.pulsemaster.us/pef-pulsemaster/faq(accessed on 23 December 2015).

85. Declercq, B.; Delarue, E.; D'haeseleer, W. Impact of the economic recession on the European power sector's CO2 emissions. *Energy Policy* 2011, *39*, 1677–1686.

86. Mithá, O. *Análise de Projectos de Investimento*; Escolar Editora: Lisboa, Portugal, 2004. (In Portuguese)

CHAPTER 5

AUTO-THERMAL REFORMING USING MIXED ION-ELECTRONIC CONDUCTING CERAMIC MEMBRANES FOR A SMALL-SCALE H2 PRODUCTION PLANT

Vincenzo Spallina, Tommaso Melchiori, Fausto Gallucci, and Martin van Sint Annaland

Chemical Process Intensification, Chemical Engineering and Chemistry, Eindhoven University of Technology, 5600MB Eindhoven, The Netherlands

ABSTRACT

The integration of mixed ionic electronic conducting (MIEC) membranes for air separation in a small-to-medium scale unit for H_2 production (in the range of 650–850 Nm^3/h) via auto-thermal reforming of methane has been investigated in the present study. Membranes based on mixed ionic electronic conducting oxides such as $Ba_{0.5}Sr_{0.5}Co_{0.8}Fe_{0.2}O_{3-\delta}$ (BSCF) give sufficiently high oxygen fluxes at temperatures above 800 °C with high purity (higher than 99%). Experimental results of membrane permeation tests are presented and used for the reactor design with a detailed reactor model. The assessment of the H_2 plant has been carried out for different operating conditions and reactor geometry and an energy analysis has been carried out with the flowsheeting software Aspen Plus, including also the turbomachines required for a proper thermal integration. A micro-gas turbine is integrated in the system in order to supply part of the electricity required in the system. The analysis of the system shows that the reforming efficiency is in the range of 62%–70% in the case where the temperature at the auto-thermal reforming membrane reactor (ATR-MR) is equal to 900 °C. When the electric consumption and the thermal export are included the efficiency of the plant approaches 74%–78%. The design of the reactor has been carried out using a reactor model linked to the Aspen flowsheet and the results show that with a larger reactor volume the performance of the system can be improved, especially because of the reduced electric consumption. From this analysis it has been found that for a production of about 790 Nm^3/h pure H_2, a reactor with a diameter of 1 m and length of 1.8 m with about 1500 membranes of 2 cm diameter is required.

INTRODUCTION

Currently approximately 95% of all the hydrogen used is produced from fossil fuels as primary feedstock, where technologies such as natural gas reforming and coal gasification are applied at very large scale [1]. Hydrogen's share in the energy market is increasing with the implementation of fuel cell systems and the growing demand for zero-emission fuels. The potential of hydrogen as a future energy carrier has been reviewed by different authors [2,3], who concluded that the costs associated with building the infrastructure and with transporting hydrogen over large distances are not sustainable from an economic point of view and in particular for the

Citation: Vincenzo Spallina, Tommaso Melchiori, Fausto Gallucci, and Martin van Sint Annaland, Auto-Thermal Reforming Using Mixed Ion-Electronic Conducting Ceramic Membranes for a Small-Scale H2 Production Plant, Molecules 2015, 20, 4998-5023; doi:10.3390/molecules20034998

transport sector. Additionally, large-scale hydrogen production plants are dependent on the market volumes and subjected to the cost evolution of the products. An attractive option is a small stationary unit for on-site hydrogen production via steam methane reforming (SMR) or methane partial oxidation (POX) [4]. Auto-thermal reforming (ATR) combines both SMR and POX so that the heat required for the endothermic reaction is supplied by using a proper amount of oxidant and no external furnace and indirect heat exchangers are required, thus making the system more compact. All these systems also require water-gas-shift reactor(s) to enhance the H_2 yield and some form of gas purification such as a pressure swing adsorption (PSA) system. Using air as oxidant for the auto-thermal reaction leads to higher volume flow rates and a high fuel consumption due to the intrinsic dilution with N_2. When using pure oxygen, cryogenic air separation is currently the most efficient and mature technology for producing large quantities of O_2 (>0.75 t/h of pure O_2) with a specific energy consumptions in the range of 160–250 kWh/t_{O2} depending on the O_2 purity [5], but inefficient for small scale applications, where different other systems have recently been proposed based on N_2adsorption, molten salt absorption and polymeric membranes [6].

The study of inorganic membrane (also called mixed ionic-electronic conducting membrane or MIEC) materials and processes for O_2 production has increased rapidly in recent years due to the potential to improve the overall efficiency and environmental footprint, while reducing the cost [7,8,9,10,11] In a MIEC membrane, O_2 permeates through the membrane after applying to the system a driving force which can be either an electrical potential gradient or a chemical potential gradient. The MIEC membrane can also be integrated in a membrane reactor in case the permeated O_2 reacts at the permeate side, as shown in Figure 1. The use of MIEC membranes for integrated O_2 separation has been studied for different applications, where most of the proposed ones relate to the production of syngas and oxy-combustion systems [12,13]. Praxair Inc. is working on the integration of oxygen transport membranes in a coal gasification unit [13] using multi-stage oxygen membrane units operated at different pressures to gradually convert the syngas into exhaust gases and separating the by-products downstream. For a natural gas fired plant for pre-combustion CO_2 capture [14], Chiesa et al. [15] proposed integrating the oxygen membranes with an auto-thermal reforming unit with CO_2 separation by amine absorption. H_2 production using oxygen perm-selective membranes has also been studied, amongst others by Gallucci et al. [16,17]. They proposed a multi-stage fluidized bed membrane reactor, where oxygen membranes are used to permeate O_2 that reacts with methane and steam to form reformed syngas in the bottom part and where H_2 selective membranes are used in the top part of the reactor to separate pure H_2 and simultaneously convert the remaining syngas into a CO_2-rich stream that can be separated and stored. MIEC membranes have also been studied for the oxidative coupling of methane to produce C_2hydrocarbons, which is a highly exothermic reaction that requires a reaction temperature of 700–900 °C and low oxygen partial pressures to achieve high C_2 yields [18]. O_2 membrane reactors have also been studied to control the temperature of the system to avoid problems with hot-spot formation, which may occur in conventional auto-thermal reforming reactors, where the combustion zone is separated from the catalytic zone [19].

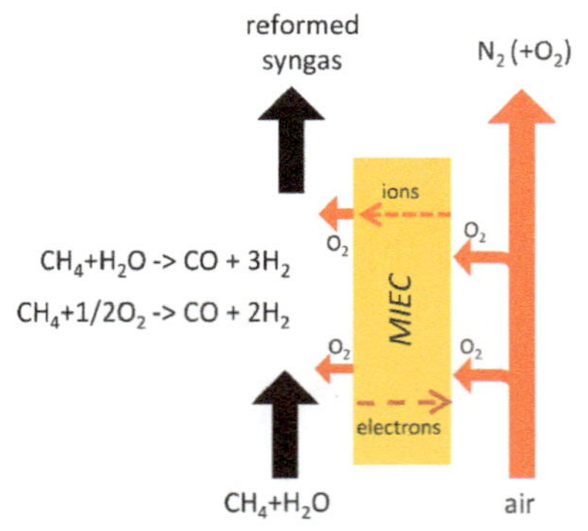

Figure 1: Schematic of an ATR-MR based on oxygen transport membrane.

In this paper MIEC membranes are integrated in an auto-thermal reforming system (ATR-MR) for H_2 production at small-to-medium scale, where the ATR-MR system is fully integrated with the other components of the plant including the micro-gas turbine. Micro-gas turbines have been studied in recent years as valid alternatives to internal combustion engines in the 10–1000 kW_e range, which differ from the widely used commercial heavy duty gas-turbines (mostly used in large scale combined cycles) in their different size, design and performance [20,21,22,23]. To carry out plant simulations, a model has been developed in the flowsheeting software program Aspen. The Aspen flowsheet was coupled with an external executable file that via an Excel interface imports the values of the variables related to the streams entering the reactor block (such as flow rates, compositions, temperatures, etc.). The specific reactor model developed in Matlab solves mass and energy balance for the membrane reactor according to the specific design and returns the values of the variables at the outlet. After that, the Aspen model performs the calculations of the downstream units. A detailed analysis of the plant is discussed in the paper with a specific focus on the design of the membrane reactor. The membrane flux, membrane dimension and scale-up is based on experimental results that have been obtained with permeation tests of a $Ba_{0.5}Sr_{0.5}Co_{0.8}Fe_{0.2}O_{3-\delta}$ (BSCF) membrane, which is shortly described first.

MEMBRANE TESTING

The oxygen permeation properties of BSCF membranes were studied in a high-temperature oxygen permeation setup as shown in Figure 2.

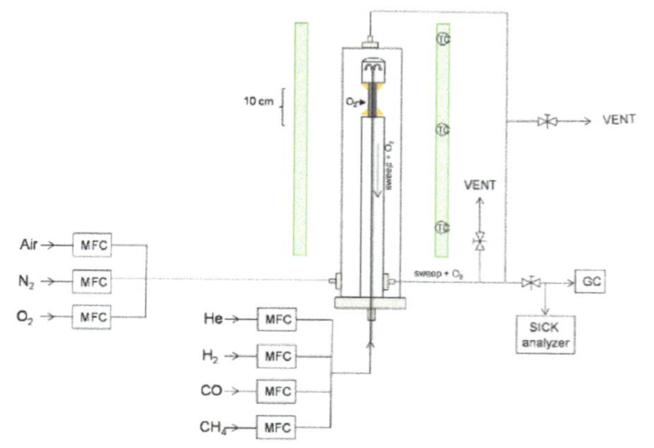

Figure 2. Layout of the experimental setup used for the permeation test. Membrane length: 10 cm, membrane tube outer diameter: 3.5 mm, membrane thickness: 0.5 mm.

Description of the Experimental Setup

Air is fed from the bottom in an oven (type Eli-tube 250/26, ELicra, Nijkerk, The Netherlands) which can reach temperatures up to 1150 °C. Due to the size, the oven has three parts that are heating the entire reactor and each part is controlled by a thermocouple. The BSCF perovskite membranes were prepared by the Flemish Institute of Technological Research (VITO, Mol, Belgium). Gold paste and Schott glass were used as sealants to fix the membrane between a dense alumina tube and alumina cup [24]. A capillary was placed inside the tube (and the membrane) through which sweep gas (He or reducing gases) is fed. During the permeation experiments, a maximum heating rate of 3 °C/min was used (to prevent the sealing and membrane from cracking due to thermal stresses). Once the membrane has been activated inside the reactor, the temperatures were kept above 800 °C and below 1000 °C to retain the chemical structure of the membrane and prevent thermal stresses in the reactor. Permeation flux experiments have been conducted by adding an air flow through the shell and a sweep gas through the capillary. During non-reducing conditions, He (99.995%) was used as sweep gas.

The oxygen permeation flux was measured as a function of the He flow rate (from 50 to 350 SmL/min) at different temperatures in the range of 850–1000 °C (Figure 3). The results clearly show that at higher temperatures the O_2 permeation rate through the membrane increases, because of the higher O_2 bulk diffusion and surface exchange kinetics. More details about these membranes can be found in [24]. As expected, increasing the sweep gas flow rate also increased the O_2permeation as a result of the higher partial pressure difference across the membrane. The experimental results from these permeation tests have been correlated to develop a permeation expression that has been implemented in the reactor model, as discussed in the next section.

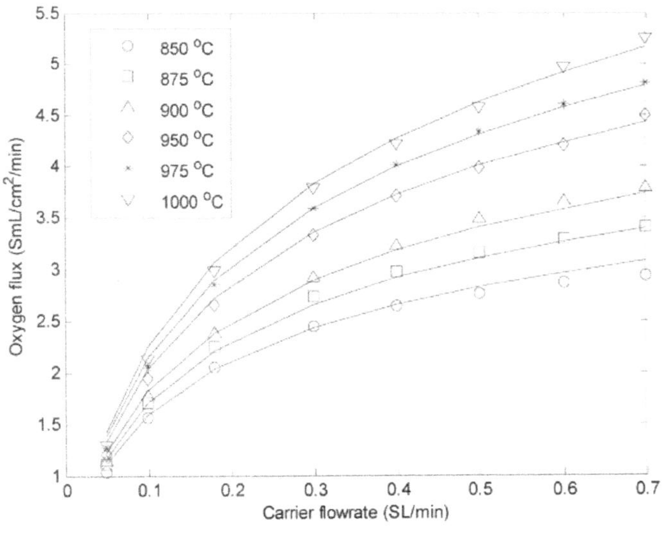

Figure 3: Comparison between experimental (symbols) and calculated (continuous lines) oxygen fluxes using BSCF membrane at different temperatures and carrier flow rates.

REACTOR MODEL

Description of the Model

The model describes a catalytic bed reactor in which a number of membrane tubes are immersed, and the corresponding model equations have been listed in Table 1. The catalyst is supposed to be placed in the shell side, where the partially reformed methane is fed, and where the reactions of reforming oxidation and water-gas shift take place. Inside the tubes, on the other hand, no particles are present. Air is fed here, co-current with the fuel flowing in the shell. Oxygen partially permeates through the membranes surface, and reacts with methane when it reaches the catalytic section. No back permeation of gas is assumed inside the tubes, so no reaction occurs in the air section. The model solves the chemical species balances in both the retentate (air) and permeate (fuel) side, assuming a 1D geometry for the two zones. The equations include transport by convection, production or consumption due to reaction (only in the fuel side), and production and consumption due to mass transfer from the air side to the fuel side. This term is actually present only in the equation for oxygen, because the membrane is assumed to be impermeable to all the other components because of the very high perm-selectivity measured for the BSCF membrane. The mass transfer terms are expressed as a function of the oxygen concentration difference between the bulk and the membrane surface. The mass transfer coefficients are calculated with the correlation by McCabe et al. [25].

Table 1: Equations for the membrane reactor model

Equations for the Fuel Side

$$\frac{dc_i^f v^f}{dz} = r_i + h_i^f a^f (c_i^{m,f} - c_i^f) \qquad (1)$$

$$\frac{dc^f v^f}{dz} = \sum_{j=1}^{N} \left(r_j + h_j^f a^f (c_j^{m,f} - c_j^f)\right) \qquad (2)$$

$$c^f Cp^f \frac{dT^f}{dz} = -\sum_{j=1}^{N} [H_j(T^f) r_j] - U a^f (T^f - T^a) + \sum_{j=1}^{N} [h_j^f a^f (c_j^{m,f} - c_j^f)(H_j(T^a) - H_j(T^f))] \qquad (3)$$

$$\frac{dP^f}{dz} = -\frac{\rho^f v^f}{d_p} \frac{(1-\varepsilon)}{\varepsilon^3} \left(\frac{150\mu^f (1-\varepsilon)}{\rho^f d_p} + 1.75 v^f\right) \qquad (4)$$

$$r_i = \sum_{j=1}^{NR} v_{ij} R_j \, \rho_{cat} (1-\varepsilon) \qquad (5)$$

Equations for the Air Side

$$\frac{dc_i^a v^a}{dz} = -h_i^a a^a (c_i^a - c_i^{m,a}) \qquad (6)$$

$$\frac{dc^a v^a}{dz} = -\sum_{j=1}^{N} \left(h_j^a a^a (c_j^a - c_j^{m,a})\right) \qquad (7)$$

$$c^a Cp^a \frac{dT^a}{dz} = U a^a (T^f - T^a) \frac{S^f}{S^a} \qquad (8)$$

Equations at the Membrane Interface

$$h_i^f (c_i^{m,f} - c_i^f) = J^m \qquad (9)$$

$$-h_i^a (c_i^{m,a} - c_i^a) = J^m \frac{S^f}{S^a} \qquad (10)$$

$$k^f (T^f - T^{m,f}) = U (T^f - T^a) \qquad (11)$$

$$k^a (T^{m,a} - T^a) = U (T^f - T^a) \frac{S^f}{S^a} \qquad (12)$$

Equation for the Oxygen Flux Through the Membrane

$$J^m = \frac{D_v k_r \left((p_i^{m,a})^n - (p_i^{m,f})^n\right)}{2\delta k_f (p_i^{m,a} p_i^{m,f})^n + D_v \left((p_i^{m,a})^n + (p_i^{m,f})^n\right)} \qquad (13)$$

$$n = 0.5$$

Heat and Mass Transfer Coefficients

$$h_i^f = 0.0096 \frac{D_i}{d^{eq,f}} Sc_f^{0.346} Re_f^{0.913} \qquad (14)$$

$$h_i^a = 0.0096 \frac{D_i}{d^{eq,a}} Sc_a^{0.346} Re_a^{0.913} \qquad (15)$$

$$k^f = 0.17 \frac{\lambda^f}{d_p} \left(\frac{Pr_f}{0.7}\right)^{1/3} Re_p^{0.79} \qquad (16)$$

$$k^a = 0.023 \frac{\lambda^a}{R^m} (\mathrm{Pr}_a)^{1/3} Re_a^{0.8} \qquad (17)$$

$$U = \frac{1}{\frac{1}{k^f} + \frac{1}{k^a}\frac{S^f}{S^a} + \frac{R^m}{\lambda^m} \ln\left(\frac{R^m + \delta}{R^m}\right)} \qquad (18)$$

Reaction Rate Expressions

$$R_1 = \frac{k_{1a} p_{CH4} p_{O2}}{(1 + K_{CH4}^{ox} p_{CH4} + K_{O2}^{ox} p_{O2})^2} + \frac{k_{1b} p_{CH4} p_{O2}}{(1 + K_{CH4}^{ox} p_{CH4} + K_{O2}^{ox} p_{O2})} \qquad (19)$$

$$R_2 = \frac{k_2 (p_{CH4} p_{H2O} - p_{H2}^3 p_{CO} / K_2)}{p_{H2O}^{1.596}} \qquad (20)$$

$$R_3 = \frac{k_3 (p_{CO} p_{H2O} - p_{H2} p_{CO2} / K_3)}{p_{H2O}} \qquad (21)$$

A kinetic model based on the Ni/Al_2O_3 catalyst was considered, including three reactions: methane total combustion, methane steam reforming and water-gas-shift reaction, as follows:

$$R_1 : CH_4 + 2O_2 \rightarrow CO_2 + 2H_2O$$

$R_2 : CH_4 + H_2O \leftrightarrow CO + 3H_2$

$R_3 : CO + H_2O \leftrightarrow CO_2 + H_2$

The direct reaction of partial oxidation of methane is not included in the mechanism: as a matter of fact, for the considered catalyst, and indirect route of formation of syngas is usually assumed in the literature, as shown by the study by Dissanayake et al. [26], followed by other authors [27,28]. According to this, steam must be formed by total combustion (when it is not directly provided), before consecutively reacting with methane to form CO and H_2. Water gas shift is also important to determine the final ratio between the products.

The kinetics of the methane total combustion is described by the equations proposed by Trimm and Lam [29], while the methane steam reforming and water-gas-shift by the corrections by Numaguchi and Kikuchi [30]. It must be pointed out that, while Numaguchi and Kikuchi [30] actually derived their parameters for the Ni based catalyst, the model by Trimm and Lam [29] is valid for the Pt/Al_2O_3 system. For this reason, the equation and adsorption constants for the methane combustion were corrected for the Ni/Al_2O_3 catalyst, as proposed by de Smet et al. [28]. The corresponding kinetic parameters used in this work have been reported in Table 2.

Table 2: Kinetic parameters for methane combustion, reforming and water gas shift reaction

	Pre exponential factor	Activation energy (kJ/mol)
$k_{1a}(mol\,s^{-1}kg_{cat}^{-1}bar^{-2})$	8.11×10^5	86.0
$k_{1b}(mol\,s^{-1}kg_{cat}^{-1}bar^{-2})$	6.82×10^5	86.0
$k_2(mol\,s^{-1}kg_{cat}^{-1}bar^{-0.404})$	2.62×10^5	106.9
$k_3(mol\,s^{-1}kg_{cat}^{-1}bar^{-1})$	2.45×10^2	54.5
$K^{ox}_{CH4}(bar^{-1})$	1.26×10^{-1}	−27.3
$K^{ox}_{O2}(bar^{-1})$	7.87×10^{-7}	−92.8

In the catalytic section, the pressure drop over the bed is taken into account, using the Ergun equation in its differential form. A pseudo-homogeneous energy balance was also added for both the fuel and air compartments, to describe the axial temperature profiles in the tubes and shell. These equations include the source terms due to the heat of reaction, and heat exchange through the membranes. The global heat transfer coefficient was calculated considering the internal and external heat transfer resistances, as well as the heat conduction resistance through the membrane, where the heat transfer coefficients were calculated from correlations from the literature [31,32]. In the equation for the fuel side, a further source term is present that describes the effect of mixing of the reacting gas with the oxygen being transported through the membrane at a different temperature. This term is related to the enthalpy associated to the mass flux. The evaluation of the heat capacity and enthalpies of the single species is performed through the Shomate equation, with the thermodynamic parameters from the National Institute of Standards and Technology (NIST) database [33].

Finally, a total mass balance is considered both for the tubes and for shell side. These two equations are used to evaluate the local gas velocity that can change due to changes in the temperature, pressure, or as a result of a reaction occurring with a variation in the total number of moles, and mass transfer between the different compartments, where the ideal gas law is used as equation of state. The oxygen membrane permeation depends on the values of the oxygen concentration at the internal and external membrane surfaces, evaluated with Equations (9) and (10), while for the oxygen permeation flux the generalized Equation (13) is used [34]. The permeability parameters were evaluated by comparing experimental results for the BSCF membrane, as shown in the next section. These parameters are also function of the membrane temperature, so the temperature profiles at the tube-shell interface are also calculated using Equations (11) and (12). Note that the equation for the oxygen permeation flux depends non-linearly on the interface oxygen partial pressures, so that the final problem to solve is a mixed differential and algebraic system (DAE). The model was implemented and compiled in Matlab.

Determination of Membrane Permeability Parameters

The previously described membrane reactor model was first used to estimate the values of the permeability parameters in the permeation flux expression (Equation (13)). In their studies, Xu and Thomson [34], and Hong et al. [35] reported values for the pre-exponential factors and activation energies to calculate D_v, k_r and k_f for a $La_{0.6}Sr_{0.4}Co_{0.2}Fe_{0.8}O_{3-\delta}$ (LSCF)-type perovskite membrane. The experimental results reported for the BSCF membranes show higher oxygen fluxes compared to their data (as expected for this kind of membranes), and the parameters were fitted to the presented experimental data.

In order to calculate the parameters, a modified version of the model described previously was used. The experimental data to fit were obtained by fixing the reactor temperature in each test. For this reason, the energy balance could be ignored and the temperature was fixed in the model. In addition, in the presented tests, the air flow rate was very high compared to the permeated oxygen flux, so that the oxygen fraction at the retentate side could be considered constant and only the permeate side needed to be modeled. Finally, the model was adapted to the different geometry of the experimental setup (single tube and permeate at the inner side). The model was coupled with an optimization solver that minimizes the sum of square errors, defined as the difference between the calculated and experimental oxygen fluxes for each test, by changing the values of the parameters in Equation (13). The results of the fitting are shown in Figure 3, and the obtained parameters are reported in Table 3. The figure shows that the agreement between experimental and calculated values is quite good with a mean relative error equal to ±2.5%.

Table 3: Permeability parameters obtained from the data fitting

-	Pre exponential Factor	Activation Energy (kJ/mol)
$D_v(m^2s^{-1})$	9.823	91.8
$k_r(mol\,m^{-2}s^{-1})$	15.36	56.3
$k_f(m\,s^{-1}Pa^{-0.5})$	308.5	267.0

DESCRIPTION OF THE PLANT

Plant Design

The schematic layout of the H_2 production plant used in this analysis is presented in Figure 4. The main assumptions for the calculation are in Table 4. The different points (the symbol # stands for point) in Figure 4 are summarized in Table 5. The methane stream (#1) is first pre-heated up to 400 °C and mixed with steam (#23) at 400 °C. The feed is pre-heated to 600 °C and sent to the Heat Exchanger Reforming (HER). In the HER, the feed gas passes through a catalytic bed in which both gas heating and steam methane reforming take place while the hot flue gases from the ATR Membrane Reactor are passing through the tubes immersed in the catalytic reactor and are cooled by supplying the heat required for the reaction. Despite the fact expensive metal dusting resistant materials are required compared to conventional adiabatic primary reformer units, the HER has been commercialized [36] thanks to the optimal heat recovery accomplished with this system. The HER is also widely studied especially for ATR units, where the use of pre-reforming boosts the overall efficiency [37]. The syngas is fed to the ATR-MR at 650 °C (#4) partially reformed. The methane conversion into reformed syngas is enhanced by the addition of the pure oxygen which is permeating via the MIEC membranes. The temperatures of the gases (at the permeate and retentate sides) increases and the CH_4 conversion is almost complete. The hot gasses are leaving the HER at around 700 °C and are afterwards cooled down by producing part of the required steam for the process at 400 °C (and the desired pressure). After two stages of water gas shift (WGS), the gasses are further cooled down and the H_2O in the stream is condensed and separated. Two WGS stages at high and low temperatures are beneficial for the system because they increase the amount of H_2 produced while reducing the heat content of the PSA-offgas. Part of the H_2-rich stream is burnt (#9) to pre-heat the air used in the MIEC and the remaining part (#7) is fed to the PSA where pure H_2 is produced and then compressed to the desired pressure (#8). The inlet air (#11) is compressed in an air compressor (#12) and it is pre-heated in two steps: in the first step air is heated up to 600 °C in a regenerative heat exchanger using high temperature exhausts coming from the post-combustor downstream the gas turbine (#17). This temperature is limited to about 650–700 °C when stainless steel or advanced austenitic components are used, otherwise very expensive alloy materials are required with an increase of the regenerative heat exchanger cost of 3.5 up to 5.5 times [38]. The air stream is finally heated up to the operating inlet temperature of the membranes (800–900 °C) by burning part of the H_2-rich stream (#9).

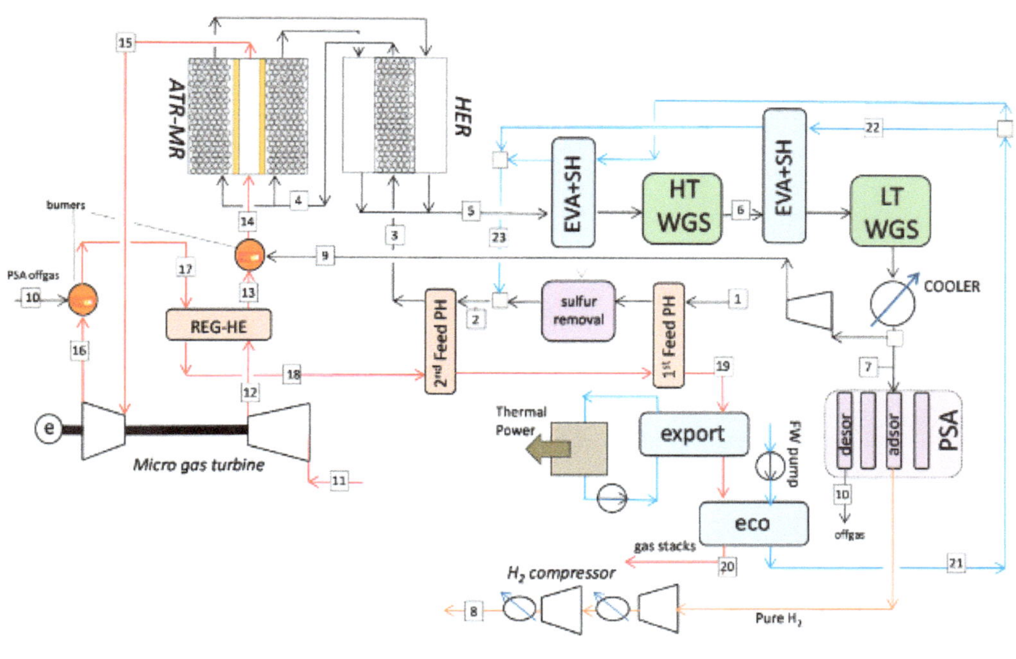

Figure 4: Plant design of the auto-thermal process with MIEC membranes for small-scale H_2 production.

At the membranes outlet the O_2-depleted air (#10) is expanded in the gas turbine and afterwards the stream #16 is used to burn the PSA-offgas. It should be noted that the PSA offgas could be burnt before the expansion of the gas to increase the efficiency of the gas turbine with a gain in the net power production from the micro gas turbine. However, this last solution requires a PSA-offgas compressor and a more complicated gas turbine layout, which also includes a specific cooling system to avoid excessive overheating of the blades. The O_2-depleted air is finally cooled down (#19) by pre-heating the fuel feed, and used for combined heat production of intermediate pressure (20 bar) steam at 220 °C and to supply the required heat for the economizer that is used to heat up the H_2O for process (#21).

Main Assumptions and Methodology

The analysis of the plant and the ATR-MR design has been studied for different configurations, where the main assumptions used for the calculations have been reported in Table 4 (the main assumptions are taken from [39]). The operating pressure of the ATR-MR has been varied from 5 bar to 20 bar. Although conventional micro-gas turbines are typically operated with lower compression ratios (from 2 up to 10 for the high pressure system), a higher compression ratio has been selected in order to investigate a wider range of operating conditions for the reactor. For the micro gas turbine no cooling system has been considered and the main assumptions are taken from [40] assuming that the compressor is a centrifugal turbomachine and the turbine is a radial inflow expander.

With the studied configuration, the amount of steam produced is sufficient to meet a steam-to-carbon-ratio (S/C) in the range of 2–2.8. This number is relatively high for ATR reactors that usually operated with a S/C in the range of 1.5 to 2. However, due to the high amount of heat available in the system the high temperature (HT) syngas coolers have been used for the production of steam for the process. The H_2 separation efficiency of the PSA has been assumed equal to 80% (the main assumption is that this efficiency is not function of the operating pressure).

Table 4: List of assumptions used in the analysis

Main Assumptions	
Fuel feedstock composition (% Vol.)	100% CH_4
LHV_{NG} (MJ/kg)	50 MJ/kg
Ambient air conditions	1 bar, 15 °C
Water feed conditions	1 bar, 15 °C
Air composition (%vol)	79% N_2; 21% O_2
Process conditions	
Pre-reforming inlet temperature, °C	650
Maximum reforming temperature, °C	900–1200 °C
Reforming pressure, bar	5–20
steam-to-carbon ratio	2–3
HT-WGS inlet temperature °C	400
LT-WGS inlet temperature, °C	250
Pressure drops, % of inlet pressure	1
Heat Exchangers	
ΔT_{min} gas-gas	20
ΔT_{min} gas-liquid	10
Pressure drops, % of inlet pressure	2
H_2 compressor and PSA	
PSA H_2 Separation purity	100%
PSA H_2 separation efficiency*	80%
H_2 separation process, bar	5–20
Number of intercooled compression stages	depending on the permeate pressure
Final H_2 pressure for plant export, bar	20
H_2 outlet temperature, °C	30
Polytropic efficiency for compression stages,%	80%
pump/compressors mech-electric efficiency, %	95%
Gas Turbine	
Air compressor isentropic efficiency	75%
Gas expander isentropic efficiency	70%
Electric mechanic efficiency	95%
Steam cycle parameters	
pressure drops economizers, % of inlet pressure	25%
pressure drops superheaters, % of inlet pressure	8%
Max. steam temperature, °C	400
steam export at 20 bar	saturated
pump hydraulic efficiency	80%
pumps mech-electric efficiency	94%

The overall plant has been simulated with Aspen Plus 7.3 and the membrane reactor mass and energy balances have been simultaneously solved with the dedicated 1D model implemented in Matlab as described before. In this way, after fixing the geometry of the membrane reactor and the main operating conditions, simulation is able to return the overall plant performance and reactor characteristics, such as the permeated O_2 flow rate, gas velocity, pressure drops, gas composition and temperature profiles along the membrane reactors. This characteristic of the model permits to simultaneously perform the process and reactor design and optimization.

Table 5: Thermodynamic conditions of the streams in Figure 4

Stream	T	p	m	N	$N*LHV_{mol}$	Composition (% vol.)							LHV
#p	°C	bar	kg/h	kmol/h	MW	CH_4	CO	CO_2	H_2	H_2O	N_2	O_2	MJ/kmol
#1	15.0	10.0	252.0	15.7	3.50	1.000	0.000	0.000	0.000	0.000	0.000	0.000	802.3
#2	400.0	10.0	934.4	53.6	3.50	0.293	0.000	0.000	0.000	0.707	0.000	0.000	235.2
#3	600.0	10.0	934.4	53.6	3.50	0.293	0.000	0.000	0.000	0.707	0.000	0.000	235.2
#4	650.0	10.0	934.4	66.8	3.84	0.136	0.031	0.068	0.366	0.399	0.000	0.000	207.0
#5	700.0	9.8	1157.3	85.1	3.45	0.000	0.133	0.051	0.444	0.370	0.000	0.001	146.1
#6	433.0	9.7	1157.3	85.0	3.34	0.000	0.034	0.151	0.542	0.273	0.000	0.000	141.7
#7	35.0	9.2	709.4	58.3	3.03	0.000	0.010	0.236	0.754	0.000	0.000	0.000	186.9
#8	30.0	20.0	70.9	35.2	2.38	0.000	0.000	0.000	1.000	0.000	0.000	0.000	244.0
#9	44.5	10.1	69.0	5.7	0.29	0.000	0.010	0.236	0.754	0.000	0.000	0.000	186.9
#10	35.0	1.1	638.5	23.1	0.64	0.000	0.025	0.594	0.381	0.000	0.000	0.000	100.0
#11	15.0	1.0	2642.8	91.6	0.00	0.000	0.000	0.000	0.000	0.000	0.790	0.210	0.00
#12	367.3	10.3	2642.8	91.6	0.00	0.000	0.000	0.000	0.000	0.000	0.790	0.210	0.00
#13	600.0	10.3	2642.8	91.6	0.00	0.000	0.000	0.000	0.000	0.000	0.790	0.210	0.00
#14	900.0	10.0	2711.8	95.1	0.00	0.000	0.000	0.015	0.000	0.045	0.761	0.179	0.00
#15	863.7	10.0	2489.0	88.1	0.00	0.000	0.000	0.016	0.000	0.049	0.821	0.115	0.00
#16	522.9	1.1	2489.0	88.1	0.00	0.000	0.000	0.016	0.000	0.049	0.821	0.115	0.00
#17	1000.2	1.0	3127.5	106.6	0.00	0.000	0.000	0.147	0.000	0.123	0.679	0.051	0.00
#18	835.2	1.0	3127.5	106.6	0.00	0.000	0.000	0.147	0.000	0.123	0.679	0.051	0.00
#19	618.4	1.0	3127.5	106.6	0.00	0.000	0.000	0.147	0.000	0.123	0.679	0.051	0.00
#20	107.1	1.0	3127.5	106.6	0.00	0.000	0.000	0.147	0.000	0.123	0.679	0.051	0.00
#21	15.0	1.0	682.4	37.9	0.00	0.000	0.000	0.000	0.000	1.000	0.000	0.000	0.00
#22	175.0	11.0	213.3	11.8	0.00	0.000	0.000	0.000	0.000	1.000	0.000	0.000	0.00
#23	400	10	682.4	37.9	0.00	0.000	0.000	0.000	0.000	1.000	0.000	0.000	0.00

Since the reforming efficiency, given by Equation (22), only takes into account the amount of produced H_2, a different index has been used in order to compare the plant performance including also the electricity and heat produced/consumed as proposed by Martinez et al. [41]. To consider the contribution of the electricity and the heat flows exchanged with the environment, the equivalent natural gas (NG) thermal input has been calculated according to the equation (23), in which the reference thermal efficiency (η_{th} equal to 90%) corresponds to the conventional industrial boiler used for steam production and the electric efficiency represents the typical performance of a NG-fired combined cycle (η_{el} assumed equal to 58.1%):

$$\eta_{ref} = \frac{\dot{m}_{H_2,pure} LHV_{H_2}}{\dot{m}_{GN} LHV_{GN}} \tag{22}$$

$$\eta_{ref,eq} = \frac{\dot{m}_{H_2,pure} LHV_{H_2}}{\dot{m}_{GN,eq} LHV_{GN}} \text{ where } \dot{m}_{GN,eq} = \dot{m}_{GN} - \frac{E_{el}}{\eta_{el} LHV_{GN}} - \frac{Q_{th}}{\eta_{th} LHV_{GN}} \tag{23}$$

RESULTS AND DISCUSSION

Detail of the Plant Flowsheet

The design of the system has been optimized by varying the main parameters that affect the plant performance. As far as the membrane reactor is concerned, the sensitivity analysis has been focused on the size of the reactor (reactor diameter and length) and the number of membranes. The optimization is carried out in order to maximize the equivalent H_2 efficiency. The analysis of the results also includes the main effects on the plant design, which are subsequently further discussed. The detailed mass balance of the reference case is reported in Table 5 and it is referring to an ATR-MR module operated at 10 bar, fuel fed at 650 °C after pre-reforming using a mixture of methane and H_2O (S/C = 2.4) and (depleted) air fed at 900 °C with an O_2 concentration of 17.9% after the pre-burner. At the permeate side, pre-reformed syngas is produced at 1127 °C: the CH_4 is completely converted, and the CO content is 13.3%. The overall CO conversion is 94.4%, where after the first WGS stage 75% of the CO is already converted and after the low temperature Water Gas Shift (LT-WGS) a final CO concentration of 1% (dry basis) is obtained. The amount of H_2-rich syngas used in a pre-combustor to pre-heat air from 600 °C up to 900 °C is 8.9% of the total syngas produced, while the main stream is sent to the PSA unit. At the retentate side, the O_2-depleted air is released at 863 °C with 11.5% of O_2 and expanded to 1.05 bar. The gas leaving the expander is heated up to about 1000 °C after the post-combustor that is used to convert the PSA-offgas coming from the desorption beds.

The exhaust streams are cooled down to 620 °C by pre-heating the fuel feed streams and after that they are used to produce saturated steam at 20 bar (0.19 kg/s). The exhaust gases are finally cooled down by pre-heating the process water and released to the stack at 107 °C. The total amount of pure H_2 produced is 0.02 kg/s, which corresponds to about 790 Nm^3/h. Looking at the quality of conversion of the syngas (in terms of energy), it is important to note that the energy of the fuel increases after the pre-reforming (from 3.5 to 3.84 MW), as expected after an endothermic reaction, and slightly decreases at the outlet of the ATR-MR, explained by the combination of the endothermic reaction of SMR and the reaction with O_2 where an oxygen-to-carbon ratio (O/C) of 0.89 is achieved, indicating that a large part of the chemical energy is converted into heat inside the system. After that, the energy of the fuel decreases because of the WGS reaction and, after the PSA the amount of energy in the pure H_2 stream is 2.38 MW which corresponds to a reforming efficiency of 68%.

The presence of HER which works at high temperature (650 °C) impacts significantly with the overall efficiency of the system. More than 57% of the CH_4 at the feed side is converted in the pre-reforming reactor and the remaining providing about 65% of the H_2 produced in the reforming reactor (and 53% of the total amount of H_2 produced).

In terms of electricity requirements, the net power consumption is 28.34 kW_e. The low Turbine Inlet Temperature (TIT) which results from the ATR-MR model significantly reduces the efficiency of the thermodynamic cycle and the overall power production is almost zero (+1.4 kW) since the power produced from the expander is balanced by the compressor consumption. The power requirement to bring the pure H_2 from the PSA separation pressure (9.2 bar) up to 20 bar is 28.5 kW, while the auxiliary consumptions of the syngas fan and feedwater pump are about 0.8 kW. The reactor has been designed with a shell diameter of 1 m and a shell (and membrane) length of 1.8 m with 1500 membranes with a diameter of 2 cm. With this design about 60% of the reactor volume is occupied by the membranes and the feasibility of this design must be further verified from an engineering point of view.

The results show that the reforming efficiency is approaching 70% (the equivalent reforming efficiency is about 76%–77%) which is slightly lower than the system operated with fired tubular reforming as presented in [41]. however this configuration is designed for small-to-medium scale applications and therefore the operating conditions, such as the steam production and the turbomachines are not as efficient as in the case of a large scale H_2 plant.

Sensitivity Analysis

A sensitivity analysis has been carried out on the effect of the operating pressure and the temperature of the oxidant at the inlet and the results are summarized in the Table 6. An additional analysis includes the effect of the reactor design on the performance of the plant. The O_2 concentration at the retentate side has been kept constant (in the range of 11.4%–11.9%) in order to ensure enough O_2 to incinerate the PSA-offgas in the post-combustor.

The micro-gas turbine power is strictly dependent on the gas temperature at the ATR-MR retentate side. In order to increase the temperature, the reactor size has been increased. Increasing the reactor volume (about 10 times and 20 times), the outlet temperature at the retentate side increases from 863.7 °C to 925 °C and consequently the power required from the exterior decreases from −28.3 kW to −12.4 kW. Looking at the energy balance, this increase is caused by a higher power for the turbine. A larger reactor volume also allows for a larger available membrane surface area. Moreover, the inlet conditions of the reactor—such as S/C, gas composition and temperature—remain substantially the same, but due to the different amount of heat transfer in the system the amount of O_2 permeation changes since it strongly depends on the membrane surface temperature. This issue will be further discussed in the next section.

The sensitivity analysis has also been carried out concerning the operating pressure of the system. Increasing the operating pressure of the ATR-MR the following effects on the plant performance can be deduced:

- The fuel conversion at the pre-reforming stage is lower and therefore more heat of reaction is required in the ATR-MR, so that more oxygen has to be permeated from the air and therefore more air has to be used since the same O_2volume fraction in the depleted air exhaust was specified for this analysis. The resulting O/C at the ATR-MR increases from 0.8 to 0.94.
- Due to the lower heat required at the pre-reforming stage the amount of steam produced for the process increases by increasing the S/C (from 2.2 to 2.9), because more heat is available from the exhaust gas cooling. However, the combination of higher pressure (which reduces the reforming conversion) and higher

S/C (which improves the methane conversion) in the pre-reforming section the performance has slightly decreased.

- The amount of pure H_2 decreases (with a reforming efficiency drop from 70.5% to 66.6%) due to the higher O/C used which increases the CH_4 conversion by the methane partial oxidation instead of steam methane reforming which is particularly high at the pre-reforming stage.
- In terms of electricity production, the micro-gas turbine (micro-GT) net electricity production decreases because a low compressor ratio ($\beta_{compressor}$, in the range of 5) is optimum to maximize the efficiency of the thermodynamic cycle operated with such TIT; the electricity consumption for the H_2 compressor is significantly lower (from 56.2 kW at 5 bar to 3.3 kW at 20 bar) due to the higher H_2 separation pressure; it must be pointed out that in this analysis the efficiency of the PSA in terms of H_2 purity has been kept constant, however, it is known that at low pressure the purity of H_2 decreases unless some systems adopting vacuum PSA are applied [42]. The overall power consumption increases by increasing the operating pressure.
- At higher pressure the amount of heat for export increases due to more heat available in the plant at lower temperature which also results from a lower thermal input (from CH_4) converted into H_2 and/or electricity.
- The equivalent reforming efficiency slightly decreases from 77% to 74%.

Another sensitivity analysis has been carried out by increasing the O/C ratio in the system in order to have the streams at the reactor outlet at higher temperature. The main consequence of an increased O/C ratio is a lower amount of pure H_2produced with a drop in the reforming efficiency of about 5.4 percentage points. Due to the higher temperature resulting from the ATR-MR and especially at the retentate side, the overall electricity consumption significantly drops in particular because the micro-GT net electricity increases with about 24 kW_e with a corresponding decrease in the total power consumption from 28.6 to 4.3 kW_e. At higher O/C ratios also the thermal power for export significantly increases (more than 20%) due to the higher temperature of the gasses. In terms of equivalent reforming efficiency, the increase in the O/C reduces the performance of the system from 76% to 73.6%, however, the decreases in the H_2 production is partly balanced by the lower power consumption and thermal production. At higher O/C ratios the membrane area of the ATR-MR increases with 35% compared to the reference case due to higher amount of O_2 to be separated.

Finally, the inlet temperature of the oxidant has been considered to be 800 °C. Compared to the same case operated with air fed at 900 °C it can be seen that a higher O/C ratio is required in order to reach similar outlet temperatures. The O_2volume fraction at the feed side of the ATR-MR is 18.9% instead of 17.9%. In this case, less fuel is required after the WGS reactors to be used in the first burner (and therefore the consumption of the syngas fan is about 30% less). The high amount of heat available in the syngas coolers (also because the syngas flow rate is higher) allows the system to work with higher S/C ratios (in the range of 3–4). Less power is consumed for H_2 compression since less pure H_2 is produced (the reforming efficiency drops from 62%–63% to 59%–60%) with a gain in the electricity export of about 8 kW_e. The reactor volume (small case) is higher when operating at 800 °C in order to ensure a high O_2 conversion at the permeate side and the membrane area is also higher especially because of (i) the average temperature of the membrane is lower, especially at the entrance of the reactor, with a correspondingly lower O_2 permeation flux, while (ii) a higher amount of O_2 that has to be separated.

Reactor Design

To further elucidate the effect of the reactor design and behavior on the plant performance, in this section the axial concentration and temperature profiles in the ATR-MR are discussed in more detail. It must be noticed that the membrane diameter adopted in this analysis is 2 cm instead of 3.5 mm as for the experimental tests: the reason we used a bigger diameter is related to the required upscaling in case of small/ medium size application.

Table 6: Results of the sensitivity analysis

Summary of Performance											
Oxidant Inlet Temperature, °C	900									800	
Case	10 bar			5 bar		20 bar		10 bar high O/C		10 bar high O/C	
	Reference Small	Medium	Big	Small	Medium	Small	Medium	Small	Medium	Small	Medium
Reactor design											
N membrane	1500	1400	750	1370	1350	1500	1430	1520	1480	1750	1600
Diameter, m	1	2	2	1.1	2	1	2	1.2	2	1.5	2
Length, m	1.8	5	10	2.1	5	1.8	5	2.4	5	3	5
Main parameters											
pure H_2, kg/h	70.93	71.72	71.47	73.44	73.51	69.35	69.34	65.26	65.11	61.59	63.40
air flow rate, kg/h	2642.8	2642.8	2642.8	2360.6	2360.6	2875.2	2875.2	3257.2	3257.2	3444.0	3444.0
O_2 separated, kg/h	222.88	213.24	215.73	201.85	199.08	235.86	234.16	265.13	268.36	332.39	313.79
O_2 ret fraction, %vol.	11.5%	11.8%	11.7%	11.4%	11.5%	11.6%	11.7%	11.7%	11.6%	11.4%	11.9%
steam for process, kg/h	682.43	643.11	642.20	634.80	581.02	841.97	780.06	722.89	799.08	1101.30	926.44
O/C	0.89	0.85	0.86	0.80	0.79	0.94	0.93	1.05	1.07	1.32	1.25
S/C	2.41	2.27	2.27	2.24	2.05	2.98	2.76	2.55	2.82	3.89	3.27
$T_{out\ ret}$, °C	863.7	899.9	924.9	871.1	915.1	863.9	908.4	928.6	994.0	978.4	1012.0
$T_{out\ perm}$, °C	1127.0	1065.3	1054.1	1103.3	1092.7	1107.2	1064.8	1238.4	1205.8	1231.7	1243.4
Energy Balance											
compressor, kW_e	−272.58	−272.58	−272.58	−153.58	−153.58	−429.02	−429.02	−335.95	−335.95	−355.22	−355.22
turbine, kW_e	273.61	283.61	289.70	184.33	191.82	360.49	375.59	358.87	379.08	384.54	397.20
net micro-GT, kW_e	1.03	11.02	17.12	30.75	38.25	−68.53	−53.43	22.91	43.12	29.32	41.97
H_2 compressor, kW_e	−28.57	−28.89	−28.79	−56.19	−56.24	−3.31	−3.31	−26.28	−26.22	−24.81	−25.54
FW pump, kW_e	−0.30	−0.29	−0.29	−0.13	−0.12	−0.79	−0.73	−0.32	−0.36	−0.49	−0.41
syngas fan, kW_e	−0.49	−0.49	−0.49	−0.53	−0.53	−0.66	−0.66	−0.62	−0.62	−0.44	−0.43
net electric power, kW_e	−28.34	−18.64	−12.45	−26.09	−18.64	−73.28	−58.13	−4.31	15.92	3.58	15.59
Q_{export}, kW_{th}	376.08	364.25	367.31	280.93	308.92	430.70	448.98	477.15	422.48	363.79	397.99
reforming efficiency	68.1%	68.9%	68.6%	70.5%	70.6%	66.6%	66.6%	62.7%	62.5%	59.1%	60.9%
equivalent reforming efficiency	76.1%	77.1%	77.1%	76.3%	77.5%	74.0%	75.1%	73.6%	72.8%	67.0%	70.3%
$A_{mem}/A_{mem,ref}$	1.00	2.59	2.78	1.07	2.50	1.00	2.65	1.35	2.74	1.94	2.96
O_2 flux, $mol_{O2}/(s\text{-}m^2)$	0.0114	0.0042	0.0040	0.0097	0.0041	0.0121	0.0045	0.0100	0.0050	0.0087	0.0054

For the reference case the gas composition at the permeate side is depicted in Figure 5: the CH_4 conversion proceeds along the reactor length via SMR by reacting with H_2O which decreases in the first part of the reactor, and POX by reacting with the permeated O_2 which is always at very low concentration (below 10^{-3} in mole fraction). H_2 is mostly produced in the first part of the reactor while, after the CH_4 conversion is approaching 100%, H_2 and CO are consumed. Based on the proposed kinetics mechanism the O_2 reacts only with the CH_4 (R1) which is virtually instantaneously produced (due to the methanation and reverse WGS reactions which are at the equilibrium) and burnt by reacting with the remaining O_2, thus overall resulting in H_2 and CO combustion. In the same way, two different temperature rise velocities along the reactor can be distinguished, in particular, the first part where the cooling effect of the SMR is prominent and, the last part where the combustion reactions prevail, as shown in Figure 6.

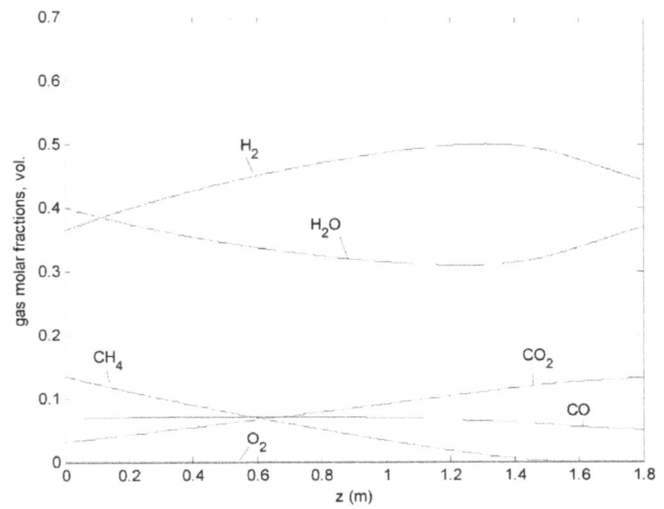

Figure 5: Axial mole fraction profiles at the permeate side of the ATR-MR (reference case).

Based on the gas species profile, after the CH_4 is completely converted (around 1.4 m) the maximum H_2 production is achieved. However, as shown in Figure 6a, the final temperature of the gas at the permeate and retentate sides would be 800 °C and 900 °C respectively. The lower gas temperatures at the reactor outlet will have two main effects on the thermal balance of the plant: (i) the gas turbine will be operated with a lower TIT, reducing the electrical efficiency and therefore increasing the overall electric consumption; (ii) a lower permeate gas temperature will reduce the heat available at pre-reforming unit, and (iii) less steam will be produced for the process which affect the CH_4 conversion in the pre-reforming section.

A temperature cross-over can be observed for all cases (Figure 6): at the beginning, the temperature of the air side is higher than the temperature at the fuel side, so the air is partially cooled down. At the same time, oxygen starts permeating, combustion occurs, and this creates a further temperature increase in the fuel side, until its temperature exceeds the temperature at the retentate side. From this point on, the retentate is heated up.

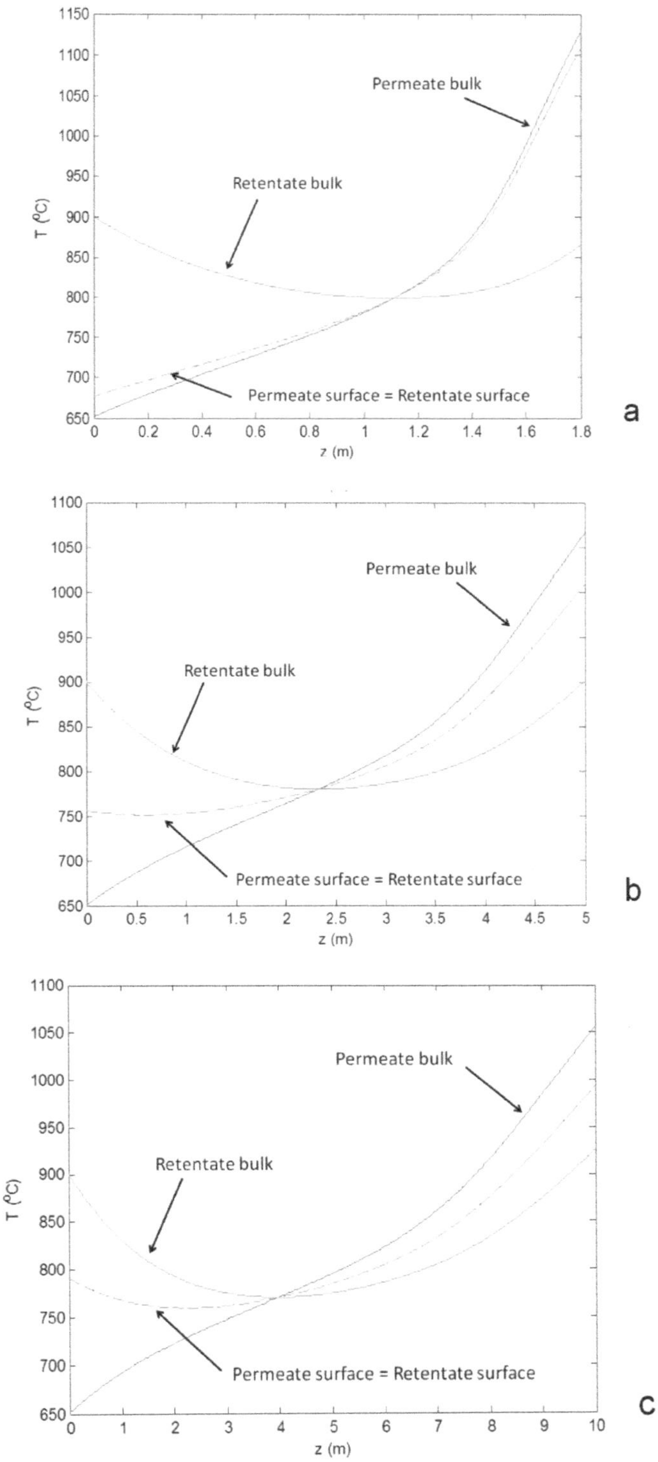

Figure 6: Axial temperature profiles of the gas streams and the membrane surface: (**a**) small size reactor; (**b**) medium size; (**c**) big size reactor. Continuous lines are related to the bulk phase and dashed lines are related to the membrane surface at retentate (green) and permeate (blue) side.

Both temperatures increase because the thermal contribution of combustion is always prevailing in the last section of the reactor. The temperatures at the surface of the permeate and retentate sides coincide (and cannot be distinguished in Figure 6) because the membrane thickness is small (500 μm) and therefore the heat transfer resistance is negligible compared to the others. When the methane concentration is low, methanation occurs (reverse reforming), and the produced CH_4 is immediately burned by the oxygen permeating from the membrane tubes. In this sense, oxygen is shifting the equilibrium of steam reforming towards the reagents. The global effect is the oxidation of syngas to steam and carbon dioxide. With respect to Figure 5 and Figure 6a, for instance, at z = 1.6 m, where the CO (13.4%) and H_2 (49.1%) concentrations are decreasing and the temperature is increasing, the methane concentration in the catalytic area is almost zero (0.08%). In these conditions the reverse reforming reaction rate (methanation) is equal to 2 mol/m^3/s. The reverse water gas shift rate is 5.7 mol/m^3/s. Additionally, the methane combustion rate is 2.9 mol/m^3/s, according to the kinetic mechanism considered in this paper [28]. The combination of these reaction rates results in an overall decrease in the CO and H_2 concentrations and heat production.

It is important to note that the axial temperature profile does not show a typical drop in the temperature at the beginning of the reactor which occurs in case of co-feeding CH_4 and H_2O. In fact, within this configuration, 40% of the CH_4 has been already converted in the HER system and therefore the heat management of the ATR-MR system is much less complicated, especially in view of the oxygen transport membrane (OTM) which requires high temperatures to achieve acceptable O_2permeation.

A comparison between the performance of the ATR-MR with different reactor sizes shows that:

- For small reactors the axial temperature profile at the membrane surface follows closely the permeate bulk temperature profile. When increasing the reactor size, the axial temperature profile at the membrane surface tends to be closer to the average temperature of both reactor compartments (permeate and retentate), as a result of the increased residence time which increases the heat transfer from both sides of the reactor (Figure 6a vs. Figure 6b,c). Consequently, the resulting reactor outlet temperatures of the streams are also closer when increasing the reactor size.
- The heat transfer coefficient at the shell side (the permeate side) is higher in the case of a small reactor size due to the higher gas velocity which reduces the difference between the temperature of the bulk flow and the membrane surface at the permeate side.
- Despite the larger membrane area by increasing the reactor volume and length, the permeated oxygen is substantially the same because of mass transfer limitations prevailing in the shell (permeate side): in the small reactor the difference of the O_2 concentration between the retentate and permeate side is higher (Figure 7a) than in the other cases (Figure 7b,c) which corresponds to a higher O_2 flux. In fact, at the permeate side the gas velocity decreases by increasing the reactor diameter and therefore the mass transfer coefficient decreases because of the decreased Reynolds number. A small difference occurs between the medium and big size of the reactor due to the similarities in the reactor geometry and total membrane area.

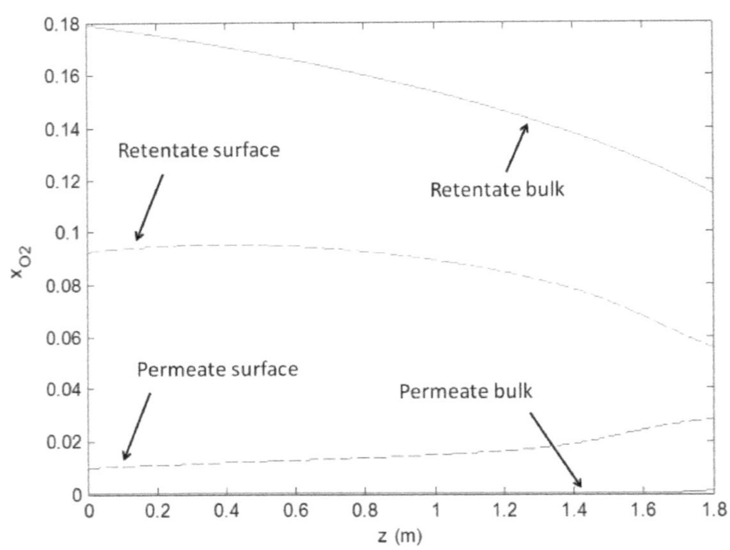

a

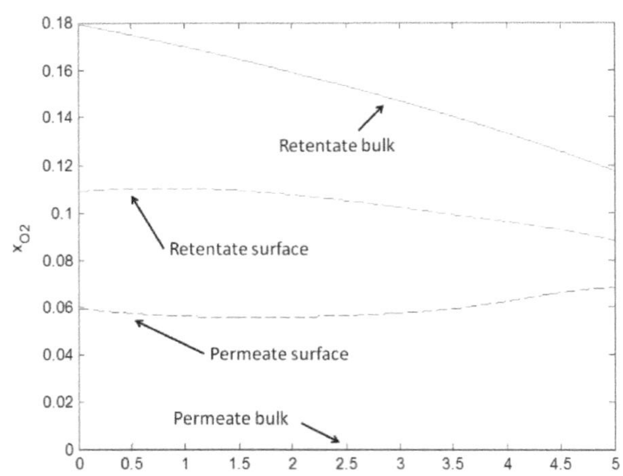

b

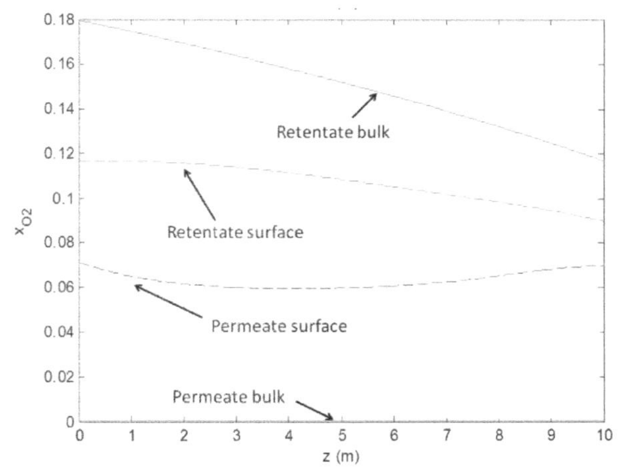

Figure 7: Oxygen volume fraction profiles along the reactor of the gas streams and the membrane surface. (**a**) small size reactor; (**b**) medium size; (**c**) big size reactor. Continuous lines are related to the bulk phase and dashed lines are related to the membrane surface at retentate (green) and permeate (blue) side.

The system operated with a small reactor shows that the O_2 conversion at the reactor outlet is not complete (Figure 7a), because the rate of the O_2 consumption depends on the amount of catalyst in the reactor which is apparently insufficient for the small reactor design. In this respect, a proper design of the system should consider a different shell and membrane length: in the last part no oxygen should permeate but the gas should be in contact with catalyst in order to enhance the conversion approaching almost to the thermodynamic equilibrium. From a thermal point of view, this configuration would also be beneficial for the outlet temperature of the retentate, because of the increased contact time so that part of the produced heat can be exchanged with the O_2 depleted air to increase the TIT of the micro-GT.

This analysis shows the influence of the reactor design on cycle performance. It is worthy to note that from an engineering point of view, increasing the reactor size would increase the cost of the plant due to the material required, the catalyst and membrane area. In case the cost of the ATR-MR impacts significantly to the overall capital cost then the improvement in the overall efficiency (1–3 percentage points depending on the operating conditions as shown in Table 6) would not be beneficial for the economics of the system. For a more precise insight into this issue, a separate techno-economic analysis of this plant should be performed considering the interaction between the size of reactor and duty of the downstream units.

CONCLUSIONS

The integration of an ATR-MR with oxygen transport membranes in a small-to-medium scale H_2 production plant has been studied in this work. The OTM's have been tested in the lab, exhibiting a relatively high O_2 permeability, and the main transport parameters for the oxygen permeation through the BSCF membrane have been determined from the experimental data. A detailed membrane reactor model has been developed to describe the axial gas compositions and temperature profiles on the retentate and permeate sides and has subsequently been used for the design of the membrane reactor and the calculation of the required membrane area. The results show that in the configuration with a bigger reactor the heat transfer is increased, thereby decreasing the temperature difference between the permeate and retentate sides with a corresponding improvement in the electric performance and reduced thermal gradients along the reactor. In terms of mass transfer, the O_2flux is reduced when using a bigger reactor, but the conversion of the permeated O_2 increases according to the kinetic model. The assessment of the plant performance and a sensitivity analysis of the main parameters has been carried out with Aspen Plus software and the reactor model has been included so that the results of the model were simultaneously updated during the simulation. The results show that the reforming efficiency is approaching 70% (the equivalent reforming efficiency is about 76%–77%). The combination of the ATR-MR with HER allows to recover efficiently the heat of the syngas and reduces the heat duty for the reforming, however, the cost of the entire system must be verified. From an economic point of view also the reactor size must be carefully considered: in particular, due to the difference in the electricity import required for different reactor sizes, it could be beneficial for the system to have a bigger reactor (with the associated capital cost) and less operating costs for auxiliaries and in particular for the H_2 compressor. Another important aspect is the possibility to couple the system with a CO_2 capture and utilization. In fact CO_2-rich stream from the PSA-offgas can be fully oxidized in a dedicated OTM module and separated afterwards by water condensation instead of using chemical absorption processes which are high capital cost units. However, the advantages in terms of performance and the economic feasibility of an H_2 production plant with reduced CO_2 emissions at small medium scale must be verified.

ACKNOWLEDGMENTS

The research leading to these results has received funding from the European Union Seventh Framework Programme (FP7/2007-2013) under grant agreement n° NMP3-LA-2011-262840 (DEMCAMER project).

AUTHOR CONTRIBUTIONS

All the authors designed the research, Fausto Gallucci and Martin van Sint Annaland designed the membrane reactor system, the experimental part of the work and are project leaders of the whole project, Tommaso Melchiori developed the membrane reactor model and Vincenzo Spallina designed the integrated system. All the authors contributed to write the paper, read and approval the final manuscript.

REFERENCES

1. Rostrup-Nielsen, J. New aspects of syngas production and use. Catal. Today 2000, 63, 159–164.
2. Dunn, S. Hydrogen futures: Toward a sustainable energy system. Int. J. Hydrogen Energy 2002, 27, 235–264.
3. Ball, M.; Wietschel, M. The future of hydrogen–opportunities and challenges. Int. J. Hydrogen Energy 2009, 34, 615–627.
4. Ogden, J. Review of small stationary reformers for hydrogen production; IEA: Princeton, NJ, USA, 2001; p. 52.
5. Beysel, G. Enhanced Cryogenic Air Separation A Proven Process Applied to Oxyfuel: Future Prospects.

In Presented at 1st Oxyfuel Combustion Conference, Cottbus, Germany, 8 September 2009.

6. Smith, A.; Klosek, J. A review of air separation technologies and their integration with energy conversion processes.Fuel Process. Technol. 2001, 70, 115–134.
7. Sunarso, J.; Baumann, S.; Serra, J.M.; Meulenberg, W.A.; Liu, S.; Lin, Y.S.; Diniz da Costa, J.C. Mixed ionic–electronic conducting (MIEC) ceramic-based membranes for oxygen separation. J. Membr. Sci. 2008, 320, 13–41.
8. Bose, A.C. Inorganic Membranes for Energy and Environmental Applications; Springer: Pittsburgh, PA, USA, 2009; p. 318.
9. Han, J.; Zeng, Y.; Lin, Y.S. Oxygen permeation through fluorite-type bismuth-yttrium-copper oxide membranes. J. Membr. Sci. 1997, 132, 235–243.
10. Engels, S.; Beggel, F.; Modigell, M.; Stadler, H. Simulation of a membrane unit for oxyfuel power plants under consideration of realistic BSCF membrane properties. J. Membr. Sci. 2010, 359, 93–101.
11. Pinacci, P.; Louradour, E.; Wimbert, L.; Gindrat, M.; Jarligo, M.O.; Vassen, R.; Comite, A.; Serra, J.M.; Jewulski, J.; Mancuso, L.; et al. Dense Membranes for Oxygen and Hydrogen Separation (DEMOYS): Project Overview and First Results. Energy Procedia 2013, 37, 1030–1038.
12. IEA GHG. Oxy Combustion Processes for CO_2 Capture from Power Plant. Report Number 2005/9; Mitsui Babcock: Renfrew, UK, 2005.
13. Wilson, J.; Christie, M.; Degenstein, N.; Shah, M.; Li, J. OTM Based Oxy-fuel Combustion for CO_2 Capture. In Proceedings of the 34th International Technical Conference on Clean Coal & Fuel Systems, Clearwater, FL, USA, 31 May–4 June 2009.
14. Romano, M.C.; Chiesa, P.; Lozza, G. Pre-combustion CO_2 capture from natural gas power plants, with ATR and MDEA processes. Int. J. Greenh. Gas. Control. 2010, 4, 785–797.
15. Chiesa, P.; Romano, M.C.; Spallina, V.; Turi, D.M.; Mancuso, L. Efficient low CO_2 emissions power generation by mixed conducting membranes. Energ. Procedia 2013, 37, 905–913.
16. Gallucci, F.; Annaland, M.; Kuipers, J. Autothermal reforming of methane with integrated CO_2 capture in a novel fluidized bed membrane reactor. Part 1: Experimental demonstration. Top. Catal. 2008, 51, 133–145.
17. Gallucci, F.; Annaland, M.; Kuipers, J. Autothermal reforming of methane with integrated CO_2 capture in a novel fluidized bed membrane reactor. Part 2 comparison of reactor configurations. Top. Catal. 2008, 51, 146–157.
18. Liu, S.; Tan, X.; Li, K.; Hughes, R. Methane Coupling Using Catalytic Membrane Reactors. Catal. Rev. 2001, 43, 147–198.
19. Hüppmeier, J.; Barg, S.; Baune, M.; Koch, D.; Grathwohl, G.; Thöming, J. Oxygen feed membranes in autothermal steam-reformers—A robust temperature control. Fuel 2010, 89, 1257–1264.
20. Kang, J.Y.; Kang, D.W.; Kim, T.S.; Hur, K.B. Comparative economic analysis of gas turbine-based power generation and combined heat and power systems using biogas fuel. Energy 2014, 67, 309–318.
21. Pilavachi, P.A. Mini- and micro-gas turbines for combined heat and power. Appl. Therm. Eng. 2002, 22, 2003–2014.
22. Technology Characterization: Microturbines; EPA report. EPA: Arlington, VA, USA, 2008.
23. Mansouri Majoumerd, M.; Nikpey Somehsaraei, H.; Assadi, M.; Breuhaus, P. Micro gas turbine configurations with carbon capture—Performance assessment using a validated thermodynamic model. Appl. Therm. Eng. 2014, 73, 172–184.
24. Di Felice, L.; Middelkoop, V.; Anzoletti, V.; Snijkers, F.; van Sint Annaland, M.; Gallucci, F. New high temperature sealing technique and permeability data for hollow fiber BSCF perovskite membranes. Chem. Eng. Process. Process. Intensif. 2014.
25. McCabe, W.L.; Smith, J.; Harriott, P. Unit operations of Chemical Engineering, 5th ed.; McGraw-Hill: Singapore, 1993; p. 668.
26. Dissanayake, D.; Rosynek, M.; Kharas, K.C.C.; Lunsford, J.H.J. Partial Oxidation of Methane to Carbon Monoxide and Hydrogen over a Ni/Al_2O_3 Catalyst. J. Catal. 1991, 132, 117–127.
27. De Groote, A.M.; Froment, G.F. Simulation of the catalytic partial oxidation of methane to synthesis gas. Appl. Catal. A Gen. 1996, 138, 245–264.
28. De Smet, C.R.H.; de Croon, M.H.J.M.; Berger, R.J.; Marin, G.B.; Schouten, J.C. Design of adiabatic fixed-bed reactors for the partial oxidation of methane to synthesis gas. Application to production of methanol and hydrogen-for-fuel-cells. Chem. Eng. Sci. 2001, 56, 4849–4861.
29. Trimm, D.L.; Lam, C.W. The combustion of methane on platinum-alumina fibre catalysts—I. Kinetics and mechanism.Chem. Eng. Sci. 1980, 35, 1405–1413.
30. Numaguchi, T.; Kikuchi, K. Intrinsic kinetics and design simulation in a complex reaction network: Steam-methane reforming. Chem. Eng. Sci. 1988, 43, 2295–2301.
31. Li, C.; Finlayson, B. Heat transfer in packed beds—A reevaluation. Chem. Eng. Sci. 1977, 32, 1055–1066.
32. Perry, R.H.; Green, D.W.; Maloney, J.O. Perry's Chemical Engineers' Handbook, 7th ed.; McGraw-Hill: New York, NY, United States, 1997; pp. 5–16.
33. NIST Chemistry WebBook. Available online: http://www.nist.gov (accessed on 6 August 2014).
34. Xu, S.J.; Thomson, W.J. Oxygen permeation rates through ion-conducting perovskite membranes. Chem. Eng. Sci. 1999,54, 3839–3850.
35. Hong, J.; Kirchen, P.; Ghoniem, A.F. Numerical simulation of ion transport membrane reactors: Oxygen permeation and transport and fuel conversion. J. Membr. Sci. 2012, 407–408, 71–85.
36. Olsson, H.; Rudbeck, P.; Andersen, K. Adding Hydrogen Production Capacity by Heat Exchange Reforming. In

Proceedings of the Conference XIV Refinery Technology Meet (RTM) on Energy & Environment Challenges for the Hydrocarbon Sector, Kovalam, India, 20–22 September 2007.

37. Manzolini, G.; Macchi, E.; Binotti, M.; Gazzani, M. Integration of SEWGS for carbon capture in Natural Gas Combined Cycle. Part B: Reference case comparison. Int. J. Greenh. Gas. Control. 2011, 5, 214–225.

38. Rosfjord, T.; Manager, P.; Chen, A.; Mulugeta, J.; Bhatia, T. Advanced Microturbine Systems Final Report DOE/CH/11060–1; DOE: East Hartford, CT, USA, 2007; p. 155.

39. European best practice guidelines for assessment of CO_2 capture technologies. Available online: http://www.energia.polimi.it/news/D%204_9%20best%20practice%20guide.pdf (accessed on 11 September 2014).

40. Visser, W.P.J.; Shakariyants, S.A.; Oostveen, M. Development of a 3 kW Microturbine for CHP Applications. J. Eng. Gas. Turbines Power 2011, 133, 042301.

41. Martínez, I.; Romano, M.C.; Chiesa, P.; Grasa, G.; Murillo, R. Hydrogen production through sorption enhanced steam reforming of natural gas: Thermodynamic plant assessment. Int. J. Hydrogen Energ. 2013, 38, 15180–15199.

42. Khunpolgrang, J.; Yosantea, S.; Kongnoo, A.; Phalakornkule, C. Alternative PSA process cycle with combined vacuum regeneration and nitrogen purging for CH_4/CO_2 separation. Fuel 2015, 140, 171–177.

CHAPTER 6

ANALYSIS OF GENETIC VARIATION AND POTENTIAL APPLICATIONS IN GENOME-SCALE METABOLIC MODELING

João G. R. Cardoso[1], Mikael Rørdam Andersen[2], Markus J. Herrgård[1], and Nikolaus Sonnenschein[1]

[1] The Novo Nordisk Foundation Center of Biosustainability, Technical University of Denmark, Hørsholm, Denmark

[2] Department of Systems Biology, Technical University of Denmark, Lyngby, Denmark

Genetic variation is the motor of evolution and allows organisms to overcome the environmental challenges they encounter. It can be both beneficial and harmful in the process of engineering cell factories for the production of proteins and chemicals. Throughout the history of biotechnology, there have been efforts to exploit genetic variation in our favor to create strains with favorable phenotypes. Genetic variation can either be present in natural populations or it can be artificially created by mutagenesis and selection or adaptive laboratory evolution. On the other hand, unintended genetic variation during a long term production process may lead to significant economic losses and it is important to understand how to control this type of variation. With the emergence of next-generation sequencing technologies, genetic variation in microbial strains can now be determined on an unprecedented scale and resolution by re-sequencing thousands of strains systematically. In this article, we review challenges in the integration and analysis of large-scale re-sequencing data, present an extensive overview of bioinformatics methods for predicting the effects of genetic variants on protein function, and discuss approaches for interfacing existing bioinformatics approaches with genome-scale models of cellular processes in order to predict effects of sequence variation on cellular phenotypes.

INTRODUCTION

Genetic engineering has been used for several decades to manipulate microorganisms in order to allow production of valuable products, including primary metabolites (e.g., amino-acids and organic acids), secondary metabolites (e.g., antibiotics), and enzymes or other recombinant proteins (Adrio and Demain, 2010). Genetic engineering is thus a central part in the quest to establish sustainable and efficient processes for the production of fuels, chemicals, food ingredients, and pharmaceutical products.

Most of these achievements would not been possible without sequencing technologies that allowed us to identify the genetic sequences and validate the genetic manipulations in microorganisms. More recently, Next-Generation Sequencing (NGS) technologies have provided us with the capability of fast and cheap sequencing of DNA at an unprecedented scale. NGS has allowed *de novo* assembly of the genomes of thousands of organisms for which no genome sequences were previously available, ranging from complex multicellular organisms

Citation: Cardoso JGR, Andersen MR, Herrgård MJ and Sonnenschein N (2015) Analysis of genetic variation and potential applications in genome-scale metabolic modeling. Front. Bioeng. Biotechnol. 3:13. doi: 10.3389/fbioe.2015.00013

(Li et al., 2010; Nakamura et al., 2013; Pegadaraju et al., 2013; Kelley et al., 2014) to microorganisms (Soares-Castro and Santos, 2013; Yamamoto et al., 2014). NGS technologies also provide us with the means to re-sequence organisms (Atsumi et al., 2010; Wang et al., 2014), i.e., the sequencing of genetically distinct strains that are close enough to a reference strain with a sequenced genome. Re-sequencing is used to determine genetic variants ranging from single nucleotide variants (SNV) to more complex structural variants such as large deletions, inversions, and translocations. The falling cost of sequencing allows routine re-sequencing of strains isolated from the wild, monitoring the genetic stability of production strains during genetic engineering and fermentation processes, and determining the genetic basis of adaptive laboratory evolution (ALE) (Herrgård and Panagiotou, 2012). In addition to biotechnological applications, re-sequencing of microbial strains plays also a key role in other areas such as epidemiology of infectious diseases caused by bacterial and fungal pathogens, and in understanding the effects of human activity on microbial diversity and evolution in the environment.

Genome-scale metabolic models (GSMs), consisting of biochemical reactions and their relations to the genome and proteome of a cell [through gene–protein-reaction (GPR) associations], are a proven framework for the *in silico* analysis of the metabolic physiology of microbes. Genome-scale metabolic models have also been used successfully for the design of metabolically engineered strains with improved production of commercially valuable proteins and metabolites: recombinant antibodies, food additives (e.g., vanillin), organic acids, ethanol, among others (Tepper and Shlomi, 2009; Brochado et al., 2010). These models have become increasingly popular over the past decade, and more than 100 models for different organisms have been published up to this date (http://optflux.org/models). The greatest strength of GSMs lie in their simplicity and computational efficiency; new GSMs can be readily built from genomic annotations complemented with limited experimental data, and predictions from GSMs can be obtained using standard mathematical optimization methods (Varma and Palsson, 1993; Segrè et al., 2002; Shlomi et al., 2005) allowing phenotypic predictions within minutes.

Genetic variation that entails a complete loss of function – commonly referred to as gene knockout – has been successfully used to tailor GSMs to a specific genotype to improve the production of valuable compounds [e.g., biobutanol (Lee et al., 2008), sesquiterpene (Asadollahi et al., 2009), vanillin (Brochado et al., 2010), polyhydroxyalkanoates (Puchałka et al., 2008), or L-valine (Park et al., 2007)], but so far no methodological framework has been developed that would allow the incorporation of other types of genetic variants systematically. In this work, we review existing tools for analyzing genetic variants that capture more subtle changes such as synonymous and non-synonymous SNVs in coding regions or variants in promoter or other regulatory regions. We will focus on outlining the challenges of combining more subtle genetic variant information with GSMs in order to use models to predict strain-specific phenotypes.

UNVEILING THE EFFECTS OF GENETIC VARIATION

Genetic Variability

Genetic variants, including SNVs and larger structural variants are commonly seen when natural or engineered strains are re-sequenced (Figure 1). SNVs can be found across the genome in different functional regions: (i) protein coding sequences, (ii) promoters and other regulatory elements such as ribosome binding sites, (iii) splice sites and other regions affecting transcript structures, and (iv) other genomic regions with unknown direct connections to any given protein function. Moreover, insertions or deletions of nucleotides (indels) within a coding region can cause a shift in the open reading frame usually denoted as frameshift mutations (Figure 1A). At the genome structure level, chromosomal rearrangements, e.g., swaps, inversions, deletions, and insertions, can affect the function of one or more proteins (Figure 1B).

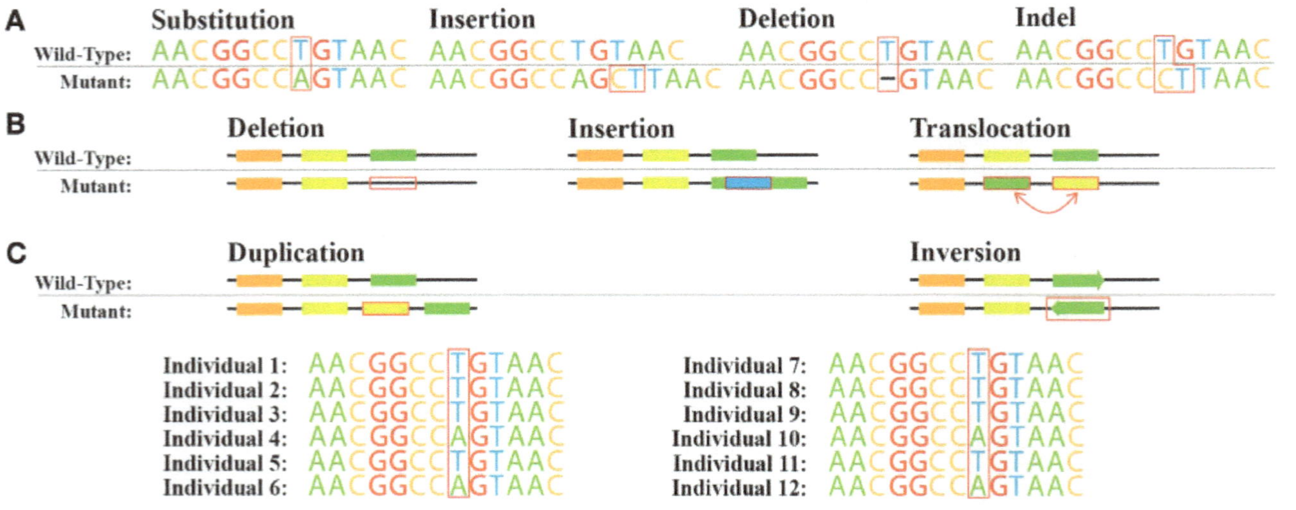

Figure 1: Common genetic variations. Variations at the (A) nucleotide level and (B)structural level. (C) Single nucleotide polymorphism A/T across a population.

The spectrum of the resulting effects caused by these genetic variations on individual gene or protein function or expression is very broad. Non-synonymous SNVs or in-frame indels in protein coding sequences can disrupt, enhance, or modify the activity of the protein depending on the exact amino-acid change introduced. Introduction or removal of a stop codon by specific

SNVs or out-of-frame indels would be expected to result in more drastic changes of protein function. For example, the appearance of a stop codon might lead to the separation of a multi-domain protein to multiple individual single-domain proteins. The removal or replacement of a stop codon could cause translational read-through leading to an elongated protein with potential new functions (Long et al., 2003). SNVs and indels in regulatory regions such as promoters can affect the transcription or translation processes giving rise to variation in expression levels in specific proteins. In eukaryotes, variants within introns can also affect transcript structures by introducing new exons or removing existing ones. Some variations can also be completely silent with no change of phenotype, for example, a change in a stop codon location might not change the protein activity. Ideally, we should be able to predict the degree in which single and multiple genetic variants within or near a coding locus affect the relevant protein function or expression. This would allow us to rapidly make sense of the vast quantities of re-sequencing data that is becoming available without having to test the effects of all variants experimentally.

Larger-scale structural variations, such as duplications, deletions, translocations, and inversions, can have significant effects on the expression or activity of individual proteins. For example, there can be a complete loss of one or more genes, or a duplication of genomic regions can modify the expression of multiple genes within or nearby these regions (Blount et al., 2012). Very large-scale genomic changes, such as duplication of entire chromosomes, can change the activity of hundreds of proteins at once and have been reported in both natural microbial strains (Gordon et al., 2009) and in strains created by ALE (Caspeta et al., 2014). The effects of structural genomic variation are often more systemic than the effects of smaller scale variations, but any framework attempting to predict the phenotypic effects of genetic variation needs to consider both small- and large-scale variation.

In silico: Predicting the Effect of Genetic Variants

A major challenge to understanding the phenotypic consequences of genetic variation lies in our ability to predict the mechanistic consequences of mutations. Proteins are very complex structures that fall into different functional categories and can be characterized by many distinct properties. For example, how protein activities are measured depends on their functional category: transcription factors can be characterized by their binding strength to a certain promoter region while metabolic enzymes would typically be characterized by their catalytic activity and specificity for a certain substrate. Moreover, proteins do not operate in isolation but interact with each other and with metabolites, and these interactions have consequences on the activities of proteins. Here, we provide a non-exhaustive review of the types of methods that are commonly used to predict the effects of genetic variants on protein function.

The study of single nucleotide polymorphisms (SNP) that affect human health is one of the major focus areas of modern medical research. In human genetics, SNPs are single nucleotide substitutions found in more than 1% of a population. Several algorithms were implemented to determine the effect of SNPs, mostly specialized to the analysis of human genotyping data (see Table 1 and Figure 2). One limitation of most of these algorithms is that they are binary classifiers – deleterious or neutral, disease-causing or neutral, and tolerant or intolerant. This means that the genetic changes will either be predicted to have no effect or to cause some measurable, negative impact on the phenotype. This may not be an issue in the context of human diseases as SNP data are primarily used in diagnostics. However, fine tuning engineered microbial strains requires more than a black and white approach for predicting variant effects on protein function. This is because many genetic variants can yield proteins with either increased or decreased activity, requiring methods that are able to predict also potential gains or modifications of functions. In particular, when mutagenesis and selection or ALE methods are applied, one commonly sees gain of function mutations of specific genes that are crucial for the adaptation to, for example, new carbon sources (Conrad et al., 2011).

Table 1: A summary of the available software tools for predicting the effect of the genetic variants

Tool	Description	Reference
AUTO-MUTE	Uses the "4-Body Statistical Potential" to compute a set of features – based on protein 3D structure – used to train a Random Forest model to predict *neutral* or *disease*-associated SNPs.	Masso and Vaisman (2010)
Align-GVGD	This algorithm is based on multiple sequence alignment and Grantham distance to identify missense SNPs. The authors propose a measure to calculate how much the substitution changes the Grantham distance.	Tavtigian (2005)
CADD	A machine-learning approach that uses a SVM model to predict deleterious phenotypes caused by SNPs.	Kircher et al. (2014)
Chasman and Adams (2001)	A probabilistic approach to identify which SNPs have an effect on the protein function using structural and evolutionary features that compare the variation against a dataset of mutations of lac repressor and T4 lysozyme.	Chasman and Adams (2001)
CONDEL	**Con**sensus **del**eteriousness provides a score computed based on the weighted average of the normalized scores of five different tools: LogR.E-value, MAPP, mutation assessor, polyphen, and STIF.	González-Pérez and López-Bigas (2011)
Evolutionary action	Evolutionary action is a function that links genotype with phenotype using evolutionary information, by quantifying the impact of SNPs on the fitness of a population; it correlates with disease-associated mutations.	Katsonis and Lichtarge (2014)
FATHMM	Uses Hidden Markov Models (HMMs) to obtain position-specific information. The prediction is based on the probability change of the HMM between wild-type and mutant.	Shihab et al. (2012)
FunSAV	A random forest classifier for predicting deleterious SNPs. It combines properties of the mutated protein with other tools (i.e., nsSNPAnalyzer, PANTHER, PhD-SNP, PolyPhen2, SIFT, and SNAP).	Wang et al. (2012)
FuzzySnps	A machine-learning approach that uses a Random Forest model trained by combining "4-Body Statistical Potential" and sequence-based features to identify tolerant and intolerant SNPs.	Barenboim et al. (2008)

Goldgar et al. (2004)	A probabilistic approach to determine if a SNP is disease-causing, which is achieved by computing the likelihood of the protein to be similar to previously classified mutated proteins in a dataset.	Goldgar et al. (2004)
HANSA	It is a machine-learning classifier that uses a SVM model to predict whether a SNP will be neutral or disease-causing.	Acharya and Nagarajaram (2011)
LogR.E-value	Uses the *E*-value computed by the HMMER algorithm using PFAM motifs to distinguish between deleterious and neutral SNPs.	Clifford et al. (2004)
LS-SNP	A workflow/database that uses predefined rules and machine-learning (SVN) approach to systematically characterize known SNPs.	Karchin et al. (2005)
Krishnan and Westhead (2003)	Two machine-learning approaches – using SVM and Decision Trees models – are used to predict the "effect" or "no-effect" of a SNP.	Krishnan and Westhead (2003)
MAPP	**M**ultivariate **A**nalysis of **P**rotein **P**olymorphism uses statistical analysis to predict the deleterious effect of SNPs.	Stone (2005)
Mutation assessor	Predicts the degree of impact in a protein by scoring the mutation based on the impact it causes regarding the properties of a multiple sequence alignment of homologous sequences.	Reva et al. (2011)
Mutation taster 2	Uses a Bayes classifier to predict disease associated effects caused by SNPs or Indels. The classifier uses a set of features that includes splicing site and polyadenylation signal information along with structural and evolutionary properties.	Schwarz et al. (2014)
MutPred	Uses a machine-learning approach to predict disease or neutral SNPs. The features used refer to a probability of loss or gain of function regarding several functional and structural properties of the encoded protein. The authors trained SVM and Random Forest models in this work.	Li et al. (2009)
nsSNPAnalyzer	Uses a Random Forest model trained with features (consisting of SIFT score and information from multiple sequence alignment and protein 3D structures) to identify disease associated SNPs.	Bao et al. (2005)
Papepro	A SVM prediction model is used by the authors to separate deleterious from neutral SNPs.	Tian et al. (2007)
Panther	Using an internal database of HMM, an evolutionary score is computed and the method predicts deleterious or neutral effects with a probability attached. The cutoff can be defined by the user (default is 3).	Thomas and Kejariwal (2004)
PhD-SNP	This approach uses one of two SVM models: one is trained using sequence profile features and the other is trained using sequence features. The choice of which model to use is based on a preliminary decision: if the mutation exists in the homology profile, the first model is used, otherwise the prediction is done using the second model.	Capriotti et al. (2006)
PMut	Predicts pathological or neutral effects of amino-acid substitutions. The prediction model is a neural network using structural-, physicochemical-, and evolutionary-based features, all calculated using sequence information only (without requiring a3D protein structure).	Ferrer-Costa et al. (2005)
Polyphen	A set of rules defined by the authors is used to predict the effect of a SNP. These rules are built based on three properties: PSIC score, substitution site properties, and substitution type properties. If one of the rules matches, the output can be deleterious or benign, otherwise the substitution is classified as neutral.	Ramensky (2002)
PolyPhen2	The follow up version of Polyphen, uses a naive Bayes predictor to predict damaging, benign, or neutral effects of SNPs. It uses structural information if available.	Adzhubei et al. (2010)
PROVEAN	**Pro**tein **V**ariation **E**ffect **AN**alyzer computes a score based on evolutionary information to predict if a genetic variant (i.e., SNP or Indel) is neutral or deleterious.	Choi et al. (2012)
RCOL	Applies a Bayes' formula to calculate the probability of a SNP to be deleterious. The likelihood is tested using 20 structural and physicochemical parameters.	Terp et al. (2002)
SAPRED	Using a SVM prediction model, the authors combine features computed from evolutionary, structural, and physicochemical properties to predict disease associated SNPs.	Ye et al. (2007)
SIFT	Using a PSSM, SIFT determines the probability of a substitution being tolerated in a given position.	Ng and Henikoff (2001)
SNAP	Identifies non-neutral SNPs using machine-learning approaches that combines a battery of Neural Network models.	Bromberg et al. (2008)
SNPs3D	Combines a set of features obtained from protein 3D structure and evolutionary information to predict deleterious effects using a SVM model.	Yue et al. (2006)
SNPs&GO	A machine-learning approach that includes GO annotations as features in a SVM model to predict whether a SNP is neutral or disease associated.	Calabrese et al. (2009)
SNPs&GO^{3D}	It is the successor of SNPs&GO. It includes new features obtained from protein 3D structure.	Capriotti and Altman (2011)
Sunyaev (2001)	This approach uses a set of seven rules empirically defined by the authors to identify nsSNPs. If one of the rules is matched, then the SNP is likely to be deleterious.	Sunyaev (2001)
SuSPect	A SVM model implementation to predict disease phenotypes caused by SNPs. The authors started with a high number of features until they identified nine that provided best performance.	Yates et al. (2014)
VarMode	A machine-learning approach using a SVN model to predict the effect of SNPs that includes information regarding known protein–protein interactions. It predicts non-synonymous SNPs.	Pappalardo and Wass (2014)

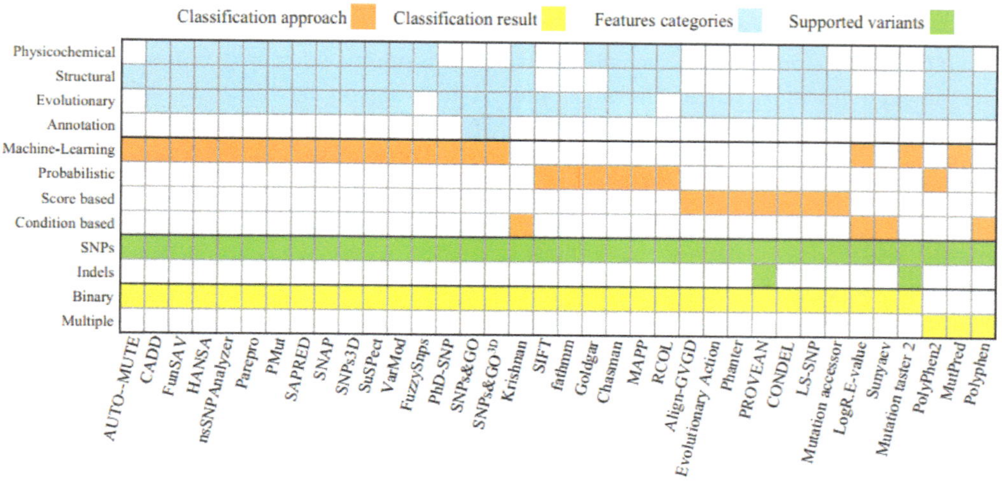

Figure 2: Summary of properties and approaches for software listed in Table 1

The approaches found fall into four different categories: *Machine-Learning*, *Probabilistic*, *Score* (calculating a summarizing score of a set of hand-picked statistics), and *Rule* (using a set of empirically derived rules). These approaches provide one of two types of classifications each: a binary classification (e.g., neutral or deleterious) or a multi-classification (e.g., benign, neutral, and deleterious). The features used by those approaches can be computed based on properties of the following five categories: (i) physicochemical properties (e.g., solvent accessibility, polarity, charge, disorder, and Grantham), (ii) structural information about the primary, secondary, and tertiary structure of a protein (e.g., α-helices, β-sheets, and coil), (iii) evolutionary properties (multiple sequence alignments, position-specific scoring matrices, and Hidden Markov models), and (iv) genome annotation (GO terms or other protein function annotations). The supported variants were determined either by accessing the tools' websites or by the description of the approach itself.

Of the existing algorithms (Table 1), *SIFT* (**S**orting **I**ntolerant **f**rom **T**olerant) (Ng and Henikoff, 2001) is often used as a gold standard to compare the performance of new algorithms or as a foundation for novel prediction strategies. SIFT and related approaches are based on the notion that evolutionary conservation can be used to predict the functional importance of each amino-acid in a protein and the impact of specific amino-acid substitutions. These methods typically use multiple sequence alignments of related proteins to determine a probabilistic description of what amino-acid substitutions are allowed in specific sites within the target protein. These descriptions can be used to determine the probability that non-synonymous coding SNPs observed in a re-sequencing data set will be tolerated by the protein; substitutions with a probability score smaller than a threshold are assumed to be deleterious (Kumar et al., 2009).

Sorting intolerant from tolerant provides only a binary deleterious/non-deleterious classification, and other methods have been developed to allow predicting cases where SNPs improve protein function. The *Polyphen* (Ramensky, 2002) and *PolyPhen2* (Adzhubei et al., 2010) approaches provide the means to discriminate three states when analyzing the effect of a SNP: benign, neutral, or deleterious. *Polyphen* uses a list of predetermined rules that combine the output of multiple algorithms using combinations of structural and sequence-based measures of mutation impact. *PolyPhen2* uses a machine-learning approach (a naive Bayes model) to predict an overall score for the variant effect, and the classification to three categories is based on thresholds. Although the algorithm is trained with human datasets, similar methods could potentially be used to build predictive models for variant effects in microorganisms. The overall variant effect score could also be exploited in more advanced methods that combine scores from different variants affecting different proteins to make phenotypic predictions.

Most studies on genetic variation focus on SNPs and disregard indels, which are also commonly observed when related microbial strains are compared to each other. The *PROVEAN* (Choi et al., 2012) and *Mutation taster 2* (Schwarz et al., 2014) approaches are capable of analyzing both SNPs and indels. *PROVEAN* uses substitution matrix scores (i.e., BLOSUM62) with gap and extension penalties to compute a variation score between the wild-type and mutant. More recently, *Mutation taster 2* computes several features (structural and evolutionary properties) for the mutated sequence using a Bayes classifier.

One possible approach for improving our ability to predict variant effects on protein function would be to predict effects of amino-acid changes on protein stability and folding (Khan and Vihinen, 2010). There are a number of tools available for these tasks (Khan and Vihinen, 2010), and stability predictions could be used to predict variant effects on protein function, as strongly destabilizing mutations would result in complete loss of function for the protein. Methods for predicting variant effects on protein stability have only been found to be moderately accurate in independent evaluation studies (Khan and Vihinen, 2010). For this reason, stability predictors should be combined with other variant effect prediction approaches to improve their predictive power for general variant effect analysis. The application of these types of stability prediction methods will be discussed in Section 3.2 in more detail together with the applications of metabolic modeling.

The majority of algorithms (53%) for variant effect prediction listed in Table 1 rely on machine-learning approaches [e.g., AUTO-MUTE (Masso and Vaisman, 2010), FunSAV (Wang et al., 2012), or HANSA (Acharya and Nagarajaram, 2011)], which is a practical strategy given the huge amount of data available for human diseases. Regarding the selection of features, most methods use evolutionary conservation information (92%) and more than half rely on structural properties (69%). The selection of sufficient features is a challenge in itself; no matter what approach is used, it is necessary to define which properties and attributes of proteins are capable of discriminating the phenotypes of interest. The improvements in the prediction capabilities provided by sequence-, evolution-, or structural-based features has been previously studied, and these studies have shown that the inclusion of structural properties leads to significant improvements in predictive power (Saunders and Baker, 2002). This has been recently confirmed by a benchmark performance test that includes several of the existing algorithms (Thusberg et al., 2011). Another effort to benchmark and improve different approaches is the Critical Assessment of Genome Interpretation (CAGI) community, which organizes a benchmark competition on predicting the effect of genetic variants on known disease phenotypes.

While the majority of algorithms aim to predict variant effects on individual proteins, a different objective is followed by the SNP-IN method that predicts how protein–protein interactions (PPIs) are affected by a SNP (Zhao et al., 2014). This is achieved by a set of features that includes the relative free energy change between wild-type and mutant PPI, the energy of all interactions in a protein complex, and other physicochemical properties, e.g., hydrophobic solvation or water bridges. Using these features, supervised and semi-supervised machine-learning approaches are used to predict how deleterious SNPs are. This approach is a very interesting, as changes in PPIs could be used to explain epistatic interactions between multiple variants. Like some previously mentioned prediction algorithms, SNP-PI requires an existing 3D model of the protein structure and, in addition, knowledge of the PPIs a given protein is involved in.

At a larger scale, genome-wide association studies are used to identify how differences between hundreds of thousands of individuals and make genotype to phenotype consequences. This approaches work as black boxes and make use of statistical and machine-learning approaches that require huge datasets. The current work and applications (e.g., clinical risk assessment) have been recently reviewed (Okser et al., 2014).

In vivo: Deep Mutational Scanning and TN-SEQ

Next-generation sequencing has enabled studying the effects of genetic variation on individual proteins or regulatory elements *in vivo* and *in vitro*. Deep mutational scanning (DMS) is an effective high-throughput method to measure the effects of mutations on protein stability and function (Fowler and Fields, 2014). The space of all possible amino-acid substitutions in a protein is exhaustively screened by first constructing a library of sequence variants using standard techniques like error prone PCR, then by using a high-throughput assay to select variants based on a fitness measure (e.g., growth rate, ligand binding, or product fluorescence), and finally by applying deep sequencing to the selected and unselected sequence variant pools. This approach results in a matrix that contains fitness values for each amino-acid substitution discovered in the selected pool. Depending on the method used for creating sequence diversity and sequencing depth, DMS can also be used to measure epistatic effects between substitutions at different sites.

The applicability of DMS is primarily limited by the lack of high-throughput functional assays for most proteins and, so far, DMS has not been applied to metabolic enzymes. When DMS can be applied at a broader scale, the results obtained from the assay could increase the predictive power of bioinformatic tools for genetic variation analysis by providing more complete training datasets for the types of predictive methods discussed in the previous section. Methods similar to DMS can also be used to systematically study effects of genetic variation in regulatory regions on protein expression using fluorescence protein-based assays.

Here, we will highlight a few case studies using DMS and related methods to study protein or regulatory element function. In the analysis of *Saccharomyces cerevisiae* poly(A)-binding protein (Melamed et al., 2013), strong epistatic effects between substitutions at specific sites were discovered. Although epistasis was not widespread, this is worrying from a computational modeling perspective, as modeling approaches usually do not account for epistasis. Another important highlight is the identification of alternative start codons. Although analyzed in previous studies, the DMS has shown that some amino-acids can be replaced by methionine and yield functional proteins (Kim et al., 2013). This biological information can be extrapolated to other studies and is highly relevant when developing strategies to understand the effect of mutations, either *in vivo* or*in silico*. Strategies similar to DMS have also been used to systematically study the effects of variation in transcription factor binding sites and other regulatory elements such as ribosomal binding sites (Kosuri et al., 2013). These studies will build the foundation for predicting effects of non-coding sequence variants on protein expression.

The methods described above allow us to systematically study the effects of a large number of variants in individual proteins or regulatory regions. In microorganisms, it is also possible to use a next-generation sequencing-based method called Tn-seq to systematically study the effect of disruption of a large number of genomic loci on cellular phenotypes (van Opijnen and Camilli, 2013). Transposons are mobile DNA elements that can disrupt a genetic locus by integrating themselves into it (Figure 1B). Tn-seq, using high density transposon insertion libraries, can be used to interrogate the function of, for example, regulatory elements and specific protein domains in a single genome-wide assay (van Opijnen and Camilli, 2013). Tn-seq has found many applications in microbiology, and it has been used for the identification of gene function, understanding genome organization, mapping genetic interactions, or assessing gene essentiality (van Opijnen and Camilli, 2013; Yang et al., 2014). Tn-seq does not offer a resolution on the single base-pair level, but the method can be rapidly used to generate sub-gene-level information relating, for example, to the essentiality of specific domains in a protein. This information in turn could be used to improve variant effect predictions, as variants in essential domains of a protein would be more likely to be predicted to be deleterious than variants in non-essential domains of the same protein.

PREDICTING PHENOTYPES FROM GENOTYPES AT THE GENOME-SCALE

Statistical and Network-Oriented Approaches for Predicting Phenotypes from Genotypes

Section 2 focused on the task of predicting the effects of genetic variation on individual protein function or expression. However, this is only a small part of a much larger problem, which of predicting cellular or organism phenotypic effects of all the genetic variants present in a genome. This requires combing the effects of variation on the function and expression of all proteins. So far, there have been surprisingly few efforts to take all genetic variants discovered in an individual (either a human or a microbial strain) and attempt to predict how certain phenotypes would be affected by all these variants together (Burga and Lehner, 2013; Lehner, 2013).

One of the first systematic attempts toward this goal was the pioneering study by Jelier et al. in *S. cerevisiae*, where growth phenotypes of selected yeast strains under different conditions were predicted from genetic differences between a reference strain and the strain of interest (Jelier et al., 2011). This was achieved by first predicting effects of coding and regulatory variants on protein function and expression using approaches similar to the one outlined in the previous section. These variant effect predictions were then combined into a single phenotypic prediction for the strain, using published single gene deletion growth phenotyping data for a yeast reference strain under the same condition. This approach can be considered to be highly simplistic, as the effects of multiple genetic variants acting on separate proteins were treated cumulative. Despite this, the approach still allowed accurate prediction of growth phenotypes across a broad range of conditions. There have also been a number of other approaches for predicting broader phenotypic consequences of single variants by mapping the variant data onto biological networks such as PPI or genetic networks (Carter et al., 2013). However, these approaches have typically not attempted to use the whole genotype of an individual (i.e., more than one variant at a time) to predict specific phenotypes.

Using Genome-Scale Metabolic Models for Interpreting Genetic Variants

The phenotype prediction methods described above are data-driven and use statistical models to predict the effects of genetic variants in the context of biological networks. However, for metabolic networks we can go beyond statistical models and graph-based descriptions to constraint-based models that are scalable to the genome-level and incorporate physicochemical, flux capacity, and reaction directionality constraints [see Price et al. (2004) for a review of constraint-based modeling]. This type of mechanistic modeling approach is very useful for understanding genetic changes that affect specific metabolic phenotypes. For example, the study of SNPs that affect mitochondrial metabolism (Jamshidi and Palsson, 2006) is a good example of how variant data can be mapped onto metabolic networks in order to explain the mechanistic basis of disease phenotypes.

A genome-scale metabolic models are composed of biochemical reactions, collected from literature and the genome annotation of an organism. This system of reactions is encoded as a matrix of stoichiometric coefficients that is usually referred to as stoichiometry matrix[1]. Assuming metabolism is in a steady-state, i.e., metabolite concentrations do not change over time, all fluxes have to balance each other. These flux-balances constitute linear constraints that can easily be analyzed using methods from linear algebra.

Furthermore, after inclusion of further constraints, e.g., known uptake and secretion rates and knowledge about reaction directionality, linear optimization methods can compute biologically relevant flux vectors that maximize defined objective functions. For example, growth can be simulated by maximizing the consumption of biomass precursors in empirically determined proportions. This type of analysis is usually referred to as flux balance analysis [FBA; see Orth et al. (2010) for a comprehensive introduction to this method].

Global optimal solutions to this linear optimization problems can be calculated very efficiently using linear programing (computation times are on a millisecond to second range for genome-scale models). Thus, one can compute thousands of phenotypes in a few minutes, simply by changing the constraints of the problem [see Lewis et al. (2012) for a comprehensive list of available *in silico*methods and (Bordbar et al., 2014) for a review of their applications].

Since the relationship between reactions, enzymes, and genes (usually referred to as GPR associations) is usually known and encoded in these models, the effect of a gene knockout can readily be mapped to the associated reactions by constraining their fluxes to be zero or by removal from the model. This way FBA can be used to compute the metabolic phenotype associated with a metabolic gene deletion, making it suitable for the analysis of genetic variation data that involves deletions or other mutations that lead to the complete loss of function of enzymes.

Flux balance analysis assumes that knockout strains can recover to an optimal growth phenotype, which might be unrealistic in cases where regulatory mechanisms – not modeled explicitly in these models – might not be able to accommodate the desired state. Other methodologies [e.g., ROOM (Shlomi et al., 2005), MoMA (Segrè et al., 2002), MiMBl (Brochado et al., 2012), and RELATCH (Kim and Reed, 2012)] employ more plausible assumptions and have been shown to improve the accuracy of knockout predictions. For example, MoMA minimizes the euclidean distance of the wild-type and mutant flux distributions, assuming that a mutant reaches the closest feasible flux distribution that is not necessarily optimal. The predictive power of FBA and these other approaches have been extensively assessed using genome-wide gene knockout assays (Snitkin et al., 2008) and transposon insertion libraries (Yang et al., 2014) and have resulted generally in a high degree of accuracy (Monk and Palsson, 2014).

Constraint-based models have also been applied to predict epistatic interactions by simulating effects of pairwise gene deletions, but with a significantly reduced accuracy in comparison to single deletions (Szappanos et al., 2011). Furthermore, simulations of multiple gene deletions have been successfully applied in developing design strategies for metabolic engineering by redirecting flux to desired products (Milne et al., 2009; Blazeck and Alper, 2010). A number of limiting factors can diminish the ability of constraint-based models to predict phenotypic effects of loss of function mutations: (i) missing reactions and erroneous GPRs, (ii) erroneous flux constraints due to the lack of thermodynamic or regulatory information, and (iii) the assumption of a fixed biomass composition that is known to change across growth conditions. Even with these limitations, constraint-based models still outperform statistical models in predicting consequences of gene deletions (Szappanos et al., 2011).

Since constraint-based models have demonstrated good ability to predict phenotypic outcomes of single and multiple gene deletions, these models should also be useful for predicting effects of other genetic variants. A SNV or indel that is predicted to reduce the maximal flux rate of an enzyme can be used to constrain the upper bound of a flux. FBA and similar methods can be used to compute the effects of these variations on the phenotype, providing a system-wide overview of the effects caused by the substitution (Jamshidi et al., 2007). This is a fast and effective way of predicting phenotypes, but it requires that one can estimate the effect the variant has on the maximum flux rate. Nevertheless, cases of complete loss of function fall into the same category as gene knockouts, and combining the bioinformatic prediction tools discussed in Section 2.2 with modeling capabilities can be used to integrate variant data. This approach can also be extended to any number of variants and genes, with the caveat that epistatic interactions are currently not captured accurately by the models.

There is currently only a limited number of studies that use GSMs to systematically explore the effects of genetic variants on phenotypes. Chang et al. (2013) conducted a study where GSMs coupled with protein structures of metabolic enzymes (GEM-PRO[2]) were used to interpret genetic variant data of *Escherichia coli* strains evolved to tolerate high temperatures (Chang et al., 2013). In this study, a GSM of *E. coli* was constrained using experimentally or bioinformatically determined thermostabilities of metabolic enzymes. Since the maximum flux capacity of a reaction is proportional to the concentration of active enzyme, temperature changes can be modeled by varying the flux constraints accordingly. This enables the prediction of enzymatic steps that are disproportionately temperature sensitive. For the evolved strains, flux balance analysis was used to explore the adaptation of the mutated enzymes; constraints associated with mutated proteins were relaxed to explain the experimentally measured growth rates (Chang et al., 2013). The study did not include separate predictions of variant effects on protein function, but rather treated all variants observed in a protein as potentially affecting its activity.

A more recent study by Nam et al. (2014) describes the use of GSMs for understanding the metabolic effects of cancer mutations. In particular, Nam et al. use genetic mutation information, gene expression profile data, and a human GSM (Thiele et al., 2013) to construct context-specific models for different cancer types. Loss and gain of function were systematically analyzed. Loss of function was modeled as described above (i.e., constraining affected reactions' fluxes to 0). Gain of a function, on the other hand, was modeled by adding novel promiscuous activities as predicted by chemoinformatic approaches. This approach allowed the prediction of potential oncometabolites.

Kinetic Modeling of Genetic Variants

As mentioned in the previous section, constraint-based modeling does not provide any information about the dynamic behavior of a metabolic system. A full kinetic description of a biochemical reaction network can be formulated using ordinary differential equations (Heinrich and Schuster, 1996). The major advantage of using kinetic models to study effects of genetic variation lies in their ability to account for mutations affecting catalytic or regulatory sites of an enzyme, causing either a gain or loss of catalytic activity, or binding sites of allosteric regulators.

Previous studies of red blood cell metabolism provide an overview on how SNPs can alter kinetic parameters and how kinetic models can be used to explain metabolic syndromes caused by enzyme deficiencies (Jamshidi, 2002; Jamshidi and Palsson, 2009). A disadvantage of using kinetic models is that kinetic parameters are not available for most enzymes and measuring the parameters can be challenging. For this reason, building predictive genome-scale kinetic models remains a challenge (Stanford et al., 2013). Kinetic models are a viable tool for interpreting genetic variant data only in specific cases like, for example, the red blood cell that harbors a relatively simple metabolism.

CONSIDERATIONS AND FUTURE DIRECTIONS

Methods and Tools to Predict the Effect of Genetic Variants

Many approaches have been explored in the past decade to understand and analyze the effects of genetic variation. In particular, the most active field has been the application of NGS techniques to characterize of genetic variation in the context of human disease. The amount of disease related information makes machine-learning approaches very suitable for the purpose of predicting effects of single genetic variants. Since most prediction methods have been trained and tested with human data, many of the existing methods do not perform as well or are simply not suited for the analysis of microbial genetic variants.

The other area where the study of microbial genetic variation lags behind human genetics is the systematic collection of variant and phenotyping data. Efforts to collect human genotype and phenotype data in a standardized way are currently underway with databases such as dbSNP and European Variation Archive. The UniProt database also collects variants found in the proteins sequences when this information is available. Every day thousands of new environmental or pathogenic isolates and laboratory developed microbial strains are sequenced around the world, but there is no centralized repository for this data in common use. We argue that it is of utmost importance to collect genetic variant data together with associated phenotypic data in a standard way for microbes as well.

All the existing algorithms for variant effect prediction are used to classify variants to preassigned categories (for example deleterious or non-deleterious). The approaches that predict deleterious effects can already be handled as knockouts in modeling their phenotypic effects using GSMs, but more subtle effects of mutations are missed by this approach. In order to improve our ability to predict phenotypes, there is a need to move beyond classification toward quantitative measures of variant effects on individual protein function. There are numerous features related to protein function that may be relevant for predicting variant effects: evolutionary and conservation, physicochemical (e.g., charge, polarity, or free energy), and structural (e.g., secondary structures, spatial distances between amino-acids or B-factors).

Existing methods for predicting variant effects have been primarily focused on generic predictors for all proteins irrespective of their function (e.g., enzymes, transcription factors, transporters, chaperons, etc.) and how do they behave in their environment (i.e., interaction with other elements: proteins, metabolites, DNA, etc.). This limits the predictive power of the methods in cases where additional information is readily available such as the relatively well studied field of microbial metabolism. For example, for metabolic enzymes, information on how kinetic parameters are affected by mutations and how these parameters vary between enzymes from different species is systematically collected in databases such as BRENDA. This type of information could be used to build improved variant effect predictors specifically for metabolic enzymes.

Modeling and High-Throughput Data Analysis

Improvements in genome-wide variant effect prediction can also come from improving or extending genome-scale modeling approaches. Recent innovations like GEM-PRO, as discussed in Section 3.2, fulfill the requirement of 3D protein structures to predict the effects of genetic variation at the protein level and could be used to systematically analyze the effect of genetic variation on a genome-scale for metabolism.

Approximately 10–30% of the genes encoded in a microbial genome are represented in metabolic GSMs, limiting the utility of these models for interpreting genomic variant data. Metabolic GSMs can be extended in a number of ways to increase coverage of the overall set of genes. The transcriptional regulatory network represented as interactions between transcription factors and target genes, can help extend the coverage of predictive models and can be integrated with metabolic GSMs in a number of ways (Covert et al., 2004; Chandrasekaran and Price, 2010). These integrated models have been successfully used to make phenotypic predictions.

Another recent extension of GSMs is ME-Models[3]. These models account for the entire machinery needed for gene and protein expression, providing a higher coverage of cellular functions and a higher resolution of cellular composition (O'Brien et al., 2013). ME-models have also been extended further to incorporate protein translocation from the cytoplasm to the periplasm (Liu et al., 2014). Currently, most of these extensions of GSMs have only been developed for *E. coli* and significant efforts will be required to build these extended models for other bacteria as well as eukaryotic model organisms such as *S. cerevisiae*.

The development of accurate kinetic models of metabolism, which could be useful for investigating the effects of mutations on allosteric regulation and catalytic activity, is still a tedious process. These models are usually limited to small parts of metabolism focusing on central carbon metabolism (Chassagnole et al., 2002; Peskov et al., 2012; Machado et al., 2014). There are two main reasons for these limitations: the models become huge in size and kinetic information of many enzymes is still unknown. Protocols (Stanford et al., 2013) and methodologies (Chowdhury et al., 2014) are being developed to bring kinetic modeling to the genome-scale, but the resulting models have not yet reached sufficiently mature stage for use in variant effect prediction.

In comprehensive level, a strategy for building whole-cell models by combining multiple individual models of different cellular processes including cell cycle, metabolism, transcription, and transport has been proposed (Karr et al., 2012). This strategy that also allows combining models using different representations (constraint-based, kinetic, and stochastic) was used to build a functioning whole-cell model of one of the simplest prokaryotes, *Mycoplasma genitalium*. Efforts toward building more complete genome-scale models of microbes will continue as more and more information is collected and computing power increases. These models will bring us closer to the goal of genome-wide prediction of phenotypes from genotyping data.

Opportunities

Genetic engineering tools, such as MAGE (Wang et al., 2009) or CRISPR/Cas9 (Xu et al., 2014), already allow us to quickly edit genomes in a precise and accurate fashion at the single base-pair resolution level at multiple loci simultaneously. These methods will allow us to map epistatic interactions of variants within a single gene and between multiple genes more comprehensively than before. On the other hand, new *in silico* tools for predicting variant effects on phenotypes outlined above open the way to a new style of modeling at the scale of single nucleotides. These new modeling tools will greatly benefit from better training datasets that can be obtained using MAGE, CRISPR/Cas9 or other genome editing methods systematically to map epistatic interactions. The application of these novel strategies provides a way to fine tune activities of proteins in the context of complete cellular networks. For example, we envision that in the future we will have predictive models of how engineering of multiple enzymes at the single amino-acid level would affect the production of a desired metabolite.

To achieve the maximum potential of genome-scale biochemical network modeling and genetic variant analysis, a link must be created between these two fields. The necessary information to connect both worlds is already there: we know the genes, the proteins, and the reactions. The major limitations are in the current methods and data sources. On the one hand, we must overcome the limitations of the tools available to predict variant effects by allowing more fine grained predictions of how a variant may affect any given protein function or expression. The usage of protein folding predictions, for example, has already been established in metabolic modeling (Chang et al., 2013), and it should be possible to use tools that predict variant effects on protein stability together with genome-scale models. On the other hand, we need to improve biochemical network modeling techniques: this is a evolving field and in the past decade there have been efforts to standardize the construction of models (Thiele and Palsson, 2010) and improving prediction methods by including high-throughput data (Machado and Herrgård, 2014).

Finally, it should be acknowledged that there will always be limitations in using solely genomic variant data as the basis for making phenotypic predictions for specific strains. We may also need to measure intermediate phenotypes such as transcript, protein, or metabolite levels for these strains in order to make predictions of how a given genotype affects a specific phenotype (Burga and Lehner, 2013). Fortunately enough comprehensive multi-omic datasets are currently being collected for wild-type microbial strains, allowing refinement of modeling and bioinformatic approaches for phenotypic prediction (Ishii et al., 2007; Skelly et al., 2013). Hopefully, systematizing such datasets and a concerted action between modelers, geneticists, microbiologists, and bioinformaticians will allow us to achieve the prediction of changed and novel metabolic capabilities of a microbial strain from genomic re-sequencing data.

CONFLICT OF INTEREST STATEMENT

The authors declare that the research was conducted in the absence of any commercial or financial relationships that could be construed as a potential conflict of interest.

ACKNOWLEDGMENTS

JC, MH, and NS acknowledge support by the Novo Nordisk Foundation through the Novo Nordisk Foundation Center for Biosustainability. MA acknowledges funding from a Biotechnology-based Synthesis and Production Research grant from the Novo Nordisk Foundation.

REFERENCES

1. Acharya, V., and Nagarajaram, H. A. (2011). Hansa: an automated method for discriminating disease and neutral human nsSNPs.*Hum. Mutat.* 33, 332–337. doi: 10.1002/humu.21642
2. Adrio, J.-L., and Demain, A. L. (2010). Recombinant organisms for production of industrial products. *Bioeng. Bugs* 1, 116–131. doi:10.4161/bbug.1.2.10484
3. Adzhubei, I. A., Schmidt, S., Peshkin, L., Ramensky, V. E., Gerasimova, A., Bork, P., et al. (2010). A method and server for predicting damaging missense mutations. *Nat. Methods* 7, 248–249. doi:10.1038/nmeth0410-248
4. Asadollahi, M. A., Maury, J., Patil, K. R., Schalk, M., Clark, A., and Nielsen, J. (2009). Enhancing sesquiterpene production in*Saccharomyces cerevisiae* through in silico driven metabolic engineering. *Metab. Eng.* 11, 328–334. doi:10.1016/j.ymben.2009.07.001
5. Atsumi, S., Wu, T.-Y., Machado, I. M. P., Huang, W.-C., Chen, P.-Y., Pellegrini, M., et al. (2010). Evolution genomic analysis, and reconstruction of isobutanol tolerance in *Escherichia coli*. *Mol. Syst. Biol.* 6, 449. doi:10.1038/msb.2010.98
6. Bao, L., Zhou, M., and Cui, Y. (2005). nsSNPAnalyzer: identifying disease-associated nonsynonymous single nucleotide polymorphisms. *Nucleic Acids Res.* 33, W480–W482. doi:10.1093/nar/gki372
7. Barenboim, M., Masso, M., Vaisman, I. I., and Jamison, D. C. (2008). Statistical geometry based prediction of nonsynonymous SNP functional effects using random forest and neuro-fuzzy classifiers. *Proteins* 71, 1930–1939. doi:10.1002/prot.21838
8. Blazeck, J., and Alper, H. (2010). Systems metabolic engineering: genome-scale models and beyond. *Biotechnol. J.* 5, 647–659. doi:10.1002/biot.200900247
9. Blount, Z. D., Barrick, J. E., Davidson, C. J., and Lenski, R. E. (2012). Genomic analysis of a key innovation in an experimental*Escherichia coli* population. *Nature* 489, 513–518. doi:10.1038/nature11514
10. Bordbar, A., Monk, J. M., King, Z. A., and Palsson, B. O. (2014). Constraint-based models predict metabolic and associated cellular functions. *Nat. Rev. Genet.* 15, 107–120. doi:10.1038/nrg3643
11. Brochado, A. R., Andrejev, S., Maranas, C. D., and Patil, K. R. (2012). Impact of stoichiometry representation on simulation of genotype-phenotype relationships in metabolic networks. *PLoS Comput. Biol.* 8:e1002758. doi:10.1371/journal.pcbi.1002758
12. Brochado, A. R., Matos, C., Møller, B. L., Hansen, J., Mortensen, U. H., and Patil, K. R. (2010). Improved vanillin production in baker's yeast through in silico design. *Microb. Cell Fact.* 9, 84. doi:10.1186/1475-2859-9-84
13. Bromberg, Y., Yachdav, G., and Rost, B. (2008). SNAP predicts effect of mutations on protein function. *Bioinformatics* 24, 2397–2398. doi:10.1093/bioinformatics/btn435
14. Burga, A., and Lehner, B. (2013). Predicting phenotypic variation from genotypes phenotypes and a combination of the two.*Curr. Opin. Biotechnol.* 24, 803–809. doi:10.1016/j.copbio.2013.03.004
15. Calabrese, R., Capriotti, E., Fariselli, P., Martelli, P. L., and Casadio, R. (2009). Functional annotations improve the predictive score of human disease-related mutations in proteins. *Hum. Mutat.* 30, 1237–1244. doi:10.1002/humu.21047
16. Capriotti, E., and Altman, R. B. (2011). Improving the prediction of disease-related variants using protein three-dimensional structure. *BMC Bioinformatics* 12(Suppl. 4):S3. doi:10.1186/1471-2105-12-S4-S3
17. Capriotti, E., Calabrese, R., and Casadio, R. (2006). Predicting the insurgence of human genetic diseases associated to single point protein mutations with support vector machines and evolutionary information. *Bioinformatics* 22, 2729–2734. doi:10.1093/bioinformatics/btl423
18. Carter, H., Hofree, M., and Ideker, T. (2013). Genotype to phenotype via network analysis. *Curr. Opin. Genet. Dev.* 23, 611–621. doi:10.1016/j.gde.2013.10.003
19. Caspeta, L., Chen, Y. P., Ghiaci, A. F., Buskov, S., Hallstrom, B. M., Petranovic, D., et al. (2014). Altered sterol composition renders yeast thermotolerant. *Science* 346, 75–78. doi:10.1126/science.1258137
20. Chandrasekaran, S., and Price, N. D. (2010).

Probabilistic integrative modeling of genome-scale metabolic and regulatory networks in *Escherichia coli* and *Mycobacterium tuberculosis*. *Proc. Natl. Acad. Sci. U.S.A.* 107, 17845–17850. doi:10.1073/pnas.1005139107

21. Chang, R. L., Andrews, K., Kim, D., Li, Z., Godzik, A., and Palsson, B. Ø (2013). Structural systems biology evaluation of metabolic thermotolerance in *Escherichia coli*. *Science* 340, 1220–1223. doi:10.1126/science.1234012
22. Chasman, D., and Adams, R. (2001). Predicting the functional consequences of non-synonymous single nucleotide polymorphisms: structure-based assessment of amino acid variation. *J. Mol. Biol.* 307, 683–706. doi:10.1006/jmbi.2001.4510
23. Chassagnole, C., Noisommit-Rizzi, N., Schmid, J. W., Mauch, K., and Reuss, M. (2002). Dynamic modeling of the central carbon metabolism of *Escherichia coli*. *Biotechnol. Bioeng.* 79, 53–73. doi:10.1002/bit.10288
24. Choi, Y., Sims, G. E., Murphy, S., Miller, J. R., and Chan, A. P. (2012). Predicting the functional effect of amino acid substitutions and indels. *PLoS ONE* 7:e46688. doi:10.1371/journal.pone.0046688
25. Chowdhury, A., Zomorrodi, A. R., and Maranas, C. D. (2014). k-OptForce: integrating kinetics with flux balance analysis for strain design. *PLoS Comput. Biol.* 10:e1003487. doi:10.1371/journal.pcbi.1003487
26. Clifford, R. J., Edmonson, M. N., Nguyen, C., and Buetow, K. H. (2004). Large-scale analysis of non-synonymous coding region single nucleotide polymorphisms. *Bioinformatics* 20, 1006–1014. doi:10.1093/bioinformatics/bth029
27. Conrad, T. M., Lewis, N. E., and Palsson, B. Ø (2011). Microbial laboratory evolution in the era of genome-scale science. *Mol. Syst. Biol.* 7, 509–509. doi:10.1038/msb.2011.42
28. Covert, M. W., Knight, E. M., Reed, J. L., Herrgård, M., and Palsson, B. Ø (2004). Integrating high-throughput and computational data elucidates bacterial networks. *Nature* 429, 92–96. doi:10.1038/nature02456
29. Ferrer-Costa, C., Gelpi, J. L., Zamakola, L., Parraga, I., de la Cruz, X., and Orozco, M. (2005). PMUT: a web-based tool for the annotation of pathological mutations on proteins. *Bioinformatics* 21, 3176–3178. doi:10.1093/bioinformatics/bti486
30. Fowler, D. M., and Fields, S. (2014). Deep mutational scanning: a new style of protein science. *Nat. Methods* 11, 801–807. doi:10.1038/nmeth.3027
31. Goldgar, D., Easton, D., Deffenbaugh, A., Monteiro, A., Tavtigian, S., and Couch, F. (2004). Integrated evaluation of DNA sequence variants of unknown clinical significance: application to BRCA1 and BRCA2. *Am. J. Hum. Genet.* 75, 535–544. doi:10.1086/424388
32. González-Pérez, A., and López-Bigas, N. (2011). Improving the assessment of the outcome of nonsynonymous SNVs with a consensus deleteriousness score, condel. *Am. J. Hum. Genet.* 88, 440–449. doi:10.1016/j.ajhg.2011.03.004
33. Gordon, J. L., Byrne, K. P., and Wolfe, K. H. (2009). Additions, losses, and rearrangements on the evolutionary route from a reconstructed ancestor to the modern *Saccharomyces cerevisiae* genome. *PLoS Genet.* 5:e1000485. doi:10.1371/journal.pgen.1000485
34. Heinrich, R., and Schuster, S. (1996). *The Regulation of Cellular Systems*. Boston, MA: Springer. doi:10.1007/978-1-4613-1161-4
35. Herrgård, M., and Panagiotou, G. (2012). Analyzing the genomic variation of microbe cell fact in the era of new biotechnology.*Comput. Struct. Biotechnol. J.* 3, 1–8. doi:10.5936/csbj.201210012
36. Ishii, N., Nakahigashi, K., Baba, T., Robert, M., Soga, T., Kanai, A., et al. (2007). Multiple high-throughput analyses monitor the response of *E. coli* to perturbations. *Science* 316, 593–597. doi:10.1126/science.1132067
37. Jamshidi, N. (2002). In silico model-driven assessment of the effects of single nucleotide polymorphisms (SNPs) on human red blood cell metabolism. *Genome Res.* 12, 1687–1692. doi:10.1101/gr.329302
38. Jamshidi, N., and Palsson, B. Ø. (2006). Systems biology of SNPs. *Mol. Syst. Biol.* 2, 38. doi:10.1038/msb4100077
39. Jamshidi, N., and Palsson, B. Ø. (2009). Using in silico models to simulate dual perturbation experiments: procedure development and interpretation of outcomes. *BMC Syst. Biol.* 3:44. doi:10.1186/1752-0509-3-44
40. Jamshidi, N., Vo, T. D., and Palsson, B. Ø. (2007). In silico analysis of SNPs and other high-throughput data. *Methods Mol. Biol.* 366, 267–285. doi:10.1007/978-1-59745-030-0_15
41. Jelier, R., Semple, J. I., Garcia-Verdugo, R., and Lehner, B. (2011). Predicting phenotypic variation in yeast from individual genome sequences. *Nat. Genet.* 43, 1270–1274. doi:10.1038/ng.1007
42. Karchin, R., Diekhans, M., Kelly, L., Thomas, D. J., Pieper, U., Eswar, N., et al. (2005). LS-SNP: large-scale annotation of coding non-synonymous SNPs based on multiple information sources. *Bioinformatics* 21, 2814–2820. doi:10.1093/bioinformatics/bti442
43. Karr, J. R., Sanghvi, J. C., Macklin, D. N., Gutschow, M. V., Jacobs, J. M., Bolival, B., et al. (2012). A whole-cell computational model predicts phenotype from genotype. *Cell* 150, 389–401. doi:10.1016/j.cell.2012.05.044
44. Katsonis, P., and Lichtarge, O. (2014). A formal perturbation equation between genotype and phenotype determines the Evolutionary Action of protein-coding variations on fitness. *Genome Res.* 24, 2050–2058. doi:10.1101/gr.176214.114
45. Kelley, J. L., Peyton, J. T., Fiston-Lavier, A.-S., Teets, N. M., Yee, M.-C., Johnston, J. S., et al. (2014). Compact genome of the Antarctic midge is likely an adaptation to an extreme environment. *Nat. Commun.* 5, 4611. doi:10.1038/ncomms5611
46. Khan, S., and Vihinen, M. (2010). Performance of protein stability predictors. *Hum. Mutat.* 31, 675–684.

doi:10.1002/humu.21242

47. Kim, I., Miller, C. R., Young, D. L., and Fields, S. (2013). High-throughput analysis of in vivo protein stability. *Mol. Cell. Proteomics* 12, 3370–3378. doi:10.1074/mcp.O113.031708

48. Kim, J., and Reed, J. L. (2012). RELATCH: relative optimality in metabolic networks explains robust metabolic and regulatory responses to perturbations. *Genome Biol.* 13, R78. doi:10.1186/gb-2012-13-9-r78

49. Kircher, M., Witten, D. M., Jain, P., O'Roak, B. J., Cooper, G. M., and Shendure, J. (2014). A general framework for estimating the relative pathogenicity of human genetic variants. *Nat. Genet.* 46, 310–315. doi:10.1038/ng.2892

50. Kosuri, S., Goodman, D. B., Cambray, G., Mutalik, V. K., Gao, Y., Arkin, A. P., et al. (2013). Composability of regulatory sequences controlling transcription and translation in *Escherichia coli*. *Proc. Natl. Acad. Sci. U.S.A.* 110, 14024–14029. doi:10.1073/pnas.1301301110

51. Krishnan, V. G., and Westhead, D. R. (2003). A comparative study of machine-learning methods to predict the effects of single nucleotide polymorphisms on protein function. *Bioinformatics* 19, 2199–2209. doi:10.1093/bioinformatics/btg297

52. Kumar, P., Henikoff, S., and Ng, P. C. (2009). Predicting the effects of coding non-synonymous variants on protein function using the SIFT algorithm. *Nat. Protoc.* 4, 1073–1081. doi:10.1038/nprot.2009.86

53. Lee, J., Yun, H., Feist, A. M., Palsson, B. Ø, and Lee, S. Y. (2008). Genome-scale reconstruction and in silico analysis of the*Clostridium acetobutylicum* ATCC 824 metabolic network. *Appl. Microbiol. Biotechnol.* 80, 849–862. doi:10.1007/s00253-008-1654-4

54. Lehner, B. (2013). Genotype to phenotype: lessons from model organisms for human genetics. *Nat. Rev. Genet.* 14, 168–178. doi:10.1038/nrg3404

55. Lewis, N. E., Nagarajan, H., and Palsson, B. O. (2012). Constraining the metabolic genotype–phenotype relationship using a phylogeny of in silico methods. *Nat. Rev. Microbiol.* 10, 291–305. doi:10.1038/nrmicro2737

56. Li, B., Krishnan, V. G., Mort, M. E., Xin, F., Kamati, K. K., Cooper, D. N., et al. (2009). Automated inference of molecular mechanisms of disease from amino acid substitutions. *Bioinformatics* 25, 2744–2750. doi:10.1093/bioinformatics/btp528

57. Li, R., Fan, W., Tian, G., Zhu, H., He, L., Cai, J., et al. (2010). The sequence and de novo assembly of the giant panda genome.*Nature* 463, 311–317. doi:10.1038/nature08696

58. Liu, J. K., O'Brien, E. J., Lerman, J. A., Zengler, K., Palsson, B. Ø, and Feist, A. M. (2014). Reconstruction and modeling protein translocation and compartmentalization in *Escherichia coli* at the genome-scale. *BMC Syst. Biol.* 8:110. doi:10.1186/s12918-014-0110-6

59. Long, M., Betrán, E., Thornton, K., and Wang, W. (2003). The origin of new genes: glimpses from the young and old. *Nat. Rev. Genet.* 4, 865–875. doi:10.1038/nrg1204

60. Machado, D., and Herrgård, M. (2014). Systematic evaluation of methods for integration of transcriptomic data into constraint-based models of metabolism. *PLoS Comput. Biol.* 10:e1003580. doi:10.1371/journal.pcbi.1003580

61. Machado, D., Rodrigues, L. R., and Rocha, I. (2014). A kinetic model for curcumin production in *Escherichia coli*. *BioSystems*125, 16–21. doi:10.1016/j.biosystems.2014.09.001

62. Masso, M., and Vaisman, I. I. (2010). Knowledge-based computational mutagenesis for predicting the disease potential of human non-synonymous single nucleotide polymorphisms. *J. Theor. Biol.* 266, 560–568. doi:10.1016/j.jtbi.2010.07.026

63. Melamed, D., Young, D. L., Gamble, C. E., Miller, C. R., and Fields, S. (2013). Deep mutational scanning of an RRM domain of the *Saccharomyces cerevisiae* poly(A)-binding protein. *RNA* 19, 1537–1551. doi:10.1261/rna.040709.113

64. Milne, C. B., Kim, P.-J., Eddy, J. A., and Price, N. D. (2009). Accomplishments in genome-scale in silico modeling for industrial and medical biotechnology. *Biotechnol. J.* 4, 1653–1670. doi:10.1002/biot.200900234

65. Monk, J., and Palsson, B. Ø (2014). Predicting microbial growth. *Science* 344, 1448–1449. doi:10.1126/science.1253388

66. Nakamura, Y., Mori, K., Saitoh, K., Oshima, K., Mekuchi, M., Sugaya, T., et al. (2013). Evolutionary changes of multiple visual pigment genes in the complete genome of Pacific bluefin tuna. *Proc. Natl. Acad. Sci. U.S.A.* 110, 11061–11066. doi:10.1073/pnas.1302051110

67. Nam, H., Campodonico, M., Bordbar, A., Hyduke, D. R., Kim, S., Zielinski, D. C., et al. (2014). A systems approach to predict oncometabolites via context-specific genome-scale metabolic networks. *PLoS Comput. Biol.* 10:e1003837. doi:10.1371/journal.pcbi.1003837

68. Ng, P. C., and Henikoff, S. (2001). Predicting deleterious amino acid substitutions. *Genome Res.* 11, 863–874. doi:10.1101/gr.176601

69. O'Brien, E. J., Lerman, J. A., Chang, R. L., Hyduke, D. R., and Palsson, B. Ø (2013). Genome-scale models of metabolism and gene expression extend and refine growth phenotype prediction. *Mol. Syst. Biol.* 9, 693–693. doi:10.1038/msb.2013.52

70. Okser, S., Pahikkala, T., Airola, A., Salakoski, T., Ripatti, S., and Aittokallio, T. (2014). Regularized machine learning in the genetic prediction of complex traits. *PLoS Genet.* 10:e1004754. doi:10.1371/journal.pgen.1004754

71. Orth, J. D., Thiele, I., and Palsson, B. Ø (2010). What is flux balance analysis? *Nat. Biotechnol.* 28, 245–248. doi:10.1038/nbt.1614

72. Pappalardo, M., and Wass, M. N. (2014). VarMod: modelling the functional effects of non-synonymous variants. *Nucleic Acids Res.* 42, W331–W336. doi:10.1093/nar/gku483

73. Park, J. H., Lee, K. H., Kim, T. Y., and Lee, S. Y. (2007). Metabolic engineering of *Escherichia coli* for the production of L-valine based on transcriptome analysis and in silico gene knockout simulation. *Proc.*

Natl. Acad. Sci. U.S.A. 104, 7797–7802. doi:10.1073/pnas.0702609104

74. Pegadaraju, V., Nipper, R., Hulke, B., Qi, L., and Schultz, Q. (2013). De novo sequencing of sunflower genome for SNP discovery using RAD (restriction site associated DNA) approach. *BMC Genomics* 14:556. doi:10.1186/1471-2164-14-556

75. Peskov, K., Mogilevskaya, E., and Demin, O. (2012). Kinetic modelling of central carbon metabolism in *Escherichia coli*. *FEBS J.*279, 3374–3385. doi:10.1111/j.1742-4658.2012.08719.x

76. Price, N. D., Reed, J. L., and Palsson, B. Ø (2004). Genome-scale models of microbial cells: evaluating the consequences of constraints. *Nat. Rev. Microbiol.* 2, 886–897. doi:10.1038/nrmicro1023

77. Puchałka, J., Oberhardt, M. A., Godinho, M., Bielecka, A., Regenhardt, D., Timmis, K. N., et al. (2008). Genome-scale reconstruction and analysis of the *Pseudomonas putida* KT2440 metabolic network facilitates applications in biotechnology.*PLoS Comput. Biol.* 4:e1000210. doi:10.1371/journal.pcbi.1000210

78. Ramensky, V. (2002). Human non-synonymous SNPs: server and survey. *Nucleic Acids Res.* 30, 3894–3900. doi:10.1093/nar/gkf493

79. Reva, B., Antipin, Y., and Sander, C. (2011). Predicting the functional impact of protein mutations: application to cancer genomics. *Nucleic Acids Res.* 39, e118–e118. doi:10.1093/nar/gkr407

80. Saunders, C. T., and Baker, D. (2002). Evaluation of structural and evolutionary contributions to deleterious mutation prediction. *J. Mol. Biol.* 322, 891–901. doi:10.1016/S0022-2836(02)00813-6

81. Schwarz, J. M., Cooper, D. N., Schuelke, M., and Seelow, D. (2014). MutationTaster2: mutation prediction for the deep-sequencing age. *Nat. Methods* 11, 361–362. doi:10.1038/nmeth.2890

82. Segrè, D., Vitkup, D., and Church, G. M. (2002). Analysis of optimality in natural and perturbed metabolic networks. *Proc. Natl. Acad. Sci. U.S.A.* 99, 15112–15117. doi:10.1073/pnas.232349399

83. Shihab, H. A., Gough, J., Cooper, D. N., Stenson, P. D., Barker, G. L. A., Edwards, K. J., et al. (2012). Predicting the functional molecular, and phenotypic consequences of amino acid substitutions using hidden Markov models. *Hum. Mutat.* 34, 57–65. doi:10.1002/humu.22225

84. Shlomi, T., Berkman, O., and Ruppin, E. (2005). Regulatory on/off minimization of metabolic flux changes after genetic perturbations. *Proc. Natl. Acad. Sci. U.S.A.* 102, 7695–7700. doi:10.1073/pnas.0406346102

85. Skelly, D. A., Merrihew, G. E., Riffle, M., Connelly, C. F., Kerr, E. O., Johansson, M., et al. (2013). Integrative phenomics reveals insight into the structure of phenotypic diversity in budding yeast. *Genome Res.* 23, 1496–1504. doi:10.1101/gr.155762.113

86. Snitkin, E. S., Dudley, A. M., Janse, D. M., Wong, K., Church, G. M., and Segrè, D. (2008). Model-driven analysis of experimentally determined growth phenotypes for 465 yeast gene deletion mutants under 16 different conditions. *Genome Biol.*9, R140. doi:10.1186/gb-2008-9-9-r140

87. Soares-Castro, P., and Santos, P. M. (2013). Towards the description of the genome catalogue of *Pseudomonas sp.* strain M1.*Genome Announc.* 1, e146–e112. doi:10.1128/genomeA.00146-12

88. Stanford, N. J., Lubitz, T., Smallbone, K., Klipp, E., Mendes, P., and Liebermeister, W. (2013). Systematic construction of kinetic models from genome-scale metabolic networks. *PLoS One* 8:e79195. doi:10.1371/journal.pone.0079195

89. Stone, E. A. (2005). Physicochemical constraint violation by missense substitutions mediates impairment of protein function and disease severity. *Genome Res.* 15, 978–986. doi:10.1101/gr.3804205

90. Sunyaev, S. (2001). Prediction of deleterious human alleles. *Hum. Mol. Genet.* 10, 591–597. doi:10.1093/hmg/10.6.591

91. Szappanos, B., Kovács, K., Szamecz, B., Honti, F., Costanzo, M., Baryshnikova, A., et al. (2011). An integrated approach to characterize genetic interaction networks in yeast metabolism. *Nat. Genet.* 43, 656–662. doi:10.1038/ng.846

92. Tavtigian, S. V. (2005). Comprehensive statistical study of 452 BRCA1 missense substitutions with classification of eight recurrent substitutions as neutral. *J. Med. Genet.* 43, 295–305. doi:10.1136/jmg.2005.033878

93. Tepper, N., and Shlomi, T. (2009). Predicting metabolic engineering knockout strategies for chemical production: accounting for competing pathways. *Bioinformatics* 26, 536–543. doi:10.1093/bioinformatics/btp704

94. Terp, B. N., Cooper, D. N., Christensen, I. T., Jøgensen, F. S., Bross, P., Gregersen, N., et al. (2002). Assessing the relative importance of the biophysical properties of amino acid substitutions associated with human genetic disease. *Hum. Mutat.* 20, 98–109. doi:10.1002/humu.10095

95. Thiele, I., and Palsson, B. Ø. (2010). A protocol for generating a high-quality genome-scale metabolic reconstruction. *Nat. Protoc.* 5, 93–121. doi:10.1038/nprot.2009.203

96. Thiele, I., Swainston, N., Fleming, R. M., Hoppe, A., Sahoo, S., Aurich, M. K., et al. (2013). A community-driven global reconstruction of human metabolism. *Nat. Biotechnol.* 31, 419–425. doi:10.1038/nbt.2488

97. Thomas, P. D., and Kejariwal, A. (2004). Coding single-nucleotide polymorphisms associated with complex vs. Mendelian disease: evolutionary evidence for differences in molecular effects. *Proc. Natl. Acad. Sci. U.S.A.* 101, 15398–15403. doi:10.1073/pnas.0404380101

98. Thusberg, J., Olatubosun, A., and Vihinen, M. (2011). Performance of mutation pathogenicity prediction methods on missense variants. *Hum. Mutat.* 32, 358–368. doi:10.1002/humu.21445

99. Tian, J., Wu, N., Guo, X., Guo, J., Zhang, J., and Fan, Y. (2007). Predicting the phenotypic effects of non-synonymous single nucleotide polymorphisms based on support vector machines. *BMC Bioinformatics* 8:450.

doi:10.1186/1471-2105-8-450

100. van Opijnen, T., and Camilli, A. (2013). Transposon insertion sequencing: a new tool for systems-level analysis of microorganisms. *Nat. Rev. Microbiol.* 11, 435–442. doi:10.1038/nrmicro3033

101. Varma, A., and Palsson, B. Ø (1993). Metabolic capabilities of *Escherichia coli* II. optimal growth patterns. *J. Theor. Biol.* 165, 503–522. doi:10.1006/jtbi.1993.1203

102. Wang, H. H., Isaacs, F. J., Carr, P. A., Sun, Z. Z., Xu, G., Forest, C. R., et al. (2009). Programming cells by multiplex genome engineering and accelerated evolution. *Nature* 460, 894–898. doi:10.1038/nature08187

103. Wang, L., Han, X., Zhang, Y., Li, D., Wei, X., Ding, X., et al. (2014). Deep resequencing reveals allelic variation in Sesamum indicum. *BMC Plant Biol.* 14:225. doi:10.1186/s12870-014-0225-3

104. Wang, M., Zhao, X., Takemoto, K., Xu, H., Li, Y., Akutsu, T., et al. (2012). FunSAV: predicting the functional effect of single amino acid variants using a two-stage random forest model. *PLoS One* 7:e43847. doi:10.1371/journal.pone.0043847

105. Xu, T., Li, Y., Nostrand, J. D. V., He, Z., and Zhou, J. (2014). Cas9-based tools for targeted genome editing and transcriptional control. *Appl. Environ. Microbiol.* 80, 1544–1552. doi:10.1128/AEM.03786-13

106. Yamamoto, K., Tamaki, H., Cadillo-Quiroz, H., Imachi, H., Kyrpides, N., Woyke, T., et al. (2014). Complete genome sequence of*Methanoregula formicica* SMSPT a mesophilic hydrogenotrophic methanogen isolated from a methanogenic upflow anaerobic sludge blanket reactor. *Genome Announc.* 2, e870–e814. doi:10.1128/genomeA.00870-14

107. Yang, H., Krumholz, E. W., Brutinel, E. D., Palani, N. P., Sadowsky, M. J., Odlyzko, A. M., et al. (2014). Genome-scale metabolic network validation of *Shewanella oneidensis* using transposon insertion frequency analysis. *PLoS Comput. Biol.* 10:e1003848. doi:10.1371/journal.pcbi.1003848

108. Yates, C. M., Filippis, I., Kelley, L. A., and Sternberg, M. J. E. (2014). SuSPect: enhanced prediction of single amino acid variant (SAV) phenotype using network features. *J. Mol. Biol.* 426, 2692–2701. doi:10.1016/j.jmb.2014.04.026

109. Ye, Z.-Q., Zhao, S.-Q., Gao, G., Liu, X.-Q., Langlois, R. E., Lu, H., et al. (2007). Finding new structural and sequence attributes to predict possible disease association of single amino acid polymorphism (SAP). *Bioinformatics* 23, 1444–1450. doi:10.1093/bioinformatics/btm119

110. Yue, P., Melamud, E., and Moult, J. (2006). SNPs3D: candidate gene and SNP selection for association studies. *BMC Bioinformatics* 7:166. doi:10.1186/1471-2105-7-166

111. Zhao, N., Han, J. G., Shyu, C.-R., and Korkin, D. (2014). Determining effects of non-synonymous SNPs on protein-protein interactions using supervised and semi-supervised learning. *PLoS Comput. Biol.* 10:e1003592. doi:10.1371/journal.pcbi.1003592

CHAPTER 7

SEPARATE HYDROLYSIS AND CO-FERMENTATION FOR IMPROVED XYLOSE UTILIZATION IN INTEGRATED ETHANOL PRODUCTION FROM WHEAT MEAL AND WHEAT STRAW

Borbála Erdei[1], Balázs Frankó[2], Mats Galbe[1], and Guido Zacchi[1]

[1]Lund University, Department of Chemical Engineering, SE-221 00 Lund, Sweden.

[2]Budapest University of Technology and Economics, Department of Applied Biotechnology and Food Science, Szt Gellért tér 4, 1111 Budapest, Hungary

ABSTRACT

Background

The commercialization of second-generation bioethanol has not been realized due to several factors, including poor biomass utilization and high production cost. It is generally accepted that the most important parameters in reducing the production cost are the ethanol yield and the ethanol concentration in the fermentation broth. Agricultural residues contain large amounts of hemicellulose, and the utilization of xylose is thus a plausible way to improve the concentration and yield of ethanol during fermentation. Most naturally occurring ethanol-fermenting microorganisms do not utilize xylose, but a genetically modified yeast strain, TMB3400, has the ability to co-ferment glucose and xylose. However, the xylose uptake rate is only enhanced when the glucose concentration is low.

Results

Separate hydrolysis and co-fermentation of steam-pretreated wheat straw (SPWS) combined with wheat-starch hydrolysate feed was performed in two separate processes. The average yield of ethanol and the xylose consumption reached 86% and 69%, respectively, when the hydrolysate of the enzymatically hydrolyzed (18.5% WIS) unwashed SPWS solid fraction and wheat-starch hydrolysate were fed to the fermentor after 1 h of fermentation of the SPWS liquid fraction. In the other configuration, fermentation of the SPWS hydrolysate (7.0% WIS), resulted in an average ethanol yield of 93% from fermentation based on glucose and xylose and complete xylose consumption when wheat-starch hydrolysate was included in the feed. Increased initial cell density in the fermentation (from 5 to 20 g/L) did not increase the ethanol yield, but improved and accelerated xylose consumption in both cases.

Citation: Erdei et al.: Separate hydrolysis and co-fermentation for improved xylose utilization in integrated ethanol production from wheat meal and wheat straw. Biotechnology for Biofuels 2012 5:12, doi:10.1186/1754-6834-5-12

Conclusions

Higher ethanol yield has been achieved in co-fermentation of xylose and glucose in SPWS hydrolysate when wheat-starch hydrolysate was used as feed, then in co-fermentation of the liquid fraction of SPWS fed with the mixed hydrolysates. Integration of first-generation and second-generation processes also increases the ethanol concentration, resulting in a reduction in the cost of the distillation step, thus improving the process economics.

BACKGROUND

The use of bioethanol is beneficial for several reasons. One is that it can be easily integrated into the current fuel distribution system, and another is that it will reduce the production of greenhouse gases as the raw material used is biomass. Bioethanol can be produced from lignocellulosic biomass in three main steps: pretreatment, enzymatic hydrolysis, and fermentation. Enzymatic hydrolysis and fermentation can be carried out either separately, using separate hydrolysis and fermentation (SHF), or simultaneously, using simultaneous saccharification and fermentation (SSF).

Over the past decade SSF has become the preferred process, since end-product inhibition of the enzymes can be avoided by performing fermentation in the same vessel at the same time as hydrolysis [1]. The capital investment cost of the plant is also reduced as fewer tanks are required [2]. However, a disadvantage of SSF is that the operating temperature must be a compromise between the optimal temperatures for hydrolysis and fermentation, whereas these can be optimized independently in SHF. Furthermore, the yeast produced during the SHF process can be recycled after fermentation of the hydrolysate, which is not possible in SSF. The yeast thus represents a yield loss as it is difficult to separate it from the solid residue (lignin) [3].

Wheat straw and other agricultural residues generally consist of high amounts of hemicelluloses [4]. Xylose fermentation is therefore very important for these lignocellulosic raw materials in order to effectively convert all the sugars into ethanol, to increase the concentration in the fermentation broth. However, wild-type *Saccharomyces cerevisiae*, which is the most commonly used yeast in ethanol fermentation due to its attractive properties, such as high yield of ethanol, high specific rate of fermentation [5], and its high tolerance to the end product, ethanol [6], is not able to metabolize xylose. Therefore, naturally xylose-fermenting yeasts, such as *Candida shehatae* or *Pichia stipitis* [7, 8] have been widely studied since their ability to ferment xylose was discovered the early 1980s. Unfortunately, their tolerance to inhibitors [9] and ethanol [10] is limited, and they also require a very low and well-controlled supply of oxygen for effective xylose-fermentation [11, 12], which makes them difficult to use in large-scale production.

The ideal solution, combining the robustness and tolerance of *S. cerevisiae* with the ability to ferment xylose has been sought by metabolic engineering. In principle, genes from bacteria and fungi encoding xylose isomerase (XI) [13, 14], or genes from fungi encoding xylose reductase (XR) and xylitol dehydrogenase (XDH) [15] can be introduced into *S. cerevisiae*. The endogenous gene *XKS1* encoding xylulokinase (XK) must also be overexpressed for the strain to be able to utilize xylose for growth and ethanol production [16]. TMB3400 [17] is an industrial strain of *S. cerevisiae* containing genes that encode for XR/XDH/XK, which is able to co-ferment xylose and glucose in non-detoxified lignocellulose hydrolysate of spruce [18, 19], as well as various pretreated raw materials [20–23] in simultaneous saccharification and co-fermentation (SSCF).

In *S. cerevisiae* TMB3400, xylose and glucose are competitively transported by the same transport protein [24, 25], but xylose has an approximately 200-fold lower affinity [26]. To avoid the inhibition of xylose uptake by glucose, the glucose concentration in the medium must be low. It has been reported that low, but non-zero, concentration of glucose enhances xylose utilization [25], which means that a slow release or feeding of glucose is required. This is one of the reasons why SSCF has become an interesting process option, as glucose is released during hydrolysis.

Co-fermentation of glucose and xylose in wheat straw hydrolysate has been investigated by Olofsson *et al.* [21]. Their results showed that almost complete xylose fermentation can be obtained if a controlled glucose feed is applied during fermentation. However, during SSF the release of sugar is not controlled, as all the cellulase enzymes are added at once. Thus prefermentation [20] and enzyme feeding strategies combined with fed-batch fermentation have recently been studied [27] as a means of controlled glucose release. The best result so far, 80% xylose uptake, has been achieved by Olofsson *et al.* using a yeast concentration of 4 g/L and controlled feeding of cellulases giving a glucose release rate of 2 g/L h [28]. Simultaneous glucose and xylose uptake has been modelled in a study by Bertilsson *et al.*, indicating that a glucose feed rate between 5 and 10 g/L h would be suitable to obtain maximum xylose uptake rate, with a yeast cell concentration of 5 g/L [29]. The effect of controlled glucose feeding has also been studied on barley straw hydrolysate by Linde *et al.* [30], however, only 74% and 51% of the xylose was consumed by the yeast TMB3400 at glucose feed rates of 0.21 g/L h and 0.45 g/L h, respectively.

Wheat hydrolysate is derived from wheat meal, following the first generation ethanol production process through liquefaction and saccharification. The hydrolysate obtained is a glucose-rich solution as well as a potential source of nutrients, which has several advantages in the process. The use of wheat-starch hydrolysate, as a complex nutrient source, has been shown to be a potential supplement for lignocellulosic hydrolysates [31]. It also has a positive effect on glucose fermentation in SSF, increasing both the ethanol yield and concentration [32], which is important for the process economy [33]. Therefore including wheat-starch hydrolysate in the feed in SHF, which is a first-generation glucose source, provides considerable possibilities for improvement of xylose uptake and hence increase the ethanol yield and concentration.

To the best of our knowledge, no study has yet been performed to evaluate separate hydrolysis and co-fermentation (SHCF) of steam-pretreated wheat straw combined with wheat-starch hydrolysate feed as a means of improving the co-fermentation of glucose and xylose. In the present work, two process configurations for SHCF were investigated using steam-pretreated wheat straw combined with wheat-starch hydrolysate feed, employing the xylose-fermenting yeast TMB3400.

RESULTS AND DISCUSSION

Enzymatic hydrolysis

Enzymatic hydrolysis of the whole SPWS slurry was performed at 7.5% WIS in Config. 1 to supply fermentation with hydrolysate. Enzymatic hydrolysis was first performed using SPWS 1 material supplemented with Cellic CTec and Cellic Htec enzyme preparations at a cellulase loading of 20 FPU/g glucan. The final glucose and xylose concentrations of 55.6 g/L and 30.6 g/L, respectively, were reached after 48 h, corresponding to yields of 96% and 91% of the theoretical, respectively. Since the hemicellulose sugars are mostly released during steam pretreatment, the xylose concentration only increased slightly during hydrolysis. Enzymatic hydrolysis of wheat straw has been investigated by Jorgenssen *et al.* [34], who obtained lower cellulose and xylose conversion yield at similar WIS loading, 75% and 65%, respectively, probably due to the different pretreatment conditions (steam pretreatment at 195°C or 205°C for 12 min with no chemical addition) and a lower enzyme mixture loading (13.3 FPU/g glucan) of a less effective enzyme cocktail (Celluclast 1.5 FG L and Novozym 188).

Enzymatic hydrolysis was then performed on the SPWS 2 material with a lower enzyme loading of Cellic Ctec2 (14 FPU/g glucan) when using wheat-starch hydrolysate feed. This experiment resulted in lower sugar concentrations due to the slightly lower WIS content (7.0%), and lower yields. The glucose and xylose concentrations reached their maxima of 32.3 g/L and 15.6 g/L, respectively, after 72 h, corresponding to yields of 63% and 62% of the theoretical, respectively.

Although the same pretreatment method was applied to both materials used in this study, SPWS 1 and SPWS 2 had quite different characteristics. This appears to be due to the different temperature control systems used for SPWS 2. Better temperature control reduced the temperature overshoot during the first few minutes of pretreatment and decreased the amount of steam admitted into the reactor, resulting in slightly milder pretreatment conditions and higher WIS.

Hydrolysis of the solid fraction of the SPWS was performed at 18.5% WIS supplemented with Cellic Ctec2 at an enzyme loading of 14 FPU/g glucan. The design of the solid-state fermentor allowed adequate mixing from the very beginning of hydrolysis and the slurry became liquefied already after 5 h allowing homogeneous samples to be taken. The final glucose and xylose concentrations were 101.7 g/L and 19.9 g/L, respectively, after 120 h, corresponding to 75% and 69% of the theoretical yield. These yields are higher than those obtained by Jorgenssen*et al.* [34], which would have been about 60% and 55% for glucose and xylose, respectively, recalculated assuming a linear correlation between cellulose and hemicellulose conversion and solid loading (WIS). The supernatant of the hydrolyzed SPWS solid fraction was used as a sugar-rich component in the feed for fermentation in Config. 2. Since the optimization of enzymatic hydrolysis was not the aim of this study, further investigations will be required to find the optimum temperature, WIS, and enzyme loading.

FERMENTATION

Glucose-fed fermentation

Fermentation of the SPWS hydrolysate (Config. 1) or the liquid fraction of SPWS (Config. 2) was first carried out in a simplified way, using a glucose solution as the feed instead of wheat-starch hydrolysate (Config. 1_G and Config. 2_G). The results obtained are presented in Table 1.

Table 1: Results obtained from fermentation fed with glucose solution (different concentrations in Config. 1G and Config. 2_G[a])

Expt Config.	Yeast load (g/L)	Xylose consumption[b] (%)	Xylose conc. (g/L)	Xylitol formation[c] (%, g/L)	Glycerol conc. (g/L)	Ethanol conc. (g/L)	Ethanol yield (g/g)	Ethanol yield[d] (%)
Config. 1_G	5	98	0.4	19 (4.3)	6.2	53.1	0.48	93
Config. 2_G	5	93	1.6	12 (2.6)	5.4	60.5	0.53	104

The raw material used for the experiments was SPWS 1.
[a]See Methods.
[b]Related to the total amount of available xylose.
[c]Related to consumed xylose.
[d]Corresponding to the maximum theoretical yield from available glucose and xylose.
G, glucose feed.

In Config. 1_G, fermentation was carried out on SPWS hydrolysate obtained after enzymatic hydrolysis, using the whole SPWS slurry with a WIS content of 7.5%. Figure 1 shows the concentration profiles of the substrates and the ethanol during fermentation. The glucose concentration was 50 g/L at the start of fermentation, corresponding to a glucose:xylose ratio of about 2 in the supernatant, which is very high. This sugar ratio could have a negative effect on the xylose uptake, but the glucose was consumed rapidly during the first 8 h, after which xylose uptake took place. More than 90% of the glucose was consumed in the first 5 h, resulting in a volumetric ethanol productivity of 4.4 g/L h during this time. Surprisingly, some xylose was consumed during this period, although has been claimed that the transport of xylose into the cells of genetically modified *S. cerevisiae* is inhibited by glucose [35]. When the glucose concentration fell below 5 g/L, the xylose uptake increased to 0.11 g/g cell h. Although, after 8 h there was no glucose in the medium, the xylose uptake reached 0.14 g/g cell h.

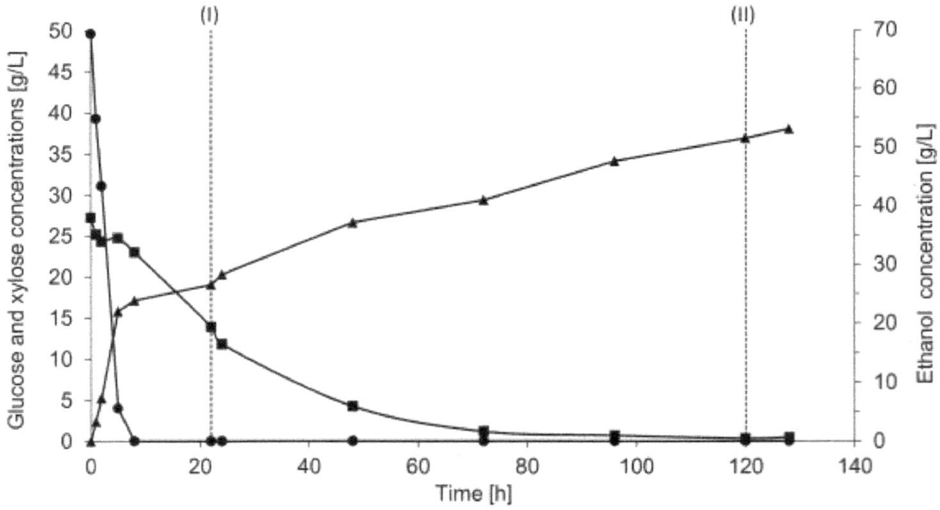

Figure 1: Concentrations of glucose (circles), xylose (squares), and ethanol (triangles) during fermentation in Config. 1_G with 5 g/L yeast. Glucose solution was fed to the fermentor for 96 h between (I) and (II) at a constant flow rate of 2.1 mL/h. (T: 32°C, pH 5, 300 rpm).

The volumetric ethanol productivity decreased from 3.0 to only 0.3 g/L h after the first 8 h, since xylose was the only available sugar in the medium. The low glucose feed rate (0.6 to 0.5 g/L h) started after 22 h (labeled I in Figure 1) and continued for 96 h. This corresponds to the glucose uptake as all the glucose was immediately consumed, that is, the glucose concentration in the medium was below the detection level. After 24 h the xylose uptake gradually declined, but almost all the xylose was consumed after 72 h, and 98% by the end of the fermentation period (120 h). These results are similar to that observed by Olofsson [21], who reported almost complete xylose consumption when glucose feed was applied during fermentation of wheat straw hydrolysate, although the initial glucose:xylose ratio in their study was one order of magnitude lower. The xylose consumption rate was lower in the present study (0.11 g xylose/g cell h *vs.* 0.15 g/g cell h), which is probably due to the initial glucose:xylose ratio. A final ethanol concentration of 53 g/L was obtained in the present study, corresponding to 93% of the theoretical yield, which is significantly higher than that obtained by Olofsson, 71% [21]. Besides ethanol, 4.3 g/L xylitol and 6.2 g/L glycerol were also formed in our experiment; mostly during the first 24 h. Small amounts of lactic acid (5 g/L) were also produced after 72 h of fermentation.

From this experiment, it can be concluded that glucose present from the beginning is consumed already in 5 h. The aim was to start feeding when most of the glucose was consumed and to provide a small amount of glucose during the fermentation. As a result the batch-fermentation phase could be shortened to only 5 h. Complete xylose consumption already after 72 h made it possible to shorten the feeding period, and lower the residence time required for fermentation.

In Config. 2_G, lignocellulose hydrolysate following pretreatment was fermented with an initial yeast concentration of 5 g/L. Figure 2 shows the concentrations of the products and substrates during fermentation. The initial glucose concentration in the medium was very low, 2.1 g/L. Glucose solution feeding was started already after 1 h (labeled I in Figure 2). Cellulose enzymes were also added at that time to ensure the breakdown of the oligomer sugars in the hydrolysate. The flow rate of the feed was constant, but since the volume was increasing, the rate of glucose feeding changed from 2.2 g/L h to 1.8 g/L h linearly during the 48-h feeding time. This feed rate corresponds to an average of 0.44 g glucose/g cell h, and is in the same range as the glucose release rate (0.5 g glucose/g cell h, calculated by the authors) that showed the best results in the study by Olofsson [21] on SSCF with TMB3400. The xylose uptake during the first 8 h was much faster, than in Config. 1_G (0.17 g/g cell h vs. 0.11 g/g cell h), probably due to the lower initial glucose concentration. However, xylose consumption had almost ceased after 24 h, when the xylose concentration fell below 4 g/L. At this point the glucose:xylose ratio in the medium became too high, and glucose utilization was prioritized over xylose, since they are competitively transported by the same transport protein [25]. At the end of the fermentation period (120 h) 1.5 g/L xylose remained in the broth, resulting in 93% xylose consumption. The xylose consumption in this experiment was sufficient to apply a similar feeding strategy also when the experiments were fed with real hydrolysates.

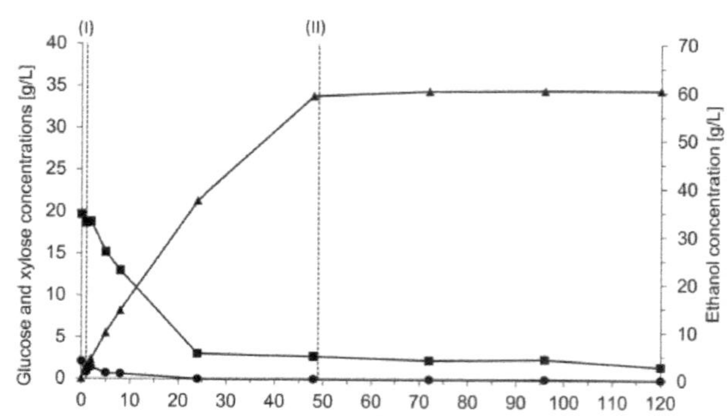

Figure 2: Concentrations of glucose (circles), xylose (squares), and ethanol (triangles) during fermentation in Config. 2_G using 5 g/L yeast. Glucose solution was fed to the fermentor for 48 h between (I) and (II) at a constant flow rate of 4.2 mL/h. (T: 32°C, pH 5, 300 rpm).

Wheat-starch-hydrolysate-fed fermentation

Wheat-starch hydrolysate feed was used in these experiments as an alternative glucose source. The effect of increased cell mass concentration was also investigated. All the results after 120 hours' fermentation are given in Table 2. Figures 3 and 4 show the concentrations of the sugars and the ethanol during fermentation in Config. 1 and Config. 2, respectively.

Table 2: Results obtained after 120 h of fermentation in Config.

Expt Config.	Yeast load (g/L)	Xylose consumption[b] (%)	Xylose conc. (g/L)	Xylitol formation[c] (%, g/L)	Glycerol conc. (g/L)	Ethanol conc. (g/L)	Ethanol yield (g/g)	Ethanol yield[d] (%)
Config. 1	5 (fresh)	100	0.0	20 (2.5)	6.9	43.1	0.49	97
Config. 1	5 (fresh)	98	0.0	20 (2.6)	5.7	43.0	0.48	93
Config. 1	5 (2w)[e]	100	0.0	21 (2.8)	5.6	41.4	0.45	88
Config. 1	20 (2w)[e]	100	0.0	13 (1.6)	6.9	41.5	0.46	91
Config. 2	5 (fresh)	76	6.5	10 (1.9)	6.6	53.3	0.44	87
Config. 2	5 (fresh)	58	11.1	9 (1.4)	5.5	52.0	0.42	83
Config. 2	5 (2w)[e]	72	6.6	9 (1.4)	5.6	49.8	0.44	87
Config. 2	20 (2w)[e]	90	2.4	13 (2.7)	7.6	49.0	0.43	84

1 or Config. 2a

[a]The raw material used for the experiments was SPWS 2.

[b]Related to the total amount of available xylose.

[c]Related to consumed xylose.

[d]Corresponding to the maximum theoretical yield from both glucose and xylose.

[e]Yeast stored in a refrigerator for two weeks (see Methods).

Config. 1, wheat-starch hydrolysate feed; Config. 2, mixed hydrolysates feed.

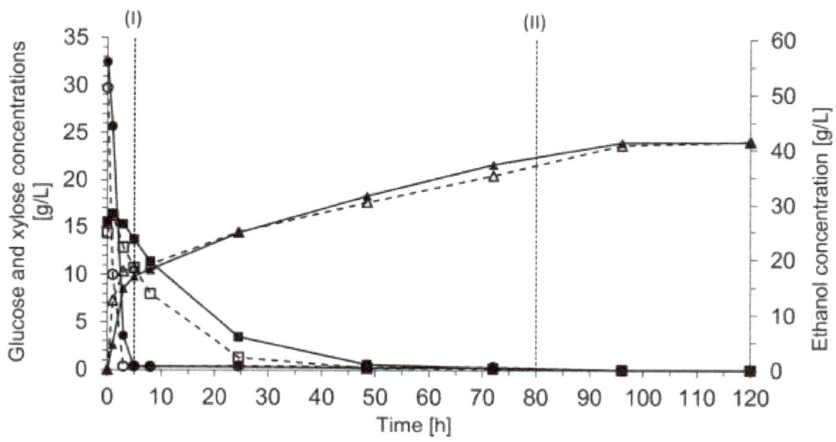

Figure 3: Concentrations of glucose (circles), xylose (squares), and ethanol (triangles) during fermentation in Config. 1 using 5 g/L yeast (filled symbols) or 20 g/L yeast (open symbols). Wheat starch hydrolysate was fed to the fermentor for 75 h between (I) and (II) at a constant flow rate of 2.1 mL/h. (T: 32°C, pH 5, 300 rpm).

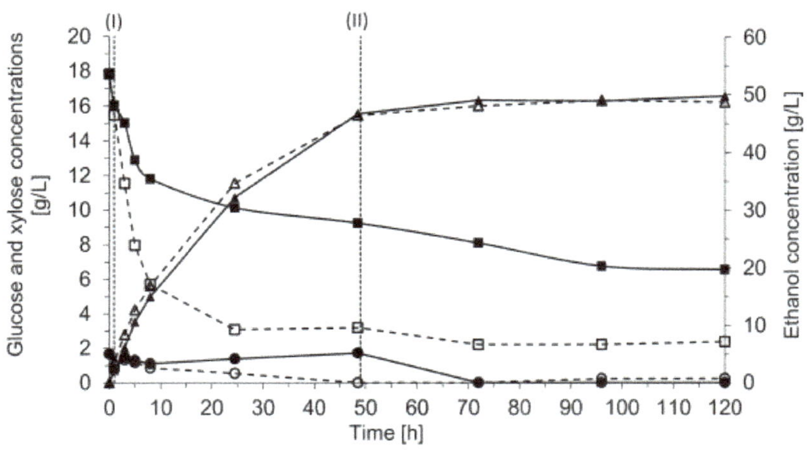

Figure 4: Concentrations of glucose (circles), xylose (squares), and ethanol (triangles) during fermentation in Config. 2 with a yeast concentration of 5 g/L (filled symbols) or 20 g/L (open symbols). The mixed hydrolysates of SPWS hydrolysate and wheat-starch hydrolysate was fed to the reactor for 48 h between (I) and (II) at a constant flow rate of 11.2 mL/h. (T: 32°C, pH 5, 300 rpm).

Config. 1 with wheat-starch hydrolysate feed

Wheat-starch hydrolysate feed was used in Config. 1 and was fed to the fermentors at a constant rate for 75 h after the glucose concentration had fallen to 0 g/L (Figure 3). The ethanol yield after fermentation reached an average of 95% of the theoretical based on available glucose and xylose, corresponding to a concentration of about 43 g/L ethanol, when using a yeast concentration of 5 g/L. The xylose was taken up completely; however, approximately 20% of the consumed xylose resulted in xylitol formation (Table 2). Xylitol production commenced when xylose consumption started, and it was produced until there was no xylose available in the medium; after which the xylitol concentration decreased somewhat until the end of fermentation. This phenomenon has been observed previously in defined media, suggesting that the glucose feed is insufficient to provide ATP for maintenance and glycerol production [21]. Xylitol is usually produced in *S. cerevisiae* strains that harbour the eukaryotic xylose utilization pathway (XR/XDH), due to the redox imbalance caused by the inability of the cell to provide enough NAD^+ for the XDH reactions [18]. The xylitol formation was in the same range as that observed in a study by Olofsson *et al.* on wheat straw hydrolysate at 30°C [21], but much lower than the yield obtained on defined medium in the same study. This is due to external electron acceptors such as furfural or hydroxyl-methyl-furfural, which are present in the lignocellulose hydrolysate, which reduce xylitol formation [17]. However, the concentration of the inhibitors was very low, and the furfural present was assimilated very quickly. It may be advantageous to apply somewhat more severe pretreatment to generate slightly more sugar degradation products, which, up to a certain concentration, have been shown to increase the ethanol yield [32, 36].

Effect of increased yeast concentration in Config. 1

Increasing the yeast concentration to 20 g/L gave similar results regarding the ethanol yield, the ethanol concentration and xylose uptake, but the xylitol formation decreased to 12% of the consumed xylose (Table 2). The glycerol concentration increased from 5.6 g/L to 6.9 g/L, which is probably due to the increased cell concentration. The xylose was completely consumed after 48 h, thus a feeding rate 0.8 g glucose/L h is sufficient to enhance xylose consumption. A higher glucose feed rate could be applied but this may lead to a decrease in the xylose uptake. This is exactly what can be seen when the glucose feed rate was increased in the experiments in Config. 2. Whether a shorter residence time would be more beneficial, apart from the complete xylose uptake can only be determined by an economic evaluation.

Config. 2 with mixed hydrolysates feed

Config. 2 is a possible alternative for controlled glucose feeding, by separating some of the SPWS liquid fraction from the fibre residue (without washing) before enzymatic hydrolysis with a high WIS loading. This would produce a glucose-rich supernatant, which could be used as feed in the fermentation. In the present study the wheat-starch hydrolysate was also mixed with the feed in order to investigate the integrated process. The liquid fraction of SPWS was used in the fermentation and then fed with the mixed hydrolysates for 48 h (Figures 4 and 5B). One of the drawbacks of this configuration is that a considerable amount of xylose is present in the feed, which in this case led to remarkably decreased xylose consumption during the second 24-h period of feeding (Figure 4). The ethanol yield with an initial yeast concentration of 5 g/L was between 83% and 87% (Table 2), corresponding to an ethanol concentration of about 50 g/L. The xylose consumption varied between 58% and 76% with very low xylitol formation in relation to the xylose uptake (< 10%), corresponding to only 1.4 g/L xylitol after 120 h of fermentation. The glycerol concentration was 5.6 g/L during the last hour. These results are in the same range as those reported for barley straw hydrolysate [30], although the best xylose uptake was observed with a much lower glucose feed rate of 0.21 g/L h.

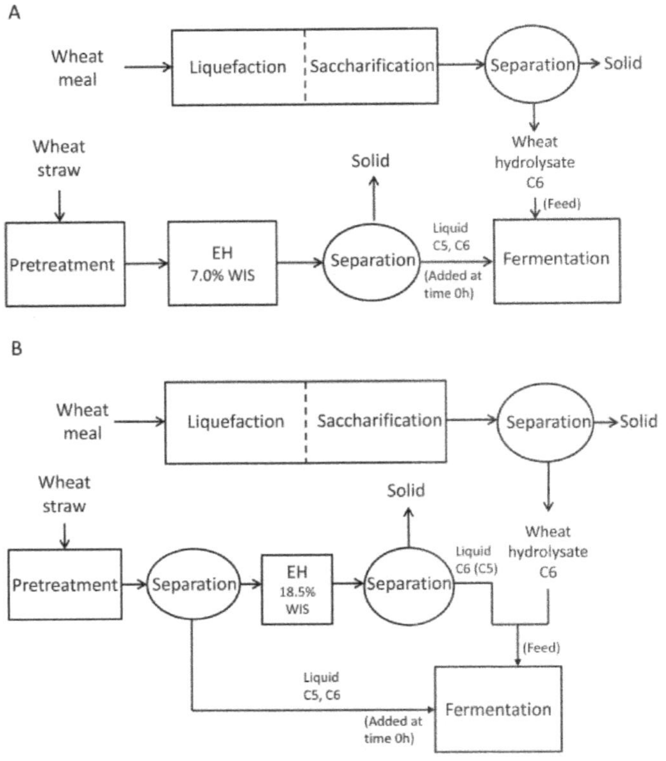

Figure 5: Experimental configurations used to investigate the fermentation of steam-pretreated wheat straw. (A) Config. 1, separate hydrolysis of the whole SPWS slurry and co-fermentation of glucose and xylose in the SPWS hydrolysate fed with wheat-starch hydrolysate (or with glucose solution). **(B)** Config. 2, co-fermentation of glucose and xylose in the liquid fraction of SPWS, which was fed with the mixed hydrolysates of SPWS hydrolysate and wheat-starch hydrolysate, (or with glucose solution).

Increased yeast concentration in Config. 2

High, yeast concentration (20 g/L) was also investigated, but this did not improve the ethanol yield, despite the fact that the xylose uptake increased to 90%. However, the xylitol and the glycerol concentrations increased to 2.7 g/L (13% of the consumed xylose) and 7.6 g/L, respectively.

Comparison of the two configurations

Despite the high initial feed rate in Config. 2, 4.4 g/L h, corresponding to 0.88 g glucose/g cell h, the glucose:xylose ratio was still low enough to result in a faster xylose uptake than that obtained in Config. 1 with 5 g/L cell mass, with a much higher initial glucose concentration. After 8 h as the xylose concentration decreased; however, to maintain the glucose:xylose ratio the feed rate should also have been adjusted, thought it would have been rather difficult as there was also xylose present in the feed. Accumulation of glucose, which was observed (Figure 4), should be avoided, indicating that a lower glucose feed rate should be applied. Nevertheless, the xylose was completely assimilated in Config. 1, which indicates that the initial glucose:xylose ratio has an influence on the initial xylose uptake, but not on the final uptake. As a higher ethanol yield was reached Config. 1 is a more advantageous process configuration in case of co-fermentation of xylose and glucose with the *S. cerevisiae* TMB3400. It is also less demanding from a technical point of view.

CONCLUSIONS

Complete xylose utilization could be achieved with TMB3400 co-fermentation of glucose and xylose in SPWS hydrolysate when the glucose-rich wheat-starch hydrolysate was fed to the fermentor. A high ethanol concentration, up to 5% (wt/wt), and a high ethanol yield, above 0.42 g/g, were achieved with both configurations (Config. 1 and 2), based on the utilization of both glucose and xylose; however, the yields and the xylose consumption were higher in Config. 1. Although increased cell concentration did not result in improved ethanol yield in fermentation, an improved xylose uptake rate was observed due to the increased fermentation capacity. Integration of the first-generation process into the second-generation (2 G) process also increases the ethanol concentration, which will result in a decrease in the cost of distillation, thus improving process economy. In this way, introduction of the 2 G bioethanol production may be considerably facilitated.

METHODS

Two configurations for SHCF of glucose and xylose were used to investigate the effect of controlled wheat-starch hydrolysate feed on the xylose uptake. Both are illustrated in Figure 5. The first configuration (Figure 5A, denoted Config. 1) was used for assessment of glucose and xylose co-fermentation in the SPWS hydrolysate. The whole slurry of SPWS 1 or 2 was diluted to 7.5% or 7.0% water insoluble solids (WIS) prior to enzymatic hydrolysis, which was followed by separation of SPWS hydrolysate and the solid fraction. SPWS hydrolysate was then subjected to fermentation. The supernatant from the liquefied and saccharified wheat meal (wheat-starch hydrolysate) was used as feed to the fermentor, commencing when the glucose present in the SPWS hydrolysate had been consumed. The extensive amount of C6 sugars present in the SPWS hydrolysate, resulting in a high glucose:xylose ratio, may have a negative effect on the xylose fermentation.

In order to investigate the glucose and xylose consumption and to obtain information about the feeding strategy, the method was simplified so that wheat-starch hydrolysate was replaced by glucose solution at a glucose concentration corresponding to that of the hydrolysate. Using glucose solution in the first trials (Config. 1_G and Config. 2_G) was also motivated by the simpler preparation procedure and the more exact dosing of glucose.

Batch fermentation in Config. 1_G was run for 24 h, and the glucose solution was then fed into the fermentor for 96 h. Based on the results of this experiment, it was decided to feed wheat-starch hydrolysate in Config. 1 for 75 h after 5 h of batch fermentation.

Another configuration, Config. 2 (Figure 5B), was used where the SPWS slurry was pressed to 18.5% dry matter (DM), enzymatically hydrolyzed and then separated into SPWS hydrolysate and a solid fraction. SPWS hydrolysate was mixed with the wheat-starch hydrolysate (mixed hydrolysates) and used as feed for fermentation. In the fermentation step the liquid fraction of SPWS 2 slurry was fermented for 1 h in order to ferment the glucose already present and then fed with glucose solution (Config. 2_G) or with the mixed hydrolysates (Config. 2) for 48 h. As opposed to Config. 1, a small amount of C6 and a large amount of C5 sugars were present in the broth in Config. 2 from the beginning of the fermentation. More C6 sugars were present in the feed in this case, which could be advantageous for the xylose fermenting *S. cerevisiae* TMB3400; however, xylose in the feed may have a counterproductive effect. Both configurations were investigated in order to assess fermentation of xylose.

Raw materials and pretreatment

Two batches of wheat straw (Straw 1 and Straw 2) were used in the study. Both were kindly provided by Johan Håkansson Lantbruksprodukter (Lunnarp, Sweden). Straw 1 and Straw 2 consisted of about 93% and 92% DM, respectively. Their compositions are given in Table 3.

Table 3: Composition of the two batches of wheat straw, expressed as a percentage of dry matter

	Straw 1 (%)	Straw 2 (%)
Glucan	38.6	34.2
Xylan	25.8	25.2
Arabinan	3.9	3.3
Lignin	20.4	21.9
Ash	4.4	4.1

The straw was milled with a knife-mill and then sieved to obtain particles in the range of 2 to 10 mm. The milled straw was stored in plastic bags at room temperature prior to impregnation. Impregnation was performed with 0.2% sulphuric-acid solution at a liquid:dry matter ratio of 20 for 1 h, and the straw was then pressed to an average DM content of 50% and stored in sealed plastic buckets overnight until pretreated. The impregnated straw was pretreated in a steam pretreatment unit, described previously [37], at 190°C for 10 min, as it was previously optimized by Linde *et al.* [38]. The steam pretreated wheat straw (SPWS) slurry was thoroughly mixed and stored at 4°C. Total DM and WIS contents were determined from three representative samples. Soluble sugars, degradation products, and total soluble sugars after acid hydrolysis were determined from the liquid fraction, according to the standardized analytical procedures of the National Renewable Energy Laboratory (NREL) [39]. The solid fraction of SPWS was extensively washed with hot water in order to remove all soluble substances, dried, and milled before compositional analysis. The structural carbohydrates and lignin content of this solid residue were determined according to the NREL method [40]. The composition of the SPWS from the two batches (SPWS 1 and 2) is given in Table 4. Straw 2 was pretreated after modification of the control system providing improved temperature control. This might explain the increased WIS content in the batch of SPWS 2. SPWS 1 was used for the experiments run with glucose feed only (Config. 1_G and Config. 2_G), while the SPWS 2 material was used for all the trials run with the real wheat-starch hydrolysate feed (Config. 1 and Config. 2).

Table 4: Composition of the two batches of steam-pretreated wheat straw (SPWS)

			SPWS 1	SPWS 2
Total solids (%)			12.3	18.7
WIS content (%)			8.2	13.5
Solid fraction (% of WIS)		Glucan	62.2	58.6
		Xylan	5.8	6.3
		Arabinan	0.6	0.4
		Lignin	23.5	27.7
		Ash	3.4	3.0
Liquid fraction (g/L)	Sugars[a]	Glucose	5.0	9.8
		Xylose	30.4	43.7
		Arabinose	3.9	5.8
	By-products	Furfural	2.0	1.7
		HMF	0.1	0.1
		Acetic acid	2.9	2.4

[a]Total sugars containing both monomer and oligomer sugars.

Cell cultivation

Inoculum

The inoculum culture of *Saccharomyces cerevisiae* TMB3400 was prepared on xylose-supplemented YPD plates (10 g/L yeast extract, 20 g/L peptone, 20 g/L glucose, 20 g/L xylose, and 15 g/L agar). The cells were added to a 300-mL Erlenmeyer flask with 70 mL aqueous solution containing 23.8 g/L glucose, 23.8 g/L xylose, 10.8 g/L $(NH_4)_2SO_4$, 5.0 g/L KH_2PO_4, and 1.1 g/L $MgSO_4$ $7H_2O$. 14.4 mL/L trace metal solution and 1.4 mL/L vitamin solution was prepared according to Taherzadeh *et al.* [41] and also added to the solution. The pH was adjusted to pH 5 with 0.25 M NaOH. The Erlenmeyer flask was sealed with a cotton plug and incubated on a rotary shaker at 30°C for 24 h at 180 rpm.

Batch cultivation

Batch cultivation was performed in a 2-L LABFORS fermentor (Infors AG, Bottmingen, Switzerland) with a working volume of 0.5 L. Cultivation was started by adding 70 mL inoculum to a medium containing 20.0 g/L glucose, 22.5 g/L $(NH_4)_2SO_4$, 10.5 g/L KH_2PO_4, 2.2 g/L $MgSO_4$ $7H_2O$, 60.0 mL/L trace metal solution, and 6.0 mL/L vitamin solution. The pH was adjusted to 5 with 2.5 M NaOH. The stirrer speed was set to 700 rpm and the aeration rate was 1 vvm, corresponding to an average of 0.5 L/min. The dissolved oxygen concentration was continuously measured with an oxygen sensor. Fed-batch cultivation was started when the oxygen concentration showed a rapid increase, indicating that all the sugar and the ethanol had been consumed.

Fed-batch cultivation

Batch cultivation was followed by a 24-h fed-batch phase, since short-term adaptation of the cells used for fermentation of lignocellulosic hydrolysate has been shown to improve fermentation performance [42]. Fed-batch cultivation was performed by the continuous addition of 847 mL SPWS liquid supplemented with glucose and salt solution to a total volume of 1 L, resulting in a final volume of 1.5 L. The glucose concentration in the SPWS liquid solution was adjusted to 70.0 g/L. Salts were added to the solution to concentrations of 11.3 g/L$(NH_4)_2SO_4$, 5.3 g/L KH_2PO_4, and 1.1 g/L $MgSO_4$ $7H_2O$. The final concentration of diluted SPWS liquid was equivalent to that obtained when the slurry from pretreatment had been diluted to 7.5% WIS. The pH of the hydrolysate was adjusted to pH 5 with solid NaOH. The SPWS liquid solution was added to the fermentor at a constant rate of 125 mL/h for 24 h. The pH was continuously adjusted to pH 5 with 2.5 M NaOH. The stirrer speed was increased to 900 rpm and the rate of aeration to 1.5 vvm. When all the sugar and the ethanol produced had been consumed, cultivation was stopped and the culture broth was centrifuged at 4,000 rpm for 10 min. The supernatant was discarded and the DM of the harvested cells was determined before use in fermentation. The cells used for starch-hydrolysate-fed fermentation with 20 g/L cell concentration were stored in a refrigerator for two weeks (it has previously been found that the performance of this yeast does not change when stored in a refrigerator for up to a month, data not shown). However, reference fermentation experiments supplemented with the 5 g/L stored yeast were run in parallel.

Enzymatic hydrolysis

Config. 1

Enzymatic hydrolysis of the whole SPWS slurry in Config. 1_G was performed in a 20-L bio-lab reactor (Bioengineering AG, Wald, Switzerland) with a total working weight of 10 kg. The WIS concentration of the SPWS 1 material was adjusted to 7.5% and hydrolysis was carried out at 40°C, pH 4.8 for 120 h, at 850 rpm. Cellic CTec cellulase (95 FPU/g, 590 β-glucosidase IU/g, 712 xylanase IU/g, and 63 β-xylanase IU/g) and Cellic HTec xylanase (7998 xylanase IU/g and 99 β-xylanase IU/g) enzyme preparations were added at a ratio of 10:1 based on the glucan content of the WIS in the steam pretreated material (20 FPU/g glucan). Both enzyme preparations were kind gifts from Novozymes A/S (Bagsværd, Denmark).

Enzymatic hydrolysis in the starch-hydrolysate-fed experiments (Config. 1) was performed using SPWS 2 at a WIS content of 7.0%, using the newly released cellulase enzyme, Cellic CTec2 (Novozymes A/S) cellulase (128 FPU/mL, 4465 β-glucosidase IU/mL) at an enzyme loading of 14 FPU/g glucan. The reason for the different enzyme loading is that the enzyme preparation has a high concentration of glucose (280 g/L), which we were not aware of when the experiments were designed. The FPU activity determined before the experiments was actually 24% higher than the actual value corrected for the glucose content. The parameters of the experiments were therefore recalculated afterwards, resulting in an FPA dosage of 14 FPU/g glucan instead of the intended 20 FPU/g. The liquid fraction (SPWS hydrolysate) after this step was separated from the solid fraction with a press-filter and then subjected to fermentation.

Config. 2

Enzymatic hydrolysis of the solid fraction of SPWS 2 slurry with 18.5% WIS was performed in two batches, 3.6 and 3.1 kg, in a 12-L solid-state fermentor (Terrafors, Infors AG). The temperature in the fermentor was maintained at 40°C with the help of an external water bath. The pH was set to 5 at the beginning of hydrolysis and when the slurry became more liquefied the pH was readjusted with 50% NaOH in conjunction with sampling. Cellic CTec2 enzyme preparation was used at a dosage of 14 FPU/g glucan. After hydrolysis, the liquid fraction (SPWS hydrolysate) was separated from the solid fraction. The SPWS hydrolysate was mixed with the wheat-starch hydrolysate (mixed hydrolysates) before being fed to the fermentation vessel.

Wheat hydrolysate

In order to produce wheat-starch hydrolysate, wheat meal was enzymatically treated using two-step hydrolysis, consisting of liquefaction and saccharification. The hydrolysis of 1.2 or 5 kg batches was performed using a 2- or 10-L evaporator (Büchi Labortechnik AG, Flawil, Switzerland). In the first step, wheat meal slurry (35% DM) was liquefied by thermostable α-amylases (Termamyl SC; Novozymes A/S, Bagsværd, Denmark) at 85°C, pH 5.5 for 3 h. Termamyl was added at a dosage of 0.5 g enzyme/kg DM meal. In the following step, amyloglucosidase (Spirizyme Fuel; Novozymes A/S) was loaded at a level of 1 mL/kg DM after the pH had been adjusted to 4.2 and the temperature reduced to 60°C. Saccharification was performed at 60°C for 24 h to ensure complete breakdown of the oligomers into monomer sugars. The hydrolysis broth was centrifuged twice, first at 4,000 rpm, for 10 min in 500-mL flasks. The supernatant was then transferred to 50-mL centrifuge tubes and centrifuged at 4,500 rpm for 10 min. The supernatant was saved and frozen prior to further use. The glucose concentration in the starch hydrolysate was determined to be between 316 and 340 g/L in different batches.

Ethanol fermentation

The experimental set-up for each fermentation performed, working volumes, feed volumes, and the total amount of sugar added in the different fermentation experiments are presented in Table 5. The sugar concentrations in the different fermentation experiments varied due to the different amounts of sugar after enzymatic hydrolysis, but the yields were always calculated based on the actual amount of sugar added.

Table 5: Experimental set-up and the amount of sugars[a] added at 0 h and with the feed[b] in the various fermentation experiments

Feed	Expt Config	Yeast load (g/L)	Feeding start; stop (h)	$G_{feed\ rate}$ (g/h)	V_{final} (L)	V_{feed}		$Glucose_{added\ at\ 0h}$ (g)	$Xylose_{added\ at\ 0h}$ (g)	$Glucose_{feed}$		$Xylose_{feed}$ (g)
						SPWS hydr. (mL)	Starch hydr. (mL)			SPWS hydr. (g)	Starch hydr. (g)	
Glucose (G)	Config. 1_G	5	22; 120	0.6	1.2	180[e]		49.6	27.3	56.3[f]		-
	Config. 2_G	5	1; 49	2.2	1.2	180[e]		4.4	27.2	104.7[f]		-
Wheat-starch hydrolysate	Config. 1	5	5; 80	0.8	1.2		180[d]	31.7	15.4	-	57.0	-
	Config. 1	5	5; 80	0.8	1.2		180[d]	31.7	15.4	-	60.6	-
	Config. 1	5[c]	5; 80	0.7	1.0		150[d]	27.6	13.2	-	50.5	-
	Config. 1	20	5; 80	0.7	1.0		150[d]	25.3	12.2	-	50.5	-
	Config. 2	5	1; 49	2.3	1.2	492	180[d]	4.8	21.5	50.0	57.0	9.8
	Config. 2	5	1; 49	2.3	1.2	492	180[d]	4.8	21.5	50.0	60.6	9.8
	Config. 2	5[c]	1; 49	1.8	1.0	367	150[d]	3.4	15.2	35.2	50.5	7.6
	Config. 2	20	1; 49	1.8	1.0	367	150[d]	3.4	15.2	35.2	50.5	7.6

[a]Glucose and xylose
[b]Glucose, SPWS hydrolysate, and wheat-starch hydrolysate.
[c]Reference for the 20 g/L run with the yeast stored in the refrigerator for two weeks.
[d]Wheat-starch hydrolysate in the feed mixture (17% of V_{final}).
[e]Volume of glucose solution feed
[f]Glucose added with the glucose solution feed.
G, glucose.

Fermentation with glucose feed

Glucose-fed fermentations were performed in 2-L LABFORS bioreactors (Infors AG) with a working volume of 1.2 L. The SPWS hydrolysate obtained from the enzymatic hydrolysis of material containing 7.5% WIS in Config. 1_G, or the liquid fraction of SPWS (diluted with water to a concentration equivalent to that which would have been obtained if the slurry from pretreatment had been diluted to 7.5% WIS) in Config. 2_G was added to the fermentor and autoclaved at 120°C for 20 min. Nutrients were dissolved separately in water at final concentrations of 1 g/l $(NH_4)_2HPO_4$, 0.05 g/L $MgSO_4$ $7H_2O$, and 2 g/L yeast extract, sterilized, and added to the bioreactor before inoculation. The nutrient concentration was doubled when the cell concentration of 20 g/L was used. The pH of the hydrolysates was set to 5 using solid NaOH before autoclaving and then controlled with 10% NaOH solution during fermentation. Fermentation was carried out at 300 rpm and 32°C for 120 h.

In Config. 1_G fermentation was allowed to continue for another 8 h after feeding had ceased (120 h) before it was stopped, so that all the glucose could be consumed. Nitrogen gas was flushed through the bioreactors before fermentation to ensure an anaerobic environment. Yeast cell suspensions at a dry cell concentration of 5 g/L were added to the fermentors. Then 180 mL glucose solution was fed to each fermentor (Table 5). The solutions had different concentrations in each configuration, such that the amount of glucose added would be the same as when using wheat-starch hydrolysate.

After 22 h, a glucose solution at a concentration of 323 g/L was fed to the fermentation vessel of Config. 1_G for another 96 h at a constant flow rate. The glucose addition of 0.6 g/h decreased from 0.6 to 0.5 g/L h linearly as the volume increased due to the feed. The solution prepared had approximately the same glucose concentration as the wheat hydrolysate.

In Config. 2_G the feed had a higher concentration of glucose (600 g/L). It was also added at a constant flow rate but for only 48 h, starting after 1 h. The extra glucose added corresponded to the amount of glucose theoretically available in the form of glucan in the WIS fraction, which would have been hydrolyzed separately. Due to the volume increase, the rate of glucose addition changed linearly from 2.2 to 1.8 g/L h during the 48 h. One hour after starting fermentation Cellic CTec and Cellic HTec mixture (10:1) was added at a dosage of 5 FPU/g (theoretical) WIS to ensure the breakdown of the oligomers present in the hydrolysate.

Fermentation with wheat-starch hydrolysate feed

The experiments were performed in the 2 L Labfors fermentors for 120 h using a working volume of 1.2 or 1.0 L (Table 5). In Config. 1 SPWS hydrolysate, after the 7.5% enzymatic hydrolysis or the liquid fraction of SPWS in Config. 2 was added to the fermentor and autoclaved at 120°C for 20 min. The fermentations were initiated by adding cell suspension, which would result in a concentration of 5 g/L in the final volume. The fermentation parameters, such as pH, temperature, stirrer speed, and the nutrient concentration were the same as in the fermentations with glucose feed. The fermentation of Config. 1 was then fed with wheat-starch hydrolysate and Config. 2 with the mixed hydrolysates described above. An increased cell concentration, 20 g/L, was also used in another set of experiments, to allow for comparison with these experiments which were run with yeast that had been stored in a refrigerator for two weeks.

Sampling and analysis

Samples were withdrawn frequently after the first hour. The concentrations at 0 h were calculated from the known concentrations of the input materials. Samples taken from enzymatic hydrolysis and fermentation were centrifuged in 10-mL centrifuge tubes at 4,000 rpm for 5 min (Labofuge 200; Thermo elektron LED GmbH, Osterode, Germany). The supernatant was filtered using 0.2-μm filters, and the filtered samples were stored at -20°C prior to analysis.

Hydrolysate samples after pretreatment, samples from the NREL fiber and total soluble sugars analysis, and samples withdrawn from enzymatic hydrolysis and fermentation were analyzed using a high-performance liquid chromatograph equipped with a refractive index detector (both from Shimadzu, Kyoto, Japan). Cellobiose, glucose, xylose, galactose, arabinose, and xylitol were separated on an Aminex HPX-87P column (Bio-Rad Laboratories, Hercules, CA, USA) at 85°C with Millipore water at a flow rate of 0.6 mL/min. Ethanol, glycerol, lactic acid, acetic acid, HMF, and furfural were separated on an Aminex HPX-87H column (Bio-Rad Laboratories) at 50°C, with 5 mM H_2SO_4 at a flow rate of 0.5 mL/min. All samples were neutralized with $CaCO_3$(s) and filtered through a 0.2-μm filter before analysis.

Yield calculations

The sugar yield in enzymatic hydrolysis was calculated on the basis of the total available sugars in the liquid and the solid fraction of SPWS. The theoretical amount of glucose released during hydrolysis is 1.11 times the amount of glucan and 1.13 times the amount of xylan in the solid material (due to the addition of water in hydrolysis). The ethanol yield was calculated based on the total amount of fermentable sugars added to the fermentation stage, that is, the sum of available glucose and/or xylose in the liquid fraction of SPWS (including monomers and oligomers) and in the feed. Based on a maximum ethanol yield of 0.51 g/g (yield from hexoses and xylose), the percentage of theoretically available ($Y^*_{E/S}$, %) was calculated as $Y^*_{E/S} = (Y_{E/S}/0.51)*100$, where $Y_{E/S}$ is the ethanol yield from the sugar substrates.

ACKNOWLEDGEMENTS

We gratefully acknowledge the Swedish Energy Agency for its financial support.

AUTHORS' CONTRIBUTIONS

BE, MG, and GZ designed and coordinated the study. BE and BF carried out the experiments and analyzed the results. BE prepared the manuscript. MG and GZ reviewed the manuscript. All authors read and approved the final manuscript.

REFERENCES

1. Ballesteros M, Oliva JM, Negro MJ, Manzanares P, Ballesteros I: Ethanol from lignocellulosic materials by a simultaneous saccharification and fermentation process (SFS) with Kluyveromyces marxianus CECT 10875. Process Biochemistry. 2004, 39: 1843-1848. 10.1016/j.procbio.2003.09.011.
2. Wingren A, Galbe M, Zacchi G: Techno-economic evaluation of producing ethanol from softwood: Comparison of SSF and SHF and identification of bottlenecks. Biotechnol Prog. 2003, 19: 1109-1117.
3. Olofsson K, Bertilsson M, Liden G: A short review on SSF - an interesting process option for ethanol production from lignocellulosic feedstocks. Biotechnol Biofuels. 2008, 1: 7-10.1186/1754-6834-1-7.
4. Wyman CE: Ethanol production from lignocellulosic biomass: overview. Handbook on Bioethanol: Production and Utilization. Edited by: Wyman CE. 1996, Washington, DC: Taylor & Francis, 1-18.
5. Verduyn C, Postma E, Scheffers WA, van Dijken JP: Energetics of Saccharomyces cerevisiae in anaerobic glucose-limited chemostat cultures. J Gen Microbiol. 1990, 136: 405-412.
6. Casey GP, Ingledew WMM: Ethanol tolerance in yeasts. Crit Rev Microbiol. 1986, 13: 219-280. 10.3109/10408418609108739.
7. Ligthelm ME, Prior BA, Preez JC: The oxygen requirements of yeasts for the fermentation of d-xylose and d-glucose to ethanol. Appl Microbiol Biotechnol. 1988, 28: 63-68.
8. Prior BA, Kilian SG, Dupreez JC: Fermentation of D-xylose by the yeasts Candida shehatae and Pichia stipitis: prospects and problems. Process Biochem. 1989, 24: 21-32.
9. Roberto IC, Lacis LS, Barbosa MFS, de Mancilha IM: Utilization of sugar cane bagasse hemicellulosic hydrolysate by pichia stipitis for the production

of ethanol. Process Biochem. 1991, 26: 15-21. 10.1016/0032-9592(91)80003-8.

10. Preez JC, Bosch M, Prior BA: The fermentation of hexose and pentose sugars by Candida shehatae and Pichia stipitis. Appl Microbiol Biotechnol. 1986, 23: 228-233.

11. Skoog K, Hahn-Hagerdal B: Xylose fermentation. Enzyme Microb Technol. 1988, 10: 66-80. 10.1016/0141-0229(88)90001-4.

12. Grootjen DRJ, van der Lans RGJM, Luyben KC: Effects of the aeration rate on the fermentation of glucose and xylose by Pichia stipitis CBS 5773. Enzyme Microb Technol. 1990, 12: 20-23. 10.1016/0141-0229(90)90174-O.

13. Karhumaa K, Hahn-Hägerdal B, Gorwa-Grauslund MF: Investigation of limiting metabolic steps in the utilization of xylose by recombinant Saccharomyces cerevisiae using metabolic engineering. Yeast. 2005, 22: 359-368. 10.1002/yea.1216.

14. Kuyper M, Hartog MMP, Toirkens MJ, Almering MJH, Winkler AA, van Dijken JP, Pronk JT: Metabolic engineering of a xylose-isomerase-expressing Saccharomyces cerevisiae strain for rapid anaerobic xylose fermentation. FEMS Yeast Res. 2005, 5: 399-409. 10.1016/j.femsyr.2004.09.010.

15. Kötter P, Amore R, Hollenberg CP, Ciriacy M: Isolation and characterization of the Pichia stipitis xylitol dehydrogenase gene, XYL2, and construction of a xylose-utilizing Saccharomyces cerevisiae transformant. Curr Genet. 1990, 18: 493-500. 10.1007/BF00327019.

16. Eliasson A, Christensson C, Wahlbom CF, Hahn-Hägerdal B: Anaerobic xylose fermentation by recombinant Saccharomyces cerevisiae carrying *XYL1, XYL2*, and *XKS1* in mineral medium chemostat cultures. Appl Environ Microbiol. 2000, 66: 3381-3386. 10.1128/AEM.66.8.3381-3386.2000.

17. Wahlbom CF, van Zyl WH, Jonsson LJ, Hahn-Hägerdal B, Otero RRC: Generation of the improved recombinant xylose-utilizing Saccharomyces cerevisiae TMB 3400 by random mutagenesis and physiological comparison with Pichia stipitis CBS 6054. FEMS Yeast Res. 2003, 3: 319-326. 10.1016/S1567-1356(02)00206-4.

18. Hahn-Hägerdal B, Karhumaa K, Fonseca C, Spencer-Martins I, Gorwa-Grauslund M: Towards industrial pentose-fermenting yeast strains. Appl Microbiol Biotechnol. 2007, 74: 937-953. 10.1007/s00253-006-0827-2.

19. Sonderegger M, Jeppsson M, Larsson Cr, Gorwa-Grauslund MF, Boles E, Olsson L, Spencer-Martins I, Hahn-Hagerdal Bl, Sauer U: Fermentation performance of engineered and evolved xylose-fermenting Saccharomyces cerevisiae strains. Biotechnol Bioeng. 2004, 87: 90-98. 10.1002/bit.20094.

20. Bertilsson M, Olofsson K, Lidén G: Prefermentation improves xylose utilization in simultaneous saccharification and co-fermentation of pretreated spruce. Biotechnol Biofuels. 2009, 2: 8-10.1186/1754-6834-2-8.

21. Olofsson K, Rudolf A, Lidén G: Designing simultaneous saccharification and fermentation for improved xylose conversion by a recombinant strain of Saccharomyces cerevisiae. J Biotechnol. 2008, 134: 112-120. 10.1016/j.jbiotec.2008.01.004.

22. Rudolf A, Baudel H, Zacchi G, Hahn-Hägerdal B, Lidén G: Simultaneous saccharification and fermentation of steam-pretreated bagasse using Saccharomyces cerevisiae TMB3400 and Pichia stipitis CBS6054. Biotechnol Bioeng. 2008, 99: 783-790. 10.1002/bit.21636.

23. Öhgren K, Bengtsson O, Gorwa-Grauslund MF, Galbe M, Hahn-Hägerdal B, Zacchi G: Simultaneous saccharification and co-fermentation of glucose and xylose in steam-pretreated corn stover at high fiber content with Saccharomyces cerevisiae TMB3400. J Biotechnol. 2006, 126: 488-498. 10.1016/j.jbiotec.2006.05.001.

24. Kilian SG, Uden N: Transport of xylose and glucose in the xylose-fermenting yeast Pichia stipitis. Appl Microbiol Biotechnol. 1988, 27: 545-548.

25. Meinander NQ, Boels I, Hahn-Hägerdal B: Fermentation of xylose/glucose mixtures by metabolically engineered Saccharomyces cerevisiae strains expressing XYL1 and XYL2 from Pichia stipitis with and without overexpression of TAL1. Bioresour Technol. 1999, 68: 79-87. 10.1016/S0960-8524(98)00085-6.

26. Kötter P, Ciriacy M: Xylose fermentation by Saccharomyces cerevisiae. Appl Microbiol Biotechnol. 1993, 38: 776-783. 10.1007/BF00167144.

27. Olofsson K, Palmqvist B, Liden G: Improving simultaneous saccharification and co-fermentation of pretreated wheat straw using both enzyme and substrate feeding. Biotechnol Biofuels. 2010, 3: 17-

28. Olofsson K, Wiman M, Liden G: Controlled feeding of cellulases improves conversion of xylose in simultaneous saccharification and co-fermentation for bioethanol production. J Biotechnol. 2010, 145: 168-175. 10.1016/j.jbiotec.2009.11.001.

29. Bertilsson M, Andersson J, Lidén G: Modeling simultaneous glucose and xylose uptake in *Saccharomyces cerevisiae* from kinetics and gene expression of sugar transporters. Bioprocess Biosyst Eng. 2008, 31: 369-377. 10.1007/s00449-007-0169-1.

30. Linde M: Process development of bioethanol production from wheat and barley residues. Department of Chemical Engineering Volume PhD. 2007, Lund: Lund University

31. Brandberg T, Karimi K, Taherzadeh MJ, Franzen CJ, Gustafsson L: Continuous fermentation of wheat-supplemented lignocellulose hydrolysate with different types of cell retention. Biotechnol Bioeng. 2007, 98: 80-90. 10.1002/bit.21410.

32. Erdei B, Barta Z, Sipos B, Réczey K, Galbe M, Zacchi G: Ethanol production from mixtures of wheat straw and wheat meal. Biotechnol Biofuels. 2010, 3: 16-10.1186/1754-6834-3-16.

33. Galbe M, Sassner P, Wingren A, Zacchi G: Process engineering economics of bioethanol production. Biofuels. Edited by: Olsson L. 2007, Berlin: Springer, 303-327.

34. Jorgensen H, Vibe-Pedersen J, Larsen J, Felby C: Liquefaction of lignocellulose at high-solids concentrations. Biotechnol Bioeng. 2007, 96: 862-870. 10.1002/bit.21115.
35. Meinander NQ, Hahn-Hagerdal B: Influence of cosubstrate concentration on xylose conversion by recombinant, XYL1-expressing Saccharomyces cerevisiae: a comparison of different sugars and ethanol as cosubstrates. Appl Environ Microbiol. 1997, 63: 1959-1964.
36. Palmqvist E, Hahn-Hägerdal B, Galbe M, Zacchi G: The effect of water-soluble inhibitors from steam-pretreated willow on enzymatic hydrolysis and ethanol fermentation. Enzyme Microb Technol. 1996, 19: 470-476. 10.1016/S0141-0229(95)00234-0.
37. Palmqvist E, Hahn-Hägerdal B, Galbe M, Larsson S, Stenberg K, Szengyel Z, Tengborg C, Zacchi G: Design and operation of a bench-scale process development unit for the production of ethanol from lignocellulosics. Bioresour Technol. 1996, 58: 171-179. 10.1016/S0960-8524(96)00096-X.
38. Linde M, Jakobsson EL, Galbe M, Zacchi G: Steam pretreatment of dilute H_2SO_4-impregnated wheat straw and SSF with low yeast and enzyme loadings for bioethanol production. Biomass Bioenergy. 2008, 32: 326-332. 10.1016/j.biombioe.2007.09.013.
39. Sluiter A, Hames B, Ruiz R, Scarlata C, Sluiter J, Templeton D: Determination of sugars, byproducts, and degradation products in liquid fraction process samples. 2006, Golden, CO: National Renewable Energy Laboratory
40. Sluiter A, Hames B, Ruiz R, Scarlata C, Sluiter J, Templeton D, Crocker D: Determination of structural carbohydrates and lignin in biomass. 2008, Golden, CO: National Renewable Energy Laboratory
41. Taherzadeh MJ, Liden G, Gustafsson L, Niklasson C: The effects of pantothenate deficiency and acetate addition on anaerobic batch fermentation of glucose by *Saccharomyces cerevisiae*. Appl Microbiol Biotechnol. 1996, 46: 176-182. 10.1007/s002530050801.
42. Alkasrawi M, Rudolf A, Liden G, Zacchi G: Influence of strain and cultivation procedure on the performance of simultaneous saccharification and fermentation of steam pretreated spruce. Enzyme Microb Technol. 2006, 38: 279-286. 10.1016/j.enzmictec.2005.08.024.

CHAPTER 8

A MACROPOROUS TIO_2 OXYGEN SENSOR FABRICATED USING ANODIC ALUMINIUM OXIDE AS AN ETCHING MASK

Chih-Cheng Lu [1], Yong-Sheng Huang [1], Jun-Wei Huang [1], Chien-Kuo Chang [2], and Sheng-Po Wu [1]

[1]Institute of Mechtronic Engineering, National Taipei University of Technology, Taipei 106, Taiwan

[2]Graduate Institute of Mechanical and Electrical Engineering, National Taipei University of Technology, Taipei 106, Taiwan

ABSTRACT

An innovative fabrication method to produce a macroporous Si surface by employing an anodic aluminium oxide (AAO) nanopore array layer as an etching template is presented. Combining AAO with a reactive ion etching (RIE) processes, a homogeneous and macroporous silicon surface can be effectively configured by modulating AAO process parameters and alumina film thickness, thus hopefully replacing conventional photolithography and electrochemical etch methods. The hybrid process integration is considered fully CMOS compatible thanks to the low-temperature AAO and CMOS processes. The gas-sensing characteristics of 50 nm TiO_2 nanofilms deposited on the macroporous surface are compared with those of conventional plain (or non-porous) nanofilms to verify reduced response noise and improved sensitivity as a result of their macroporosity. Our experimental results reveal that macroporous geometry of the TiO_2 chemoresistive gas sensor demonstrates 2-fold higher (~33%) improved sensitivity than a non-porous sensor at different levels of oxygen exposure. In addition, the macroporous device exhibits excellent discrimination capability and significantly lessened response noise at 500 °C. Experimental results indicate that the hybrid process of such miniature and macroporous devices are compatible as well as applicable to integrated next generation bio-chemical sensors.

INTRODUCTION

With the rapid evolution of microelectronic and microelectromechanical systems (MEMS) technologies, recent industrial products show a clear trend towards miniaturization and neatness. In addition to the traditional CMOS process, MEMS has become a key technique for shrinking device and system dimensions, and the most emerging and cutting-edge field in sensor research during the past two decades. It has effectively reduced device dimensions, lowered material volume and element cost, and more importantly, lessened operational power consumption if necessary. Besides, it also demonstrates the capability of batch processing to produce numerous microsensors and microsystems, which ensures the functionality and availability of CMOS- or silicon-based devices. Relevant studies on the use of SOI technology platform as a full CMOS-MEMS approach for low-power gas detection [1–3] were carried out. These are quite advantageous in the

Citation: Chih-Cheng Lu , Yong-Sheng Huang, Jun-Wei Huang, Chien-Kuo Chang, and Sheng-Po Wu, A Macroporous TiO_2 Oxygen Sensor Fabricated Using Anodic Aluminium Oxide as an Etching Mask, Sensors 2010, 10(1), 670-683; doi:10.3390/s100100670

gas sensors development track. However, one still can see that these micro-scale gas sensors may face poorer sensitivity or response characteristics compared to bulk ones, due to significant reduction of the surface reaction area in chemically sensitive materials. Hence the use of porous or macroporous materials to enhance the deficient sensitivity and response speed of MEMS gas sensors is very popular because the sensor sensitivity is proven to be proportional to the specific surface area [4].

Of all gas sensors, chemoresistive gas sensors deposited with metal oxide thin films have been well developed and are widely employed in industrial applications and environmental control. A number of metal oxide semiconducting materials such as ZnO, SnO_2, TiO_2, In_2O_3 and WO_3 have been studied to detect hazardous and combustible gases [5–10]. Metal oxide bulk films were mainly deposited by physically dispersion or physical vapour deposition (PVD) on the active areas of miniature sensors. However, considerable power losses due to high-temperature operation and poor sensitivity resulting from reduced active area of the gas-sensitive films are still problematic issues. Recently, along with widespread applications of nanostructural materials such as nanotubes, nanoparticles, nanobelts and nanowires, their significant nanoporosity and extensive surface areas could provide promising solutions to improve sensor sensitivity and response characteristics. Alternatively, one may use mesoporous particles as the gas-sensitive film with dielectrophoresis (DEP) processes [11]. However there was room for improvement of the immobilisation method and adhesion strength between the active material and device. These nanostructural particles may separate from the sensor surface because of external force or environmental factors and this can deteriorate a sensor's lifetime. On the other hand, conventional poration approaches to fabricate a porous surface mainly include lithographic patterning [12,13] and electrochemical hydrofluoric acid etch [14]. However, the former is limited by lithographic resolution and expensive cost; in the latter it is still difficult to regulate uniform pore size and density, though it was able to produce nanoporous materials.

To date, there have been a number of continuous investigations on the anodic aluminium oxide (AAO) process since Keller *et al.* [15] discovered porous aluminium oxide layers in 1953 by immersing aluminium metal in electrochemical acid solutions. The AAO layer, essentially a cluster of alumina nanopores, features macroporous (*i.e.*, pore diameter >50 nm), self-organized, highly ordered and homogeneous pore arrays [16–21] and can be prepared directly on aluminium metal sheets, or on silicon substrates by coating high-quality aluminium metal films [22–25]. It has been known that the adoption of acid electrolytes including $C_2H_2O_4$ [17,18], H_3PO_4 [19,20], H_2SO_4 [23] is able to produce AAO nanopores with more satisfactory poration rates and pore uniformity than alkali electrolytes. AAO nanopores normally range from tens of nanometers to hundreds of nanometers in diameter, depending on the electric or chemical parameters modulated and thus have larger specific surface areas than plain thin films with the same dimensions. It is considered a simple, low-temperature, cost-effective and MEMS-compatible process, which makes it possible to act as a platform for thin film-based sensors, solar cells, fuel cells or quantum dot applications.

Herein we present a novel approach to produce macroporous metal oxide gas sensors which combines the AAO thin film and microfabrication processes. As can be seen, the AAO layer features highly ordered and homogeneous alumina nanopores that make it suitable as a dry-etching mask on the silicon substrate to generate a macroporous surface. The geometric features such as pore diameter, pore density and pore depth can be experimentally controlled by modulating AAO process parameters such as aluminium film quality, applied potential and electrolyte conditions. Experimental work including preparation of AAO process and fabrication flow of this gas sensor is introduced first, followed by a description of the characteristics of the macroporous surface and as-deposited TiO_2 thin films. Response characteristics and sensitivity of macroporous and non-porous TiO_2 films to different oxygen concentrations at room temperature (31 °C) and a high temperature (500 °C) have been provided and compared. All these experimental results verify the promising feasibility of this class of macroporous TiO_2 thin film gas sensors fabricated using such well-developed techniques.

EXPERIMENTAL

Preparation of the AAO Process

Before the beginning of the AAO process, the silicon wafer was deposited with an aluminium thin film of 1 μm RF sputtered onto the front side of the silicon substrate. The AAO process was then carried out by using 0.3 M oxalic acid solution at 4 °C in an electrochemical cooling tank for 90 min. The electrochemical reaction was conducted under dc 40–60 volts by a Keithley 2000 sourcemeter and the corresponding anodic current during the AAO process was closely monitored by a computer connected to a LabVIEW® data acquisition adaptor. The anodic I-V curve exhibited a significant drop to very low current within 100 s during when the aluminium oxidation procedure started, and ended at various time points under different voltage, as shown in Figure 1. After the oxidation process, the alumina thin film was immersed in a 5% phosphoric acid solution for 4 h at room temperature. This step was relatively vital to produce a homogeneous pore diameter distribution and pore-widening effect in the alumina nanopore array. In addition, the alumina layer and the barrier layer located at the base of the alumina layer were further thinned using a 30% phosphoric acid solution at room temperature to make it suitable to act as an etch mask for later use. Compared with the AAO methods developed earlier, the AAO process developed in our laboratory was adjusted and simplified so that merely one pore-widening step was required and the barrier layer in the bottom part could be easily removed. The as-fabricated AAO films in Figure 2 exhibit anodic oxidation results before and after the alumina pore-widening step.

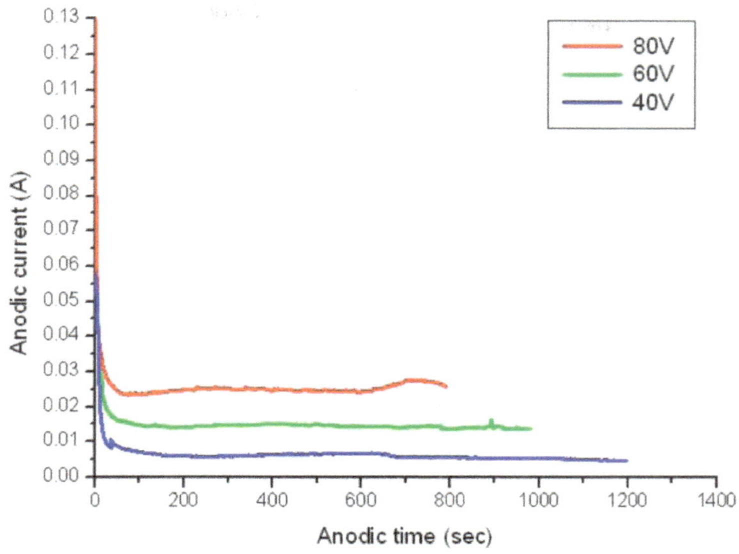

Figure 1: Anodic current monitoring during the AAO process under various voltage.

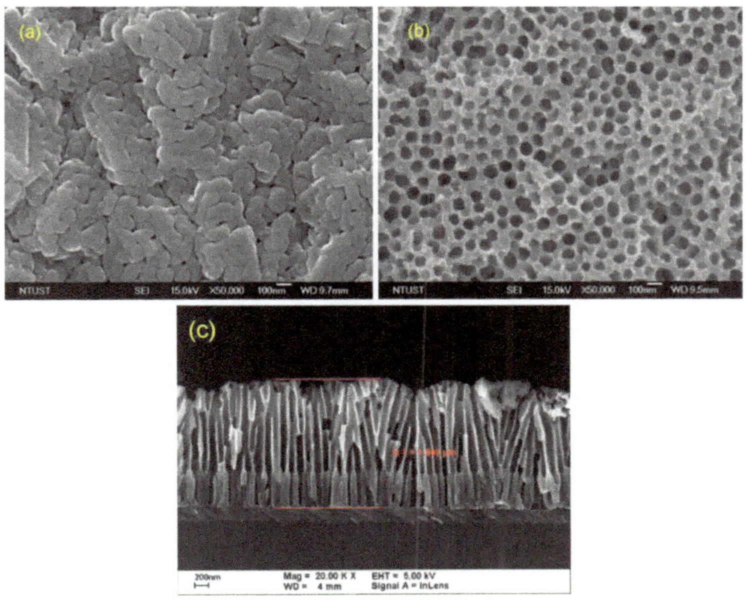

Figure 2: SEM images of the AAO layer: (a) processed under 60 V, 10 min, (b) processed under 60 V, 40 min (the pore-widening step), (c) illustrated with cross-section before the pore-widening step.

Manufacture of Macroporous Silicon

With thinned AAO films, it is very feasible to create porous silicon, silicon oxide or any other treatable materials. The thinned layer of anodic aluminium oxide readily acts as an dry-etch mask for an isotropic reactive ion etch (RIE) process, enabling the silicon surface under the alumina mask to generate macroporous pores estimated 100–700 nm in diameter. The alumina nanopore array works as a competent etch mask due to the good selectivity relative to silicon in an RIE process with CF_4:Ar:O_2 = 5:0.5:0.5 in sccm and a pressure of 90 mTorr. The thin barrier layer at the base of the alumina mask can be removed first. On completion of RIE process, the alumina mask was removed by a 30% H_3PO_4 solution at 60 °C for 1 h. A scanning electron microscopic (SEM) micrograph of the macroporous silicon surface after an RIE process is shown in Figure 3.

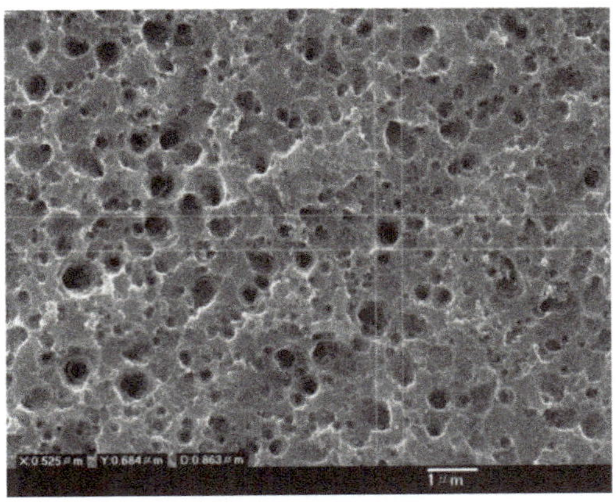

Figure 3: SEM image of the macroporous Si surface after RIE process and removal of the alumina layer.

Design and Fabrication of Macroporous TiO_2 *Gas Sensors*

To fabricate macroporous TiO_2 gas sensors, the integration of AAO and RIE etch processes was performed. The sensor was manufactured by our revised AAO process mentioned before and microfabrication techniques. The process flow for the macroporous TiO_2 sensor is shown in Figure 4. A similar concept of sensor fabrication was also reported by Lu and Chen [26] in 2009; however, the AAO layer was not removed and the gas-sensitive TiO_2 film was directly grown onto it. During the AAO electrochemical process, according to our experimental practice, the alumina layer was inclined to peel off or get damaged by other oxidation steps, which may not afford sufficient adhesion to the substrate. It is our belief that the AAO layer as the dry-etch mask can produce a porous silicon surface, and after its removal, a reliably adhesive TiO_2 macroporous thin film deposited onto the silicon surface can be highly guaranteed. The RIE process readily etched the silicon substrate surface to form nano-scale pores ranging from 100 nm to 750 nm in diameter. After an ultrasonic cleaning step, the gas-sensitive layer of polycrystalline TiO_2 was deposited onto the porous silicon surface to configure a macroporous TiO_2 thin film by a RF sputtering process. In consideration of the nanopore diameter distribution on the silicon surface, the TiO_2 thin film with a thickness of 50 nm was selected to verify its sensitivity and response properties. Finally, silver electrodes were deposited on the macroporous TiO_2 thin film by a lift-off technique. The material characterization and geometric profile of the porous TiO_2 thin film will be shown and discussed later.

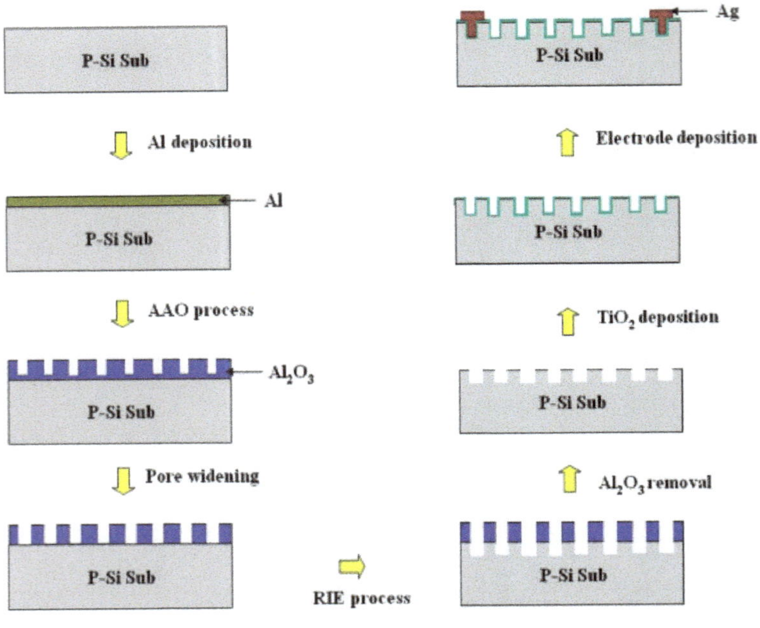

Figure 4: The processing schematic of a macroporous TiO_2 gas sensor.

Gas Characterization of the Macroporous TiO_2 *Sensor*

A gas test system was utilised to characterise various responses to toxic or hazardous gases. This system basically includes two-channel gas pipelines, a gas mixer and a test chamber with high-precision mass flow controllers (MFCs). Other facilities such as a mechanical vacuum pump and a vacuum gauge, a multi-channel thermometer, a Keithley 2400 sourcemeter and a LabVIEW® data acquisition card were also appropriately linked to this system. A configuration diagram of the gas test system is shown in Figure 5.

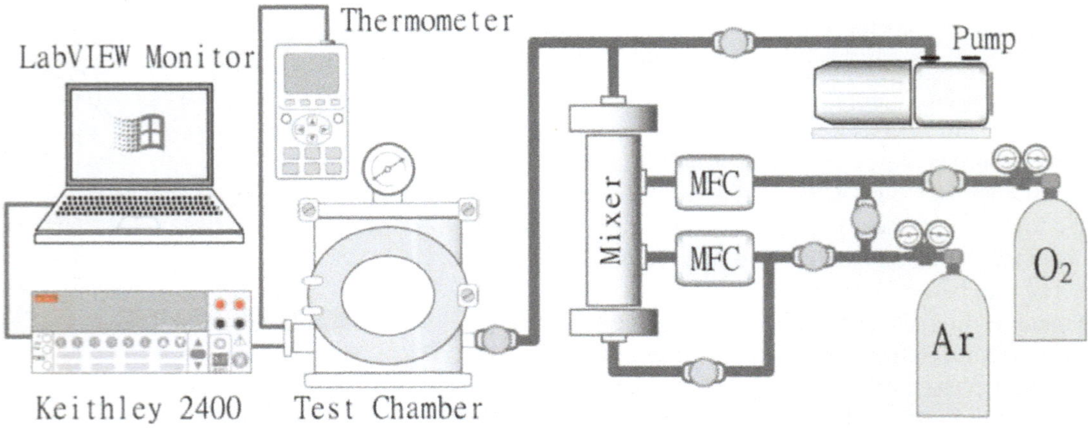

Figure 5: A configuration schematic of the gas test system.

Initially, argon carrier gas was introduced into the system to evacuate or dilute the air inside the system as oxygen in air may affect the accuracy of the response to an analyte gas. The fabricated TiO_2 sensor was placed in the test chamber and characterised under exposure to a mixture of analyte and carrier gases in atmosphere pressure (gas adsorption phase), then evacuated to a low pressure of 10^{-3} Torr or heated up by a heater (gas desorption phase), and simultaneously measured by*in situ* resistance readout from the electrodes using a LabVIEW® software interface.

RESULTS AND DISCUSSION

The surface geometry of the macroporous TiO_2 sensor was verified by a SEM image, and its X-ray diffraction (XRD) pattern was analyzed to examine the thin film crystalline structure. Furthermore, discussions on AAO process experiments and parametric operation were reported. Atomic force microscopy (AFM) was also employed to investigate the aspect ratio of pore depth to pore diameter in the fabricated nanopores that were not clearly shown in the SEM images. Next, to explore the gas-sensitive effect enhanced by the macroporous TiO_2 surface, plain and AAO-based porous thin films of 50 nm thick were deposited and made into gas sensors. Oxygen analyte gas mixtures of different concentrations mixed with an argon carrier gas was prepared and supplied to measure chemoresistive response and sensitivity due to analyte gas exposures at the room temperature of 31 °C and an elevated temperature of 500 °C, respectively. All gas tests were conducted at ambient atmosphere pressure, room temperature of 31 °C and 65% RH. As usual, the sensor sensitivity to an analyte gas can be defined as $S = \Delta R/R_0 \times 100\%$, where R_0 is the baseline resistance in the presence of the carrier gas without an analyte gas and ΔR is the difference between R_0 and the resultant resistance after the analyte gas appears in the test chamber and an equilibrium is reached. All test results were compared, analyzed and summarized.

Characterization of the Porous TiO_2 *Thin Film*

Figure 6 exhibits an SEM diagram of the macroporous geometry of the 50-nm TiO_2 thin film after AAO process and sputtering deposition.

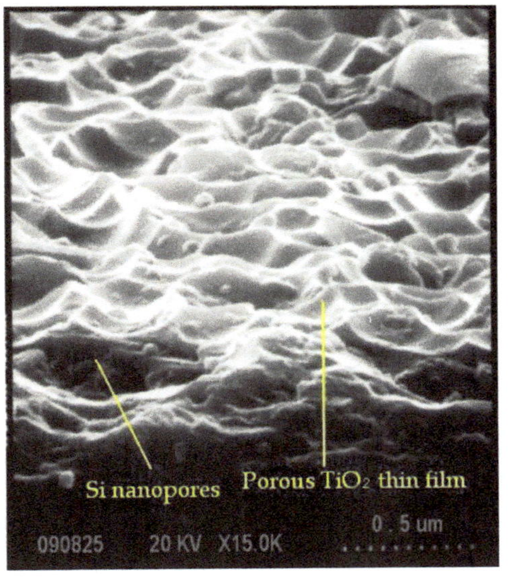

Figure 6: SEM image of the macroporous TiO_2 thin film deposited on the silicon substrate.

As one can see, the TiO_2 layer successfully covers the porous silicon substrate and duplicates the wavy geometry underneath. From this evidence, it is shown that the specific surface area of the TiO_2 thin film is significantly increased compared to a conventionally deposited plain (or non-porous) thin film. Next, X-ray diffraction (XRD) pattern analysis was carried out to characterize the composition of as-deposited TiO_2 thin film. The XRD pattern was obtained using a Bede/D1 diffractometer with Cu Kα radiation (1.541 Å) made by MAC science Ltd. Usually, it is noted that small-angle XRD analysis is typically employed for thin films while large-angle XRD analysis is for bulk materials. As one can see a full-range angle XRD scan was employed that confirmed prominent intensity peaks at 2-theta degree of 24.97° and 43.85°, as shown in Figure 7, which can be explained as the crystalline structure of anatase instead of rutile after film deposition based on JCPDS database and considered suitable for gas detection [27]. No annealing process was applied to the TiO_2 layer deposited in our experiments.

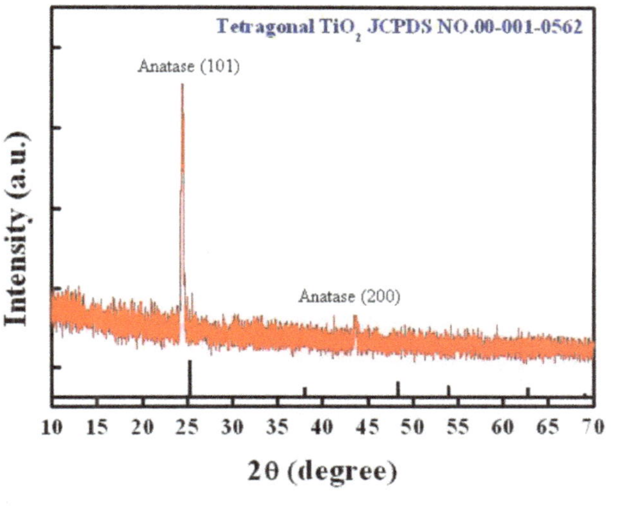

Figure 7: XRD pattern analysis of the porous TiO_2 thin film.

Modulation and Analysis of AAO Porosity

In the anodic aluminium oxidation process, the deposition of quality aluminium films can mostly dominate the orderly geometry of the alumina nanopore array. Some operating steps such as wafer cleaning, particle removal, deposition parameter setup and substrate materials are closely associated with the successful implementation of AAO layers. A common failure encountered is AAO film peel-off in the beginning or half way through the process. Our experimental results indicate that higher anodic voltages, for instance, greater than 100 V tend to cause peel-off or deionization of the aluminium film in oxalic acid solution. Therefore, the hard anode mode with anodic voltage ranging from 40 V to 80 V is commonly employed. The higher the working voltage, the shorter the AAO reaction period, as shown in Figure 1. Such an operation may result in a larger aspect ratio of depth to width (>16) in the alumina nanopore array, which is not advantageous for use as a mask for an RIE etching process. However, the thickness of the alumina layer can be further reduced by a 30% H_3PO_4 solution at room temperature and this makes a suitable RIE etch mask with a lower aspect ratio (<5). Also, appropriate manipulation of anodic voltage and current has a vital impact on determination of nanopore diameters. It is noted that nanopores with larger diameters can be induced at higher anodic voltages and *vice versa*. Moreover, the Joule heat generated in the anodic oxidation is able to raise the solution temperature and thus cause unstable electrochemical AAO reactions, leading to peel-off of the AAO film. A magnetic stirring rotator is placed in the solution to diminish the undesired thermal effect. To characterize the as-etched silicon surface, an atomic force microscope was employed to describe the surface geometry. Figure 8 illustrates the AFM geometry showing the 2-D porous silicon surface and a cross-sectional profile along Y-direction (not to scale). The profile indicates that a nanopore etched on the silicon surface has an estimated aspect ratio of 0.35 at pit A in Figure 8, which is suitable for thin film deposition and coverage.

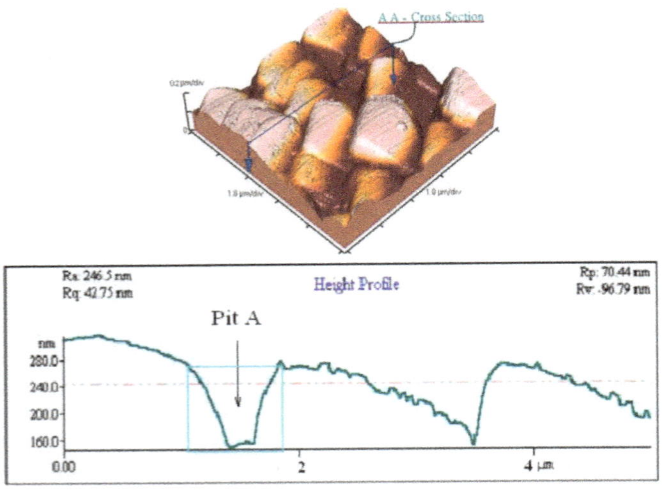

Figure 8: The porous TiO_2 thin film of AFM measurement.

Oxygen Response at Room Temperature (31°C)

It's been recognized that the resistance change of metal oxide thin films such as TiO_2 or SnO_2 behaves as an *n-type*semiconductor in an oxygen-contained environment [28]. At high temperatures the temporal resistance change under gas exposure is caused by analyte gas molecules that react irreversibly with electrons or holes of the TiO_2 lattice vacancies. A surface depletion region referred to as the Debyé length is thus formed on TiO_2 surface and affects the Schottky barrier, which is dominated by charge carrier concentrations. The Schottky barrier is exponentially associated with material conductance and the conductance variation is proportional to gas concentrations as well. In oxygen or other oxidizing gases, it was shown that the resistance in TiO_2 can be increased by exposing to rising concentrations of oxygen, and vice versa in reducing gases. The macroporous TiO_2 thin film, as a result, represents a higher specific surface area and contributes higher charge carrier concentrations that can enhance the conductance, and thus demonstrates decreased resistance when compared with a plain TiO_2 thin film. A number of previous studies disclosed that most TiO_2-based sensors must operate over 400 °C for better stability and performance [4–6]. However, to examine the impact of macroprorosity to TiO_2 sensors, a series of gas tests addressing gas response to different oxygen concentrations for porous and non-porous TiO_2 sensors were carried out. These TiO_2 gas sensors were tested at room temperature of 31 °C, 65% RH and 1 atm in a completely sealed chamber with a gas inlet from the MFCs and the mixer, and a gas outlet to ambient air. With precise MFCs one can modulate various gas flow rates to produce expected concentrations of the analyte gas. Figure 9 shows the resistance change of the sensor in response to 500 ppm, 1,500 ppm and 2,500 ppm of oxygen gas. It is observed that resistance signals were highly unstable and fluctuated, despite the increase of oxygen concentrations. Higher oxygen concentrations up to 1% even did not help diminish the phenomenon. It was also shown that three temporal response curves of different oxygen concentrations shown in Figure 9 were very close to each other and discrimination ability was limited at room temperature regardless of TiO_2 macroporous films. However, the ascending trend of resistance variation with elapsed time was clearly verified and still consistent with oxidized characteristics of the TiO_2 thin film. These indistinct resistance responses and significant noise are not unexpected because experiments were operated at room temperature and could be attributed to wide-gapped Ag electrodes in part.

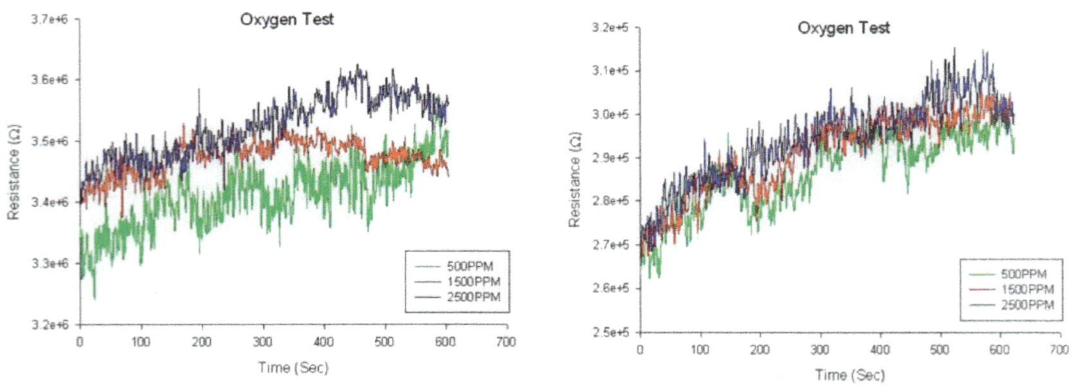

(a) (b)

Figure 9: Temporal resistance response *vs.* different oxygen concentrations at room temperature of 31 °C with (a) a plain (non-porous) TiO_2 film, and (b) a macroporous TiO_2 film. The film thickness is 50 nm.

Oxygen Response at High Temperature (500 °C)

Applying most test conditions of the room-temperature experiment, porous and non-porous sensors were characterized instead at a high temperature of 500 °C. The experiments were carried out to reveal resistance change of the sensors *vs.* 4,000 ppm, 5,000 ppm and 6,000 ppm of oxygen gas, as illustrated in Figure 10. It was shown that, unlike the unstable response in room-temperature tests, both temporal resistance changes to a constant oxygen concentration for both porous and non-porous devices became more stable and less noisy, particularly prominent for the porous type sensor. A significant high temperature effect on improving the resistance response of TiO_2 thin films has been demonstrated when compared to the room-temperature operation. Our experimental results also showed that the TiO_2 resistance at 500 °C is lower than that at 31 °C. Two things may explain this phenomenon. One is the *n-type* semiconducting model at high temperatures; *i.e.*, the higher the temperature, the lower the resistance. The second is that TiO_2 resistance can be also affected by humidity factor and it was revealed that the resistance of TiO_2 thin film lessens while relative humidity falls [29], particularly when operating at high temperatures. In addition, the porous TiO_2 sensor obviously proved better discrimination ability for oxygen with different concentrations than the non-porous one.

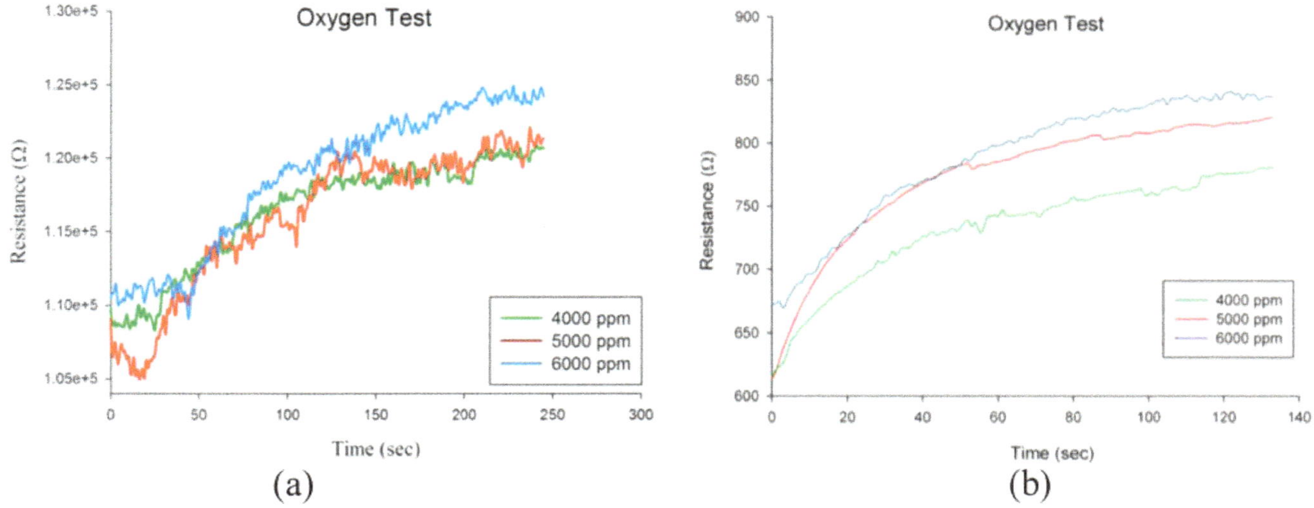

Figure 10: Temporal resistance response *vs.* different oxygen concentrations at a high temperature of 500 °C with (a) a plain (non-porous) TiO_2 film, and (b) a macroporous TiO_2 film. The film thickness is 50 nm.

In summary, the response behaviors suggested that TiO_2 gas-sensing films should be considered an *n-type* semiconductor when exposed to an oxidizing gas. More interestingly, we converted the resistance response characteristics into charts expressing sensitivity *versus* oxygen concentration.

In Figure 11 it is clearly observed that the macroporous TiO_2 sensor can effectively elevate sensor sensitivity by more than 2-fold (~33%) when compared to the non-porous TiO_2 sensor at different levels of oxygen exposure, both at room temperature and high temperatures. Also, one can clearly see that both resistance response and sensitivity of the sensor are increasingly proportional to rising oxygen concentrations, delivering an informative and favourable behaviour for analyte calibration.

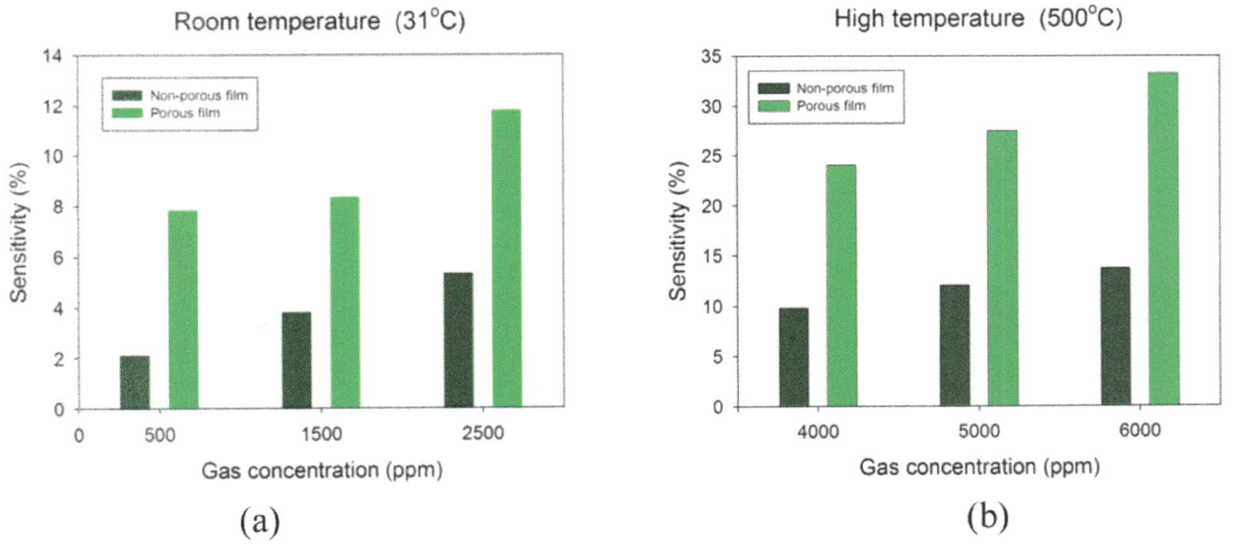

Figure 11: Comparative charts of sensitivity *vs.* oxygen concentrations with respect to non-porous and porous TiO_2 thin film sensors at (a) room temperature of 31 °C and (b) a high temperature of 500 °C.

CONCLUSIONS

A new approach to fabricate macroporous TiO_2 oxygen sensors utilizing anodic aluminium oxide nanostructures as an effective dry-etch mask for production of a porous silicon substrate is described. This new method is mainly developed for generating a higher specific surface area that is believed to reduce sensor dimensions and enhance sensor sensitivity by integrating low-temperature AAO and microfabrication process. With positive experimental results, it is noteworthy that the AAO-fabricated TiO_2 porous gas sensor can not only lessen response noise, but also significantly enhance sensor sensitivity by a significant 2-fold increase (~33%) compared to that of a non-porous TiO_2 sensor at 500 °C for oxygen detection. The macroporous TiO_2 sensors are able to achieve successful oxygen discrimination in terms of distinct resistance response and low measurement noise at high temperatures. We could attribute these advantages to an increased specific surface area in macroporous TiO_2 thin films fabricated using the proposed hybrid process.

ACKNOWLEDGMENTS

We are grateful to CIC and NCHC, NARL for chip implementation service and the use of computer facilities. This research was partially supported by NSC, Taiwan, ROC under the grant of 98-2221-E-027-066.

REFERENCES

1. Suehle, J.S.; Cavicchi, R.E.; Gaitan, M.; Semancik, S. Tin oxide gas sensor fabricated using CMOS micro-hotplates and in-situ processing. *IEEE Electron Dev. Lett* **1993**, *14*, 118–120.
2. Lu, C.C.; Liao, K.H.; Udrea, F.; Covington, J.A.; Gardner, J.W. A CMOS-MEMS smart sensory platform using SOI technology for high temperature gas detection. *J. Micromech. Microeng* **2008**, *18*. No. 075010..
3. Barrettino, D.; Graf, M.; Song, W.H.; Kirstein, K.U. Hotplate-based monolithic CMOS microsystems for gas detection and material characterization for operating temperatures up to 500 °C. *IEEE J. Solid-State Circ* **2004**, *39*, 1202–1207.
4. Rothschild, A.; Komem, Y. The effect of grain size on the sensitivity of nanocrystalline metal-oxide gas sensors. *J. Appl. Phys* **2004**, *95*, 6374–6380.
5. Mardare, D.; Iftimie, N.; Luca, D. TiO_2 thin films as sensing gas materials. *J. Non-Crystal. Solids* **2008**, *354*, 4396–4400.
6. Devi, G.S.; Hyodo, T.; Shimizu, Y.; Egashira, M. Synthesis of mesoporous TiO_2-based powders and their gas-sensing properties. *Sens. Actuat. B: Chem* **2002**, *87*, 122–129.
7. Carney, C.M.; Yoo, S.H.; Akbar, S.A. TiO_2-SnO_2 nanostructures and their H_2 sensing behavior. *Sens. Actuat. B: Chem***2005**, *108*, 29–33.
8. Meng, D.; Toshinari, Y.; Yanbai, S.; Liu, Z.; Toshio, K. Preparation of WO_3 nanoparticles and application to NO_2sensor. *Appl. Surf. Sci* **2009**, *256*, 1050–1053.
9. Ali, M.; Wang, C.Y.; Rohlig, C.; Stauden, C.C.; Ambacher, V.O. NO_x sensing properties of In_2O_3 thin films grown by MOCVD. *Sens. Actuat. B: Chem* **2008**, *129*, 467–472.
10. Ryu, H.W.; Park, B.S.; Akbar, S.A.; Lee, W.S.; Lee, K.J.; Seo, Y.J.; Shin, D.C.; Park, J.S.; Choi, G.P. ZnO sol-gel derived porous film for CO gas sensing. *Sens. Actuat. B: Chem* **2003**, *96*, 717–722.
11. Lu, C.C.; Liao, K.H. Microfabrication and chemoresistive characteristics of SBA-15-templated mesoporous carbon gas sensors with CMOS compatibility. *Sens. Actuat. B: Chem* **2009**.
12. Calvert, J.M. Lithographic patterning of self-assembled films. *J. Vac. Sci. Technol. B* **1993**, *11*, 2155–2158.
13. Marrian, C.R.K.; Colton, R.J. Low-voltage electron beam lithography with a scanning tunneling microscope. *Appl. Phys. Lett* **1990**, *56*, 755–757.
14. Connolly, E.J.; Timmer, B.; Pham, T.M.; Groeneweg, J.; Sarro, P.M. A porous SiC ammonia sensor. *Sens. Actuat. B: Chem* **2005**, *109*, 44–46.
15. Keller, F.; Hunter, M.S.; Robinson, D.L. Structural features of oxide coating on aluminium. *Electrochem. Soc* **1953**, *100*, 411–419.
16. Masuda, H.; Fukuda, K. Ordered metal nanohole arrays made by a 2-step replication of honeycomb structure of anodic alumina. *Science* **1995**, *268*, 1466–1468.
17. Wang, Y.D.; Chua, S.J.; Sander, M.S.; Chen, P.; Tripathy, S.; Fonstad, C.G. Fabrication and properties of nanoporous GaN films. *Appl. Phys. Lett* **2004**, *85*, 816–818.
18. Hidetaka, A.; Mamoru, M.; Megumi, Y.; Sachiko, O. Transfer of nanoporous pattern of anodic porous alumina into Si substrate. *Appl. Phys. Lett* **2003**, *83*, 4408–4410.
19. Masuda, H.; Satoh, M. Fabrication of gold nanodot array using anodic porous alumina as an evaporation mask. *J. Appl. Phys* **1996**, *35*, 126–129.
20. Sander, M.S.; Tan, L.S. Nanoparticle arrays on surfaces fabricated using anodic alumina films as templates. *Adv. Funct. Master* **2003**, *13*, 393–397.
21. Almawlawi, D.; Bosnick, K.A.; Osika, A.; Moskovits, M. Fabrication of nanometer-scale patterns by ion-milling with porous anodic alumina masks. *Adv. Mater* **2000**, *12*, 1252–1256.
22. Hu, M.; Yu, R.; MacManus-Driscoll, J.L.; Robinson, A.P. Large-area silica nanotubes with controllable geometry on silicon substrates. *Appl. Surf. Sci* **2009**, *255*, 3563–3566.
23. Nasirpouri, F.M.; Abdollahzadeh, M.J.; Almasi, N. A comparison between self-ordering of nanopores in aluminium oxide films achieved by two- and three-step anodic oxidation. *Curr. Appl. Phys* **2009**, *9*, 91–94.
24. Lee, W.; Ji, R.; Ulrich, G.; Kornelius, N. Fast fabrication of long-range ordered porous alumina membranes by hard anodization. *Nat. Mater* **2006**, *5*, 741–743.
25. Hatice, U.; Tae, B.H.; Kimb, H.K. Thin alumina nanoporous membranes for similar size biomolecule separation. *J. Membr. Sci* **2009**, *343*, 1–6.
26. Lu, C.; Chen, Z. High-temperature resistive hydrogen sensor based on thin macroporous rutile TiO_2 film on anodic aluminium oxide. *Sens. Actuat. B: Chem* **2009**, *140*, 109–115.

27. Tang, H.; Prasad, K.; Sanjines, R.; Levy, F. TiO_2 anatase thin films as gas sensors. *Sens. Actuat. B: Chem* **1995**, *26*, 71–75.

28. Morrison, S.R. Mechanism of semiconductor gas sensor operation. *Sens. Actuat* **1987**, *11*, 283–287.

29. Buso, D.; Post, M.; Cantalini, C.; Mulvaney, P.; Martucci, A. Gold nanoparticle-doped TiO_2 semiconductor thin films:gas sensing properties. *Adv. Funct. Mater* **2008**, *18*, 3843–3849.

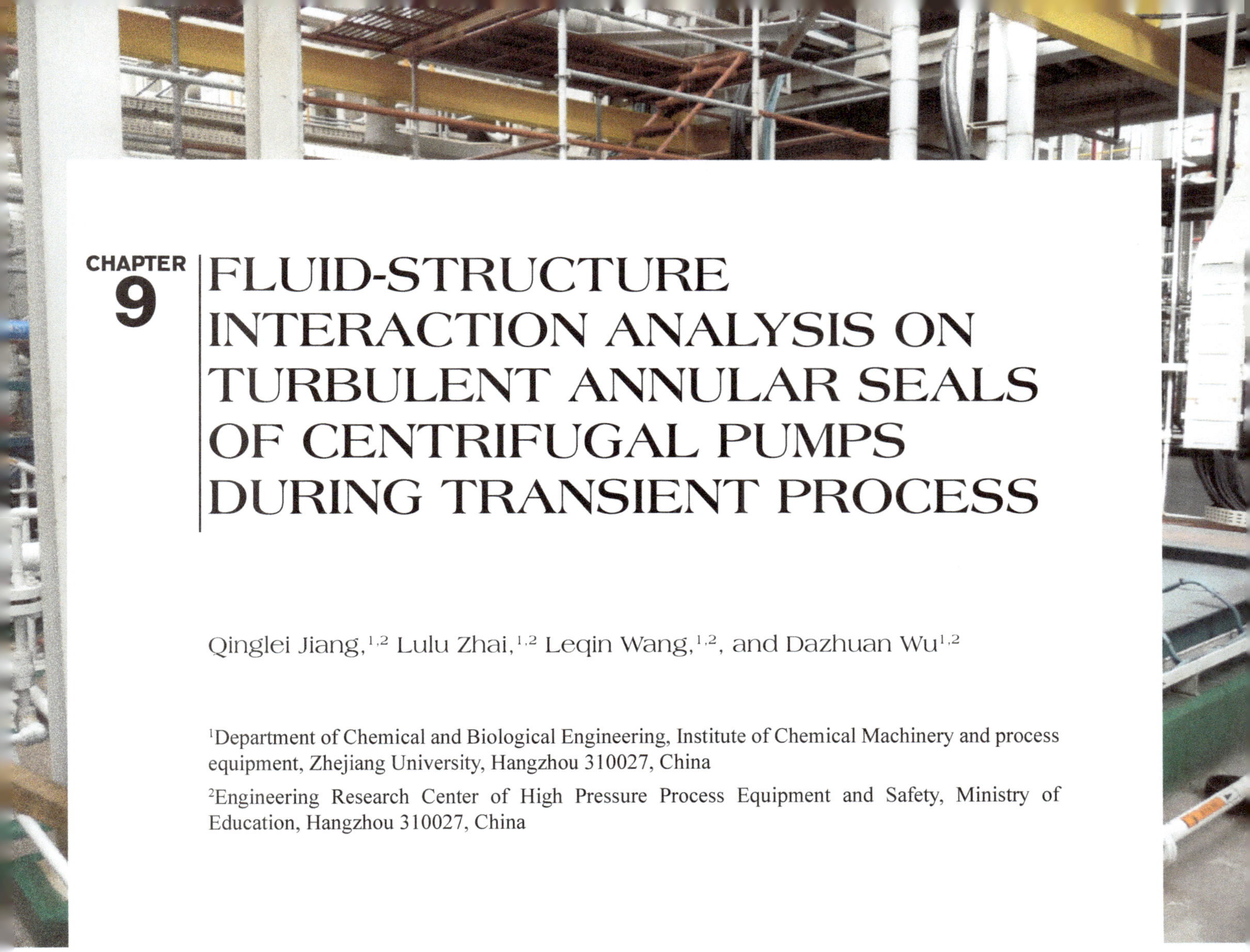

CHAPTER 9

FLUID-STRUCTURE INTERACTION ANALYSIS ON TURBULENT ANNULAR SEALS OF CENTRIFUGAL PUMPS DURING TRANSIENT PROCESS

Qinglei Jiang,[1,2] Lulu Zhai,[1,2] Leqin Wang,[1,2], and Dazhuan Wu[1,2]

[1]Department of Chemical and Biological Engineering, Institute of Chemical Machinery and process equipment, Zhejiang University, Hangzhou 310027, China

[2]Engineering Research Center of High Pressure Process Equipment and Safety, Ministry of Education, Hangzhou 310027, China

ABSTRACT

The current paper studies the influence of annular seal flow on the transient response of centrifugal pump rotors during the start-up period. A single rotor system and three states of annular seal flow were modeled. These models were solved using numerical integration and finite difference methods. A fluid-structure interaction method was developed. In each time step one of the three annular seal models was chosen to simulate the annular seal flow according to the state of rotor systems. The objective was to obtain a transient response of rotor systems under the influence of fluid-induced forces generated by annular seal flow. This method overcomes some shortcomings of the traditional FSI method by improving the data transfer process between two domains. Calculated results were in good agreement with the experimental results. The annular seal was shown to have a supportive effect on rotor systems. Furthermore, decreasing the seal clearance would enhance this supportive effect. In the transient process, vibration amplitude and critical speed largely changed when the acceleration of the rotor system increased.

INTRODUCTION

In multistage centrifugal pumps, annular seal passages, such as sealing rings and balance drums, can provide fluid-induced forces, which play an important role in determining the rotor dynamic stability of a centrifugal pump [1, 2]. The flow in these annular seal passages is generally called the annular seal flow. The annular seal flow in pumps can generate reaction forces to support rotating shafts similar to journal bearings. A completely developed turbulent flow usually exists in annular seals than a laminar flow in journal bearings. In the past, the influence of the annular seal flow on rotor systems was usually studied indirectly in terms of dynamic coefficients. In the 1970s, Hirs presented a turbulent lubrication theory called the "bulk flow theory" in his doctoral thesis [3]. This result was published in the Journal of the American Society of Mechanical Engineers [4]. Childs [5] developed a method to obtain dynamic coefficients (K, k, M, C, c) based on the "bulk flow theory" using perturbation analysis. This method has been widely used in numerical and experimental studies of the pump shaft stability analysis [6, 7]. The start-up process of the centrifugal pump has gained increasing attention recently. In many cases, centrifugal pumps must experience

Citation: Qinglei Jiang, Lulu Zhai, Leqin Wang, and Dazhuan Wu, "Fluid-Structure Interaction Analysis on Turbulent Annular Seals of Centrifugal Pumps during Transient Process," Mathematical Problems in Engineering, vol. 2011, Article ID 929574, 22 pages, 2011. doi:10.1155/2011/929574

a quick-start process due to its function demand [8, 9]. In this process, the rotor response involves simultaneous free and forced vibrations, a typical transient response very much different from the normal steady response [10]. The dynamic coefficients method mentioned above is not applicable in such situations. First, this method disregards the impeller difference eccentricity, whereas, in the quick-start process, the eccentricity rapidly increases. Second, the amplitude and phase angle of the impellers change rapidly with the increase in rotating speed. The dynamic coefficients method cannot consider nonlinear characteristics of fluid-induced forces and inertia effects of rotor systems due to its linear simplification [5]. Given these disadvantages of the traditional methods, studying the influence of the annular seal on rotor systems in the quick-start process using other methods is beneficial. The fluid-structure interaction method is a good choice because it can directly transfer data between two domains. Richardet and Lornage [11, 12] studied shaft-fluid coupling problems using two similar FSI methods. The finite element method and finite difference method were used to solve structural analysis and fluid film analysis. Afterwards, the data of the structural and fluid meshes were transferred between each other based on an interfacing grid concept.

To conduct the FSI analysis, a proper annular seal model must be selected first. Although Richardet and Lornage [11, 12] used the classical laminar model to simulate a small clearance flow in the FSI analysis, the laminar model is not applicable in the present study. The reason is that in centrifugal pumps, the annular seal flow is not maintained in the laminar state but grows from a laminar state to transition regime state and finally a turbulent state with the increase of the rotor rotating speed. A similar phenomenon was presented by Taylor in 1922 [13]. He found that the axial vertex (Taylor vertex) appears between two concentric cylinders when the internal cylinder starts to rotate. This vertex will break if the angular velocity continues to increase, indicating that turbulent flow occurs in the angular clearance. Laminar flow can be easily simulated by classical Reynolds equations [14]; thus, much attention has been given to the transition regime and turbulent flow. Con et al. [3,15, 16] proposed three turbulent theories on the angular clearance flow in the 1970s. These three theories found a way to express turbulent shear stress using the Prandtl mixing length theory, law of wall, and experimental results. These expressions of turbulent stress were then submitted to the momentum equations of clearance flow, leading to turbulent lubrication equations, including turbulent factors. The angular clearance flow is usually in the transition regime state between the laminar and turbulent states when the Reynolds number is between 700 and 2000. The transition regime flow cannot be directly simulated by classical Reynolds equations or turbulent theories. In the late 1980s [17, 18], numerical and experimental studies were carried out for the transition regime flow. Roberts and Mason studied these characteristics in 1986 [19] and then proposed a transition flow model. The transition flow model is similar in form to the turbulent model except for the difference in turbulent factor expressions.

In traditional FSI methods, fluid mesh is usually much finer than the structure mesh. Thus data transfer between two domains is usually difficult and inaccurate due to missing data. In the current paper, the rotor system was modeled based on the node-element method instead of the traditional FE method. The deformation of the impeller itself is ignored in the node-element method. Because only the radial displacement and the deflection angle of the impeller are useful in the rotor dynamic analysis and the deformation of the impeller is usually not necessary in FSI calculation, this simplification is accurate enough. Moreover, this simplification can overcome problems on missing data and remarkably reduce computing time by replacing the impeller with a node. In the current paper, a typical single rotor system was modeled and coupled with an annular seal flow model. The transient response in the start-up process was then obtained using the FSI method presented. Fluid-induced forces showed great influence on the rotor dynamic analysis. The results can be helpful in the improvement of the vibration characteristics of the centrifugal pump during the start-up period.

MODELS OF ROTORS AND TURBULENT FLOW

Dynamic Model of a Rotor System

The dynamic model of a single rotor system was first established. System equations of motion were obtained under the following assumptions. (a) A rotating shaft is supported simply at both ends and carries one disk at the middle of span. The mass of the shaft is negligible. (b) The angular acceleration of the shaft is constant with time. Figure 1 shows the cross-section of the shaft at the disk position, including the geometrical center S, the gravitational center of the disk, eccentricity of the disk e, static steady position of geometrical center O□, and deflection of shaft r.

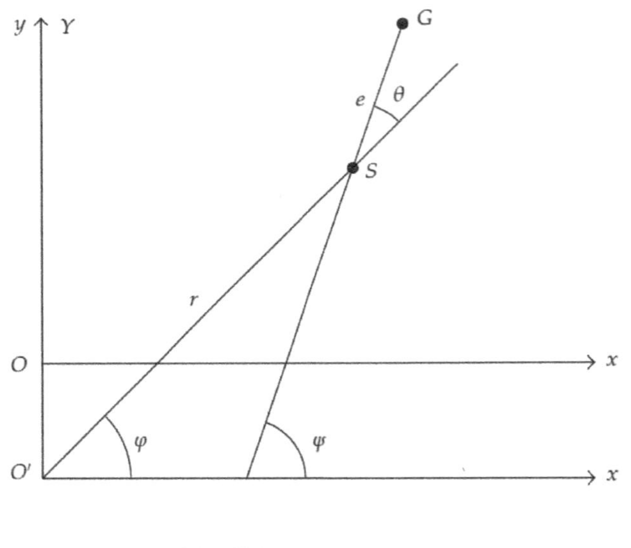

Figure 1: Position of the disk.

The geometrical center S and gravitational center have the following relations:

$$x_c = x + e\cos\psi,$$

$$y_c = y + e\sin\psi. \tag{2.1}$$

After derivation,

$$\dot{x}_c = \dot{x} - e\dot{\psi}\sin\psi,$$
$$\dot{y}_c = \dot{y} + e\dot{\psi}\cos\psi,$$
$$\ddot{x}_c = \ddot{x} - e\ddot{\psi}\sin\psi - e\dot{\psi}^2\cos\psi,$$
$$\ddot{y}_c = \ddot{y} + e\ddot{\psi}\cos\psi - e\dot{\psi}^2\sin\psi. \quad (2.2)$$

The reaction force and damping force generated by the shaft and applied on the disk can be expressed as follows:

$$F_x = -kx, \qquad F_y = -ky,$$
$$R_x = -c_1\dot{x}_c, \qquad R_y = -c_1\dot{y}_c, \quad (2.3)$$

where k is the stiffness of shaft and c_1 is the lateral motion damping coefficient of the disk, which is usually approved by the experiments.

Based on the theorem of motion of centre of mass

$$m\ddot{x}_c = F_x + R_x,$$
$$m\ddot{y}_c = F_y + R_y - mg. \quad (2.4)$$

The equations of motion of the geometrical center S can be written as follows:

$$m\ddot{x} + c_1(\dot{x} - e\dot{\psi}\sin\psi) + kx = me\dot{\psi}^2\cos\psi + me\ddot{\psi}\sin\psi,$$
$$m\ddot{y} + c_1(\dot{y} + e\dot{\psi}\cos\psi) + ky = me\dot{\psi}^2\sin\psi - me\ddot{\psi}\cos\psi - mg. \quad (2.5)$$

The system equations can be simplified as follows after substituting $x=X$, $y=Y-mg/k$:

$$m\ddot{X} + c_1(\dot{X} - e\dot{\psi}\sin\psi) + kX = me\dot{\psi}^2\cos\psi + me\ddot{\psi}\sin\psi,$$
$$m\ddot{Y} + c_1(\dot{Y} + e\dot{\psi}\cos\psi) + kY = me\dot{\psi}^2\sin\psi - me\ddot{\psi}\cos\psi. \quad (2.6)$$

Suppose $X=r\cos\varphi$, $Y=r\sin\varphi$, $\psi=\varphi+\theta$; then

$$\ddot{r} - r\dot{\varphi}^2 + \frac{k}{m}r + \frac{c_1}{m}(\dot{r} - e\dot{\psi}\sin\theta) = e\dot{\psi}^2\cos\theta + e\ddot{\psi}\sin\theta,$$
$$r\ddot{\varphi} + 2\dot{r}\dot{\varphi} + \frac{c_1}{m}(r\dot{\varphi} + e\dot{\psi}\cos\theta) = e\dot{\psi}^2\sin\theta - e\ddot{\psi}\cos\theta. \quad (2.7)$$

The dimensionless form of system equations can be obtained after substituting $\omega_n^2 = k/m$, $\rho=r/e$, $D^*=c/(2m\omega_n)$:

$$\ddot{\rho} - \rho\dot{\varphi}^2 + 2D^*\left(\dot{\rho} - \frac{\dot{\psi}\sin\theta}{\omega_n}\right) + \rho = \omega_n^2\left(\dot{\psi}^2\cos\theta + \ddot{\psi}\sin\theta\right),$$
$$\rho\ddot{\varphi} + 2\dot{\rho}\dot{\varphi} + 2D^*\left(\rho\dot{\varphi} + \frac{\dot{\psi}\cos\theta}{\omega_n}\right) = \omega_n^2\left(\dot{\psi}^2\sin\theta - \ddot{\psi}\cos\theta\right). \quad (2.8)$$

For convenience of calculation, system equations can be expressed as follows:

$$\ddot{\rho} = -\left(1-\dot{\varphi}^2\right)\rho - 2D^*(\dot{\rho} - \dot{\psi}\sin\theta) + \left(\dot{\psi}^2\cos\theta + \ddot{\psi}\sin\theta\right),$$
$$\ddot{\varphi} = -\frac{2\dot{\rho}\dot{\varphi}}{\rho} - \frac{2D^*(\rho\dot{\varphi} + \dot{\psi}\cos\theta)}{\rho} + \frac{(\dot{\psi}^2\sin\theta - \ddot{\psi}\cos\theta)}{\rho}. \quad (2.9)$$

The solving process of these system equations of motion is shown as follows.

(a) Using the relationships $\kappa = \omega_0/\omega_n$, $\tau = \omega_n t$, $\lambda = \ddot{\psi}/2\omega_n^2$ and supposing the single rotor system is initially rotating at a uniform speed when $\tau = 0$, $\dot{\psi} = \omega_0$, the equations of motion of the disk can be expressed by the following equation:

$$m\ddot{r} + c_1\dot{r} + kr = F_0 e^{i\omega_0 t}, \quad (2.10)$$

where $F_0 = me\omega_0^2$.

The steady response (amplitude and phase angle) of this single rotor system can be obtained [20]:

$$r_0 = \frac{F_0/k}{\sqrt{(1-s^2)^2 + (2\xi s)^2}},$$
$$\theta_0 = \arctan\frac{2\xi s}{1-s^2}, \quad (2.11)$$

where $\sqrt{\xi} = c/2\sqrt{km}$, $s = \omega_0/\omega_n$.

The dimensionless form of response is

$$\rho_0 = \frac{\kappa^2}{\sqrt{(1-\kappa^2)^2 + (2\kappa D^*)^2}},$$
$$\theta_0 = \arctan\frac{2\kappa D^*}{1-k^2}. \quad (2.12)$$

Initially, $\dot{\psi}_0 = \kappa$, $\dot{\varphi}_0 = \kappa$, $\dot{\rho}_0 = 0$, $\psi_0 = 0$, $\varphi_0 = -\theta_0$.

(b) Equation (2.9) is solved after substituting the initial values. Then, $\ddot{\rho}_0$ and $\ddot{\varphi}_0$ are obtained.

(c) After all initial conditions become available, the next time step starts by calculating the variables as follows:

$$\dot{\rho}_1 = \dot{\rho}_0 + \Delta\tau\ddot{\rho}_0, \qquad \rho_1 = \rho_0 + \frac{\Delta\tau(\dot{\rho}_0 + \dot{\rho}_1)}{2},$$
$$\dot{\varphi}_1 = \dot{\varphi}_0 + \Delta\tau\ddot{\varphi}_0, \qquad \varphi_1 = \varphi_0 + \frac{\Delta\tau(\dot{\varphi}_0 + \dot{\varphi}_1)}{2}.$$

(2.13)

(d) Solving (2.9) by substituting $\dot{\psi}_i = \kappa + 2\lambda\tau$, $\psi_i = \kappa\tau + \lambda\tau^2$, and $\theta_i = \psi_i - \varphi_i$, then $\ddot{\rho}_1$ and $\ddot{\varphi}_1$ are obtained.

(e) Repeat steps (b)–(d).

Equation (2.9) is solved using the numerical time-marching method. For linear systems, the time step verifies <1/10f, where f is the highest frequency of the shaft rotation. However, due to the highly nonlinear property of the fluid-induced force, the time step used in the numerical computation is =10–6 s, which is significantly lower than the critical limit (10–3 s). This indicates that about 2000 calculations are made per rotation.

Turbulent Flow Model of the Annual Seal

With the increase of the rotor angular velocity, the small clearance flow in the centrifugal pump will grow from a laminar state to a transition regime and finally to a turbulent state. Proper models that simulate all three states flow should be selected to study the influence of the annular seal flow on the transient response of rotor systems through the start-up process. Based on the three turbulent theories and transition regime model presented by con and Roberts et al. [3, 15, 16, 19] numerical results were obtained and then compared with the experimental results.

Figure 2 is the schematic of the annular seal flow channel cross-section. The internal boundary has eccentricity e. Figure 2 shows the angular velocity of the internal and external boundaries ωj, ωb; the radius of the internal and external boundaries Rb and Rj, the external boundary center Ob, the internal boundary center Oj, the eccentricity of internal boundary e', and the local clearance of flow channel h. Suppose there is a laminar flow in this channel; then the Reynolds laminar equation can be used to simulate the flow [14]:

$$\frac{\partial}{\partial x}\left(\frac{h^3}{\mu}\frac{\partial p}{\partial x}\right)+\frac{\partial}{\partial z}\left(\frac{h^3}{\mu}\frac{\partial p}{\partial z}\right)=6(U_j+U_b)\frac{\partial h}{\partial x}+12\frac{\partial h}{\partial t}. \tag{2.14}$$

For the turbulent flow, the following general equation refers to the turbulent theories [3, 15, 16]:

$$\frac{\partial}{\partial x}\left(\frac{G_x h^3}{\mu}\frac{\partial p}{\partial x}\right)+\frac{\partial}{\partial z}\left(\frac{G_z h^3}{\mu}\frac{\partial p}{\partial z}\right)=\frac{U_j+U_b}{2}\frac{\partial h}{\partial x}+\frac{\partial h}{\partial t}, \tag{2.15}$$

where $h=(R_b-R_j)(1+\varepsilon\cos\varphi), U_j=R_j\omega_j\simeq R_b\omega_j, U_b=R_b\omega_b$. G_x and G_z are the turbulent factors and expressions given in (4.1)–(4.3).

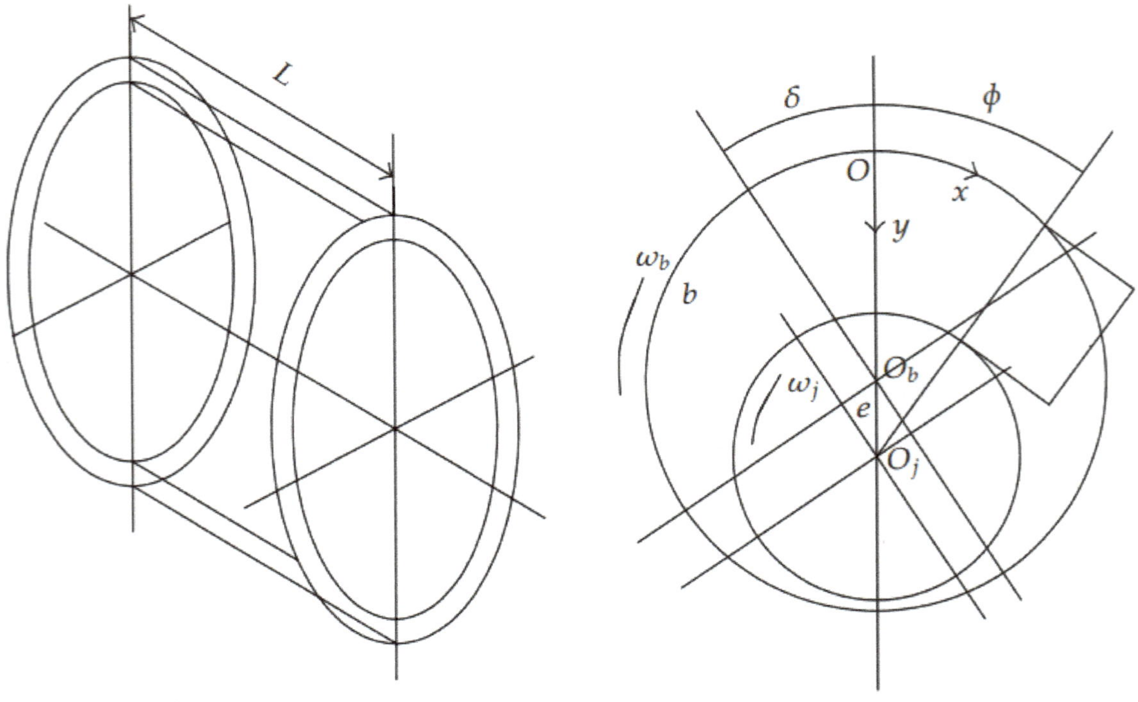

Figure 2: Schematic of the small clearance flow cross-section.

By using the dimensionless expression as follows, the dimensionless form of (2.15) can be obtained (refer to document) [21]:

$$\psi=\frac{R_b-R_j}{R_b},\quad \varepsilon=\frac{e}{R_b\psi},\quad \varphi=\frac{x}{R_b},\quad \bar{z}=\frac{2z}{B},$$

$$\bar{h}=1+\varepsilon\cos\varphi,\quad \bar{p}'=\frac{\psi^2}{6\mu\omega}p,\quad \lambda=\frac{B}{2R_b},$$

$$\frac{\partial}{\partial\varphi}\left(6G_x\bar{h}^3\frac{\partial\bar{p}'}{\partial\varphi}\right)+\frac{6G_z\bar{h}^3}{\lambda^2}\frac{\partial^2\bar{p}'}{\partial\bar{z}^2}=\frac{\partial\bar{h}}{2\partial\varphi}+\frac{1}{\omega}\frac{\partial\bar{h}}{\partial t}, \tag{2.16}$$

where $\omega=\omega_j+\omega_b$ and B is the length of the annular seal channel. Add $\bar{h}=1+\varepsilon\cos\varphi$ to (2.16):

$$\frac{\partial}{\partial\varphi}\left(G_x\bar{h}^3\frac{\partial\bar{p}'}{\partial\varphi}\right)+\frac{G_z\bar{h}^3}{\lambda^2}\frac{\partial^2\bar{p}'}{\partial\bar{z}^2}=-\frac{\varepsilon\sin\varphi}{12}+\frac{\varepsilon}{6\omega}\sin\varphi\frac{d\delta}{dt}+\frac{\cos\varphi}{6\omega}\frac{d\varepsilon}{dt}. \tag{2.17}$$

Suppose

$$\omega^*=\left(\omega-2\frac{d\delta}{dt}\right),\quad \bar{p}=\frac{\psi^2}{6\mu\omega^*}p; \tag{2.18}$$

then (2.17) can be further simplified:

$$\frac{\partial}{\partial\varphi}\left(G_x\bar{h}^3\frac{\partial\bar{p}}{\partial\varphi}\right)+\frac{G_y\bar{h}^3}{\lambda^2}\frac{\partial^2\bar{p}}{\partial z^2}=-\frac{\varepsilon\sin\varphi}{12}+\frac{\cos\varphi}{6\omega^*}\frac{d\varepsilon}{dt}. \tag{2.19}$$

Add $q=(2/\omega^*)(d\varepsilon/dt), k_{x1}=1/12G_x, k_{z1}=1/12G_z$; then the final form of the turbulent equation can be obtained:

$$\frac{\partial}{\partial\varphi}\left(\frac{\bar{h}^3}{k_{x1}}\frac{\partial\bar{p}}{\partial\varphi}\right)+\frac{\bar{h}^3}{k_{z1}\lambda^2}\frac{\partial^2\bar{p}}{\partial z^2}=-\varepsilon\sin\varphi+q\cos\varphi. \tag{2.20}$$

Equation (2.20) has a form similar to the dimensionless equation of the laminar model [21]. However, there is no turbulent factor k_{x1} and k_{z1} exists in laminar model. Thus, solving the laminar and turbulent models using one program is convenient. Furthermore, this program can also be used in solving the transition regime flow, except for the difference between the expressions of these models. The turbulent factors (k_{x1}, k_{z1}) presented here are different from the turbulent factors (k_x, k_z) in (3.3)–(4.3) and those presented by Roberts ($k_x=k_z=6c_fR_{em}$). The relationships between them are $k_{x1}=k_x/12$, $k_{z1}=k_z/12$.

FSI ANALYSIS METHOD

To accomplish the FSI, motion equations of a single rotor system and the turbulent lubrication equation must be solved separately. The deflection of the shaft under the influence of the fluid force can then be obtained through data transfer in each time step. Figure 3 shows the annular seal flow channel when considering fluid-induced forces. Fluid forces generated by the annular seal flow require the decomposition of the coordinate system to be fixed on the wall. Figure 4 shows the process of data transfer. Details of the calculation process are as follows.

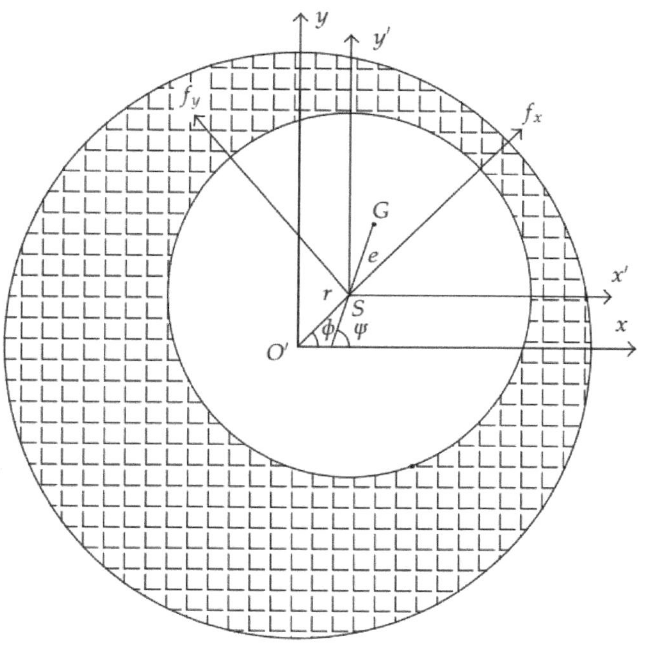

Figure 3: Schematic of the FSI theory.

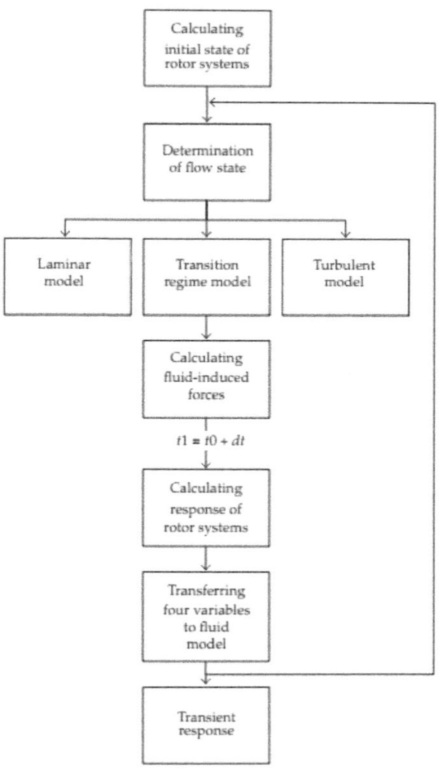

Figure 4: Comparison of the numerical results with the experimental results in the transition regime.

(a) Solve (2.12) to obtain the initial state of rotor systems based on boundary conditions.

(b) Select a proper model to simulate the annular seal flow based on the initial state of rotor systems. The fluid-induced forces f_x and f_y can then be obtained from the integration of pressure applied on the annular seal channel. These forces are introduced in (2.11):

$$m\ddot{x} + c_1(\dot{x} - e\dot{\psi}\sin\psi) + kx = me\dot{\psi}^2\cos\psi + me\ddot{\psi}\sin\psi + f_x\cos\varphi - f_y\sin\varphi,$$
$$m\ddot{y} + c_1(\dot{y} + e\dot{\psi}\cos\psi) + ky = me\dot{\psi}^2\sin\psi - me\ddot{\psi}\cos\psi - mg + f_x\sin\varphi + f_y\cos\varphi. \tag{3.1}$$

The dimensionless form of (3.1) is as follows:

$$\ddot{r} - r\dot{\varphi}^2 + \frac{k}{m}r + \frac{c_1}{m}(\dot{r} - e\dot{\psi}\sin\theta) = e\dot{\psi}^2\cos\theta + e\ddot{\psi}\sin\theta + \frac{f_x}{m},$$
$$r\ddot{\varphi} + 2\dot{r}\dot{\varphi} + \frac{c_1}{m}(r\dot{\varphi} + e\dot{\psi}\cos\theta) = e\dot{\psi}^2\sin\theta - e\ddot{\psi}\cos\theta + \frac{f_y}{m}. \tag{3.2}$$

From (3.2) we obtain

$$\ddot{\rho} = -1\left(1 - \dot{\varphi}^2\right)\rho - 2D^*(\dot{\rho} - \dot{\psi}\sin\theta) + \left(\dot{\psi}^2\cos\theta + \ddot{\psi}\sin\theta\right) + \frac{f_x}{me\omega_n^2},$$
$$\ddot{\varphi} = -\frac{2\dot{\rho}\dot{\varphi}}{\rho} - \frac{2D^*(\rho\dot{\varphi} + \dot{\psi}\cos\theta)}{\rho} + \frac{(\dot{\psi}^2\sin\theta - \ddot{\psi}\cos\theta)}{\rho} + \frac{f_y}{me\omega_n^2}. \tag{3.3}$$

Equation (3.3) can then be solved as (2.9).

(c) Eccentricity of the disk e, angular velocity of rotation $d\psi/dt$, angular velocity of whirling motion $d\varphi/dt$, and deflection velocity of the whirling motion dr/dt can be obtained from the results ρ, $\dot{\psi}$, $\dot{\varphi}$, and $\dot{\rho}$ calculated from (3.2). The four variables $e, d\psi/dt, d\varphi/dt$, and dr/dt will be introduced into the turbulent lubrication equations. Hirs' turbulent lubrication theory is used in this paper. By solving (2.6), fluid forces f_x and f_y are obtained.

(d) By repeating (b) and (c), the curves of the numerical solutions are obtained.

RESULTS AND DISCUSSION

Verification of the Rotor Dynamic Model

The dynamic model was verified based on the experimental results presented by Yanabe and Tamura [10] in 1971. Structural parameters of the experimental apparatus are shown in Figure 5. A rotating shaft supported on the ball bearings carries a single disk midway between them. The shaft with a diameter $d1$ of 17 mm has a span $L1$ of 520 mm. The disk has a diameter $D1$ of 120 mm, thickness T of 50 mm, and weight of 4.7 kg. This rotor is driven by a 0.5 sp motor with variable speed and is given a speed varying at 0–90 1/s.

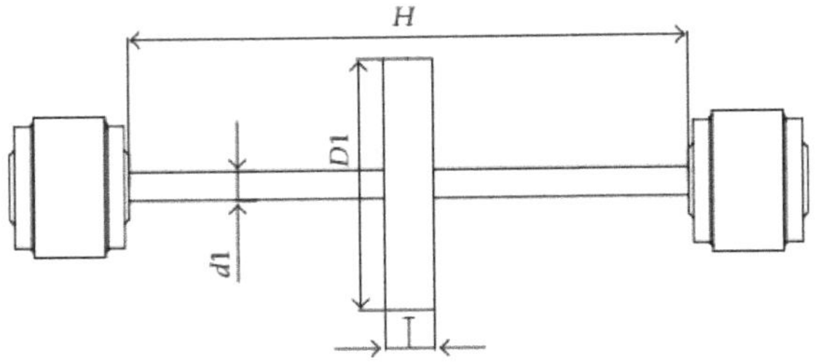

Figure 5: Schematic of the experimental apparatus.

The experimental results of the response were compared with the numerical solutions in Figure 6. There is a slight discrepancy between the experimental value and the calculated one. However, the periodicity of the beats was determined to be in good agreement. Figure 7 compares the rotating speed and the whirling angular velocity of the shaft. When acceleration is constant, the rotating speed increases linearly. Figure 7 also shows that the whirling angular velocity is divided into three regions. In the first region, the rotating speed is almost the same with the whirling angular velocity. However, in the second and third regions, the whirling angular velocity fluctuates around the rotating speed. In the second region, the whirling angular velocity is always smaller than the rotating speed, whereas it is always bigger in the third region at the peak value position. The second region starts from the rotating speed close to the critical speed of the rotor system. The peak value increases with the acceleration of the rotating speed, indicating that the backwards whirling motion may occur when rotor systems pass through the critical speed. Based on the results shown in Figure 7, the phase delay results in Figure 8 are also divided into three regions. In the second region, the phase delay starts to increase steadily because the whirling angular velocity is always smaller than the rotating speed and then fluctuates in the third region. The phase delay angle mainly depends on the damping ratio of the shaft materials. The damping ratio used in this paper was obtained from empirical data, which might not be identical to the damping ratio of the materials used in the experiment apparatus. There is a 4π phase angle difference between the numerical results and the experiment results.

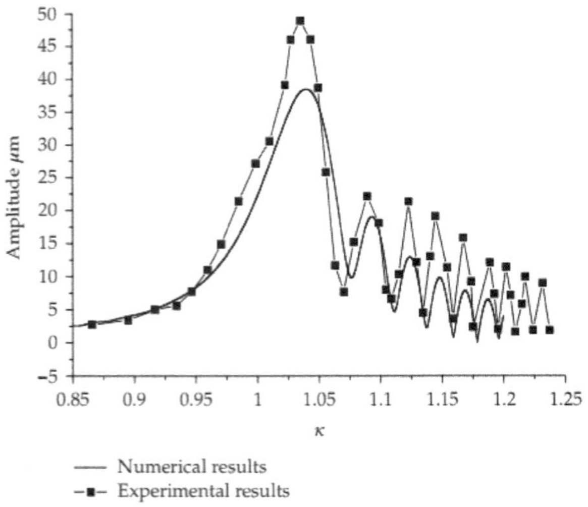

Figure 6: Comparison between the numerical and experimental results.

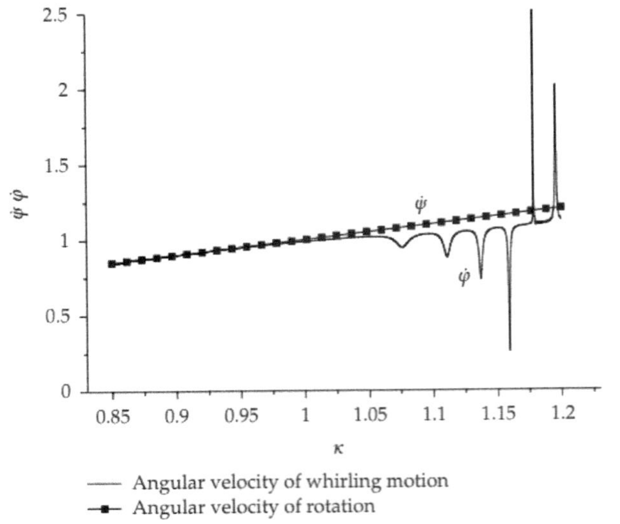

Figure 7: Comparison between the angular velocity of whirling motion and rotation.

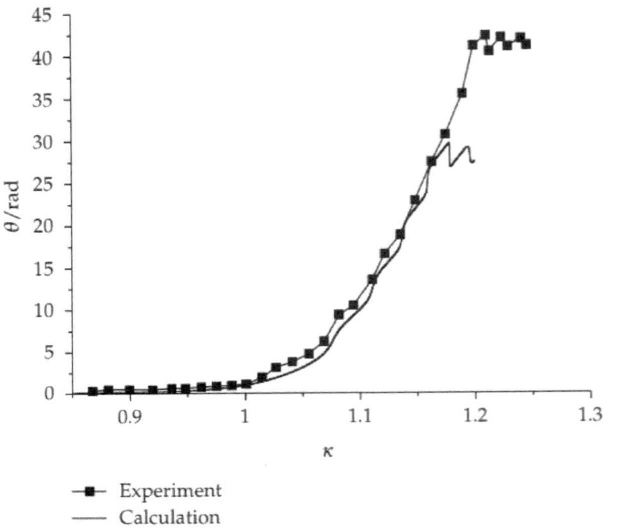

Figure 8: Comparison of the phase delay angle between the numerical and experimental results.

Verification of the Turbulent Flow Model

The turbulent flow model was verified based on the experimental apparatus built in the 1980s by Roberts and Hinton [22]. The annular seal passages used in this investigation were formed by a shaft and a plain cylinder around it. There is a central circumferential groove in the annular seal through which the medium is supplied. The shaft is made of mild steel, has a fixed axis of rotation, and is mounted on two ball bearings. Using this apparatus, Roberts carried out two experiments, namely, experiment a and experiment b. The parameters of the annular seal channel used in these two experiments were different. The structural parameters of this apparatus are shown in Figure 9. The radius of shaft D, radial clearance C, and length of clearance channel L in experiment a are 100, 0.55, and 33 mm, respectively. In experiment b, these three parameters are 120, 0.186, and 30 mm, respectively.

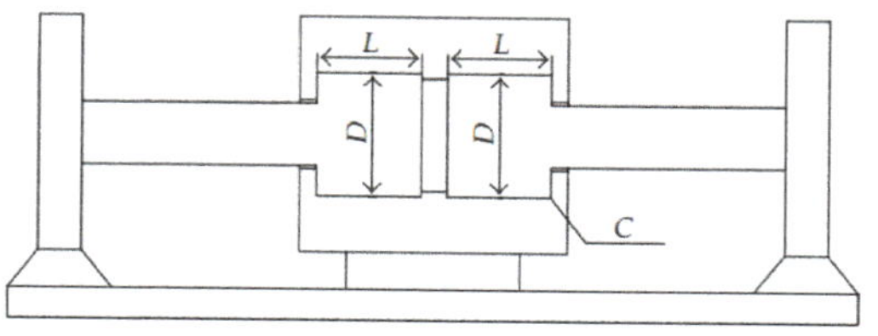

Figure 9: Structural parameters of the experimental apparatus.

First, a laminar model was used to simulate the annular seal flow in four different conditions. The numerical results were compared with the experiment results in Figures 10, 11, and 12. These four conditions correspond to the mean Reynolds numbers 17, 663, 1076, and 48600, respectively. Results were chosen from the pressure distribution at the center position of the annular seal channel in the circumferential direction. Boundary conditions corresponding to the numerical results shown in Figures 10–12 are provided in Table 1.

Table 1: Boundary conditions used in Figures 10–12

Figure no.	Experiments no.	Mediums	Rotating speed r/min	Eccentricity ratio ε	Normal Reynolds number R_{em}
10	b	tellus22	320 r/min	0.49	/
11	b	water	/	0.52	1076
12	a	tellus23	/	0.77	663
13	a	Water	/	0.91	48600

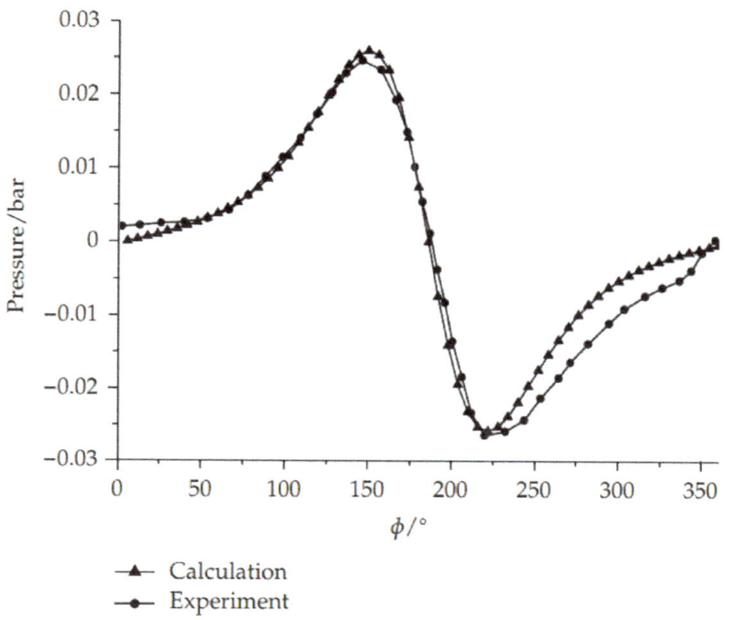

Figure 10: Comparison of the laminar results with the experimental results in the laminar state.

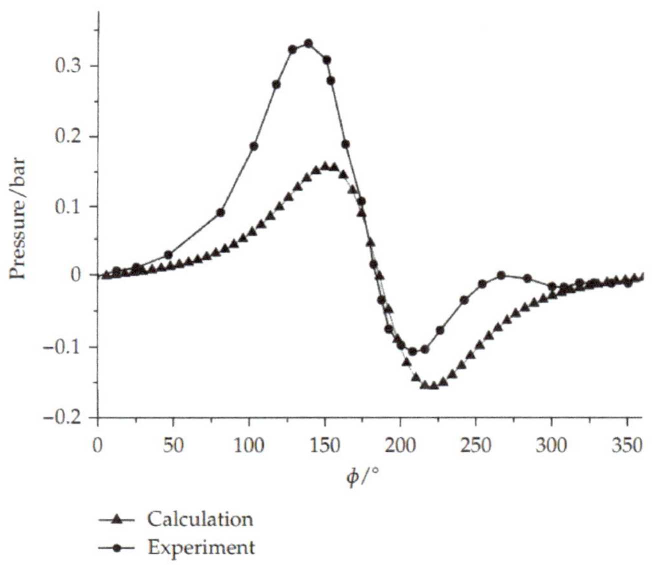

(a)

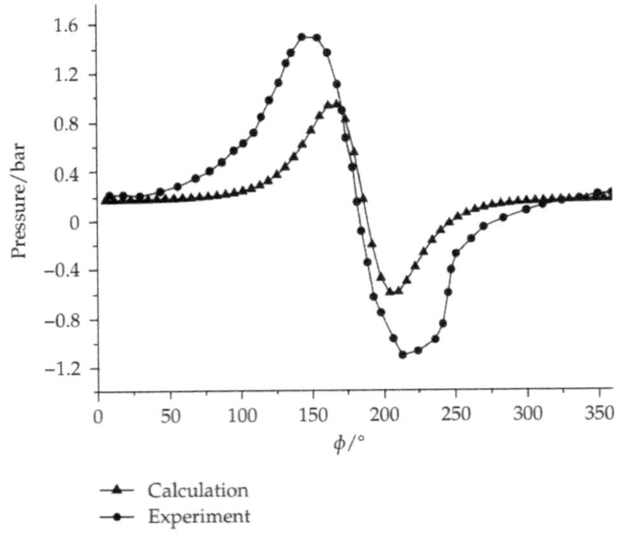

(b)

Figure 11: Comparison of the laminar results with the experimental results in the transition regime.

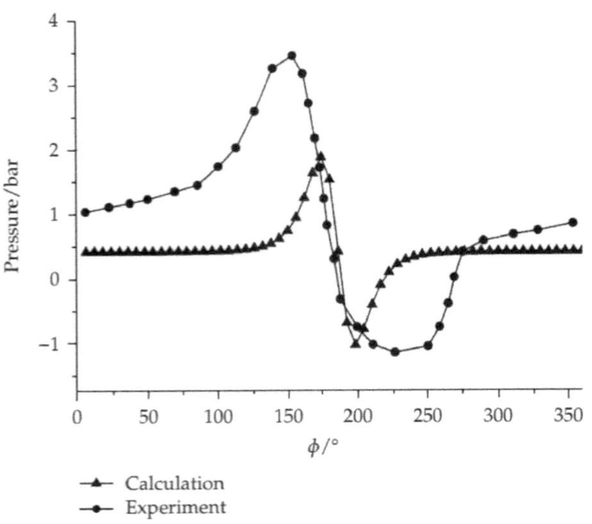

Figure 12: Comparison of the laminar results with the experimental results in the turbulent state.

The numerical results are shown to be in good agreement with the experiment results when the annular seal flow is at the laminar state. However, in the transition and turbulent states, the numerical results are no longer correct. This indicates that the laminar model is only applicable to the laminar state.

The turbulent theories presented by Con et al. [15] and the transition regime model presented by Roberts were then used to simulate the annular seal flow in the regime and turbulent states. Based on the finite difference method, the numerical results obtained by solving the equations of the three turbulent theories were compared with the experiment results in Figure 13. The numerical results have the same boundary conditions as those shown in Figure 12. Turbulent coefficients $G_{x\,z}$ of the three turbulent lubrication theories developed by Con, Ng-PAN, and Hirs, respectively, are presented in (4.1)–(4.3) [12]. The numerical results show better agreement with the experimental results compared with Figure 12, except that the absolute values of the numerical results are more or less still larger than those of the experimental results. Compared with the other two turbulent theories, Con's theory is clearly the best in simulating the annular seal flow:

$$k_x = \frac{1}{G_x} = 12 + 0.026R_e^{0.8265},$$

$$k_z = \frac{1}{G_z} = 12 + 0.0198R_e^{0.741}, \tag{4.1}$$

$$k_x = \frac{1}{G_x} = 12 + 0.0136R_e^{0.9},$$

$$k_z = \frac{1}{G_z} = 12 + 0.0043R_e^{0.96}, \tag{4.2}$$

$$k_x = \frac{1}{G_x} = 12 + 0.0687R_e^{0.75},$$

$$k_z = \frac{1}{G_z} = 12 + 0.0392R_e^{0.75} \tag{4.3}$$

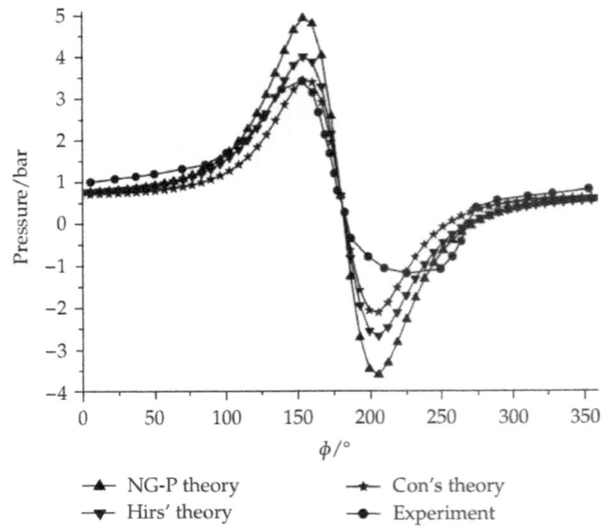

Figure 13: Comparison of the different turbulent theory results with the experimental results.

The annular seal flow is maintained in the transition regime flow state when the Reynolds number is between 500 and 2000. In this regime called the transition regime flow, Roberts presented the turbulent expression of the friction factor $c_f k_x = k_z = 6c_f R_{em}$, to simulate the transition flow. The c_f was then plotted against the Reynolds number based on the experimental results [19]. The turbulent factors k_x, k_z of the transition regime used in this paper were obtained by consulting Roberts' plots. The numerical results of the laminar theory, turbulent theory, and transition regime model were compared with the experimental results, respectively, in Figure 13. The experimental results and the boundary conditions of the numerical computation used in Figure14 are the same as those used in Figure 11(a). The results of the laminar and turbulent theories have greater absolute values at the hump position in either positive or negative region. By introducing the friction factor $c_f$in simulating the transition regime flow, the numerical results clearly improved compared with the other results based on the laminar or turbulent theories. However, there are still discrepancies between these results, indicating that the friction factor c_f presented by Roberts may need to be revised before it is used here.

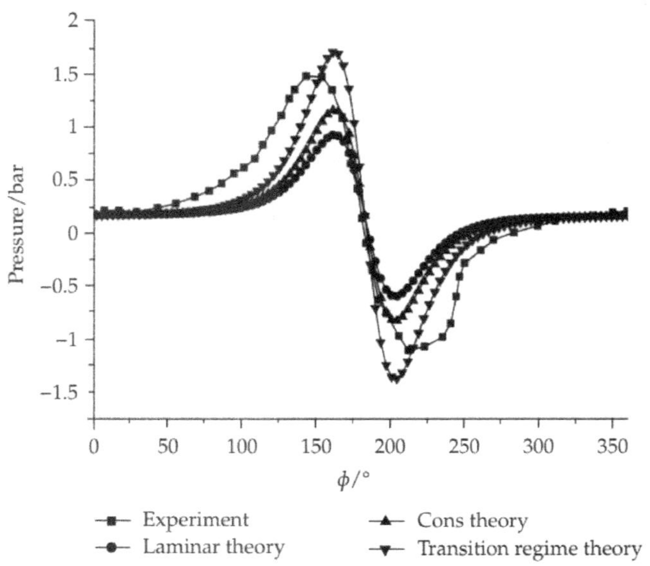

Figure 14: Comparison of the numerical results with the experimental results in the transition regime.

Results of the FSI

Based on the FSI method and related equations presented above, a single rotor system and annular clearance flow were modeled and solved. The transient response of the rotor system in the start-up process was obtained when considering the fluid-induced forces. Considering the nonlinear characteristics of the fluid-induced forces, the time step was set to 1e-6s to guarantee the accuracy of the results.

Unless specifically explained, the rotor system has the same structural parameters and boundary conditions as those in Section 2. Length of the annular seal is 33 mm, inner radius is 50 mm, outer radius is 50.5 mm, and the medium is water with viscosity of 0.001 Pa.s.

Figure 15 provides the transient response of the single rotor system. The amplitude of the response decreases, whereas the critical speed increases under the influence of the fluid-induced forces. This proves the existence of the supporting effects of the annular seal flow. Fluid-induced forces change with the increase in rotating velocity. Figure 16 shows that forces in the x direction are smaller by one order of magnitude than those in the y direction. Fluid forces in the y direction clearly have a shape similar to the curve in Figure 15, illustrating that the forces are mainly affected by the radial displacement of disk. The larger shaft deformation leads to higher fluid-induced forces, thus changing the critical speed corresponding to the largest shaft deformation of rotor systems. Moreover, the magnitude of fluid-induced forces is about 1 N, which leads to small effects of the annular seal flow on the vibration characteristics of rotor systems.

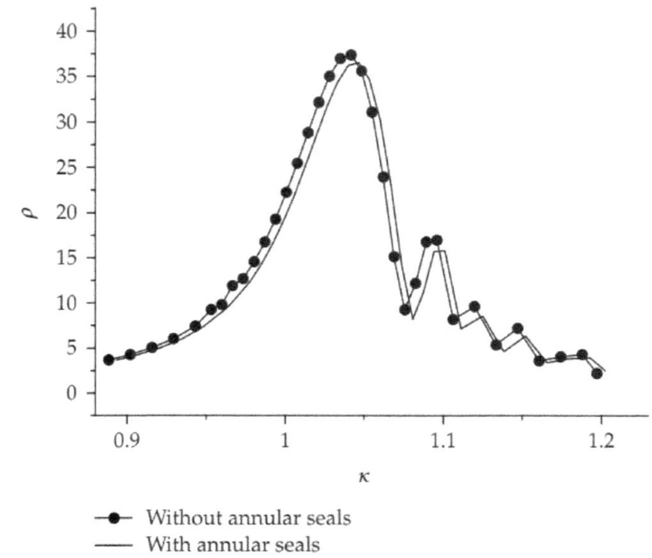

Figure 15: Transient response of the rotor systems under the influence of the annular seal flow.

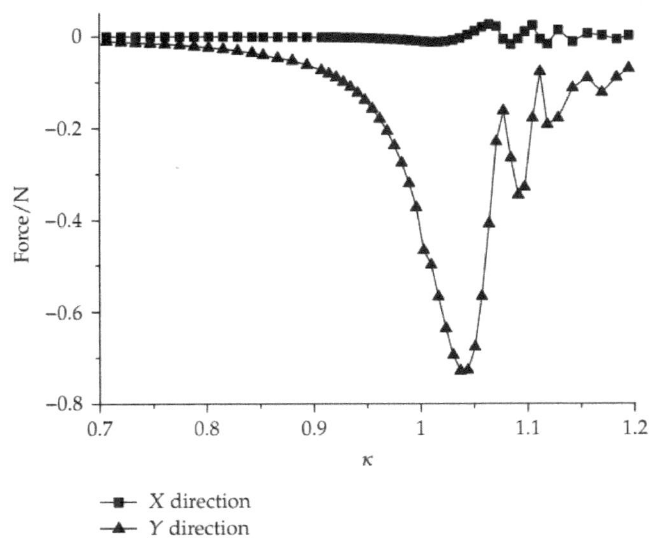

Figure 16: Fluid-induced forces in the x and y directions.

According to the study of Childs et al. [5, 6], annular seals play a supporting role similar to journal bearing in rotor systems. This can explain the increase of critical speed and the decrease of response in Figure 15. However, supporting effect is directly bound up with the annular clearance size and pressure drop of the annular seals. Because pressure drop between the two ends of the annular seals is relatively small in this paper, the increase of critical speed and the decrease of response are not very obvious.

Figures 17 and 18 show the variation of the whirling angular velocity, the phase delay of the whirling motion with the increase in rotating velocity under the influence of fluid-induced forces, and the comparison between these results when considering the annular seal flow or not. Figure 18 clearly shows that the whirling angular velocity, which considers the fluid-induced forces, is slightly smaller in most regions but larger in several peak value positions than that which does not consider fluid-induced forces. This result leads to two groups of phase delay, as shown in Figure 18, which are almost the same, indicating that the annular seal flow does not affect the damping ratio of rotor systems.

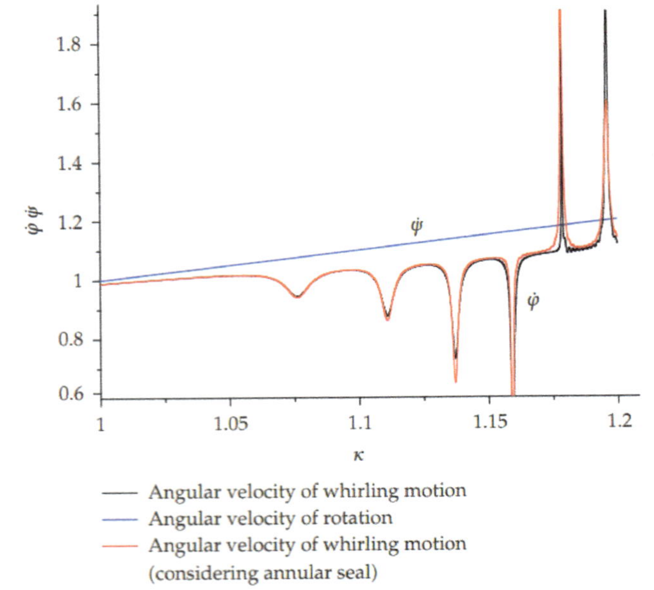

Figure 17: Influence of the annular seal flow on the angular velocity of the whirling motion.

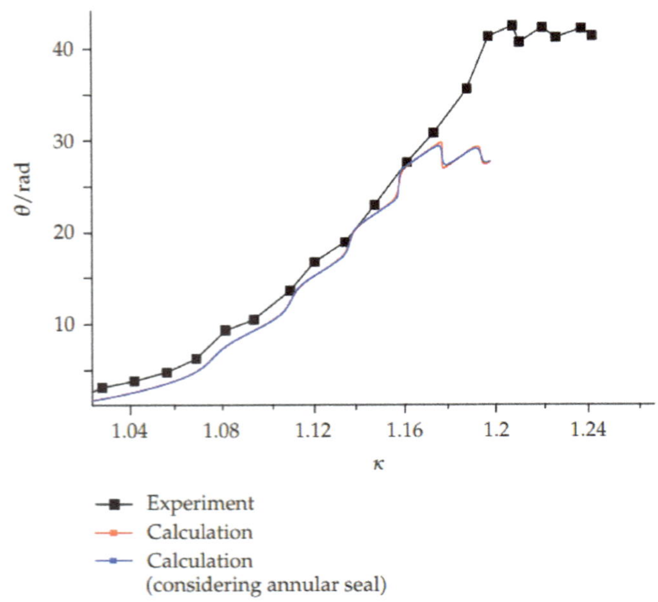

Figure 18: Influence of the annular seal flow on the phase delay angle.

Transient Effects of the Rotor System

To study the influence of the annular seal flow on a rotor system response in the transient process, three different accelerations of the rotating velocity and size of the annular clearance were solved.

First, three different accelerations (λ=0.00026, 0.00052, 0.00104) were selected to determine the influence of starting time on the vibration characteristics of rotor systems. Figure 19 shows that the amplitude clearly decreases when acceleration increases. The amplitude decreases from 37 μm to 18 μm when starting time increases from 0.184 s to 1.84 s, showing a decrease of 51%. The critical speed increases from 268.83 1/s to 303.73 1/s when acceleration increases from 0.00026 to 0.00104. This proves that acceleration increase can clearly postpone the rotating speed corresponding to the critical speed. Here, we apply the term critical speed to the rotating speed at which the maximum deflection occurs in a transient motion.

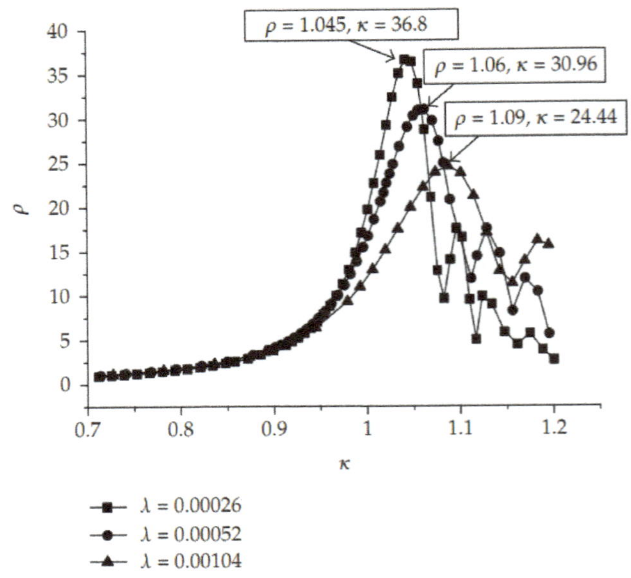

Figure 19: Comparison of the transient response under different accelerations.

Subsequently, the influence of fluid-induced forces on the single rotor system was studied by selecting three different clearances while setting the accelerating time to 0.184 s. Clearances are 0 (regardless of the fluid-induced forces), 0.45, and 0.85 mm, respectively. The numerical results are shown in Figure 20. The annular seal can clearly reduce the amplitude of a single rotor system. The smaller clearance results in a larger amplitude decrease. The amplitude decreases from 40 μm to 28 μm when the clearance is 0.45 mm compared with the results that disregard the fluid-induced forces. Figure 20 also shows that along with the decrease in clearance size, critical speed markedly increases as well. When the clearance was set to 0.45 mm, the critical speed increased to 314 1/s, which is a 10% growth rate compared with that not considering the annular seal flow.

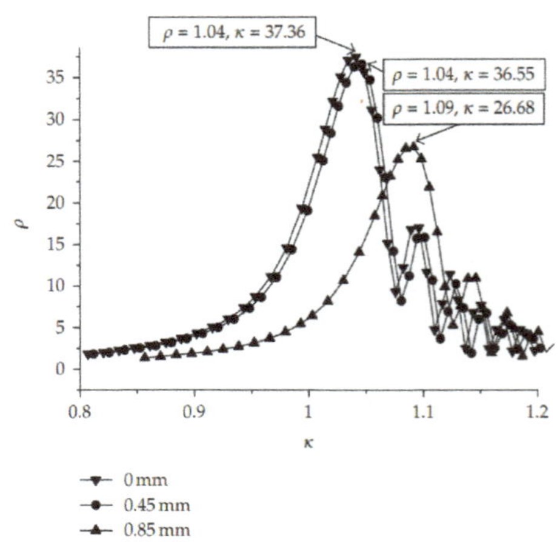

Figure 20: Comparison of the transient response of different seal clearances.

CONCLUSIONS

A single rotor system and an annular seal flow were modeled and verified. Based on these models, an FSI method was developed in this paper. The influence of the annular seal flow on rotor systems during the start-up process was studied through an FSI analysis. Compared with the traditional rotor dynamic analysis, the amplitude of the transient response decreases, whereas the critical speed increases when considering the annular seal flow. This indicates a supportive effect of the annular seal similar with bearings on the rotors. Moreover, the decrease in clearance size enhances the supportive effects. The amplitude and critical speed of the transient response were greatly influenced by the acceleration of the rotor system in the start-up process. The following are the useful results obtained in this paper.

- Con's turbulent theory and Roberts' transition regime model can be used to simulate the turbulent and transition regime flows in the annular seal of the centrifugal pump. By considering three different flow states existing in the annular seal, a whole start-up process of the centrifugal pump can be simulated in the calculating process.
- The transient response of rotor systems when considering the fluid-induced forces can be obtained by solving two domain equations and then transferring the data between these two domains. In the FSI calculation, rotor systems can be reduced to the node-element model, which does not affect the accuracy of the results.
- The fluid-induced forces generated by the annular seal flow can lead to a decrease in the response amplitude and increase in the critical speed of the centrifugal pump. Furthermore, these results can be enhanced by reducing the clearance size. The vibration characteristics of rotor systems improve when acceleration of the rotating speed is increased.

ACKNOWLEDGMENT

The support for this work by the Fundamental Research Funds for the Central Universities (no. KYJD09016) is gratefully acknowledged.

REFERENCES

1. W. Diewald and R. Nordmann, "Dynamic analysis of centrifugal pump rotors with fluid-mechanical interactions," Journal of Vibration, Acoustics, vol. 111, no. 4, pp. 370–378, 1989.
2. Z. X. Lu, "Numerical calculation of dynamic coefficients of angular seal ring," Chinese Journal of Applied Mechanics, vol. 12, no. 1, pp. 81–86, 1995.
3. G. G. Hirs, Fundamentals of a Bulk-Flow Theory for Turbulent Lubricant Films, University of Technology Delft, Delft, The Netherlands, 1970.
4. G. G. Hirs, "A bulk-flow theory for turbulence in lubricant films," Journal of Lubrication Technology, vol. 95, pp. 137–146, 1973.
5. D. W. Childs, "Dynamic analysis of turbulent annular seals based on Hirs' lubrication equation," Journal of lubrication technology, vol. 105, no. 3, pp. 429–436, 1983.
6. D. W. Childs and H. K. Chang, "Analysis and testing for rotordynamic coefficients of turbulent annular seals with different, directionally-homogeneous surface-roughness treatment for rotor and stator elements," Journal of Tribology, vol. 107, no. 3, pp. 296–306, 1985.
7. E. A. Baskharone, A. S. Daniel, and S. J. Hensel, "Rotordynamic effects of the shroud-to-housing leakage flow in centrifugal pumps," Journal of Fluids Engineering, vol. 116, no. 3, pp. 558–563, 1994.
8. L. Q. Wang, Z. F. Li, D. Z. Wu, et al., "Transient flow around an impulsively started rotating and translating circular cylinder using dynamic mesh method," International Journal of Computational Fluid Dynamics, vol. 21, no. 3-4, pp. 127–135, 2007.
9. Z. F. Li, D. Z. Wu, L. Q. Wang, and B. Huang, "Numerical simulation of the transient flow in a centrifugal pump during starting period," Journal of Fluids Engineering, vol. 132, no. 8, pp. 081102-1–081102-8, 2010.
10. S. Yanabe and A. Tamura, "Vibration of a shaft passing through a critical speed," Bulletin of the JSME, vol. 14, no. 76, pp. 1050–1057, 1971.
11. J. Richardet and G. A. Rieutord, "A three-dimensional fluid-structure coupled analysis of rotating flexible assemblies of turbo machines," Journal of Sound and Vibration, vol. 219, no. 1, pp. 61–76, 1998.
12. D. Lornage, E. Chatelet, and G. Jacquet-Richardet, "Effects of wheel-shaft-fluid coupling and local wheel deformations on the global behavior of shaft lines," Journal of Engineering for Gas Turbines and Power, vol. 124, no. 4, pp. 953–957, 2002.
13. G. I. Taylor, "Stability of viscous flow contained between two rotating cylinders," Philosophical Transactions of Royal Society, vol. 223, pp. 289–343, 1923.
14. C. M. Taylor and D. Dowson, "Turbulent lubrication theory-application to design," Journal of Lubrication Technology, vol. 96, no. 4, pp. 36–47, 1974.
15. V. N. Constantinescu, "On turbulent lubrication," Proceedings of The Institution of Mechanical Engineers, vol. 173, no. 38, pp. 881–889, 1959.
16. C. W. Ng and C. H. T. Pan, "A linearised turbulence to calculation of lubrication flows," Journal of Basic Engineering, vol. 87, no. 4, pp. 675–688, 1965.
17. H. F. Black and M. H. Walton, "Theoretical and experimental investigation of a short 360° journal-bearing in the transition superlaminar regime," Journal of Mechanical Engineering Science, vol. 16, no. 5, pp. 286–297, 1974.
18. P. J. Mason, An Investigation of Journal-Bearing Behavior in the Superlaminar Flow Regime, University of Sussex, Bridlington, UK, 1982.
19. J. B. Roberts and P. J. Mason, "An experimental investigation of pressure distributions in a journal bearing operating in the transition regime," Proceedings of Institution Mechanical Engineers, vol. 200, no. 4, pp. 251–264, 1986.
20. Y. Z. Liu, W. L. Chen, and L. Q. Chen, "Vibration

mechanics," High Education Press, pp. 31–32, 1998.

21. B. X. Chen, Z. G. Qiu, and H. S. Zhang, Fluid Lubrication Theories and Application, China Machine Press, Beijing, China, 1991.

22. J. B. Roberts and R. E. Hinton, "Pressure distributions in a superlaminar journal bearing," Journal of Lubrication Technology, vol. 104, no. 2, pp. 187–195, 1982.

CHAPTER 10

ON THE INTEGRATION OF LEADERSHIP IN ENERGY AND ENVIRONMENTAL DESIGN (LEED)®ND PROTOCOL WITH THE ENERGY PLANNING AND MANAGEMENT TOOLS IN ITALY: STRENGTHS AND WEAKNESSES

Giuliano Dall'O' [1], Annalisa Galante [1], Nicola Sanna [1], and Karin Miller [2]

[1]Architecture, Building Environment and Construction Engineering (A.B.C.) Department, Polytechnic of Milan, Via E. Bonardi 9, Milano 10133, Italy

[2]YR&G Sustainability Consulting, Education, and Analysis, Chicago, IL 60661, USA

ABSTRACT

Owners and municipalities face the choice to renovate or rebuild buildings in order to improve energy efficiency and sustainability. The process of upgrading the existing building stock can be supported by land management tools, *i.e.*, municipal energy plans (MEP) or sustainable building codes (BC), that municipalities use to plan, check and monitor actions taken. Many local and regional authorities are involved in the Covenant of Mayors supported by the European Union (EU): the sustainable energy action plan (SEAP) is the key document in which the Covenant signatory outlines how it intends to reach CO_2 reduction target by 2020. Sustainability at the urban scale is also supported by voluntary certification schemes such as Leadership in Energy and Environmental Design (LEED)® for neighbourhood development proposed by the United States Green Building Council (USGBC), Building Research Establishment Environmental Assessment Method (BREEAM) Communities proposed by UK Building Research Establishment (BRE) and Comprehensive Assessment System for Built Environment Efficiency (CASBEE) for Urban Development proposed by the Japan GreenBuild Council. The fact that there are so many tools to manage urban sustainability is positive, but what happens when multiple tools are applied to the same territory? Overlap, redundancy, and conflicts in management may be critical elements. The purpose of this article is to analyze these critical issues, highlighting the elements of integration and possible synergies for effective management of sustainability at the local level.

INTRODUCTION

The environmental quality of urban spaces is a necessity and at the same time, a great opportunity. In recent decades, the state of the environment has become a dominant theme, and many nations have signed international commitments to reducing green house gas (GHG) emissions, starting with the Kyoto Protocol [1]. Human-induced environmental change is not only a global issue but also a local issue; environmental problems become more critical in cities, as those spaces hold the greatest concentration of population. The latest edition of Demographia World Urban Areas [2] estimated that urban areas account for approximately 48% of the world's population. In some countries, as in the U.S., population density in city cores are decreasing [3], but the higher concentration of population

Citation: Giuliano Dall'O', Annalisa Galante, Nicola Sanna, and Karin Miller, On the Integration of Leadership in Energy and Environmental Design (LEED)® ND Protocol with the Energy Planning and Management Tools in Italy: Strengths and Weaknesses, Energies 2013, 6, 5990-6015; doi:10.3390/en6115990 in suburban areas brings about its own set of issues. The major causes of environmental impacts in urban areas can be linked to local traffic patterns (especially private transport) and building energy consumption related to heating ventilation and air-conditioning (HVAC) systems, domestic hot water (DHW) systems and electrical purposes. With regards to the European Union (EU), the Green Paper [4] estimated that the residential and tertiary sectors, the major part of which are buildings, account for more than 40% of final energy consumption in the EU and are expanding, a trend which is bound to increase energy consumption and hence carbon dioxide emissions.

In the EU, energy policies cover both new buildings and existing buildings. The European Directive 2002/91/EC [5], named Energy Performance Building Directive (EPBD), highlights the fact that buildings will have an impact on long-term energy consumption and new buildings should therefore meet minimum energy performance requirements, tailored to the local climate. The transposition of the above mentioned Directive in the Member States generated new regulations and laws aimed to significantly increase the energy performances of new construction. The more recent Directive 2010/31/EC [6] goes further, in terms of energy performances for buildings; by 31 December 2020, all new buildings must be nearly zero energy buildings, and after 31 December 2018, new buildings occupied and owned by Public Authorities must be nearly zero energy buildings.

A critical issue concerns the existing building stock. The energy performance of existing buildings is very low and an energy policy that focuses on turn-over (*i.e.*, replacement of existing buildings with new buildings) is not always realistic since this process would take too many years. EPBD states that "Major renovations of existing buildings above a certain size should be regarded as an opportunity to take cost-effective measures to enhance energy performance". A new effective strategy for improving the sustainability of urban areas should include actions aimed at retrofits on existing buildings and new rules for the design and construction of new buildings that should be, according to the above mentioned Directive 2010/31/EC nearly zero energy and, in the near future, zero energy. Actions to improve the environmental sustainability at the urban scale not only affect the buildings but also the urban context. The most appropriate means to act in a rational way are planning policies and programming.

Following the first energy crisis of 1973, many countries have adopted energy planning tools to support more traditional urban planning. At that time, energy issues were primarily concerns about energy security. Countries that have approved an energy strategy, through national energy plans (NEP) and regional energy plans (REP), are in most cases characterized by a reliance on imported energy from fossil fuels. The oil embargo thus had an undeniably positive effect of promoting energy efficiency and the use of renewable energy sources.

Since 1992, when the United Nations Conference on Environment and Development (UNCED) was held in Rio de Janeiro (Brazil), and up to the above-mentioned Kyoto Protocol, the environmental issue has become better defined. Countries are faced with both energy and environmental issues; the need to reduce energy consumption, and in many cases the import of fossil fuels, and the need to counter the effects of climate-change generated by the inefficient use of energy. Energy and environmental planning, starting with national and regional strategies such as the NEP and the REP, in many cases also concern the urban or municipal scale.

The municipal energy plans (MEP) are the tools that, supporting the more traditional town planning schemes (TPS), ensure that the development of urban areas contributes to improving the quality of urban life through a planned reduction of energy consumption. In many cases, energy planning is also supported by two important tools: the integrated plan of action (IPA), required managing land at the neighborhood level; and building codes (BC), tools that, among other things, define the rules for building activities and the minimum energy and environmental performance of new buildings and those subject to renovation. These instruments reflect the Italian situation, however many countries, not only European, adopt a similar structure.

The above mentioned energy planning instruments are managed by the public administration (the state, the regions, and the municipalities); rules and strategies, once approved and adopted, must be complied with by stakeholders (planners, designers, builders and citizens). Energy sustainability at the local level (e.g., municipal level), however, can be encouraged by initiatives that promote and stimulate a market that values the energy and environmental quality not only of buildings, but also of neighborhoods.

The paradigm in this case changes; the new rules are not followed because there is a law or regulation governed by the public authority, but as a voluntary choice, knowing that a building or neighborhood with greater energy and environmental quality has a higher market value. The rules of these voluntary approaches are defined by reference protocols developed by associations or non-governmental bodies. At the international level, the most diffused protocols are the Leadership in Energy and Environmental Design (LEED)® for Neighborhood Development rating system, or LEED® ND, proposed by the United States Green Building Council (USGBC) [7], the Building Research Establishment Environmental Assessment Method (BREEAM)® Communities rating system proposed by the Building Research Establishment (BRE) [8,9] and Comprehensive Assessment System for Built Environment Efficiency (CASBEE)® for Urban Development [10] proposed by the Japan GreenBuild Council.

The municipal framework for promoting sustainability in the EU is enriched by an important initiative, the Covenant of Mayors, adopted by many municipalities. The Covenant of Mayors is a voluntary commitment, made by local and regional authorities, to increasing energy efficiency and use of renewable energy sources within their territories. Through their commitment, Covenant signatories aim to meet and exceed the EU's objective to reduce CO_2 by 20% by 2020.

A sustainable energy action plan (SEAP) is the key document in which the Covenant signatory outlines how they intend to reach their CO_2 reduction target by 2020. The document defines the activities and measures set up to achieve the targets, together with time frames and assigned responsibilities.

The direction of development is moving towards an improvement in environmental sustainability. This change, almost unanimously considered necessary, covers all areas responsible for energy consumption, and more generally speaking, resources; the main areas being transport, industry and construction (residential and tertiary sectors) [11].

Energy and environmental planning tools were developed in order to provide concrete responses and approaches to issues related to the use of energy from fossil fuels and the reduction in GHG. The intent of the latest tools, such as LEED®ND and BREEAM® Communities, is to encourage a change in the housing market by stimulating an economic interest in the value of choices that lead to greater energy efficiency and improved sustainability. The fact that there are so many tools to manage urban sustainability is positive, but what happens when multiple tools are applied to the same territory? Overlap, redundancy, and conflicts in management between LEED and other tools may be critical elements. The purpose of this article is to analyze these critical issues, highlighting the elements of integration and possible synergies for effective management of sustainability at the local level.

In order to better emphasize the synergies and potential problems of use of LEED with the other methods that can arise from the application of several instruments in the same territory, a case study including an operational and practical approach will be used. The case study addresses Baranzate, a municipality of about 10,000 inhabitants, located on the outskirts of Milan (Italy), in which most of these tools have been applied or are in the implementation stage. This paper takes into account the framework of Italy, a nation where these environmental issues are beginning to emerge, the results and discussions of this work, however, have international value. The movement of the Covenant of Mayors, which generates the SEAP tool, is a movement promoted by the European Commission. The LEED® ND protocol, considered as reference point to evaluate the possible integration of the energy and sustainability planning tools, is an international rating system. The decision to analyze this instrument, instead of on BREEAM® Communities, is due to its greater flexibility for application outside the UK and due to the fact that the Italian GBC has developed GBC Neighborhoods, a version of LEED®ND tailored for a local application [12]. This paper provides an important contribution as it investigates the integration of instruments contributing to a new approach to energy and environmental management of the land at the urban scale.

ENERGY PLANNING, SUSTAINABLE AND CERTIFICATION PROTOCOLS FOR DISTRICTS: STATE OF THE ART

Energy Planning Tools at the Regional and Municipal Level

The topic dealt with in this paper is current and therefore, an object of study, research and analysis. Many of the considerations in this paper are inspired by works of several authors that, although within specific sectors, have contributed to the advancement of state of the art tools and methods for energy planning, as well as the role of protocols for the sustainable certification of neighborhoods.

A very detailed analysis of integrated planning in cities and territories is offered by the work of review published in the article by Mirakyan and De Guio [13]. The authors state that "although the integrated energy and environmental planning processes of cities and territories with more than 50,000 inhabitants differ, previous studies suggest that long-term, model-based energy planning processes have a common scheme that can also be used as a framework for reviewing the methods and the tools that are used in The Integrated Energy Planning (IEP) of these cities and territories".

The concept of IEP was proposed, during the 1970s by the International Energy Agency (IEA) in response to the oil crisis, to increase energy diversity and decrease dependence on foreign oil. Different IEP methodologies, including integrated resource planning (IRP), integrated assessment of supply and demand-side options (IASDO) and least-cost planning (LCP), have typically been practiced at the national level. The implementation of the IEP at a lower level (territory or city) in several countries has increased in conjunction with the liberalization of energy markets [14]. The European Commission has recognized the importance of using integrated approaches in the sustainable development of cities and territories [15]. From a methodological point of view, Mirakyan and De Guio [13] stated that the planning processes can be divided into four phases: preparation and orientations, model design and detailed analysis, prioritization and decision, implementation and monitoring. The authors analyzed other important aspects of the IEP, *i.e.*, participatory and governance levels, methods and tools, software resources, implementation and monitoring. Climate change, urban energy and planning practices in the Italian context are the subjects of research carried out by Zanon and Venores [16]. Energy planning and spatial planning in many cases are managed separately despite a logical potential for integration: the authors deal with this problem in a paper that focuses on the Italian case and analyses the possibility of integrating energy planning with spatial planning, the effectiveness of plan implementation mechanisms, and the prospect of integrating public-led interventions with market tools. The thesis of the authors is that public authorities need to make use of all the policy tools at their disposal, because energy efficiency and GHG emissions abatement can be attained only when the market incorporates some of the benefits (which are short-term for energy savings but long-term for climate change) and when the behaviors of consumers and citizens change so as to reduce consumption and contain mobility (mainly connected with residential well-being and urban space quality). We agree with the authors when they argue that these goals can be pursued only partially "by regulation". The results of the case studies presented, the MEPs of three cities located in northern Italy, highlight the partial inefficiency and ineffectiveness of the authoritative command-and-control instruments, as well as of the financial implementation strategies.

Brandoni and Polonara [17] investigated the role of MEP in the REP process with results derived from 12 MEP developed for urban areas located in Marche Region, in the centre of Italy. The results note that decentralized viewpoints, with respect to regional and municipal governments, are important in the energy planning process, contributing to the transition to a low-carbon society and increasing citizens' awareness of feasible alternatives to the current fossil fuel based energy system. The study highlights the importance of consistency between the different levels of energy planning, in this case, regional and municipal. The definitions of the strategies within the MEPs require knowledge of the energy and environmental quality of existing buildings stock. Several papers focus on this important aspect of energy planning, providing tools and methods of analysis. Dall'O', Galante and Torri [18] propose a methodology for the energy performance classification of residential building stock on an urban scale. The method, which is based mainly on information that municipalities already own, is tested on a medium-sized municipality located in the Lombardy Region (Italy). Within a few years, the market for high energy-performance buildings grew substantially; to date, nearly 7500 energy performance certificates for buildings of Class A and Class A+ have been issued [19].

The study developed by Caputo, Costa e Ferrari [20] describes a methodology able to characterize the energy performance of the built environment in a city or neighborhood and to evaluate the effects of different energy strategies. The methodology was designed to be a useful tool for energy policies at the urban level. More technical is the paper of Manfren, Caputo and Costa [21] that provides their contribution to the debate on urban-scale distributed power generation.

Planning for Sustainability and Certification Protocols for Neighborhoods

The topic of sustainability planning at the urban scale has been the subject of many works that highlight interest in an open debate. Most research addresses the issue of sustainability with reference to buildings and not the neighborhood scale (the issue of the sustainability of neighborhoods is more recent). Considerable interest in the topic of sustainability in buildings has generated many, perhaps too many, certification protocols and this can sometimes become a critical element. Haapio and Viitaniemi [22] present a review of building environmental assessment tools. The 16 assessment tools are analyzed as a group rather than as individuals considering all the possible aspects (*i.e.*, environmental, economic and social). The authors state that since the field of building environmental assessment tools is vast, the comparison of the tools and their results is difficult, if not impossible. The very detailed, accurate and complete work provides an interesting perspective for reflection. Insights and analysis of the environmental assessment tools, regarding buildings, are provided by other authors as Ding [23], Kyrkou and Karthaus [24], Alyami and Rezgui [25], Goldstein, Herbøl and Figueroa [26].

The issue of the classification systems of sustainability at the neighborhood scale is covered by Haapio [27]. The author makes a comparison between the international assessment tools for urban communities: LEED® for Neighborhood Development, BREEAM® Communities, and CASBEE® for Urban Communities. Table 1, taken from the author's paper, summarizes the categories of the three assessment tools. The author confirms that the interest in certification systems is increasing amongst authorities, and especially amongst global investors. Sustainable Neighborhood Development is the theme of research work presented by Miller [28], a thesis for a Master in Urban Planning and Policy Design, comparing U.S. and European approaches. According to the author, the U.S. approaches urban sustainability issues with a different perspective than most European countries. Wherein European planners are tasked with adaptive reuse of previously developed, historic sites and a very limited opportunity for entirely new developments, planners in the U.S. face a temporal, cultural challenge where new development is valued, space is not limited, and buildings have a short lifespan. The research work analyses and compares the latest case studies of sustainable neighborhoods, made in recent years in the U.S. and Europe, in which the rating systems LEED® ND and BREEAM®Communities were applied.

In our work, the comparison between U.S. and European approaches is useful in order to understand some critical aspects of the U.S. LEED® ND protocol that has been designed for a local context sometimes very different from the European one, although the Italian version of the protocol [12], partly takes these differences into account and makes it more suitable for application in the Italian context. Since GBC neighborhoods is perfectly compatible with LEED® ND, for convenience we will refer to it in the following as LEED® ND.

Table 1: Categories of the urban assessment tools (Haapio, [27]). LEED: Leadership in Energy and Environmental Design; BREEAM: Building Research Establishment Environmental Assessment Method; CASBEE: Comprehensive Assessment System for Built Environment Efficiency.

LEED for Neighbourhood Development	BREEAM Communities	CASBEE for Urban Communities
USA	UK	Japan
Smart location and linkage—favours development of cities and suburban areas. Development, revitalisation and services are important aspects. Protects areas, populations and water bodies. Neighbourhood pattern and design—emphasizes public transportation and reduction of auto dependency.Reaches for rich neighbourhood by increasing social interaction. Green infrastructure and buildings—devotes to decreasing environmental impact caused by construction and maintenance of buildings and infrastructure. Energy and water efficiency are emphasised. Additional categories: Innovation and design process; Regional priority (LEED, 2009); Criteria: 53 criteria. They are evaluated differently—some are worth 10 points, some only 1 point. From the main categories, the total is 100 points, and from additional categories it is possible to earn 10 extra points.	Climate and energy—focuses on reducing the project's contribution to climate change; Community—supports vibrant communities and encourages to integrate with surrounding areas; Place shaping—provides a framework for the design and layout of the local area; Ecology and biodiversity—aims at conserving the ecological value of the site; Transportation—focuses on sustainable transportation options, and encouraging walking and cycling; Resources—emphasize sustainable and efficient use of resources; Business—aims at providing opportunities for local businesses and creating jobs in the region; Buildings—focus on the overall sustainability performance of buildings. (BREEAM, 2009); Criteria: 51 criteria. All credits are equal. Values criteria from1 to 3 points.	$Q_{UD}1$—Natural environment (microclimates and ecosystems); $Q_{UD}2$—Service functions for the designated area; $Q_{UD}3$—Contribution to the local community (history, culture, scenery and revitalization); $LR_{UD}1$—Environmental impact on microclimates, façades and landscape; $LR_{UD}2$—Social infrastructure; $LR_{UD}3$—Management of the local environment (CASBEE, 2007); Criteria: 80 criteria. All credits are equal. Uses five-step scale based on the ratio of achieved and maximum points.

ENERGY PLANNING AT AN URBAN AND NEIGHBOURHOOD LEVEL IN ITALY

General Framework

In this section, we analyze the energy planning tools at an urban and neighborhood level in the Italian context. The purpose of the analysis is to verify understand how these tools interact with each other, both on a technical level and on the level of governance. Through this comparison, we can verify the synergy, the potential but also the weaknesses of the LEED® ND protocol when placed into a comprehensive framework for energy planning and sustainability. The diagram inFigure 1 shows the relationship between LEED® ND Protocol and other tools for energy and urban planning, and summarizes the concepts that we want to express through our analysis.

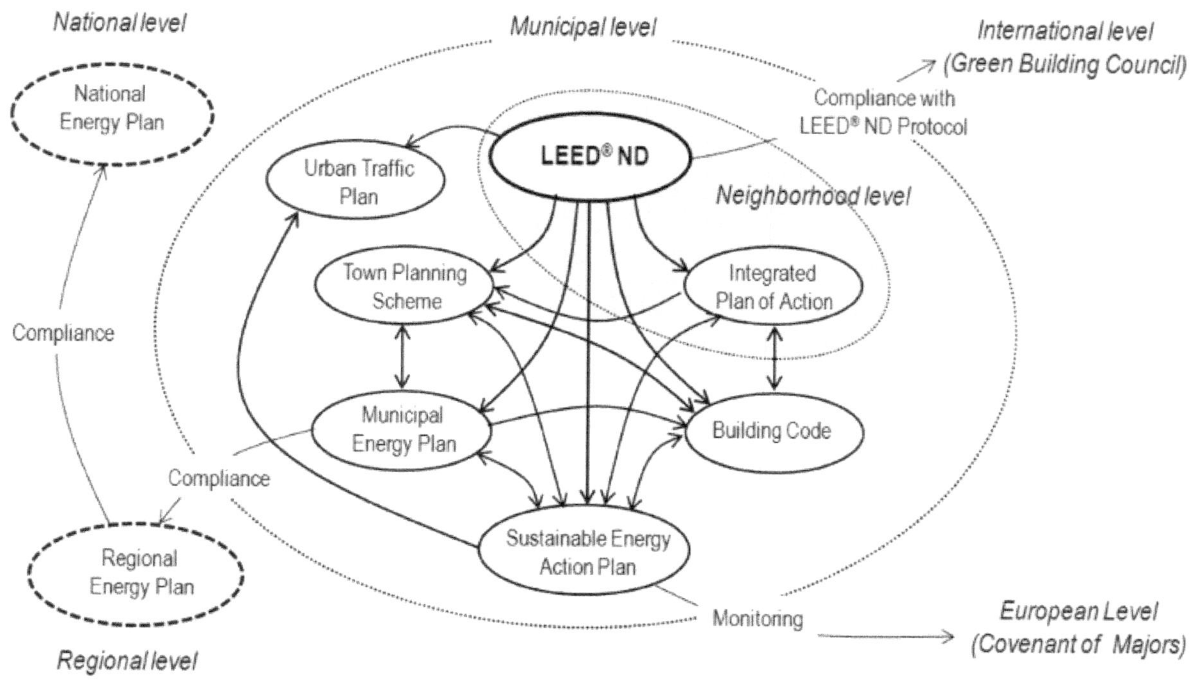

Figure 1: Relationship between LEED ND protocol and other tools for energy and urban planning. Arrows indicate the direction of the interaction.

Italian National energy and environmental policies are closely linked to strategies fostered through the EU directives to be transposed by the Member States and of course international agreements (e.g., the Kyoto Protocol). This general premise is very important in understanding the interactions between the various instruments of energy planning and programming.

Energy planning in Italy follows a top-down logic. Article 7 of Decree-law 112 Article 7 of Decree-Law No. 112 [29] introduces the concept of a "National Energy Strategy (SEN)" as a means of orientation and programming of national energy policy. The most recent document approved by the Italian Parliament is the SEN [30]. General objectives of the plan are to reduce energy costs, fully achieve and surpass all the European objectives for the environment, and promote a greater security of supply and industrial development of the energy sector.

The Italian Constitution considers energy as a competing matter between State and Regions: Regions, therefore, have autonomy in transposing European directives on energy issues; however, regional strategies must ensure compliance with the general principles established at national level. The REP have been introduced by law 10/91 [31], and are therefore mandatory. The same law requires municipalities with populations over 50,000 to develop a MEP, providing that there should be consistency between the MEP and the TPS.

At the Municipal level, in addition to MEP and TPS, there are other planning tools: the IPA, implementation tools that municipalities use to manage the development of suburban areas (neighborhood level), and the BC, rules defined and approved by the municipality for the construct buildings and urban infrastructure. Municipalities are allowed to propose BC more restrictive than those set by national or regional laws with regards to energy and sustainability issues. Some municipalities approved (by the town council) and adopted the SEAP within the European Covenant of Majors project: the SEAP must be consistent with other tools, in particular with the MEP and the BC. Since LEED® ND also considers aspects of the sustainability of traffic, it is important to compare the protocol with the planning tool for traffic management at the urban scale, in this case, the Urban Traffic Plan (UTP).

The scheme of Figure 1 shows that the LEED® ND Protocol cannot be considered an independent planning tool, since the choices made to obtain the desired level of sustainability interact with other energy planning and urban planning tools. As for the governance of the processes, Figure 1 highlights the many public institutions and organizations that are involved: the State for NEP, the Region for MEP, the Municipality for MEP, TPS, IPA, BC and SEAP, the European Commission with the technical support of the Joint Research Center (JRC) of The Institute for Environmental Protection and Research (ISPRA—Istituto Superiore per la Protezione e la Ricerca Ambientale) for SEAP and Green Building Council (GBC) for LEED® ND.

In the following section, we explore aspects of the tools of energy and environmental planning at the municipal scale, leaving LEED® ND for a more detailed discussion in the next section.

Municipal Energy Plans (MEP)

Article 5 of the Italian Law 10/91 [31] states that the TPS of municipalities with a population greater than 50,000 must include a specific plan at the municipal level concerning the use of renewable sources of energy, namely a Municipality Energy Plan. It is interesting to note that the legislature does not consider a MEP as independent tools but elements that are tightly integrated with urban planning (TPS). The typical objectives of a MEP are summarized as follows:

- acquire systematic data on the flows of energy within the municipal area, considered a snapshot of the energy situation to highlight any critical elements;
- define and plan actions for energy efficiency on the basis of a clear framework for evaluating each action, as economic resources but also as benefits for energy saving and reduction of environmental impact;
- monitoring, by means of dynamic indicators, the impact of the introduced measures through benefits: monitoring allows one to change the strategies in the event that the expected benefits are not achieved.

In Italy, there are 136 municipalities with a population greater than 50,000 inhabitants, constituting a population of about 21 million inhabitants, corresponding to 36% of the Italian population. The Italian municipalities that have approved a MEP, however, are many more. According to Cittalia [32], 3318 municipalities have approved a MEP, or about 41% of the 8093 Italian municipalities [33]. This demonstrates that other small municipalities that are not obligated by law also use the instrument.

Building Codes (BC)

The first genuine "sustainable" Italian BC was endorsed by the City of Carugate, a municipality near Milan (Italy), on 2003 [34,35]. The BC of Carugate introduced mandatory rules in order to guarantee high energy performance for new buildings and buildings subject to major renovation works for the first time. Among the new rules for energy and environmental performance of buildings, innovative at that time (the EPBD Directive had just been approved and not yet implemented in Italy), the most advanced were the following:

- optimization of the orientation of buildings in neighborhoods in order to take advantage of solar radiation for heating and lighting;
- maximum level of U-values for the building envelope (*i.e.*, external walls, windows, roofs and basement) were very low if compared with the standard used at that time;
- mandatory use of solar thermal for DHW (minimum integration of 50% on consumption required;
- mandatory efficient technologies for HVAC and DHW systems (e.g., condensing boiler, local thermal controls, *etc.*);
- recovery and reuse of stormwater.

The mandatory use of solar thermal systems for DHW, within the BC of Carugate, was inspired by the first European regulation, the Solar Ordinance of Barcelona (Spain) that came into force in 2000 [36]. Following the example of Carugate, many other municipalities in the Province of Milan adopted a "sustainable" BC. In order to define a similar standard, the Province of Milan published guidelines for sustainable building regulations [37]. In order to promote the implementation of the sustainable BC at national level, Dall'O' and Galante [38] defined a guideline taking into account the results of the most interesting case studies.

The BC, strongly oriented to the implementation of energy efficiency, renewable energy and, generally, the sustainability of the urban spaces, have found much success in Italy; the National Observatory Building Regulations Energy Saving in the recent report [39] estimated that in Italy, 1003 municipalities (about 10% of the total municipalities) have changed their BC in order to insert new energy and environmental criteria and objectives for improving the energy performance of buildings and the sustainability of the urban landscapes.

Sustainable Energy Action Plans (SEAP)

A SEAP is the key document in which the Covenant signatory (of the Covenant of Mayors) outlines how it intends to reach its CO_2 reduction target by 2020. It defines the activities and measures set up to achieve the targets, together with timeframes and assigned responsibilities. Covenant signatories are free to choose the format of their SEAP, as long as it is in line with the general principles set out in the Covenant SEAP guidelines. There are 4628 committed signatory cities involving a population of 167,145,782 citizens, there are 2861 currently adopted SEAP. The level of involvement of Italian municipalities and Italian communities is very high; 2446 committed signatory cities (representing 29,910,055 citizens), which corresponds to 51% of the Italian population. Italian submitted SEAPs are 63% of the whole project [40]. While the SEAP refers to a European dimension, and the MEP refers to the national dimension, the objectives of the two planning tools are similar: the reduction of GHG emissions through an action plan focused on the reduction of the use of energy from fossil fuels. The political approach however is different: it is worth remembering that in Italy, the MEP is mandatory for municipalities with a population higher of 50,000 inhabitants, while joining the Covenant of Mayors is voluntary. After more than twenty years from the publication of law 10/91 [31], 3318 municipalities have adopted a MEP, after only five years, 2446 cities are committed signatories of the Covenant of Mayors and that number is increasing significantly. The Covenant of Mayors project is a demonstration that actions from below, even on a voluntary basis, can sometimes be more successful than top-down actions. From the technical point of view, there may be significant overlapping elements between the MEP and the SEAP and the two instruments and the planned actions should be carefully assessed, to avoid redundancy or contradictions.

LEED ND® CERTIFICATION SCHEME

Structure of the Certification Scheme

The USGBC defines their vision of neighborhoods as "In basic terms, a neighborhood is an area of dwellings, employment, retail, and civic places and their immediate environment that residents and/or employees identify with in terms of social and economic attitudes, lifestyles, and institutions. By itself the neighborhood is a village, but combined with other neighborhoods it becomes a town or a city. Similarly, several neighborhoods with their centers at transit stops can constitute a transit corridor. The neighborhood, as laid out in LEED® ND, is in contrast to sprawl development patterns, which create pod-like clusters that are disconnected from surrounding areas" [7].

LEED® ND emphasizes the creation of compact, walkable, vibrant, mixed-use neighborhoods with good connections to nearby communities. In addition to neighborhood morphology, pedestrian scale, and mix of uses, the rating system also emphasizes the location of the neighborhood and the performance of the infrastructure and buildings within it. Together, well-located and well-designed green neighborhood developments will play an integral role in reducing GHG emissions and improving quality of life.

GBC Italia in 2013 has defined a certification protocol named GBC Neighborhoods [12]. The effort of GBC Italia was to verify the compatibility of the regulatory and legislative of the protocol which is strongly inspired by LEED® ND, in the Italian context. The structure of GBC Neighborhoods is basically identical to that of LEED® ND.

GBC neighborhoods considers a certifiable area an area that possesses a minimum of two buildings which constitutes a set of relationships that will contribute to creating a functional and social mix, and the characteristics of a permanent settlement. Sustainable neighborhoods are places where there are buildings, infrastructure and residents, where the service functions and workplaces are walkable. The projects being evaluated may also be single-purpose small interventions to complement existing neighborhoods that integrate relationships and quality of services to the residents.

GBC Neighborhoods is broken down into the three categories of Smart Location & Linkage (SLL), Neighborhood Pattern and Design (NPD), and Green Infrastructure & Buildings (GIB), plus additional points for Innovation & Design Process (IDP) and Regional Priority Credits (RPC).

The following sections analyses the Prerequisites and Credits provided by GBC Neighborhoods and related to their possible applicability to existing planning instruments. Specifically, we identified, based on the experience of the authors, the following types of interactions:

- "Binding Rule—Direct influence" or the Prerequisite or Credit can be entered as a binding rule, therefore directly affecting the planning instrument;
- "Binding Rule—Indirect influence" or the Prerequisite or Credit cannot be inserted into the planning instrument, having an indirect impact on it;
- "Optional potentially incentivized rule—Direct influence", Credit can be entered as an optional rule, therefore having a direct bearing on the planning instrument, municipalities to decide whether to apply an incentive scheme, e.g., the acquisition of more points for the effective achievement of the objectives laid down by specific Credit;
- "Optional potentially incentivized rule—Indirect influence", Credit is not entered directly into the planning instrument, but its application affects it indirectly.

To better explain, let's take an example of the Credit "Location with Reduced Automobile Dependence". This can certainly be entered as a binding rule with direct influence on IPA, as it can be required by municipalities for a given neighborhood. For the TPS, the optional rule can be applied, since it can be extended to the whole territory, but if it is applied (for example with an incentive scheme related to the acquisition of one to seven points) it has a direct influence on everything within the TPS from the urban point of view. Similarly, it can be considered an optional rule for both the UTP and the SEAP since it has direct influence on both transport and the built environment. When applied to the MEP, it becomes an Optional Rule indirect influence because the criteria cannot be inserted in the Plan, but it is possible application has an indirect influence on the instrument, through the positive effects on the reduction of emissions.

Smart Location & Linkage (SLL): Where to Build

SLL is characterized by five prerequisites and nine additional credits—the prerequisites include Smart Location, Imperiled Species and Ecological Communities, Wetland and Water Body Conservation, Agricultural Land Conservation, and Floodplain Avoidance. The credits in this category are generally based on smart growth design principles, focusing on site selection. Emphasis within the credits is placed on Preferred Locations and Locations with Reduced Automobile Dependence [28].

Neighborhood Pattern and Design (NPD): What to Build

Neighborhood Pattern and Design is highly influenced by New Urbanist principles, with the credits of Walkable Streets, Mixed-Income Diverse Communities, and Compact Development being given the highest point allotment and also being prerequisites for the category [28].

Green Construction and Technology (GCT): How to Manage Environmental Impacts

GCT is the last standard LEED® ND category. The prerequisites in this category are a Certified Green Building, Minimum Building Energy Efficiency, Minimum Building Water Efficiency and Construction Activity Pollution Prevention. Credits in this category most closely align with the other LEED® programs, emphasizing especially, Certified Green Buildings, Stormwater Management, and On-site Renewable Energy Sources [28].

Innovation & Design Process (IDP) and Regional Priority Credits (RPC)

Additional points can be achieved for both IDP and RPC. Innovation & Design (I&D) points are awarded to projects implementing a sustainable practice that doesn't fall under the three primary categories, for example, one point of I&D is designated for the inclusion of a LEED® Accredited Professional on the project design team. The RPC are defined by regional committees to place emphasis on particular credits falling under the three primary categories that are particularly relevant to that region. A project will get additional points for choosing to implement a credit that has been selected for regional importance in the project area [28].

ANALYSIS OF THE CASE STUDY

Description of the Urban Context

The case study analysed is the town of Baranzate, a medium-small municipality located north-east of the Province of Milan (Italy) (Lat. 45°31'37" Nord, Long. 9°06'51" East). Baranzate has a population of 11,448 inhabitants [41], the average population of a typical Italian municipality, and a total area of 2.78 km^2, with a population density of 4117.99 4 inhabitant/km^2, about twice the density of the Province of Milan that is 2023.84 inhabitants/km^2. The climate is temperate [2404 Winter Degree Days (DDH)], sub-continental, is characterized during the year as follows: six months temperate (from March to June and from September to October), four months cold and wet (from November to February) and two hot and humid months (July and August). The average annual air temperature is equal to 13.2 °C with an average temperature range (difference between the mean temperature of the warmest month, July, and the most cold-January) of 21.3 °C. The design air temperature for the heating systems is −5 °C while the average value of the external air temperature during the winter season is 8.4 °C.

The urban context is characterized by the presence of many industrial and service sectors areas, some of which, partly because of the economic crisis, are no longer used. From the social point of view, the Municipality of Baranzate is an interesting and unique case, considering the Italian context. The population is characterized by a strong presence of foreign nationals, average to lower-middle class workers, who have created communities through which integration and social cohesion are sought. The municipal government has proposed, in recent years, a policy strongly oriented towards social cohesion but also to the improvement of energy and environmental sustainability. All the tools of sustainable planning have been activated:

- the approval of a TPS that aims to recover and enhance unused urban areas;
- the approval of a BC which requires high energy quality standards for new buildings (energy class A) and for buildings subject to major renovation;
- the implementation of a UTP that addresses sustainable mobility;
- signatory to the Covenant of Mayors and the approval of an SEAP;
- the implementation of a MEP in order to reduce energy consumption and GHG emissions.

The political and administrative context of Baranzate seemed ideal to test a possible implementation of the LEED Protocol and its integration with the energy planning and management tools approved or in the approval process.

Possible Implementation of LEED® *ND Protocol*

Figure 2 shows the general layout of the town highlighting the areas in which urban redevelopment is scheduled within the TPS. The first analysis step is to identify areas that may be certified using the LEED® ND Protocol.

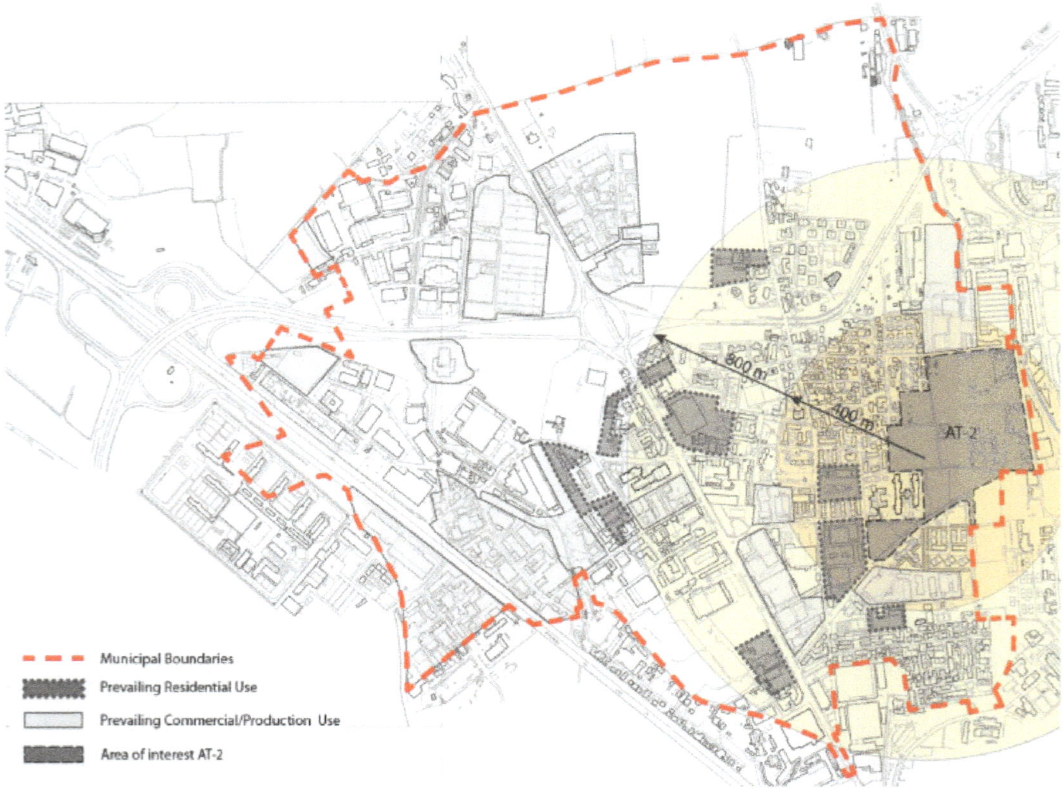

Figure 2: General layout of the town highlighting the areas in which the urban redevelopment is scheduled within the Town Planning Scheme (TPS).

Within the TPS, that is the framework for the urban programming and development, it is possible to define particular areas in which to apply the IPA. Figure 2 highlights the IPA of Baranzate by means of grayscale areas.

The areas highlighted in light gray are characterized by the presence of industries and commercial areas; these areas are not compatible for LEED® ND certification. The areas highlighted in dark gray, however, have a strong presence of residential and therefore could be considered. An area located to the East of the city was chosen for the test of a possible application of the LEED® ND Protocol, shown in larger scale in Figure 3.

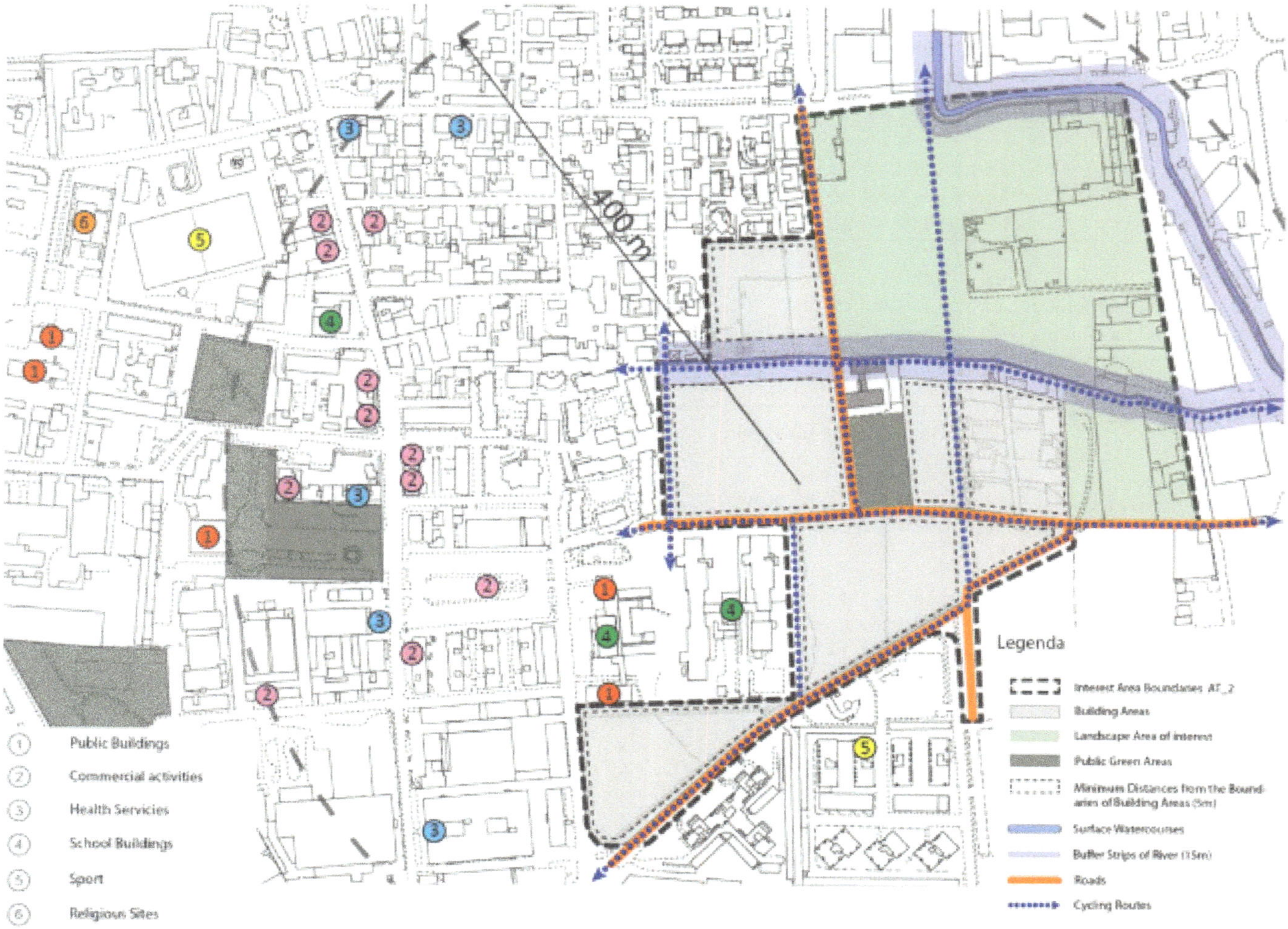

Figure 3: Characteristics of the urban area covered by the application of the LEED®ND Protocol.

For the area under consideration, all the provisions contained in the planning and sustainable tools already approved or in the approval phase (TPS, IPA, MEP, UTP and SEAP) were carefully analyzed in order to verify the compatibility and compliance with the prerequisites and credits of the LEED® ND certification scheme. This analysis and comparison, based on a concrete case, was very useful for processing Table 2, Table 3 and Table 4.

Table 2: Relationship between prerequisites and credits of "Smart Location & Linkage (SLL)" category and tools for energy and urban planning. Legend: ■ Binding Rule—Direct Influence; □ Binding Rule—Indirect Influence; • Optional potentially incentivized rule—Direct Influence; ○ Optional potentially incentivized rule—Indirect Influence. IPA: Integrated Plan of Action; MEP: Municipal Energy Plan; BC: Building Code; UTP: Urban Traffic Plan; SEAP: Sustainable Energy Action Plan.

SLL—Pre-Requisites and Credits	Tools for Energy and Urban Planning					
	TPS	IPA	MEP	BC	UTP	SEAP
PR01—Smart Location	■	■	-	□	□	-
PR02—Imperiled Species and Ecological Communities	■	■	-	-	■	-
PR03—Wetland and Water Body Conservation	■	■	-	□	-	-

SLL—Pre-Requisites and Credits	Tools for Energy and Urban Planning					
	TPS	IPA	MEP	BC	UTP	SEAP
PR04—Agricultural Land Conservation	■	□	-	-	□	-
PR05—Floodplain Avoidance	■	□	-	-	-	-
CR01—Preferred Location	•	•	-	-	○	-
CR02—Brownfield Redevelopment	•	■	-	-	-	○
CR03—Location with Reduced Automobile Dependence	•	■	○		•	•
CR04—Bicycle Network and Storage	□	■	-	■	■	□
CR05—Housing and Job Proximity	■	■	-	-	□	□
CR06—Steep Slope Protection	•	■	-	-	-	-
CR07—Site Design for Habitat or Wetland and Water Body Conservation	•	■	-	-	-	-
CR08—Restoration of Habitat or Wetlands and Water Bodies	■	■	-	□	-	□
CR09—Long-Term Conservation Management of Habitat or Wetlands and Water Bodies	■	□	-	■	-	□

Table 3: Relationship between prerequisites and credits of "Neighborhood Pattern and Design (NPD)" category and tools for energy and urban planning. Legend: ■ Binding Rule—Direct Influence; □ Binding Rule—Indirect Influence; • Optional potentially incentivized rule—Direct Influence; ○ Optional potentially incentivized rule—Indirect Influence.

NPD—Pre-Requisites and Credits	Tools for Energy and Urban Planning					
	TPS	IPA	MEP	BC	UTP	SEAP
PR01—Walkable Streets	-	■	-	■	□	-
PR02—Compact Development	■	■	-	□	-	-
PR03—Connected and Open Community	■	■	-	□	■	□
CR01—Walkable Streets	□	■	-	■	■	-
CR02—Compact Development	•	■	-	-	○	○
CR03—Mixed-Use Neighborhood Centers	•	■	-	-	-	•
CR04—Mixed Income Diverse Communities	•	•	-	•	-	•
CR05—Reduced Parking Footprint	□	■	-	■	■	□
CR06—Street Network	•	■	-	-	■	-
CR07—Transit Facilities	■	■	-	■	■	□
CR08—Transportation Demand Management	○	○	○	-	•	○
CR09—Access to Civic and Public Spaces	•	•	-	○	○	-
CR10—Access to Recreation Facilities	•	•	-	○	○	-
CR11—Visitability and Universal Design	•	■	-	•	-	-
CR12—Community Outreach and involvement	-	•	-	-	-	•
CR13—Local Food production	•	■	-	-	○	•
CR14—Tree-Lined and Shadowed Streets	•	•	-	•	•	-
CR15—Neighborhood Schools	•	■	-	•	○	-
CR16—Acoustic Climate (Only GBC Neighborhood)	-	■	-	■	□	-

Table 4: Relationship between prerequisites and credits of "Green Construction and Technology (GCT)" category and tools for energy and urban planning. Legend: ■ Binding Rule—Direct Influence; □ Binding Rule—Indirect Influence; • Optional potentially incentivized rule—Direct Influence; ○ Optional potentially incentivized rule—Indirect Influence.

GCT—Pre-Requisites and Credits	Tools for Energy and Urban Planning					
	TPS	IPA	MEP	BC	UTP	SEAP
PR01—Certified Green Buildings	□	■	■	■	-	□
PR02—Minimum Building Energy Efficiency	□	■	■	■	-	■
PR03—Minimum Building Water Efficiency	-	■	■	■	-	-
PR04—Construction Activity Pollution Prevention	□	■	-	■	□	-
CR01—Certified Green Buildings	○	•	○	○	-	○
CR02—Building energy Efficiency	○	•	•	•	-	•
CR03—Building Water Efficiency	-	•	•	•	-	-
CR04—Water-Efficient Landscaping	-	■	-	■	-	□
CR05—Existing Building Use	○	■	•	•	-	•
CR06—Historic Resource Preservation and Adaptive Reuse	■	■	-	■	-	-
CR07—Minimized Site Disturbance in Design Construction	□	■	-	■	-	□
CR08—Stormwater Management	-	■	-	■	-	-
CR09—Heat Island Reduction	□	■	■	■	-	□
CR10—Solar Orientation	-	•	•	•	-	-
CR11—On-Site Renewable Energy Sources	-	■	■	■	-	■
CR12—District Heating and Cooling	-	•	•	•	-	•
CR13—Infrastructure Energy Efficiency	-	■	■	-	■	■
CR14—Wastewater Management	-	•	•	•	-	-
CR15—Recycled Content in Infrastructure	-	•	-	•	•	•
CR16—Solid Waste Management Infrastructure	-	•	•	•	-	•
CR17—Light Pollution Reduction	□	■	■	■	□	□

In the area covered by the study, shown in Figure 2, three scenarios were considered:

- *Baseline scenario*: the prerequisites and credits in the LEED® ND were compared with the actions already provided for within the energy and environmental planning tools;
- *Silver scenario*: the energy and environmental planning tools already implemented were integrated with additional rules (prerequisites and credits) in order to achieve the Silver certification of the LEED® ND Protocol;
- *Gold scenario*: the energy and environmental planning tools already implemented were integrated with additional rules (prerequisites and credits) in order to achieve the Gold certification of LEED® ND Protocol.

The results of the three simulations are shown in Table 5, which highlights the scores achieved in the three areas of evaluation (SLL, NPD and GCT). The baseline scenario shows the impossibility of obtaining the LEED® ND certification, as the rules contained in the actual planning tools do not comply with the prerequisites that are mandatory in the Protocol. By implementing the existing planning tools with a few additional rules, and limited effort for the public administration, it is possible to aspire to the Silver certification. There is still a margin, and the need for other additional rules, to get certified Gold.

Table 5: Results of application of LEED ND to Baranzate municipality. SLL: Smart Location & Linkage; NPD: Neighbourhood Pattern and Design; GCT: Green Construction and Technology.

Rating scenario	Rating				Additional effort
	SLL	NPD	GCT	Total (*)	
Baseline	10	13	3	26	NA
Silver	11	32	9	52	low
Gold	13	34	15	62	medium

(*): The credits for Innovation & Design Process (IDP) and Regional Priority Credits (RPC) were not considered in this simulation.

RESULTS AND DISCUSSION

The aim of the LEED® ND Protocol is to become an effective tool for the transition of a neighborhood or a part of the city towards an ideal situation in which the concept of sustainability becomes a fact and not a fascinating theory. Raimi + Associates and the Natural Resources Defense Council [42] defines a neighborhood as a place with its own unique character and function, where people can live, work, shop, and interact with their neighbors. The most sustainable neighborhoods tend to exhibit high levels of walkability, a sense of place, social cohesion and stability, and neighborhood resiliency amidst changing economic and sociopolitical conditions.

As the LEED-ND program is market driven, local governments have the opportunity to influence new development in a way that was only previously possible with zoning and permitting. Local governments can prepare themselves for the implementation of LEED® ND developments and shape the future growth of a city by considering a number of aspects related to the program [28].

An important added value of this protocol is guaranteed by GBCS governance, non-profit-association charged to promote the protocols in specific areas, so considering local specificities, but within a general framework of world reference (WGBC—World Green Building Council).

The role of GBC is a key point for the success of the strategy proposed. Sedlacek and Maier [43] discuss the role that such organizations can play in the respective construction and real estate industry and under what circumstances a GBC can contribute positively to the development of a “greener” or “more sustainable” stock of buildings.

The authors state that activities of GBC can broadly be categorized into three areas:

- promotion of sustainable construction and building awareness of the issues of sustainability;
- lobbying for BC and policies that support sustainable buildings and sustainable development in general;
- identifying best practice examples through the application of sustainable building rating systems.

The first area targets the general public. The GBCs aim to alter behavior by providing information about issues of sustainability and about sustainable materials and construction techniques.

The second area takes on the opposite position. It assumes that sustainable buildings can only be implemented through governmental regulation. Unless a public policy body requests these standards in BC they will not be implemented.

The third area takes on an intermediate position. It encompasses more than just general information, but less than mandatory regulations for everyone. This area attempts to identify particularly good examples of sustainable buildings by measuring them against the yardstick of a sustainable building rating system. This measurement is voluntary and usually paid for by the building developer. This approach combines two policy elements: on the one hand it implicitly informs the general public about sustainable buildings and sustainable construction techniques, on the other hand it identifies individual buildings to be of particular quality with respect to sustainability [43].

To better understand the role and potential of tools such as LEED® ND, it is necessary to make a clarification.

LEED® ND is a voluntary tool that was created with the aim of breaking the “circle of blame”, so defined by David Cadman; constructors don’t build environmentally friendly buildings because developers don’t commission them;

developers don't commission them because investors don't require them; investors don't require them because they don't believe that occupiers demand them; occupiers don't demand them because they say that they are not offered to them, and round and round it goes [42].

This argument basically states that although there is a potential demand for "environmentally friendly buildings", the market is either too complex for this potential demand to become visible or the market participants are too dumb to see it. If these were really the basic problem of sustainable building and construction, it would be easy to solve for GBCs [43].

The other energy and environmental planning tools (TPS, IPA, MP, UTP and SEAP), are defined and approved by the Public Administrations (State, Region, Municipality) in order to protect the environment and citizens against environmental, economic and social critical or as a concrete response to external demands (*i.e.*, Kyoto Protocol, European Directives, *etc.*). While LEED® ND is voluntary, the other tools are in most cases mandatory.

Analyzing Table 2, Table 3 and Table 4 it is possible to observe that the relationships between LEED® ND and the BC are very strong. The structure of the Protocol is very similar to that of the BC, both contain rules on how to make choices in the process of construction.

An Italian BC normally contains rules that are mandatory and in some cases, rules that are suggested in order to increase the energy performance and the sustainability of the buildings. The Protocol contains prerequisites and credits that allow individuals to obtain different levels of global certification, up to LEED® Platinum. In this case, it is possible to choose a strategy considering a cost-effective approach. The added value of this Protocol is that there is a single reference document, in other words, a unique standard that theoretically could be the same for the entire Country. BC are usually proposed by the Municipalities; so theoretically, but unfortunately also in practice, in Italy there could be 8092 (the number of municipalities in Italy) BC.

It is important to have a national reference that has an international matrix, a standard that could be periodically updated in order to integrate the advantages in technology innovation in the fields related to energy and sustainability.

This protocol could be the real driver for a policy strongly oriented to the sustainability of land use, for a "greening" of our neighborhoods and of our cities, and could be, a uniting element for all the other planning tools around energy and sustainability. Foreign investors, in addition, would be better ensured in the case where land planning references an international standard.

The application of a LEED® ND standard, however, presents a critical issue. A rating system for a building or multiple buildings can be considered a private matter, and easy to manage. When the subject of the certification is a neighborhood, inevitably public areas and public infrastructure must be considered, *i.e.*, spaces and objects related to planning choices operated by Public Authorities (primarily municipalities). For this reason the comparison with the existing planning tools and the involvement of the Public Authorities is crucial.

The case study of Baranzate, discussed in this paper, demonstrates that in some cases, in order to guarantee compliance with the prerequisites of the Protocol, it is necessary to modify some rules contained in the existing planning tools.

LEED® ND was born in a territorial context (U.S.) in which new construction operations and new urbanization are favored. As highlighted by Miller [28], the situation in Europe is quite different, many of the interventions of urban redevelopment are implemented inside the city and could consider new urbanization but with urban constraints that cannot be exceeded or that could be exceeded but in a process that could take several years.

The prerequisite that makes it mandatory to have at least a LEED®-certified building seems restrictive and too commercial, a protocol for a sustainable neighborhood might consider, alternatively, other recognized certification standards for sustainable buildings or green buildings although developed outside of the GBC.

Another critical aspect is the cost of the certification. Who has fiscal responsibility for the LEED certification, and who maintains the certification in following years?

Some remarks must be made about the procedure [28]. The first of these regards the definition of neighborhood characteristics and boundaries; LEED® ND doesn't define the size of neighborhoods. This can create issues such as larger projects experiencing difficulty in achieving some credits while smaller projects may not adequately represent the idea of a neighborhood.

Projects located in dense urban neighborhoods that consist of a single building, partial block or block have an advantage over larger stand-alone projects in that they can achieve rating without much effort. The location and access to public transportation are the main areas of benefit and given the emphasis on these areas, where the advantage is located. The problem with this advantage is that innovation may not be inspired if a project can succeed in an existing context [28].

In order to overcome the weaknesses of the integration of LEED® ND Protocol with the energy planning and management tools, we suggest a path of possible actions:

- municipal administration should take the lead in actively promoting LEED® ND certification protocol;
- the prerequisites of the LEED® ND Protocol should be carefully considered in the implementation phase of the energy and environmental tools, the simulation of the Baranzate case study shows that additional efforts to comply with the LEED® ND Protocol may be limited;
- once this decision has been made, municipalities should verify that there are no inconsistencies between the protocol and the existing planning tools, that is the guarantee that the environmental parameters applied in certified areas become standard parameters for the entire city;
- energy and sustainable planning tools should be consistent with the baseline LEED® ND certification; property developers could be encouraged to achieve higher ratings (silver, gold and platinum) by means of incentives (e.g., higher volumes or lower costs for the public infrastructure works);
- since the GBC are non-profit associations open to external contributions, municipalities should consider

the possibility of being an active part in defining the criteria (for example by participating in working groups), in all cases an agreement protocol between the local GBC and the local municipalities is advisable;

- for municipalities, it is difficult and in some cases impossible, to assume the costs of certification; the costs should be paid by property developers of the area, the standard to be achieved should be included in the "development agreement" to be signed between the municipality and property developers.

CONCLUSIONS

The scope of this paper was to verify the relationship, and also the possible synergies between international and non-governmental environmental rating systems for neighborhoods (*i.e.*, LEED® ND, BREEAM® Communities and CASBEE® for Urban Development) and energy and environmental planning management tools in Italy. The Italian GBC recently proposed GBC Neighborhoods, the Italian version of LEED® ND, and for this reason, LEED® ND was considered the reference protocol in this paper even though many of considerations emerging from this study may also apply to other protocols.

The premise of GBC Neighborhoods states that "GBC Neighborhoods not born as an instrument of urban planning, but it can be a useful tool for municipalities that could adopt for promoting, encouraging and sustainable redevelopment of urban land, or for the preparation of guidelines for development".

LEED® ND is an excellent rating system for sustainability at the urban scale and can fit well with existing tools if its use is limited to promote the sustainability market. However, we believe that LEED® ND is an instrument capable of expressing a greater potential to support planning strategies at the urban or neighborhood scale.

This paper is not intended as a criticism of LEED® ND but, on the contrary, a valuable contribution towards the implementation of this tool until it becomes effective for the planning of sustainable energy systems.

The great interest in energy planning and sustainable land use has stimulated a series of tools that often are likely to be inconsistent and redundant in certain aspects. In this paper we discussed the Italian context but in many countries, particularly European ones, the complexity of the situation is not different. There is a level of scale (national, regional, municipal, neighborhood), but there is also a level of governance. The supply chain management of these tools starts from the European level (EU is responsible for the directives), goes through the national and regional level and ends at the municipal level. Within this difficult and complex governance, LEED® ND is inserted, a rating system that comes with a clear non-governmental connotation, a private initiative based on the involvement of stakeholders in the interest for the community.

The potential of LEED® ND is remarkable, first of all, we have an international rating system protocol that stimulates a real market of sustainability, no longer limited to the scale of individual buildings but extended to the neighborhood level, then an effective operational framework that can feed the need for a common growth comparison, compatible with international strategies. From the technical point of view, the tool is flexible enough to be adapted to local situations that can be very different.

Since LEED® ND does not apply to a single building but a neighborhood, we need to compare the compatibility of its rules with the rules contained in local planning instruments (see Figure 1) as the governance of these tools is charged to the Public Administration (*i.e.*, Central Government, Region, Municipality), one of the objectives of this paper was to verify this compatibility. Another important issue to be resolved is the governance of the tool. Who will be responsible for the application of LEED® ND? A private organization, such as a real estate developer, or a public body such as the municipality, or a private-public consortium.

In this paper, the LEED® ND Protocol has been tested on a real case. The potential, but also the weaknesses, were highlighted with suggestions in order to overcome these weaknesses. The study shows that, if the goal is to use LEED® ND not only as a tool to enhance the energy and environmental sustainability of small parts of the city (*i.e.*, neighborhoods), but also to lay the foundation for improving the sustainability of the whole city, the integration of LEED® ND Protocol with the energy planning and management tools must consider significant involvement of the Public Administration with the local GBC.

ACKNOWLEDGMENTS

This study could not have been achieved without the technical and organizational support of the Municipality of Baranzate and in particular the city councilors Veronica Banfi and Pietro Prisciandaro, the head of the technical office Sonia Pagnacco and Antonio Federico who contributed to the collection of the technical documentation available.

REFERENCES

1. United Nations. Kyoto Protocol to the United Nations Framework Convention on Climate Change. 1998. Available online: http://unfccc.int/resource/docs/convkp/kpeng.pdf (accessed on 30 September 2013).
2. Demographia World Urban Areas. World Urban Areas Population and Density: A 2012 Update. Available online: http://www.newgeography.com/content/002808-world-urban-areas-population-and-density-a-2012-update (accessed on 30 September 2013).
3. Fee, K.; Hartley, D.A. Urban Growth and Decline: The Role of Population Density at the City Core. Available online: http://www.clevelandfed.org/research/commentary/2011/2011-27.cfm (accessed on 30 September 2013).
4. Commission of the European Communities. *Green Paper: A European Strategy for Sustainable, Competitive and Secure Energy*; COM 105 Final; European Commission: Brussels, Belgium, 2006.
5. European Community. Directive 2002/91/EC of the European Parliament and of the Council of 16 December 2002 on the energy performance of buildings. *Off. J. Eur. Commun.* 2002, *L1*, 65–71.
6. European Parliament. Directive 2010/31/EU of the European Parliament and of the Council of 19 May 2010 on the energy performance of buildings (recast). *Off. J.*

Eur. Union 2010, *L153*, 13–35.

7. United States Green Building Council (USGBC). LEED for Neighborhood Development. Available online: http://www.usgbc.org/neighborhoods (accessed on 30 September 2013).
8. BREEAM for Communities. *BREEAM Communities Assessor Manual: Development Planning Application Stage*; Building Research Establishment Ltd.: Watford, UK, 2009.
9. BREEAM Communities 2012—An Introduction for International Use. 2012. Available online: http://www.breeam.org/filelibrary/BREEAM%20Communities/BREEAM_Communities_-_An_introduction_for_international_use_%28KN5260%29.pdf (accessed on 30 September 2013).
10. CASBEE. *CASBEE for Urban Development (2007 Edition)*; Institute for Building Environment and Energy Conservation (IBEC): Tokyo, Japan, 2007.
11. Dall'O', G. *Green Energy Audit of Buildings—A Guide for a Sustainable Energy Audit of Buildings*; Springer-Verlag: London, UK, 2013.
12. Green Building Council (GBC) Italia. *GBC Quartieri*, Available online: http://www.gbcitalia.org/page/show/gbc-quartieri (accessed on 30 September 2013).
13. Mirakyan, A.; de Guio, R. Integrated energy planning in cities and territories: A review of methods and tools. *Renew. Sustain. Energy Rev.* 2013, *22*, 289–297.
14. Cormio, C.; Dicorato, M.; Minoia, A.; Trovato, M. A regional energy planning methodology including renewable energy sources and environmental constraints. *Renew. Sustain. Energy Rev.* 2003, *7*, 99–130.
15. *Thematic Strategy on the Urban Environment—Interim Communication*; COM 60; European Commission: Brussels, Belgium, 2004.
16. Zanon, B.; Verones, S. Climate change, urban energy and planning practices: Italian experiences of innovation in land management tools. *Land Use Policy* 2013, *32*, 343–355.
17. Brandoni, C.; Polonara, F. The role of municipal energy planning in the regional energy-planning process. *Energy*2012, *48*, 323–338.
18. Dall'O', G.; Galante, A.; Torri, M. A methodology for the energy performance classification of residential building stock on an urban scale. *Energy Build.* 2011, *48*, 211–219.
19. Dall'O', G.; Belli, V.; Brolis, M.; Mozzi, I.; Fasano, M. Nearly zero-energy buildings of the Lombardy region (Italy), a case study of high-energy performance buildings. *Energies* 2013, *6*, 3506–3527.
20. Caputo, P.; Costa, G.; Ferrari, S. A supporting method for defining energy strategies in the building sector at urban scale. *Energy Policy* 2013, *55*, 261–270.
21. Manfren, M.; Caputo, P.; Costa, G. Paradigm shift in urban energy systems through distributed generation: Methods and models. *Appl. Energy* 2011, *88*, 1032–1048.
22. Haapio, A.; Viitaniemi, P. A critical review of building environmental assessment tools. *Environ. Impact Assess. Rev.*2008, *28*, 469–482.
23. Ding, G.K.C. Sustainable construction—The role of environmental assessment tools. *J. Environ. Manag.* 2008, *86*, 451–464.
24. Kyrkou, D.; Karthaus, R. Urban sustainability standards: Predetermined checklists or adaptable frameworks? *Procedia Eng.* 2011, *21*, 204–211.
25. Alyami, D.H.; Rezgui, Y. Sustainable building assessment tool development approach. *Sustain. Cities Soc.* 2012, *5*, 52–62.
26. Goldstein, B.P.; Herbøl, M.; Figueroa, M.J. Gaps in tools assessing the energy implications of renovation *versus*rebuilding decisions. *Curr. Opin. Environ. Sustain.* 2013, *5*, 244–250.
27. Haapio, A. Towards sustainable urban communities. *Environ. Impact Assess. Rev.* 2012, *32*, 165–169.
28. Miller, K. Sustainable Neighborhood Development 2010. A Toolkit for Understanding US and European Approaches. Master's Thesis, Polytechnic University of Milan, Milan, Italy, October 2010.
29. Italian Parliament. Conversione in Legge, con Modificazioni, del Decreto-Legge 25 Giugno 2008, n. 112, Recante Disposizioni Urgenti per lo Sviluppo Economico, la Semplificazione, la Competitività, la Stabilizzazione della Finanza Pubblica e la Perequazione Tributaria. (in Italian). Available online: http://www.parlamento.it/parlam/leggi/08133l.htm (accessed on 30 September 2013).
30. Italian Parliament. Strategia Energetica Nazionale: Per Un'energia più Competitiva e Sostenibile. (in Italian). 2013. Available online: http://www.mise.gov.it/images/stories/normativa/20130314_Strategia_Energetica_Nazionale.pdf(accessed on 30 September 2013).
31. Italian Parliament. Norme per L'attuazione del Piano Energetico Nazionale in Materia di Uso Razionale Dell'energia, di Risparmio Energetico e di Sviluppo delle fonti Rinnovabili di Energia. 1991. Available online: http://www.autorita.energia.it/it/docs/riferimenti/legge_10_91.htm (accessed on 30 September 2013).
32. Cittalia, Fondazione ANCI Ricerche. Politiche di Rispetto Ambientale, Il Piano Energetico dei Comuni. 2013. Available online: http://www.cittalia.it/index.php?option=com_content&view=article&id=4796:politiche-di-rispetto-ambientale-il-piano-energetico-dei-comuni&catid=3:notizie&Itemid=14 (accessed on 30 September 2013).
33. ISTAT. Strumenti di Pianificazione per L'energia per i Comuni Capoluogo di Provincial. 2013. Available online: http://dati.istat.it/Index.aspx?DataSetCode=DCCV_STPIANEN&Lang= (accessed on 30 September 2013).
34. Dall'O', G.; Galante, A.; Scansani, S. *Valutazione del Potenziale di Risparmio Energetico nel Controllo del Processo di Riqualificazione Edilizia: Simulazione di un Caso*; (in Italian). Congresso Nazionale ATI: Padova, Italy, 2003.
35. Dall'O', G.; Solaini, G.; Galante, A. *Il Regolamento*

Edilizio come Strumento di Integrazione delle Scelte Progettuali per la Sostenibilità: Il Caso di Carugate; (in Italian). 44° Convegno Internazionale AICARR: Milano, Italy, 2004.

36. European Solar Thermal Industry Federation (ESTIF). Best Practice Regulation for Solar Thermal. Available. 2007. Available online: http://www.estif.org/fileadmin/estif/content/policies/STAP/Best_practice_solar_regulations.pdf(accessed on 30 September 2013).
37. *Linee Guida per la Definizione di un Regolamento Edilizio Tipo Provinciali, Assessorato all'Ambiente, Settore Energia*; (in Italian). Provincia di Milano: Milan, Italy, 2005.
38. Dall'O', G.; Galante, A. *Efficienza Energetica e Rinnovabili nel Regolamento Edilizio Comunale*; (in Italian). Edizioni Ambiente: Milano, Italy, 2009.
39. CRESME Ricerche; Legambiente. Rapporto Onre 2013, Edilizia Sostenibile in Crescita. (in Italian). Available online: http://www.legambiente.it/contenuti/dossier/rapporto-onre-2013-edilizia-sostenibile-crescita (accessed on 30 September 2013).
40. European Commission. Covenant of Mayors—Sustainable Energy Action Plans. 2013. Available online: http://www.eumayors.eu/actions/sustainable-energy-action-plans_en.html (accessed on 30 September 2013).
41. ISTAT. Censimento ISTAT 2011. Available online: http://censimentopopolazione.istat.it/ (accessed on 30 September 2013).
42. Raimi + Associates and the Natural Resources Defense Council (NRDC). A Citizen's Guide to LEED for Neighborhood Development: How to Tell if Development is Smart and Green. 2013. Available online: https://www.nrdc.org/cities/smartgrowth/files/citizens_guide_LEED-ND.pdf (accessed on 30 September 2013).
43. Sedlacek, S.; Maier, G. Can green building councils serve as third party governance institutions? An economic and institutional analysis. *Energy Policy* 2012, *49*, 479–487.

CHAPTER 11

MULTI-STEP USAGE OF IN VIVO MODELS DURING RATIONAL DRUG DESIGN AND DISCOVERY

Charles H. Williams [1], and Charles C. Hong [1,2]

[1]Division of Cardiovascular Medicine, Department of Medicine, Vanderbilt University School of Medicine, Nashville, TN 37232, USA

[2]Research Medicine, Veterans Affairs TVHS, Nashville, TN 37212, USA

ABSTRACT

In this article we propose a systematic development method for rational drug design while reviewing paradigms in industry, emerging techniques and technologies in the field. Although the process of drug development today has been accelerated by emergence of computational methodologies, it is a herculean challenge requiring exorbitant resources; and often fails to yield clinically viable results. The current paradigm of target based drug design is often misguided and tends to yield compounds that have poor absorption, distribution, metabolism, and excretion, toxicology (ADMET) properties. Therefore, an *in vivo* organism based approach allowing for a multidisciplinary inquiry into potent and selective molecules is an excellent place to begin rational drug design. We will review how organisms like the zebrafish and *Caenorhabditis elegans* can not only be starting points, but can be used at various steps of the drug development process from target identification to pre-clinical trial models. This systems biology based approach paired with the power of computational biology; genetics and developmental biology provide a methodological framework to avoid the pitfalls of traditional target based drug design.

INTRODUCTION

Cancer, Alzheimer, diabetes; all are leading causes of death in the US. Unlike exogenous factors like HIV/AIDS and influenza, they are the result of endogenous developmental programming behaving in an aberrant fashion. As the average lifespan of human increases, certain biological machineries in our bodies start to break down. Even if all exogenous infections were obliterated from the face of the globe we would still face these diseases. Pharmaceutical companies face the challenge of modulating these developmental processes with small molecules. The human genome consists of approximately 25,000 genes [1]. Of these genes only 3000 of which are thought to be druggable and 50% of those thought to be disease causing [2]. A list of ~1500 potentially druggable disease causing biomolecules is now the center of focus in the pharmaceutical industry. The current paradigm of drug design revolves around these biomolecule targets and designing and identifying small molecules that modulate the activity of them *in vitro* or *in silico;* this is called a "target centered" approach (Figure 1a). Let us look at the track record of this approach. Only 1 of 5000 discoveries makes it to market from

Citation: Charles H. Williams, and Charles C. Hong, Multi-Step Usage of in Vivo Models During Rational Drug Design and Discovery, Int. J. Mol. Sci. 2011, 12(4), 2262-2274; doi:10.3390/ijms12042262

the bench side. The average time it takes a drug to reach the bedside from discovery is 12 years, and a single pharmaceutical agent costs from 500 million to 2 billion dollars to bring to market [3]. How could all but one of 5000 discoveries end up being useless as a therapeutic? The answer may lie in the "target centered" paradigm that has driven drug design for the past 50 years.

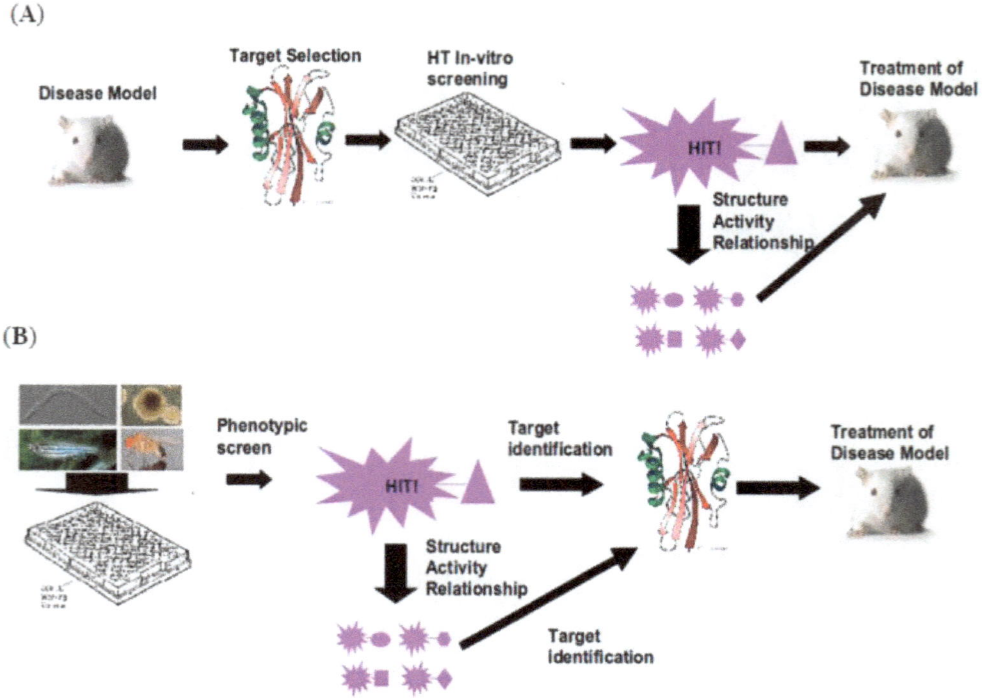

Figure 1: Workflow for two paradigms of drug discovery. (**A**) Conventional "Target centered" drug discovery; (**B**) *In vivo* model based drug discovery.

This paradigm is not without its successes. The anti-hypertensive agent Captopril produced by Bristol Meyer-Squibb is a potent and reversible inhibitor of Angiotensin-converting enzyme. Approved by the FDA in 1981 it rapidly became an effective treatment for hypertension and heart failure [4]. And Novartis produced an FDA approved drug Aliskiren, in 2007 with a combination of X-ray crystallography and computer aided design [5]. However, most of these target centric designs often fail to meet the standard when ADMET (absorption, digestion, metabolism and toxicity) is evaluated after years of research and millions of dollars. In place of this paradigm, a systems biology approach is emerging using a phenotypic screen that inherently takes into account certain ADMET properties. The following article will present a systematized development method for rational drug design based on phenotype driven discovery.

WORKFLOW

The first step in *in vivo* phenotypic discovery is the defining the desired phenotype of a "hit" compound. Currently, there are two major types of phenotypic screens. First is a forward chemical genetic screen, which consists of inducing a desired phenotype in a wild typesetting in your model organism (Figure 1B). The second is a therapeutic screen, taking a disease model and reversing it with a compound. However, before either screen can be done a decision must be made about which model system to use. There is also an emerging third type of screen, known as a pathway screen. This screen looks for a change in a particular signaling pathway *in vivo*. The second decision that must be made is which model will be used for the screen. The model organisms are, namely, *Drosophila*, *C. elegans*, zebrafish, or stem cells, which are all discussed below and summarized (Table 1). Further, strategies to identify the molecular target of the hit ligand must be considered; since a compound with completely unknown mechanism of action is unlikely to gain ready acceptance. Various techniques for target identification exist and are not mutually exclusive. Broadly, they are transcriptome profiling, affinity pull down, affinity response target screening, and yeast 3 hybrid screening. At this juncture, it is important to consider whether a "hit" compound is bioactive in live animals, depending on what model was used for initial screening. Assuming the ultimate goal of a chemical screen is to discover novel therapeutics for humans, it would be important to test whether the small molecule intervention robustly elicits desired effects without toxicities in inexpensive model organisms, prior to advancing the small molecule toward much more expensive clinical testing. Indeed, it has been estimated that just 10% improvements in predicting failures before clinical trials could save $100 million per drug [6].

Table 1: Comparison of *in vivo* small molecule discovery models

	C. elegans	*D. melanogaster*	*D. rerio*	*M. musculus*
Generation Time	3–5 days	10–14 days	3–4 months	6–8 weeks
Media	Solid or liquid	Solid	Liquid	N/A
Ease of Obtaining Embryos	+++++	+++++	++++	N/A
Number of Genes	~19,000	~13,000	~25,000	~25,000
Homology to Human Genome	>50%	>60%	>70%	>90%
Annual Cost	+	+	++	++++

ORGANISMAL MODELS

Invertebrates are small, low cost and tend to have high fecundity; as such, they are one possible avenue for screening models. One such invertebrate worming its way into drug discovery is *C. elegans*; it is a nematode with a short life cycle (approximately 3.5 days) and can be raised in liquid media while consuming *E. coli*. Each adult hermaphrodite consists of 959 cells and can produce ~300 larvae by self-fertilization. Since it started being used by Sydney Brenner in 1960s [7], it has been used to study cancer [8], neuronal cell death [9], and cilia [10]. *C. elegans* has also been established as a disease model of Alzheimer's disease [11], Parkinson's disease [12], Friedrich ataxia [13], and diabetes mellitus [14]. Recently the small nematode has made progress as a screening tool, in part due to a HTS method of arraying larvae [15]. For example, Kwok and colleagues identified a novel calcium antagonist that targets egl-19, the l-type calcium channel alpha1-subunit [16]. Additionally, an automated image analysis based high-throughput screen utilizing transgenic worms identified known autophagy enhancers that could be used to treat human liver diseases caused by protein misfolding [17]. Despite the versatility of *C. elegans* as a screening model, it has a few important drawbacks. First, its homology to the human genome is relatively low, with approximately 50% of human genes having orthologues [18]. Many organs in the human body do not have corresponding structures in *C. elegans*. The *C. elegans* body is also covered by a thick cuticle that is hard to penetrate [19]. Many compounds will not penetrate it unless a special solvent containing naphthalene and para-dichlorobenzene is added to the media [20]. Overall, *C. elegans* is promising model for *in vivo* small molecule screening to yield tools for simple biologically conserved pathways.

Another invertebrate that shows potential for small molecule screening is the fruit fly *Drosophila melanogaster*. It has been used for over a century in genetics research. *Drosophila* is anatomically more similar to mammals and has more than 60% genomic homology to humans [21]. As with *C. elegans* there are many mutants and even a fair number of disease models ranging from muscular dystrophy [22], to Alzheimers [23]. Recently this model has successfully been used by Chang and colleagues in a therapeutic screen. Briefly, a screen of 2000 compounds resulted in identification of nine molecules that rescue the *Drosophila* model of fragile X syndrome [24]. This screen, in addition to providing new small molecule tools, uncovered a novel function for muscarinic cholinergic receptors acting in parallel to the GABAergic pathway in rescuing fragile X syndrome phenotypes. Despite these impressive results, the drawbacks of using *Drosophila* as a small molecule screening model are three-fold. First, the organism is covered with a cuticle that is hard to penetrate. Secondly, *Drosophila*does not grow in a liquid media so precise dosing of small molecules in a high throughput manner is difficult. Thirdly, as with *C. elegans*, the *Drosophila* model lacks some anatomical (e.g., closed circulatory system) and genetic components that are present in humans.

As a vertebrate model of human disease, mouse has long been the gold standard. However, the size, labor and time requirements for mice make them cost prohibitive for conducting small molecule screens. However, there has been some limited success in chemical screening using mice. A proof of principle comparison of anti-TB drugs has been conducted and shows promise as a methodology for conduction large scale screens for TB therapeutics [25]. In recent years, another vertebrate model has emerged at the forefront of small molecule screening. Over the past 20 years the zebrafish has made a substantial impact in biological research. It has been used to study multiple areas ranging from vascular development [26,27] and neural development [28], to disease models for cancer [29] and melanocyte development [30]. The embryos are roughly 1 mm in size and 3–6 embryos can comfortably be arrayed in the wells of a 96-well plate [31]. Its size, low cost, and fecundity make zebrafish an attractive model for basic research. In addition, the liquid media, genetic homology to humans (over 70%), and rapid development of most organs within 48 h post fertilization, make it an ideal model for small molecule screening. With these advantages, the small teleost is swimming its way into the field of chemical genetics. There are four major models for phenotypic screening in zebrafish; they are chemical genetic, therapeutic, transgene assisted and pathway based screens (Table 2). We have, in a chemical genetic screen, used perturbation of dorso-ventral (front to back) polarity in zebrafish embryos to discover dorsomorphin, the first selective Bone morphogenic protein (BMP) type I receptor [32]. Dorso-ventral patterning in zebrafish is established primarily through a functional antagonism between Wnt and BMP, but perturbations in numerous other pathways can distort the overall embryonic patterning. In a slightly more focused, transgene assisted, approach screening for anti-angiogenic agents, transgenic fish expressing GFP under a vasculature specific promoter were used in an automated and quantitative screen. This identified two known anti-angiogenic agents along with one other novel compound [33]. Zebrafish have also been used to screen for modifiers of specific pathways such as Fgf through the use of a florescent reporter [34].

Table 2: Summary of types of phenotypic screens in zebrafish.

Embryo	Type of Screen	Readout
 Wild type	Chemical genetic	 Morphological defect
 Disease model (eg. Blood pooling)	Therapeutic	 Restored to wild type
 Trangenic Embryo Tg(Flk:dsRed)	Transgene assisted	 Altered anatomy visualized through transgenic marker
 Transgenic Reporter Line Tg(Top:dGFP)	Pathway Reporter inhibitor /enhancer	 Down regulated Reporter gene

Table 3: Overview of molecular target identification technologies.

Strategy	Advantages	Disadvantages
Affinity Chromatography	Traditionally used and readily accepted	Requires sophisticated equipment Low throughput Requires chemical modification
Expression Profiling	No chemical modifications High throughput	Data can be noisy Requires sophisticated bioinformatics Imprecise
Yeast 3 Hybrid	High throughput	Non native environment Not suitable for membrane bound proteins Requires chemical modifications
DARTS	No chemical modifications Does not require high affinity	Low throughput

In addition to organism based phenotypic screens, chemical screen using pluripotent stem cells has emerged as a viable alternative. A major benefit of screening in human derived stem cells is that one doesn't need to worry about translatability in terms of conservation and orthology. Stem cells are grown in liquid media and can be arrayed in a 96-well format to form uniform sized embryoid bodies [35]. Stem cells can give rise to all three germ layers and any cell type should be derived. Small molecules have successfully been used to create numerous cell types including cardiomyocytes [36,37]. Because of this attractive therapeutic potential, for directed differentiation, a small molecule screens using stem cells (R1 cells) was conducted by Zhu and colleagues for small molecules that promoted differentiation into definitive endoderm by assaying for expression of a sox17-rfp [38]. Additionally, ectoderm derived neurospheres were also used for assay for neurogenic compounds in a phenotypic screen [39]. However, as with many cell based assays this model for phenotypic screening could also potentially lead to identification of compounds that have poor ADME properties. This methodology could be applied to the generation of any number of cell types for which specific markers are available, such as Beta-cells and could substantially help the field of regenerative medicine.

TARGET IDENTIFICATION

Traditional drug design starts with the identification of a possible therapeutic target. For phenotypic screens target identification is the bottle neck for drug development. However, major advances are being made in the field for more efficient and rapid identification. These techniques can for our purposes be broken into two broad categories; techniques requiring modification of the small molecule and techniques that can use the native molecule.

The first two methods, that don't require modification of the ligand, are based on comparative transcriptome profiling. By utilizing a network systems biology approach one can identify the central nodes of affected gene clusters. This can yield a broad view of the mechanism of action [40]. One can also use hierarchical clustering which has been successful in both yeast and rat tissue [41]. This however provides documentable cell-type specific effects, and also microarray specific effects. An alternative approach is through the use of the Connectivity Map [42]. This takes a rank based pattern matching strategy applied to a database of over 7000 expression profiles representing 1309 compounds to identify similarities and thus potential target pathways [43].The problem with the listed techniques thus far, has been that they are unable to pinpoint specific binding partners. There is a new technique called Drug Affinity Response Target Screening (DARTS) [44]. When a small molecule binds a molecule there is a physical response. This response can be proteolytically protective, and the resultant protected peptide sequence can yield the target protein. The efficacy of this method has been tested with both high and low binding affinity molecules. This technique does not depend on an *in vivo* response and gives large amount of flexibility.

If one has a good understanding of the structure activity relationship for the compound and the phenotype one can alter the small molecule so it can be affixed to a linker. Once a linker is affixed to the small molecule, there are two approaches that can be taken. The first is the chemical proteomic approach; this approach is based on affinity chromatography and utilizes the pull down of candidate proteins. This has been the golden standard in the field for many years. With advances in MS technology and protein technologies there are new flavors of this approach that use isotope labeled proteins (a significant review of these methods was written by Rix and Superti-Furga [45]). Another technique that can be used is a variation of the yeast two hybrid screen. The yeast 3 hybrid screen requires the small molecule to be linked to a methotrexate moiety (the anchor). This molecule will attach itself to the hybrid dihydrofolate reductase-LexA DNA binding domain via the methotrexate. The small molecule bait will bring any target molecules (from a cDNA library of Gal4 activation domain fusions) within range of the DNA binding domain to elicit a His-3 reporter [46]. This is a versatile system and the cDNA libraries are commercially available from a number of sources for a number of different model systems.

SAR AND COMPOUND OPTIMIZATION

Most SAR studies are done with *a priori* knowledge of the target molecule or pathway. Recent advances in computer technology have made molecular modeling based on crystallized protein structure feasible for even small laboratories [47]. Traditionally, an *in vitro* assay is used for testing analogs to determine SAR. However, it is possible to conduct SAR *in vivo*using a phenotypic model as a read out. Hao and colleagues used the zebrafish model to perform the first *in vivo* SAR of dorsomorphin analogues. With this system they were able to derive compounds with differing specificities for Vegf and Bmp receptors [48]. Although the target for dorsomorphin was known, this new paradigm of *in vivo* SAR could be used prior to any knowledge of target pathways as it only requires a simple read-out. This *in vivo* approach also helps in identifying not just the most potent forms of the compound; but also takes into account the *in vivo* permeability of the compound.

VERTEBRATE TOXICITY

One of the major reasons potential drugs fail before they reach the market is because of the off target effects that manifest during clinical trials. The International Conference on Harmonization (ICH) S7A guidelines state that prior to clinical trial in humans, that pharmacophores must be evaluate on the vital functions, namely the circulatory system, CNS, G.I., and skeletal systems [49]. Of the models discussed above, only the zebrafish and mouse have all these systems (Figure 2). Traditionally, toxicological data is obtained from mouse. This is however, as mentioned before, a costly model system. In recent years zebrafish have emerged as a viable, low cost alternative for determining compound affects on these major organ systems and could be a useful tool for identifying liabilities of pharmacophores during early stages of drug development.

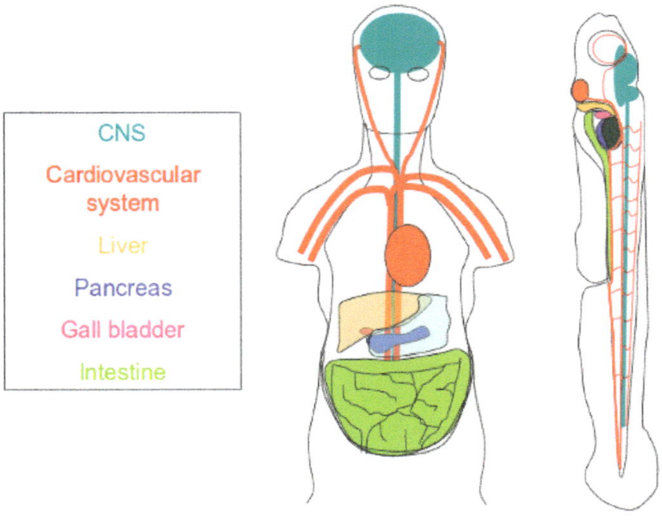

Figure 2: Conservation of organ systems between zebrafish and humans. The zebrafish is a versatile model that is useful not only drug discovery but rapid development of organ systems makes it ideal for assessing biochemical safety and toxicity

The zebrafish circulatory system forms and is functional by 23 h post fertilization (hpf), and by 48 hpf the heart has undergone looping to form a distinct atrium and ventricle. The transparency of the zebrafish embryo allows for easy visualization of both the heart and blood circulation. A screen was done with 23 drugs known to cause QT prolongation and*torsades de pointes* in man, to identify if zebrafish would phenocopy the results in 3 days post fertilization (dpf) zebrafish embryos. The results identified 18 that caused brachycardia [50]. Furthermore, a transgenic line that expresses GFP under a cardiac specific promoter has been developed and an automated method of determining heart rate in a high throughput manner has been developed [51]. For further cardiovascular effects, zebrafish blood can be visualized for hemorrhages and an image capture analysis of single erythrocytes can be conducted to assess the contractility of the heart [52]. Within the first 48 h post fertilization zebrafish develop a touch response, at 68 hpf they have a visual startle response, are capable of free swimming at 96 hpf, and develop an auditory startle response by 5 dpf. Screens have been conducted on all of these nervous system responses and have shown remarkable predictability of identifying problems also caused in humans [53–56]. For example, in a study of 8 compounds that can cause visual impairment in man, 6 inhibited the optokinetic motor response correctly in zebrafish [57]. At 36 hpf the zebrafish digestive tract begins to form, and it becomes fully functional and zebrafish are fed exogenously by 5 dpf [58]. As with blood cells, the optical clarity of the zebrafish allows for easy visualization of the peristaltic contractions of the intestine. In a study of compound effects on gut mobility 8 of 10 compounds showed corresponding decreased contractility in zebrafish [58]. Finally, as a vertebrate the zebrafish model allows one to assess bone mineralization by 10 dpf. This is a proof of principle as the study was conducted with prednisolone which substantially decreased bone mineralization in zebrafish [59]. The flexibility and anatomical conservation between man and zebrafish make it ideal for assessing toxicity of various organ systems.

AFTERWORD

Rational drug design has traditionally depended on a thorough understanding of the disease to be treated so a target protein can be selected and ligands can be screened against that molecular target. However, many diseases are complex, and regardless of the progress made in understanding the disease a coherent model cannot be established. This is particularly true for psychological and neurological diseases such as Alzheimer's and Parkinson's disease. In the treatment of many psychological disorders the only pharmaceuticals available target serotonin, dopamine, or norepinephrine signaling. These diseases are diagnosed based on behavior; therefore a model with quantifiable behaviors is necessary for *in vivo* screening. Recently, a screen for chemical modulators of wake/rest cycle was conducted using zebrafish [60]. This study implicated a novel pathway involving ERG (ether-a-go-go) potassium channel proteins. Another potential avenue for *in vivo* screening is cancer. Zebrafish can be injected with human cancer cells at 2 dpf and assessed for both angiogenic response (1 day post implantation) and metastatic behavior (5 days post implantation) [61]. Conducting a small molecule screen on these fish could yield novel therapeutics that target the cancer cells specifically in an *in vivo* environment. Furthermore different cancer cell lines respond differently and as such this screen could be conducted on different lines of cancer cells and yield different results. Whether a disease is well understood or not, phenotypic screens using *in vivo* small animal models like the zebrafish show great promise for aiding drug discovery and development at many steps of the drug development process.

REFERENCES

1. International Human Genome Sequencing Consortium. Finishing the euchromatic sequence of the human genome.*Nature* 2004, *431*, 931–945.
2. Hopkins, AL; Groom, CR. The druggable genome. *Nat. Rev. Drug Discov* 2002, *1*, 727–730.
3. Adams, CP; Brantner, VV. Spending on new drug development. *Health Econ* 2010, *19*, 130–141.
4. Talele, TT; Khedkar, SA; Rigby, AC. Successful applications of computer aided drug discovery: Moving drugs from concept to the clinic. *Curr. Top Med. Chem* 2010, *10*, 127–141.
5. Zon, LI; Peterson, RT. *In vivo* drug discovery in the zebrafish. *Nat. Rev. Drug Discov* 2005, *4*, 35–44.
6. Cohen, NC. Structure-based drug design and the discovery of aliskiren (Tekturna): Perseverance and creativity to overcome a R&D pipeline challenge. *Chem. Biol. Drug Des* 2007, *70*, 557–565.
7. Brenner, S. In the beginning was the worm. *Genetics* 2009, *182*, 413–415.
8. Poulin, G; Nandakumar, R; Ahringer, J. Genome-wide RNAi screens in *Caenorhabditis elegans*: Impact on cancer research. *Oncogene* 2004, *23*, 8340–8345.
9. Putcha, GV; Johnson, EM. Men are but worms: Neuronal cell death in *C. elegans* and vertebrates. *Cell Death Differ* 2004,*11*, 38–48.
10. Smith, KR; Kieserman, EK; Wang, PI; Basten, SG; Giles, RH; Marcotte, EM; Wallingford, JB. A role for central spindle proteins in cilia structure and function. *Cytoskeleton* 2010, *68*, 112–124.
11. Ewald, CY; Li, C. Understanding the molecular basis of Alzheimer's disease using a *Caenorhabditis elegans* model system. *Brain Struct. Funct* 2010, *214*, 263–283.
12. Harrington, AJ; Hamamichi, S; Caldwell, GA; Caldwell, KA. *C. elegans* as a model organism to investigate molecular pathways involved with Parkinson's disease. *Dev. Dyn* 2010, *239*, 1282–1295.
13. Puccio, H. Multicellular models of Friedreich ataxia. *J Neurol* 2009, *256*(Suppl. 1), 18–24.
14. Morcos, M; Hutter, H. The model Caenorhabditis elegans in diabetes mellitus and Alzheimer's disease. *J. Alzheimers Dis* 2009, *16*, 897–908.
15. Burns, AR; Kwok, TCY; Howard, A; Houston, E; Johanson, K; Chan, A; Cutler, SR; McCourt, P; Roy, PJ. High-throughput screening of small molecules for bioactivity and target identification in Caenorhabditis elegans. *Nat. Protoc* 2006, *1*, 1906–1914.
16. Kwok, TCY; Ricker, N; Fraser, R; Chan, AW; Burns, A; Stanley, EF; McCourt, P; Cutler, SR; Roy, PJ. A small-molecule screen in *C. elegans* yields a new calcium channel antagonist. *Nature* 2006, *441*, 91–95.
17. Gosai, SJ; Kwak, JH; Luke, CJ; Long, OS; King, DE; Kovatch, KJ; Johnston, PA; Shun, TY; Lazo, JS; Perlmutter, DH; *et al.* Automated high-content live animal drug screening using *C. elegans* expressing the aggregation prone serpin α1-antitrypsin Z. *PLoS One* 2010, *5*, e15460.
18. Harris, TW; Chen, N; Cunningham, F; Tello-Ruiz, M; Antoshechkin, I; Bastiani, C; Bieri, T; Blasiar, D; Bradnam, K; Chan, J; *et al.* WormBase: A multi-species resource for nematode biology and genomics. *Nucleic Acids Res* 2004,*32*(Suppl. 1), D411–D417.
19. Wells, WA. High-throughput worms. NemaPharm, Inc. *Chem. Biol* 1998, *5*, R147–R148.
20. Kokel, D; Li, Y; Qin, J; Xue, D. The nongenotoxic carcinogens naphthalene and para-dichlorobenzene suppress apoptosis in *Caenorhabditis elegans*. *Nat. Chem. Biol* 2006, *2*, 338–345.
21. Bier, E. *Drosophila*, the golden bug, emerges as a tool for human genetics. *Nat. Rev Genet* 2005, *6*, 9–23.
22. van der Plas, MC; Pilgram, GSK; de Jong, AW; Bansraj, MR; Fradkin, LG; Noordermeer, JN. *Drosophila* Dystrophin is required for integrity of the musculature. *Mech. Dev* 2007, *124*, 617–630.
23. Nichols, CD. *Drosophila* melanogaster neurobiology, neuropharmacology, and how the fly can inform central nervous system drug discovery. *Pharmacol. Ther* 2006, *112*, 677–700.
24. Chang, S; Bray, SM; Li, Z; Zarnescu, DC; He, C; Jin, P; Warren, ST. Identification of small molecules rescuing fragile X syndrome phenotypes in *Drosophila*. *Nat. Chem. Biol* 2008, *4*, 256–263.
25. Nikonenko, BV; Samala, R; Einck, L; Nacy, CA. Rapid, simple *in vivo* screen for new drugs active against Mycobacterium tuberculosis. *Antimicrob. Agents*

Chemother 2004, *48*, 4550–4555.

26. Williams, C; Kim, S; Ni, TT; Mitchell, L; Ro, H; Penn, JS; Baldwin, SH; Solnica-Krezel, L; Zhong, TP. Hedgehog signaling induces arterial endothelial cell formation by repressing venous cell fate. *Dev. Biol* 2010, *341*, 196–204.
27. Hong, CC; Kume, T; Peterson, RT. Role of crosstalk between phosphatidylinositol 3-kinase and extracellular signal-regulated kinase/mitogen-activated protein kinase pathways in artery-vein specification. *Circ. Res* 2008, *103*, 573–579.
28. Takada, N; Kucenas, S; Appel, B. Sox10 is necessary for oligodendrocyte survival following axon wrapping. *Glia* 2010,*58*, 996–1006.
29. Goessling, W; North, TE; Zon, LI. New waves of discovery: Modeling cancer in zebrafish. *J. Clin. Oncol* 2007, *25*, 2473–2479.
30. Richardson, J; Lundegaard, PR; Reynolds, NL; Dorin, JR; Porteous, DJ; Jackson, IJ; Patton, EE. Mc1r Pathway regulation of zebrafish melanosome dispersion. *Zebrafish* 2008, *5*, 289–295.
31. Hao, J; Williams, CH; Webb, ME; Hong, CC. Large scale zebrafish-based *in vivo* small molecule screen. *J. Vis. Exp* 2010,*46*, 2243.
32. Yu, PB; Hong, CC; Sachidanandan, C; Babitt, JL; Deng, DY; Hoyng, SA; Lin, HY; Bloch, KD; Peterson, RT. Dorsomorphin inhibits BMP signals required for embryogenesis and iron metabolism. *Nat. Chem. Biol* 2008, *4*, 33–41.
33. Tran, TC; Sneed, B; Haider, J; Blavo, D; White, A; Aiyejorun, T; Baranowski, TC; Rubinstein, AL; Doan, TN; Dingledine, R; *et al.* Automated, quantitative screening assay for antiangiogenic compounds using transgenic zebrafish. *Cancer Res* 2007, *67*, 11386–11392.
34. Molina, G; Vogt, A; Bakan, A; Dai, WX; de Oliveira, PQ; Znosko, W; Smithgall, TE; Bahar, IB; Lazo, JS; Day, BW; *et al.* Zebrafish chemical screening reveals an inhibitor of Dusp6 that expands cardiac cell lineages. *Nat. Chem. Biol* 2009, *5*, 680–687.
35. Ao, A; Williams, CH; Hao, J; Hong, CC. Modified Mouse Embryonic Stem Cell based Assay for Quantifying Cardiogenic Induction Efficiency. *J Vis Exp* 2011. in press.
36. Zhang, X. Modulation of embryonic stem cell fate and somatic cell reprogramming by small molecules. *Reprod. Biomed. Online* 2010, *21*, 26–36.
37. Wang, H; Hao, J; Hong, CC. Cardiac Induction of Embryonic Stem Cells by a Small Molecule Inhibitor of Wnt/β-Catenin Signaling. *ACS Chem. Biol* 2010, *6*, 192–197.
38. Zhu, S; Wurdak, H; Wang, J; Lyssiotis, CA; Peters, EC; Cho, CY; Wu, X; Schultz, PG. A small molecule primes embryonic stem cells for differentiation. *Cell Stem Cell* 2009, *4*, 416–426.
39. Saxe, JP; Wu, H; Kelly, TK; Phelps, ME; Sun, YE; Kornblum, HI; Huang, J. A phenotypic small-molecule screen identifies an orphan ligand-receptor pair that regulates neural stem cell differentiation. *Chem. Biol* 2007, *14*, 1019–1030.
40. Arrell, DK; Terzic, A. Network systems biology for drug discovery. *Clin. Pharmacol. Ther* 2010, *88*, 120–125.
41. Waring, JF; Ciurlionis, R; Jolly, RA; Heindel, M; Ulrich, RG. Microarray analysis of hepatotoxins *in vitro* reveals a correlation between gene expression profiles and mechanisms of toxicity. *Toxicol. Lett* 2001, *120*, 359–368.
42. Lamb, J; Crawford, ED; Peck, D; Modell, JW; Blat, IC; Wrobel, MJ; Lerner, J; Brunet, JP; Subramanian, A; Ross, KN; *et al.* The Connectivity Map: using gene-expression signatures to connect small molecules, genes, and disease. *Science*2006, *313*, 1929–1935.
43. Rho, SB; Kim, B; Kang, S. A gene signature-based approach identifies thioridazine as an inhibitor of phosphatidylinositol-3′-kinase (PI3K)/AKT pathway in ovarian cancer cells. *Gynecol. Oncol* 2011, *120*, 121–127.
44. Lomenick, B; Hao, R; Jonai, N; Chin, RM; Aghajan, M; Warburton, S; Wang, J; Wu, RP; Gomez, F; Loo, JA; *et al.* Target identification using drug affinity responsive target stability (DARTS). *Proc. Natl. Acad. Sci. USA* 2009, *106*, 21984–21989.
45. Rix, U; Superti-Furga, G. Target profiling of small molecules by chemical proteomics. *Nat. Chem. Biol* 2009, *5*, 616–624.
46. Kley, N. Chemical dimerizers and three-hybrid systems: Scanning the proteome for targets of organic small molecules. *Chem. Biol* 2004, *11*, 599–608.
47. Senderowitz, H; Marantz, YG. Protein-coupled receptors: Target-based *in silico* screening. *Curr. Pharm. Des* 2009, *15*, 49–68.
48. Hao, J; Ho, JN; Lewis, JA; Karim, KA; Daniels, RN; Gentry, PR; Hopkins, CR; Lindsley, CW; Hong, CC. *In vivo*structure-activity relationship study of dorsomorphin analogues identifies selective VEGF and BMP inhibitors. *ACS Chem. Biol* 2010, *5*, 245–253.
49. FDA ICH Guidance for industry: S7A Safety Pharmacology Studies for Human Pharmaceuticals: US Department of Health and Human Services. Available online: http://www.fda.gov/downloads/Drugs/GuidanceComplianceRegulatoryInformation/Guidances/ucm074959.pdf (accessed on 28 March 2011).
50. Milan, DJ; Jones, IL; Ellinor, PT; MacRae, CA. *In vivo* recording of adult zebrafish electrocardiogram and assessment of drug-induced QT prolongation. *Am. J. Physiol. Heart Circ. Physiol* 2006, *291*, H269–H273.
51. Burns, CG; Milan, DJ; Grande, EJ; Rottbauer, W; MacRae, CA; Fishman, MC. High-throughput assay for small molecules that modulate zebrafish embryonic heart rate. *Nat. Chem. Biol* 2005, *1*, 263–264.
52. Schwerte, T; Pelster, B. Digital motion analysis as a tool for analysing the shape and performance of the circulatory system in transparent animals. *J. Exp. Biol* 2000, *203*, 1659–1669.
53. Airhart, MJ; Lee, DH; Wilson, TD; Miller, BE;

Miller, MN; Skalko, RG. Movement disorders and neurochemical changes in zebrafish larvae after bath exposure to fluoxetine (PROZAC). *Neurotoxicol. Teratol* 2007, *29*, 652–664.

54. Richards, FM; Kimber, G; Butler, P; Waldron, G; Knight, T; Goldsmith, P; Fleming, AL. Prediction of ophthalmic effects in man by screening drugs for the optokinetic and optomotor responses in zebrafish. *Invest. Ophthalmol. Vis. Sci*2006, *2622*, 123.

55. Berghmans, S; Hunt, J; Roach, A; Goldsmith, P. Zebrafish offer the potential for a primary screen to identify a wide variety of potential anticonvulsants. *Epilepsy Res* 2007, *75*, 18–28.

56. Best, JD; Berghmans, S; Hunt, JJ; Clarke, SC; Fleming, A; Goldsmith, P; Roach, AG. Non-associative learning in larval zebrafish. *Neuropsychopharmacology* 2008, *33*, 1206–1215.

57. Pack, M; Solnica-Krezel, L; Malicki, J; Neuhauss, SC; Schier, AF; Stemple, DL; Driever, W; Fishman, MC. Mutations affecting development of zebrafish digestive organs. *Development* 1996, *123*, 321–328.

58. Berghmans, S; Butler, P; Goldsmith, P; Waldron, G; Gardner, I; Golder, Z; Richards, FM; Kimber, G; Roach, A; Alderton, W; *et al.* Zebrafish based assays for the assessment of cardiac, visual and gut function-potential safety screens for early drug discovery. *J. Pharmacol. Toxicol. Methods* 2008, *58*, 59–68.

59. Barrett, R; Chappell, C; Quick, M; Fleming, A. A rapid, high content, *in vivo* model of glucocorticoid-induced osteoporosis. *Biotechnol. J* 2006, *1*, 651–655.

60. Rihel, J; Prober, DA; Arvanites, A; Lam, K; Zimmerman, S; Jang, S; Haggarty, SJ; Kokel, D; Rubin, LL; Peterson, RT; *et al.* Zebrafish behavioral profiling links drugs to biological targets and rest/wake regulation. *Science* 2010, *327*, 348–351.

61. Snaar-Jagalska, BE. ZF-CANCER: Developing high-throughput bioassays for human cancers in zebrafish. *Zebrafish*2009, *6*, 441–443.

CHAPTER 12

APPLICATION OF 2K FULL FACTORIAL DESIGN IN OPTIMIZATION OF SOLVENT-FREE MICROWAVE EXTRACTION OF GINGER ESSENTIAL OIL

Mumtaj Shah[1] and S. K. Garg[2]

[1]Department of Chemical Engineering, SOET, ITM University, Gwalior 475001, India

[2]Department of Chemical Engineering, HBTI, Kanpur 208002, India

ABSTRACT

The solvent-free microwave extraction of essential oil from ginger was optimized using a 2^3 full factorial design in terms of oil yield to determine the optimum extraction conditions. Sixteen experiments were carried out with three varying parameters, extraction time, microwave power, and type of sample for two levels of each. A first order regression equation best fits the experimental data. The predicted values calculated by the regression model were in good agreement with the experimental values. The results showed that the extraction time is the most prominent factor followed by microwave power level and sample type for extraction process. An average of 0.25% of ginger oil can be extracted using current setup. The optimum conditions for the ginger oil extraction using SFME were the extraction time 30 minutes, microwave power level 640 watts, and sample type, crushed sample. Solvent-free microwave extraction proves a green and promising technique for essential oil extraction.

INTRODUCTION

The principal aim of green chemistry and engineering is to reduce chemical related impact on human health and to search alternative, environmentally friendly and energy efficient production methods. Green and clean extraction methods can offer more natural products, free from toxic solvents. The search for such green extraction methods is highly emphasized in essential oils industries since last decade because of consumer's preference towards natural products. Essential oils are volatile extract of the spices, medicinal and aromatic plants. The history of essential oil extraction and their use for various purposes is very old.

Zingiber officinale Roscoe, commonly known as ginger, is a member of Zingiberaceae family. Most Zingiberaceae family spices are fibrous rooted perennial herb which is cultivated in many tropical and subtropical areas, India, North East Asia, Australia, and Japan. The use of ginger as spice and medicine is very old and is mentioned in earliest Chinese and Sanskrit literature [1].

Ginger species possesses aromatic properties and has a commercial importance. There are two valuable extracts of ginger, essential oil which varies as 0.8–4.2% and oleoresin in the range of about 7% depending on its origin habitat and agronomic treatment of culture [2]. Ginger oil possesses

Citation: Mumtaj Shah and S. K. Garg, "Application of 2k Full Factorial Design in Optimization of Solvent-Free Microwave Extraction of Ginger Essential Oil," Journal of Engineering, vol. 2014, Article ID 828606, 5 pages, 2014. doi:10.1155/2014/828606

the natural aroma of crude ginger and is globally used in flavour, perfumer, and pharmaceutical and liqueur industry [3]. The therapeutic properties of ginger oil are antiseptic, antispasmodic, carminative, cephalic, expectorant, febrifuge, laxative, and stomachic [4, 5].

The general methods used for extraction of ginger oil are hydrodistillation, steam distillation, H-S distillation, SFE-CO_2, solvent extraction, and microwave extraction as a recent technique; the oil yield, extraction time, and quality of oil extracted from each method differ significantly and dried rhizome was used in all the methods [6,7]. Steam distillation is still the principal method of essential oil extraction in industry. All the conventional extraction methods possess the common characteristic of boiling the plant material with water or with organic solvents. Longer extraction time of these conventional methods may degrade the oil quality at the same time leaving the toxic solvent residue in essential oils.

Today, microwave technology for extraction of essential oils and natural extracts has got much attention from scientist community. In 1992, Pare [8] was the first who demonstrated the use of microwave energy for the extraction of naturally produced compounds from plant tissues; extraction has advantages in terms of yield and selectivity, with better extraction time and essential oil composition free from residual solvents, contaminants, or artifacts, and is also environmentally friendly. Solvent-free microwave extraction (SFME) is a combination of microwave heating and distillation to extract essential oils from plant materials. SFME involves placing the plant material in a microwave reactor without adding any solvent and distillation is performed at atmospheric pressure. The internal heating of the in situ water within the plant material increases the internal pressure and makes the oil cells burst. This process thus frees essential oil which is evaporated by the in situ water of the plant material. SFME has been used for the extraction of essential oils from aromatic and medicinal plants [9–12].

In this study, solvent-free microwave extraction was adopted for oil extraction of green ginger rhizome. The extraction process was optimized in terms of essential oil yield using full factorial design, determining those variables that most influenced essential oil yield. A 2^3 full factorial plan was carried out to model the process. The simple first degree model was used which gave a representation of the response function according to the variables.

MATERIALS AND METHODS

Raw Material

The ginger rhizome used in this research was purchased from local market nearby H.B.T.I., Kanpur, India. The rhizome was somewhat yellowish brown in color, branched, and laterally flattened. Pretreatment of ginger sample was done to ensure maximum yield of essential oil. Raw ginger was first washed with clean water and partially peeled to remove excess bark. Two types of sample were used: sliced and crushed. Slicing was done longitudinally to an average thickness of 0.20 cm and soaked in water for 18 hours to increase the moisture content for better yield [13].

Experimental Setup

In the present experimental work the process adopted was solvent-free microwave assisted extraction. Extraction was performed at atmospheric pressure. This was a domestic multimode microwave reactor with frequency of 2450 MHz with a maximum delivered output power of 800 W and input power of 1200 W, having the voltage supply of 230 volts and dimensions of the oven cavity are 206 mm (H) × 300 mm (W) × 302 mm (D), with total capacity of 18.5 liters.

The microwave oven was mechanically modified from its original condition to collect the water and oil vapors coming from the ginger sample once it was heated. Standard procedure was followed in modification of microwave oven to prevent the leakage of microwaves, published elsewhere [14]. The extraction vessel was a 1000 mL flat bottomed flask connected to a condensing system. The hole and the microwave containment choke were centered above the turntable and approximately in the center of the microwave chamber. Schematic diagram of the microwave extraction apparatus used for essential oil extraction is shown in Figure 1.

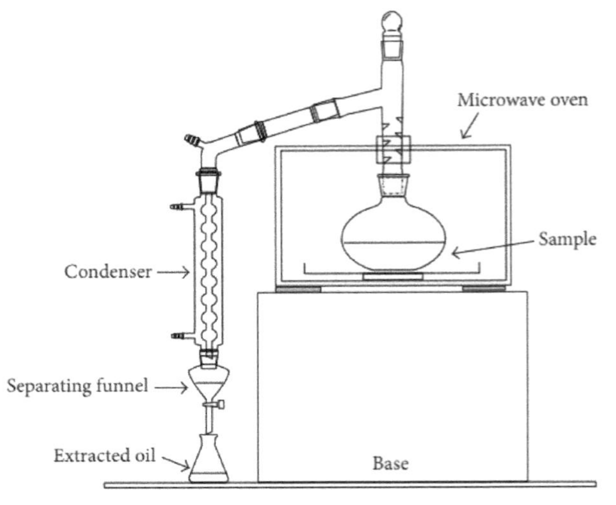

Figure 1: Microwave extraction apparatus.

Extraction of Essential Oil

The fresh ginger rhizomes were washed, cleaned, partially peeled, and thinly grated or crushed, as per requirement prior to extraction. A 100 g of sample was placed in reactor. The runs were taken at three different levels of time and microwave power and with two types of sample, crushed and sliced. Pale yellow colored oil, with a warm, spicy, lemon like odor, was obtained which was separated and dried over the minimum amount of anhydrous sodium sulfate to remove traces of moisture. The essential oil so obtained was stored at low temperature (4±2° C) in dark; the percentage oil yield is expressed as follows:

$$\text{Oil yield (\%)} = \left(\frac{\text{mass of oil extracted}}{\text{mass of sample}}\right) \times 100. \tag{1}$$

Chemical Composition of Essential Oil

A standard gas liquid chromatography was used for the analysis of essential oil of ginger obtained from SFME. The analysis was carried out at Fragrance and Flavors Development Centre

(FFDC), Kannauj, Uttar Pradesh, India. GLC analysis of the ginger oil was performed on split mode HP (Hewlett-Packard) (make HP 5890) having detector flame ionization with a carbowax column (30 m × 0.25 m i.d., film thickness 0.325 μm). The oven temperature was programmed from 50° C (initial time 5 min) to 230° C with 50° C/min and injector and detector have the temperatures 230° C and 240° C, respectively. The identification of compounds was done by comparing with the retention times of available standard. Area percentage in the chromatogram was used to know the percentage of each compound in the oil. α-zingiberene (30%), α-curcumene (9%), β-sesquiphellandrene (4%), and bisabolene (3.2%) were identified as major constituents in extracted ginger essential oil.

Mathematical Treatment

To investigate the efficiency of extraction process on ginger essential oil yield, a two-level, 2^3 full factorial design was constructed with two replications of experimental runs. Three variables were chosen, namely, extraction time (X_1), microwave power (X_2), and sample type (X_3). Each independent variable had 2 levels which were coded as −1 and +1.The coded values of independent variables were found from equation and given in Table 1:

$$\mathbf{x}_i = \frac{x_i - x_0}{\Delta x}, \tag{2}$$

where x_0 is the base value at the center of experimental domain and x_i is original variable and Δx is the average value of the difference between highest and lowest values. The sixteen runs in design matrix of 2^3 full factorial designs are set up by randomization. A multiple regression, first degree model was used to express the response as a function of all three factors, which are centered and reduced variables:

$$y = \beta_0 + \beta_1 x_1 + \beta_2 x_2 + \beta_3 x_3 + \beta_{12} x_1 x_2 + \beta_{13} x_1 x_3 + \beta_{23} x_2 x_3 + \beta_{123} x_1 x_2 x_3, \tag{3}$$

where β_0, β_i, β_{ij}, and β_{ijk} represents the average effect, main effects, and two way and three way interactions effects, respectively. Design Expert version 8.0.7.1 (trial version) was applied for performing the experimental design and the data analysis [15].

Table 1: Coded and natural variables in 2^3 factorial design

Levels	X_1 [min.]	X_2 [watt]	X_3
Basic level (0)	20	464	—
High level (+1)	30	640	Sliced
Low level (−1)	10	288	Crushed
Interval	10	176	—

RESULTS AND DISCUSSION

Statistical Analysis

Table 2 shows the design matrix of experimental outcome, as carried out above specified levels as shown in Table 1 and expressed as extraction yield (y) of essential oil extracted from ginger sample.

Table 2: The 2^3 factorials design including the corresponding responses

Run	Code variables			Oil yield
	x_1 (extraction time)	x_2 (microwave power)	x_3 (sample type)	y (%)
1	-1	-1	-1	0.10
2	1	-1	-1	0.26
3	-1	1	-1	0.14
4	1	1	-1	0.35
5	-1	-1	1	0.22
6	1	-1	1	0.20
7	-1	1	1	0.28
8	1	1	1	0.44
9	-1	-1	-1	0.12
10	1	-1	-1	0.25
11	-1	1	-1	0.14
12	1	1	-1	0.32
13	-1	-1	1	0.24
14	1	-1	1	0.17
15	-1	1	1	0.31
16	1	1	1	0.46

It is observed from response table that an interaction model best represents the microwave extraction process in terms of coded variables:

$$Y = 0.25 + 0.056X_1 + 0.055X_2 + 0.040X_3 + 0.031X_1X_2 + 0.028X_2X_3 - 0.029X_1X_3 + 0.019X_1X_2X_3. \quad (4)$$

This regression equation shows that the optimum yield should be located in the experimental domain or very close to it, as the values of β's are not very high. An average yield of essential oil we can extract from the current experimental setup is β_0 = 0.25% for the current levels of the factors. To assess the goodness of fit of the empirical model and to check the adequacy of model that has been generated by the factorial experiment, analysis of variance (ANOVA) was conducted at 95% confidence level ($P = 0.05$) given in Table 3. Results showed adjusted-R^2 (0.9783), an adequately high degree of correlation between experimental and predicted values. Coefficient of determination R^2 (0.9885) was desirably high (close to 1), adequate precision (AP) 30.411, standard deviation (SD) 0.016, and coefficient of variation (CV) 6.32%. Based on the ANOVA result, it is clear that the model is good simulation of extraction experiment. The predicted R^2 value is also in agreement with the adjusted-R^2 value.

Table 3: Analysis of variance

Source	Sum of squares	Degree of freedom (df)	Mean square	F value	P value prob. > F	Remark
Model	0.17	7	0.024	97.83	<0.0001	Significant
A-extraction time	0.051	1	0.051	202.50	<0.0001	
B-power	0.048	1	0.048	193.60	<0.0001	
C-sample	0.026	1	0.026	102.40	<0.0001	
AB	0.016	1	0.016	62.50	<0.0001	
AC	0.013	1	0.013	52.90	<0.0001	
BC	0.012	1	0.012	48.40	0.0001	
ABC	5.625E − 003	1	5.625E − 003	22.50	0.0015	
Pure error	2.000E − 003	8	2.500E − 004			
Cor total	0.17	15				

Figure 2 shows the normal probability plot of the residuals for percentage of oil yield. The normal distribution of error can be seen as the value of residuals fit on a straight line to a major extent. The ANOVA results for degradation produced the Pareto chart of main and interaction effects shown in Figure 3, bar lengths being proportional to the absolute value of the estimated effects, which not only help compare the relative importance of effects but also decide the ranking of factors.

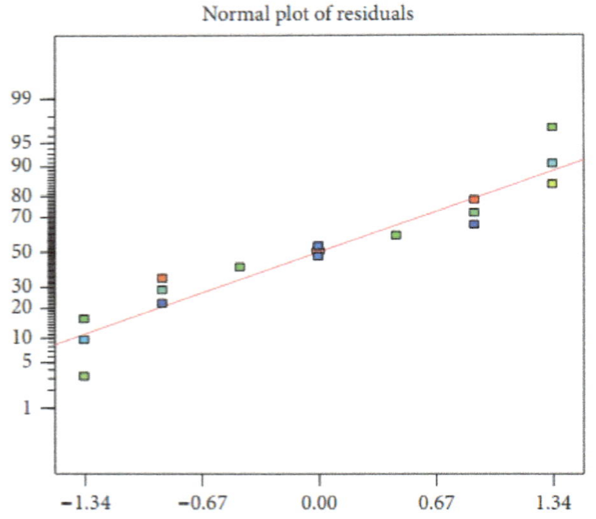

Figure 2: Normal probability plot of the residuals.

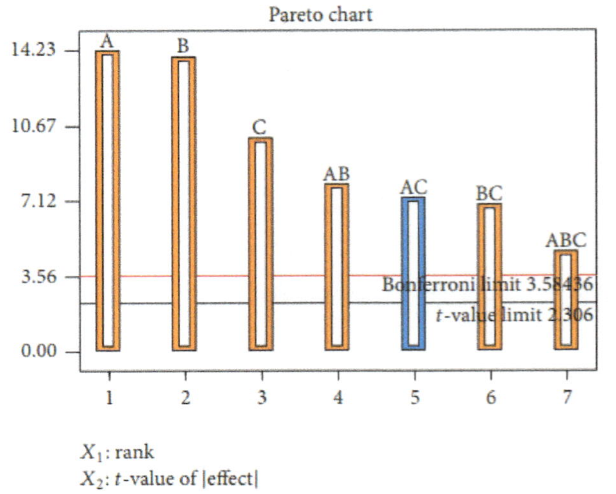

Figure 3: Pareto chart for significance of parameters.

The interpretation of this chart indicates that the influence of the extraction time is important in the response, followed by power level and both the factors have similar effect on extraction process. Sample type is less effective in comparison to other two factors; uneven crushing of socked ginger sample was favored and extraction using sliced sample resulted in lower yield. The combined effect of duration of extraction and type of sample was not favorable, as was the combined effect of duration of extraction and heating power level. However, we note the very low values for these interactions, which can therefore be ignored in practice, especially for small-scale production units.

Process Optimization

In order to optimize the conditions of extraction experiment, a desirability function (d) for the simultaneous optimization of multiple responses was used. This function can be described as follows:

$$d = \left(\prod_{i=1}^{N} m_i^{r_i}\right)^{1/\sum r_i}, \quad (5)$$

where d varies in range of $0 \le d \le 1$, and N, r_i, and m represent the number of responses, importance of a particular response, and partial desirability function for specific responses, respectively [16]. Design expert software has inbuilt optimization tool which uses (5). This study aimed to extract the highest amount of essential oil from the sample.

Numerical optimization of extraction process was done using Design Expert 8.0.7.1 and Table 4 shows the settings provided to Design Expert and the final result obtained in numerical optimization. There were 17 solutions found but solution with the highest yield was selected as final result.

Table 4: Setting goal for each factor and response for formulation of optimization and selected optimized optimum conditions

Factors/response	Goal	Selected optimum conditions	
		Coded	Actual
Extraction time (min.)	In range	1	30 min
Microwave power (w)	In range	1	640 watts
Sample type	In range	1	Crushed sample
Oil yield	Maximum	0.45%	0.45%

The optimum condition in SFME is whenever extraction was performed at high level of all three factors. In other words, the best combination of condition to extract essential oil from green ginger rhizome was extraction time 30 min., microwave power level at 640 watts, and crushed sample by using SFME. This optimum oil yield which was 0.45% carries a desirability value of 0.972.

CONCLUSIONS

Solvent-free microwave extraction was used for the extraction of useful essential oil from green ginger rhizome. It can be concluded from this study that duration of time is the most dominant factor followed by microwave power and type of sample. For green ginger, solvent-free microwave extraction proved a promising technique with high quality of essential oil. An average of 0.25% of ginger oil can be extracted using current setup.

REFERENCES

1. B. N. Korla and N. P. Dohroo, "Production technology in ginger—a review," Agricultural Reviews, vol. 12, no. 1, pp. 22–36, 1991.
2. M. Noor Azian, M. S. Sazalina, and M. R. Haira Rizan, Essential Oil and Active Ingredients Extraction from Ginger Plants, Annual Progress Report Centre of Lipids Engineering & Applied Research, Kuala Lumpur, Malaysia, 2001.
3. E. Langner, S. Greifenberg, and J. Gruenwald, "Ginger: history and use," Advances in Therapy, vol. 15, no. 1, pp. 25–44, 1998.
4. Y. Hori, T. Miura, Y. Hirai et al., "Pharmacognostic studies on ginger and related drugs—part 1: five sulfonated compounds from Zingiberis rhizome (Shokyo)," Phytochemistry, vol. 62, no. 4, pp. 613–617, 2003.
5. B. H. Ali, G. Blunden, M. O. Tanira, and A. Nemmar, "Some phytochemical, pharmacological and toxicological properties of ginger (Zingiber officinale Roscoe): a review of recent research," Food and Chemical Toxicology, vol. 46, no. 2, pp. 409–420, 2008.
6. Y. Yonei, H. Ohinata, R. Yoshida, Y. Shimizu, and C. Yokoyama, "Extraction of ginger flavor with liquid or supercritical carbon dioxide," The Journal of Supercritical Fluids, vol. 8, no. 2, pp. 156–161, 1995.
7. M. J. Alfaro, J. M. R. Bélanger, F. C. Padilla, and J. R. J. Paré, "Influence of solvent, matrix dielectric properties, and applied power on the liquid-phase microwave-assisted processes (MAP™) extraction of ginger (Zingiber officinale)," Food Research International, vol. 36, no. 5, pp. 499–504, 2003.
8. J. R. J. Pare, "Microwave assisted process for extraction and apparatus therefore," Canadian patent, CA 2055390, 1992.
9. M. E. Lucchesi, J. Smadja, S. Bradshaw, W. Louw, and F. Chemat, "Solvent free microwave extraction ofElletaria cardamomum L.: a multivariate study of a new technique for the extraction of essential oil,"Journal of Food Engineering, vol. 79, no. 3, pp. 1079–1086, 2007.
10. M. A. Ferhat, B. Y. Meklati, F. Visinoni, M. A. Vian, and F. Chemat, "Solvent free microwave extraction of essential oils: green chemistry in the teaching laboratory," Chemistry Today, vol. 26, no. 2, pp. 48–50, 2008.
11. M. E. Lucchesi, F. Chemat, and S. Jacqueline, "Solvent free microwave extraction: an innovative tool for rapid extraction of essential oil from aromatic herbs and spices," Journal of Microwave Power and Electromagnetic Energy, vol. 39, no. 3-4, pp. 137–140, 2004.
12. X. Chen, Y. Zhang, Y.-G. Zu, X.-Y. Yu, and J.-L. Li, "Optimization of solvent-free microwave extraction of essential oil from the fruits of Schisandra chinensis and its DPPH radical scavenging activity," Food Science, vol. 32, no. 14, pp. 85–89, 2011.
13. M. N. Azian, A. A. M. Kamal, and M. N. Azlina, "Changes of cell structure in ginger during processing,"Journal of Food Engineering, vol. 62, no. 4, pp. 359–364, 2004.

14. S. S. Chen and M. Spiro, "Study of microwave extraction of essential oil constituents from plant materials," Journal of Microwave Power and Electromagnetic Energy, vol. 29, no. 4, pp. 231–241, 1994. ·

15. D. C. Montgomery and C. R. George, Applied Statistics and Probability for Engineers, John Wiley & Sons, Singapore, 3rd edition, 2003.

16. G. Derringer and R. Suich, "Simultaneous optimization of several response variables," Journal of Quality Technology, vol. 12, no. 4, pp. 214–219, 1980.

CHAPTER 13

D-FNN BASED MODELING AND BP NEURAL NETWORK DECOUPLING CONTROL OF PVC STRIPPING PROCESS

Shu-zhi Gao,[1] Jing Yang,[2] and Jie-sheng Wang[3]

[1]College of Information and Engineering, Shenyang University of Chemical Technology, Shenyang 110142, China

[2]Multi-Functional Design and Research Academy, Zhengzhou University, Zhengzhou 450001, China

[3]School of Electronic and Information Engineering, University of Science and Technology Liaoning, Anshan 114044, China

ABSTRACT

PVC stripping process is a kind of complicated industrial process with characteristics of highly nonlinear and time varying. Aiming at the problem of establishing the accurate mathematics model due to the multivariable coupling and big time delay, the dynamic fuzzy neural network (D-FNN) is adopted to establish the PVC stripping process model based on the actual process operation datum. Then, the PVC stripping process is decoupled by the distributed neural network decoupling module to obtain two single-input-single-output (SISO) subsystems (slurry flow to top tower temperature and steam flow to bottom tower temperature). Finally, the PID controller based on BP neural networks is used to control the decoupled PVC stripper system. Simulation results show the effectiveness of the proposed integrated intelligent control method.

INTRODUCTION

With VCM as a raw material, the suspension method to produce polyvinyl chloride (PVC) resin is a kind of typical intermittent chemical production process [1]. The traditional polymerization monomer conversion ratio is controlled to the scope 85%–90%. Unreacted vinyl chloride monomer is recovered by self-pressure, and 1%-2% of the monomers are still remained in PVC slurry. In the production, vinyl chloride monomer must be getting a further removal and recovery due to its some toxicity. The removal of residual vinyl chloride in polyvinyl chloride can improve the quality of polyvinyl chloride, reduce the cost of production, and solve the problem of environmental pollution caused in plastic processing and usage. At present, the stripping technique is very effective among all the removal methods of redundant VCM monomers in PVC slurry [2]. Now the research on the control methods in PVC industry mostly is in the stage of polymerization and distillation, where many advanced control techniques, such as neural network, fuzzy control, expert system, and predictive control methods, have been widely used [3–5]. Li [3] proposed a BP neural network decoupling controller applied to the temperature control of rectifying column. The simulation results show that the algorithm has strong robustness. Xiao et al. [4] established steady-state optimization model of expanded bed energy consumption and designed feedback control system

Citation: Shu-zhi Gao, Jing Yang, and Jie-sheng Wang, "D-FNN Based Modeling and BP Neural Network Decoupling Control of PVC Stripping Process," Mathematical Problems in Engineering, vol. 2014, Article ID 681259, 13 pages, 2014. doi:10.1155/2014/681259

of multivariable state based on artificial neural network for the drying process of fluidized bed, effectively removing the decoupling association between the variables and achieving the dynamic optimization control of bed temperature. The actual operation results show that this system has strong robustness and achieve the minimum energy consumption control indicators of expanded bed. Khairiyah et al. [5] proposed the predictive control system based on neural network applied to the control of the rectifying tower, which makes the system have a smaller overshoot and shorter response time.

However, the research on the advanced control methods for the PVC stripping process is lack. The stripping process has characteristics of highly nonlinear, strong coupling, and time varying. It is difficult to establish a precise mathematical model and achieve good control effect by using traditional control methods.

Dynamic fuzzy neural network (DFNN) is a hybrid model of fuzzy theory and neural network method, whose function is equivalent to the TSK fuzzy system [6–10]. Based on the problem complexity and precision demand, D-FNN model can be constructed combining the system prior knowledge. An improved fuzzy neural network modeling method [11] was proposed to realize the soft sensor of molten index in the polymerization process, in which a hybrid learning method based on LM algorithm and a steepest descend method to train the network parameters exist. Simulation results show that the model has the merits of being not sensitive to initial values, fast convergence velocity, and high forecasting precision. The particle swarm algorithm (PSO) is adopted to optimize the parameters of the fuzzy neural network to set up a soft-sensor model of acrylonitrile yield [12]. Li and Lee [13] proposed an interval type-2 fuzzy system integrated with an observer-based hierarchical fuzzy neural controller, which can greatly reduce the number of adjusted parameters. Wu et al. proposed the maximum likelihood estimators used in nonparametric maximum likelihood fuzzy neural networks (MFNNs) for nonlinear regression problems and simple weight updating rules based on gradient descent and iteratively reweighted least squares (IRLS) will be derived [14]. Nasr and Chtourou [15] proposed a self-organizing map-based initialization for hybrid training based on a two-stage learning approach for feedforward neural networks. The weights between input and hidden layers are firstly adjusted by Kohonen algorithm with fuzzy neighborhood, whereas the weights connecting hidden and output layers are adjusted using gradient descent method. Hung-Ching proposed an adaptive self-constructing fuzzy neural network (ASCFNN) controller for a real inverted pendulum system, where the Mahalanobis distance (M-distance) method in the structure learning is also employed to determine if the fuzzy rules are generated/eliminated or not [16]. D-FNN is also successfully applied in many fields, such as function approximation, permanent-magnet synchronous motor drive control [17, 18].

In this paper, the D-FNN model of the PVC stripping process has been established firstly. Then the neural network decoupling modules and BP neural network PID controller are applied in the stripping temperature control. The effectiveness of the proposed intelligent control strategy is verified by the simulation experiments. The paper is organized as follows. In Section 2, the technique flowchart of PVC stripping process is introduced. The stripping process model based on dynamic fuzzy neural networks is presented in Section 3. In Section 4, the neural network decoupling controller of stripping process is summarized. In Section 5, simulation results are introduced in detail. Finally, the conclusion illustrates the last part.

TECHNIQUE FLOWCHART OF PVC STRIPPING PROCESS

In the polymerization reaction process of vinyl chloride, according to the different techniques and resin grades, the polymerization conversion rate of vinyl chloride is generally controlled within the scope of (80%, 90%). Even the unreacted monomer is disposed by the autoclave recovery and precipitated groove blowing, and there still remains small amount of vinyl chloride monomer in resin products, which will influence the products quality and increase the hazards when using the produced plastic. In addition, vinyl chloride emission into atmosphere per year not only causes a huge waste of raw materials, but also causes serious environment pollution. Currently, among all the approaches to the removal of excess VCM monomer in PVC slurry, the stripping technique is very effective.

Firstly, the stripping mechanism of PVC slurry is introduced in detail. In the PVC suspension to be stripping treated in the off chute of PVC polymerization, the mass fraction of PVC is about 25%–30%, the VCM is 2%–4%, and the rest is water. The mass approximation allocation ratio of VCM in gassed is about 1 : 1000 : 100. Because of the different amount distribution of VCM in the three phases, the slurry itself has the characteristic of the diffused concentration gradient. In PVC slurry, the VCM in the solid PVC either by parsing or by spreading must pass through the film layer to a low concentration of VCM in water for spreading through the pores in PVC particles. Moreover, the solubility of VCM in water changes within the variation of temperature. In different temperature, when the VCM in water reaches to solubility, the part which diffused from PVC into water will break the static pressure of the liquid water layer and diffuse into the gas phase. With the above mechanism, the heat exchange makes the slurry's temperature increase so that the VCM will be parsed.

There are two main tower stripping devices. What we research in this paper is the sieve wear weir plate stripper. The stripping technique of the PVC slurry is shown in Figure 1.

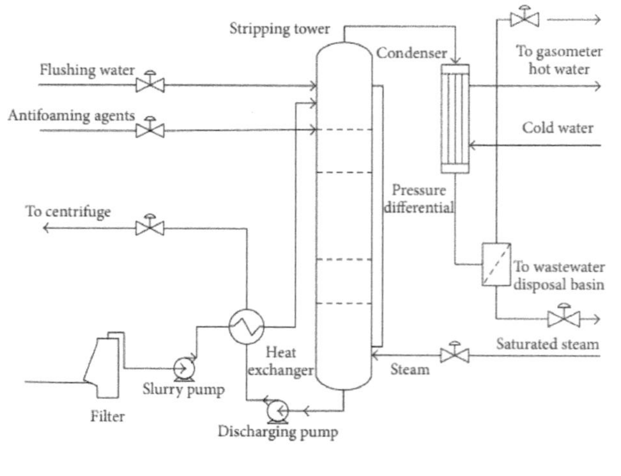

Figure 1: Flow diagram of PVC stripping process.

After the polymerization reaction, the PVC slurry is threw into the top of stripper from the polymerization vessel through the inlet slurry pump. But before entering the stripper, they need to carry through the heat exchange with the slurry stripped from the bottom of stripper and then go into the top of stripper. PVC slurry sprinkles evenly on the tray by sprinklers and flow from the top to bottom along the strays from the first layer stray. The low pressure steam enters from the stripper bottom, rises through the orifice of the tray from the bottom to top, and countercurrent contacts with the declining PVC slurry at each plate for heat and mass transfer. Finally, it will resolve internal vinyl chloride and the residue in the surface of the resin particles. Steam and parsed vinyl chloride gas are discharged from the top of the stripper. After the condensation by overhead condenser, it enters into the condensate sink, then separates condensate, and controls the level of condensate sink. The excess condensate will be discharged to the centrifuge mother liquor water recovery unit, and the separate vinyl chloride gas will be recovered to vinyl chloride gas cabinets. The key parameters influencing the PVC stripping process can mainly include temperature, steam, slurry flow rate, and pressure. Table 1 gives the parameters values in the PVC stripper controller of a certain type resin.

Table 1: Parameters of PVC string tower

Parameter	Scope	Alarm value	Normal scope	Real value
Temperature in tower top	0–150°C	Upper limit: 110°C Lower limit: 70°C	80–105°C	85°C
Temperature in tower bottom	0–150°C	Upper limit: 130°C Lower limit: 100°C	105–113°C	107°C
Pressure in tower top	0-0.1 MPa	Upper limit: 0.08 MPa Lower limit: 0.02 MPa	0.03–0.06 MPa	0.03 MPa
Pressure difference in tower	0–60 MPa	Upper limit: 30 MPa Lower limit: 5 MPa	10–25 MPa	15 MPa
Steam flow rate	0–4000 kg/h	Upper limit: 3500 kg/h Lower limit: 1000 kg/h	1500–3000 kg/h	2200 kg/h
Inlet slurry flow rate	0–80 m^3/h	Upper limit: 70 m^3/h Lower limit: 15 m^3/h	35–50 m^3/h	40–45 m^3/h
Steam pressure in tower	0–0.6 MPa	Upper limit: 0.5 MPa Lower limit: 0.15 MPa	0.35–0.4 MPa	0.38 MPa

Through the analysis of PVC stripping process, it has the characteristics of strong nonlinear and coupling. Because of lack of effective mathematical model and strong coupling between the variables, it is difficult to achieve satisfactory control effect. The integrated model and hybrid intelligent control strategies based on neural network technology are put forward for the PVC stripping process in this paper.

The data used for modeling all derived from the history data recorded in the operating process of PVC stripping tower. Each data sample includes the following elements: top tower temperature, bottom tower temperature, top tower pressure, tower differential pressure, slurry flow rate, and steam flow rate. According to the analysis of stripping technique, the key factors which influence vinyl chloride removed from PVC are the top tower temperature and the bottom tower temperature. So there are two mainly important control factors, which are the steam flow rate and the slurry flow rate during the process regulation. Therefore, the D-FNN model is built in which the steam flow rate and slurry flow rate are the input variables, and the top tower temperature and the bottom tower temperature are the output variables.

STRIPPING PROCESS MODELLING BASED ON DYNAMIC FUZZY NEURAL NETWORKS

According to the analysis of the stripping process, the changeable slurry flow or the steam flow rate not only influences the top tower temperature but also the bottom tower temperature. Consequently, the coupling exists in the PVC stripping process. The coupling object model needs to be built and the stripping process needs to be decoupled. Suppose the dynamic system model is described as the following nonlinear equation:

$$y(k) = f\left[y(k-1),\dots,y(k-n_a),u_1(k),\dots,u_1(k-n_b),\right.$$
$$\left.u_2(k),\dots,u_2(k-n_b)\right], \tag{1}$$

where y is the top tower temperature or bottom tower temperature, u_1 is the steam flow rate, u_2 is the slurry flow rate, and k is the sample time (20 s).

The D-FNN control strategy of the PVC stripping process is put forward shown in Figure 2, whose inputs variables are the steam flow rate and the slurry flow rate and output variables are the top tower temperature and the bottom tower temperature.

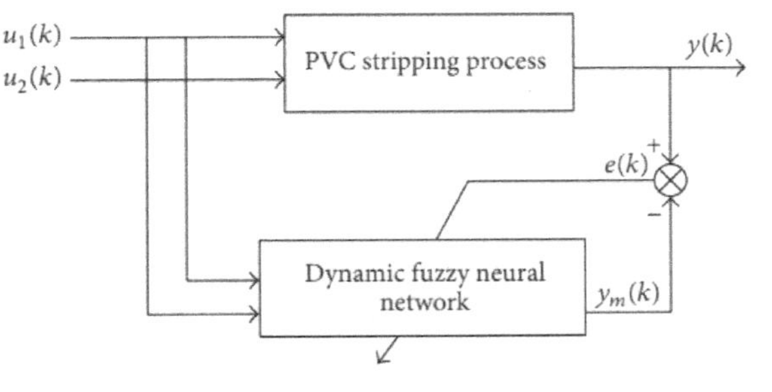

Figure 2: Modeling of stripping process based on D-FNN.

Structure of D-FNN

The structure diagram of D-FNN [19–23] is shown in Figure 3. In Figure 3, x_1, x_2,...,$_r$ is the input linguistic variables, y is the system output, MFi_j represents the jth membership function of the ith input variable, R_j is the jth fuzzy rule, N_j is the jth normalized node, w_j is the connection weight of the jth rule, and u refers to the numbers of the system rules. Then each layer of the D-FNN is described in detail.

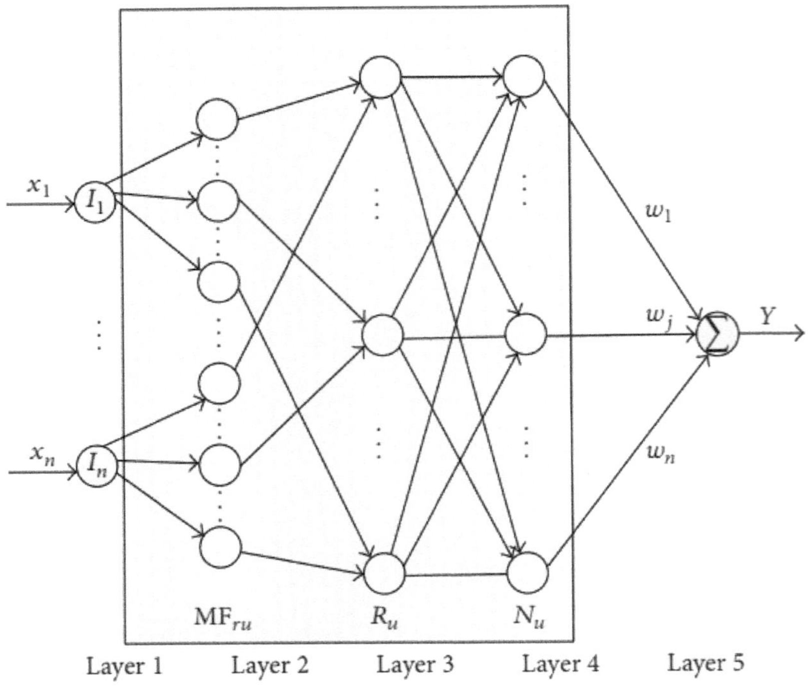

Figure 3: Structure diagram of D-FNN.

Layer 1 is the input layer, whose each node represents an input linguistic variable.

Layer 2 is the membership function layer, whose each node represents a membership function (MF) defined in the followed Gaussian function:

$$\mu_{ij}(x_i) = \exp\left[-\frac{(x_i - c_{ij})^2}{\sigma_j^2}\right], \tag{2}$$

where $i = 1, 2, \ldots, r$, $j = 1, 2, \ldots u$, μ_{ij} is the jth membership function, c_{ij} is the center of the jth Gaussian membership function, σ_j is the width of the jth Gaussian membership function of x_i, and r is the number of input variables.

Layer 3 is the T-norm layer, whose each node represents a possible IF-part of fuzzy rules. For the jth rule R_j, its output is defined as follows:

$$\phi_j = \exp\left[-\frac{\sum_{i=1}^{r}(x_i - c_{ij})^2}{\sigma_j^2}\right] = \exp\left[\frac{-\|X - C_j\|^2}{\sigma_j^2}\right], \tag{3}$$

where $j = 1, \ldots, u$, $X = (x_1,\ldots,x_r)$ and $C_j = c_{1j},\ldots,c_{rj}$ is the center of the jth RBF unit.

Layer 4 is the normalization layer, where the output of the N_j node is defined as

$$\psi_j = \frac{\varphi_j}{\sum_{k=1}^{u}\varphi_k} \quad j = 1, 2, \ldots, u. \tag{4}$$

Layer 5 is the output layer, whose each node is an output variable, a summation of input signals described as follows:

$$y(K) = \sum_{k=1}^{u} w_k \cdot \psi_k, \tag{5}$$

where y is the output variable and w_k is the weight of the kth rule.

For the TSK model, it can be represented as follows:

$$w_k = a_{k0} + a_{k1}x_1 + \cdots + a_{kr}x_r \quad k = 1, 2, \ldots, u. \tag{6}$$

Then (2)–(5) are fed into (5) to obtain

$$y(X) = \sum_{i=1}^{u}\left[(a_{i0} + a_{i1}x_1 + \cdots + a_{ir}x_r) \times \exp\left(-\frac{\|X - C_i\|^2}{\sigma_i^2}\right)\right] \times \left(\sum_{i=1}^{u}\exp\left(-\frac{\|X - C_i\|^2}{\sigma_i^2}\right)\right)^{-1}. \tag{7}$$

Learning Algorithm of D-FNN

Each node in layer 3 of the D-FNN represents an RBF unit or the IF-part for fuzzy rules. If the number of fuzzy rules is required to be identified, the D-FNN structure is not predefined. So a new learning algorithm for the D-FNN is adopted to automatically determine the fuzzy rules satisfying the required system performances.

Rule Generation Criteria

If there are only few rules, the system cannot fully represent the input-output state space. But if there are too many rules, the system can be unnecessarily complicated, and the generic ability of D-FNN can become worse because of the greatly increased calculation burden. So the system errors are the principal elements to determine whether the new rules can be added.

The error criterion can be described as follows. For the ith observation point (X_i, t_i), where X_i is the input vector and t_i is the desired output, compute the overall D-FNN output yi according to (7). Define the error as follows:

$$\|e_i\| = \|t_i - y_i\|. \tag{8}$$

If the following condition is satisfied, a new rule should be added. In (9), k_e is determined according to the desired accuracy of the D-FNN:

$$\|e_i\| > k_e. \tag{9}$$

Accommodation Boundary

Gaussian function has good locality characteristic because its output is monotonic decreasing along with the increasing distance between the output and the centers. When the membership function of input variables is described by Gaussian function, the whole input space is divided by a series of Gaussian function. If a new sample is located in some Gaussian function, that is to say within accommodation boundary, the new sample can be represented by the existed Gaussian function, without new Gaussian unit produced by D-FNN.

The accommodation boundary criterion is described as follows. For the ith observation point (X_i, t_i), calculate the distance $d(j)$ between the observation X_i and the center C_j of the existed RBF units by

$$d_i(j) = \|X_i - C_j\|, \quad j = 1, 2, \ldots, u. \tag{10}$$

Find the minimum distance by (11). If $d_{\min} > k_d$, a new fuzzy rule should be added. Otherwise, the observation X_i can be represented by the nearest existed RBF unit. Here, k_d is the effective radius of the accommodation boundary

$$d_{\min} = \arg\min(d_i(j)). \tag{11}$$

Hierarchical Learning Strategy

The concept of hierarchical learning is that the accommodation boundary of each RBF unit is not fixed but adjusted dynamically in the following style. Firstly, the accommodation boundaries are set larger for achieving rough but global learning. Then, they are gradually reduced for fine adjustment. Inspired by this idea, a simple method based on monotonically decreasing function is proposed to reduce both the effective radius of each RBF unit and error index gradually [22]. To be more specific, the discussed above k_e and k_d are not constant. They are changeable according to the following equations:

$$k_e = \max\left[e_{\max} \times \beta^i, e_{\min}\right],$$

$$k_d = \max\left[d_{\max} \times \gamma^i, d_{\min}\right], \tag{12}$$

where $e_{\max}$ is the predefined maximum error, $e_{\min}$ is the desired D-FNN accuracy, β ($0 < \beta < 1$) is the convergence constant, γ ($0 < \gamma < 1$) is the decay constant, and $d_{\max}$ and $d_{\min}$ are the largest and the smallest length of the input space, respectively.

The key idea of the hierarchical learning is to firstly find the most troublesome positions, which have large errors between the desired and the actual outputs but are not properly covered by existing fuzzy rules. This stage is called the coarse learning. When k_e and k_d reach $e_{\min}$ and $d_{\min}$, respectively, this stage is named the fine learning.

Allocation of Precedent Parameters

After a fuzzy rule is obtained, the problem is how to determine its parameters. The width of the RBF unit has great significant on its generalization. If it is less than the distance between adjacent inputs, the RBF unit does not generalize well and D-FNN will not give meaningful outputs in response to inputs. However, if the width is too large, the output of the RBF nodes may always be large irrespective of inputs. Hence, the initial parameters of the new produced rule are allocated based on the following equations:

$$C_i = X_i,$$

$$\sigma_i = k \times d_{\min}, \tag{13}$$

where k ($k > 1$) is an overlap factor.

What we needed to pay attention to is that when the first pattern (X_1, t_1) enters the D-FNN, it is selected as the first fuzzy rule: $C_1 = X_1$, $\sigma_1 = \sigma_0$, where σ_0 is a preset constant. It is worth highlighted that only when $\|e_i\| > k_e$, $d_{\min} > k_d$, one fuzzy rule needs to be added.

Determination of Decision Parameters

According to the criteria of rule generation, suppose u fuzzy rules are generated based on the n observation datum. So the output of N nodes can be obtained according to (4), which can be represented as follows:

$$\psi = \begin{bmatrix} \psi_{11} & \cdots & \psi_{1n} \\ \vdots & & \vdots & \vdots \\ \psi_{u1} & \cdots & \psi_{un} \end{bmatrix}. \tag{14}$$

Then, for any input X_j (x_{1j}, x_{2j},...,$_{rj}$), the system output y_j can be calculated by (5), which can be described in more compact form:

$$W\Psi = Y. \tag{15}$$

Suppose the ideal output is $T = (t_1, t_2, \ldots, _n) \in R^n$. The relation between $\Psi \in \mathfrak{R}^{(r+1)u\times n}$ and $T \in \mathfrak{R}^n$ is represented as follows:

$$Y = W \times \Psi,$$

$$\tilde{E} = \|T - Y\|. \tag{16}$$

Find an optimal coefficient vector based on

$$W^* \times \Psi = T. \tag{17}$$

The optimal W^* is in the following form:

$$W^* = T\left(\Psi^T\Psi\right)^{-1}\Psi^T, \tag{18}$$

where Ψ^T is the transposition of Ψ and $\Psi^+ = (\Psi^T\Psi)^{-1}\Psi^T$ is the pseudoinverse of Ψ.

Simulation Results

The samples are derived from the actual operation datum of the certain chemical corporation, in which 250 groups of input-output data have been selected to identify the system randomly. The other 80 groups of input-output data are used to test the identification effect. Then the D-FNN models are obtained, in which the steam flow rate and slurry flow rate are inputs variables, and the top tower temperature and the bottom tower temperature are output variables separately. The identification effect and test results are shown in Figures 4 and5. It can be seen form Figures 4 and 5 that the output curves of the D-FNN model and the actual curves of the tower temperature match very well, and the errors are within ±0.1, which can meet the requirement of the process optimization control.

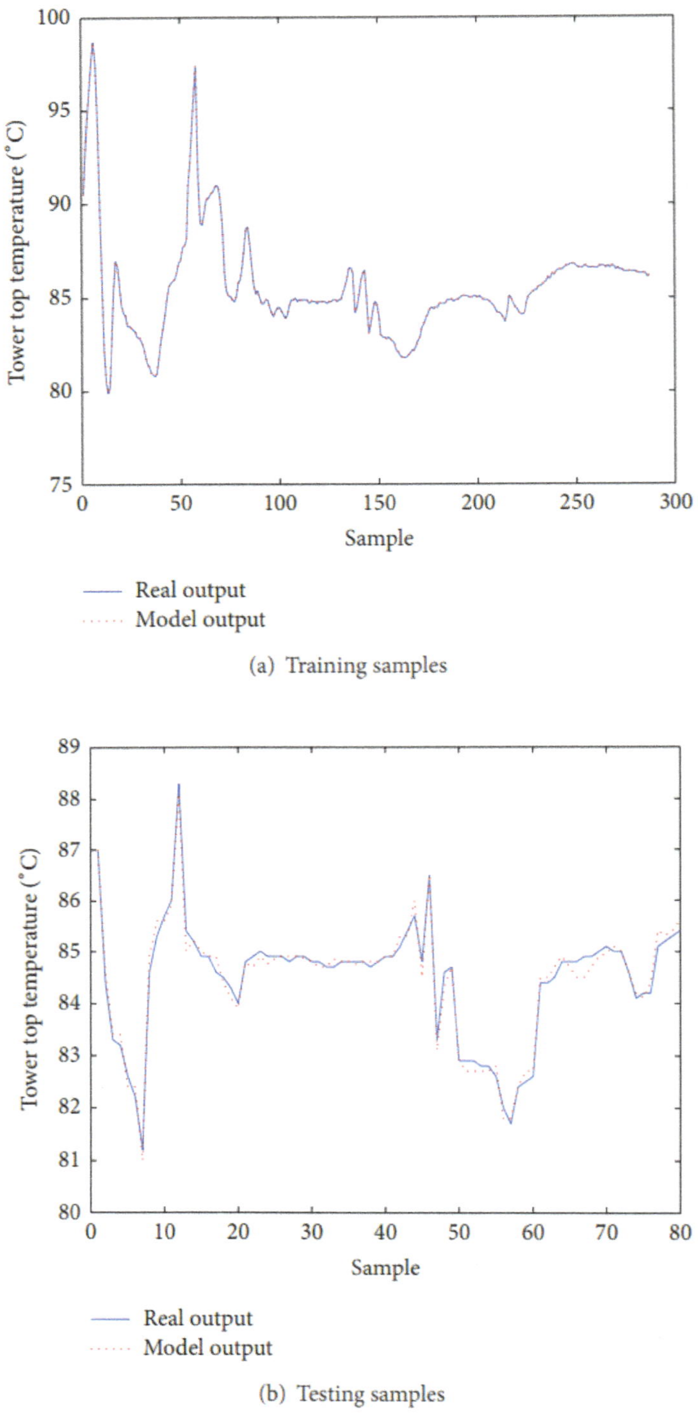

Figure 4: Simulation results of D-FNN modeling of tower top temperature.

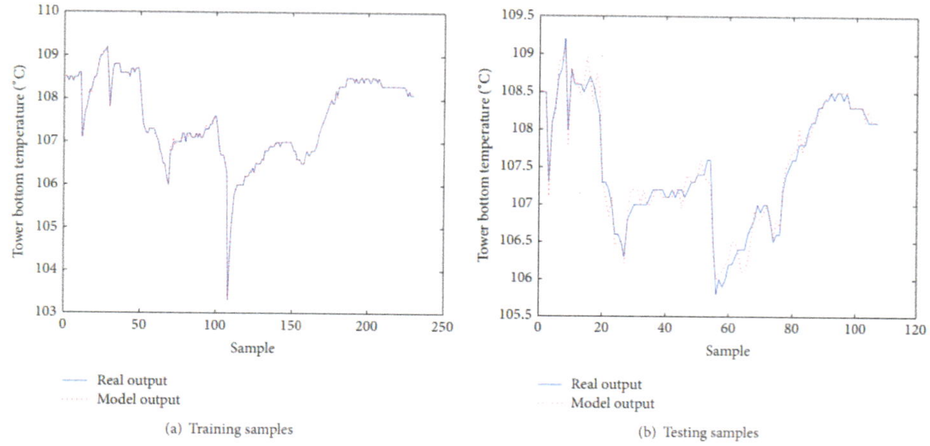

Figure 5: Simulation results of D-FNN modeling of tower bottom temperature.

NEURAL NETWORK DECOUPLING CONTROLLER OF STRIPPING PROCESS

Structure of Intelligent Decoupling Controller

The intelligent decoupling control strategy based on neural network is put forward aiming at the above-established PVC stripping process model. The system diagram is shown in Figure 6. The control system consists of PVC stripper object (slurry flow rate to top tower temperature and steam flow rate to tower bottom temperature), two neural network open-loop decoupling modules ND_1 and ND_2, and two PID controllers online optimized by two BP neural networks NN_1 and NN_2.

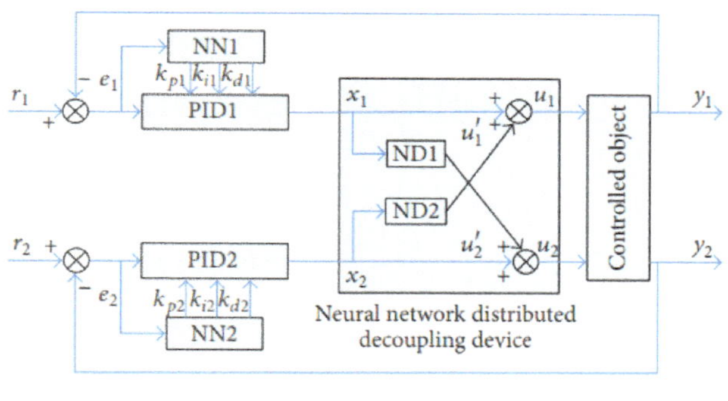

Figure 6: Decoupling controller structure of PVC stripping process.

In Figure 6, r_i is the system input, y_i is the system output, x_i is the controller output vector, u_i is the output vector of the neural network decoupling module, and u_i' is the control input vector (i = 1, 2). The intelligent decoupling control system of the stripping process can be divided into two stages.

Training Stage

The neural network PID controller is disconnected and the system is in open-loop training state. Firstly, the neural decoupling module ND1 is trained and ND2 is maintained as no input. If decoupling is successful, the channel between the slurry flow rate and the tower top temperature will not have output (y_1 is 0). Then y_1 is set as the training index function of ND1. The connection weights of ND1's each layer are adjusted. Similarly, ND2 is trained in the same way. Finally, the whole stripper controlled object is decoupled into two SISO systems (slurry flow rate, the tower top temperature, and the steam flow rate, the tower bottom temperature).

Control Stage

NN_1 and NN_2 are used to online optimize the PID controller parameters for realizing the closed-loop control for the two SISO systems. After training of ND_1 and ND_2, the weights are maintained unchanged. At this time, the entire stripper controlled object is decoupled into two SISO systems. Then two BPNN PID controllers ($NN_1$and NN_2) are, respectively, adopted to realize the closed-loop control of two SISO systems. NN_1 and NN_2 will be used to online adjust the PID controller's parameters K_p, K_i, and K_d.

Structure of Neural Network Decoupling Module

The neural network decoupling module consists of $n \times (n - 1)$ SISO neural network (n is the number of the object inputs), whose structure is shown in Figure 7. Each neural network decoupling module is single-input-single-output (SISO) subsystem, which is responsible for only one channel decoupling. So its structure and indicator function are relatively simple because the total objective function has been decomposed into single objective functions and the decoupling modules' interconnection problem needs not be considered. At the same time, the dispersion of the objective function is also easy to achieve parallel distributed processing in real time. In addition, the decoupling structure does not change the dynamic characteristics of the controlled process main channel. It will make the design problem easy in engineering. As long as the reasonable structure and parameters of controller c_i are chosen, the required dynamic characteristics can be obtained.

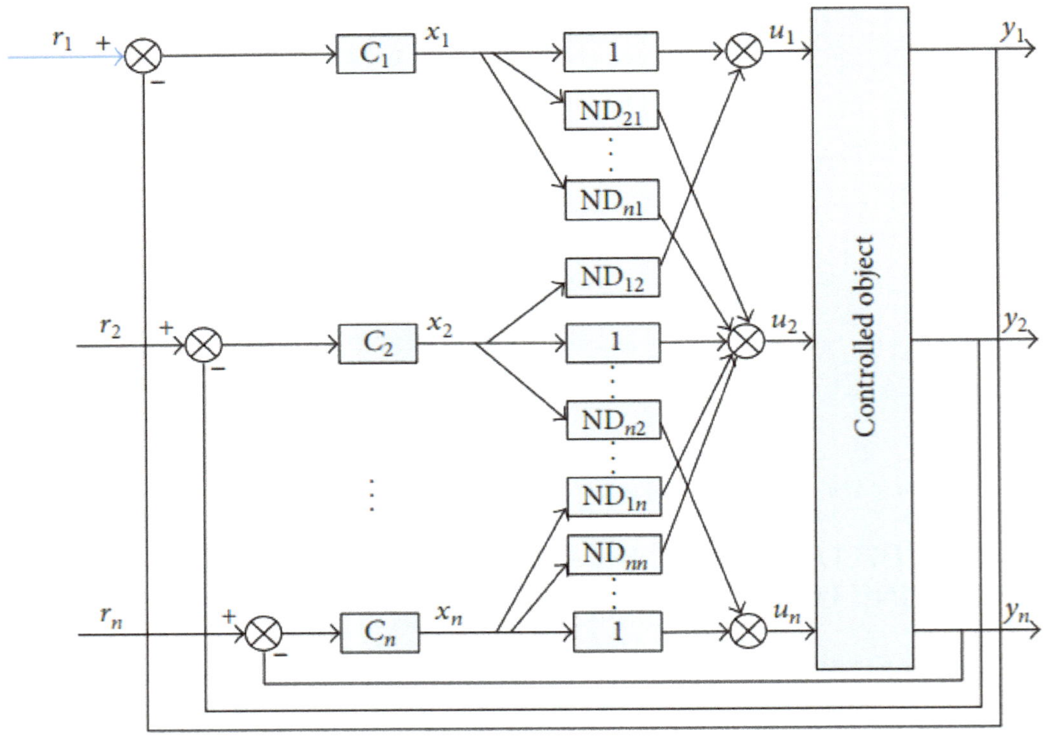

Figure 7: Structure diagram of decoupling module.

The PVC stripping process is double input and double output coupling system. The structure of the neural decoupling module is shown in Figure 7.

The decoupling module is composed of two different neural networks ND_1 and ND_2, which are dynamic time delay TDNN networks. Each network has an input terminal and an output terminal. The decoupling module outputs can be expressed as

$$u_2'(k) = f_1\left[x_1(k), x_1(k-1), \wedge, x_1(k-n_1), u_2'(k-1),\right.$$
$$\left. u_2'(k-2), \wedge, u_2'(k-m_2)\right],$$
$$u_1'(k) = f_2\left[x_2(k), x_2(k-1), \wedge, x_2(k-n_2), u_1'(k-1),\right.$$
$$\left. u_1'(k-2), \wedge, u_1'(k-m_1)\right], \qquad (19)$$

where u is the output of neural network decoupling module, k is sampling frequency, and m_1, m_2, n_1, and n_2 are the positive integers, which are determined according to the controlled process' order.

The decoupling module in this paper adopts a three-layer neural network. The neural network's excitation function is the hyperbolic tangent function $(x) = (1 - e^{-x})/(1 + e^{-x})$. Its derivative is $(x) = 1/(2 \times (1 - f^2(x)))$. $u_1(k)$ and $u_2(k)$ are the outputs of the decoupling module. After superimposing the controller outputs x_1 and x_2, the system control output is obtained by $u_1 = u_1' + x_1$ and $u_2 = u_2' + x_2$.

Training Algorithm of Neural Network Decoupling Module

For the neural decoupling module shown in Figure 8, theoretically speaking, the expected output of the neural decoupling module should be viewed as the target function. But the NN expected output is unknown. In order to avoid this contradiction, the neural decoupling module and the controlled object are viewed as a generalized object. Thus the neural network training can adopt the system output as the performance function. For neural decoupling modules ND_1 and ND_2, the performance index functions are defined as follows:

$$J_1 = \frac{1}{2}\sum_{K=0}^{N}\left[y_2(k)\right]^2,$$
$$J_2 = \frac{1}{2}\sum_{K=0}^{N}\left[y_1(k)\right]^2. \qquad (20)$$

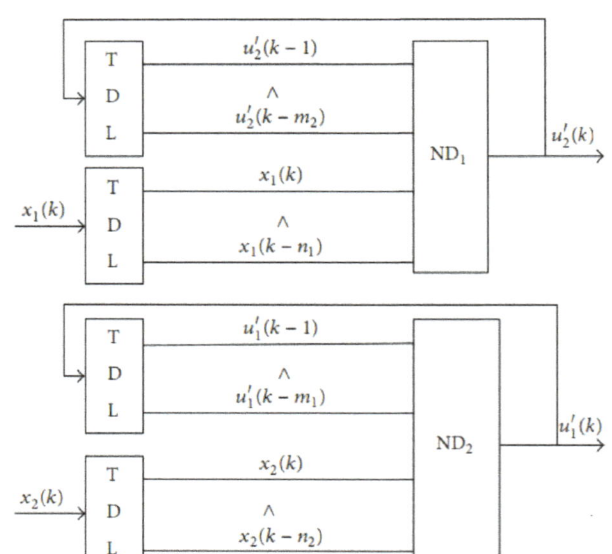

Figure 8: Schematic diagram of distributed decoupling.

According to the BP learning algorithm [24, 25], the network weights are updated based on the following equation:

$$\Delta w(k) = -\eta \frac{\partial J}{\partial w} + \alpha * \Delta w(k-1), \tag{21}$$

where η is the learning rate, α is the momentum coefficient, is the weights of the corresponding neural network, and is the number for training samples.

For ND_1 and ND_2, differentiate the performance index function to obtain

$$\Delta w(k) = -\eta \sum^{N} y_2 \frac{\partial y_2}{\partial u_2'} \frac{\partial u_2'}{\partial w} + \alpha * \Delta w(k-1),$$

$$\Delta w(k) = -\eta \sum^{N} y_1 \frac{\partial y_1}{\partial u_1'} \frac{\partial u_1'}{\partial w} + \alpha * \Delta w(k-1). \tag{22}$$

For the convenience of discussion, the subscripts are omitted to get the universal formula:

$$\Delta w(k) = -\eta \sum^{N} y \frac{\partial y}{\partial u'} \frac{\partial u'}{\partial w} + \alpha * \Delta w(k-1), \tag{23}$$

where Δw is the change of weight.

In order to adopt BP learning algorithm, the partial derivative $\partial y / \partial u$ needs to be calculated. However, the learning rate is an adjustable parameter. The partial derivatives multiply the learning rate's product deciding the change of weights. Therefore, the absolute value of the partial derivatives is not very important. If the partial derivative's symbols are known, it is recorded as sign $(\partial y / \partial u)$. Adjust the learning rate to obtain

$$\eta' = \eta \left| \frac{\partial y}{\partial u'} \right|. \tag{24}$$

Therefore, (23) can be written as

$$\Delta w(k) = -\eta' \operatorname{sign}\left(\frac{\partial y}{\partial u'}\right) \sum^{N} y \frac{\partial u'}{\partial w} + \alpha * \Delta w(k-1). \tag{25}$$

In the actual industrial production, there are many controlled processes which have monotone increasing or decreasing characteristics. That is to say, y is the monotonically increasing or decreasing function of u'. When y is the monotone increasing of u', $\operatorname{sign}(\partial y / \partial u) = 1$, or it is −1.

The relationship between network input and output can be expressed as

$$u' = w^{(2)} u^{(1)},$$

$$u^{(1)} = f\left(w^{(1)} u^{(0)}\right), \tag{26}$$

$$u^{(0)} = \left[x(k), x(k-1), \wedge, x(k-n), u'(k-1),\right.$$

$$\left. u'(k-2), \wedge, u'(k-m)\right]^T, \tag{27}$$

where u' is the output signal of the network output layer, $u^{(1)}$ is the output signal of the network middle layer, $u^{(1)} = [u_1^{(1)}, u_2^{(1)}, u_3^{(1)}, \wedge]$ and $u^{(0)}$ is the input signal of the network input layer, and $w^{(2)} = [w_0^{(2)}, w_1^{(2)}, w_2^{(2)}, \ldots]$ is the connection weight vector between network middle layer and output layer. $w^{(1)} = \begin{bmatrix} w_{11}^{(1)} & w_{12}^{(1)} & \cdots \\ w_{21}^{(1)} & w_{22}^{(1)} & \cdots \\ \cdots & \cdots & \ddots \end{bmatrix}$ is the connection weight vector between network input layer and middle layer.

For the output layer,

$$\frac{\partial u'}{\partial w_i^{(2)}} = \frac{1}{2}\left(1 - u'^2\right) u_i^{(1)}. \tag{28}$$

For the middle layer,

$$\frac{\partial u'}{\partial w_{ji}^{(1)}} = \frac{\partial u'}{\partial u_j^{(1)}} \frac{\partial u_j^{(1)}}{\partial w_{ji}^{(1)}} = \frac{1}{2}\left(1 - u'^2\right) w_j^{(2)} \frac{1}{2}\left(1 - u_j^{(1)2}\right) u_i^{(0)}. \tag{29}$$

Put (28) and (29) into (23); the weight of the output layer and middle layer is updated by the following equations.

For the output layer,

$$\Delta w_i(k) = -\frac{\eta'}{2} \operatorname{sign}\left(\frac{\partial y}{\partial u'}\right) \sum^{N} y\left(1 - u'^2\right) u_i^{(1)}$$

$$+ \alpha * \Delta w_i(k-1). \tag{30}$$

For the middle layer

$$\Delta w_{ji}(k) = -\frac{\eta}{4} \operatorname{sign}\left(\frac{\partial y}{\partial u'}\right) \sum^{N} y\left(1 - u'^2\right) w_j^{(2)}\left(1 - u_j^{(1)2}\right) u_i^{(0)}$$

$$+ \alpha * \Delta w_{ji}(k-1). \tag{31}$$

The algorithm procedure is described as follows.

(1) Connect the neural network decoupling module ND_1 and ND_2 with the controlled process.

(2) Initialize the neural decoupling module weights with the smaller random values.

(3) Give the input signals and make the controller output signal correspond to the input signal of the neural decoupling module.

(4) According to the selected samples and time interval, adopt (26) to calculate the output of the neural network u', $u^{(1)}$, $u^{(0)}$, and N values of the process variables changed with time (set u' = 0 at the begin of calculation).

(5) According to (30) and (31), calculate the weight variances of the output layer and middle layer and update weights $^{(1)}$ and $w^{(2)}$.

(6) Replete steps (4) and (5) until reaching the specified iterations to determine $^{(2)}$ and $w^{(1)}$ at the end.

The above process does not involve each variable's subscript. In practice, the training of network weights can be completed individually according to the above procedure. When ND_1, ND_2,..., ND_n are determined, the whole system can be viewed as n single variable nonlinear systems. Then each loop can be designed the corresponded controller.

PID Controller Based on BP Neural Network

The structure of PID controller based on BP neural network is shown in Figure 9. According to the system's operation states, the PID controller parameters are adjusted to achieve the optimization performance. The neural units in the output layer are in correspondence with three PID controller adjustable parameters k_p, k_i, and k_d. Through self-learning of neural network, the weighted coefficients are adjusted to make the neural network output corresponding to the PID controller parameters with a certain optimal control law.

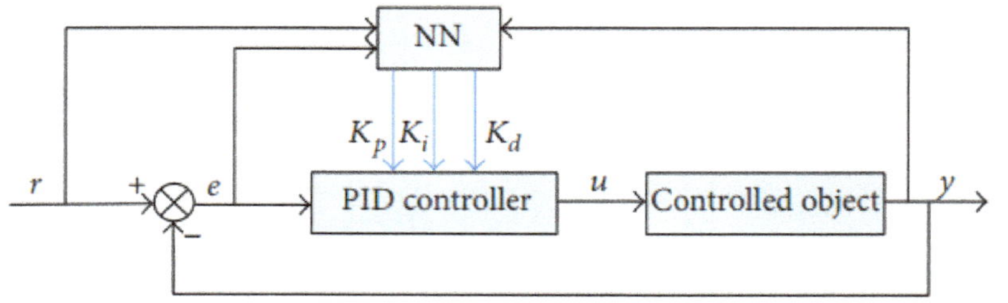

Figure 9: PID controller based on BP neural network.

Three layers BP neural network [26] is used for the PVC stripping process. The input layer have 3 neurons: input r, output y, and error error:

$$O_i = x_{i1}(i), \quad i = 1, 2, 3, \tag{32}$$

where the subscripts i, j, and l, respectively, represent the input layer, the hidden layer, and the output layer.

The hidden layer has 5 neurons. The input of the hidden layer is calculated by

$$net_j(k) = \sum_{i=1}^{3} w_{ji} O_i. \tag{33}$$

The output of the hidden layer is calculated by

$$O_j(k) = f\left(net_j(k)\right) \quad j = 1, 2, \ldots, 5. \tag{34}$$

The activation function of the hidden layer neuron adopts the symmetrical sigmoid function $(x) = \tanh(x) = (e^x - e^{-x})/(e^x + e^{-x})$.

The output layer has 3 neurons. The input of the output layer is calculated by

$$net_l(k) = \sum_{j=1}^{5} w_{lj} O_j(k). \tag{35}$$

The output of the output layer is calculated by

$$O_l(k) = g\left(net_l(k)\right), \quad l = 1, 2, 3. \tag{36}$$

The output layer nodes are in accordance with three adjustable parameters k_p, k_i, and k_d. Since k_p, k_i, and k_d cannot be negative, the output layer neuron's activation function is not negative sigmoid function: $(x) = 1/(2 \times (1 + \tanh(x))) = e^x/(e^x + e^{-x})$.

The performance index function is selected as $(k) = 1/(2 \times (r(k) - y(k))^2)$. The network weights w_{ji} and w_{lj} are revised according to the gradient descent method. That is to say, according to (k), the weighted coefficients are searched and adjusted in the negative gradient direction with an inertia which make search rapidly converge to global minima.

From the output layer to the hidden layer,

$$\Delta w_{lj}(k) = -\eta \frac{\partial E(k)}{\partial w_{lj}} + \alpha \Delta w_{lj}(k-1), \tag{37}$$

where η is the learning rate and α is the inertial coefficient:

$$\frac{\partial E(k)}{\partial w_{lj}(k)} = \frac{\partial E(k)}{\partial y(k)} \cdot \frac{\partial y(k)}{\partial u(k)} \cdot \frac{\partial u(k)}{\partial O_l(k)} \cdot \frac{\partial O_l(k)}{\partial net_l(k)} \cdot \frac{\partial net_l(k)}{\partial w_{lj}(k)},$$
$$\frac{\partial net_l(k)}{\partial w_{lj}(k)} = O_j(k). \tag{38}$$

Because $\partial(k)/\partial u(k)$ is unknown, it is approximated by the symbol function $\mathrm{sign}(\partial y(k)/\partial u(k))$. The problem of calculation accuracy can be solved by adjusting the learning rate η to compensate.

For the output layer,

$$\frac{\partial u(k)}{\partial O_1(k)} = error(k) - error(k-1),$$
$$\frac{\partial u(k)}{\partial O_2(k)} = error(k),$$
$$\frac{\partial u(k)}{\partial O_3(k)} = error(k) - 2error(k-1) + error(k-2). \tag{39}$$

Thus the learning algorithm of the output layer weights is described as

$$\Delta w_{lj}(k) = \alpha \Delta w_{lj}(k-1) + \eta \delta_l O_j(k),$$
$$\delta_l = error(k)\, sign\left(\frac{\partial y(k)}{\partial u(k)}\right) \frac{\partial u(k)}{\partial O_l(k)} g'\left(net_l(k)\right)$$
$$l = 1, 2, 3,$$
$$g'(x) = g(x)\left(1 - g(x)\right). \tag{40}$$

Similarly, the learning algorithm of the hidden layer weights is described as

$$\Delta w_{ji}(k) = \alpha \Delta w_{ji}(k-1) + \eta \delta_j O_i(k),$$
$$\delta_j = f'\left(net_j(k)\right) \sum_{l=1}^{3} \delta_l w_{lj}(k) \quad j = 1, 2, \ldots 5,$$
$$f'(x) = \frac{\left(1 - f^2(x)\right)}{2}. \tag{41}$$

The algorithm procedure is described as follows.

(1) Determine the structure of BP neural network.That is to say, determine the number of the input layer nodes and the hidden layer nodes. Given all initial weighting coefficients, select the learning rate η and the inertial coefficient α.

(2) Sample and obtain (k) and $y(k)$. Then calculate $erro(k) = r(k) - y(k)$.

(3) Calculate each layer's input and output of the neural network. The output in BPNN is the three adjustable parameters of the PID controller.

(4) Calculate the PID output according to the incremental algorithm.

(5) Train BP neural network to online adjust the weights for realizing the self-tuning of the PID controller parameters.

(6) Set k=k+1 and repeat from step (2) to step (6).

SIMULATION EXPERIMENTS

In order to verify the effectiveness of the D-FNN model and the proposed intelligent decoupling control strategy based on neural network, the D-FNN model is used in neural network decoupling control strategy. Based on the tower top temperature D-FNN model and tower bottom temperature D-FNN model, the neural network decoupling modules ND_1 and ND_2 and neural network controller NN_1 and NN_2 are designed. The structure and the parameters of all neural networks are set as follows. The structure of ND_1 and ND_2 is 5-4-1, the momentum factor η_{ND} = 0.005, and the learning rate α_{ND} = 0.9. The structure of NN_1 and NN_2 is 3-5-3, the inertia η_{NN} = 0.25, and learning rate α_{NN} = 0.05. According to the technique of the PVC stripping process, the optimum of the stripper tower top temperature is 85° C. Thus in our simulation, the tower top temperature is 85° C. The temperature should not exceed 110° C, otherwise it will make the PVC resin decomposed and metamorphic. Therefore the upper limit is 110° C. The optimum of the stripper tower bottom temperature is 107° C and its upper limit is 113° C. According to the actual possible working conditions of the stripping process, this paper considers the following situations in the simulation.

(1) Without any disturbance, the tracking performance of the tower top temperature and tower bottom temperature is verified, which is shown in Figure 10. It can be seen form Figure 10 that the control system can track the input changes and has a high response performance.

(2) The random disturbance $\Delta y1$ = 0.01 is added on the tower top temperature and the simulation results are shown in Figure 11. The random disturbance $\Delta y2$ = 0.01 is added on the tower bottom temperature and the simulation results are shown in Figure 12. By adding the disturbance signals on the tower top temperature and tower bottom temperature, the system decoupling effects are verified. It can be seen form Figures 10 and 11 that the neural network decoupling modules improve the performance of the control system and the transaction between two loops can be controlled effectively.

Based on the above simulation results, the following conclusions can be made. (1) The tracking performance of the system is better and is able to adapt to the change of set points. The response time and other parameters are satisfactory. (2) The coupling degree between the slurry flow rate, the tower top temperature, and steam flow rate, the tower bottom temperature, has been significantly reduced and the decoupling is achieved. (3) The system has certain robustness for the external disturbances. It can overcome the temperature disturbance of the tower top temperature and the tower bottom temperature.

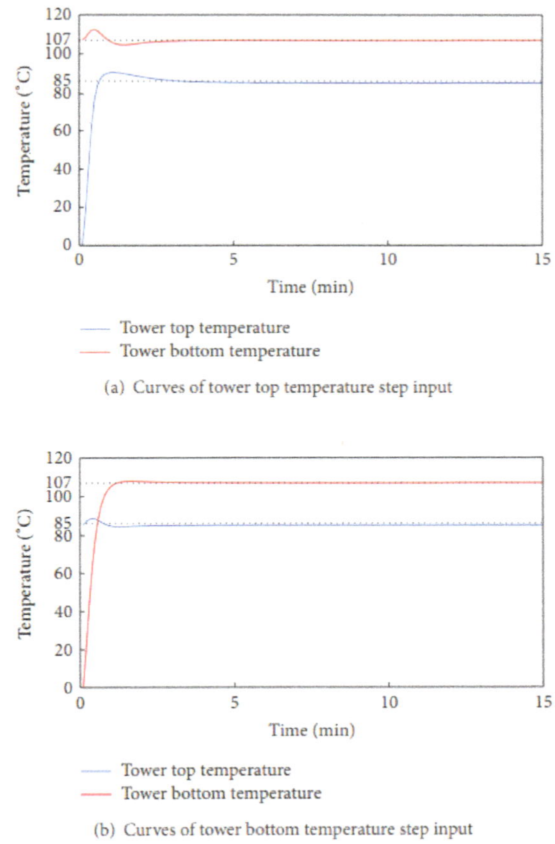

Figure 10: Simulation results of step input without disturbance.

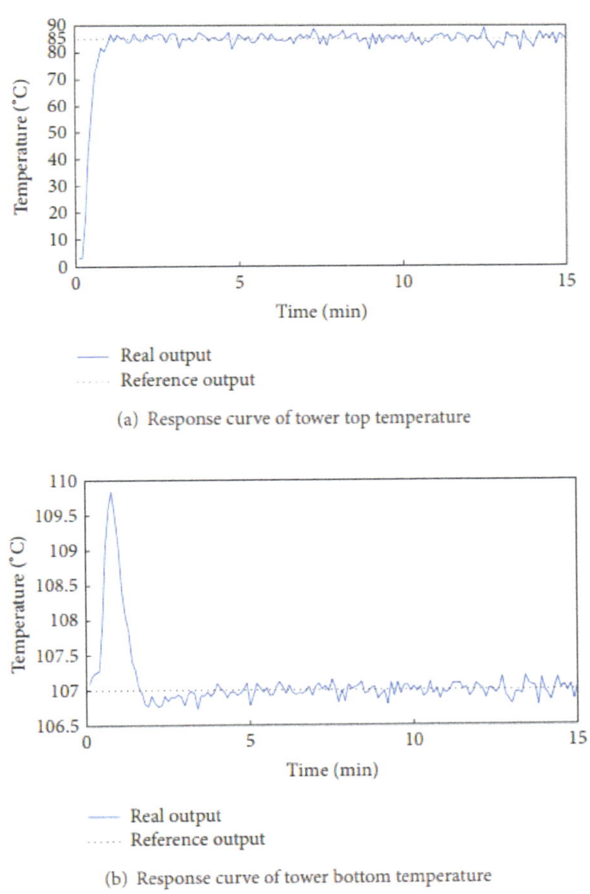

Figure 11: Simulation results with tower top temperature disturbance.

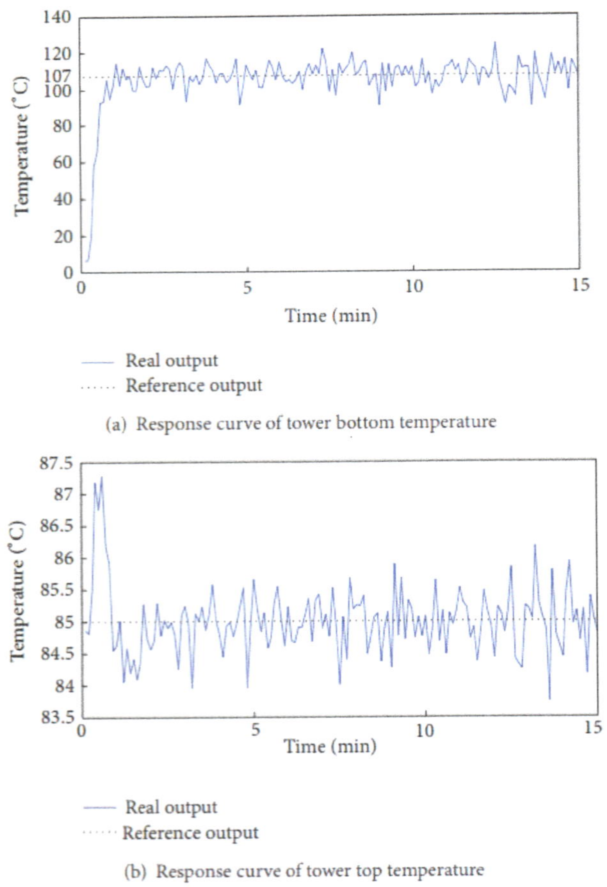

Figure 12: Simulation results with tower bottom temperature disturbance.

CONCLUSIONS

In the paper, the D-FNN was used to set up the PVC stripping process model. Then the strategy of combination of the neural network decoupling module and the BP neural network PID controller is put forward for controlling the stripping process temperature. The simulation results verify the effectiveness of the proposed PVC stripping intelligent control strategy.

ACKNOWLEDGMENTS

This work is partially supported by the Program for China Postdoctoral Science Foundation (Grant no. 20110491510), the Program for Liaoning Excellent Talents in University (Grant no. LJQ2011027), and the Program for Research Special Foundation of University of Science and Technology of Liaoning (Grant no. 2011ZX10).

REFERENCES

1. J. S. Wang, S. Han, and Q. P. Guo, "Echo state networks based predictive model of vinyl chloride monomer convention velocity optimized by artificial fish swarm algorithm," Soft Computing, vol. 18, no. 3, pp. 457–468, 2014. ·
2. Z. G. Hui, "The application of stripping technology in the production of PVC," Polyvinyl Chloride, vol. 7, no. 2, pp. 8–12, 2007.
3. H. Li, "Design of Multivariable Fuzzy-neural Network Decoupling Controller," Control and Decision, vol. 21, no. 5, pp. 593–596, 2006.
4. W. G. Xiao, Z. M. Wang, Y. F. Guo, et al., "PVC boiling bed drying process optimization control system,"Automation and Instruments in Chemical Industry, vol. 21, no. 5, pp. 12–15, 1994.
5. M. Y. Khairiyah, C. E. Boo, and T. S. Amy, Formulation of Model Predictive Control Algorithm For Nonlinear Processes, University of Technology Malaysia, 2006.
6. S. Q. Wu and J. Xu, Dynamic Fuzzy Neural Network-Design and Application, Tsinghua University Press, Beijing, China, 2008.
7. J. S. Wang and Q. P. Guo, "D-FNN based soft-sensor modeling and migration reconfiguration of polymerizing process," Applied Soft Computing, vol. 13, no. 4, pp. 1892–1901, 2013.
8. C. K. Kwong, T. C. Wong, and K. Y. Chan, "A methodology of generating customer satisfaction models for new product development using a neuro-fuzzy approach," Expert Systems with Applications, vol. 36, no. 8, pp. 11262–11270, 2009.
9. K. Y. Chan, C. K. Kwong, and Y. C. Tsim, "Modelling and optimization of fluid dispensing for electronic packaging using neural fuzzy networks and genetic algorithms," Engineering Applications of Artificial Intelligence, vol. 23, no. 1, pp. 18–26, 2010.
10. C. K. Kwong, K. Y. Chan, and Y. C. Tsim, "A genetic algorithm based knowledge discovery system for the design of fluid dispensing processes for electronic packaging," Expert Systems with Applications, vol. 36, no. 2, pp. 3829–3838, 2009.
11. R. L. Liu, H. Y. Su, and J. Chu, "A soft sensor modeling algorithm based on modified fuzzy neural network," Information and Control, vol. 32, no. 4, pp. 367–370, 2003.
12. G.-C. Chen, Y.-F. Xu, and J.-S. Yu, "Soft-sensor modelling of acrylonitrile yield based on particle swarm optimization fuzzy neural networks," Journal of System Simulation, vol. 19, no. 23, pp. 5370–5372, 2007. ·
13. I. Li and L. W. Lee, "Interval type 2 hierarchical FNN with the H-infinity condition for MIMO non-affine systems," Applied Soft Computing, vol. 12, no. 8, pp. 1996–2011, 2012. ·
14. H.-K. Wu, J.-H. Hsieh, Y.-L. Lin, and J.-H. Jeng, "On maximum likelihood fuzzy neural networks," Fuzzy Sets and Systems, vol. 161, no. 21, pp. 2795–2807, 2010. ·
15. M. B. Nasr and M. Chtourou, "A self-organizing map-based initialization for hybrid training of feedforward neural networks," Applied Soft Computing Journal, vol. 11, no. 8, pp. 4458–4464, 2011.
16. H. C. Lu, M. H. Chang, and C. H. Tsai, "Adaptive self-constructing fuzzy neural network controller for hardware implementation of an inverted pendulum system," Applied Soft Computing, vol. 11, no. 5, pp. 3962–3975, 2011.
17. S. Wu and M. J. Er, "Dynamic fuzzy neural networks—a novel approach to function approximation,"IEEE Transactions on Systems, Man, and Cybernetics B:

Cybernetics, vol. 30, no. 2, pp. 358–364, 2000. · ·

18. F. J. Lin, C. H. Lin, and P. H. Shen, "Self-constructing fuzzy neural network speed controller for permanent-magnet synchronous motor drive. Fuzzy Systems," IEEE Transactions on Fuzzy Systems, vol. 9, no. 5, pp. 751–759, 2001.
19. X. Deng and X. Wang, "Incremental learning of dynamic fuzzy neural networks for accurate system modeling," Fuzzy Sets and Systems, vol. 160, no. 7, pp. 972–987, 2009.·
20. C.-F. Juang and C.-D. Hsieh, "A locally recurrent fuzzy neural network with support vector regression for dynamic-system modeling," IEEE Transactions on Fuzzy Systems, vol. 18, no. 2, pp. 261–273, 2010.
21. A. Subasi, "Automatic detection of epileptic seizure using dynamic fuzzy neural networks," Expert Systems with Applications, vol. 31, no. 2, pp. 320–328, 2006. · ·
22. H. Adeli and X. Jiang, "Dynamic fuzzy wavelet neural network model for structural system identification," Journal of Structural Engineering, vol. 132, no. 1, pp. 102–111, 2006.
23. C.-H. Lee and C.-C. Teng, "Identification and control of dynamic systems using recurrent fuzzy neural networks," IEEE Transactions on Fuzzy Systems, vol. 8, no. 4, pp. 349–366, 2000.
24. B. H. M. Sadeghi, "BP-neural network predictor model for plastic injection molding process," Journal of Materials Processing Technology, vol. 103, no. 3, pp. 411–416, 2000.
25. Z. Xiao, S.-J. Ye, B. Zhong, and C.-X. Sun, "BP neural network with rough set for short term load forecasting," Expert Systems with Applications, vol. 36, no. 1, pp. 273–279, 2009.
26. J. Yi, Q. Wang, D. Zhao, and J. T. Wen, "BP neural network prediction-based variable-period sampling approach for networked control systems," Applied Mathematics and Computation, vol. 185, no. 2, pp. 976–988, 2007.

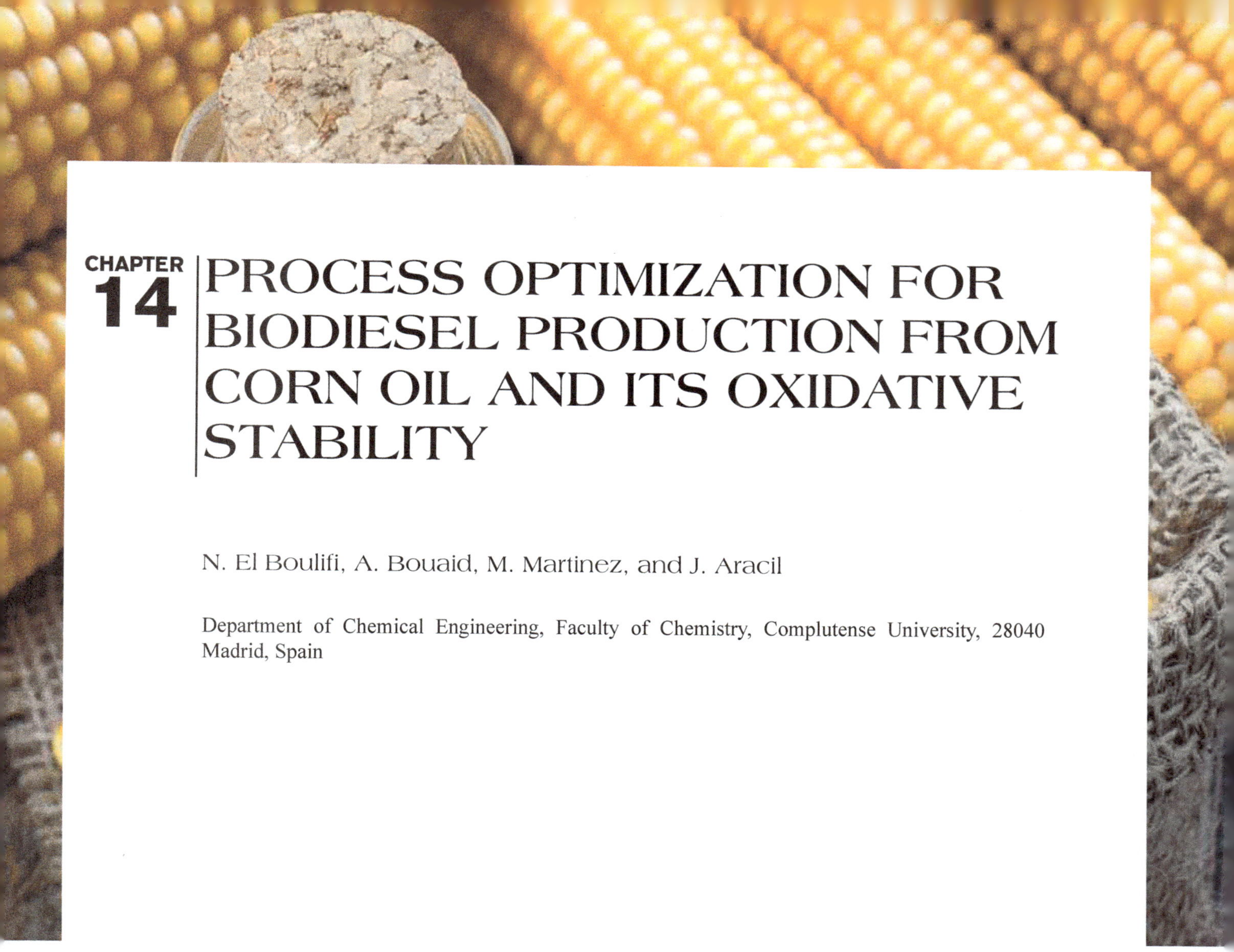

CHAPTER 14

PROCESS OPTIMIZATION FOR BIODIESEL PRODUCTION FROM CORN OIL AND ITS OXIDATIVE STABILITY

N. El Boulifi, A. Bouaid, M. Martinez, and J. Aracil

Department of Chemical Engineering, Faculty of Chemistry, Complutense University, 28040 Madrid, Spain

ABSTRACT

Response surface methodology (RSM) based on central composite design (CCD) was used to optimize biodiesel production process from corn oil. The process variables, temperature and catalyst concentration were found to have significant influence on biodiesel yield. The optimum combination derived via RSM for high corn oil methyl ester yield (99.48%) was found to be 1.18% wt catalyst concentration at a reaction temperature of 55.6◦ C. To determine how long biodiesel can safely be stored, it is desirable to have a measurement for the stability of the biodiesel against such oxidation. Storage time and oxygen availability have been considered as possible factors influencing oxidative instability. Biodiesel from corn oil was stored for a period of 30 months, and the physico-chemical parameters of samples were measured at regular interval of time. Results show that the acid value (AV), peroxide value (PV), and viscosity (v) increased while the iodine value (IV) decreased. These parameters changed very significantly when the sample was stored under normal oxygen atmosphere. However, the v, AV, and IV of the biodiesel sample which was stored under argon atmosphere were within the limit by the European specifications (EN 14214).

INTRODUCTION

In the last few years, the world's energy demand is increasing due to the needs from the global economic development and population growth. However, the most important part of this energy currently used is the fossil energy sources. The problem is fossil fuels are nonrenewable. They are limited in supply and will one day be depleted. There is an increased interest in alternative renewable fuels. As biodiesel is an environmentally friendly fuel, it is the best candidate to replace fossil-diesel, which has lower emissions than that of fossil-diesel, it is biodegradable, nontoxic, and essentially free of sulphur and aromatics [1]. However, only nitrogen oxides increase using biodiesel as fuels [2, 3].

Renewable feedstocks such as vegetable oils and animal fats have been used as raw materials for biodiesel production [4]. The general way to produce biodiesel fuels is transesterification of fat or oil triacylglycerols with short-chain alcohol such as methanol or ethanol in presence of alkaline or acid catalysts [5–7].

Vegetable oils are promising feedstocks for biodiesel production since they are renewable origin and can be

Citation: N. El Boulifi, A. Bouaid, M. Martinez, and J. Aracil, "Process Optimization for Biodiesel Production from Corn Oil and Its Oxidative Stability," International Journal of Chemical Engineering, vol. 2010, Article ID 518070, 9 pages, 2010. doi:10.1155/2010/518070

produced on a large scale. More than 95% of biodiesel production feedstocks come from edible oils since they are varying considerably with location according to climate and availability. In the United States, soybean oil is the most common biodiesel feedstock, whereas in Europe and in tropical countries, rapeseed oil and palm oil are the most common source for biodiesel, respectively, [8]. However, some of these oil sources are commodities whose prices are strongly influencing biodiesel cost, generally in the proportion of 70–80% [9]. In order to reduce the biodiesel cost, many researchers are interested in waste edible oils [10] and nonedible oils like karanja, mahua and jatropha [7, 11, 12].

Another alternative comes into play when looking to other industries. That is the case of ethanol, whose primary feedstock is corn. Plants ethanol from corn gives the integrated biorefineries a hydrocarbon-based source of renewable carbon for the production of fuels and chemicals. Ethanol is formed when starch is subjected to hydrolysis, followed by glucose fermentation. During this process, also some by-products including corn gluten meal, gluten feed and corn oil are formed. Therefore, corn oil can be extracted as a by-product by using new technology that will make ethanol production more efficient. This corn oil can be converted into a biofuel, such as biodiesel [13].

Biodiesel is being commercialized as a substitute or blending stock of fossil-diesel. However, biodiesel is less resistant to oxidation than typical fossil fuel and therefore doping of biodiesel in fossil-diesel will affect the stability of fuel significantly [14]. Thus, as the demand and the production of biodiesel grows fast, the development of methods to ensure the quality of the biodiesel industry and standardization becomes an urgent topic for market introduction of biodiesel.

As corn biodiesel chemically is a mixture of long-chain fatty acid methyl esters (FAMEs), it is more susceptible to autoxidation and thus has a higher level of chemical reactivity than fossil-diesel. This oxidation instability is dependant on the number and location of methylene-interrupted double bounds in the FAMEs. Thus carbons that are simultaneously allylic to two olefinic groups will be extremely susceptible to the initiation of peroxidation. An early study [15] measured the relative rate of oxidation for the methyl esters of oleic (18 : 1), linoleic (18 : 2) and linolenic (18 : 3) acids to be 1 : 12 : 25.

A number of reports have appeared on the storage and oxidative stability of biodiesel synthesized from different vegetable and frying oils (including sunflower, soybean and rapeseed oil) [16–18]. However, there is no available information about storage stability of biodiesel from corn oil. These stability studies were carried out under different storage condition, such as elevated temperature and exposure to light, air, water, and other contaminants. In other studies biodiesel was subjected to accelerated methods of oxidation, including EN 14112 (Rancimat method) and pressurized differential scanning calorimetry [19, 20]. The oxidation stability does not only depend on the raw material on the production process, but also on how the oil pressed and refined [21].

In a previous work [16], long-term storage tests on biodiesel from different sources have been conducted under different storage condition. Some chemicals properties have been severely affected by oxidation of the FAMEs, when the biodiesel was exposed to air and daylight. In order to understand whether normal oxygen atmosphere availability can adversely affect these properties, the study of the change in chemical and physical proprieties of biodiesel in absence of oxygen (under argon atmosphere) becomes necessary.

In this paper, the storage stability of biodiesel made from corn oil was investigated over a storage time of 30 months under argon atmosphere conditions, whose properties were compared to a corn biodiesel sample which stored under normal oxygen atmosphere. The focus of the study was on the influence of storage time on the biodiesel properties, such as PV, AV, IV and v. This work also discusses the results of experiments carried out to evaluate the different variables affecting the alkaline methanolysis of corn oil. The optimum value for the variables affecting the process will be determined by application of factorial design and response surface methodology.

Factorial design of experiments gives more information per experiment than unplanned approaches; it allows to see interactions among experimental variables within the range studied, leading to better knowledge of the process and therefore reducing research time and costs [22].

EXPERIMENTAL SECTION

Equipment

Reactions were carried out in a batch stirred reactor of 500 cm^3 volume, equipped with a reflux condenser and a mechanical stirrer. The impeller speed was set at 600 rpm to avoid external mass transfer limitation [6].

Materials

Corn oil was supplied by Koipe Spain. The quality control of the corn oil used in this study is presented in Table1. The characteristics were determined according to AOCS official method. Methanol of 99.8% purity was supplied by Panreac (Spain). The catalyst used was potassium hydroxide purchased from Merck (Barcelona, Spain).

Table 1: Composition analysis results of corn oil.

PV (meq/Kg)	AV (mg KOH/g)	ν (cst at 40°C)	IV (mg I_2/g)	Main fatty acid composition (%)					
				Palmitic (C16 : 0)	Stearic (C18 : 0)	Oleic (C18 : 1)	Linoleic (C18 : 2)	Linolenic (C18 : 3)	other
2.26	0.23	39.28	125.4	12	2.4	27.3	55.8	1	1.5

PV: peroxide value; AV: acid value; ν: viscosity; IV: iodine value.

PRODUCTION OF CORN OIL FAMES

The corn oil FAMEs were produced via a transesterification process in which a strong alkali catalyst was used. This latter is frequently used in the transesterification reaction, primarily due to its significant advantages in terms of the smaller quantity of catalyst used and the shorter reaction time required [6].

Corn oil was used as the raw oil to be transesterified with methanol in a reacting tank. The temperature values are below the boiling point of methanol (63°C), to prevent the methanol in the reactant mixture from evaporating. The molar ratio of the methanol and corn oil was set at 6 : 1. The potassium hydroxide was stirred with methanol for 10 minutes using an electric-magnetic stirrer to form potassium methoxide, which was then poured into the reacting tank and mixed with the corn oil. The total reaction time was 60 minutes. Almost total conversion to corn oil FAMEs was achieved quickly after a few minutes from the start of the reaction, depending on the reaction conditions.

Fatty Acid Methyl Esters (FAMEs) Purification

At the end of a run the reaction mixture was allowed to cool down. The upper phase consisted of FAME, and the lower phase was glycerol.

Once the glycerol and FAME phases have been separated, the last one was purified by gentle washing with distilled water to remove residual catalyst, glycerol, and soaps. The pH of washing water was initially very high 10.22 due to the dissolved KOH. After 3 successive rinses with water, the washing water became clear and its pH was 7.9. The washing process was continued (twice more) until a pH of about 7 was achieved. Finally, the methyl ester phase was distilled to remove the residual water.

The final water content of the corn oil FAME was less than 0.01%. Water in the sample can promote microbial growth, lead to tank corrosion, participate in the formation of emulsions, as well as cause hydrolysis or hydrolytic oxidation [23].

Analytical Method

Reaction products were monitored by capillary column gas chromatography, using a Hewlett-Packard 5890 series II equipped with a flame ionization detector (FID). The injection system was split-splitless. The carrier gas was helium at a flow rate of 1 mL/minutes. The analytical procedures and operating conditions have been described in detail in a previous work [24]. The internal standard technique was used to quantify the amount of the chemical species.

Preparation the FAME Samples

Two biodiesel samples, three litres each of were stored at ambient temperature for 30 months at two different storage conditions: the samples were stored in closed glass exposed to daylight, one under normal oxygen conditions and the other under atmosphere argon. During storage, samples were taken out periodically and different quality parameters (PV, AV, IV, and ν) were monitored. The analyses were done according to the following procedures: PV (AOCS Cd 8-53), AV (AOCS-Ca 5a-40), ν (ISO 3104), and IV (AOCS cd 1-25).

RESULTS AND DISCUSSION

Statistical Analysis

The synthesis of biodiesel was studied using factorial design of experiments. The experimental design applied to this study was a full two-level factorial design 2^2 (two factors each, at two-levels) and extended to response surface methodology (RSM). The response selected, Y, was the yield of methyl ester. The factors chosen were reaction temperature, X_T, and initial catalyst concentration, X_C. Initial alcohol/oil molar ratio was fixed at 6:1. Selection of the levels was carried out based on the results obtained in preliminary studies [25]. Temperature levels were selected according to reactants properties, thus the lower value was set at 34.4°C and the higher at 55.6°C. The levels of catalyst concentration were chosen on the basis of preliminary experiments where the amount of catalyst was progressively increased, and the ester yield monitored versus time. The levels chosen were 0.86 and 1.18 wt.%, based on the total mass reaction. The statistical analysis was thereafter applied. The experimental matrix for the factorial design is shown in Table2. All the runs were performed at random. Four experiments were carried out at the centre point level, coded as "0", for experimental error estimation.

Table 2: Experimental matrix and experimental results.

Run	X_C	X_T	C (%)	T (°C)	Y(%)
1	−1	−1	34.4	0.86	94.40
2	+1	−1	55.6	0.86	98.25
3	−1	+1	34.4	1.18	96.50
4	+1	+1	55.6	1.18	99.48
5	0	0	45	1.02	98.25
6	0	0	45	1.02	97.85
7	0	0	45	1.02	98.30
8	0	0	45	1.02	98.40
9	$-\alpha$	0	30	1.02	97.50
10	$+\alpha$	0	60	1.02	98.70
11	0	$-\alpha$	45	0.80	96.9
12	0	$+\alpha$	45	1.24	96.4

Note: T: temperature; C: catalyst concentration; X: coded value; Y: biodiesel yield.

The use of analysis and factorial design of experiments allowed us to express the amount of ester produced as a polynomial model (if the levels of the factors are equally spaced, then orthogonal polynomials may be used). The response, yield of ester, may be thus expressed as a function of the significant factors.

Linear Stage

A linear stage was considered in the first step. Table 2 shows the experiments corresponding to the 2^2 factorial design (experiments 1–4) and four experiments in the centre point to evaluate the experimental error (experiments 5–8). A statistical analysis was carried out on these experimental values, and the main effects and interaction effects of the variables were calculated. The analysis of the main effects and its interaction for the chosen response and the test of statistical significance are given in Table 3.

Table 3: 2^2 factorial design: statistical analysis.

Response:		Y(%)
Number of experiments:		4
Degree of freedom:		3
Results of statistical analysis		
Y (average)		97.16
	Main effects and interactions	
	$X_T = 1.66$, $X_C = 3.41$, $X_{TC} = -0.43$	
	Statistical Significance of t-test	
Confidence level: 95%,		Standard deviation, $S = 0.24$
Student's t value, $t = 3.182$,		Confidence interval: ± 0.38
	Significant effects and Interactions: $X_T(+)$, $X_C(+)$, $X_{TC}(-)$	
	Statistical significance of curvature	
$Y_C = 98.2$, $Y = 97.16$		Curvature: $Y_C - Y- = 1.04$
$S = 0.11$, $t = 3.182$		Confidence curvature interval: ± 0.54
		Curvature: significant
	Response equation:	
$Y = 97.16 + 1.7X_C + 0.83X_T - 0.21X_{TC}$		$r = 0.98$

Temperature (X_T), catalyst concentration (X_C), and catalyst concentration-temperature interaction effects (X_{TC}) were fitted by multiple regression analysis to a linear model. The response function for the significant main effects and interactions is

$$Y = 97.16 + 1.7X_C + 0.83X_T - 0.21X_{TC}, \quad r = 0.98. \tag{1}$$

The statistical analysis of experimental results revealed that the most significant factor is the catalyst concentration, while it also shows a significant value for curvature for the chosen responses. These data indicate the nonlinearity of the model and thus justifies planning a more complex design to fit the data to a second-order model.

Nonlinear Stage

To better predict the effect of variables, a quadratic model was investigated. Here, the 2^2 experiment design was expanded to a circumscribed central composite design by the addition of 4 new experiments (run 9–12 in Table 2), called start points and coded $\pm\alpha$. The value of α, which is the distance from the center point to the start point, is $2^{n/4}$, where n is the number of factors (for two factors, $\alpha = 1.414$). The corresponding model is the complete quadratic surface between response and the factors, given by the equation

$$Y = \beta_0 + \sum_{i=1}^{2} \beta_i X_i + \sum_{i=1}^{2} \beta_{ii} X_i^2 + \sum_{i \neq j}^{2} \beta_{ij} X_i X_j, \tag{2}$$

where Y is the response (methyl ester yield), X_i and X_j are the uncoded independent variables, and β_0, β_i, β_{ii}, and β_{ij} are intercept, linear, quadratic and interaction constant coefficients, respectively.

The coefficients of (2) were determined by multiple regression analysis. This analysis includes all the independent variables and their interactions, regardless of their significance levels. The best-fitting response surfaces found can be written as follows:

The statistical model was obtained from coded levels. Equation (3) can be represented as dimensional surfaces and contour plots, as shown in Figure 1. These show the ester yield predicted for the experimental range of temperature and initial catalyst concentration. The influence of these variables on the ester yield will now be discussed. The influence of the main factors and interactions will be derived from (1) and (3).

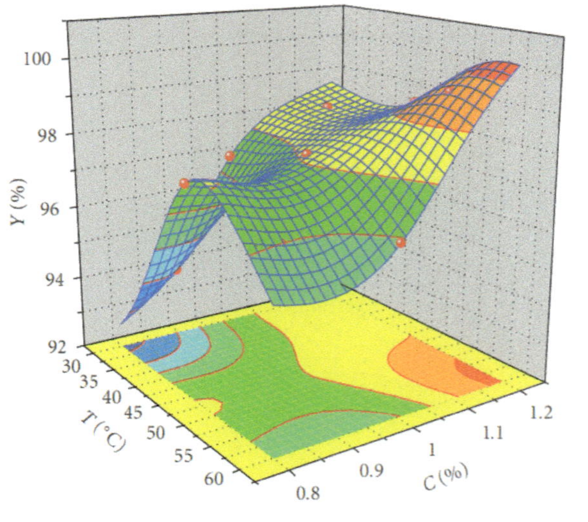

Figure 1: Experimental yield versus temperature and catalyst concentration.

Influence of Temperature

For both linear and nonlinear models, the temperature influence is statistically significant in the range studied. This effect has a positive influence on the response. As the temperature increases, the solubility of methanol in the oil increases and so does the speed of reaction. As a matter of fact, at low temperatures, methanol is not soluble at all in the oil; when the stirring is started an emulsion appears. The reaction takes place at the interface of the droplets of alcohol in the oil and then as soon as the first FAMEs are formed, the alcohol solubilizes progressively because the esters are mutual solvents for the alcohol and the oil.

Influence of Catalyst Concentration

From the statistical analysis it can be concluded that, within the experimental range, initial catalyst concentration is the most important factor on the transesterification process. It has a positive influence on the response; that is, ester yield increases with increasing catalyst concentration.

Influence of Interaction

The nonlinear model (Central Composite Design) gives the binary influences of all the factors used in the design. Interaction of significant main effects temperature and catalyst concentration (T-C) is significant and has a negative influence on the process, this effect may be due to the formation of by-products (soaps).

Analysis of Response: Yield of Ester

The ester yield generally increases with increasing catalyst concentration and temperature, but progressively decreases at higher level of temperature and lower level of catalyst concentration. This finding may be explained by the formation of by-products, possibly due to triglycerides saponification processes, a side reaction which is favoured at high temperatures. This side reaction produces potassium soaps and, thus, decreases the ester yield.

The surface plot and contours of ester yield versus temperature and catalyst concentration obtained when individual experimental data are plotted are shown in Figure 1. A comparison among these plots shows that the maximum ester yield is achieved at the higher level for both temperature and catalyst concentration. Figure 2present a normal probability plot of methyl ester yield which confirmed the significant influence of catalyst and temperature in the range studied. Figure 3 present a plot of the residual distribution, defined as the difference between calculated and observed values, over the observed values for the response studied FAME. The quality of the fit is good because the residual distribution does not follow a trend with respect to the predicted variables. All the residuals are smaller than 1.7% for FAME yield, which indicate that the model adequately represent the methyl ester yield over the experimental range studied.

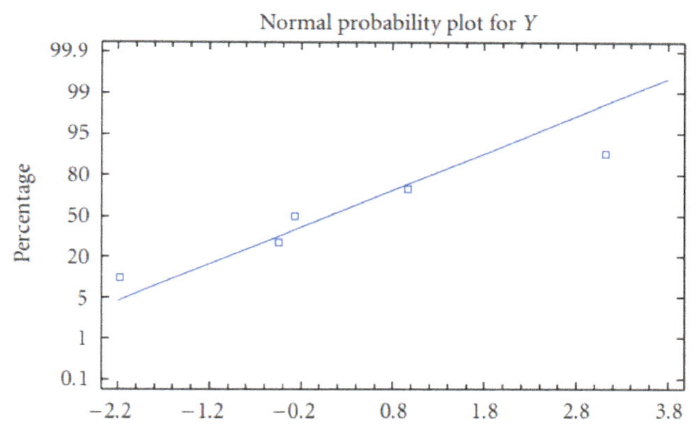

Figure 2: Normal probability plot for the methyl esters yield.

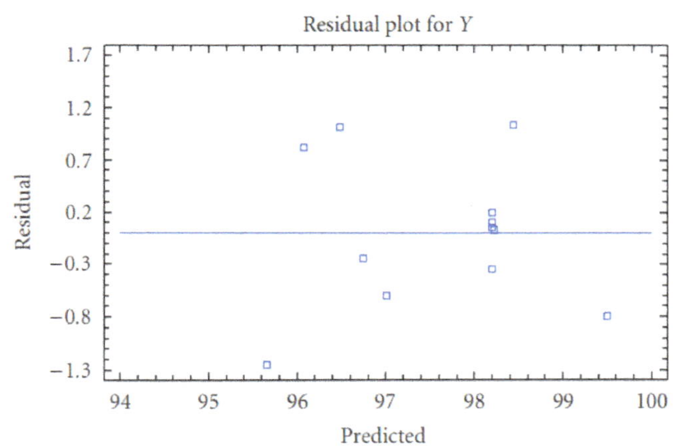

Figure 3: Residual plot of methyl esters yield for the second-order model.

Figure 4 presents the plot of experimental values versus predicted ones for biodiesel from Corn oil. From this figure, it can be observed that there are no tendencies in the linear regression fit, so the model explains the experimental range studied adequately.

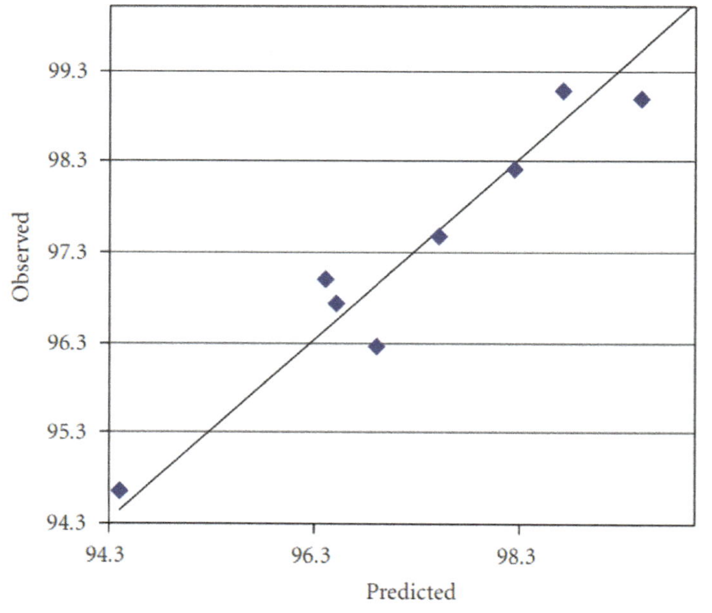

Figure 4: Experimental versus predicted values.

Long Storage Stability of Biodiesel

The oxidation reactions affect the fuel quality of biodiesel, primarily during extended storage. The oxidation stability study was conducted for a period of 30 months. At regular intervals, samples were taken to measure the following physico-chemical quality parameters: PV, AV, IV, and v.

Some of the most important qualities of biodiesel are shown before and after storage in Table 4. These parameters were compared with some of the biodiesel standards (European Union Standards, EN 14214).

Table 4: Quality control of corn oil biodiesel fuels used in this study before and after storage time of 30 months compared to EN 14214

Properties	Unit	Corn biodiesel before storage	Corn biodiesel after storage		Specification EN 14214
			Under argon atmosphere	Under normal oxygen atmosphere	
Viscosity at 40°C	mm^2/s	4.14	4.72	5.82	3.5–5.0
Density at 15°C	g/cm^3	0.865	0.871	0.895	0.86–0.90
Water content	wt%	0.15	0.21	0.29	<0.5
Biodiesel yield	wt%	99.48	98.7	96.23	>96.5
Monoglyceride content	wt%	0.40	0.27	0.3	<0.80
Diglyceride content	wt%	0.11	0.1	0.1	<0.20
Triglyceride content	wt%	0.0	0.0	0.0	<0.20
Free glycerol	wt%	0.01	0.01	0.01	<0.02
Acid value	mg KOH/g	0.047	0.21	0.7	<0.5
Iodine value	mg I_2/g	126.6	98.4	80	<120
Peroxide value	meq/Kg	3.0	14	31.8	—
Cloud point	°C	−5	−4	−3	—
Pour point	°C	−6	−6	−5	—
Cold filter plugging point	°C	−8	−8	−9	<−7

Peroxide Value (PV)

Although PV is less suitable for monitoring oxidation [26], and is not specified in the biodiesel fuel standards, this parameter influences cetane number (CN), a parameter that is specified in the fuel standard. An increase in PV involves an increase in CN, and therefore may reduce ignition delay time [27]. The variation of PV of corn oil FAME in two different storage conditions is shown in Figure 5. Sample stored under argon atmosphere showed an increase in PV from 3 to 19.6 meq/kg and then a decrease to 14 meq/kg. It can be see that the faster increase rate happens during the first 6 months. The same results were obtained by another author [28]. The reasons why such behaviours exist are not clearly resolved in prior work. However, the increasing of PV during the first 6 months may be due to the presence of the dissolved oxygen remaining in the biodiesel sample that allows the formation of compounds like hydroperoxides. Furthermore, sample stored under normal oxygen atmosphere conditions showed a continuing increase in PV over time. If oxygen is not available in sufficient abundance, the formation of ROOH can slow or even stop while ROOH decomposition continues. This will tend to cause a peak in the ROOH concentration followed by a decrease.

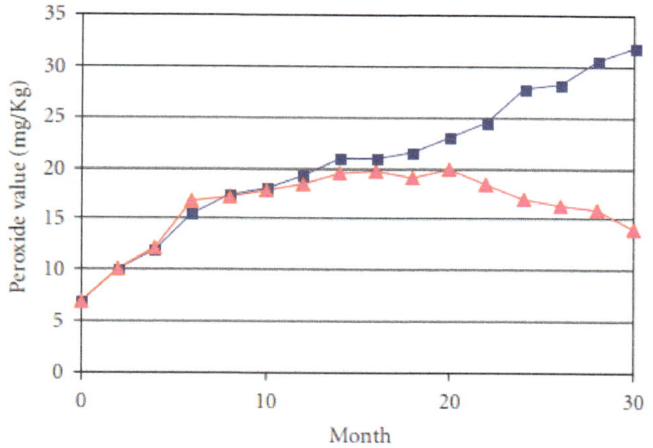

Figure 5: Variation in the peroxide value of corn oil biodiesel stored ■ under normal oxygen atmosphere and ▲ under argon atmosphere.

Acid Value (AV)

Once the peroxides have formed, they decompose and interreact to form numerous secondary oxidation products including aldehydes, which further oxidized into acids. Acids can also be formed when traces of water cause hydrolysis of the esters into alcohol and acids. Acid value measured in mg KOH/g is one of the significant indicators of oxidative degradation in lipids. The change in AV during storage for the two corn oil FAME samples is shown in Figure 6. It can be seen that the AV increase continuously in sample stored under normal oxygen atmosphere from 0.047 to 0.7 mg KOH/g. In opposition to it, in the sample stored under argon atmosphere, the AV increased fast during the first 4 months, and then remained almost constant. This behaviour may be due to the presence of the dissolved oxygen remaining in the biodiesel sample that allows the formation of shorter chain fatty acids [18]. In this case, the specification limit of 0.5 mg KOH/g was not exceeded with the corn oil FAME sample stored under argon atmosphere after a storage time of 30 months.

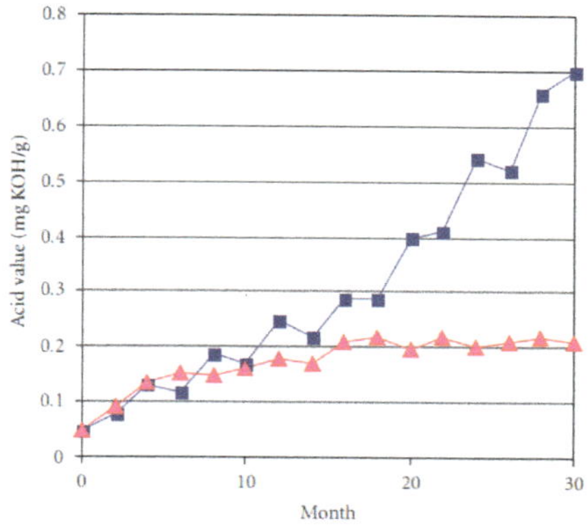

Figure 6: Variation in the acid value of corn oil biodiesel stored ■ under normal oxygen atmosphere and ▲ under argon atmosphere.

Viscosity (ν)

Due to the hydroperoxides decomposition, the oxidative linking of fatty acid methyl ester chains can occur, giving as a result higher molecular weights species. During storage one of the obvious results of polymer formation is an increase in biodiesel viscosity [29]. The formation of higher molecular weight species, with higher viscosity, is a reason why the viscosity specification in biodiesel standards can be used to assess the fuel quality status of stored biodiesel. Figure 7 shows the increase in the kinematic viscosity of the two corn oil FAME samples versus storage time. Corn oil FAME sample which was stored under argon atmosphere, was found to have lower viscosity compared to the sample which was stored under normal oxygen atmosphere. A similar result had been observed previously by other researchers [30]; they noted that the oxidation processes led to the formation of free fatty acids, double bond isomerization (usually cis to trans), saturation and products of higher molecular weight, and viscosity increases with increasing oxidation. Under conditions where oxygen is available, fatty acid moieties are joined by both C–O–C linkages and C–C linkages [31]. When ROOH decomposition occurs under an inert atmosphere, only C–C linkages in resulting polymers are observed [31].

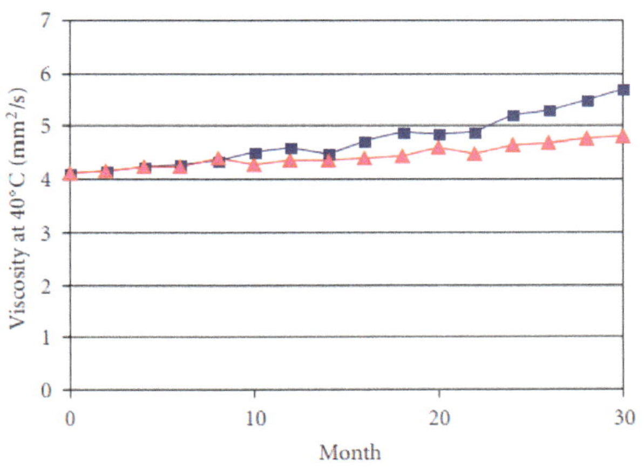

Figure 7: Variation in the iodine value of corn oil biodiesel stored ■ under normal oxygen atmosphere and ▲ under argon atmosphere.

Iodine Value (IV)

One of the oldest and most common methods of determining the level of unsaturation in a fatty oil or ester is the iodine value [32], it purportedly addresses the use of oxidative stability and it is included in the European biodiesel standard. Oxidation, which consists of a complex series of chemical reactions, is characterized by a decrease in the total unsaturated content of biodiesel due to elimination of hydrogen adjacent to a double bond and the formation of free radicals [33]. Our experiment shows that the IV of the two corn oil FAME samples fell along the storage time (Figure 8). The sample which was stored under normal oxygen atmosphere appeared to have the highest rate of reduction in IV, implying that the largest conversion of polyunsaturated fatty acids to hydroperoxides took place. In contrast to this result, the sample which was stored under argon atmosphere was observed to have low rate of decrease in IV. Corn oil FAME samples (IV = 126.6) with a relatively high IV compared to other biodiesel samples [4]; this sample contains low levels of saturated FAME and high levels of mono- and polyunsaturated FAME which make it much susceptible to oxidation.

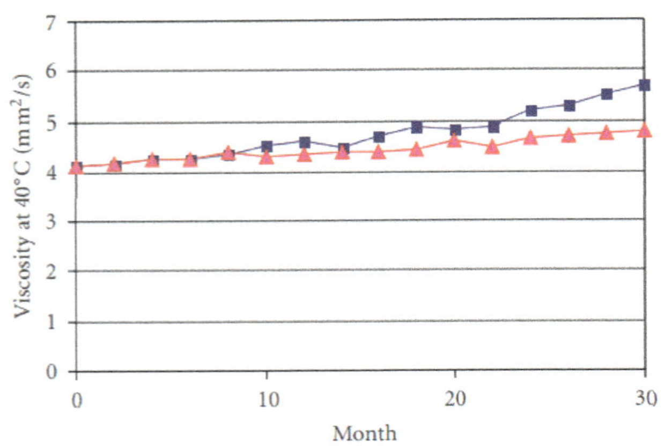

Figure 8: Variation in the kinematic viscosity of corn oil biodiesel stored ■ under normal oxygen atmosphere and ▲ under argon atmosphere.

For biodiesel samples, the peroxide value, acid values, and viscosity tended to increase, and iodine value to decrease over time. The sample stored under argon atmosphere never exceeded the specification limit of the studied parameters along the 30 months of storage. Biodiesel from corn oil stored under argon atmosphere did not demonstrate rapid increase in peroxide value, acid value and viscosity, compared to the one stored under normal oxygen atmosphere.

When storage is done in argon atmosphere condition, at ambient temperature without oxygen availability, PV, AV, and v increases slowly; while PV peaks and then decreases, AV and v continue to increase, but at a lower rate. Besides IV decreases during the storage.

Biodiesel kept in an argon atmosphere could increase its stability and also show positive effects retarding oxidative degradation of the biodiesel produced from corn oil.

CONCLUSIONS

In this work, a fully central composite design has been applied to optimize the synthesis process of FAME from corn oil using potassium hydroxide as catalyst. The study of the factors (temperature and catalyst concentration) affecting the response shows that, within, the experimental range considered, the most important factor is the initial catalyst concentration. For the yield of ester, this factor has a positive influence. The temperature has a positive influence in both responses. The T-C interaction is small and negative, due to the formation of by-products (soaps). According to this study, and from an economical point of view, the maximum yield of ester (98.75%) can be obtained, working with an initial catalyst concentration (1.1%) and operation temperature (55°C). These models are useful to determine the optimum operating conditions for the industrial process using a minimal number of experiments with the consequent economical benefit.

Biodiesel synthesised, which consists of long-chain FAME, generally suffers from lower oxidation stability. Results from this study suggest that for a remarkably stable biodiesel and in order to avoid oxidation, special precautions must be taken during long storage, such as storage under argon atmosphere. Nevertheless this action delays oxidation but it doesnot prevent it. The results of this work have allowed develop a methodology to overcome the obvious problems of long storage stability of biodiesel. Long-term storage study gives us a better understanding of the effect of the normal oxygen atmosphere on the stability of biodiesel. The use of corn oil as an alternative raw material and renewable feedstock to produce biodiesel which fulfills the specification of EU standards for biodiesel (EN 14214) is of great interest to build an integrated and self-sustained biorefinery.

ACKNOWLEDGMENT

Financial support from the (CICYT), Spanish Project CTQ 2006-10467/PPQ is gratefully acknowledged.

REFERENCES

1. F. Staat and E. Vallet, "Vegetable oil methyl ester as a diesel substitute," Chemistry & Industry, vol. 21, pp. 856–863, 1994.
2. M. P. Dorado, E. Ballesteros, J. M. Arnal, J. Gomez, and F. J. L. Gimenez, "Testing waste olive oil methyl ester as a fuel in a diesel engine," Energy & Fuels, vol. 17, no. 6, pp. 1560–1565, 2003.
3. N. Usta, "An experimental study on performance and exhaust emissions of a diesel engine fuelled with tobacco seed oil methyl ester," Energy Conversion and Management, vol. 46, no. 15-16, pp. 2373–2386, 2005.
4. S. P. Singh and D. Singh, "Biodiesel production through the use of different sources and characterization of oils and their esters as the substitute of diesel: a review," Renewable and Sustainable Energy Reviews, vol. 14, no. 1, pp. 200–216, 2010.
5. J. M. Marchetti, V. U. Miguel, and A. F. Errazu, "Possible methods for biodiesel production," Renewable and Sustainable Energy Reviews, vol. 11, pp. 1300–1311, 2007.
6. G. Vicente, M. Martínez, and J. Aracil, "Integrated biodiesel production: a comparison of different homogenous catalysts systems," Bioresource Technology, vol. 92, pp. 297–305, 2004.

7. Y. C. Sharma and B. Singh, "Development of biodiesel from karanja, a tree found in rural India," Fuel, vol. 87, no. 8-9, pp. 1740–1742, 2008.

8. G. Knothe, "Current perspectives on biodiesel," Information, vol. 13, no. 12, pp. 900–903, 2002. ·

9. H. Shi and Z. Bao, "Direct preparation of biodiesel from rapeseed oil leached by two-phase solvent extraction," Bioresource Technology, vol. 99, no. 18, pp. 9025–9028, 2008.

10. F. Ferella, G. Mazziotti, I. De Michelis, V. Stanisci, and F. Vegliò, "Optimization of the transesterification reaction in biodiesel production," Fuel, vol. 89, pp. 36–42, 2010.

11. S. V. Ghadge and H. Raheman, "Process optimization for biodiesel production from mahua (Madhuca indica) oil using response surface methodology," Bioresource Technology, vol. 97, no. 3, pp. 379–384, 2006.

12. A. K. Tiwaria, A. Kumara, and H. Raheman, "Biodiesel production from jatropha oil (Jatropha curcas) with high free fatty acids: an optimized process," Biomass and Bioenergy, vol. 31, no. 8, pp. 569–575, 2007.

13. GS AgriFuels to Convert Corn Oil into Biodiesel at Ethanol Facilities, GS AgriFuels Corporation, New York, NY, USA, 2006.

14. R. O. Dunn and G. Knothe, "Oxidative stability of biodiesel in blends with jet fuel by analysis of oil stability index," Journal of the American Oil Chemist's Society, vol. 80, no. 10, pp. 1047–1048, 2003. ·

15. F. D. Gunstone and T. P. Hilditch, "The union of gaseous oxygen with methyl oleate, linoleate, and linolenate," Journal of the Chemical Society, pp. 836–841, 1945. ·

16. A. Bouaid, M. Martínez, and J. Aracil, "Long storage stability of biodiesel from vegetable and used frying oils," Fuel, vol. 86, no. 16, pp. 2596–2602, 2007.

17. A. Bouaid, M. Martínez, and J. Aracil, "Production of biodiesel from bioethanol and Brassica carinata oil: oxidation stability study," Bioresource Technology, vol. 100, no. 7, pp. 2234–2239, 2009.

18. P. Bondioli, A. Gasparoli, L. D. Bella, and T. Silvia, "Evaluation of biodiesel storage stability using reference methods," European Journal of Lipid Science and Technology, vol. 104, no. 12, pp. 777–784, 2002. ·

19. B. R. Moser, "Comparative oxidative stability of fatty acid alkyl esters by accelerated methods," Journal of the American Oil Chemist's Society, vol. 86, no. 7, pp. 699–706, 2009.

20. R. O. Dunn, "Effect of oxidation under accelerated conditions on fuel properties of methyl soyate (biodiesel)," Journal of the American Oil Chemist's Society, vol. 79, no. 9, pp. 915–920, 2002. ·

21. M. H. Chahine and R. F. Macneill, "Effect of stabilization of crude whale oil with tertiary-butylhidroquinone and other antioxidants upon keeping quality of resultant deodorized oil. A feasibility study," Journal of the American Oil Chemist's Society, vol. 51, no. 3, pp. 37–41, 1974.

22. G. Box and J. Hunter, "Response surface methods," in Statistics for Experiments Part IV: Building Models and Using Them, chapter 5, John Wiley & Sons, New York, NY, USA, 1978.

23. G. Knothe, "Analyzing biodiesel: standards and other methods," Journal of the American Oil Chemist's Society, vol. 83, no. 10, pp. 823–833, 2006.

24. T. Garcia, A. Coteron, M. Martínez, and J. Aracil, "Optimization of the enzymatic synthesis of isopropyl palmitate using a central composite design," Transactions of Chemical Engineers, vol. 73, pp. 140–144, 1995.

25. G. Vicente, A. Coteron, M. Martínez, and J. Aracil, "Application of the factorial design of experiments and response surface methodology to optimize biodiesel production," Industrial Crops and Products, vol. 8, no. 1, pp. 29–35, 1998.

26. A. Monyem, M. Canakci, and J. H. Van Gerpen, "Investigation of biodiesel thermal stability under simulated in-use conditions," Applied Engineering in Agriculture, vol. 16, pp. 373–378, 2000.

27. G. Knothe, "Structure indices in FA chemistry. How relevant is the iodine value?" Journal of the American Oil Chemist's Society, vol. 79, no. 9, pp. 847–854, 2002. ·

28. P. Bondioli, A. Gasparoli, L. D. Bella, S. Tagliabue, and G. Toso, "Biodiesel stability under commercial storage conditions over one year," European Journal of Lipid Science and Technology, vol. 105, no. 12, pp. 735–741, 2003.

29. H. A. Moser, P. C. Cooney, C. D. Evans, and J. C. Cowan, "The stability of soybean oil: effect of time and temperature on deodorization," Journal of the American Oil Chemist's Society, vol. 43, no. 11, pp. 632–634, 1966.

30. G. Knothe and K. R. Steidley, "Kinematic viscosity of biodiesel fuel components and related compounds. Influence of compound structure and comparison to petrodiesel fuel components," Fuel, vol. 84, no. 9, pp. 1059–1065, 2005.

31. M. W. Formo, E. Jungermann, F. Noris, and N. O. V. Sonntag, "Bailey&s Industrial Oil and Fat Products," in Bailey's Industrial Oil and Fat Products, D. Swern, Ed., vol. 1, pp. 698–711, John Wiley & Sons, New York, NY, USA, 4th edition, 1979.

32. G. Knothe and R. O. Dunn, "Dependence of oil stability index of fatty compounds on their structure and concentration and presence of metals," Journal of the American Oil Chemist's Society, vol. 80, no. 10, pp. 1021–1026, 2003. ·

33. S. Naz, H. Sheikh, R. Siddiqi, and S. A. Sayeed, "Oxidative stability of olive, corn and soybean oil under different conditions," Food Chemistry, vol. 88, no. 2, pp. 253–259, 2004.

CHAPTER 15

SYNTHESIZING CONFIGURABLE BIOCHEMICAL IMPLEMENTATION OF LINEAR SYSTEMS FROM THEIR TRANSFER FUNCTION SPECIFICATIONS

Tai-Yin Chiu[1,2], Hui-Ju K. Chiang[2,3], Ruei-Yang Huang[2] , Jie-Hong R. Jiang[2,4], and François Fages[3]

[1] Department of Physics, National Taiwan University, Taipei, Taiwan,

[2] Graduate Institute of Electronics Engineering, National Taiwan University, Taipei, Taiwan,

[3] EPI Lifeware, Inria Paris-Rocquencourt, Rocquencourt, France,

[4] Department of Electrical Engineering, National Taiwan University, Taipei, Taiwan

ABSTRACT

The ability to engineer synthetic systems in the biochemical context is constantly being improved and has a profound societal impact. Linear system design is one of the most pervasive methods applied in control tasks, and its biochemical realization has been proposed by Oishi and Klavins and advanced further in recent years. However, several technical issues remain unsolved. Specifically, the design process is not fully automated from specification at the transfer function level, systems once designed often lack dynamic adaptivity to environmental changes, matching rate constants of reactions is not always possible, and implementation may be approximative and greatly deviate from the specifications. Building upon the work of Oishi and Klavins, this paper overcomes these issues by introducing a design flow that transforms a transfer-function specification of a linear system into a set of chemical reactions, whose input-output response precisely conforms to the specification. This system is implementable using the DNA strand displacement technique. The underlying configurability is embedded into primitive components and template modules, and thus the entire system is adaptive. Simulation of DNA strand displacement implementation confirmed the feasibility and superiority of the proposed synthesis flow.

INTRODUCTION

Advancements in synthetic biology have resulted in the development of biochemical systems of increasing complexity that are capable of using living cells as well as cell-free systems. Synthetic biology holds promise for biotechnology, biomedicine, bio-environmental, bioenergy, and other applications. Many computation and control design examples have been demonstrated either in vivo or in vitro. For instance, oscillators [1], toggle switches [2], logic gates [3], band pass filters [4], and analog circuits [5] have been designed and implemented in living cells, while digital circuits [6], neural networks [7], and switchable memories [8] have been demonstrated in cell-free systems. These developments show a clear growing trend in design complexity. Indeed, the more sophisticated systems we can build, the better skills we will have to comprehend biology and to build even more sophisticated systems.

The increasing system complexity renders the necessity of design automation tools. Constructing biochemical systems bottom-up using pre-characterized parts in synthetic biology is

Citation: Chiu T-Y, Chiang H-JK, Huang R-Y, Jiang J-HR, Fages F (2015) Synthesizing Configurable Biochemical Implementation of Linear Systems from Their Transfer Function Specifications. PLoS ONE 10(9): e0137442. doi:10.1371/journal.pone.0137442

analogous to designing electronic systems using pre-designed standard modules. While electronic design automation (EDA) has largely enabled system design under exponential capacity growth over the past five decades thanks to the driving force of Moore's law, design automation in synthetic biology may play a similar key role in the construction of complex biochemical systems [9]. Computer-aided modeling, simulation, synthesis, and verification are crucial because a biochemical design is often intended to function in a biological context that is often too complex to be fully characterized.

Depending on its target application, a system is often designed for a specific application rather than as a general purpose computing machine. In control applications, linear systems are pervasive due to their simplicity of design and analysis. Any linear control system can be realized with three primitive components: integration, gain, and summation. Realizing linear control with biochemical reactions has been proposed by Oishi and Klavins [10], who showed that three types of chemical reactions (catalysis, degradation, and annihilation) are sufficient to implement the three primitive components. In principle, any polynomial ordinary differential equation can be approximated by chemical reaction networks (CRNs) [11]. Therefore, any linear control system can be built using CRNs.

To realize CRNs, nucleic acids have been exploited as a universal tool for biomolecular computation [12]. By appropriately changing their nucleotide sequences, the interaction between nucleic acids can be precisely controlled and programmed in vitro. Specifically, toehold-mediated or enzyme-free DNA strand displacement (DSD) is a promising approach to perform biological computations. Although the DSD mechanism has been studied since the 1970s [13–17], until recently it was systematically used to build a molecular machine made of DNA and RNA [18]. The DSD method is attractive for several reasons. First, the kinetics of DSD devices can be engineered by controlling toehold binding rates, which may range from 1 to 6×10^6 M^{-1} s^{-1} [19–21]. Second, composability [22] can be achieved. In typical implementation schemes, single-stranded DNAs play the role of signals and double-stranded DNAs act as gates, and they work together with the mechanism of toehold-mediated DNA displacement [19]. DNA signals of different domain lengths [23–25] have been proposed. For example, gates can be composed to realize complex systems using a 2-domain signal architecture [25]. Third, the DSD mechanism works autonomously [26] as long as the DNA or RNA fuels are supplied. With these advantages, many biomolecular devices, e.g., [27–29], have been designed using this powerful mechanism. The DSD technology allows computation and interfacing with molecular components in living organisms [30] and shows potential in bio-sensing and control, biomedicine, and other applications.

Despite the advancement of biochemical implementation of linear systems, four challenges remain to be solved. First, the construction of Oishi and Klavins's system requires that the rate constants of the underlying reactions be carefully matched to achieve the intended integration, gain, and summation functions. This requirement imposes substantial practicality restrictions because, in reality, the reaction rates of available reactions can be limited. Thus, not all gain and summation blocks can be realized. Second, once a system is constructed, its function is fixed and cannot be changed without redesign. However, this fixed functionality can be inadequate for a system reacting to its biochemical environment, which is intrinsically stochastic and often full of uncertainty. Thus, designing systems with dynamic adaptation capabilities is crucial, especially in the biochemical context. Third, the CRN implementations of the gain and summation components are approximative. In essence, the transfer functions of the gain and summation components contain extra poles rather than just the ideal scalar and summation. When a complex system is built from these components, these additional poles might cause system behavior that deviates from its specification and can even lead to unwanted instability. Fourth, an automated flow of the synthesis of CRN from the transfer function specification of a linear system is lacking. Although a linear system can be compiled into a CRN with a direct block-by-block conversion from a block diagram, the block diagram may not be available in the first place and the CRN implementation may only approximate the specification. Specifying a linear system using transfer functions can be more natural than using block diagrams. However, converting a transfer function into a block diagram suitable for CRN realization can be a nontrivial optimization task.

In this paper, we tackle the above challenges. To address the first two challenges, we devise a mechanism to make linear control systems configurable by adding auxiliary species to CRN as control knobs. The concentrations of the auxiliary species can be adjusted not only to compensate for reaction rate mismatch but also to reconfigure a control system to dynamically adapt to environmental changes. Hence, implementing linear control systems in biochemistry can be made more practical. To resolve the last two challenges, we apply parallel decomposition on a given transfer function and express it as a summation of elementary modules, and propose a CRN solution that achieves the exact implementation of the elementary modules. The CRN implementation of a transfer function can be further mapped and realized using the DSD method. The proposed method lends itself to an automated design flow (as depicted in Fig 1) where the linear system to be synthesized is specified by a transfer function, which is decomposed for CRN implementation and further mapped into the DSD reactions. Through simulation using Visual DSD [31, 32], we show the feasibility and superiority of our proposed design automation flow for synthesizing a specified linear system into DSD reactions.

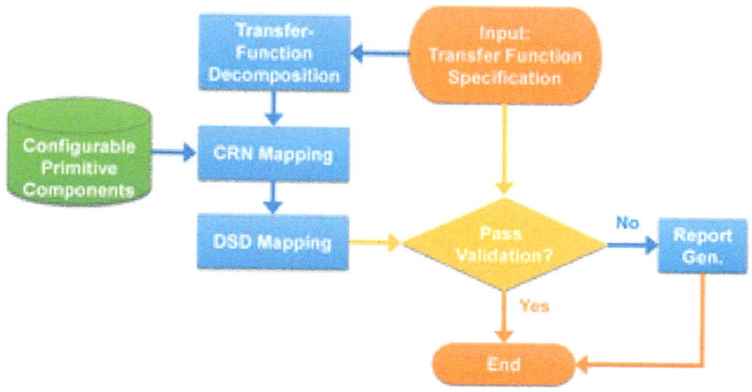

Figure 1: Automated flow of linear system synthesis.

METHODS AND RESULTS

Although synthetic biology shows promising potential in biotechnology applications, it poses grand challenges to complex system design. A typical reactive system may involve sensing, actuation, information processing, and other tasks, while a control unit is one of the fundamental ingredients that constitute a reactive system. In this work, we address the design automation problem of linear control systems that are to be realized by biochemical reactions.

Any linear time-invariant (LTI) system can be fully specified by means of a transfer function, which corresponds to the Laplace transform of the impulse response of the system, with all initial conditions set to zero. In essence, once the impulse response of a linear system is known, the output y(t) of the system with any input u(t) can be characterized using the transfer function. However, for a multiple-input multiple-output (MIMO) linear system, the principle of superposition suggest that the total effect on any output due to all the inputs acting simultaneously is obtained by summing up the output effects due to the individual inputs acting alone. Consequently any MIMO system can be specified by a matrix of transfer functions. For simplicity, in this paper we focus on single-input single-output (SISO) systems without loss of generality.

Configurable Primitive Components

A transfer function can be realized by building a network of primitive components consisting of integration, gain, and summation blocks; therefore, implementing an LTI system in biochemical systems amounts to realizing these three basic components with chemical reactions [10]. However, the CRN realizations of these components proposed by Oishi and Klavins [10] are deficient in that rate constants have to be carefully matched and the designed control system cannot be reconfigured. These deficiencies impose a practicality issue that restricts system realization and a flexibility issue that prevents dynamic system adaptation. We can overcome these issues by introducing configurable primitive components as discussed below.

Following Oishi and Klavins [10], we represent a real signal x by the difference ($x^+ - x^-$) between the concentrations of two complementary molecular species x^+ and x^-. In this paper, we do not distinguish notationally between a species and its concentration. Moreover, we abbreviate a pair of chemical reactions of complementary species

$$x^+ + \cdots \xrightarrow{r^+} y^- + \cdots,$$
$$x^- + \cdots \xrightarrow{r^-} y^+ + \cdots,$$

with rate constants r^+ and r^-, to

$$x^\pm + \cdots \xrightarrow{r^\pm} y^\mp + \cdots,$$

where the plus and minus signs in the superscripts $\pm$ and $\mp$ are ordered.

Table 1 shows the chemical reactions and transfer functions of the proposed configurable primitive components, which we will elaborate on in the following sections.

Table 1: Three primitive components, their chemical reactions, and their transfer functions.

Type	Reactions	Transfer Function
Integration	$\begin{cases} x^\pm + u^\pm \xrightarrow{r_1^\pm} x^\pm + u^\pm + y^\pm \\ y^+ + y^- \xrightarrow{r_{\text{int}}} \emptyset \end{cases}$	$Y(s) = \frac{k}{s}U(s)$, $k \equiv r_1^+ x^+ = r_1^- x^-$
Gain & Summation	$\begin{cases} x_i^\pm + u_i^\pm \xrightarrow{r_i^\pm} x_i^\pm + u_i^\pm + y^\pm, i = 1, \ldots, n \\ z^\pm + y^\pm \xrightarrow{r_0^\pm} z^\pm \\ y^+ + y^- \xrightarrow{r_{gs}} \emptyset \end{cases}$	$Y(s) = \sum_{i=1}^{n} \frac{k_i}{s + k_0} U_i(s)$, $\begin{cases} k_i \equiv r_i^+ x_i^+ = r_i^- x_i^-, i = 1, \ldots, n \\ k_0 \equiv r_0^+ z^+ = r_0^- z^- \end{cases}$

doi:10.1371/journal.pone.0137442.t001

Integration Block

An integration block takes a signal u(t) as input, and produces a signal $y(t) = k\int_0^t u(\tau)\mathrm{d}\tau + y(0)$, for some constant $k \in \mathbb{R}$, as output. The chemical realization of an integration block for $k \geq 0$ consists of a pair of catalytic reactions

$$x^\pm + u^\pm \xrightarrow{r_1^\pm} x^\pm + u^\pm + y^\pm$$

and an annihilation reaction

$$y^+ + y^- \xrightarrow{r_{\text{int}}} \emptyset,$$

where $r_1^+, r_1^-, r_{\text{int}}$ are the rate constants. The reactions differ from those of Oishi and Klavins [10] in that the auxiliary species $x^\pm$ are newly added to the catalytic reactions. Both the auxiliary species $x^\pm$ and the input species $u^\pm$ serve as catalysts.

With the definition that $r_1^+ x^+ = r_1^- x^- \equiv k$, the signal y is exactly the integration of signal u as described by the ordinary differential equation

$$\dot{y} = \dot{y}^+ - \dot{y}^- = r_1^+ x^+ u^+ - r_1^- x^- u^- = ku,$$

for $\dot{y}^\pm = r_1^\pm x^\pm u^\pm - r_{\text{int}} y^+ y^-$. Taking the Laplace transform of the above equation, we obtain the transfer function $\frac{Y}{U} = \frac{k}{s}$. Because the concentrations of x^+ and x^- can be controlled externally, in theory it is always possible to design a reaction network to meet any required k. For $k < 0$, the reactions are the same except that the catalytic reactions should be modified by reversing the complementary species of the output signal y as

$$x^{\pm} + u^{\pm} \xrightarrow{r_1^{\pm}} x^{\pm} + u^{\pm} + y^{\mp},$$

where $y^{\mp}$ replaces the original $y^{\pm}$.

Gain and Summation Blocks

A weighted summation block takes a number of input signals $u_i(t)$, i = 1, 2, …, n and produces an output signal $y(t) = \sum_{i=1}^{n} k_i u_i(t)$ for $k_i \in \mathbb{R}$. A gain block is a special weighted summation block with only one input u(t) that produces output $y(t) = k_1 u(t)$ for $k_1 \in \mathbb{R}$. The gain block with $k_1 \geq 0$ can be realized by one pair of catalytic reactions,

$$x^{\pm} + u^{\pm} \xrightarrow{r_1^{\pm}} x^{\pm} + u^{\pm} + y^{\pm},$$

one pair of degradation reactions,

$$z^{\pm} + y^{\pm} \xrightarrow{r_0^{\pm}} z^{\pm},$$

and an annihilation reaction

$$y^{+} + y^{-} \xrightarrow{r_{gs}} \emptyset,$$

where $x^{\pm}$ and $z^{\pm}$ are auxiliary species. These reactions induce the following kinetic equations

$$\dot{y}^{\pm} = r_1^{\pm} x^{\pm} u^{\pm} - r_0^{\pm} z^{\pm} y^{\pm} - r_{gs}\, y^{+} y^{-}.$$

Let $k_1 \equiv r_1^{+} x^{+} = r_1^{-} x^{-}$ and $k_0 \equiv r_0^{+} z^{+} = r_0^{-} z^{-}$. The mass action of y becomes

$$\dot{y} = k_1(u^{+} - u^{-}) - k_0(y^{+} - y^{-}) = k_1 u - k_0 y. \quad (1)$$

When the steady state is reached, the changing rate of y (i.e., $\dot{y}$) equals zero, which implies $y = \frac{k_1}{k_0} u$. Note that the concentrations of the auxiliary species $x^{\pm}$ and $z^{\pm}$ are controlled externally. Therefore, it becomes theoretically possible to meet any required k_1 by tuning the concentrations of these auxiliary species. For $k_1 < 0$, the plus and minus signs of the superscripts of y in the pair of catalytic reactions should be swapped.

For the weighted summation block, the reactions are the same as those for the gain block except that the pair of catalytic reactions becomes

$$x_i^{\pm} + u_i^{\pm} \xrightarrow{r_i^{\pm}} x_i^{\pm} + u_i^{\pm} + y^{\pm}, \text{ for } i = 1, \ldots, n.$$

By setting $k_i \equiv r_i^{+} x_i^{+} = r_i^{-} x_i^{-}$ we obtain the equation

$$y = \sum_i \frac{k_i}{k_0} u_i \quad (2)$$

in the steady state. Similarly, if a scaling factor $k_i < 0$, we swap the signs in the superscript of yin the reaction corresponding to input u_i.

It is worth noting that in contrast to the integration block, the weighted summation block only approximates the intended weighted summation when the steady state is not yet reached; this is often the case in practice. This approximation can be clearly seen from its transfer function $\frac{Y}{U} = \sum_i \frac{k_i}{s + k_0}$, rather than the ideal $\frac{Y}{U} = \sum_i \frac{k_i}{k_0}$. Although this approximation may seem to inevitably cause a chemical reaction implementation to deviate from its specification, we will later show methods that can be used to circumvent this imperfection.

Case Study

To assess the benefit of the auxiliary species, we perform a proof-of-concept case study on the mass-spring-damper (MSD) system as shown in Fig 2A. The system can be modeled by the equation $M\ddot{x} + b\dot{x} + kx = F$. Using M = 1 kg, b = 10 N s/m, k = 20 N/m, and F = 1 N, by Laplace transform we derive the transfer function

$$G = \frac{1}{s^2 + 10s + 20} \approx 0.2236\left(\frac{1}{s + 2.764} - \frac{1}{s + 7.236}\right)$$

The transfer function can be implemented with the block diagram shown in Fig 2B. The proportional-integral (PI) controller to the MSD system is shown in Fig 2C.

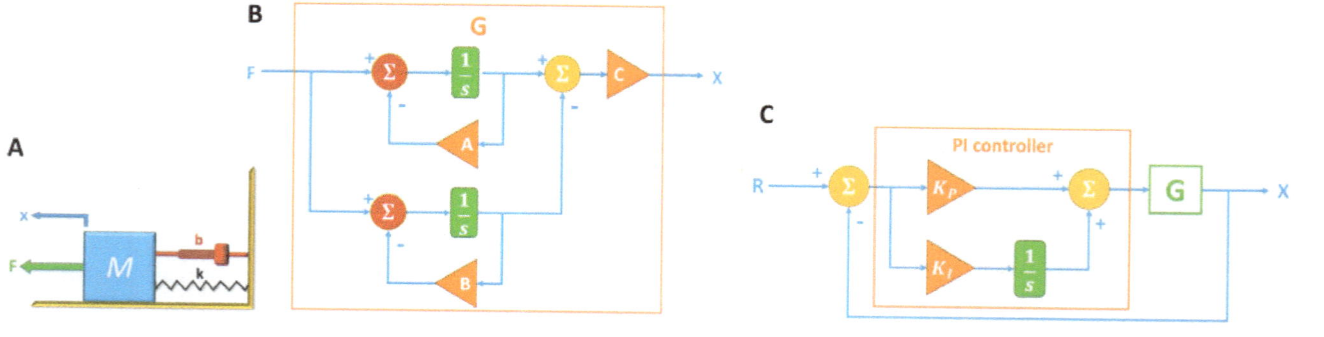

Figure 2: Mass-spring-damper system and its block diagrams. (A) MSD system. (Let F = 1 N, M = 1 kg, b = 10 N s/m, and k = 20 N/m.) (B) Block diagram of the MSD model, where the triangular blocks denote gain functions with their corresponding weights, the rectangular blocks denote integrators, and the circle blocks denote mixers for summation and/or subtraction. (Let A = 2.764, B = 7.236 and C = 0.2236.) (C) Block diagram of the PI-controlled MSD model, where G is the plant shown in (B). (Assume the values of K_p and K_I given in Table 2.)

The block diagrams are constructed with $k_0 = 10$ for all the summation and gain blocks, with the exception of the summation blocks shown in red and gain blocks A and B with $k_0 = 50$. The values of k_i's are set to αk_0, where the values of α are equal to the weights specified in the corresponding gain blocks. Additionally, we assume q_0^+ and q_0^- have the same values as k_0, and q_1^+ and q_1^- are mismatched to k_1 by 10%, where q_1 and q_0 are the rate constants of the catalytic and degradation reactions, respectively, formulated in the prior work [10].

Fig 3A1, 3A2, 3A3, and 3B show the step, impulse, and sinusoidal responses of the MSD system, and the step responses of the PI-controlled MSD system, respectively. Our method achieves better approximation to the ideal cases than the prior method [10]. One of the advantages of our method is that we can match the weight $\frac{k_1}{k_0}$ by tuning the concentrations of $x^{\pm}$ and $z^{\pm}$. In contrast, no tuning is possible in the prior method [10] to avoid the inexact gain $\frac{q_1}{q_0}$, where q_1 and q_0 are the rate constants of the catalytic and degradation reactions formulated in the prior work [10], respectively, due to the mismatch of the rate constants. (Note that the biochemical implementations have their own optimal K_P and K_I values, shown in Table 2, to approximate the ideal system.)

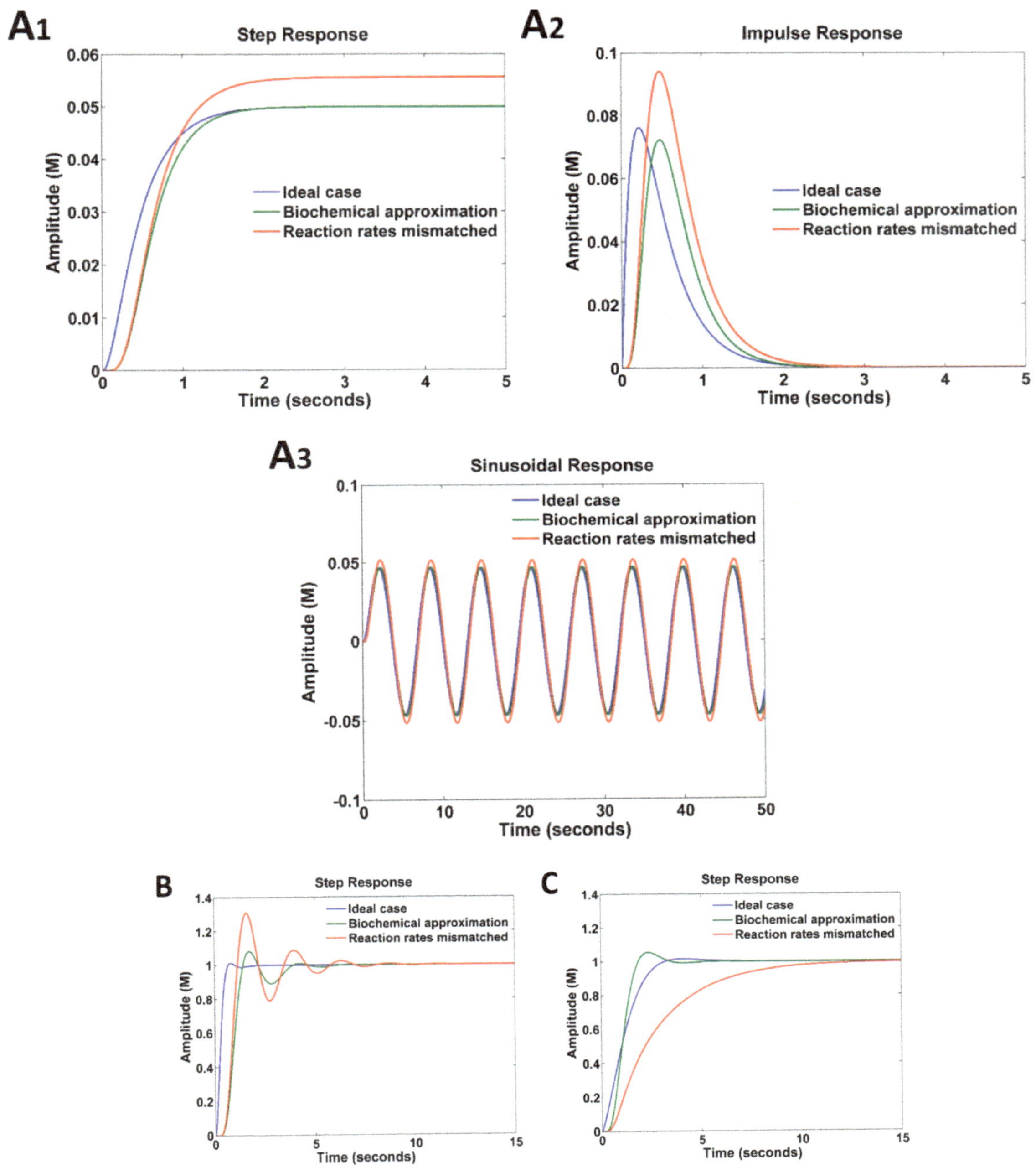

Figure 3: Responses of MSD systems with and without a PI controller. The blue, green, and red curves represent the responses in ideal, configurable biochemical implementation, and nonconfigurable biochemical implementation cases, respectively. (A) Step, impulse, and sinusoidal responses of the MSD. (B) Step responses of PI-controlled MSD. (Assume 10% rate mismatch in the MSD system.) (C) Step responses of PI-controlled MSD, where the MSD undergoes a parameter change with b = 40 N s/m and k = 60 N/m, respectively, to induce a gain change of A = 1.561, B= 38.44 and C = 0.0271.

Table 2: Values of (K_p, K_I) in ideal, configurable and non-configurable implementations of the original and new MSD systems.

	Ideal	Config. Imp.	Non-config. Imp.
Original MSD	(30, 70)	(15, 20)	(15, 20)
New MSD	(15, 55)	(40, 60)	(15, 20)

doi:10.1371/journal.pone.0137442.t002

Suppose that the spring and damper of the above MSD system are now replaced with new ones for b = 40 N s/m and k = 60 N/m. Our method can still adapt the PI-controller to the new MSD system without redesigning the PI-controller, whereas the prior method has no such capability. Because we can tune the concentrations of $x^{\pm}$ and $z^{\pm}$ in biochemical implementation, it is possible for us to adapt (K_p, K_I) to optimal values (40, 60) for the new PI-controlled MSD system, which is in contrast to the original system (15, 20). Fig 3C compares the results with and without such reconfigurability.

DSD Realization of Configurable Primitive Components

We exploit the DNA strand displacement (DSD) technique [24, 33] as an experimental chassis for our synthesis flow, and map the synthesized CRNs to DSD reactions. The simulation is conducted using the Visual DSD tool [34] for validation.

To implement each primitive component, we map the reactions listed in Table 1 to DSD reactions considering the compatibility among the components. We adopt the two-domain DSD method [25] to compile CRNs into nucleic acid-based chemistry to achieve flexible composability. In this method, any free (i.e., unbound) single-stranded DNA species consist of two domains: one toehold domain and one recognition domain. A toehold may initiate the binding between a single-stranded DNA species and a double-stranded DNA species. The two-domain DSD method supports compositionality, which indicates that a block implemented with DSD reactions can be cascaded with other blocks implemented with DSD reactions. Thus, the output species of one block fits the input species required by its downstream blocks.

As a notational convention, we indicate a domain t to be a toehold by a hat "^" as $\hat{t}$. We also use superscript * to indicate the Watson-Crick complement of a domain DNA sequence (e.g., domain t* is sequence CTAG if domain t is sequence GATC). For a single-stranded species, we indicate the 5′ end using a vertical bar "|" and the 3′ end by an angle bracket "⟩" or "⟨." We call a single-stranded species of the form | ⟩ (5′ end on the left and 3′ end on the right) and ⟨ | (5′ end on the right and 3′ end on the left) an upper strand and a lower strand, respectively. A double-stranded species with an upper strand |t u⟩ (and a lower strand ⟨t* u*|) is denoted as [[t u]]. A double-stranded species with one or more dangling (or exposed) toeholds is called a gate, and is used to help transduce one or more species into other species. Double-stranded species without any dangling (exposed) toeholds or single-stranded species containing no toeholds are regarded as waste because they lack the ability to react with other species. The gates and species separately used to construct catalysis, degradation, and annihilation CRNs are calledfuel.

To ignore the effects of fuel depletion, we maintain all of the separately used species at constant concentrations. Moreover, we set the binding rates of all toeholds to 0.05 nM^{-1} s^{-1}, which is consistent with the prior work [35]. We assume infinite unbinding rates [32, 36] to ignore the unproductive reactions [32] that result from the situation where a single-stranded species binds to a gate but does not further displace the strand with which it competes for the same recognition domain. Under these settings, we map the catalytic chemical reactions $x^{\pm} + u^{\pm} \xrightarrow{r_1^{\pm}} x^{\pm} + u^{\pm} + y^{\pm}$ to the DSD reactions shown in Fig 4, where the domains highlighted in red and green are toeholds. Reactions $x^{+} + u^{+} \xrightarrow{r_1^{+}} x^{+} + u^{+} + y^{+}$ and $x^{-} + u^{-} \xrightarrow{r_1^{-}} x^{-} + u^{-} + y^{-}$ can be realized by the same set of DSD reactions except for species renaming (on recognition domains). For brevity, we simply use the same set of species $|\hat{t}\ x\rangle$ (catalysis_7), $|\hat{t}\ u\rangle$ (sp_0), and $|\hat{t_2}\ y\rangle$ (sp10) to denote the species x^{+}, u^{+}, and y^{+}, respectively, to implement the former reaction and to denote x^{-}, u^{-}, and y^{-}, respectively, for the latter reaction.

The set of DSD reactions outlined in Fig 4 consists of two stages. Gate catalysis initiates the first stage of the reactions by reacting with $|\hat{t}\ x\rangle$. At the end of the first stage, the species $|a\ \hat{t}\rangle$ will be produced, and this species will react with gate catalysis_1 to initiate the second stage. During the second stage reactions, the species $|\hat{t}\ x\rangle$ and $|\hat{t}\ u\rangle$ will be produced to compensate for their consumption in the first stage.

We note that if the recognition domains of u and x were exchanged within the seven gates and one waste in the four first stage reactions, the entire set of reactions would still represent an implementation of the intended catalytic chemical reactions. However, the new arrangement of $[[u\ \hat{t}]]$ followed by $[[x\ \hat{t}]]$ would accelerate input $|\hat{t}\ u\rangle$ (effectively, u^{+} or u^{-} in the catalytic chemical reactions) consumption because the input $|\hat{t}\ u\rangle$ would directly react with the newcatalysis gate. This is in contrast to the original catalysis structure, where $[[x\ \hat{t}]]$ is followed by $[[u\ \hat{t}]]$. Therefore, this consumption acceleration of $|\hat{t}\ u\rangle$ would amplify the bias between the production and consumption rates of $|\hat{t}\ u\rangle$. Because the concentration of $|\hat{t}\ u\rangle$ could not remain constant, the input species $|\hat{t}\ u\rangle$ could not be treated as a catalyst as expected. As a result, the new set of DSD reactions will not be as effective as the example presented in Fig 4.

To calibrate the values of the rate constants $r_1^{\pm}$, we conduct an experiment on the catalytic reaction $x^{\pm} + u^{\pm} \xrightarrow{r_1^{\pm}} x^{\pm} + u^{\pm} + y^{\pm}$. Let $x^{\pm}$ be mapped to $|\hat{t}\ x\rangle$, $u^{\pm}$ to $|\hat{t}\ u\rangle$ and $y^{\pm}$ to $|\hat{t_2}\ y\rangle$ at the concentrations summarized in Table 3. Additionally, we set and maintain the concentrations of all fuel to 2000 nM. However, the concentration of auxiliary species $|\hat{t}\ x\rangle$ can be obtained by the weight $k \equiv r_1^{+}x^{+} = r_1^{-}x^{-}$ after $r_1^{\pm}$ are determined. Based on the relationship $r_1^{\pm}x^{\pm}u^{\pm} = \dot{y}^{\pm} = y_{\text{final}}^{\pm}/\tau$, where

$y^{\pm}_{\text{final}}$ are the concentrations of $|\hat{t_2}\ y\rangle$ after an elapsed time $\tau = 1000$ seconds, we derive the rate constants $r_1^{\pm}$. As seen from Table 3, the reaction rates $r_1^{\pm}$ hold almost the same value when the concentrations of $|\hat{t}\ u\rangle$ and $|\hat{t}\ x\rangle$ are smaller than 100 nM, but slightly decrease when one of the concentrations approaches 100 nM. Because $|\hat{t}\ u\rangle$ and $|\hat{t}\ x\rangle$ will be low concentrations (approximately 10^{-6} to 1 nM) in our later experiments, we set $r_1^{\pm} = 0.0249$ $nM^{-1}\ s^{-1}$. Fig 5A and 5B show the concentration of $|\hat{t_2}\ y\rangle$ over time under the concentration (in nM) settings of $(|\hat{t}\ x\rangle, |\hat{t}\ u\rangle)$ at (0.001, 0.001) and (100, 100), respectively. These results suggest that the DSD reactions produce the output $y^{\pm}$ (i.e., $|t_2\ y\rangle$) at a constant rate. The linearity of the DSD reactions indicates their adequacy in realizing our proposed configurable primitive components.

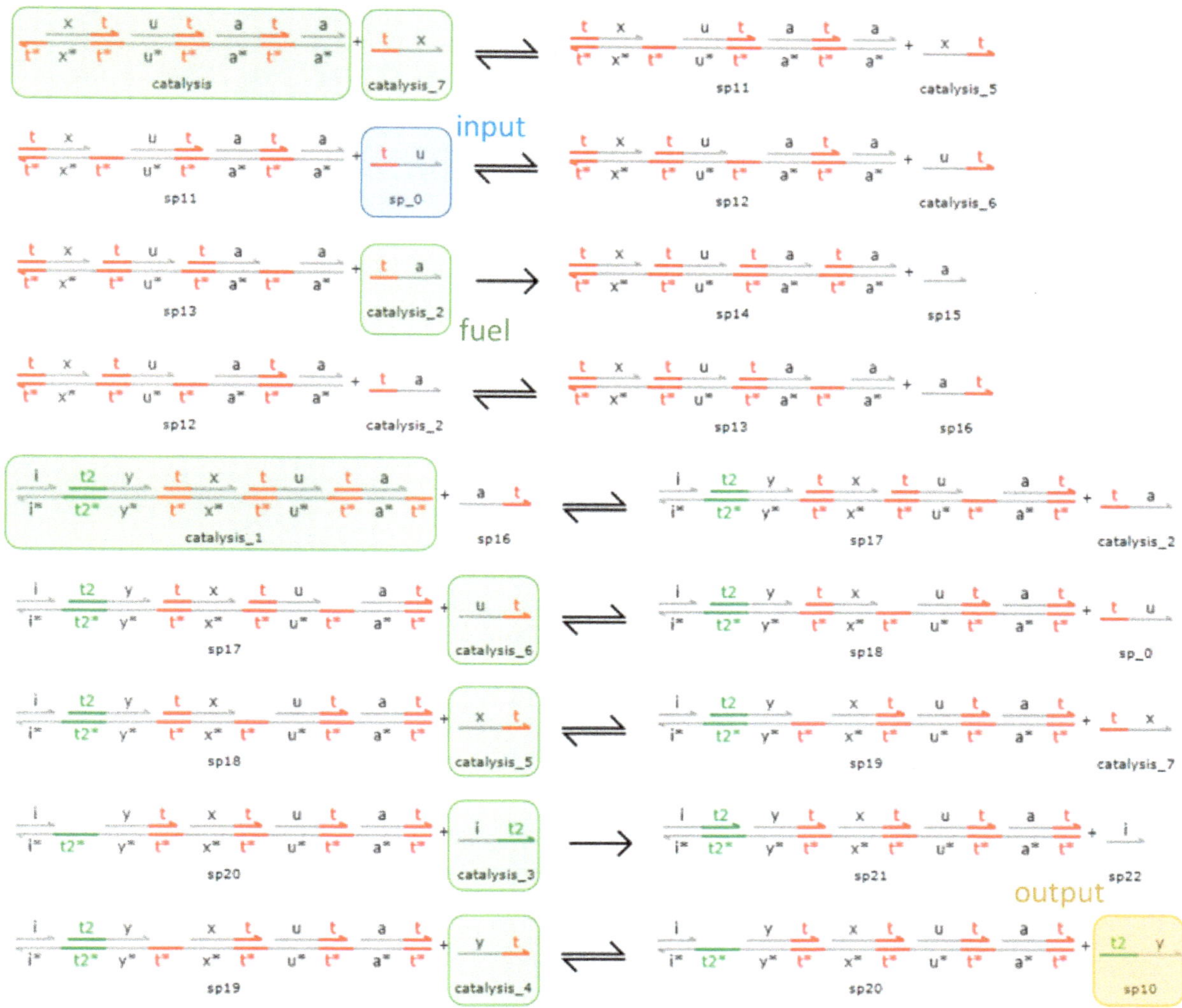

Figure 4: DSD realization of catalytic reactions. The species in green are fuels whose concentrations are fixed to 2000 nM to avoid the effect of fuel depletion. The species sp_0, sp10, and catalysis_7 represent the inputu^+ or u^-, output y^+ or y^-, and catalyst x^+ or x^-, respectively. Species catalysis_7reacts with the gate catalysis to begin the first stage of the reactions followed by a series of reactions involving species sp_0 and catalysis_2. At the end of the first stage, species sp16 is produced and reacts with the gate catalysis_1 to begin the second stage of the reactions. In this stage, species catalysis_7 and sp_0 are yielded to compensate for their consumption in the first stage. Finally, the output species sp10 is generated.

Table 3: Experimental data used for measuring the rate constant of DSD catalysis.

$(\lvert\hat{t}\ u\rangle, \lvert\hat{t}\ x\rangle, \lvert\hat{t_2}\ y\rangle)$ **(nM, nM, nM)**	$r_1^{\pm}$ **($nM^{-1}\ s^{-1}$)**
(1, 1, 24.97)	0.0250
(1, 100, 2272)	0.0227
(10, 10, 2469)	0.0247
(100, 1, 2436)	0.0244
(100, 100, 220492)	0.02205
(10^{-3}, 10^{-3}, 2.49×10^{-3})	0.0249

doi:10.1371/journal.pone.0137442.t003

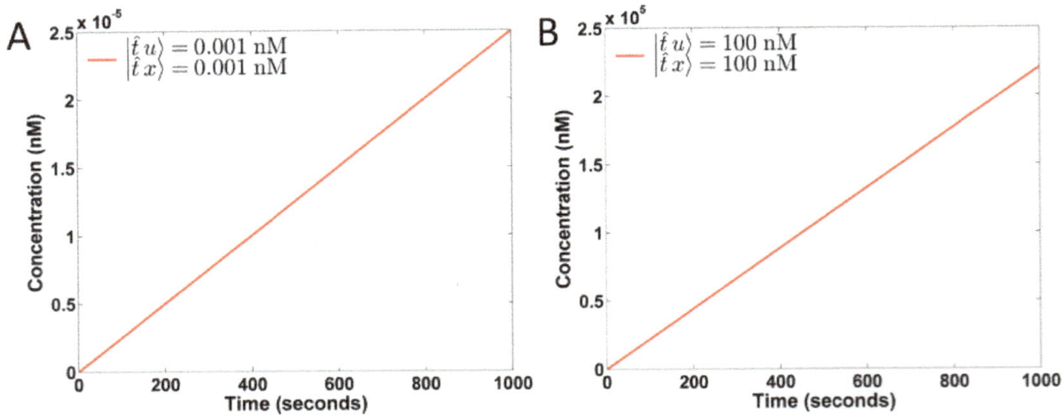

Figure 5: Concentration of $|\hat{t_2}\ y\rangle$ **over time in DSD realized catalytic reactions.** (A) $(|\hat{t}\ x\rangle, |\hat{t}\ u\rangle) = (0.001, 0.001)$ in nM and (B) $(|\hat{t}\ x\rangle, |\hat{t}\ u\rangle) = (100, 100)$ in nM.

Similar to the construction proposed by Yordanov et al. when mapping chemical reactions $y^{\pm} \xrightarrow{q_0^{\pm}} \emptyset$ to DSD reactions [37], we map the degradation chemical reactions $z^{\pm} + y^{\pm} \xrightarrow{r_0^{\pm}} z^{\pm}$ to the DSD reactions shown in Fig 6. For brevity, the DSD reactions shown only correspond to one of the reactions $z^{+} + y^{+} \xrightarrow{r_0^{+}} z^{+}$ and $z^{-} + y^{-} \xrightarrow{r_0^{-}} z^{-}$. The species $\langle \hat{t_2}^{*}[[y]]$, labeled deg, is the only gate or fuel used in this network, while the species $|\hat{t_2}\ y\rangle$ represents $y^{\pm}$. The productssp3 and sp4 are waste products. In contrast to the DSD reactions proposed by Yordanov et al.[37] where the fuel concentrations are not maintained as constants, we maintain the fuel deg at a constant concentration. Therefore, the reaction can be effectively regarded as

$$deg + |\hat{t_2}\ y\rangle \xrightarrow{0.05} deg,$$

which is the same form as the above degradation chemical reactions. Thus, the species degcorresponds to z^{+} or z^{-}, and the rate constants $r_0^{\pm}$ equal 0.05 $nM^{-1}\ s^{-1}$, which is the binding rate constant of a toehold. By solving the differential equation of the kinetics of $y^{\pm}$, we derive that its concentration at time t is $y(t)^{\pm} = y(0)^{\pm}\exp[-(0.05z^{\pm})t]$, where $y(0)^{\pm}$ are the initial concentrations of $y(t)^{\pm}$. Fig 7A and 7B plot $y(t)^{\pm}$ for $z^{\pm} = 1$ nM and 100 nM, respectively. In each case, there are three curves corresponding to different initial concentrations of $y(0)^{\pm} = 10$, 50 and 100 nM. These six curves are shown in Fig 7A and 7B and exactly fit the ideal waveforms.

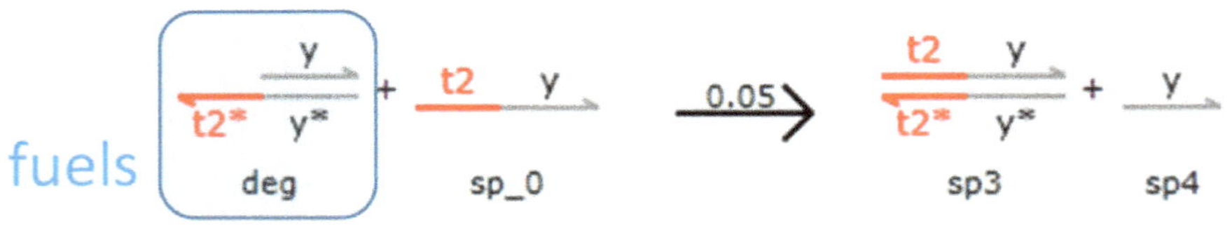

Figure 6: DSD realization of the degradation reaction. Here, deg serves as fuel with its concentration maintained at a fixed value and corresponds to catalysts $z^{\pm}$. In the reaction, the rate constant equals the toehold binding rate 0.05 $nM^{-1}\ s^{-1}$.

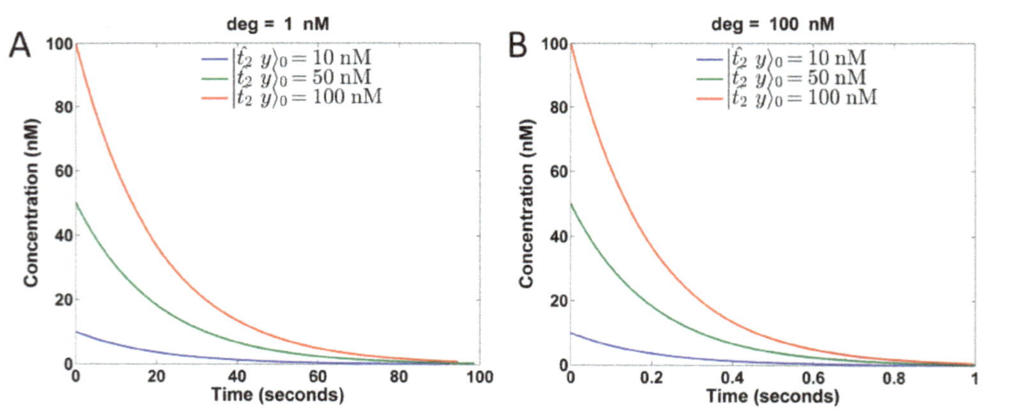

Figure 7: Simulation of DSD realized degradation reactions. (A) Concentration of $y^{\pm} = |\hat{t_2}\ y\rangle$ over time under deg = $z^{\pm}$ = 1 nM. (B) Concentration of $y^{\pm} = |\hat{t_2}\ y\rangle$ over time under deg = $z^{\pm}$ = 100 nM. The three curves in each case correspond to different initial concentrations $y(0)^{\pm}$ = 10, 50, and 100 nM.

Finally, following the construction scheme of Yordanov et al. [37], we map the annihilation chemical reaction

$$y^{+} + y^{-} \xrightarrow{r_{\text{int}} \text{ or } r_{\text{gs}}} \emptyset \tag{3}$$

to the DSD reactions provided in Fig 8, where species $|\hat{t_2}\ yp\rangle$ and $|\hat{t_2}\ ym\rangle$ represent y^+ and y^-, respectively. These species correspond to the output species of the primitive components. Two fuels with similar structures are used to make the entire DSD reaction symmetric and balance the consumption rates of $|\hat{t_2}\ yp\rangle$ and $|\hat{t_2}\ ym\rangle$. As shown in Fig 8, the fuel ann first reacts with $|\hat{t_2}\ yp\rangle$ and then the product sp11 reacts with $|\hat{t_2}\ ym\rangle$. In contrast, the fuel ann_1first reacts with $|\hat{t_2}\ ym\rangle$ followed by the reaction between the product sp15 and $|\hat{t_2}\ yp\rangle$. Therefore, the consumption rates of $|\hat{t_2}\ yp\rangle$ and $|\hat{t_2}\ ym\rangle$ are balanced and achieve effective annihilation.

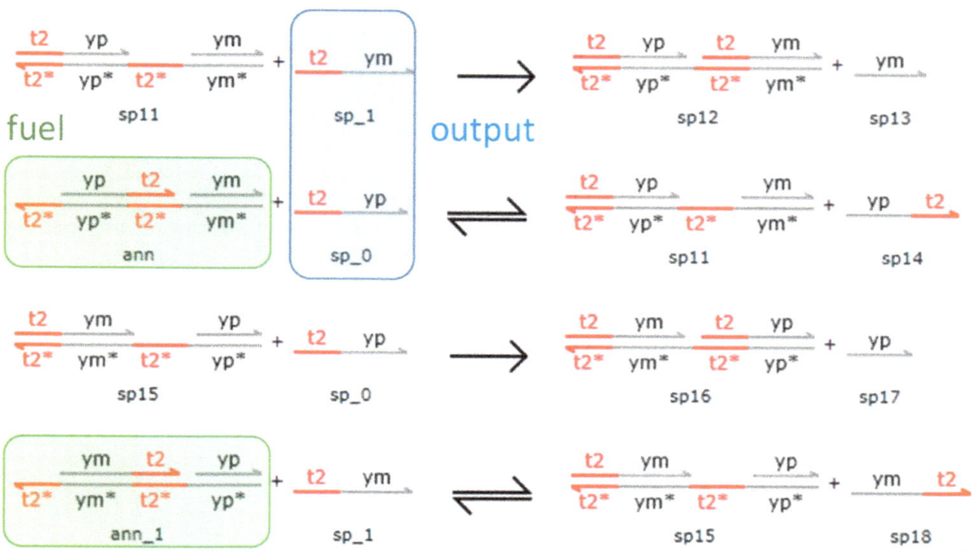

Figure 8: DSD realization of the annihilation reaction. Here, $|\hat{t_2}\ ym\rangle$ and $|\hat{t_2}\ yp\rangle$ correspond to the output species y^+ and y^-, respectively. Two fuels (ann and ann_1) with similar structures are used to ensure that the consumption rates of y^+ and y^- are balanced.

The above DSD reactions for catalysis, degradation, and annihilation form a set of basic elements that enable the construction of primitive components and any proper linear system. We built the proposed configurable primitive components of integration and weighted summation with DSD reactions. The simulation results are shown in Fig 9. Fig 9A shows the output response of the integration component computing $y = \int_0^t u(\tau)d\tau$ with respect to fixed inputs $u^+ = 2$ nM and $u^- = 1$ nM. The concentration of y^- (green curve) remains at approximately 8 nM due to the effect of the annihilation reaction [Eq (3)], while the concentration of y^+ (blue curve) grows to 105 nM after 100 seconds. As expected, the output signal $y = y^+ - y^-$ grows linearly and reaches 97 nM after 100 seconds. Fig 9B shows the simulation results of the weighted summation component computing $y = 2u_1 + u_2$. Two sets of inputs $(u_1^+, u_1^-) = (5,3)$ and $(u_2^+, u_2^-) = (2,1)$ in nM are observed. As expected, the component functions correctly with $y = 12 - 7 = 5$ (for $y^+ = 2 \times 5 + 2 = 12$ and $y^- = 2 \times 3 + 1 = 7$) nM.

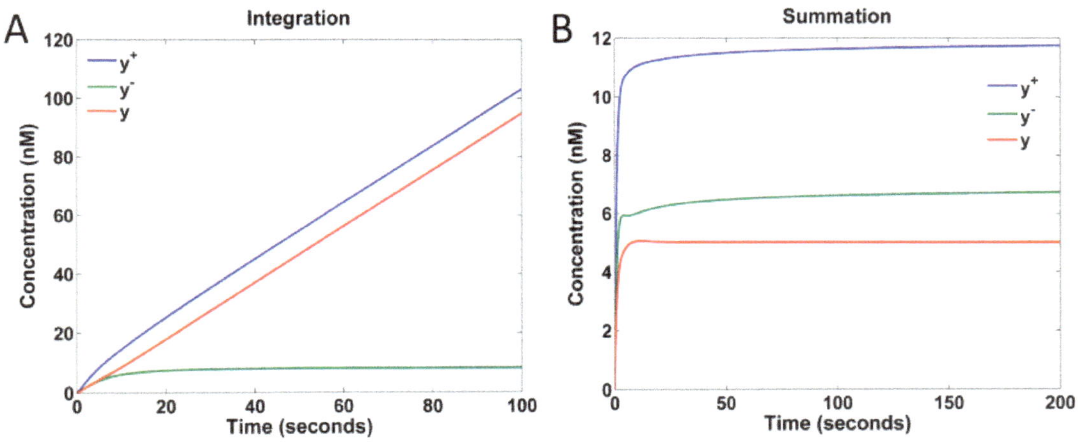

Figure 9: Simulation of DSD realized primitive components. (A) Concentrations of y^+, y^-, and $y = y^+ - y^-$ of the integration component computing $y = \int_0^t u(\tau)d\tau$ under constant inputs $u^+ = 2$ and $u^- = 1$ nM. (B) Concentrations of y^+, y^-, and $y = y^+ - y^-$ of the weighted summation component computing $y = 2u_1 + u_2$ under constant inputs $(u_1^+, u_1^-) = (5,3)$ and $(u_2^+, u_2^-) = (2,1)$ in nM.

The above simulation results confirm the successful realization of the integration and weighted summation components using DSD reactions. Because good compatibility is observed among the blocks constructed using the two-domain DSD method, these primitive DSD reactions can be composed to realize sophisticated linear systems. Later we will show that any proper transfer functions can be realized using the DSD realized primitive components.

Implementation of Transfer Functions in Biochemistry

Transfer Function Decomposition

The transfer function G(s) of an LTI system can always be written as

$$G=\frac{Y(s)}{U(s)}=\frac{b_m s^m+b_{m-1}s^{m-1}+\cdots+b_0}{a_n s^n+a_{n-1}s^{n-1}+\cdots+a_0}, \tag{4}$$

where Y(s) and U(s) are the Laplace transforms of the output y(t) and input u(t) signals, respectively. A transfer function is called strictly proper, as we shall assume, if $m < n$ (i.e., the degree of the numerator polynomial is smaller than the denominator polynomial).

Using the fundamental theorem of algebra, U(s) can be rewritten as

$$U(s)=(s+\alpha_1)^{n_1}\cdots(s+\alpha_l)^{n_l}(s^2+\beta_1 s+\gamma_1)^{m_1}\cdots(s^2+\beta_p s+\gamma_p)^{m_p},$$

where α_i, β_i and γ_i are real numbers satisfying $\beta_i^2-4\gamma_i<0$ for stable LTI systems, l and p are non-negative integers, and n_i and m_i are positive integers. By partial fraction expansion, Eq (4)can be

$$G(s)=\sum_{i=1}^{l}\sum_{j=1}^{n_i}\frac{A_{ij}}{(s+\alpha_i)^j}+\sum_{i=1}^{p}\sum_{j=1}^{m_i}\frac{B_{ij}s+C_{ij}}{(s^2+\beta_i s+\gamma_i)^j},$$

factorized as

where A_{ij}, B_{ij}, and C_{ij} are real coefficients. Because the terms of $\frac{1}{(s+\alpha_i)^j}$ and $\frac{1}{(s^2+\beta_i+\gamma_i)^j}$ can be realized by cascading j times the elementary modules of $\frac{1}{(s+\alpha_i)}$ and $\frac{1}{(s^2+\beta_i s+\gamma_i)}$, respectively, we only need to implement the set

$$\left\{\frac{D_1}{s+\alpha},\frac{D_2}{s^2+\beta s+\gamma},\frac{D_3 s}{s^2+\beta s+\gamma}\right\}, \tag{5}$$

of elementary modules, where D_1, D_2 and D_3 are real numbers. We refer to the three elementary modules as the "degree-(1, 0)," "degree-(2, 0)," and "degree-(2, 1)" modules, respectively, according to the degrees of their denominator and numerator polynomials.

Below we investigate how to implement these three elementary modules to construct any LTI system with CRNs. First, we provide a naive construction to illustrate inexactness, and then present a refined exact solution.

Naive Implementation of Elementary Modules.

To build a degree-(1, 0) module $J(j_1, j_2)$ with parameters j_1, j_2 with real values, one might use the negative feedback loop consisting of a forward integration block of $\frac{j_1}{s}$ and a negative feedback of gain j_2 (as shown in Fig 10A). The transfer function of this realization equals $\frac{j_1}{s+j_1 j_2}$, which exactly represents the target form. Recall that the introduction of auxiliary species enables the configurability of primitive components, and thus the values of j_1 and j_2 can be set as desired.

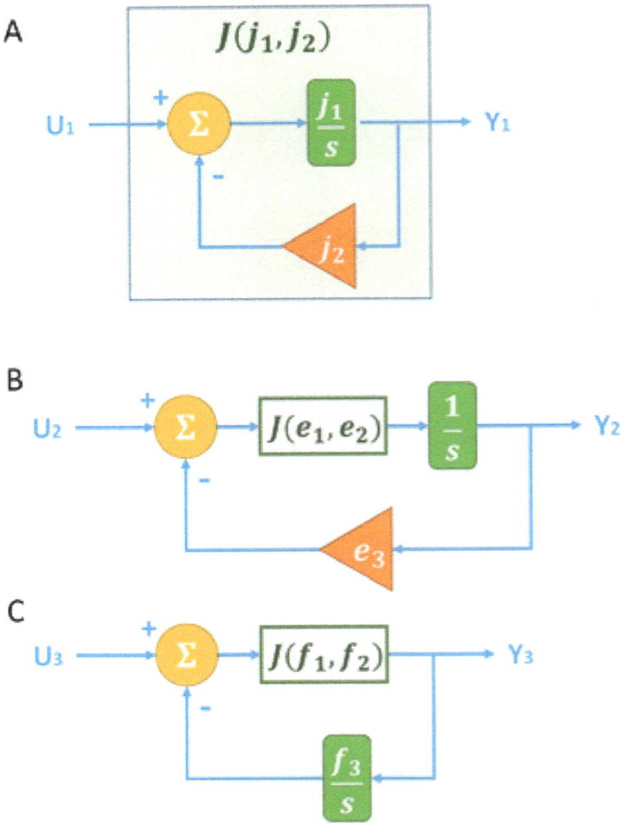

Figure 10: Block diagrams of naive implementation of elementary modules. (A) Block diagram of the degree-(1, 0) module with parametric weights j_1 and j_2 to match the transfer function coefficients. (B) Block diagram of the degree-(2, 0) module with parametric weights e_1, e_2, and e_3 to match the transfer function coefficients. (C) Block diagram of the degree-(2, 1) module with parametric weights f_1, f_2, and f_3 to match the transfer function coefficients.

To build a degree-(2, 0) module, one has to increase the degree of the denominator polynomial. Therefore, one might try to use a degree-(1, 0) module for a compositional construction (as shown in Fig 10B). Let the degree-(1, 0) module be $J(e_1, e_2)$ fed to a forward integration block of $\frac{1}{s}$, and let the negative feedback have gain e_3. Then, the transfer function of the block diagram equals $\frac{e_1}{s^2+e_1e_2s+e_1e_3}$ as desired.

To build a degree-(2, 1) module, one might also try to use a degree-(1, 0) module for compositional construction (as shown in Fig 10C). Let the degree-(1, 0) module be $J(f_1, f_2)$ fed forward to the output, and place an integration block of $\frac{f_3}{s}$ on the negative feedback path. Then, the transfer function of the block diagram equals $\frac{f_1 s}{s^2+f_1f_2s+f_1f_3}$ as desired.

Although the block diagrams of Fig 10 yield the desired transfer functions of the elementary modules, they are problematic due to the imperfection of the weighted summation components realized by the chemical reactions (with transfer function

$\sum_i \frac{k_i}{s+k_0}$ instead of $\sum_i \frac{k_i}{k_0}$ as discussed earlier in the primitive component construction section). The following example demonstrates that a system constructed in this fashion may deviate drastically from the specification and can even be unstable.

Consider the LTI system specified by transfer function

$$G = \frac{3s^4 + 19s^3 + 47s^2 + 51s + 22}{s^5 + 10s^4 + 36s^3 + 67s^2 + 66s + 30}. \quad \textbf{(6)}$$

By partial fraction expansion, G can be expressed as

$$G = \frac{1}{s^2 + 3s + 3} + \frac{s}{s^2 + 2s + 2} + \frac{2}{s+5}. \quad \textbf{(7)}$$

By defining

$$\begin{cases} G_1 \equiv \dfrac{1}{s^2 + 3s + 3}, \\ G_2 \equiv \dfrac{s}{s^2 + 2s + 2}, \\ G_3 \equiv \dfrac{2}{s+5}, \end{cases} \quad \textbf{(8)}$$

they correspond to the basic elements in Eq (5) and can be implemented with the block diagrams of Fig 10 by setting $(j_1, j_2) = (2, 2.5)$, $(e_1, e_2, e_3) = (1, 3, 3)$ and $(f_1, f_2, f_3) = (1, 2, 2)$. Assume that $k_0 = a$ for some positive real constant a for all weighted summation primitive components; then, assume that k_i, i = 1, …, n, of a weighted summation primitive component are determined according to the required weights. The transfer functions after the chemical reaction realization become

$$\begin{cases} G_1 \to \dfrac{a^2(s+a)}{s^2(s+a)^3 + 3a^2 s(s+a) + 3a^3} \equiv G_1', \\ G_2 \to \dfrac{a^2 s}{s^2(s+a)^2 + 2a^2 s + 2a^2} \equiv G_2', \\ G_3 \to \dfrac{2a(s+a)}{s(s+a)^2 + 5a^2} \equiv G_3'. \end{cases} \quad \textbf{(9)}$$

Based on (asymptotic) stability analysis, the implemented system is unstable when a pole of the transfer functions (i.e., a root of the denominator polynomials) falls in the right-half of the s-plane. To avoid instability, the value of a has to be carefully determined. However, choosing a proper value for a can be tedious and sometimes even impossible.

The responses of G_1', G_2', and G_3' to the step input with an amplitude of 10^{-6} are shown in Fig 11. Specifically, Fig 11A and 11B show the responses of G_1', C and D show the responses of G_2', and E and F show the responses of G_1'. As seen from the plots of Fig 11A, 11C, and 11E, the responses are unstable under the chosen small a values, while for plots of B, D, and F, the responses are stable under the chosen large a values. However, even the stable responses may deviate from the ideal transfer function responses to some extent. When a in G_1', G_2' and G_3' increases to 6, the maximal concentrations of the auxiliary species used in each case are 6 × 3/0.0249 ≈ 720, 6 × 2/0.0249 ≈ 480, and 6 × 2.5/0.0249 ≈ 600 nM, respectively, where the factors 3, 2 and 2.5 are the weights of the gain blocks in each system. As seen in Fig 11, the naive implementations cannot approximate the ideal responses well even under such high concentrations.

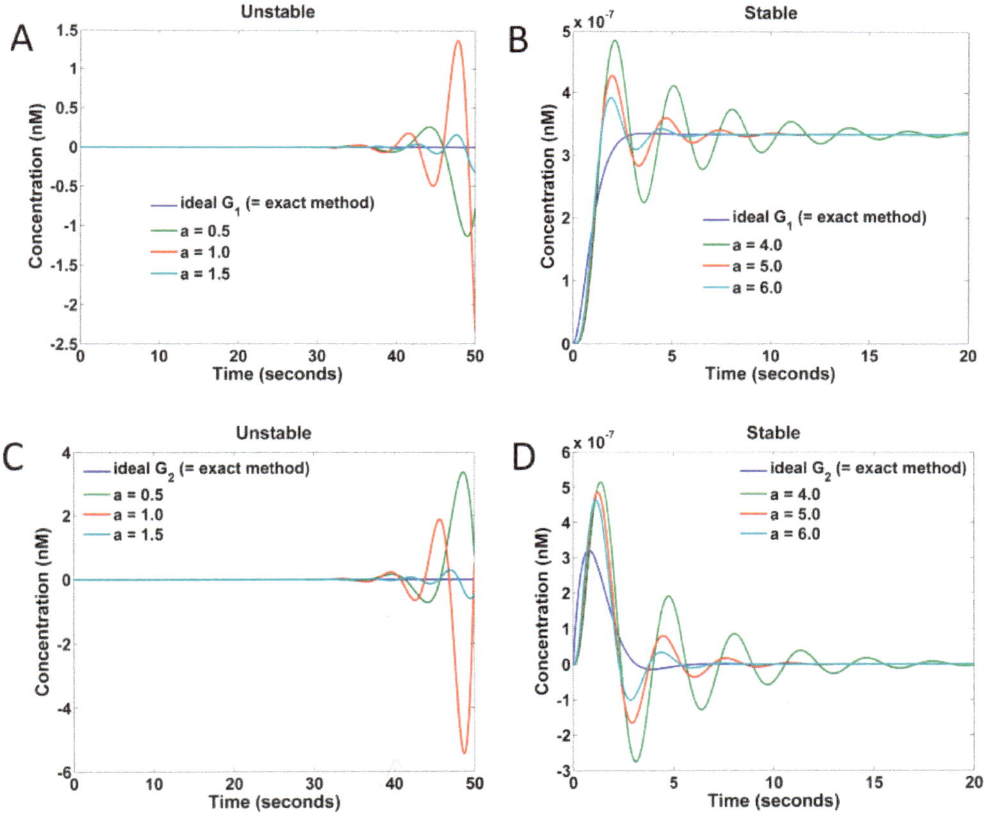

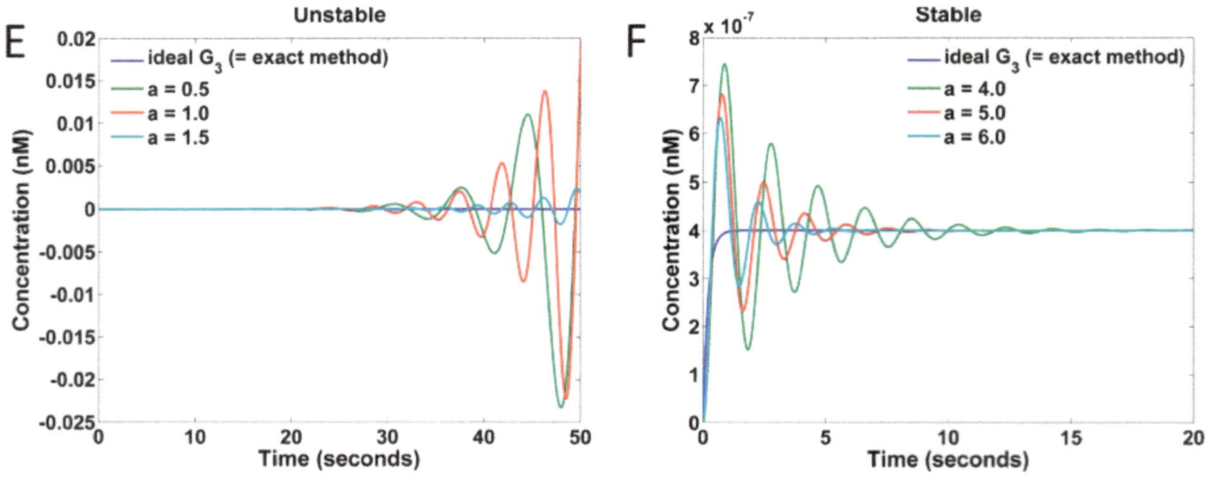

Figure 11: Responses of naive implementations to a step input of amplitude 10^{-6} nM under different values of parameter a. (A) Responses of G_1' under a = 0.5, 1.0, 1.5. (B) Responses of G_1' under a = 4.0, 5.0, 6.0. (C) Responses of G_2' under a = 0.5, 1.0, 1.5. (D) Responses of G_2' under a = 4.0, 5.0, 6.0. (E) Responses of G_3' under a = 0.5, 1.0, 1.5. (F) Responses of G_3' under a = 4.0, 5.0, 6.0.

Exact Implementation of Elementary Modules

To prevent the non-ideal effect of the weighted summation components, we present a refined method that exactly implements the three elementary modules with chemical reactions.

The degree-(1, 0) module can be realized directly by the CRN transfer function of the primitive gain block. For the other two modules, we take advantage of the non-ideal form of the weighted summation block and devise their corresponding block diagrams, whose CRN transfer functions match the specification. The block diagrams, normal transfer functions, and CRN transfer functions for the three elementary modules are summarized in Fig 12. The CRN transfer function of the degree-(2, 0) module can be calculated from

$$\begin{cases} X_1 = \dfrac{k_1}{s+k_0}U - \dfrac{k_2}{s+k_0}Y, \\ Y = \dfrac{k_3}{s}X_1. \end{cases} \quad \textbf{(10)}$$

Therefore, the CRN transfer function $\frac{Y}{U}$ equals $\frac{Y}{U} = \frac{k_1 k_3}{s^2 + k_0 s + k_2 k_3}$. Note that if the gain and summation blocks are ideal, the normal transfer function should be $\frac{k_1 k_3}{k_0 s + k_2 k_3}$. Similarly, the CRN transfer function of the degree-(2, 1) module can be derived as $\frac{Y}{U} = \frac{k_1 s}{s^2 + k_0 s + k_2 k_3}$. As a result, the responses can be made exact through the control of auxiliary species concentrations.

Block Diagram	Normal Transfer Function	CRN Transfer Function
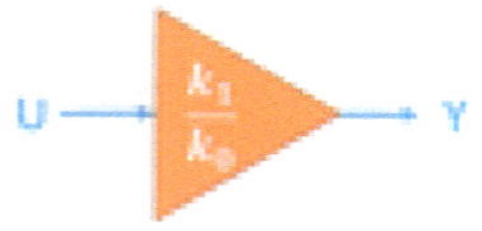	$\dfrac{k_1}{k_0}$	$\dfrac{k_1}{s+k_0}$
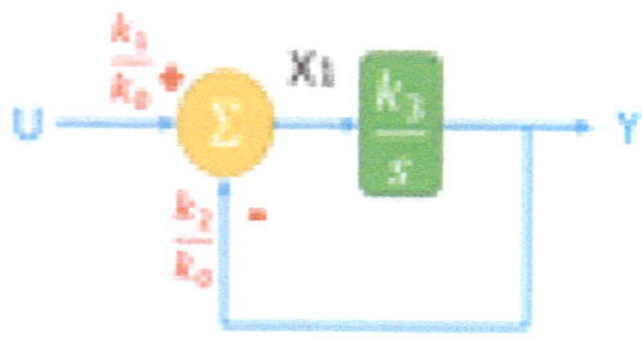	$\dfrac{k_1 k_3}{k_0 s + k_2 k_3}$	$\dfrac{k_1 k_3}{s^2 + k_0 s + k_2 k_3}$
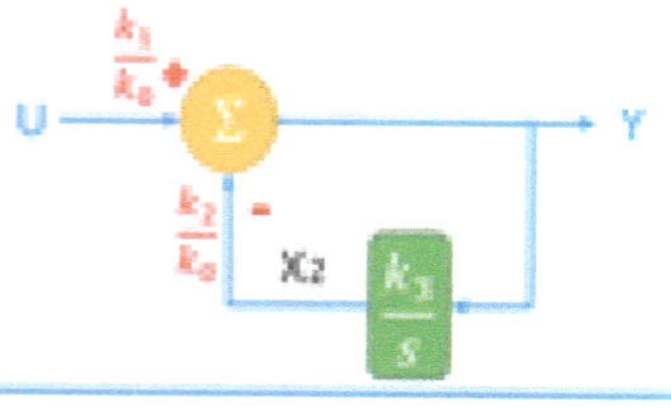	$\dfrac{k_1 s}{k_0 s + k_2 k_3}$	$\dfrac{k_1 s}{s^2 + k_0 s + k_2 k_3}$

Figure 12: Block diagrams and their transfer functions for the exact implementation of the elementary modules. The first, second, and third rows in the table correspond to the implementations of the degree-(1, 0), degree-(2, 0), and degree-(2, 1) modules, respectively. The first, second, and third columns in the table show the block diagrams, normal transfer functions, and CRN transfer functions, respectively. A normal transfer function is obtained directly from its block diagram, whereas a CRN transfer function is obtained when the gain and summation blocks in the block diagram are realized using the gain and summation CRNs in Table 1.

There are two special cases that need attention. First, if the coefficient β vanishes in the denominators of the transfer functions $\frac{D_2}{s^2+\beta s+\gamma}$ and $\frac{D_3 s}{s^2+\beta s+\gamma}$ of the degree-(2, 0) and degree-(2, 1) modules, respectively, we have to make $k_0 = 0$ in the CRN transfer functions. Therefore, we need to remove the degradation reactions $z^{\pm} + y^{\pm} \xrightarrow{r_0^{\pm}} z^{\pm}$ from the summation blocks. Second, for the negative parameter k_1 or k_2 in Fig 12, as mentioned previously the plus and minus signs of the superscripts of y in the pair of catalytic reactions of the weighted summation CRN should be swapped; for negative k_3, the superscript signs of y in the pair of catalytic reactions of the integration CRN should be swapped. Notice that k_0 can always be non-negative for stable linear systems, whose poles should all be on the left-half of the s-plane. Therefore, with these modifications our method is sufficient to implement any stable linear system.

DSD Realization of Transfer Functions

We map the naive and exact implementations of the transfer function specification of Eq (6) into DSD reactions for comparison. In the simulation, we assume the rate constants $r_0^{\pm} = 0.05$ nM^{-1} s^{-1} and $r_i^{\pm} = 0.0249$ nM^{-1} s^{-1} for i = 1, …, n.

The responses of DSD that realized naive implementation of the transfer function specification are shown in Fig 13. Specifically, Fig 13A and 13B show the results realizing the transfer function G_1 in Eq (8), C and D show those realizing G_2, and E and F show those realizing G_3. In the plots of Fig 13A, 13C, and 13E, the realized systems are unstable under the chosen small a values, while in the plots of B, D, and F the realized systems are stable under the chosen large a values. These results are consistent with the CRN responses of Eq (9) in Fig 11.

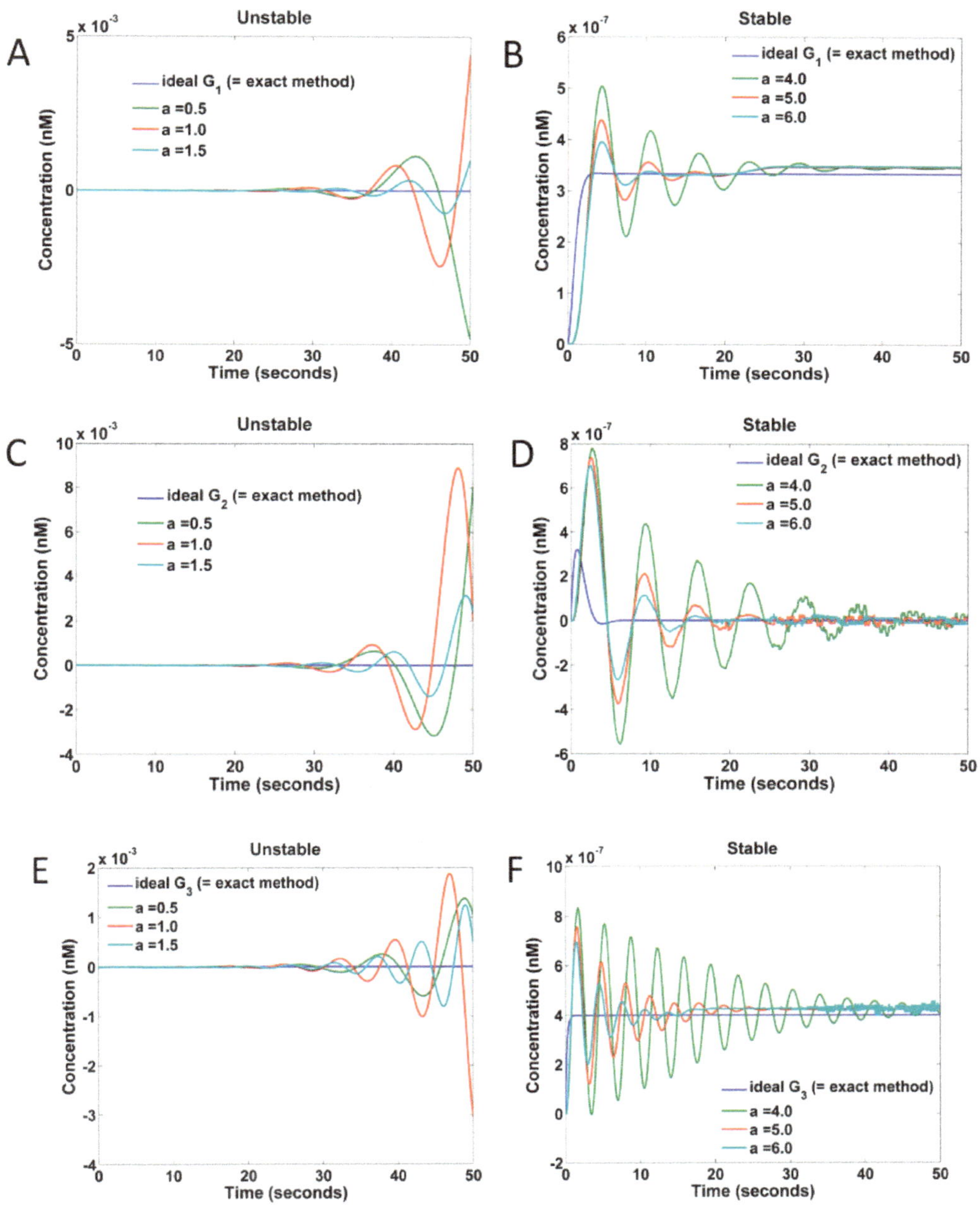

Figure 13: Responses of DSD realized naive implementations to a step input of amplitude 10^{-6} nM under different values of parameter a. (A) Responses of G_1' under a = 0.5, 1.0, 1.5. (B) Responses of G_1' under a = 4.0, 5.0, 6.0. (C) Responses of G_2' under a = 0.5, 1.0, 1.5. (D) Responses of G_2' under a = 4.0, 5.0, 6.0. (E) Responses of G_3' under a = 0.5, 1.0, 1.5. (F) Responses of G_3' under a = 4.0, 5.0, 6.0.

The maximum k_i's in the DSD realizations of G'_1, G'_2, and G'_3 are 6.0 × 3 = 18.0, 6.0 × 2 = 12.0, and 6.0 × 2.5 = 15.0, respectively. Because the rate constants $r_i^{\pm}$, i = 1, …, n are set to 0.0249 $nM^{-1}\ s^{-1}$, the corresponding concentrations of the catalysts $x^{\pm}$ in the gain and summation blocks in Fig 10 of G'_1, G'_2, and G'_3 are approximately 18.0/0.0249 ≈ 720, 12.0/0.0249 ≈ 480, and 15/0.0249 ≈ 600 nM, respectively. Not only are such high concentrations undesirable, but it is also difficult to find a proper a value to fit the ideal response.

To compare the naive and exact implementations of transfer function specifications under DSD realization, the responses of the targeted transfer functions G_1, G_2 and G_3 are plotted in Fig 14A, 14B, and 14C, respectively. The red curves in Fig 14A, 14B, and 14C correspond to the naive implementations with a = 6; the involved concentrations of catalysts are approximately 720, 480, and 600 nM, respectively. The species with the largest concentrations used in the exact implementations of G_1 and G_2 are both the catalysts $x^{\pm}$ in the integration CRNs. Because the weights of the integration blocks in the exact implementations of G_1 and G_2 are both equal to 2.0, the corresponding concentrations of $x^{\pm}$ in the catalytic reactions are approximately 80 nM. In contrast, in the exact implementation of G_3 the catalysts $z^{\pm}$ in the gain block (where k_0 equals 5.0) have the highest concentration (5.0/0.05 = 100 nM). Compared to the naive implementations involving concentrations as large as 720 nM, the exact implementations require much lower concentrations, and yet fit the ideal responses much better.

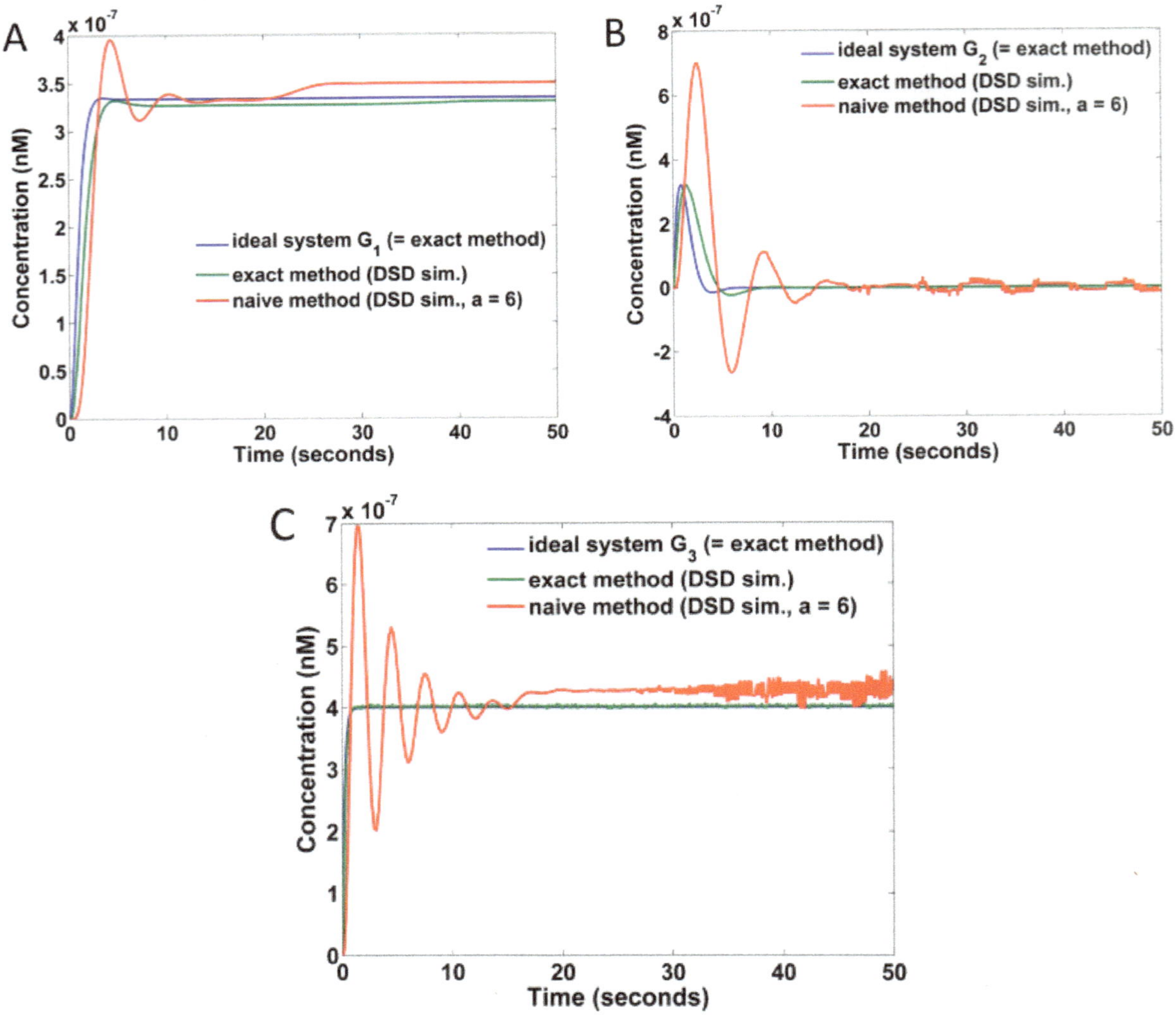

Figure 14: Response comparisons on DSD realized systems under the naive and exact implementations of transfer functions G_1, G_2, and G_3. (A) Responses of G_1 implementations. (B) Responses of G_2 implementations. (C) Responses of G_3 implementations. The curves in red corresponds to naive implementations (under the parameter setting a = 6 in G'_i), while the curves in green correspond to exact implementations.

For the overall transfer function $G = G_1 + G_2 + G_3$, the responses of the DSD reactions realized via the naive and exact implementations are compared in Fig 15 against the ideal response. There are two approaches combining the subsystems G_1, G_2, and G_3. One approach is to use a summation block to sum up the outputs Y_1, Y_2, and Y_3 of G_1, G_2, and G_3, respectively; the other approach is to keep Y_1, Y_2, and Y_3 intact without summation by assuming that the output species of Y_1, Y_2, and Y_3 share the same piece of the DNA segment (rather than the entire DNA segment) that corresponds to the intended final product. The result of the former approach is shown in Fig 15A and the latter approach is shown in Fig 15B, which corresponds to the direct superposition of individual G_1, G_2, and G_3 responses. We observe that summing Y_1, Y_2, and Y_3 with an additional summation block may result in slight distortions in comparison with direct superposition. In either case, the exact implementation fits the ideal response much better than the naive implementation.

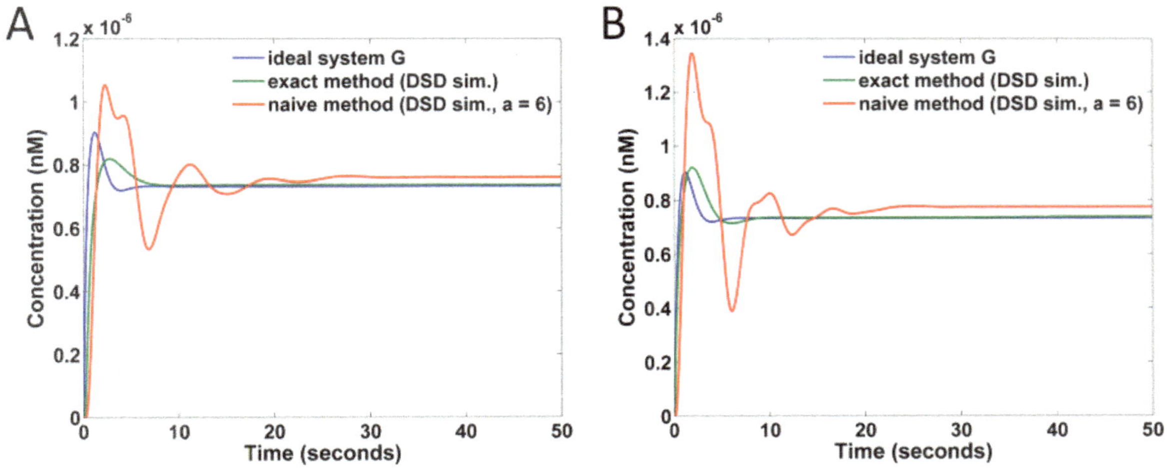

Figure 15: Response comparisons on DSD realized systems for the naive and exact implementations of transfer function G. (A) G implemented by a summation block adding G_1, G_2, and G_3, assuming a = 6 for this summation block in both the naive and exact methods. (B) G implemented by superposing G_1, G_2, and G_3, assuming that the DSD output species of G_1, G_2, and G_3 have a common sub-domain that is recognized as the output of G.

DISCUSSION

Rate Matching and Configurability

The CRN of the integration block proposed by Oishi and Klavins contains the catalytic reactions $u^{\pm} \xrightarrow{q_c^{\pm}} u^{\pm} + y^{\pm}$. The integration is made possible under the assumption that $q_c^+ = q_c^- \equiv w_{\text{int}}$. Accordingly $\dot{y} = w_{\text{int}}\dot{u}$ (i.e., output y is a weighted integration of input u). However, the CRNs of the gain block contain one pair of additional degradation reactions $y^{\pm} \xrightarrow{q_d^{\pm}} \emptyset$. The gain function with weight w_{gn} is realized by assuming $\frac{q_c^+}{q_d^+} = \frac{q_c^-}{q_d^-} = w_{\text{gn}}$. Thus, not only $q_c^+ = q_c^-$ but also $q_d^+ = q_d^- = q_c^+ / w_{\text{gn}}$ must be satisfied.

To cope with the problem of matching q_c^+ and q_c^- in the integration CRN, Oishi and Klavins provided a solution by maintaining a large rate constant q_{ua} in the annihilation reaction $u^+ + u^- \xrightarrow{q_{ua}} \emptyset$ so that q_c^+ and q_c^- do not need to be exactly the same, as long as they are close enough to w_{int}. The reason is that one of the concentrations of u^+ and u^- would approach zero under fast annihilation, and thus $\dot{y}$ equals either w_{int} u^+ or w_{int} u^-. Because q_c^+ and q_c^- are close to w_{int}, the integration can be reasonably approximated. Similarly, the matching problem in the gain CRN is overcome by asserting not only a large q_{ua}, but also a large q_{ya} in $y^+ + y^- \xrightarrow{q_{ya}} \emptyset$. Thus, by requiring that $q_c^{\pm}$ are close enough and the ratios $\frac{q_c^+}{q_d^+}$ and $\frac{q_c^-}{q_d^-}$ are close to w_{gn}, the gain function can be well approximated.

However, the above assumptions of fast annihilations $u^+ + u^- \xrightarrow{q_{ua}} \emptyset$ (and $y^+ + y^- \xrightarrow{q_{ya}} \emptyset$) may not be easily satisfied, especially when these reactions are achieved by composing multiple reactions (e.g., in the DSD realization). Two undesirable situations might occur in composite reactions even if the rate constant q_{ua} is high. One is that u^+ and u^- both degrade, but the species with the lower concentration converges to a non-zero concentration value due to the restoration of the reactants by other reactions. The other is that although the species with the lower concentration vanishes, the remaining species converges to a concentration value less than $|u^+ - u^-|$ because some fuels can react with the remaining species. The approximations may be unsound when these situations occur. Although the effect of inappropriate assumptions might not be serious for a single primitive component, for a complex system composed of many primitive components the approximation would be crude due to error accumulations.

In contrast, because our primitive components are configurable due to the addition of the auxiliary species, we can match not only the reaction rates but also the required weights by tuning the concentrations of auxiliary species. Another advantage provided by the use of auxiliary species is the ability for dynamic system adaptation. As discussed in the case study of the PI-controller, the parameters of a PI-controller may be modified by simply tuning the concentrations of auxiliary species to react to environmental changes. Thus, by using auxiliary species we may fulfill a new system with the same set of reactions by tuning only the concentrations of the auxiliary species. In contrast, without auxiliary species the system would have to be redesigned by finding new catalysts and degradations to match the new weights and rate constants.

Transfer Function Decomposition

A transfer function can be decomposed in many different ways. Two common choices are through parallel decomposition ($G = G_1 + G_2$) and serial decomposition ($G = G_1 \cdot G_2$). In this paper, we adopt the former for the following reasons. First, given a strictly proper transfer function it is not always possible to decompose it as a product of the strictly proper transfer functions. For instance, the transfer function $\frac{2s^2+3s+4}{(s+1)(s+2)(s+3)}$ cannot be decomposed serially into a system involving the cascade of two strictly proper subsystems. Because only proper systems where the degree of the numerator polynomial does not exceed that of the denominator polynomial are physically realizable, the only hope for implementing the system physically is to have a product of bi-proper subsystems, where the degrees of the numerator and denominator polynomials are the same. For example, it can be decomposed as $\frac{2s^2+3s+4}{(s+1)(s+2)} \cdot \frac{1}{(s+3)}$. However, if we implement the system with CRNs or the primitives introduced previously, the bi-proper subsystem might be an approximation due to $\frac{2s^2+3s+4}{(s+1)(s+2)} = 2 - \frac{3s}{s^2+3s+2}$, whose realization might involve a gain and a summation block, both of which are approximations. Second, by cascading subsystems using serial decomposition a system may be susceptible to error accumulation due to realization approximation. In contrast, parallel decomposition avoids the issue of error accumulation and amplification. Third, a strictly proper transfer function can always be decomposed as a summation of strictly proper transfer functions, which can be modeled by the three elementary modules described in Eq (5). Furthermore, we show that the three elementary modules allowexact CRN implementation using our configurable primitive components. For these reasons, parallel decomposition may be preferable to serial decomposition.

Nevertheless, applying parallel decomposition exclusively may not always be the best choice. The reasons are twofold. First, parallel decomposition requires one final summation block to sum up all elementary modules. This summation block, when it is implemented with a CRN, is approximative. Second, a hybrid decomposition strategy may achieve more effective implementation than applying a parallel decomposition. For instance, consider the system of transfer function $T = \frac{3s^2+8s+6}{(s+1)(s+2)(s^2+2s+2)}$. If only parallel decomposition is applied, then T is decomposed into $\frac{2}{s^2+2s+2} + \frac{1}{s+1} - \frac{1}{s+2}$. Nevertheless, if serial decomposition is also exploited, then $\frac{1}{s+1} - \frac{1}{s+2}$ can be rewritten as $\frac{1}{(s+1)(s+2)}$, which can be implemented by cascading two gain blocks with $(k_0, k_1) = (1, 1)$ and $(2, 1)$, respectively. Therefore, with this rewriting a final summation block only needs to sum up the outputs of the two subsystems rather than three subsystems. Furthermore, the hybrid parallel and serial decomposition strategy may be more cost effective than using parallel decomposition alone. Future work should exploit a good hybrid decomposition strategy and remove the approximative effect due to the final summation block in parallel decomposition.

DSD Reaction Rates

In this paper, we realize all CRNs with DSD reactions. Typically, the rate constants in DSD reactions are approximately 10^{-3} nM^{-1} s^{-1} [33]. For example, if one requires the parameter $k_0 = r_0^{\pm} z^{\pm}$ in the degradation reaction to be 100 s^{-1}, then the concentrations of $z^{\pm}$ should be of magnitude 10^5 in nM. However, such a high concentration might be impractical. To alleviate this high concentration requirement, one may try to increase the DSD reaction rates. Zhang et al. constructed and characterized DNA catalytic circuits driven by entropic gains [28]. Based on the entropy effects, a variant called the tethered entropy driven catalytic circuits was introduced [35, 38] to shorten the catalytic cycle and improve the reaction kinetics using localized hybridization reactions, which are achieved by tethering the key species in DSD reactions to increase their local concentrations. The influence of tethering is effectively presented in a speed-up factor λ, whose value can be feasibly reach approximately 10^5 [35]. Thus, the toehold binding rate constants can be scaled up to $\lambda \times 5 \times 10^{-5}$ nM^{-1} s^{-1}, where 5×10^{-5} nM^{-1} s^{-1} is the short toehold binding rate constant in diffusion-based systems [6].

In our DSD simulation, we assume the factor $\lambda = 1000$ is available to increase the toehold binding rate constants to approximately 0.05 nM^{-1} s^{-1}. Effectively, the concentrations of the auxiliary species can be reduced by a factor of 25. Under the toehold binding rate increase, a catalysis has a rate constant approximately 0.025 nM^{-1} s^{-1} in our interested range of concentrations under the intended system operation. Thus, the kinetics of the products $y^{\pm}$ will now become $\dot{y}^{\pm} = 0.025 x^{\pm} u^{\pm}$. If $x^{\pm}$ could be tuned to 1000 nM, then we can have an integration block with weight as high as 25.

Although we have reduced the required concentrations of auxiliary species using the speed-up factor λ, there are still some limitations. First, if β is set to 500 in the elementary modules $\frac{D_2}{s^2+\beta s+\gamma}$ and $\frac{D_3 s}{s^2+\beta s+\gamma}$ shown in Fig 12, then we have to prepare the catalysts $z^{\pm}$ with large concentrations (approximately 500/0.05 = 10000 nM due to the relation $\beta = r_0^{+} z^{+} = r_0^{-} z^{-}$). In contrast, if $D_3 = 500 = r_1^{+} x^{+} = r_1^{-} x^{-}$, then the concentrations of $x^{\pm}$would be 500/0.025 = 20000 nM. Thus, although we have reduced the concentrations of the auxiliary species to increase the range of coefficients in the transfer functions, further reduction may be needed.

Fuel Supply

DSD is a relatively mature technology for reliable in vitro construction of biochemical systems, and is more practical than in vivo experiments. However, one of the main obstacles for practical DSD experiment is to provide constant supply of fuels [37]. This paper assumes the fuels are large in quantity and ignore the situations where the fuels become deficient. In reality, the fuels are converted to inert waste and one has to continually provide the fuels to the system. Providing stable fuel supply can be challenging in a cellular context. Therefore, it is easier to implement DSD reactions in vitro instead of in vivo.

AUTHOR CONTRIBUTIONS

Conceived and designed the experiments: TYC RYH JHJ. Performed the experiments: TYC. Analyzed the data: TYC HJC RYH JHJ FF. Contributed reagents/materials/analysis tools: TYC HJC RYH JHJ FF. Wrote the paper: TYC HJC RYH JHJ FF.

REFERENCES

1. Elowitz MB, Leibler S. A synthetic oscillatory network of transcriptional regulators. Nature. 2000;403:335–338. doi: 10.1038/35002125. pmid:10659856
2. Gardner TS, Cantor CR, Collins JJ. Construction of a genetic toggle switch inEscherichia coli. Nature. 2000;403:339–342. doi: 10.1038/35002131. pmid:10659857
3. Tamsir A, Tabor JJ, Voigt CA. Robust multicellular computing using genetically encoded NOR gates and chemical 'wires'. Nature. 2011;469:212–215. doi: 10.1038/nature09565. pmid:21150903
4. Sohka T, Heins RA, Phelan RM, Greisler JM, Townsend CA, Ostermeier M. An externally tunable bacterial band-pass filter. Proc Natl Acad Sci USA. 2009;106:10135–10140. doi: 10.1073/pnas.0901246106. pmid:19502423
5. Daniel R, Rubens JR, Sarpeshkar R, Lu TK. Synthetic analog computation in living cells. Nature. 2013;497:619–623. doi: 10.1038/nature12148. pmid:23676681
6. Qian L, Winfree E. Scaling up digital circuit computation with DNA strand displacement cascades. Science. 2011;332:1196–1201. doi: 10.1126/science.1200520. pmid:21636773
7. Qian L, Winfree E, Bruck J. Neural network computation with DNA strand displacement cascades. Nature. 2011;475:368–372. doi: 10.1038/nature10262. pmid:21776082
8. Padirac A, Fujii T, Rondelez Y. Bottom-up construction of in vitro switchable memories. Proc Natl Acad Sci USA. 2012;109:E3212–E3220. doi: 10.1073/pnas.1212069109. pmid:23112180
9. Densmore D, Hassoun S. Design automation for synthetic biological systems. IEEE Design & Test of Computers. 2012;29:7–20. doi: 10.1109/MDT.2012.2193370.
10. Oishi K, Klavins E. Biomolecular implementation of linear I/O systems. IET Syst Biol. 2011;5:252–260. doi: 10.1049/iet-syb.2010.0056. pmid:21823756
11. Klonowski W. Simplifying principles for chemical and enzyme reaction kinetics. Biophys Chem. 1983;18:73–87. doi: 10.1016/0301-4622(83)85001-7. pmid:6626688
12. Padirac A, Fujii T, Rondelez Y. Nucleic acids for the rational design of reaction circuits. Curr Opin Biotechnol. 2013;24:575–580. doi: 10.1016/j.copbio.2012.11.011. pmid:23265857
13. Holloman WK, Wiegand R, Hoessli C, Radding CM. Uptake of homologous single-stranded fragments by superhelical DNA: A possible mechanism for initiation of genetic recombination. Proc Natl Acad Sci USA. 1975;72:2394–2398. doi: 10.1073/pnas.72.6.2394. pmid:1094467
14. Green C, Tibbetts C. Reassociation rate limited displacement of DNA strands by branch migration. Nucleic Acids Res. 1981;9:1905–1918. doi: 10.1093/nar/9.8.1905. pmid:6264399
15. Robinson BH, Seeman NC. Simulation of double-stranded branch point migration. Biophys J. 1987;51:611–626. doi: 10.1016/S0006-3495(87)83386-6. pmid:3580487
16. Weinstock PH, Wetmur JG. Branch capture reactions: Effect of recipient structure. Nucleic Acids Res. 1990;18:4207–4213. doi: 10.1093/nar/18.14.4207. pmid:2198537
17. Panyutin IG, Hsieh P. The kinetics of spontaneous DNA branch migration. Proc Natl Acad Sci USA. 1994;91:2021–2025. doi: 10.1073/pnas.91.6.2021. pmid:8134343
18. Yurke B, Turberfield AJ, Mills AP, Simmel FC, Neumann JL. A DNA-fuelled molecular machine made of DNA. Nature. 2000;406:605–608. doi: 10.1038/35020524. pmid:10949296
19. Yurke B, Mills AP. Using DNA to power nanostructures. Genetic Programming and Evolvable Machines. 2003;4:111–122. doi: 10.1023/A:1023928811651.
20. Li Q, Luan G, Guo Q, Liang J. A new class of homogeneous nucleic acid probe based on specific displacement hybridization. Nucl Acids Res. 2002;30:e5. doi: 10.1093/nar/30.2.e5. pmid:11788731
21. Zhang DY, Winfree E. Control of DNA strand displacement kinetics using toehold exchange. J Am Chem Soc. 2009;131:17303–17314. doi: 10.1021/ja906987s. pmid:19894722
22. Gyorgy A, Del Vecchio D. Modular composition of gene transcription networks. PLoS Computational Biology. 2014;10:e1003486. doi: 10.1371/journal.pcbi.1003486. pmid:24626132
23. Cardelli L. Strand algebras for DNA computing. In: Deaton R, Suyama A, editors. DNA Computing and Molecular Programming. vol. 5877 of Lecture Notes in Computer Science. Springer Berlin Heidelberg; 2009. p. 12–24. Available from:http://dx.doi.org/10.1007/978-3-642-10604-0_2.
24. Soloveichik D, Seelig G, Winfree E. DNA as a universal substrate for chemical kinetics. Proc Natl Acad Sci USA. 2010;107:5393–5398. doi: 10.1073/pnas.0909380107. pmid:20203007
25. Cardelli L. Two-domain DNA strand displacement. Mathematical Structures in Computer Science. 2013;23:247–271. doi: 10.1017/S0960129512000102.
26. Green SJ, Lubrich D, Turberfield AJ. DNA hairpins: Fuel for autonomous DNA devices. Biophysical Journal. 2006;91:2966–2975. doi: 10.1529/biophysj.106.084681. pmid:16861269
27. Seelig G, Soloveichik D, Zhang DY, Winfree E. Enzyme-free nucleic acid logic circuits. Science. 2006;314:1585–1588. doi: 10.1126/science.1132493. pmid:17158324
28. Zhang DY, Turberfield AJ, Yurke B, Winfree E. Engineering entropy-driven reactions and networks catalyzed by DNA. Science. 2007;318:1121–1125. doi: 10.1126/science.1148532. pmid:18006742
29. Bath J, Turberfield AJ. DNA nanomachines. Nat Nanotechnol. 2007;2:275–284. doi: 10.1038/nnano.2007.104. pmid:18654284
30. Amir Y, Ben-Ishay E, Levner D, Ittah S, Abu-Horowitz A, Bachelet I. Universal computing by DNA origami robots in a living animal. Nat Nanotechnol. 2014;9:353–357. doi: 10.1038/nnano.2014.58. pmid:24705510

31. Phillips A, Cardelli L. A programming language for composable DNA circuits. J R Soc Interface. 2009;6:419–436. doi: 10.1098/rsif.2009.0072.focus.

32. Lakin MR, Youssef S, Cardelli L, Phillips A. Abstractions for DNA circuit design. J R Soc Interface. 2012;9:470–486. doi: 10.1098/rsif.2011.0343. pmid:21775321

33. Zhang DY, Seelig G. Dynamic DNA nanotechnology using strand-displacement reactions. Nat Chem. 2011;3:103–113. doi: 10.1038/nchem.957. pmid:21258382

34. Lakin MR, Youssef S, Polo F, Emmott S, Phillips A. Visual DSD: A design and analysis tool for DNA strand displacement systems. Bioinformatics. 2011;27:3211–3213. doi: 10.1093/bioinformatics/btr543. pmid:21984756

35. Chandran H, Gopalkrishnan N, Phillips A, Reif J. Localized Hybridization Circuits. In: Cardelli L, Shih W, editors. DNA Computing and Molecular Programming. vol. 6937 of Lecture Notes in Computer Science. Springer Berlin Heidelberg; 2011. p. 64–83. Available from: http://dx.doi.org/10.1007/978-3-642-23638-9_8.

36. Lakin MR, Parker D, Cardelli L, Kwiatkowska M, Phillips A. Design and analysis of DNA strand displacement devices using probabilistic model checking. J R Soc Interface. 2012;9:1470–1485. doi: 10.1098/rsif.2011.0800. pmid:22219398

37. Yordanov B, Kim J, Petersen RL, Shudy A, Kulkarni VV, Phillips A. Computational design of nucleic acid feedback control circuits. ACS Synth Biol. 2014;3:600–616. doi: 10.1021/sb400169s. pmid:25061797

38. Gopalkrishnan N, Garg S, Chandran H, Reif J. Speeding up DNA circuits using localized hybridization; 2011. In Foundations of Nanoscience Meeting (FNANO).

CHAPTER 16

THE PROCESS OF NANOSTRUCTURING OF METAL (IRON) MATRIX IN COMPOSITE MATERIALS FOR DIRECTIONAL CONTROL OF THE MECHANICAL PROPERTIES

Elena Zemtsova, Denis Yurchuk, and Vladimir Smirnov

Saint Petersburg State University, Universitetskii Pr. 26, Saint Petersburg 198504, Russia

ABSTRACT

We justified theoretical and experimental bases of synthesis of new class of highly nanostructured composite nanomaterials based on metal matrix with titanium carbide nanowires as dispersed phase. A new combined method for obtaining of metal iron-based composite materials comprising the powder metallurgy processes and the surface design of the dispersed phase is considered. The following stages of material synthesis are investigated: (1) preparation of porous metal matrix; (2) surface structuring of the porous metal matrix by TiC nanowires; (3) pressing and sintering to give solid metal composite nanostructured materials based on iron with TiC nanostructures with size 1–50 nm. This material can be represented as the material type "frame in the frame" that represents iron metal frame reinforcing the frame of different chemical compositions based on TiC. Study of material functional properties showed that the mechanical properties of composite materials based on iron with TiC dispersed phase despite the presence of residual porosity are comparable to the properties of the best grades of steel containing expensive dopants and obtained by molding. This will solve the problem of developing a new generation of nanostructured metal (iron-based) materials with improved mechanical properties for the different areas of technology.

INTRODUCTION

In the past two decades scientists have paid much attention to obtaining of composite nanomaterials based on nanoobjects with varying dimensions (nanoparticles, nanofilms, nanowhiskers, etc.) [1, 2]. During such research, for example, in the field of obtaining of structural composite nanomaterials, there was expected that the preparation and use of such materials will dramatically improve mechanical properties of these materials [3–7]. However, in most cases, the effect was not so large compared with the expected [6–8]. There are many reasons, but we can distinguish the main ones:

- necessity for more precise control of nanoobject sizes (preferably in the range of 1–10 nm), for example, the most studied 3D nanoobjects (nanoparticles) which are dispersed phase for the structural composite nanomaterials. This is due to the fact that in most of the proposed methods for the synthesis there are obtained particles characterized by a fairly broad curve of

Citation: Elena Zemtsova, Denis Yurchuk, and Vladimir Smirnov, "The Process of Nanostructuring of Metal (Iron) Matrix in Composite Materials for Directional Control of the Mechanical Properties," The Scientific World Journal, vol. 2014, Article ID 979510, 12 pages, 2014. doi:10.1155/2014/979510 particle size distribution [9–13];

- necessity of development of precise methods for synthesis of a new generation of composite nanomaterials. If staying, for example, on one type of the composite nanomaterials with the metal matrix, the existing methods for producing such materials often include introduction of nanoparticles or nanowhiskers in the molten metal. Almost always particles in the solid (usually with different size) are not evenly distributed in metal matrix volume. This leads to a slight increase in the strength characteristics of composite metal nanomaterials [14, 15].

Thus, when considering the preparation of new generation of nanostructured composite nanomaterials, we should talk about the material [15] in which it is difficult to separate the constituent components of composite materials (matrix and reinforcing filler). In this material reinforcing filler should be a continuous framework (e.g., nanowires) pervading the all volume of matrix frame. This structure consists of two interpenetrating each other continuous frames.

The solution of the above problem (synthesis of highly organized nanostructured materials) involves the creation of material with required level of structural organization. In general, we are talking about artificial "super control" in the solid as an alternative to natural regulation processes, mainly represented by the process of crystallization.

The crystal structure of a solid is characterized by two main parameters: structural units (from a single atom to big molecular objects) and ordering degree (from the amorphous state to monocrystal). Auxiliary parameters are needed for describing artificially synthesized highly organized structures and for characterizing these complex systems. As a rule, the task of any solid-phase synthesis is creation of a certain structure. There are several aspects of understanding the term "synthesis of a solid chemical compound" or its physicochemical analog "synthesis of a certain structure":

- the physicochemical aspect includes consideration of composition and crystal (sometimes amorphous) structure;
- the topological aspect includes consideration of the spatial distribution of the composition and structure (actually, the spatial distribution of structural units of an object);
- the electronic aspect concerns electronic states of a solid substance under consideration.

The notion of topology implies a complex structure of matter. Therefore, we consider in more detail the notion of topology for highly organized solid compounds. This notion is related to the possibility of various spatial atomic distributions in an artificially synthesized substance [14–16].

From the viewpoint of topology, the limiting states are, on one side, a substance uniform in composition and structure (e.g., monocrystal, glass, etc.) and, on the other side, multielement (not in a chemical, but in a structural meaning) structure where each structural unit can be arbitrarily connected with others (e.g., biological objects). Artificial structures of the simplest topological organization (uniformly laminated structures, Figure 1(a)), for example, with thin oxide layers ($L > 30$ nm), have long been prepared [14]. A more complex topological organization includes nonuniform distribution of composition and structure (layered structure) along one of the spatial coordinates both of periodic (a) and of aperiodic (b, c) character. The synthesis of such structures is well run. However, certain difficulties arise if it is necessary to obtain a small nanostructure period (L) of order 1 nm.

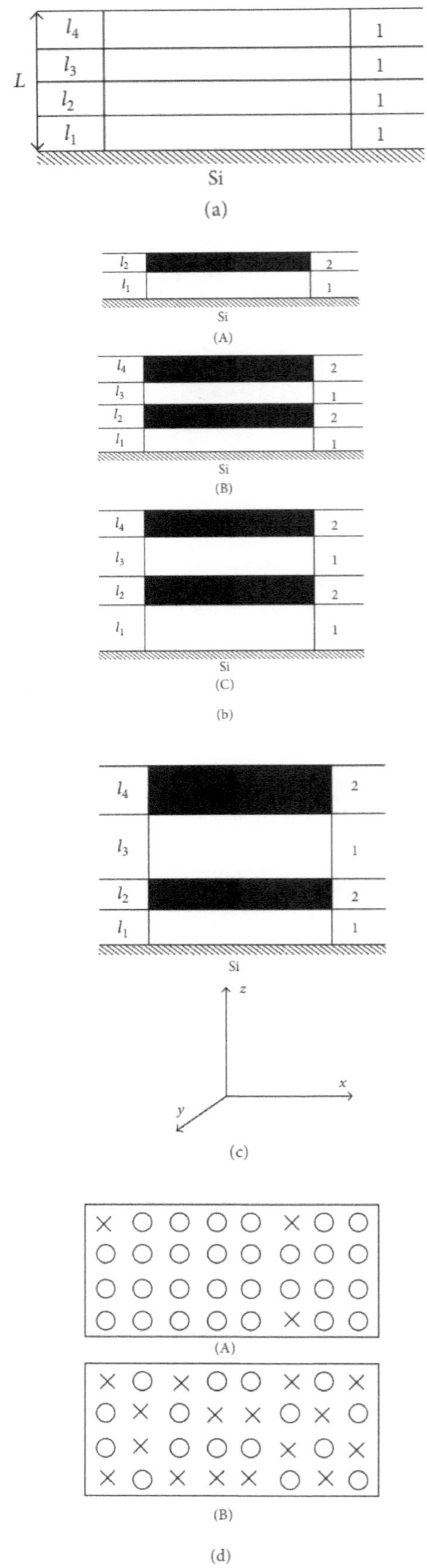

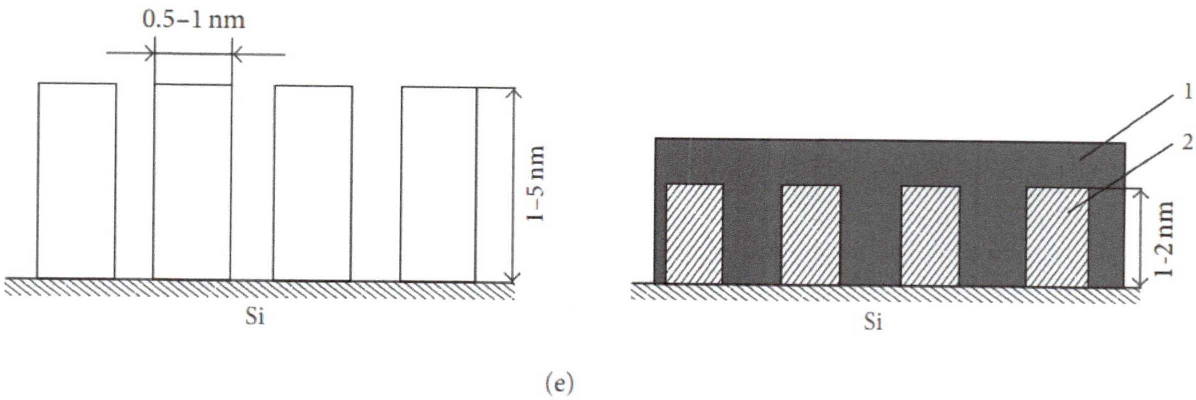

Figure 1: Scheme of two-dimensional nanostructures at silica surface. (a) Uniform distribution of chemical composition and structure, for example, monolayers of element-oxygen groups of a given composition: (l) monolayer thickness; (L) total layer thickness, $l_1 = l_2 = l_3 = l_4$; and (1) Ti-O monolayers. (b) Periodical distribution along the z-axis of element-oxygen layers containing a certain number of monolayers: (1) Fe-O groups and (2) Ti-O groups ((A) two-layer structure, $l_1 = l_2$ and (B) four-layer structure, $l_1 = l_3$ and $l_2 = l_4$). (c) Aperiodic distribution of element-oxygen layers along the E-axis (four-layer structure with $l_1 = l_2 = l_3 = l_4$). (d) Aperiodic atomic distribution in the surface monolayer plane (top view): (crosses) Fe-O groups and (circles) Ti-O groups ((A, B) different Fe-O/Ti-O ratios). (e) Aperiodic distribution of "zerodimensional" structures in the substrate plane: (1) Fe and (2) Si.

Further complication of topological organization is possible on organizing atoms also in the monolayer plane (for one or several monolayers) (Figure 1(d)). Still higher levels of organization of matter imply formation of spatial nonuniformity in a structure synthesized and differ in the degree of internal symmetry (Figure 1(e)).

In fact, we discuss the possibility of realizing the processes of superordering in a solid, that is, formation (synthesis) of ordered distributions of matter of a certain size. Let us compare the realization of natural ordering and artificial superordering in a solid. It is of note that macroscopically ordered distributions are sufficiently widespread in nature. As examples, macroscopic ordered distributions in plumeous clouds, in the radial structure of volcanoes, in ordered eutectics, and in sea sandbanks can be enumerated.

Since these examples are related to the structure of matter, the processes under consideration involve mass transfer (diffusion) and may follow the only direction to a uniform state of the system. The origin of such processes lies in that they are caused by thermodynamic driving forces and directed to extreme values of thermodynamic parameters, including the entropy maximum at equilibrium.

The closer a system to equilibrium, the more pronounced are internal self-ordering processes in the system, the easier is maintaining physicochemical conditions, and the more difficult is changing them by external action. Hence, as we approach equilibrium, the range of allowed nonuniform structures becomes narrower, while uniform structures become simpler to synthesize. The reverse situation takes place far from equilibrium: a very precise external control is needed, but almost all structures are allowed. The choice of an optimal synthetic strategy is one of the most intricate problems.

Thus, the only way to create an aperiodic order consists in changing the ordering processes described by equilibrium thermodynamics and directed to securing system stability by a synthetic program directed to obtaining highly organized solid-phase structures in the metastable state. Because of the low mobility of atoms in a solid, a certain thermodynamically unfavorable structure (e.g., diamond) can exist for any long time.

It is now firmly established that the synthesis of solid chemical compounds by crystallization can result in formation, from the same structural units, of a number of equivalent variants of composition and structure with close energies (actually, this is an unseparable mixture of variable composition). This occurs since the solidification process always proceeds at a certain supersaturation [17]. In the case of the solid-phase synthesis of highly organized structures, a solid compound to be obtained should possess the following characteristics [18]: (a) any topology including uniform and nonuniform distribution of composition and structure should be attainable; (b) aperiodic structure (periodicity is possible as a particular case); and (c) thermodynamic nonequilibrium state. We imply that the chemical potentials in the existing homogeneity regions are different under chosen conditions (temperature, pressure, etc.) and this does not lead to destruction of the resulting structure.

It is of note that chemical processes on the atomic-molecular level should play the main role in the synthesis of nanostructures with a small period and a given topology. Then all the processes are reduced to a certain sequence of surface chemical reactions between functional groups of the solid and molecules of necessary chemical nature [19]. Herewith, the only way for creating highly organized structures (of aperiodic order) is replacing crystallization by the synthesis of solid-phase structures in the metastable state. In fact, to accomplish such synthesis, one should control the activity of some chemical bonds and prevent reaction of other chemical bonds.

There is an important aspect to be used at realizing the solid-phase matrix synthesis. With microcrystal seeds, the main role is played by the surface of a solid, since physicochemical processes just develop at the surface. It is of note that thermodynamic and kinetic barriers are not eliminated but only diminished by introduction of a seed (substrate) in a system. Separate relative equilibrium is established for each substrate under given conditions. Traditional methods of epitaxial synthesis (deposition) are controllable processes. However, it is hard to speak about obtaining a nanometric substance layer of a given thickness by the epitaxial deposition method because epitaxial layers are formed by way of appearance and integration of crystal nuclei whose dimensions certainly exceed the monolayer thickness.

Additional difficulties in realizing the synthesis of solid substances arise from the fact that the overwhelming majority of reactions used in chemical science and technology belong to the type of "unorganized" reactions. We mean the reactions with particles (atoms, molecules, ions, or radicals) reacting at accidental meetings both in time and in space and by mutual orientation. In other words, there is no spatiotemporal molecular and solid-phase organization of chemical interaction in such reactions. At the same time, the possibility of such an organization is known from molecular biology, for example, the biosynthesis process [20]. For example, Merrifield has realized the solid-phase (matrix) synthesis of polypeptides [21, 22] by successive extension of the polypeptide chain on an inert polymeric substrate.

The problem of a standard solid surface remains actual in synthesis of nanostructures. Let us consider the principles of synthesis of solid substances of reproducible composition [23, 24]. They are the process irreversibility conditions, the availability of a matrix, and the program of production and accumulation of information.

When performing controlled solid-phase synthesis, a researcher actually carries out a number of operations to obtain a solid chemical substance possessing (a) a certain excess of free energy contained in the system of interatomic bonds, as compared with the initial reagents, and (b) a higher degree of ordering of interatomic bonds as compared with the initial reagent mixture at equilibrium.

Practically, the reproducible synthesis of nanostructured solid substances is realized via chemical assembling of structural units on appropriate matrices. The monolayer-by-monolayer chemical assembling of solid substances involves a number of preprogrammed surface reactions with at least bifunctional molecules of either one substance or another.

The principal significance of this method of synthesis of solid matter is "that, instead of spontaneous packing structural units in the crystallization process, their forced packing in a preset order is performed, that is, preset composition and structure of the solid substance are realized" [23]. By varying temperature and other conditions of chemical assembling, one can change the density of packing of structural units. Apart from crystalline structures, one can create other, significantly more complex structures whose number has no limit.

We now turn to chemical reactions performed in the process of synthesis. The progress in the field of precise inorganic synthesis is associated with the rapid development of surface chemistry. This branch of the inorganic chemistry of solid substances studies chemical reactions in the surface atomic layer of a solid substance and also studies the ways controlling these reactions [15]. Such reactions open the unique possibility for controlling, on the atomic level, the synthesis of new substances of reproducible composition and structure by means of reactions between functional groups of a solid substance and molecules of necessary chemical nature. Such surface reactions result in formation of surface chemical compounds.

In order to solve our problem—finding the influence of one-dimensional (nanowires, 1–50 nm) structural heterogeneities basis on TiC, amorphous or crystalline structure on the structure, and functional (mechanical) properties of bulk metallic material (metallic matrix)—it is necessary to solve the problem of substance structuring at the nanolevel. The main theoretical and experimental positions of the chemical nanostructuring are considered in [15]. Herewith surface chemical reactions of forced structuring are the most important phase of nanostructuring. Thus the basic mechanical properties of the resulting solids are provided by the creation of nanostructured metal matrix. The synthesis of nanolayers of the dispersed phase (TiC) was carried out by the method of directed synthesis developed in Russia—molecular layering (ML) [3–6, 13, 25–27], also known abroad under the name "atomic layer deposition" (ML-ALD) [8, 13, 28–34]. Currently this method is recognized by foreign experts as the main method for production of nanolayers in nanoelectronics and in a number of other related areas [9, 10, 25, 35].

Practically, the reproducible synthesis of nanostructured solid substances is realized via chemical assembling of structural units on appropriate matrices. The monolayer-by-monolayer chemical assembling of solid substances involves a number of preprogrammed surface reactions with at least bifunctional molecules of either one substance or another [36].

The main purpose of this work is to establish experimental fundamentals of the synthesis of highly organized nanostructured composite materials with the TiC dispersed phase which will directly regulate the mechanical properties of composite material.

MATERIALS AND METHODS

As noted above for the solution of our problem—finding the influence of one-dimensional (nanowires, 1–50 nm) structural inhomogeneities based on TiC on the structure of and functional (mechanical) properties of metal type (iron matrix) bulk material—we should solve the problem of structuring material at nanoscale. In this case, the most important phase of the nanostructuring of matrix is the surface chemical reaction of force structuring. Thus the basic mechanical properties of obtained material are provided by the creation of nanophase dispersion (TiC) in the volume of the metal matrix. Thus, the original metal matrix must be porous. The synthesis of nanolayers of the dispersed phase (TiC) in the pores of the matrix was based on developed in St. Petersburg State University method of directed synthesis—a method of molecular layering (ML), known abroad under the name "atomic layer deposition" (ML-ALD).

Process of preparation of nanostructured metal material includes a number of successive steps: (1) preparation of nanoparticles of iron oxide; (2) reducing iron oxides (by hydrogen or other reducing agents) in order to obtain ultrafine iron metal frame with a predetermined porosity; (3) surface nanostructuring of prepared metal porous matrix by TiC nanolayers (1–50 nm); (4) consolidation (pressing) and sintering to obtain a bulk (nonporous) material with nanostructured metal material with size of TiC nanowires (1–50 nm). The third step of the synthesis involves a fundamentally new scientific approach for the development of scientific bases of ultraprecise synthesis, that is, combined chemical and physical processes leading to the creation of a given order of the atoms in synthesized structure or nanostructured material.

Following each cycle of surface chemical reactions (SCR) carried out under strictly controlled conditions on the solid surface, there is a deposition of monolayer of new structural building units (carbides) chemically associated with the original substrate and the "thickness" of a monolayer is 2-3 Å. Carrying out a number of these reactions a layer of substance with a certain thickness can be synthesized.

Chemicals and materials were obtained from commercial suppliers: TiCl4 (99.6%), (Alfa Aesar), carbon tetrachloride (99%) (Sigma-Aldrich), $FeSO_4 \cdot 7H_2O$, NaOH, HNO_3 (Merck), and CH_4 (gas) (Lengaz).

Preparation and Reduction of Iron Oxide in Order to Obtain Ultrafine Iron Metal Frame with a Given Porosity

We presumed that for obtaining of metal particles with high dispersion need to reduce their oxides at lower temperatures [2]. This fact demands that we have to approach more carefully to the selection of the original matrix. We felt it appropriate to use the amorphous iron hydroxide (α-FeOOH). Iron hydroxide (α-FeOOH) and hydrogen with purity of at least 99.99% are used as starting substances. α-FeOOH was prepared by precipitation of salt $FeSO_4 \cdot 7H_2O$, NaOH is used as a precipitant, and distilled water is used as solvent. The resulting precipitate of hydroxides was washed with distilled water and dried. During the heating at 200°C the pyrolysis of iron oxyhydroxide (III) (FeOOH) carried out to form Fe_2O_3. Specific surface (Ssp) for iron hydroxide determined by nitrogen adsorption (BET) is 198 m²/g. Recalculation of experimental values of the specific surface S was carried out by the formula: dav = 6/ρS, (1), where ρ is density and S is surface area. The formula allows us to estimate the average particle size of powders dav. The calculated average size of FeOOH particles is 7 nm. According to the Mössbauer microscopy, obtained samples of iron (III) oxyhydroxide warmed at 25°C were the fine particles which follow from the spectrum form (doublet) of the investigated sample. Metal nanopowders are received by recovery of FeOOH (1 g) in a quartz tube furnace with nichrome heaters in a hydrogen atmosphere at a temperature range from 300 to 1000°C. Reduction was carried out by hydrogen with purity not less than 99.99%. The hydrogen flow rate was $2{,}5 \cdot 10^{-6} m^3/s$. The effect of reducing conditions on the phase composition, structure and dispersion of metallic iron powder was studied. Two forms of iron hydroxide were studied: (1) hydroxide dried at a temperature of 25°C (Sample 1), (2) hydroxide dried at 110°C (sample 2) (Table 1). The identification of the powder phases was performed by X-ray powder diffraction on a diffractometer Bruker "D2 Phaser" with a cobalt anode. XRD of metal iron powder after reduction for 90 minutes at 450°C (Figure 2) and after reconstitution for 60 minutes in the temperature range 500–1000°C showed that the samples contain only α-Fe. Comparison of the iron nanoparticle sizes of FeOOH samples, (heat-treated at 25°C and 110°C) and reduced at different temperatures, show (Figure 3) that increasing in reduction temperature from 400 to 550°C increases the size of metallic iron particles. It should be noted that for samples FeOOH heat-treated at 25°C and reduced at the temperature range 400–550°C the obtained resulting iron nanoparticles are higher than those nanoparticles obtained by reducing of samples FeOOH heat-treated at 110°C. At reduction temperatures 450–500°C, a dramatic increase in the size of nanoparticles of metallic iron with their transition to micronscale occurs. This fact we attributed to the process of sintering of metal particles. It was revealed that for reduction in the temperature range 400–450°C of initial iron hydroxides, iron nanopowders with an average particle size of 61–81 nm can be prepared, representing the phase of α-Fe.

Table 1: Specific surface and average size of aggregates of fine dispersed iron reduced at different temperatures.

Initial oxide	Reducing temperature, °C	S_{sp}, m^2/g	D_{av}, 10^{-9} m
FeOOH dried at 25°C	400	9,7	82
	450	7,2	109
	500	4,5	173
FeOOH dried at 110°C	400	8,4	91
	450	6,6	116
	500	3,7	206
	550	2,1	381

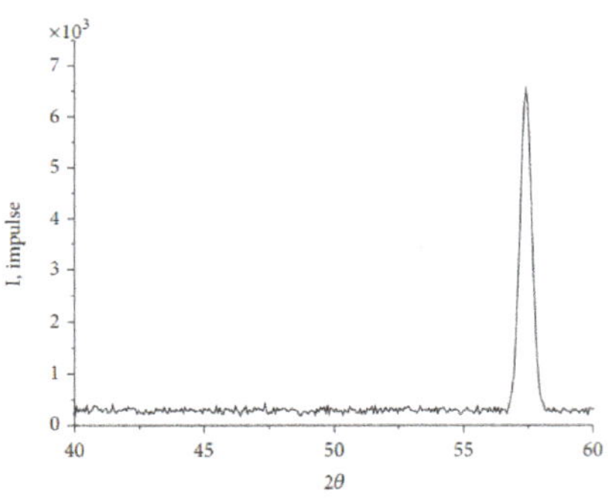

Figure 2: XRD spectrum of iron samples obtained by reduction of sample 2 in hydrogen flow at 400°C.

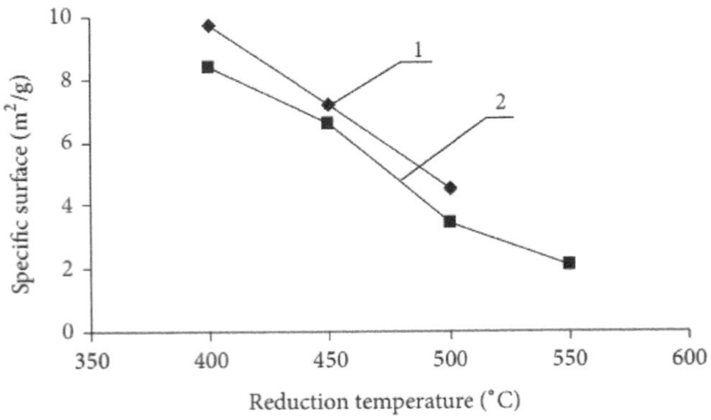

Figure 3: The dependence of the specific surface of iron on reduction temperature: (1) iron powder obtained from FeOOH dried at 25°C, (2) iron powder obtained from FeOOH dried at 110°C.

Thus, it can be assumed that the production of dispersed iron hydroxides by heat treatment of initial iron hydroxides in a reducing atmosphere is an optimal process because it allows to obtain a dispersed metal in comparison with the metal reduced from oxide.

Study of uniaxial compaction of iron powders with a particle size of 60 nm obtained by reduction of iron hydroxide in a hydrogen atmosphere was carried out in a cylindrical mold having an inner diameter 15 mm; compact height was 5–7 mm. Pressure was varied from 0.05 to 1.1 GPa. The density of compacts was determined by hydrostatic weighing with an accuracy of 2%. The most intense compaction of powders occurs at pressures of up to 0.8-0.9 GPa.

As a result, the optimal pressing conditions were determined as follows for iron nanopowders weight of powder—9 g, size of obtained sample: height (h)—4,96 mm, diameter—15 mm, ρ = 4,60 g/cm3 , applying pressure of 200 MPa for 60 seconds, holding for 180 sec, and discharge for 30 seconds.

As a result, the regulation conditions of residual porosity of metallic sample were found in the range from 1 to 35%. Such porosity is required for the carrying out of surface chemical reactions on the iron particles surface.

Synthesis of titanium-carbon groups and titanium-carbon nanostructures on the iron surface was carried out on the hydroxylated surface of samples by ML-ALD. In this paper objects of study on which we investigated the surface chemical reactions were: (1) modeling samples—monocrystalline silicon wafer (KDB-7, 5 with orientation (100)) with deposited layer of α-Fe with size 1.5 × 1.5 cm and iron film thickness of ~100 nm and (2) dispersed iron particles obtained by the method described above; dispersed iron nanoparticles are characterized by the following parameters: a particle size of 40–80 nm, specific surface area of 6 m²/g.

Since the surface of the iron particles is always oxidized and thickness of the oxide layer is not known, then all the investigated objects before the synthesis are treated (chlorinated) by vapor CCl_4 at 350°C to remove the oxide layer. Interaction reactants with the iron powder surface by ML-ALD were performed in a flow-type reactor under dry inert gas flow (helium) which simultaneously ensures removal of gaseous reaction products from the reactor. For the analysis of chemical reactions on the silicon plates with iron layer we used thermogravimetric setting in which analytical electronic scales BP 221S were used to register the change in sample weight. Quartz cup with the test substance (silicon plate with a layer of iron) is attached to scales with quartz fiber. In some experiments method of chemical vapor deposition (CVD) is used to obtain titanium carbide from the gas phase.

To study the mechanical strength of the samples of metal (iron-based) composite materials nanostructured titanium carbide, ultimate stress limit-ultimate resistance (σ_V) is determined on desktop testing machine AG-50kNXD (Shimadzu) in the resource center of innovative technologies of composite materials (St. Petersburg State University).

RESULTS AND DISCUSSION

Synthesis of titanium-carbon groups and titanium-carbide nanostructures was performed by ML-ALD on chemically prepared surface of iron sample, that is, on a surface which contains reactive functional groups (OH, OCH_3, H, Cl, or others).

Experimental Substantiation of the Synthesis of Titanium-Carbon Nanogroups and TiC Nanostructures on the Iron Particles Based on the Study of Surface Reactions of Functional Groups with Low-Molecular-Weight Reagents

In the study objects, on which the surface chemical reactions were studied are

- the model samples-plates of single-crystal silicon with deposited layer of α-Fe with thickness of ~100 nm;
- dispersed iron particles obtained by the method described above; nanoscale dispersed iron particles characterized by the following characteristics: a particle size of 40–75 nm, specific surface of the particles—6 m²/g;
- porous pressing from iron particles as starting intermediates for the synthesis of titanium-carbon nanolayers in the pores of the metal matrix.

Obtaining of Nanolayer of Titanium-Carbon Groups by ML-ALD on Silicon Sample Coated with Iron Nanolayer

To clarify the mode of deposition of titanium carbide nanolayer, model samples representing monocrystal silicon wafer with a sputtered layer of pure iron were used.

When carrying out the surface chemical reactions their behaviors are dependent on temperature and holding time in the reaction zone. Note that the reaction occurs in the range 300–700°C. Our proposed method of synthesizing of titanium carbide by ML-ALD is based on reactions of chemical condensation (1)–(4) according to the following scheme:

$$\equiv Si \cdots [Fe]_a H + CCl_4 \longrightarrow \equiv Si \cdots [Fe]_a\text{-}Cl + HCl \uparrow \quad (1)$$

$$\equiv Si \cdots [Fe]_a\text{-}Cl + CH_4 \longrightarrow \equiv Si \cdots [Fe]_a\text{-}CH_3 + HCl \uparrow \quad (2)$$

$$\equiv Si \cdots [Fe]_a{-}CH_3 + TiCl_4 \longrightarrow \equiv Si \cdots [Fe]_a{-}CH = TiCl_2 + HCl\uparrow \quad (3)$$

$$\equiv Si \cdots [Fe]_a{-}CH = TiCl_2 + CH_4 \longrightarrow \equiv Si \cdots [Fe]_a{-}CH = Ti{-}(CH_3)_2 + HCl\uparrow \quad (4)$$

where $\equiv Si\cdots[Fe]_aH$ represents surface atoms of silicon plates with a deposited film of pure iron$\cdots[Fe]_a$ with thickness of ~10 nm and containing surface hydrogen atoms (H) after treatment by hydrogen. Bond $\equiv Si\cdots[Fe]_aH$ denotes an adhesive connection with a layer of silicon with iron layer. The product of 1 cycle of reactions ML-ALD—$\equiv Si\cdots[Fe]_a{-}CH{=}Ti{-}(CH_3)_2$—contains titanium-carbon groups ($-CH{=}Ti{-}(CH_3)_2$) associated covalent bond with surface methyl sites capable to reacting with $TiCl_4$ (see scheme in Figure 7).

The synthesis was carried out on silicon wafers with a sputtered film of pure iron with thickness of ~10 nm. The sample was alternately treated at 500°C in a flow-type reactor by carbon tetrachloride and methane.

Cycle is repeated again when thickness increasing is needed. Then the reactor with sample was cooled in the hydrogen flow. To determine the coating continuity of functional groups ($-CH_3$) on the iron surface the water contact angles were studied. The study of contact angle was conducted by photomicrograph method. Based on a study of water contact angle of samples (initial silicon with hydroxyl groups, silicon with methyl groups and silicon with deposited titanium-carbon layers), regular increase of water contact angle from 43° to 76° and 94° was identified which indicates the transition to higher packing density of surface groups in the case of reactions involving methyl ($-CH_3$) group (Figure 4). Study of reactions (1)–(4) was performed in the temperature range 200–600°C.

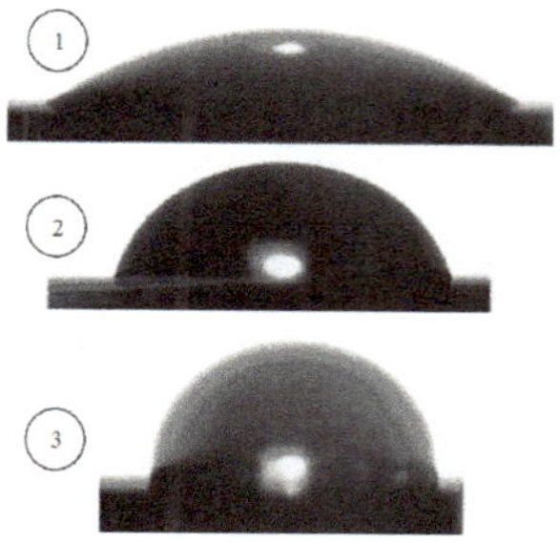

(1) Si–Fe–OH	(2) Si–Fe–CH_3	(3) Si–Fe–TiC (10 nm)
θ	θ	θ
39.8	76.8	94.8
42.8	76.1	93.1
43.1	75.4	95.5
43.5	75.4	93.8
44.6	75.3	93.2
		95
		92.9
		98
		94.6
		91.8
θav.	θav.	θav.
42.8	75.8	94.5

Figure 4: The values of contact angles by water for silicon samples with iron layer with different functional groups: (1) $\equiv$Si–Fe–OH, (2) $\equiv$Si–Fe–OCH_3, and (3) $\equiv$Si–Fe–TiC.

Data on the chemical analysis of titanium have concluded (Figure 5), that at low temperature synthesis there is observed an increase of titanium and chlorine content which reach maximum at about 350–400°C, the amount of reacted titanium and chlorine in the sample remains approximately constant and the ratio of Cl/Ti is 2,6-2,7. Above 450°C, a significant slope in the titanium content and a high chlorine concentration on the surface are observed.

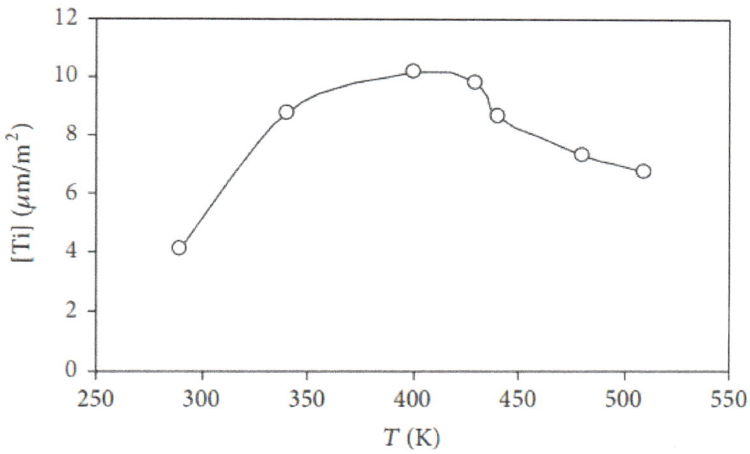

Figure 5: The dependence of Ti content in nanolayer from synthesis temperature.

Decline of the titanium content at warming temperature of 450°C can be associated with the thermal decomposition of surface groups. Apparently, in these conditions the interaction $TiCl_4$ with the surface leads to dissociative process: $TiCl_4 \rightarrow TiCl_3 + Cl_2$, resulting in chlorine chlorinating the surface to form complex C–Cl that blocks it from further reaction with $TiCl_4$.

Macrohardness study and thin section structures of the synthesized samples (Table 2) also showed that the optimal synthesis temperature is 400°C.

Table 2: Macrohardness and characterization of the microstructure of samples obtained by alternating exposures of titanium tetrachloride-methane.

T, °C	Number of cycles of ALD reactions	HV, kg/mm	Description of microstructures
200	1	84	Absence of reaction by XRD, nonuniform film
	2	82	
	4	150	
	5	144	
	6	210	
300	1	80	Absence of reaction by XRD, nonuniform film
	2	63	
	4	120	
	5	380	
	6	465	
400	1	340	Uniform fine-grained structure
	2	303	
	4	690	
	5	1302	
	6	1280	
450	1	684	Uniform fine-grained structure
	2	905	
	4	1010	
	5	1260	
	6	1200	
500	1	204	The appearance of chlorine excess which is manifested in the nonuniform structure
	2	660	
	4	785	
	5	550	
	6	930	
550	1	340	Increasing heterogeneity of structure centers
	2	280	
	4	833	
	5	535	
	6	650	

The study of dependence of titanium content on silicon samples with metallic iron layer from treatment showed that the growth dependence of the titanium content is linear and after 5-6 cycles the titanium content increasing becomes constant. Note that the optimum concentration of the hydrogen significantly increases the rate of titanium carbide formation. This is probably due to the suppression of the reaction of free carbon formation.

Production of Titanium Carbide Micron- and Nanolayer on Iron Nanoparticle Surfaces

Obtaining of titanium carbide on the surface of dispersed iron was carried out on particles with size 40–80 nm and specific surface area 6 m^2/g. Treatment of iron powder in a quartz reactor was performed in a flow of dry argon.

To study the thickness effect of the deposited layer of titanium carbide on the mechanical properties of the samples, obtaining of ultrafine micron films of titanium carbide by chemical vapor deposition (CVD) and nanolayers of titanium carbide by atomic layer deposition (ML-ALD) was investigated. Treatment of iron powder was performed in a quartz reactor in dry argon flow. In gas phase deposition of titanium carbide formation of continuous frame in the pores of metal workpiece begins with the formation of island structures on the surface. This character of the formation is associated with the feature of the reaction of chemical deposition of titanium carbide as surface is energetically heterogeneous. This leads to different rates of growth of the titanium carbide on different sites and forming island structures—three-dimensional nucleus of titanium carbide [6] (Figure 6).

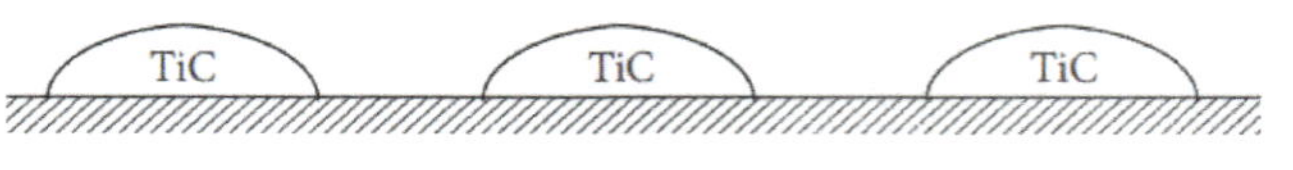

Figure 6: Structure of the samples after chemical deposition (CVD) of titanium carbide at sequential supplying of titanium tetrachloride and methane mixture.

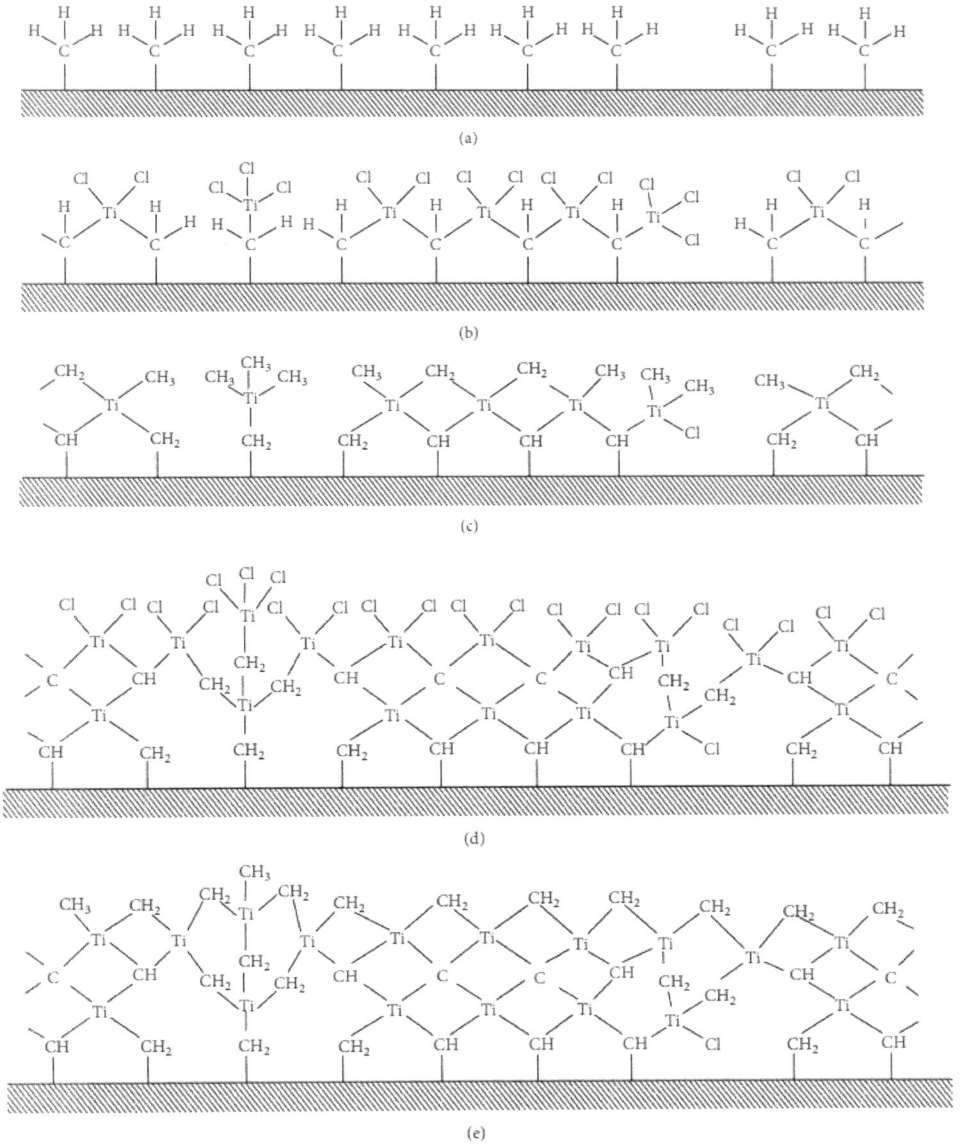

Figure 7: Diagram of the samples after process of atomic layer deposition with the cyclic supply of titanium tetrachloride and methane. (a) Initial surface of dispersed iron particles with methyl functional groups after exposure CH_4; (b) exposure of titanium tetrachloride giving titanium-chloride groups; (c) exposure of methane giving titanium-carbon groups; (d) and (e) repeated sequential exposures of titanium tetrachloride and methane, respectively (reactions (1)–(4)).

Feature of nanolayers obtained by ML-ALD is the absence of phase-formation (nucleation) under maintenance of certain synthesis conditions which allows more fine-tune of the functional properties of the resulting material.

As a result of the synthesis there were synthesized examples of iron powder with different quantity of titanium carbide deposited on the surface. The content of titanium-carbon groups in the sample was determined by the titanium content. Analysis of titanium and carbon was performed by chemical method (photometrically). Data of the chemical composition of samples are given in Table 3.

Table 3: Chemical composition of samples based on the iron particles with titanium-carbon groups.

Sample	Number of cycles processing	Composition of the sample	Estimated thickness (nm)
1	—	100% Fe	—
2	1	95,5% Fe + 1,5% TiC	0,5
3	2	97% Fe + 3% TiC	1
4	10	90,5% Fe + 12,5% TiC	5
5	CVD	84,5% Fe+ 19,5% TiC	15
6	CVD	74,5% Fe + 35,5% TiC	40

Observance of conditions of ALD reactions allows obtaining monolayers with certain chemical composition; the number of performed reactions controlled the thickness of the deposited layer of substance. Features of layers produced by chemical assembly are the absence of phase formation (nucleation) under certain synthesis conditions which allows to regulate more fine the functional properties of the obtained material.

Figure 7 shows a scheme of nanolayer structure of titanium-carbon groups. Features of the structure of nanocrystalline materials (grain size, a significant proportion of the interfaces and their state, porosity, and other structural defects) are defined by methods of their production and have a significant influence on their properties. With decreasing of grain size, strength increase with maintaining of ductility, effect of low temperature, and high superplasticity are manifested; a change in physical properties is observed.

As it follows from Figure 7 the repeated sequential exposures with titanium tetrachloride and methane, respectively, lead to the growth of amorphous nanolayer of titanium-carbon groups. Note that as following from the XRD data samples after synthesis at 400–500°C were amorphous.

From the X-ray diffraction analysis data (Figure 8) there follows that nanostructures of TiC are formed on iron particle surface after 10 cycles of surface reactions and calcination at 1100°C for 5 hours. Considerable interest to bulk nanocrystalline materials was determined from the fact that their structural and functional properties differ significantly from the properties of coarse-grained analogues.

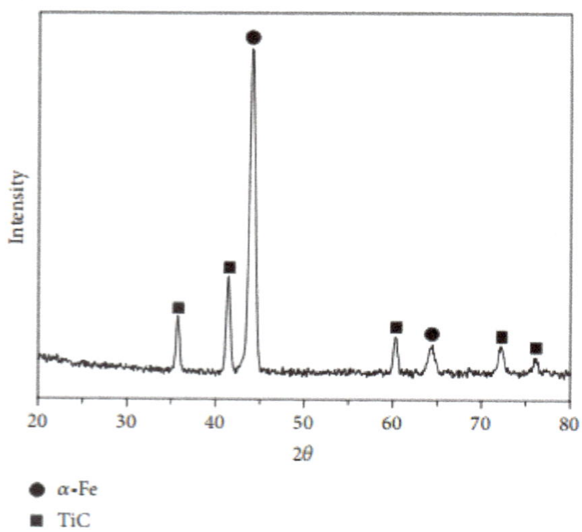

Figure 8: XRD spectrum of samples of dispersed iron with titanium-carbon nanostructures after treatment with 10 cycles of ALD reactions and sintered at 1100°C.

Analysis of the compaction and sintering of metallic iron samples nanostructured by nanowires TiC allows finding the conditions for obtaining metallic samples with low porosity. To achieve the lowest porosity (5-6%) the sintering of compacts is necessary to be carried out by heating from 500 to 800°C for 2 hours with intermediate equalizing at 800°C in hydrogen for 3 hours.

Study of Mechanical Properties of Nanostructured Bulk Metal (Iron-Based) Composite Material with Structural Nanoheterogeneities Based on TiC

Mechanical properties of nanomaterials essentially depend on the grain size. At large grain sizes a strength and hardness growth with decreasing grain size is determined by introduction of additional grain boundaries that are obstacle to dislocation motion and at small nanoscale grains a strength growth is determined by low density of existing dislocations and the difficulty of formation of new dislocations [2].

To study the mechanical strength of the samples of metal (iron-based) composite materials nanostructured titanium carbide ultimate stress limit-ultimate resistance (σV) is determined. These studies are presented in Table 4. For comparison the table shows the best mechanical properties of the steels according to [37]. Table 4 shows that the mechanical properties of the iron-based composite materials with TiC dispersed phase, despite the presence of residual porosity, are comparable with the best properties of steel grades which contain many dopants.

Table 4: The study of mechanical properties of the synthesized composite materials based on iron with dispersed phase of TiC.

Sample	Composition	Tensile strength, σ_B, MPa
	Obtained samples	
	MN-ALD method	
1	100% Fe	183
2	95,5% Fe + 1,5% TiC	800
3	97% Fe + 3% TiC	1130
4	90,5% Fe + 9,5% TiC	1370
5	84,5% Fe + 15,5% TiC	1460
	CVD method	
6	69,5% Fe + 30,5% TiC	890
7	61,5% Fe + 38,5% TiC	900
	Literature data [37]	
7	Steel AS 14CGN	1120
8	Steel 50CG	1300
9	Steel 12C2N4A	1130
10	High-strength steel	1500

*The residual porosity of the synthesized samples was 5–6 %.

It should be borne in mind that the samples have porosity, that allows to improve the mechanical properties of specimens.

Thus we can assume that the creation of metallic (iron-based) composite materials nanostructured titanium carbide using surface structuring process is promising for producing of new generation composite materials.

CONCLUSIONS

Through a nanostructuring process of metal (iron) matrix developed by authors the synthesis of samples of nanostructured metal matrix composite with nanostructural (1–50 nm) inhomogeneities on the basis of TiC with the crystal structure in the volume of the iron matrix was reformed.

The synthesis, pressing, and sintering conditions of dispersed iron nanoparticles (40–75 nm) with a surface area of 6 m^2/g were investigated.

The features of the behaviour of surface chemical reactions of functional groups on the metallic iron surface with low molecular substances ($TiCl_4$, CH_4) were studied and it was found that methyl (CH_3) groups are most suitable for the synthesis of carbide nanostructures in inert environment compared with hydroxyl (OH) or methoxy (OCH_3) groups.

A method of synthesis of titanium carbide nanostructures on the surface of dispersed iron particles using methyl (CH_3)

functional groups, $TiCl_4$, and CH_4 is developed. The sequence of synthesis steps of titanium-carbon nanostructures on the surface of the dispersed iron particles, concentration and temperature ranges at which the synthesis is carried out, and the temperature intervals of sample heating were established. Based on X-ray analysis it is shown that in a cyclic process of exposure of $TiCl_4$ vapor and CH_4 on dispersed iron titanium-carbon nanolayer with amorphous structure is formed at the surface. Crystallization and the formation of titanium carbide are carried out after calcinations at 1100°C.

Investigation of sintering and pressing conditions of dispersed iron particles with deposited titanium carbide nanostructures allows to determine the conditions for obtaining a metal (iron-based) composite material with low residual porosity with a dispersed phase based on titanium carbide.

Mechanical properties of the metal (iron-based) composite material with dispersed phase based on titanium carbide are studied and it is found that it is comparable with the best grades of steel obtained by moulding.

The results suggest that the creation of metallic (iron-based) composite materials structured by titanium carbide nanostructures using the process of surface structuring is promising for the preparation of a new generation of composite materials.

ACKNOWLEDGMENT

This work was supported by the Ministry of Education and Science of the Russian Federation, the agreement of 8575 from 13.09.2012.

REFERENCES

1. H. Gleiter, "Nanostructured materials: basic concepts and microstructure," Acta Materialia, vol. 48, no. 1, pp. 1–29, 2000.
2. C. C. Koch, I. A. Ovid'ko, S. Seal, and S. Veprek, Structural Nanocrystalline Materials: Fundamentals and Applications, Cambridge University Press, Cambridge, Mass, USA, 2007.
3. M. Hotta, H. Kita, and J. Hojo, "Nanostructured silicon carbide ceramics fabricated through liquid-phase sintering by spark plasma sintering," Journal of the Ceramic Society of Japan, vol. 119, no. 1386, pp. 129–132, 2011.
4. W.-P. Chen, M.-Y. Han, and S.-F. Yang, "Research progress of Al2O3 ceramic composites," Journal of Materials Engineering, no. 3, pp. 91–96, 2011.
5. T. M. Ulyanova, N. P. Krut'ko, P. A. Vityaz, and L. V. Titova, "Investigation of nanostructured oxides: synthesis, structure and properties," Journal of Nanoscience and Nanotechnology, vol. 11, no. 3, pp. 2107–2112, 2011.
6. V. S. Aigbodion and S. B. Hassan, "Effects of silicon carbide reinforcement on microstructure and properties of cast Al–Si–Fe/SiC particulate composites," Materials Science and Engineering A, vol. 447, no. 1-2, pp. 355–360, 2007.
7. H. Ataee-Esfahani, M. R. Vaezi, L. Nikzad, B. Yazdani, and S. K. Sadrnezhaad, "Influence of SiC nanoparticles and saccharin on the structure and properties of electrodeposited Ni–Fe/SiC nanocomposite coatings," Journal of Alloys and Compounds, vol. 484, no. 1-2, pp. 540–544, 2009.
8. M. A. Meyers, A. Mishra, and D. J. Benson, "Mechanical properties of nanocrystalline materials,"Progress in Materials Science, vol. 51, no. 4, pp. 427–556, 2006.
9. A. V. Monin, E. G. Zemtsova, N. B. Shveikina, and V. M. Smirnov, "Features of phase transitions upon the thermal treatment of Al2O3 nanoparticles," Nanotechnologies in Russia, vol. 7, no. 3-4, pp. 152–155, 2012.
10. Y. L. Dong, F. M. Xu, X. L. Shi et al., "Fabrication and mechanical properties of nano-/micro-sizedAl2O3/SiC composites," Materials Science and Engineering A, vol. 504, no. 1-2, pp. 49–54, 2009.
11. P. K. Sharma, V. V. Varadan, and V. K. Varadan, "A critical role of pH in the colloidal synthesis and phase transformation of nano size α-Al2O3 with high surface area," Journal of the European Ceramic Society, vol. 23, no. 5, pp. 659–666, 2003.
12. K. Okada, T. Nagashima, Y. Kameshima, A. Yasumori, and T. Tsukada, "Relationship between formation conditions, properties, and crystallite size of boehmite," Journal of Colloid and Interface Science, vol. 253, no. 2, pp. 308–314, 2002.
13. V. V. Tomaev, A. T. Nakusov, and E. G. Zemtsova, "Preparation and investigation of ion-conducting nanocomposite materials based on the aerosil-silver iodide system," Glass Physics and Chemistry, vol. 36, no. 1, pp. 92–99, 2010.
14. V. F. Dorfman, Synthesis of Solid-Phase Structures, Metallurgiya, Moscow, Russia, 1986, (Russian).
15. V. M. Smirnov and V. B. Aleskovskii, "Nanolevel structuring—the way to the designing of novel functional solids and materials," in Proceedings of the 1st IEEE Conference on Nanotechnology (IEEE-NANO '01), pp. 409–414, Maui, Hawaii, USA, 2001.
16. V. M. Smirnov, Chemistry of Nanostructures. Synthesis, Structure, Properties, Saint Petersburg State University, St. Petersburg, Russia, 1996, (Russian).
17. Y. D. Tretyakov and U. Lepis, Chemistry and Technology of Solid-Phase Materials, Moscow State University, Moscow, Russia, 1985, (Russian).
18. P. Glansdorff and I. Prigogine, Thermodynamic Theory of Structure, Stability and Fluctuations, John Wiley & Sons, London, UK, 1971.
19. V. M. Smirnov, Actual Problems of Chemistry of Solid Substances, Saint Petersburg State Institute of Technology, St. Petersburg, Russia, 1992.
20. J. M. Lehn, Supramolecular Chemistry: Concepts and Perspectives, Wiley-VCH, Weinheim, Germany, 1995.
21. R. B. Merrifield, "Automated synthesis of peptides," Science, vol. 150, no. 3693, pp. 178–184, 1965.
22. P. Hodge and D. C. Sherrington, Eds., Polymer-Supported Reactions in Organic Synthesis, Wiley, Chichester, UK, 1980.
23. V. B. Aleskovskii, Chemistry of Solid Substances,

Vysshaya Shkola, Moscow, Russia, 1978.

24. L. I. Maissel and R. Glang, Eds., Handbook of Thin Film Technology, McGraw-Hill, New York, NY, USA, 1970.

25. V. M. Smirnov, E. G. Zemtsova, P. E. Morozov, A. Y. Arbenin, and A. S. Vinogradov, "Special features of structural organization and magnetic properties of quasi-one-dimensional organoiron nanostructures on a silica support," Russian Journal of General Chemistry, vol. 78, no. 12, pp. 2299–2307, 2008.

26. V. B. Aleskowskii, "Several hypotheses and regularities in the chemistry of highly organized substances,"Macromolecular Symposia, vol. 136, no. 1, pp. 1–7, 1998.

27. V. M. Smirnov, E. G. Zemtsova, E. B. Ivanov, M. G. Osmolowsky, V. G. Semenov, and I. V. Murin, "Mössbauer and magneto-chemical study of solids formed by surface chemical reaction of OH–ssilica groups with iron diacetylacetonato chloride (C5H7O2)2FeCl," Applied Surface Science, vol. 195, no. 1–4, pp. 89–92, 2002.

28. R. L. Puurunen, "Surface chemistry of atomic layer deposition: a case study for the trimethylaluminum/ water process," Journal of Applied Physics, vol. 97, no. 12, Article ID 121301, pp. 52–98, 2005.

29. S. M. George, B. Yoon, and A. A. Dameron, "Surface chemistry for molecular layer deposition of organic and hybrid organic-inorganic polymers," Accounts of Chemical Research, vol. 42, no. 4, pp. 498–508, 2009.

30. V. Miikkulainen, M. Leskelä, M. Ritala, and R. L. Puurunen, "Crystallinity of inorganic films grown by atomic layer deposition: overview and general trends," Journal of Applied Physics, vol. 113, no. 2, Article ID 021301, pp. 3–96, 2013.

31. V. M. Smirnov, V. G. Povarov, G. P. Voronkov et al., "Solid-state water-mediated transport reduction of nanostructured iron oxides," Journal of Nanoparticle Research, vol. 3, no. 1, pp. 83–89, 2001.

32. V. Smirnov and N. Bobrysheva, "Magnetic properties of iron–oxygen nanostructure doped with 4f elements," Journal of Alloys and Compounds, vol. 275–277, pp. 533–536, 1998.

33. L. E. Ermakova, M. P. Sidorova, and V. M. Smirnov, "Isoelectric point of silicon oxide particles coated by monolayers of oxides of titanium and aluminum," Colloid Journal of the Russian Academy of Sciences, vol. 59, no. 4, pp. 523–525, 1997.

34. S. I. Koltsov, V. B. Kopylov, V. M. Smirnov, and V. B. Aleskovskii, "Synthesis and investigation of aluminum-oxygen layers on surface of silica," Journal of Applied Chemistry of the USSR, vol. 49, no. 3, pp. 525–532, 1976.

35. S. M. George, "Atomic layer deposition: an overview," Chemical Reviews, vol. 110, no. 1, pp. 111–131, 2010.

36. V. M. Smirnov, "Nanoscaled structuring as a way to constructing new solid substances and materials,"Russian Journal of General Chemistry, vol. 72, no. 4, pp. 590–606, 2002.

37. Y. P. Solntsev and E. I. Pryahin, Materials Science, Chempub, St. Petersburg, Russia, 2007.

CHAPTER 17

GRAPHENE-BASED NANOMATERIALS AS HETEROGENEOUS ACID CATALYSTS: A COMPREHENSIVE PERSPECTIVE

Bhaskar Garg [1], Tanuja Bisht [2] and Yong-Chien Ling [1]

[1]Department of Chemistry, National Tsing Hua University, Hsinchu 30013, Taiwan

[2]Department of Chemistry, Government Degree College, Champawat 262523, Uttarakhand, India

ABSTRACT

Acid catalysis is quite prevalent and probably one of the most routine operations in both industrial processes and research laboratories worldwide. Recently, "graphene", a two dimensional single-layer carbon sheet with hexagonal packed lattice structure, imitative of nanomaterials, has shown great potential as alternative and eco-friendly solid carbocatalyst for a variety of acid-catalyzed reactions. Owing to their exceptional physical, chemical, and mechanical properties, graphene-based nanomaterials (G-NMs) offer highly stable Brønsted acidic sites, high mass transfer, relatively large surface areas, water tolerant character, and convenient recoverability as well as recyclability, whilst retaining high activity in acid-catalyzed chemical reactions. This comprehensive review focuses on the chemistry of G-NMs, including their synthesis, characterization, properties, functionalization, and up-to-date applications in heterogeneous acid catalysis. In line with this, in certain instances readers may find herein some criticisms that should be taken as constructive and would be of value in understanding the scope and limitations of current approaches utilizing graphene and its derivatives for the same.

INTRODUCTION

The prime source of energy on Earth for the present and foreseeable future is found in chemical bonds. Catalysis affords the means of changing the rates at which chemical bonds are formed and broken making it possible to elicit a desirable product over an undesired one, the phenomenon of chemical specificity. Among catalysis of substantial interest, acid catalysis is quite prevalent and probably the most routine operation in both industrial processes and academic research worldwide. The petroleum (biodiesel production as alternative fuels), chemical (production of most chemicals and materials), plastic (production of polymers and plasticizers), and pharmaceutical industries (production of new heterocycles as "wonder drugs"), essential to a healthy economy, rely heavily on acid catalysis. Not surprisingly, therefore, it lies at the heart of our quality of life.

In industry, with foremost focus on economic benefits, a large class of liquid-phase reactions are routinely executed

Citation: Bhaskar Garg, Tanuja Bisht, and Yong-Chien Ling, Graphene-Based Nanomaterials as Heterogeneous Acid Catalysts: A Comprehensive Perspective, Molecules 2014, 19(9), 14582-14614; doi:10.3390/molecules190914582

with sulfuric acid (H_2SO_4), hydrofluoric acid (HF), and methanesulfonic acid (CH_3SO_3H) as bulk commodity catalysts. Unquestionably, the unique spatial separation (similar to that of enzymes) as well as the self-similarity of structures between the active sites in these homogeneous molecular acids not only allow consistent energetic interactions between each active site and reaction substrate (even at ambient conditions) but also make them responsive to most spectroscopic characterization techniques. Despite this, these catalysts require challenging processes for their separation from the homogeneous reaction mixture, resulting in abundant non-recyclable acid waste.

The shift towards an emphasis on catalyst recoverability and reusability has resulted in extensive research efforts devoted to the development of solid or heterogeneous acid catalysts, which are easier to separate, recover and reuse in desired reactions. In this milieu, endeavors have been made using, to name but a few, cation exchange resins (Amberlyst™-15, Nafion-H, *etc.*), solid superacids (SO_4^{2-}/ZrO_2), (supported)metal oxides/sulfides/triflates (Al_2O_3, CdS, $Yb(OTf)_3$), zeolites (H-ZSM-5), molecular sieves, silica, natural clays (bentonite, montmorillonite), and heteopolyacids to complement liquid-acid catalysts in a broad spectrum of acid-catalyzed reactions [1,2,3,4,5,6,7,8]. Although the mentioned solid catalysts certainly overcome some shortcomings, most of them are inappropriate owing to their deactivation (leaching or undesired deposits on active sites) [9,10], internal mass transfer (such as in zeolites) [11] or high mass transfer resistance (as in clay, molecular sieves *etc.*), unfavorable side reactions, high cost, and most importantly, toxicity issues associated with the presence of certain metals.

The pursuit of heterogeneous acid catalysts capable of meeting most of the abovementioned challenges has resulted in the consideration of carbon nanomaterials (CNMs), which can function as metal-free catalysts. Indeed, with the added advantage of classical nanofabrication techniques, the CNMs, which include mesoporous/amorphous carbon [12,13] and rich polytypes of carbon such as carbon nanotubes (CNTs), have produced considerable advancements in heterogeneous acid catalysis [14,15]. Notably, until the 20th century, the carbon family was only dominated by the well-known materials like graphite, diamond, fullerenes, and CNTs. However, this "aura" was lessened with the first isolation of free standing 2D graphene in 2004 [16] (in order to avoid unnecessary complexity to potential readers, we have adopted herein the common names currently in use in graphene-based catalysis research, yet we would also like to recommend the readers to consult the guest editorial published in the journal *Carbon* [17] for an exhaustive and productive account of the nomenclature of sp^2carbon nanoforms).

Graphene could be considered as a potential building block of 3D graphite, 0D fullerenes (in a broader context or to some extent, it is reasonable to declare graphene as "mother" of all carbon nanostructures, as increasingly highlighted by many research groups; certainly, however, it is a misconception that graphene can produce fullerenes– notably, fullerenes have alternate pentagonal and hexagonal arrangements, whereas in graphene only hexagons should be present), and 1D CNTs (Figure 1) [18,19] and possess many exceptional properties that have surpassed those observed with its predecessor, CNTs. Not surprisingly, therefore, graphene-based nanomaterials (G-NMs) are (and will continue to be) increasingly used in optoelectronic [18], energy conversion and storage [18,20], photocatalysis/photovoltaic [21], switching and information storage [22], bacteria killing [23], biosensors [24,25] and other technological applications [26,27,28] in the imminent carbon age. On another front, the concept of carbocatalysis within the graphene framework is a relatively new but rapidly emerging research area at the interface of organic, material, green and sustainable chemistry [29].

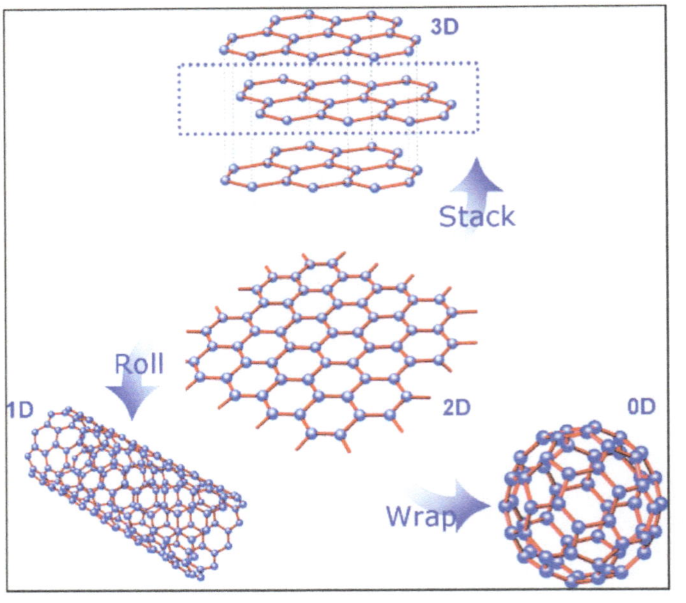

Figure 1. Graphene: the basic building block for other carbon allotropes, graphite (3D), fullerene (0D), and CNT (1D). Reprinted with permission from [18]. Copyright (2012) American Chemical Society.

It is worth commenting here that at certain instances, "carbocatalysis" can act as a bridge toward homogeneous catalysis (where both the reactants and catalysts are in a single phase) due to the existence of indefinitely persistent colloidal dispersions of carbocatalysts. In such cases, the process should be referred as "pseudo-homogeneous" because during the reaction, the catalyst is ostensibly in the same phase as reactants and products but can be recovered by means of centrifugation or membrane filtration after the reaction. Nevertheless, in order to avoid any confusion and also taking into account the recoverable features of G-NMs (lacking molecular entities), herein, we use the term "heterogeneous" throughout for graphene-based acid catalysis (G-AC).

After the seminal discovery by Bielawski and coworkers [30] of the use of graphene oxide (GO) as catalyst in alcohol oxidation, unquestionably, an appreciable range of G-NMs are currently available as catalysts for a variety of useful transformations. However, since the focus of this review article strongly lies on G-AC, interested readers are recommended to consult other review articles [29,31,32,33,34] for the corresponding unique and productive details.

The current state-of-the-art in G-AC can be largely categorized into graphite/graphene oxide (GO) and their sulfated (-O-SO_3H) or sulfonated (-SO_3H) derivatives. Whereas the former are obtained by chemical oxidation of natural graphite and/or subsequent exfoliation, the latter rely on the fundamental idea of surface functionalization of GO with sulfonating agents. Both classes exhibit unique acid activity and hold great promise in replacing traditional homogeneous and heterogeneous acid catalysts in an environmentally benign manner (Table 1). Thus far, to the best of our knowledge, G-AC has not been reviewed by any research group and we would like to be the first to shed light on this fast-growing field.

Table 1: A comparison between the characteristic features of sulfuric acid, Amberlyst™-15 and G-NMs in acid catalysis

Features	Sulfuric Acid	Amberlyst™-15	G-NMs
Physical state	*Liquid*	*Solid*	*Solid*
Acid activity	*High*	*High*	*High*
Surface Area	-	*Relatively low*	*High*
Operating conditions	*Typically harsh*	*Relatively Mild*	*Variable* [a]
Diffusion/Leaching	*None*	*Possible*	*Negligible*
Recoverability	*Difficult*	*Easy* [b]	*Easy* [b]
Reusability/cycles	*Difficult*	*Easy/Less cycles* [c]	*Easy/More cycles* [c]
Water tolerance/stability	*High*	*Relatively low*	*High*
Cost-effectiveness	*Cost-effective*	*Relatively expensive*	*Variable* [a]
Mass transfer	-	*Relatively low*	*High*
Reaction selectivity	*High*	*Variable* [a]	*Variable* [a]
Eco-friendly use	*No*	*Yes*	*Yes*
Industrial use	*Frequent*	*Selected* [d]	*Awaiting*

[a] Depending upon the types of reactants, catalyst, and nature of reactions; [b] Amberlyst™-15 can be recovered simply by filtration through a common filter paper while G-NMs are efficiently recovered either by membrane filtration or centrifugation; [c] The number of cycles, herein, should be taken as (partial)loss of acidic sites; the less cycles symbolize significant loss in acid activity while due to no or partial loss in acid activity, G-NMs can be used repeatedly; [d] Amberlyst™-15 is commercially used at industrial scale in the addition of alcohols to alkenes to form ethers for gasoline boosters.

As such, this review will begin with a brief discussion on the synthesis and structure of GO and its precursor, graphite oxide. Then, we will focus on unconventional structural models that have emerged only recently and are ascribed to be responsible for the inherent acidic properties of graphite oxide/ GO. Afterwards, we will focus on the description of various synthetic routes to obtain sulfonated and sulfated analogues of graphene. In this regard, systematization according to the type of reducing and sulfonating agents is established. An equal emphasis is placed while discussing the properties and characterization data of G-NMs based on elemental analysis (EA) and Raman spectroscopy. Lastly, the catalytic applications of G-NMs are categorized based on: (i) graphite oxide; (ii) GO; and (iii) sulfated or sulfonated derivatives and discussed in conjunction with other acid catalysts. It should be noted that in some applications, both GO and its sulfonated counterparts have been examined together and thus the organization of this review should be regarded as a flexible guide to the classification of these applications.

GRAPHITE OXIDE AND GRAPHENE OXIDE (GO)

Synthesis and Structure

Historically, graphite oxide was discovered as a way to measure the molecular weight of graphite. Indeed, this was the intention of Brodie when in 1859 [35] he heated a mixture of graphite and fuming nitric acid in the presence of potassium chlorate. The repeated oxidation (four replicates) of the as-obtained product afforded the very first sample of graphite oxide as a light yellow solid which was termed "graphic acid", based on its dispersion behavior at different pH values. In a subsequent follow-up work in 1898 [36,37], Staudenmaier improved the oxidation protocol by adding potassium chlorate in small portions and acidifying the mixture with concentrated H_2SO_4, which completely avoided the four repetitions of the oxidation steps and was thus ascertained more practical. Over the span of 60 years, in 1958 [38], Hummers introduced potassium permanganate as an oxidant in a mixture of sodium nitrate and concentrated H_2SO_4 resulting in a more heavily oxygenated form of graphite oxide. This protocol was proved safer as it not only allowed *in situ* generation of nitric acid but also avoided the use of highly corrosive fuming nitric acid. In a bid to replace *in situ* production of nitric acid with less corrosive phosphoric acid, a new oxidation protocol (Tour's method) is reported recently [39], yet Hummers method has been well appreciated among scientific community and increasingly adopted to date by many researchers worldwide.

Aside from the operative oxidation procedures, the chemical structure of graphite oxide has been the subject of extensive research over the years. Consequently, several structural models of graphite oxide have been surfaced from time to time. In this context, Hofmann-Holst [40], Ruess [41], Clauss [42], Scholz-Boehm [43], Nakajima [44], Nakajima-Matsuo [45], Lerf-Klinowski (LK) [46,47,48], Dékány [49], and Ajayan [50] models are especially notable. However, due to the variable oxidation procedures leading to sample-to sample variability, nonstoichiometric atomic composition and lack of a "hallmarked" characterization technique, the precise chemical structure of graphite oxide remains elusive and still a subject of further discussion. Overall, the LK model [48] of graphite oxide is perhaps the most popular and widely accepted among graphene research community.

As such, it is assumed that the basal plane of heavily oxygenated graphite oxide is highly populated with hydroxyl (-OH) and epoxide (C-O-C) functions, while the edge-plane consists of carbonyl (C=O) and carboxylic acid (-COOH) groups. The coverage of the oxygen-rich functional groups in graphite oxide substantiates hole defects (nanovoids and vacancies) throughout the surface and deviates it from the state of pristine graphite. Indeed, the presence of such features on the graphite oxide surface has recently been observed [51] using high-resolution transmission electron microscopy (HRTEM).

Graphite oxide is chiefly composed of carbon, oxygen, and hydrogen atoms retaining a C/O ratio in between 1.8 to 2.5, while the most common C/O ratio for different samples is ~2 [40,42,49,52]. In this context, it should be noted that the stoichiometric ratio between the constituent elements of graphite oxide is not fixed and varies depending upon the oxidation level. Consequently, regardless of the synthesis protocol followed [36,38] the composition and structure of graphite oxide do not change significantly even after the addition of excess oxidizing agent. GO, an exfoliated form of graphite oxide, is another important intermediate between graphite and graphene and usually attained via mechanical (vigorous stirring or ultrasonication), chemical (ionic liquids) or thermal treatment of graphite oxide. At this juncture, it is critical to understand the technical point of differentiation between graphite oxide and GO. Whereas graphite oxide represents the lamellar structure (consisting of layers of carbon from the original graphene lattice), GO largely exists as mono-, bi- or at most a few layers (3 to 4) of graphene sheets. Nevertheless, despite being structurally different, both are chemically similar, having a complex cocktail of oxygen functionalities (Figure 2) rendering them hydrophilic in nature. As a consequence, water molecules can readily intercalate between the layers resulting in an increase of the interlayer distance and change of hybridization state (in oxidized carbons) from sp^2 (planar) to sp^3 (tetrahedral).

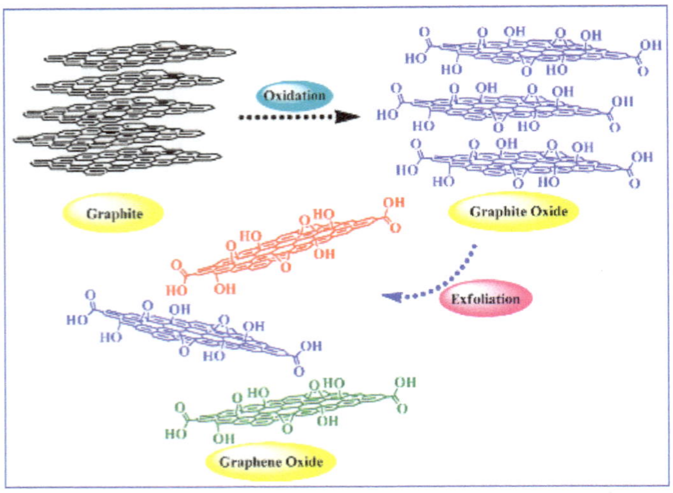

Figure 2: Synthesis of graphite oxide and graphene oxide (GO) from natural graphite.

Acidic Properties of Graphite Oxide and GO

Aqueous solutions of graphite oxide have significant acidity, with a pH in between 3 to 4. It has long been demonstrated that 100 g of graphite oxide contains 500–800 mmol of active acid sites [40,42] *i.e.*, one acid site is present for every 6–8 carbon atoms that can participate in cation exchange reactions. In order to explain such high acidity of graphite oxide, Clauss *et al.* [42] suggested the presence of enolic groups and proposed the empirical formula $C_8O_2(OH)_2$ for the composition of graphite oxide. However, the authors were unable to provide any experimental data in support of this suggestion.

In yet a more recent scenario, it is commonly assumed that graphite oxide obtained by the (modified)Hummer's method [38,48] contains lots of recoverable Brønsted acid sites (though benchmark experimental evidences are still in question) including -OH and -COOH groups on the basal plane and flake edges, respectively. Accordingly, the acidic character of graphite oxide is likely due to the synergistic effect of these functionalities. However, controlled experiments by Dhakshinamoorthy *et al.* [53] with GO, H_2SO_4, *p*-toluenesulfonic acid (*p*-TSA), and glacial acetic acid have clearly indicated that the active acid sites of GO are not indeed -COOH groups but probably HSO_4 groups as evidenced by elemental analysis (1.16 wt % of sulfur atoms). In fact, the thermal treatment of GO at 200 °C removed most of the HSO_4 groups and resulted in catalytically inactive, partially reconstituted, GO having a negligible sulfur content. A similar observation (at relatively high temperature; 400 °C) has also been reported recently while using GO as an acid catalyst in the conversion of carbohydrates into 5-ethoxymethylfurfural [54]. In an elegant work, Dimiev *et al.* [55] have recently proposed a new explanation for the graphite oxide acidity that may allow one to add critical details to existing structural models of graphite oxide with higher confidence. It is well known, as increasingly highlighted in most published work, that graphite oxide has a light yellow appearance during its first quenching with water. However, with cumulative water washings, this color gradually changes from light yellow to dark brown or even black, which indicates that significant chemical transformations occur in the course of the purification of graphite oxide. Taking this into account, the group [55] decided to avoid exposing the graphite oxide material to water. Thus, by introducing non aqueous solvents such as methanol, isopropanol, ethyl acetate and glacial/trifluoro- acetic acid during the quenching and purification steps of graphite oxide, they synthesized different samples of pristine graphite oxide (pGO) having a significant amount of vacancy defects terminated by ketone groups. The products were extensively characterized by Fourier transform infrared spectroscopy (FTIR), UV-visible and X-ray photoelectron spectroscopy (XPS), thermogravimetric analysis (TGA), ^{13}C solid state nuclear magnetic resonance (SSNMR), and scanning electron microscopy (SEM).

It was postulated that the oxidized areas (sp^3) of as-synthesized pGO are dominated by epoxide groups while covalent sulfates (primarily existing as cyclic sulfates **1**) and -OH groups are present in small amounts. When pGO was exposed to water, hydrolysis occurred in two steps. In the first step, cyclic sulfates are cleaved to monosulfate **2**, while 1,2-diol **3** and H_2SO_4 are formed in the second step of hydrolysis (Scheme 1). The as-produced **2** and H_2SO_4 were ascribed to be responsible, in part, for the highly acidic properties of graphite oxide. Furthermore, as evidenced by experimental data, it was proposed that acidic properties of graphite oxide are significantly affected by stepwise conversion of tertiary alcohols into edge-pane ketones, which remain in equilibrium with their hydrated form.

The group further examined the effect of water on GO acidity and proposed an unconventional view of GO chemistry [56], a "dynamic structural model" (DSM). As per this theory, "GO does not exist as a static structure with a given set of functional groups, but rather generates them and constantly changes its structure through interaction with water". As such, the GO acidity was brought about by CC bond cleavage, formation of vinylogous carboxylic acid **5** (rendering the -OH groups acidic) (Scheme 2), and the generation of protons. In this milieu, it should be noted that the formation of an electrical double layer at the GO interface in aqueous solutions is the main driving force for the observed GO acidity. Equally important is to understand that on prolonged exposure to water, GO flakes gradually degrade and attain a humic acid-like structure.

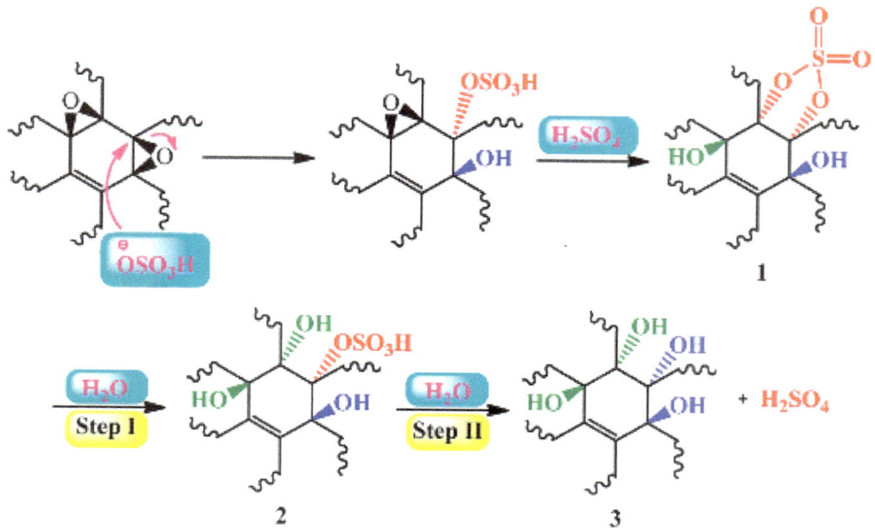

Scheme 1: Hydrolysis of cyclic sulfate **1** into hydrosulfate **2** (Step I) and sulfuric acid (Step II). Adapted with permission from [55]. Copyright (2012) American Chemical Society.

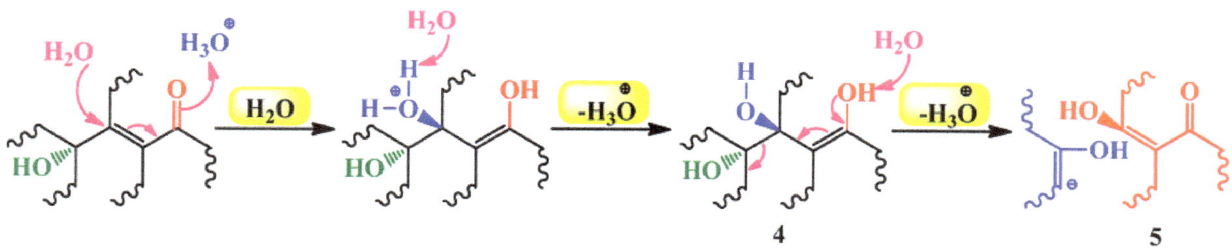

Scheme 2: A schematic representing the formation of vinylogous carboxylic acid **5** via C-C bond cleavage of enol **4**, which in turn was obtained via nucleophilic attack of water molecule on a GO fragment having one hydroxyl and ketone group. Adapted with permission from [56]. Copyright (2013) American Chemical Society.

In a more recent report, Eigler *et al.* [57] undertook a systematic study in order to determine the sulfur species in GO and its base-treated counterparts, the so called GO-NaOH, GO-$NaHSO_4$, and GO-Na_2SO_3. In contrast to the previous work on graphite oxide hydrolysis [55], the group postulated that only the first step of hydrolysis takes place, while mono/organosulfates **2** are fairly stable and remain bound on GO (see Scheme 1), a key aspect for graphite oxide/GO acidity. As evidenced by thermogravimetry-mass spectroscopy (TGA-MS), it was concluded that organosulfates are the integral part of any GO and decompose at about 250 °C. On the other hand, the weight loss step in between 700–800 °C could be attributed to the pyrolysis of inorganic sulfur as sulfates or sulfite in GO-$NaHSO_4$, and GO-Na_2SO_3, respectively (Figure 3a). Moreover, a TGA-MS analysis using sodium dodecylbenzenesulfonate (decomposition temperature of ~500 °C) as a reference completely excluded the presence of sulfonic acid groups on the GO surface. Accordingly, the structural model of GO was extended by adding organosulfate in addition to epoxy, and hydroxyl groups, which are predominantly located above and below the carbon skeleton (Figure 3b). It was estimated that, in addition to hydroxyl and epoxy groups, one organosulfate group per 20 carbon atoms is present in GO and must be located on both sides of carbon lattice as well as the edges of GO flakes. The observation brought about by this study that the basic washing procedures could apparently be effective in GO purification, especially in removing sulfur species, is quite appealing.

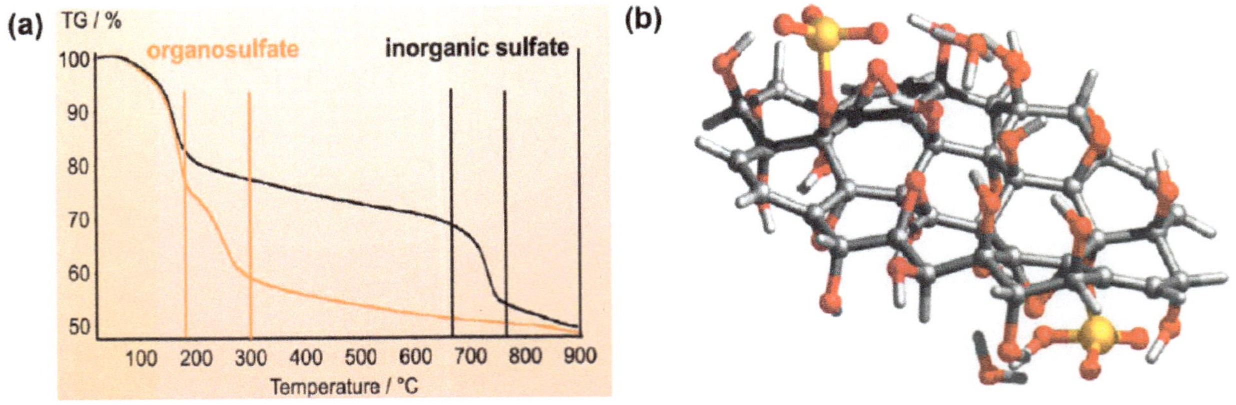

Figure 3. (**a**) The TGA data of GO-$NaHSO_4$ revealing two main decomposition steps for sulfur-containing species. (**b**) The proposed structural model of GO terminated with epoxy, hydroxyl, and organosulfate as dominant functional groups. Reproduced with permission from [57]. Copyright (2013) Wiley-VCH Verlag GmbH & Co.

In light of the above discussion, it may be concluded that -COOH groups have nothing to do with the acidic nature of graphite oxide/GO while the presence of sulfur as organosulfates, an integral part of oxidation protocols, is largely responsible for the observed acidity. As a final note, we anticipate that the exhaustive characterization of GO is a prerequisite prior to surface functionalization, especially, in sulfonation reactions as the presence of organosulfates in GO may significantly influence the characterization data of the resulting NMs and lead to misinterpretations.

G-NMS *VS.* SULFONATION AND SULFATION

The integration of sulfonic acid functional groups (-SO_3H) into CNMs is well known and emerged as a powerful strategy to produce promising solid acids for a range of acid-catalyzed reactions [12,13,14,15,58,59]. However, the sulfonation chemistry of G-NMs has developed only recently and continues to expand for diverse applications. At this juncture, it is critical to differentiate between sulfonation and sulfation reactions (Scheme 3) as the two are often confused but the structural differences are quite significant, making the use of separate terms more appropriate.

In sulfonation, sulfur trioxide (SO_3), the active species, reacts for example with an alkyl benzene and forms a C-S bond. Notably, the resultant alkyl benzenesulfonic acid is a stable molecule. Sulfation, on the other hand, involves the formation of a C-O-S bond and the resultant alcohol sulfuric acid is somewhat less stable. Unless neutralized, it decomposes to form sulfuric acid and the starting alcohol. This stability difference in the products of SO_3 reaction also has an intense impact on the choice of process used to produce sulfonates or sulfates.

In general, graphene-based sulfonation can be achieved in two steps: (a) reduction of GO to reduced graphene oxide (rGO) and (b) sulfonation of rGO. However, the direct sulfonation of GO and two-steps reduction protocols have also been undertaken at certain instances. The concrete methods, reducing and sulfonating agents are classified, summarized, and discussed below.

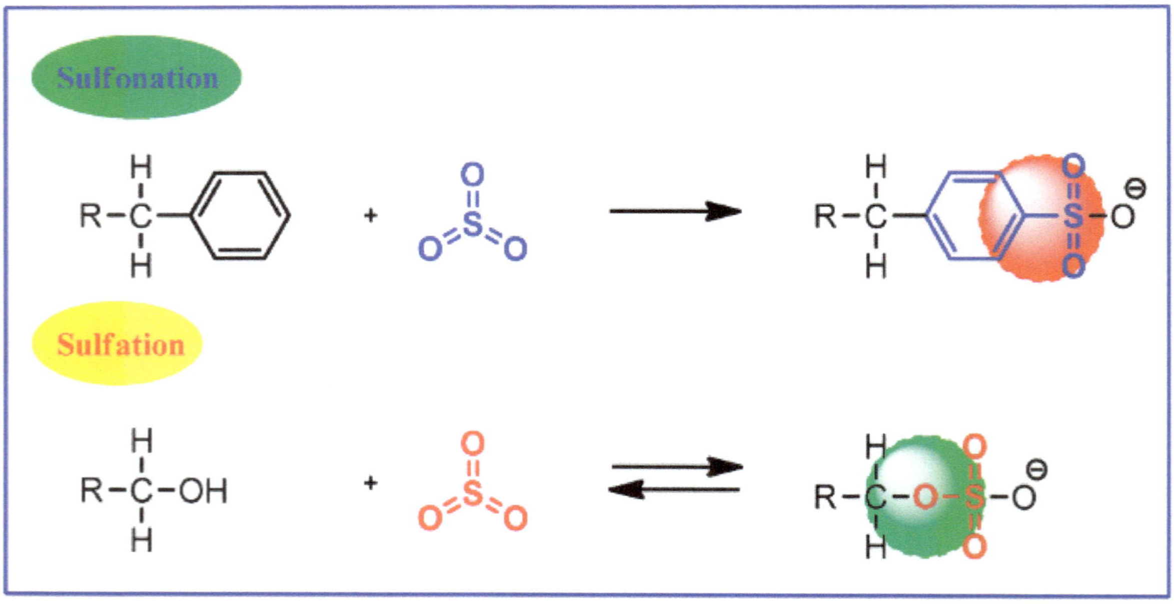

Scheme 3: Sulfonation *vs.* sulfation.

Reduction of GO

To date, one of the most important reactions of GO is its reduction leading to rGO, similar to that of pristine graphene. Although the reduction of GO can be achieved through thermal/H_2 [60,61,62], electrochemical [63], or chemical means [19,64,65,66,67,68], the latter is widely used in the graphene research and will only be discussed here. The emphasis is given to C/O ratio as well as Raman spectroscopy.

The aqueous colloidal dispersions of GO can be effectively reduced by a range of reducing agents having an established background in synthetic chemistry (see the recent review [69] for a comprehensive table of reducing agents for GO). The conversion of GO into rGO is usually achieved by an increased aggregation or hydrophobicity (due to the removal of oxygen containing groups) manifested in a color change from brown to black. Based on elemental analysis (EA), a decrease in elemental oxygen contents thereby resulting an increase of C/O ratio certainly signify the efficacy of the employed reducing agents.

Raman spectroscopy is a handy tool to characterize the structure and quality of G-NMs including graphite, graphite oxide, GO, rGO, and other carbonaceous materials [70,71]. In this context, the two most prominent features in (r)GO are the D and G vibrational bands (1100–1800 cm^{-1}). Whereas the former gives an idea about ordered or disordered structure of carbon, the latter is an indicator of the stacked carbon domain. Also of interest is the 2D band, a dominant sharp peak always observed for single layer graphene. It has been demonstrated that Raman scattering can be used as a fingerprint for single, bi, and a few layer graphene [72,73]. The ratio of intensities of two bands (I_D/I_G) can be used as a means of determining the structural changes within graphene samples. For instance, a high I_D/I_G ratio clearly indicates a certain amount of disorders/edges or exfoliation [71].

In general, graphite displays a prominent G peak at 1580 cm^{-1}, attributable to the first-order scattering of the E_{2g} mode. On the contrary, GO shows two peaks at 1363 and 1594 cm^{-1} corresponding to the D and G bands, respectively. The G band is associated with the vibrations of sp^2 carbon atoms in the 2D graphitic lattice whereas the D band is related to the vibrations of sp^3 carbon atoms of defects or disorder as mentioned above [71,73]. After the reduction, the I_D/I_G ratio changes considerably, as will be discussed below taking hydrazine (N_2H_4), sodium borohydride ($NaBH_4$), and lithium aluminum hydride ($LiAlH_4$) as representative reducing agents.

Hydrazine (N_2H_4)

N_2H_4 is known to scavenge oxygen while itself being broken down into water and nitrogen, making it an attractive option where most reducing agents exhibit moderate to strong reactivity toward water. $N_2H_4{\cdot}H_2O$ was firstly used as reducing agent by Stankovich *et al.* [74] in an effort to obtain individual sheets of graphene from GO. Under the optimized conditions, the number of polar functionalities on GO's surface was decreased significantly and resulted in aggregated layers of graphene sheets, rGO. Based on the extensive characterization data, it was proposed that the reduction of GO proceeds via a nucleophilic attack of N_2H_4 on an epoxide function. The EA results were in accord to expectations and a high C/O ratio (10.3/1) was measured for rGO over GO (2.7/1). Likewise, the Raman spectra indicated an increase in I_D/I_G ratio of as-obtained rGO which was suggested to be the result of a decrease in the size of sp^2 carbon domains (by the formation of sp^3carbons) upon reduction. Despite its good efficiency, the severe toxicity coupled with the introduction of heteroatomic impurities somewhat limits the practical use of N_2H_4.

Sodium Borohydride ($NaBH_4$)

$NaBH_4$ is one of the most used hydrides in research laboratories, as it can be readily solubilized in aqueous and alcoholic media. The first example of $NaBH_4$ usage in GO reduction was reported by Muszynski *et al.* [75] to achieve the physisorption of Au nanoparticles on amine grafted graphene. Later on, Gao *et al.* [50] performed the total reduction of GO using $NaBH_4$ as the first of a three-step protocol. The C/O ratio of as-obtained rGO was almost twice (4.78) that of GO (2.44). In addition, the Raman spectra of rGO exhibited an increased I_D/I_G ratio (1.91) compared to GO (0.95), which is consistent with what is observed with N_2H_4, as mentioned above. Unlike N_2H_4, $NaBH_4$ generates additional alcohols as principal impurities [76], a possible cause of a low C/O ratio relative to that of N_2H_4.

Lithium Aluminum Hydride ($LiAlH_4$)

$LiAlH_4$ (LAH) is another strong reducing agent of the hydride family that remained unused in GO reduction for a long time, presumably, due to its strong reactivity towards water (production of H_2 gas) and capability to reduce carbonyl, epoxy, ester, and carboxylic functions (unselective nature) to hydroxyl groups. The first LAH-triggered reduction of GO, in conjunction with N_2H_4 and $NaBH_4$, was reported by Ambrosi *et al.* [77]. The strong reducing strength of LAH was evidenced by a highest C/O ratio (12) while the reduction of GO (C/O ratio 3.4) with N_2H_4 and $NaBH_4$ provided relatively low C/O ratios of 11.5 and 9.5, respectively. The FTIR spectrum of as-obtained rGO indicated the presence of $-CH_2$ groups due to the alkene hydrogenation of unsaturated carbonyls. The GO obtained by graphite oxidation provided a considerably high I_D/I_G ratio (1.10) and was postulated to be due to the extensive disorder in the graphitc structure. On the contrary, upon reduction, the I_D/I_G ratio decreased to 0.61, 0.97, and 0.63 for $NaBH_4$, N_2H_4 and LAH, respectively. A decrease in I_D/I_Gratio clearly indicates that the disorder associated with the amorphous GO diminishes. There are a handful of other characterization techniques (XRD, XPS, EDS elemental mapping) for differentiating GO and rGO. Besides this, the AFM can be used to determine the single layer morphology of the desired sample. We recommend that readers consult the original literature for additional details, as mentioned previously.

Sulfonation of rGO and GO

As discussed above, the chemical reduction of oxygen functionalities on GO sheets produces rGO with the reintroduction of large aromatic domains. Consequently, rGO has increasingly used for the non-covalent adsorption of polymers [78,79] and aromatic species [80,81] via π–π stacking and van der Waals interactions. On the other hand, the covalent functionalization of rGO is quite uncommon and not performed significantly. In this milieu, the surface modifications of (r)GO with $-SO_3H$ groups have provided a handle to narrow down this huge gap. This section will deal the concrete methodologies used for sulfonation reactions of (r)GO along with a brief on characterization data. For the sake of clarity, sulfonation reactions are classified according to the type of sulfonating agents (Table 2).

Table 2: List of sulfonating agents for (r)GO to G-AC

Sulfonating Agents	Reducing Agents [a]	Reaction Conditions [b]	Acid Density [c]	Applications [d]	Ref.
4-Benzenediazonium sulfonate	$N_2H_4 \cdot H_2O$	$EtOH/H_2O$, H_3PO_2, 3–5 °C, 1.5 h	1.55 mmol H^+ g^{-1}	Hydrolysis of ethyl acetate	[82]
Fuming sulfuric acid	$NaBH_4$	Autoclave, 180 °C, 24 h	1.2 mmol H^+ g^{-1}	(1) Esterification reactions (2) Peckmann reaction (3) Hydration reaction	[83]
Chlorosulfonic acid	*n*-butyl lithium	$ClSO_3H$, 0 °C, RT [e], overnight, NaOH, HCl, H_2O	-	Ester exchange reactions	[84]
4-Benzenediazonium sulfonate	$NaBH_4$	0 °C, 2 h, Hydrazine, 100 °C, 24 h	-	Synthesis of xanthenes and benzoxanthenes	[85]
Chlorosulfonic acid	-	$CHCl_3$, 70 °C, 4 h	1.2 mmol H^+ g^{-1}	Chemical conversion of biomass derived carbohydrates	[86]
4-Benzenediazonium sulfonate	(i) $NaBH_4$ (ii) N_2H_4	RT [e], overnight	0.5–1.7 mmol H^+ g^{-1}	Dehydration of xylose to furfural	[87]
Ammonium sulfate	-	235 °C, 30 min, argon atmosphere	EW [f] = 725 ± 5 g/mol	As support for metal nanocatalysts	[88]
Sulfuric acid	Microwave irradiation, 190 °C	160 °C, 5 h nitrogen atmosphere	1.9 mmol H^+ g^{-1}	Production of biofuels	[89]

[a] List of reducing agents used in sulfonation methods of GO and partially reduced GO; [b] Sulfonation carried out in aqueous medium or pure solution of the sulfonating agents unless stated otherwise; [c] Determined by acid-base titrations; [d] Applications in metal-free heterogeneous acid-catalyzed reactions except ref. ([88,90], the latter is not included in the table); [e] Room temperature; [f] Equivalent weight.

Diazonium Salt of Sulfanilic Acid; 4-Benzenediazonium Sulfonate

Diazonium salts represent a class of organic compounds having a general formula of $RN_2^+ X^-$ where R may be an alkyl or aryl residue and X is an (in)organic anion. The diazonium salts having an aryl group are usually stable in aqueous solutions at temperatures below 5 °C (the N_2 group does not tend to be lost as nitrogen gas) and serve as important intermediates in the organic synthesis of azo dyes [91,92,93,94]. The nitrosation of primary aromatic amines with HNO_2, which itself generated *in situ* from $NaNO_2$ and a mineral acid, is the most common method for the preparation of diazonium salts. However, being non-nucleophilic, the diazonium tetrafluoroborate salts can be isolated and are stable at room temperature. The process of forming diazonium compounds has been given a variety of names, including diazotation, diazoniation, and diazotization.

The usage of diazonium chemistry in the graphene domain was first reported by Lomeda *et al.* [95] to achieve the high dispersibility of graphene platelets in a range of polar aprotic solvents. However, the first attempt towards graphene sulfonation was performed by Si and Samulski [96] following a three step protocol as shown in Scheme 4. It was postulated that prereduction (step 1) is crucial both to enable sulfonation reaction (step 2) and to achieve complete reduction (step 3). The EA indicated that after prereduction of GO, a S/C ratio of 1:35 could be achieved via diazotization. However, only a 1:148 ratio is possible without step 1. The as-obtained sulfonated graphene exhibited an electrical conductivity of 1250 S m^{-1}implying that an extended conjugated sp^2 network is restored in the sulfonated graphene.

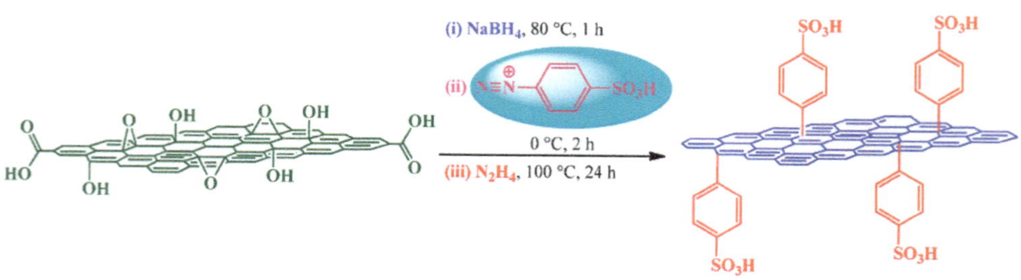

Scheme 4: Sulfonation of GO with 4-benzenediazonium sulfonate.

On another note, Ji *et al.* [82] introduced a two-step protocol avoiding the prereduction step. Specifically, GO was first reduced with $N_2H_4{\cdot}H_2O$ and dialyzed for 2 days. The diazotization of sulfanilic acid followed by sulfonation of purified rGO with 4-benzenediazonium sulfonate in the presence of hypophosphorous acid (H_3PO_2) afforded sulfonated graphene (S-graphene) having crumpling features. The solid state ^{13}C-NMR spectrum of S-graphene exhibited a small peak at 136 ppm, which was assigned to the carbons in the covalently attached phenyl-SO_3H groups. The Raman spectra of S-graphene indicated a slightly higher I_D/I_G ratio than that of rGO, which was likely a result of introduction of abundant -SO_3H groups to the sp^2 carbon network. It was further evidenced by EDS mapping where sulfur was found to be homogeneously distributed on the whole surface. The neutralization titrations of S-graphene indicated an ion exchange capacity (IEC) of 1.55 mmol H^+ g^{-1}, which was significantly higher than commercially available Nafion NR50 (0.80 mmol H^+ g^{-1}) solid acid catalyst.

Following Samulski's approach, Zhao *et al.* [97] obtained highly dispersed sheets of sulfonated graphene (SG) and applied them as adsorbents for persistent organic aromatic pollutants such as naphthalene and 1-naphthol from aqueous solutions.

A similar, three-step preparation method was also carried out by Lam *et al.* [87] during their investigations on xylose dehydration. The BET surface area of resulting NMs was measured to be 318 (GO), 680 (SGO), and 634 (SG) $m^2{\cdot}g^{-1}$; quite higher than that of Hu's work, even though the S/C ratio was nearly comparable (1:135 *vs.* 1:140) for SG. Of further interest is the clear discrepancy in the Raman data. The I_D/I_G ratio was increased from GO (0.89) to SG (1.00) in the previous work [97], whereas a significant decrease was observed in I_D/I_G ratio from GO (1.16) to SG (0.78) in the latter [87]. However, the product of step 2 (SGO; obtained after prereduction of GO followed by sulfonation) showed a slight increase in I_D/I_G ratio from GO (1.16) to SGO (1.27). Although no explanation has been given for this difference, we assume that the synthesis and purification steps of graphite oxide or GO, as mentioned previously, can significantly affect its further functionalization and experimental properties. Indeed, this assumption may be linked to the EA analysis in the same work. Surprisingly, all three GO, SGO, and SG NMs indicated the presence of 0.6% of sulfur, even though their IEC values were significantly different; 2.0, 1.7, and 0.5 mmol H^+ g^{-1} for GO, SGO, and SG, respectively. This data suggests that the presence of oxygen-functionalities, in particular, -OH groups are responsible for the acidity of G-NMs, which indeed is a misconception as mentioned above.

Zhang *et al.* [98] prepared SG sheets as described above [96] and used them as a sorbent in micro-solid-phase extraction (μ-SPE) coupled with gas chromatography-mass spectrometry for the determination of polycyclic aromatic hydrocarbons in water. Based on Samulski's pioneering work [96] on GO sulfonation, a number of other groups have also prepared SGs and used them for different applications [85,99,100].

Fuming Sulfuric Acid

Fuming sulfuric acid or oleum commonly refers to a solution of sulfur trioxide (SO_3) dissolved in sulfuric acid. It can be described by the general formula $H_2SO_4{\cdot}xSO_3$ where x is the free SO_3 molar content and is often known as 30% oleum. It is produced in the contact process in which S is oxidized to SO_3 and subsequently dissolved in concentrated H_2SO_4. Owing to its high enthalpy of hydration, it serves as a central intermediate in the production of H_2SO_4. In typical organic chemistry, oleum is used in the nitration of nitrobenzene and the manufacture of explosives. Due to its relatively low cost (~US$ 0.153 per pound of reactive SO_3), oleum sulfonation is one of the industrially preferred strategies, especially, for versatility in feedstock selection to build materials that are truly application specific. In general, oleum sulfonation is an equilibrium and hydrothermal process, which can be operated both as a batch and continuous process.

In graphene research, the oleum-assisted sulfonation was first achieved by Liu *et al.* [83] by following the route shown inScheme 5. The resulting material was termed sulfated graphene (G-SO_3H). However, we would like to rename it sulfonated graphene, which is more appropriate as there is C-S bonding rather than O-S. The TEM and AFM images indicated the sheet like structure of G-SO_3H with an average thickness of 0.8–3.0 nm. Compared with GO (0.89), G-SO_3H gave a relatively highI_D/I_G ratio (1.02), which can be attributed to a decrease in the average size of the sp^2 domain as described above. The high resolution S2p XPS spectrum of G-SO_3H displayed a peak at 168.3 eV (associated with S-O bonds); rather lower than those of SO_3H-functionalized ordered mesoporous carbon (OMC-SO_3H; 168.8 eV) and Amberlyst-15 (168.9 eV), which was likely a result of electron transfer from graphene to the -SO_3H groups [16]. In this milieu, it is important to stress that S2p binding energy is a direct measure and quite sensitive to the acidic strength [101].

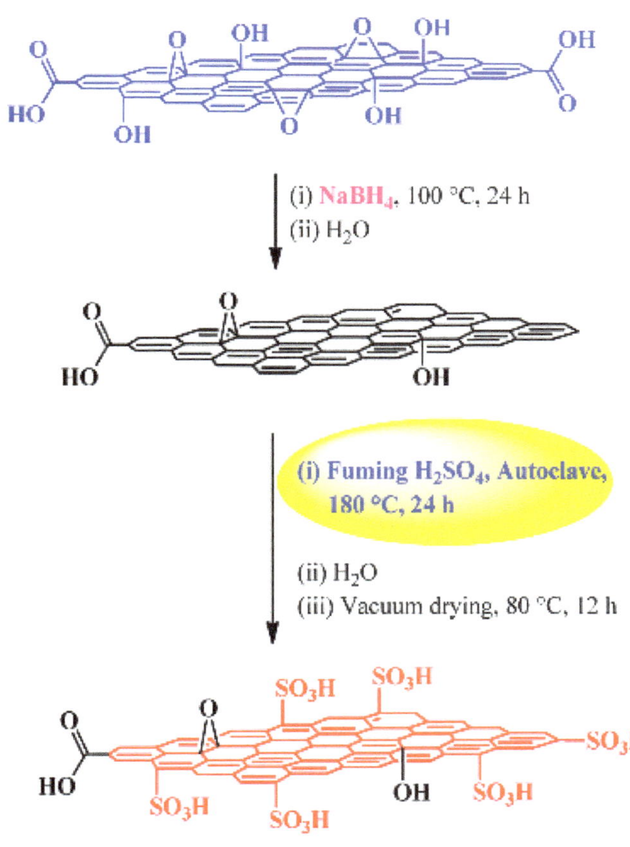

Scheme 5: Oleum-assisted sulfonation of GO into G-SO_3H.

The thermal stability of G-SO_3H was evaluated by TGA in association with OMC-SO_3H. Specifically, G-SO_3H revealed two main decomposition steps, centered at 268 and 568 °C, which were assigned to the degradation of -SO_3H groups and carbon framework, respectively. In contrast, weight loss temperatures in OMC-SO_3H were centered at 237 °C and 390 °C for -SO_3H groups and carbon framework, respectively. The better stability of carbon framework and -SO_3H groups in G-SO_3H over OMC-SO_3H was attributed to its non-amorphous characteristic leading to stronger interaction with -SO_3H groups. It would be worthwhile to remind readers again here that TGA results are in sharp contrast to Hirsch's experimental findings [57], where it is shown that -SO_3H groups dissociate at ~500 °C. Furthermore, the observed weight loss at 268 °C in G-SO_3H is quite close enough to that of organosulfates, which may be attributed to the presence of sulfur species in the parent GO. The acid content of G-SO_3H was evaluated by acid-base titration and was found closer (1.2 mmol H^+ g^{-1}) to OMC-SO_3H (1.3 mmol H^+ g^{-1}) but far below that of Amberlyst-15 (4.7 mmol H^+ g^{-1}).

While diazotization-based GO sulfonation involves tedious multi-step synthesis, oleum-assisted hydrothermal sulfonation does not, making it an attractive strategy for sulfonation of GO. Despite this, the use of autoclaves with the added disadvantage of relatively high temperature conditions (≥180 °C) make this process quite expensive as well as energy inefficient. With these realizations, we decided to reinvestigate this reaction under relatively mild conditions. Indeed, combined with a knowledge of acid functional ionic liquids [102,103,104] and their exceptional product-switching ability in tetrapyrrolic organic systems [102], we were able to synthesize sulfonated graphene with simple processing and handling, one of the important tenets of "green chemistry". In particular, the oleum-assisted hydrothermal sulfonation of rGO in a standard fume cupboard at 120 °C for 3 days afforded sulfonated graphene exhibiting an IEC of 1.55 mmol H^+ g^{-1}. The as-obtained sulfonated graphene was successfully used in the synthesis of ester plasticizers, and other industrially important chemicals. The details of these findings will be reported elsewhere.

In yet another work, Liu *et al.* [90] used oleum to develop single-layer sheets of sulfated GO (GO-OSO_3H) (though oleum-assisted sulfation is not commonly practiced) as a hole-extraction layer in high performance polymer solar cells. It was postulated that on treatment with oleum, epoxy and hydroxyl groups on the basal plane of GO are either removed or transformed into sulfate groups as shown in Scheme 6. This is quite reasonable because strong acids [50,105] such as oleum exhibit a high dehydration tendency and can lead to the partial reduction of GO. The oleum-induced basal plane reduction of GO was evidenced by comparing the XRD patterns for GO and GO-OSO_3H. In particular, a peak shift from 2θ = 11.50° (*d*= 7.65 Å) for GO to 2θ = 26.40° (*d* = 3.91 Å) was recognized for GO-OSO_3H, a characteristic feature of partially restacked graphene domain. These results were further verified by XPS, UV-visible absorption, and Raman spectroscopy, the latter of which showed a slightly higher I_D/I_G ratio for GO-OSO_3H (0.88) than that for GO (0.78).

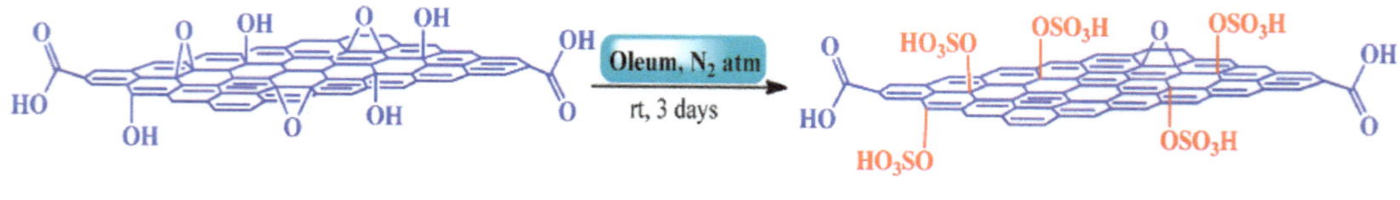

Scheme 6: Oleum-assisted sulfation of GO into GO-OSO_3H.

The acid density of GO-OSO_3H was evaluated to be 5.86 mmol H^+ g^{-1}, which was relatively higher than that of GO (3.15 mmol H^+ g^{-1}). Such a high acidity (even higher than that of Amberlyst™-15) of GO-OSO_3H was attributed to both the presence of -OSO_3H and -COOH groups and may be linked to high S content (4.49 wt %) as determined by EA. Furthermore, the acidity of GO was interpreted in terms of -COOH and -OH groups, however, no EA data was presented by the authors [90].

Chlorosulfonic Acid

Chlorosulfonic acid ($ClSO_3H$) or sulfuric chlorohydrin is another strong acid widely used to produce primarily alcohol sulfates, alcohol ether sulfates, and dye intermediates. $ClSO_3H$ is an expensive source of SO_3 (~US$ 0.255 per pound of reactive SO_3), highly toxic, corrosive, rapid, and stoichiometric reactant that can be used as a dehydrating, oxidizing, and chlorinating agent. During its reaction, it liberates hydrochloric acid as a by-product, which is normally neutralized with water or dilute basic solution. In graphene research, $ClSO_3H$ has been used as a solvent for high concentration dispersion and expansion/exfoliation of graphitic domains under ambient conditions [106,107,108].

Taking into account the excellent nucleophilicity of organolithium compounds toward epoxy and carbonyl functions, Wang *et al.* [84] introduced a novel synthesis methodology for sulfated graphene under relatively mild conditions. As such, the sulfation steps of epoxide, hydroxyl and carbonyl functions were suggested to follow the mechanism shown in Scheme 7. These include pretreatment of GO, organolithiation of rGO with *n*-butyl lithium and sulfation of as-obtained chemically converted graphene (CCG).

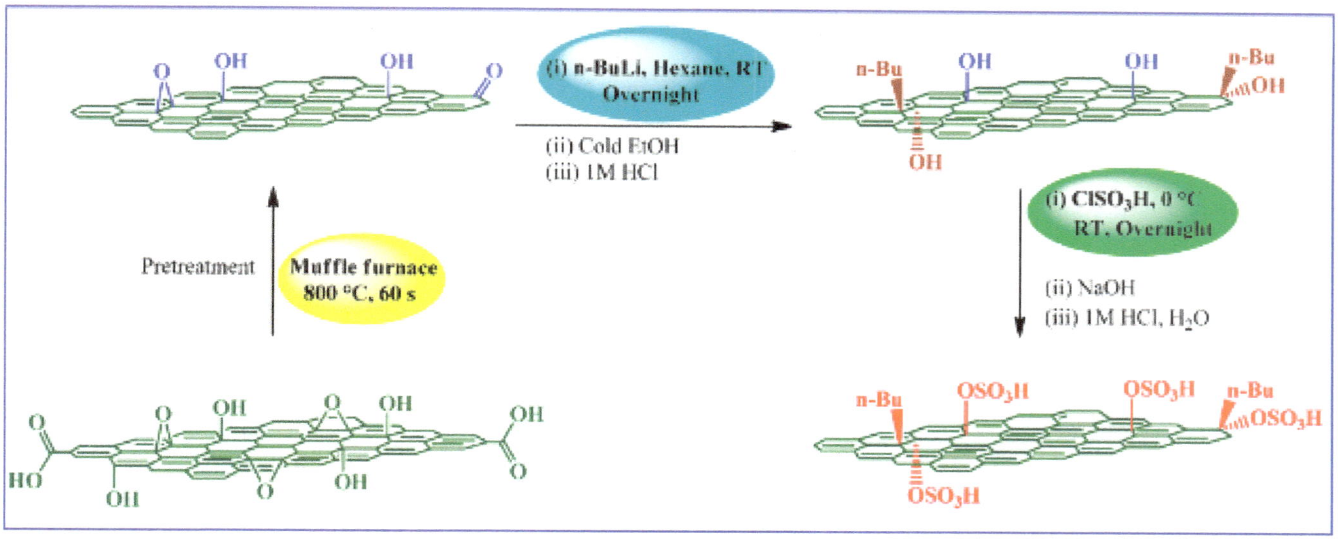

Scheme 7: $ClSO_3H$-induced sulfation of GO.

The formation of O-S bonds in the as-prepared sulfated graphene was evidenced by FTIR spectroscopy, where a peak at 3691 cm^{-1} corresponding to the O-H stretching vibrations in CCG disappeared and reappeared (at 3450 cm^{-1}) along with an additional peak at 1006 cm^{-1} (S=O stretching vibrations). The Raman spectra of rGO, CCG, and sulfated graphene indicated an I_D/I_G ratio of 0.92, 0.84, and 0.94, respectively. The thermal stability of all samples was assayed by TG analysis and determined to be: rGO > sulfated graphene > CCG. The quantitative XPS analysis of sulfated graphene showed 2.36% S by mass, however, IEC was not taken into account by the authors [84].

In another work, Upare *et al.* [86] employed $ClSO_3H$ to functionalize commercial samples (Sigma-Aldrich, Germany) of GO and activated carbon with $-SO_3H$ groups. It has been claimed that as-prepared GO-SO_3H catalyst can be used for the selective decomposition of hexose sugars into levulinic acid and other value-added derivatives. However, this report is somewhat difficult to harmonize in view of the reported experimental data. For instance, XRD pattern of GO are reported to be at $2\theta = 26°$; GO exhibits a sharp diffraction peak at $2\theta = 10°–13°$. We think that perhaps rGO is misinterpreted as GO in this report, as only the former gives diffraction patterns in the range from $2\theta = 21.5°$ to $27.5°$. Our assumption may be further linked to EA results demonstrating a very high C/O ratio of 14.12 for GO, which is quite unlikely. In fact, this value is very close to that observed (13.4) with $NaBH_4$-induced reduction of graphite oxide [76].

Combined with a knowledge of sulfonation or sulfation chemistry, we can make an educated guess that the reactivity of $ClSO_3H$ towards -OH functions of GO should even be higher than that of with oleum. Nevertheless, chlorosulfonation reactions cannot be ruled out completely.

Sulfuric Acid

Sulfuric acid (H_2SO_4), historically known as oil of vitriol, is a well-known strong mineral acid having significant oxidizing and dehydrating properties. In the chemical industry it has a central importance and is increasingly used for variety of applications such as in oil refining, fertilizer manufacturing, mineral and waste water processing, acidic drain cleaning, as well as in acid-catalyzed reactions as mentioned previously. In typical organic chemistry, it is used for sulfonation of benzene through an electrophilic aromatic substitution under reflux conditions. However, the sulfonation with H_2SO_4 can be considered as a special case of oleum sulfonaion. In present scenario, the H_2SO_4 sulfonation is principally used for production of hydrotropes by azeotropic reaction with toluene, xylene, or benzene.

Recently, Antunes *et al.* [89] have reported the use of H_2SO_4 to introduce sulfur-containing acid groups onto the graphene surface. Unlike previous studies, GO (in conjunction with CNTs and carbon black (CB)) was reduced by high temperature (190 °C) treatment with benzyl alcohol under microwave irradiation. The as-obtained rGO (coined as RGO in the original report) was subjected to H_2SO_4 (97 wt %) treatment at 160 °C for 5 h and this process was repeated twice more. The BET surface area of the as-obtained material (S-RGO) was determined to be 12 $m^2 \cdot g^{-1}$; far below than those of S-CNTs (44 $m^2 \cdot g^{-1}$) and S-CB (228 $m^2 \cdot g^{-1}$). As described above, the Raman results were consistent and an increase in I_D/I_G ratio was observed in all CNMs after H_2SO_4 treatment.

The acid strength of S-RGO in combination with rGO, Amberlyst-15, and S-CNTs was assayed by ^{31}P magic-angle spinning (MAS) NMR spectroscopy using triethylphosphine oxide (TEPO) as a basic probe. As shown in Figure 4, the appearance of signals at 55–60 ppm in both rGO and S-RGO were attributed to the presence of non-sulfur acid groups while the latter gave additional signals at higher chemical shifts (71–80 ppm), possibly due to the presence of sulfur species. As expected, the sulfonic acid-based cation exchange resin, Amberlyst™-15 (referred to as Amberlyst-15 in ref. [89]), displayed a single sharp resonance ~90 ppm demonstrating its stronger acidity than that of carbocatalysts.

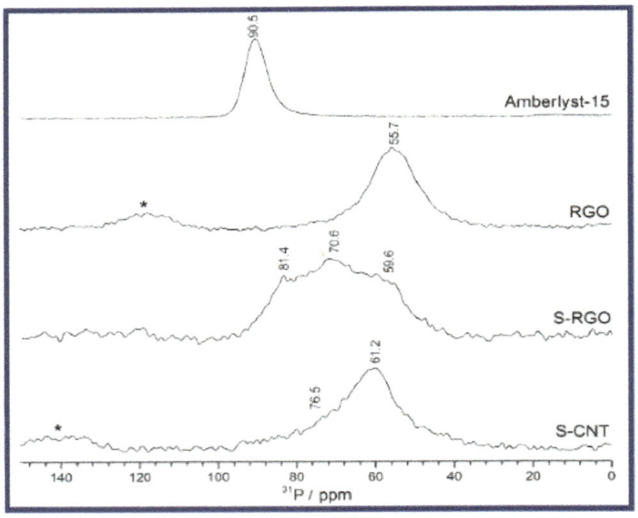

Figure 4: The ^{31}P MAS NMR spectra of TEPO adsorbed onto carbocatalysts and Amberlyst-15. Reproduced with permission from [89]. Copyright (2014) Wiley-VCH Verlag GmbH & Co.

The total acid content of S-RGO as measured by acid-base titrations was determined to be 2.2 H^+ mmol·g^{-1} and was attributed to the presence of -COOH and -OH groups. On the other hand, EA results indicated relatively low sulfur content (1.08 mmol·g^{-1}) which was credited to the presence of sulfonic acids. However, the TGA outcome is in contrast to this hypothesis. In particular, a second weight loss in TGA is observed in between 230–320 °C; far below the decomposition temperature (~500 °C) of sulfonic acid groups [57]. As such, one can assume that rather than sulfonic these are hydrosulfate groups that may be generated during oxidative treatment of rGO with H_2SO_4. In this regard, a comparison between S-RGO and GO, especially, ^{31}P MAS NMR experimental data is not only desirable but also would be praiseworthy.Aside from the above-described sulfonating agents, He *et al.* [88] have reported the use of ammonium sulfate ($(NH_4)_2SO_4$) in the simultaneous reduction and sulfonation of GO. It was demonstrated that when a mixture of GO and $(NH_4)_2SO_4$ (in mass ratio of 5:1) was dried by lyophilization and subsequently heated at 235 °C in an inert atmosphere, the sulfonic acid groups could be grafted onto GO, whereas GO could be simultaneously reduced. The as-obtained material (S-rGO) was characterized by SEM, EDX, IR and Raman spectroscopy and used as support for metal nanocatalysts. The acid-density of S-rGO in terms of equivalent weight was evaluated to be 727 ± 5 g·mol^{-1}. It is worthy of mention here that despite the efforts of many chemists since the 18th century [109,110,111,112], the exact mechanism of thermal decomposition of $(NH_4)_2SO_4$ remains elusive. Nevertheless, the use of $(NH_4)_2SO_4$ as a sulfonating agent, as far as we are aware, is not common in practice, presumably, due to its most feasible two-steps decomposition pathway [113] as shown in Equations (1) and (2):

$$2(NH_4)_2SO_4 \rightarrow (NH_4)_2S_2O_7 + 2NH_3 + H_2O \quad (1)$$

$$3(NH_4)_2S_2O_7 \rightarrow 2NH_3 + 6SO_2 + 2N_2 + 9H_2O \quad (2)$$

$$(NH_4)_2SO_4 \rightarrow 2NH_3 + SO_3 + H_2O \quad (3)$$

The ammonium pyrosulphate ($(NH_4)_2S_2O_7$), the primary condensation phase product (Equation (1)) is generally obtained at temperatures in between 100–250 °C while the reaction shown in Equation (2) occurs at relatively high temperature (>250 °C). Thus, the only possibility of generating SO_3 is the likely reaction of SO_2 with oxygen at relatively high temperature. Though a similar $(NH_4)_2SO_4$-induced sulfonation protocol was also reported previously [114], we believe that characterization data is insufficient, leading to confusion in understanding the nature of this reaction. Furthermore, the acidity is commonly presumed and interpreted from the perspective of sulfur content in -SO_3H groups. Now, if we assume that the decomposition pathway does exist in accordance with Equation (3), is this process gentle enough to leave other molecules, such as ammonia, untouched?

Indeed, the thermal decomposition of $(NH_4)_2SO_4$ within graphene domain appears to us quite interesting and it is quite reasonable to expect that some interesting materials can be formed simply by switching the experimental conditions.

APPLICATIONS OF G-NMS IN HETEROGENEOUS ACID CATALYSIS

Acid-catalyzed reactions are classified according to the type of acid catalysts. The catalytic performances of G-NMs in conjunction with other solid/liquid catalysts have been discussed in terms of catalyst loadings, products yield, recoverability, and recyclability at certain instances.

Graphite Oxide as Acid Catalyst

Michael-Type Friedel-Crafts Addition

Intrigued by the strong acidity of graphite oxide, Kumar and Rao [115] used it as catalyst in the Friedel-Crafts addition of indoles to α,β-unsaturated ketones as well as nitrostyrenes (Scheme 8). The superiority of graphite oxide over other acid catalysts including HCl, H_2SO_4, acidic Al_2O_3, nano MgO/CuO/Fe_2O_3, (sulfo)β-cyclodextrins, activated charcoal, and *p*-TSA was demonstrated by addition of indole to methyl vinyl ketone in THF-H_2O mixtures at ambient conditions. Whereas graphite oxide (20 wt %) afforded excellent yield (92%), the other catalysts only gave poorer to moderate yields (5%–67%) of addition product. For the same reaction, graphite oxide could be recycled up to five times without apparent loss in catalytic activity. Indoles with electron-donating and electron-withdrawing substituents underwent reactions smoothly in the addition to methyl vinyl ketone with no polymerized or dimerized side products. However, no reaction was observed with carboxylic acid/ester substituents, even after 24 h. Furthermore, the reaction of indole with electron deficient nitrostyrene took quite a longer time to afford products in moderate yields.

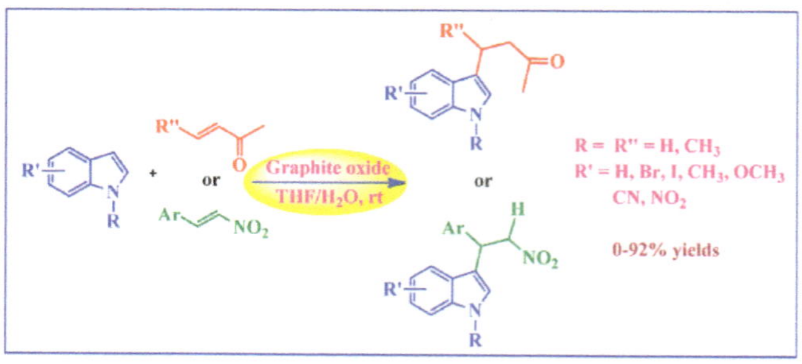

Scheme 8: Graphite oxide catalyzed Michael-type Friedel-Crafts addition of indoles.

Polymerization

Inspired by their initial discovery of GO's acidity [30], Dreyer *et al.* [116] used graphite oxide (2.5–20 wt % loadings) to catalyze the ring opening polymerization of various cyclic lactones and lactams to their corresponding carbon-reinforced, high molecular weight polyesters, or polyamide composites in good to excellent yields (39%–100%). In contrast to most other methods, which often rely on clumsy two-step protocols preparing the polymer followed by mixing of the carbon filler, the use of a graphite oxide combines the polymerization and dispersion of the filler into a single expeditious step.

Synthesis of Dipyrromethanes

Chauhan *et al.* [117] examined the reactions of pyrrole and ketones using graphite oxide as acid catalyst under a variety of conditions. Specifically, moderate to high yields of 5,5-dialkyldipyrromethanes were achieved along with linear oligomers and cyclic tetramers under ambient conditions.

Graphene Oxide (GO) as Acid Catalyst

Hydration of Alkynes

The first use of GO [30] as an acid catalyst was demonstrated in the hydration of various alkynes to their corresponding ketones (Scheme 9a). In particular, in the presence of GO, low to excellent conversions of alkynes (27%–98%) were achieved at 100 °C as monitored by ^{1}H-NMR spectroscopy. However, the need to use significantly high GO loadings (200 wt %) in these reactions somewhat limits the practical utility of this protocol.

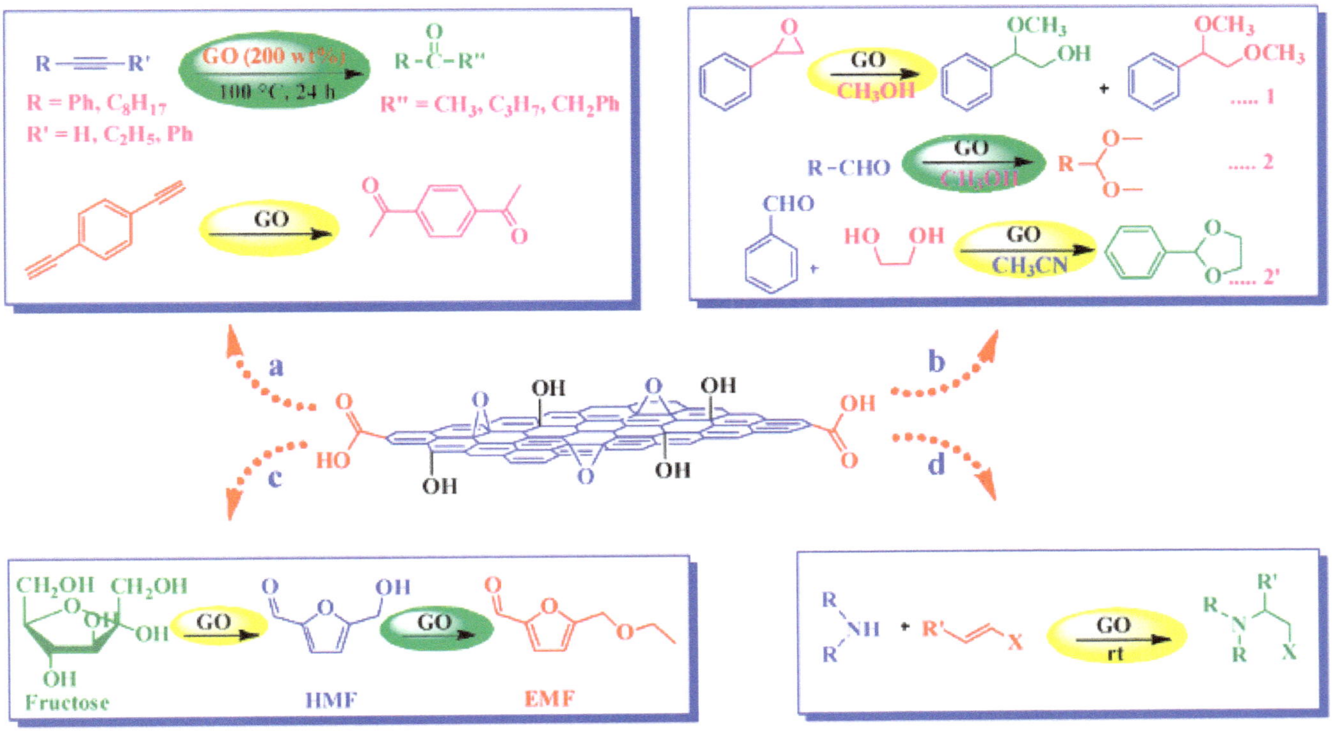

Scheme 9: GO catalyzed (**a**) Hydration of alkynes. (**b**) Ring opening of epoxides and acetalization of aldehydes. (**c**) Conversion of carbohydrates into 5-ethoxymethylfurfural (EMF). (**d**) Aza-Michael addition.

Aza-Michael Addition

The Aza-Michael addition of amines to activated alkenes in the presence of GO catalyst furnishing β-amino compounds was reported by Verma *et al.* [118] (Scheme 9d). The reactions were found to proceed under relatively mild conditions and to afford excellent product yields (70%–97%) within short reaction times. The GO was recyclable up to nine runs with the consistent catalytic activity.

Ring Opening of Epoxides and Acetalization of Aldehydes

Dhakshinamoorthy *et al.* [53] employed GO as an acid catalyst in the ring opening of epoxides with methanol and other primary alcohols as nucleophile and solvents. Under ambient conditions, the product selectivity could be switched by changing the reaction time. However, most of the reactions exhibited high regioselectivity rendering the product expected from S_N1 mechanism. Interestingly, the conversion and selectivity in the presence of GO were found comparable to those of H_2SO_4 and *p*-TSA and much higher than Norit A, activated carbon, as demonstrated by styrene oxide with methanol as a model reaction (Scheme 9b-1). The high activity of GO was attributed to the presence of hydrogen sulfate groups on GO surface. The catalyst could be easily recovered after the reaction and successfully reused for up to three cycles. In their subsequent work [119], the group further examined the activity of GO for the acetalization of aldehydes in methanol (Scheme 9b-2). The GO-induced reaction of a variety of aromatic and aliphatic aldehydes with methanol afforded the corresponding acetalized products in moderate to excellent yields as determined by GC. In the case of 4-methylbenzaldehyde, the low yield (10%) was attributed to the poisoning of GO due to the formation of toluic acid, while 4-pyridylcarboxaldehyde and acetophenone were completely unreactive. The protocol was further extended successfully to the acetalization of benzaldehyde with ethylene glycol (Scheme 9b-2').

Conversion of Carbohydrates into 5-Ethoxymethylfurfural (EMF)

Wang *et al.* [54] discovered that GO can be used as a facile acid catalyst for the one-step conversion of fructose-based carbohydrates into high-heating value liquid biofuels (Scheme 9c). In this regard, one-pot conversions of sucrose, inulin, and fructose into EMF using DMSO-ethanol mixtures (1:2.3, v/v) were achieved with yields of 34%, 66%, and 71%, respectively. Interestingly, GO-catalyzed direct etherification of 5-hydroxymethyl furfural (HMF), a dehydration product of carbohydrates, in ethanol afforded high EMF yield of 92% as determined by LC/MS. On the contrary, traditional liquid and solid acid catalysts such as Amberlyst-15, *p*-TSA, and H_2SO_4 gave only moderate yields of EMF (54%–61%) even though with 100% conversions as determined by NMR. Though a high yield of EMF (82%) was detected with $H_3PW_{12}O_{40}$, it was dissolved in the reaction mixture and couldn't be separated. The high activity of GO was attributed to the presence of hydrogen sulfate groups as mentioned above and found consistent even after several runs.

Sulfonated/Sulfated Graphene and Graphene Oxide (GO) as Acid Catalysts

Sulfonated Graphene and GO as Acid Catalysts Prepared by Diazonium Salt

Hydrolysis of Ethyl Acetate

Ji *et al.* [82] used sulfonated graphene in the water participating reaction, hydrolysis of ethyl acetate. To evaluate its activity and stability, H_2SO_4 and Nafion NR50 catalysts were employed as controls. Whereas Nafion NR50 showed poorer activity, the activity of sulfonated graphene was comparable to that of H_2SO_4 and remain unchanged even after five consecutive runs.

Synthesis of Xanthenes and Benzoxanthenes

Shaabani *et al.* [85] used sulfonated graphene in conjunction with GO for the one-pot condensation of aldehydes, 1,3-diketones, and naphthols into the corresponding xanthenes and benzoxanthenes (Scheme 10).

Unlike $InCl_3$, P_2O_5, HY zeolite, and $Sr(OTF)_2$, the G-NMs exhibited high catalytic activity in water and could be reused several times; the best results were obtained with sulfonated graphene.

Dehydration of Xylose

Lam *et al.* [87] applied sulfonated GO (SGO) and graphene in conjunction with unfunctionalized GO and graphene for the dehydration of xylose to furfural (Scheme 11). Out of the four tested catalysts (2 wt% loadings, 35 min), SGO afforded highest yield of furfural (62%) with 75% selectivity as averaged over three runs. On the other hand, yields determined in the absence of catalyst as well as in the presence of graphene, GO, and sulfonated graphene were in the following order: sulfonated graphene (55%) > GO (53%) > graphene (51%) > no catalyst (44%). Interestingly, at only 0.5 wt % loading *vs.*xylose, SGO could be reused over multiple runs maintaining an average yield of 61%, despite the presence of accumulated byproducts on its surface. It has been suggested that strongly acidic aryl sulfonic acid groups in SGO are the key active sites for the high temperature production of furfural in water.

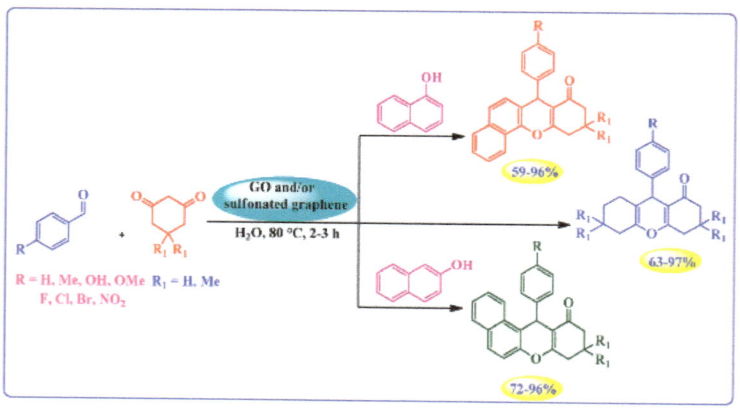

Scheme 10. GO and/or sulfonated graphene catalyzed synthesis of xanthenes and benzoxanthenes in water.

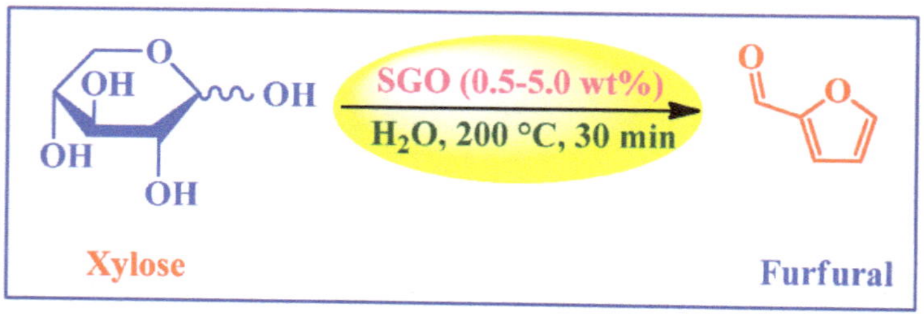

Scheme 11: Sulfonated GO catalyzed dehydration of xylose into furfural.

Sulfonated Graphene as Acid Catalyst Prepared by Fuming Sulfuric Acid or Oleum

Esterification, Peckmann, and Hydration Reactions

Liu *et al.* [83] employed sulfonated graphene (G-SO_3H) as solid catalyst in a number of acid-catalyzed reactions including esterification, hydration, and Peckmann reactions (Scheme 12). The esterification reactions of acetic acid with cyclohexanol and *n*-butanol in the presence of G-SO_3H (after pretreatment with 0.1 M H_2SO_4) afforded the corresponding esters with the highest conversions of ~79% and 89%, respectively (Scheme 12a,a'). Furthermore, G-SO_3H exhibited almost similar activity when reactions were performed under static conditions. On the contrary, the conversions were significantly low (<53% and <76% for cyclohexanol and *n*-butanol, respectively) in the presence of other solid catalysts such as OMC-SO_3H, rGO and SBA-15-SO_3H. Nevertheless, Amberlyst-15 gave second highest conversions for the same. Notably, in contrast to available reports, GO showed poorer catalytic activity with as low as <22% conversions. Surprisingly, no acid sites were detected in GO as measured by acid-base titrations and S analysis.

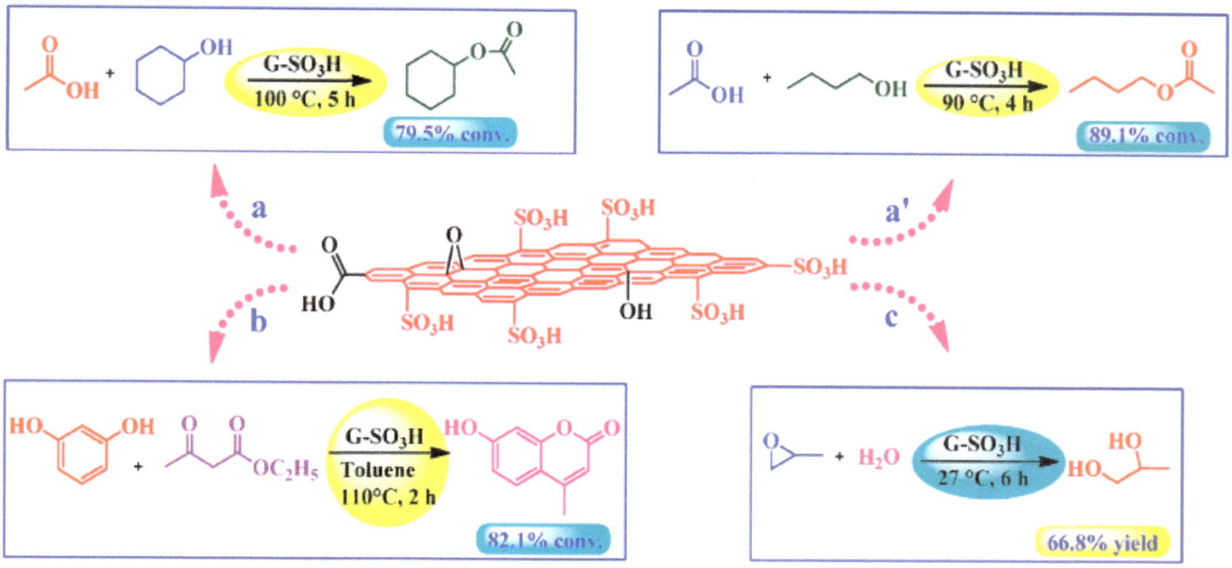

Scheme 12: G-SO_3H catalyzed (**a**) Esterification of acetic acid with cyclohexanol. (**a'**) Esterification of acetic acid with *n*-butanol. (**b**) Peckmann reaction of resorcinol with ethyl acetoacetate. (**c**) Hydration of propylene oxide.

In yet another example, G-SO_3H catalyzed Peckmann reaction of resorcinol with ethyl acetoacetate in toluene afforded corresponding coumarin with ~82% conversion (Scheme 12b). On the other hand, conversions were decreased significantly in the presence of other catalysts and determined in the following order: Amberlyst-15 (~75%) > OMC-SO_3H > SBA-15-SO_3H >> GO > rGO (~11%). A similar trend was recognized in the hydration of propylene oxide (Scheme 12c), however, GO and rGO were found completely inactive.

The recyclability of G-SO_3H was examined in association with OMC-SO_3H. After being recycled five times, G-SO_3H gave relatively higher conversions than OMC-SO_3H. The activity loss in G-SO_3H was 1.5%–2.8% whereas OMC-SO_3H had a significant reduction in activities (~7%–10%) after five cycles. The high stability and activity of G-SO_3H were attributed to the presence of stable sulfonic acid groups on its surface and unlimited mass transfer owing to its unique sheet structure.

Sulfated Graphene as Acid Catalyst Prepared by $ClSO_3H$

Ester-Exchange Reactions

Wang *et al.* [84] evaluated the catalytic efficiency of sulfated graphene for the ester-exchange reaction between ethyl acetate and a variety of alcohols. Specifically, sulfated graphene catalyzed reaction of alcohols gave corresponding esters in ≥96% isolated yield; somewhat superior to that of Dowex 50 W × 2 (50–100 mesh; 95% yield). It was postulated that nucleophilicity of alcohols play a significant role in ester-exchange reactions. The catalyst could be recycled up to three times.

Chemical Conversion of Carbohydrates to Levulinic Acid (LA)

Upare *et al.* [86] used -SO_3H functional GO catalyst in conjunction with activated carbon for the selective decomposition of glucose and fructose into levulinic acid (LA) (Scheme 13). The graphene-based catalyst gave high yield of LA (78%) and showed good reusability with reliable performance. It was postulated that chemical transformation patterns for hexose sugar decompositions are affected by the type of catalyst, the density of acid sites as well as temperature. In addition, the presence of other functional groups including -COOH and -OH groups on catalyst surface help in enhancing the adsorption of sugar for the reaction.

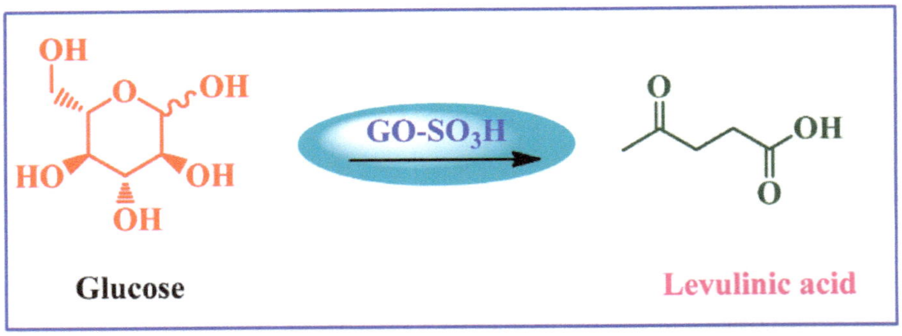

Scheme 13: Chemical conversion of glucose to levulinic acid (LA).

Sulfated rGO as Acid Catalyst Prepared by H_2SO_4

Conversion of HMF into Biofuels

Antunes *et al.* [89] used H_2SO_4-treated rGO (S-RGO) as potential catalyst in the conversion of 5-hydroxymethylfurfural (HMF) to biofuels (Scheme 14) such as 5-ethoxymethylfurfural (EMF), 5-(ethoxymethyl)furfural diethyl acetal (EMF da), and ethyl levulinate (EL). The catalytic performance of S-RGO was found superior to that of H_2SO_4 treated carbon black, CNTs as well as Amberlyst-15 and was attributed to the different acid sites and 2D structure of the catalyst. The catalyst could be reused up to three cycles without significant loss in catalytic activity.

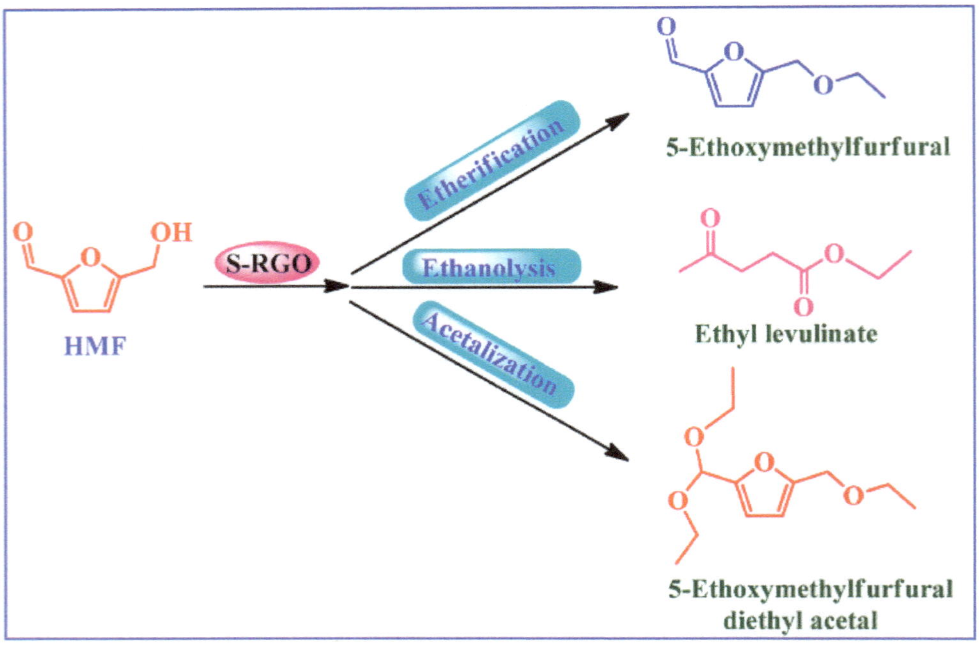

Scheme 14: Conversion of 5-hydroxymethylfurfural (HMF) into value-added biofuels.

CONCLUSIONS: FINAL REMARKS

As the building block of CNMs of all dimensions, graphene possesses many unique properties and physicochemistry. In this comprehensive review, we have shed light on graphite oxide, GO, and their sulfonated or sulfated derivatives acting as main performers in heterogeneous acid catalysis. The observation that graphite oxide and GO, prepared by a modified Hummer's method, convey inherent acidic character has motivated scientists to get new insights into the structural properties of graphite oxidation products. In this regard, new theories and approaches are constantly being explored and somewhat mystifying the commonly accepted LK model. Another important property of GO is its easy functionalization rendering it useful in preparing a library of catalytically active sulfonated or sulfated NMs. Indeed, the heterogeneous acid catalysis using these NMs has the potential to undergo a renaissance and to address many industrial processes not previously accessible to it.

Despite this, a basic understanding of working actions of sulfonating agents on (r)GO is still in its infancy, which have resulted in conceptual vagueness in many aspects. Furthermore, due to the lack of comparative yet solid experimental data of starting (GO) and final products (sulfonated or sulfated G-NMs), a more precise acid activity has not been ascertained. These shortcoming are of utmost importance since two key concerns may plague future scientists interested in using sulfonation methods, mainly whether a sulfonating agent of interest would: (i) specifically target aromatic carbon (sulfonation) or -OH functions (sulfation) in a certain mode of reaction and (ii) undergo a random mode of reaction. In this context, a fundamental understanding of sulfonation chemistry combined with routine industrial experience can only act as a guide for advancements in this area.

As a final note, while increasingly tight legislation on the control of chemical effluents currently at play, heterogeneous acid catalysis with G-NMs holds great promise to achieve the objectives of industrial catalysis following the principles of sustainable and green chemistry. We postulate that by the mutual cooperation between chemists, physicists and material scientists, graphene-based heterogeneous acid catalysis may lead to new disciplines in the 21st century. It is just a matter of patience and constant efforts to discover new catalysts or novel methodologies; after all, they are also patiently waiting to be discovered.

ACKNOWLEDGMENTS

This work was supported by National Tsing Hua University (102N1807E1) and the Ministry of Science and Technology (NSC101-2113-M-007-006-MY3) of Taiwan.

AUTHOR CONTRIBUTIONS

B. Garg and Y.-C. Ling were responsible for the subject of the review. T. Bisht performed bibliographic researches, participated in the discussions and also played an important role in preparing graphics or art work. The manuscript was designed, organized and written by B. Garg while the final editing was done by Y.-C. Ling. All authors read and approved the final manuscript. B. Garg was also responsible for the revisions and the proofs.

REFERENCES

1. Mizuno, N.; Misono, M. Heterogeneous catalysis. *Chem. Rev.* **1998**, *98*, 199–217.
2. Pal, R.; Sarkar, T.; Khasnobis, S. Amberlyst-15 in organic synthesis. *ARKIVOC* **2012**, 570–609.
3. Okuhara, T. Water-tolerant solid acid catalysts. *Chem. Rev.* **2002**, *102*, 3641–3666.
4. Harmer, M.A.; Sun, Q. Solid acid catalysis using ion-exchange resins. *Appl. Catal. A: Gen.* **2001**, *221*, 45–62.
5. Harmer, M.A.; Farneth, W.E.; Sun, Q. High surface area nafion resin/silica nanocomposites: A new class of solid acid catalyst. *J. Am. Chem. Soc.* **1996**, *118*, 7708–7715.
6. Shylesh, S.; Thiel, W.R. Bifunctional acid–base cooperativity in heterogeneous catalytic reactions: Advances in silica supported organic functional groups. *ChemCatChem* **2011**, *3*, 278–287.
7. Harmer, M.A. Industrial processes using solid acid catalysts. In *Handbook of Green Chemistry and Technology*; Clark, J.H., Macquarris, D.J., Eds.; Blackwell Publishers: London, UK, 2002; pp. 86–117.
8. Chauhan, S.M.S.; Garg, B.; Bisht, T. Syntheses of calix[4] pyrroles by amberlyst-15 catalyzed cyclocondensations of pyrrole with selected ketones. *Molecules* **2007**, *12*, 2458–2466.
9. Suwannakarn, K.; Lotero, E.; Goodwin, J.G., Jr.; Lu, C. Stability of sulfated zirconia and the nature of the catalytically active species in the transesterification of triglycerides. *J. Catal.* **2008**, *255*, 279–286.
10. Onda, A.; Ochi, T.; Yanagisawa, K. Selective hydrolysis of cellulose into glucose over solid acid catalysts. *Green Chem.***2008**, *10*, 1033–1037.
11. Suwannakarn, K.; Lotero, E.; Goodwin, J.G., Jr. A comparative study of gas phase esterification on solid acid catalysts.*Catal. Lett.* **2007**, *114*, 3–4.
12. Wang, X.; Liu, R.; Waje, M.M.; Chen, Z.; Yan, Y.; Bozhilov, K.N.; Feng, P. Sulfonated ordered mesoporous carbon as a stable and highly active protonic acid catalysts. *Chem. Mater.* **2007**, *19*, 2395–2397.
13. Nakajima, K.; Hara, M. Amorphous carbon with SO_3H groups as a solid Brønsted acid catalyst. *ACS Catal.* **2012**, *2*, 1296–1304.
14. Peng, F.; Zhang, L.; Wang, H.; Lv, P.; Yu, H. Sulfonated carbon nanotubes as a strong protonic acid catalyst. *Carbon***2005**, *43*, 2405–2408.
15. Yu, H.; Jin, Y.; Li, Z.; Peng, F.; Wang, H. Synthesis and characterization of sulfonated single-walled carbon nanotubes and their performance as solid acid catalyst. *J. Solid State Chem.* **2008**, *181*, 432–438.
16. Novoselov, K.S.; Geim, A.K.; Morozov, S.V.; Jiang, D.; Zhang, Y.; Dubonos, S.V.; Grigorieva, I.V.; Firsov, A.A. Electric field effect in atomically thin carbon films. *Science* **2004**, *306*, 666–669.
17. Guest Editorial. Nomenclature of sp^2 carbon nanoforms. *Carbon* **2012**, *50*, 741–747.
18. Wan, X.; Huang, Y.; Chen, Y. Focusing on energy and optoelectronic applications: A journey for graphene and graphene oxide at large scale. *Acc. Chem. Res.* **2012**, *45*, 598–607.
19. Compton, O.C.; Nguyen, S.T. Graphene oxide, highly reduced graphene oxide, and graphene: Versatile building blocks for carbon-based materials. *Small* **2010**, *6*, 711–723.
20. Dai, L.; Chang, D.W.; Baek, J.-B.; Lu, W. Carbon nanomaterials for advanced energy conversion and storage. *Small***2012**, *8*, 1130–1166.
21. Huang, X.; Qi, X.; Boey, F.; Zhang, H. Graphene-based composites. *Chem. Soc. Rev.* **2012**, *41*, 666–686.
22. Chen, Y.; Zhang, B.; Liu, G.; Zhuang, X.; Kang, E.-T. Graphene and its derivatives: Switching ON and OFF. *Chem. Soc. Rev.* **2012**, *41*, 4688–4707.
23. Wu, M.C.; Deokar, A.R.; Liao, J.H.; Shih, P.Y.; Ling, Y.-C. Graphene-based photothermal agent for rapid and effective killing of bacteria. *ACS Nano* **2013**, *7*, 1281–1290.
24. Guo, C.X.; Ng, S.R.; Khoo, S.Y.; Zheng, X.; Chen, P.; Li, C.M. RGD-peptide functionalized graphene biomimetic live-cell sensor for real-time detection of nitric oxide molecules. *ACS Nano* **2012**, *6*, 6944–6951.
25. Li, Y.; Wen, T.; Xue, C.; Han, Q.; Wang, Y.; Hong, J.; Zhou, X.; Jiang, H. RGO LBL modified biomimetic electrochemical sensor for detection of Sildenafil in herbal sexual health products. *Biosens. Bioelectron.* **2013**, *42*, 287–292.
26. Ganesh, G.; Chang, C.C.; Ling, Y.-C. Facile synthesis of

smart magnetic graphene for safe drinking water: Heavy metal removal and disinfection control. *ACS Sustain. Chem. Eng.* **2013**, *1*, 462–471.

27. Liu, J.Y.; Chang, H.Y.; Truong, Q.D.; Ling, Y.-C. Synthesis of nitrogen-doped graphene by pyrolysis of ionic-liquid-functionalized graphene. *J. Mater. Chem. C* **2013**, *1*, 1713–1716.
28. Ganesh, G.; Ling, Y.-C. Magnetic and fluorescent graphene for dual modal imaging and single light induced photothermal and photodynamic therapy of cancer cells. *Biomaterials* **2014**, *35*, 4499–4507.
29. Garg, B.; Ling, Y.-C. Versatilities of graphene-based catalysts in organic transformations. *Green Mater.* **2013**, *1*, 47–61.
30. Dreyer, D.R.; Jia, H.-P.; Bielawski, C.W. Graphene oxide: A convenient carbocatalyst for facilitating oxidation and hydration reactions. *Angew. Chem. Int. Ed.* **2010**, *49*, 6813–6816.
31. Su, C.; Loh, K.P. Carbocatalysts: Graphene oxide and its derivatives. *Acc. Chem. Res.* **2013**, *46*, 2275–2285.
32. Machado, B.F.; Serp, P. Graphene-based materials for catalysis. *Catal. Sci. Technol.* **2012**, *2*, 54–75.
33. Dreyer, D.R.; Todd, A.D.; Bielawski, C.W. Harnessing the chemistry of graphene oxide. *Chem. Soc. Rev.* **2014**, *43*, 5288–5301.
34. Navalon, S.; Dhakshinamoorthy, A.; Alvaro, M.; Garcia, H. Carbocatalysis by graphene-based materials. *Chem. Rev.* **2014**, *114*, 6179–6212.
35. Brodie, B.C. On the atomic weight of graphite. *Philos. Trans. R. Soc. Lond.* **1859**, *149*, 249–259.
36. Staudenmaier, L. Verfahren zür darstellung der graphitsäure. *Ber. Dtsch. Chem. Ges.* **1898**, *31*, 1481–1487.
37. Staudenmaier, L. Verfahren zür darstellung der graphitsäure. *Ber. Dtsch. Chem. Ges.* **1899**, *32*, 1394–1399.
38. Hummers, W.S., Jr.; Offeman, R.E. Preparation of graphitic oxide. *J. Am. Chem. Soc.* **1958**, *80*, 1339–1339.
39. Marcano, D.C.; Kosynkin, D.V.; Berlin, J.M.; Sinitskii, A.; Sun, Z.; Slesarev, A.; Alemany, L.B.; Lu, W.; Tour, J.M. Improved Synthesis of Graphene Oxide. *ACS Nano* **2010**, *4*, 4806–4814.
40. Hofmann, U.; Holst, R. Über die säurenatur und die methylierung von graphitoxyd. *Ber. Dtsch. Chem. Ges. A/B* **1939**, *72*, 754–771.
41. Ruess, G. Uber das graphitoxyhydroxyd (graphitoxyd). *Monatshefte Chem.* **1947**, *76*, 381–417.
42. Clauss, A.; Plass, R.; Boehm, H.-P.; Hofmann, U. Untersuchungen zür Struktur des Graphitoxyds. *Z. Anorg. Allg. Chem.* **1957**, *291*, 205–220.
43. Scholz, W.; Boehm, H.P. Untersuchungen am Graphitoxid. VI. Betrachtungen zür Struktur des Graphitoxids. *Z. Anorg. Allg. Chem.* **1969**, *369*, 327–340.
44. Nakajima, T.; Mabuchi, A.; Hagiwara, R. A new structure model of graphite oxide. *Carbon* **1988**, *26*, 357–361.
45. Nakajima, T.; Matsuo, Y. Formation process and structure of graphite oxide. *Carbon* **1994**, *32*, 469–475.
46. He, H.; Riedl, T.; Lerf, A.; Klinowski, J. Solid-state NMR studies of the structure of graphite oxide. *J. Phys. Chem.* **1996**, *100*, 19954–19958.
47. Lerf, A.; He, H.; Riedl, T.; Forster, M.; Klinowski, J. ^{13}C and ^{1}H MAS NMR studies of graphite oxide and its chemically modified derivatives. *Solid State Ion.* **1997**, *101–103 Part 2*, 857–862.
48. Lerf, A.; He, H.; Forster, M.; Klinowski, J. Structure of graphite oxide revisited. *J. Phys. Chem. B* **1998**, *102*, 4477–4482.
49. Szabó, T.; Berkesi, O.; Forgó, P.; Josepovits, K.; Sanakis, Y.; Petridis, D.; Dékány, I. Evolution of surface functional groups in a series of progressively oxidized graphite oxides. *Chem. Mater.* **2006**, *18*, 2740–2749.
50. Gao, W.; Alemany, L.B.; Ci, L.J.; Ajayan, P.M. New insights into the structure and reduction of graphite oxide. *Nat. Chem.* **2009**, *1*, 403–408.
51. Erickson, K.; Erni, R.; Lee, Z.; Alem, N.; Gannett, W.; Zettl, A. Determination of the local chemical structure of graphene oxide and reduced graphene oxide. *Adv. Mater.* **2010**, *22*, 4467–4472.
52. Szabó, T.; Tombácz, E.; Illés, E.; Dékány, I. Enhanced acidity and pH-dependent surface charge characterization of successively oxidized graphite oxides. *Carbon* **2006**, *44*, 537–538.
53. Dhakshinamoorthy, A.; Alvaro, M.; Concepción, P.; Fornés, V.; Garcia, H. Graphene oxide as an acid catalyst for the room temperature ring opening of epoxides. *Chem. Commun.* **2012**, *48*, 5443–5445.
54. Wang, H.; Deng, T.; Wang, Y.; Cui, X.; Qi, Y.; Mu, X.; Hou, X.; Zhu, Y. Graphene oxide as a facile acid catalyst for the one-pot conversion of carbohydrates into 5-ethoxymethylfurfural. *Green Chem.* **2013**, *15*, 2379–2383.
55. Dimiev, A.; Kosynkin, D.V.; Alemany, L.B.; Chaguine, P.; Tour, J.M. Pristine graphite oxide. *J. Am. Chem. Soc.* **2012**, *134*, 2815–2822.
56. Dimiev, A.; Alemany, L.B.; Tour, J.M. Graphene oxide. Origin of acidity, its instability in water, and a new dynamic structural model. *ACS Nano* **2013**, *7*, 576–588.
57. Eigler, S.; Dotzer, C.; Hof, F.; Bauer, W.; Hirsch, A. Sulfur species in graphene oxide. *Chem. Eur. J.* **2013**, *19*, 9490–9496.
58. Toda, M.; Takagaki, A.; Okamura, M.; Kondo, J.N.; Hayashi, S.; Domen, K.; Hara, M. Green chemistry: Biodiesel made with sugar catalyst. *Nature* **2005**, *438*, 178–178.
59. Liu, X.-Y.; Hunag, M.; Ma, H.-L.; Zhang, Z.-Q.; Gao, J.-M.; Zhu, Y.-L.; Han, X.-J.; Guo, X.-Y. Preparation of a carbon-based solid acid catalyst by sulfonating activated carbon in a chemical reduction process. *Molecules* **2010**, *15*, 7188–7196.
60. McAllister, M.J.; Li, J.-L.; Adamson, D.H.; Schniepp, H.C.; Abdala, A.A.; Liu, J.; Herrera-Alonso, M.; Milius, D.L.; Car, R.; Prud'homme, R.K.; *et al.* Single sheet functionalized graphene by oxidation and thermal expansion of graphite. *Chem. Mater.* **2007**, *19*, 4396–4404.

61. Schniepp, H.C.; Li, J.-L.; McAllister, M.J.; Sai, H.; Herrera-Alonso, M.; Adamson, D.H.; Prud'homme, R.K.; Car, R.; Saville, D.A.; Aksay, I.A. Functionalized single graphene sheets derived from splitting graphite oxide. *J. Phys. Chem. B*2006, *110*, 8535–8539.

62. Kaniyoor, A.; Baby, T.T.; Ramaprabhu, S. Graphene synthesis via hydrogen induced low temperature exfoliation of graphite oxide. *J. Mater. Chem.* **2010**, *20*, 8467–8469.

63. Zhou, M.; Wang, Y.; Zhai, Y.; Zhai, J.; Ren, W.; Wang, F.; Dong, S. Controlled synthesis of large-area and patterned electrochemically reduced graphene oxide films. *Chem. Eur. J.* **2009**, *15*, 6116–6120.

64. Dreyer, D.R.; Park, S.; Bielawski, C.W.; Ruoff, R.S. The chemistry of graphene oxide. *Chem. Soc. Rev.* **2010**, *39*, 228–240.

65. Amarnath, C.A.; Hong, C.E.; Kim, N.H.; Ku, B.-C.; Kuila, T.; Lee, J.H. A; Hong, C.E.; Kim, N.H; Ku, B.-C.; Kuila, T.; Lee, J.H. Efficient synthesis of graphene sheets using pyrrole as a reducing agent. *H. Efficient synthesis of graphene sheets using pyrrole as a reducing agent. Carbon* **2011**, *49*, 3497–3502.

66. Liu, J.; Tang, J.; Gooding, J.J. Strategies for chemical modification of graphene and applications of chemically modified graphene. *J. Mater. Chem.* **2012**, *22*, 12435–12452.

67. Pei, S.; Cheng, H.-M. The reduction of graphene oxide. *Carbon* **2012**, *50*, 3210–3228.

68. Liu, Y.; Zhang, Y.; Ma, G.; Wang, Z.; Liu, K.; Liu, H. Ethylene glycol reduced graphene oxide/ polypyrrole composite for supercapacitor. *Electrochim. Acta* **2013**, *88*, 519–525.

69. Chua, C.K.; Pumera, M. Chemical reduction of graphene oxide: A synthetic chemistry viewpoint. *Chem. Soc. Rev.* **2014**,*43*, 291–312.

70. Kudin, K.N.; Ozbas, B.; Schniepp, H.C.; Prud'homme, R.K.; Aksay, I.A.; Car, R. Raman spectra of graphite oxide and functionalized graphene sheets. *Nano Lett.* **2008**, *8*, 36–41.

71. Das, A.; Chakraborty, B.; Sood, A.K. Raman spectroscopy of graphene on different substrates and influence of defects.*Bull. Mater. Sci.* **2008**, *31*, 579–584.

72. Gupta, A.; Chen, G.; Joshi, P.; Tadigadapa, S.; Eklund, P.C. Raman scattering from high-frequency phonons in supported n-graphene layer films. *Nano Lett.* **2006**, *6*, 2667–2673.

73. Graf, D.; Molitor, F.; Ensslin, K.; Stampfer, C.; Jungen, A.; Hierold, C.; Wirtz, L. Spatially resolved Raman spectroscopy of single- and few-layer graphene. *Nano Lett.* **2007**, *7*, 238–242.

74. Stankovich, S.; Dikin, D.A.; Piner, R.D.; Kohlhaas, K.A.; Kleinhammes, A.; Jia, Y.; Wu, Y.; Nguyen, S.T.; Ruoff, R.S. Synthesis of graphene-based nanosheets via chemical reduction of exfoliated graphite oxide. *Carbon* **2007**, *45*, 1558–1565.

75. Muszynski, R.; Seger, B.; Kamat, P.V. Decorating graphene sheets with gold nanoparticles. *J. Phys. Chem. C* **2008**, *112*, 5263–5266.

76. Shin, H.-J.; Kim, K.K.; Benayad, A.; Yoon, S.-M.; Park, H.K.; Jung, I.-S.; Jin, M.H.; Jeong, H.-K.; Kim, J.M.; Choi, J.-Y.; *et al.* Efficient reduction of graphite oxide by sodium borohydride and its effect on electrical conductance. *Adv. Funct. Mater.* **2009**, *19*, 1987–1992.

77. Ambrosi, A.; Chua, C.K.; Bonanni, A.; Pumera, M.˙ Lithium aluminium hydride as reducing agent for chemically reduced graphene oxides. *Chem. Mater.* **2012**, *24*, 2292–2298.

78. Stankovich, S.; Piner, R.D.; Chen, X.; Wu, N.; Nguyen, S.T.; Ruoff, R.S. Stable aqueous dispersions of graphitic nanoplatelets via the reduction of exfoliated graphite oxide in the presence of poly(sodium 4-styrenesulfonate). *J. Mater. Chem.* **2006**, *16*, 155–158.

79. Bai, H.; Xu, Y.; Zhao, L.; Li, C.; Shi, G. Non-covalent functionalization of graphene sheets by sulfonated polyaniline.*Chem. Commun.* **2009**, 1667–1669.

80. Hao, R.; Qian, W.; Zhang, L.; Hou, Y. Aqueous dispersions of TCNQ-anion-stabilized graphene sheets. *Chem. Commun.* **2008**, 6576–6578.

81. Xu, Y.; Bai, H.; Lu, G.; Li, C.; Shi, G. Flexible graphene films via the filtration of water-soluble noncovalent functionalized graphene sheets. *J. Am. Chem. Soc.* **2008**, *130*, 5856–5857.

82. Ji, J.; Zhang, G.; Chen, H.; Wang, S.; Zhang, G.; Zhang, F.; Fan, X. Sulfonated graphene as water-tolerant solid acid catalyst. *Chem. Sci.* **2011**, *2*, 484–487.

83. Liu, F.; Sun, J.; Zhu, L.; Meng, X.; Qi, C.; Xiao, F.-S. Sulfated graphene as an efficient solid catalyst for acid-catalyzed liquid reactions. *J. Mater. Chem.* **2012**, *22*, 5495–5502.

84. Wang, L.; Wang, D.; Zhang, S.; Tian, H. Synthesis and characterization of sulfonated graphene as a highly active solid acid catalyst for the ester-exchange reaction. *Catal. Sci. Technol.* **2013**, *3*, 1194–1197.

85. Shaabani, A.; Mahyari, M.; Hajishaabanha, F. The synthesis of xanthenes and benzoxanthenes on graphene oxide and sulfated graphene nanosheets in water. *Res. Chem. Intermed.* **2013**.

86. Upare, P.P.; Toon, J.-W.; Kim, M.Y.; Kang, H.-Y.; Hwang, D.W.; Hwang, Y.K.; Kung, H.H.; Chang, J.-S. Chemical conversion of biomass-derived hexose sugars to levulinic acid over sulfonic acid-functionalized graphene oxide catalysts. *Green Chem.* **2013**, *15*, 2935–2943.

87. Lam, E.; Chong, J.H.; Majid, E.; Liu, Y.; Hrapovic, S.; Leung, A.C.W.; Luong, J.H.T. Carbocatalytic dehydration of xylose to furfural in water. *Carbon* **2012**, *50*, 1033–1043.

88. He, D.; Kou, Z.; Xiong, Y.; Cheng, K.; Chen, X.; Pan, M.; Mu, S. Simultaneous sulfonation and reduction of graphene oxide as highly efficient supports for metal nanocatalysts. *Carbon* **2014**, *66*, 312–319.

89. Antunes, M.M.; Russo, P.A.; Wiper, P.V.; Veiga, J.M.; Pillinger, M.; Mafra, L.; Evtuguin, D.V.; Pinna, N.; Valente, A.A. Sulfonated graphene oxide as effective catalyst for conversion of 5-(hydroxymethyl)-2-furfural into biofuels.*ChemSusChem* **2014**, *7*, 804–812.

90. Liu, J.; Xue, Y.; Dai, L. Sulfated graphene oxide as a hole-extraction layer in high performance polymer solar cells.*Phys. Chem. Lett.* **2012**, *3*, 1928–1933.

91. Garg, B.; Bisht, T.; Chauhan, S.M.S. Synthesis and anion binding properties of novel 3,12- and 3,7-bis(4'-nitrophenyl)-azo-calix[4]pyrrole receptors. *New J. Chem.* **2010**, *34*, 1251–1254.

92. Chauhan, S.M.S.; Garg, B.; Bisht, T. Synthesis and anion binding of 2-arylazo-*meso*-octamethylcalix[4]pyrroles. *Supramol. Chem.* **2009**, *21*, 394–400.

93. Chauhan, S.M.S.; Bisht, T.; Garg, B. 1-arylazo-5,5-dimethyl Dipyrromethanes: Versatile chromogenic probes for anions. *Sens. Actuators B: Chem.* **2009**, *141*, 116–123.

94. Garg, B.; Bisht, T.; Chauhan, S.M.S. 2,2'-Diaminoazobenzene, a potential scaffold for the synthesis of bis-ureas and thioureas: Solution phase anion sensing and binding studies. *Sens. Actuators B: Chem.* **2012**, *168*, 318–328.

95. Lomeda, J.R.; Doyle, C.D.; Kosynkin, D.V.; Hwang, W.-F.; Tour, J.M. Diazonium functionalization of surfactant-wrapped chemically converted graphene sheets. *J. Am. Chem. Soc.* **2008**, *130*, 16201–16206.

96. Si, Y.; Samulski, E.T. Synthesis of water soluble graphene. *Nano Lett.* **2008**, *8*, 1679–1682.

97. Zhao, G.; Jiang, L.; He, Y.; Li, J.; Dong, H.; Wang, X.; Hu, W. Sulfonated graphene for persistent aromatic pollutant management. *Adv. Mater.* **2011**, *23*, 3959–3963.

98. Zhang, H.; Low, W.P.; Lee, H.K. Evaluation of sulfonated graphene sheets as sorbent for micro-solid-phase extraction combined with gas chromatography-mass spectrometry. *J. Chromatogr. A* **2012**, *1233*, 16–21.

99. Coşkun, E.; Zaragoza-Contreras, E.A.; Salavagione, H.J. Synthesis of sulfonated graphene/polyaniline composites with improved electroactivity. *Carbon* **2012**, *50*, 2235–2243.

100. Ankamwar, B.; Surti, F. Water soluble graphene synthesis. *Chem. Sci. Trans.* **2012**, *1*, 500–507.

101. Geim, A.K. Graphene: Status and prospects. *Science* **2009**, *324*, 1530–1534.

102. Garg, B.; Ling, Y.-C. Highly efficient synthesis of N-confused *meso*-tetraspirocyclohexyl calix[4] pyrrole using Brønsted acidic ionic liquids as catalysts. *Tetrahedron Lett.* **2012**, *53*, 5674–5677.

103. Garg, B.; Lei, S.-L.; Liu, S.-C.; Bisht, T.; Liu, J.-Y.; Ling, Y.-C. Rapid identification of trimethyl and triethyl amines using sulfonic acidic ionic liquids: A TOF-SIMS study of fragmentation reactions. *Anal. Chim. Acta* **2012**, *757*, 48–55.

104. Garg, B.; Ling, Y.-C. One-pot green synthesis of azides from alcohols using Brønsted acidic ionic liquid [HMIM][BF_4] as solvent and catalyst. *J. Chin. Chem. Soc.* **2014**, *61*, 737–742.

105. Pei, S.; Zhao, J.; Du, J.; Ren, W.; Cheng, H. Direct reduction of graphene oxide films into highly conductive and flexible graphene films by hydrohalic acids. *Carbon* **2010**, *48*, 4466–4474.

106. Behabtu, N.; Lomeda, J.R.; Green, M.J.; Higginbotham, A.L.; Sinitskii, A.; Kosynkin, D.V.; Tsentalovich, D.; Parra-Vasquez, A.N.G.; Schimidt, J.; Kesselman, E.; *et al.* Spontaneous high-concentration dispersions and liquid crystals of graphene. *Nat. Nanotechnol.* **2010**, *5*, 406–411.

107. Lu, W.; Liu, S.; Qin, X.; Wang, L.; Tian, J.; Luo, Y.; Asiri, A.M.; Al-You, A.O.; Sun, X. High-yield, large-scale production of few-layer graphene flakes within seconds: Using chlorosulfonic acid and H_2O_2 as exfoliating agents. *J. Mater. Chem.***2012**, *22*, 8775–8777.

108. Mutlay, I.; Tudoran, L.B. Chlorosulfonic acid-based room temperature chemical expansion route for the bulk production of graphite nanoplatelets. *Fuller. Nanotub. Carbon Nanostruct.* **2013**, *21*, 149–157.

109. Dixon, P. Formation of sulphamic acid during the decomposition of ammonium sulphate. *Nature* **1944**, *154*, 706–706.

110. Kiyoura, R.; Urano, K. Mechanism, kinetics, and equilibrium of thermal decomposition of ammonium sulfate. *Ind. Eng. Chem. Process Des. Dev.* **1970**, *9*, 489–494.

111. Li, J.-H.; Zhang, G.-E. Investigation of the kinetics and mechanism of decomposition of ammonium hydrogen sulfate.*Acta Phys. Chim. Sin.* **1992**, *8*, 123–127.

112. Li, J.; Zhang, G.; Wang, J. The identification of the mechanism function and the kinetic investigation of the energy-storing reaction of ammonium hydrogen sulfate. *Thermochem. Acta* **1992**, *207*, 219–225.

113. Halstead, W.D. Thermal decomposition of ammonium sulphate. *J. Appl. Chem.* **1970**, *20*, 129–132.

114. Xu, Z.; Qi, Z.; Kaufman, A. Superior catalysts for proton exchange membrane fuel cells sulfonation of carbon-supported catalysts using sulfate salts. Electrochem. *Solid-State Lett.* **2005**, *8*, A313–A315.

115. Kumar, A.V.; Rao, K.R. Recyclable graphite oxide catalyzed Friedel-Crafts addition of indoles to α,β-unsaturated ketones. *Tetrahedron Lett.* **2011**, *52*, 5188–5191.

116. Dreyer, D.R.; Jarvis, K.A.; Ferreirab, P.J.; Bielawski, C.W. Graphite oxide as a carbocatalyst for the preparation of fullerene-reinforced polyester and polyamide nanocomposites. *Polym. Chem.* **2012**, *3*, 757–766.

117. Chauhan, S.M.S.; Mishra, S. Use of graphite oxide and graphene oxide as catalysts in the synthesis of dipyrromethane and calix[4]pyrrole. *Molecules* **2011**, *16*, 7256–7266.

118. Verma, S.; Mungse, H.P.; Kumar, N.; Choudhary, S.; Jain, S.L.; Sain, B.; Khatri, O.P. Graphene oxide: An efficient and reusable carbocatalyst for aza-Michael addition of amines to activated alkenes. *Chem. Commun.* **2011**, *47*, 12673–12675.

119. Dhakshinamoorthy, A.; Alvaro, M.; Puche, M.; Fornes, V.; Garcia, H. Graphene oxide as catalyst for the acetalization of aldehydes at room temperature. *ChemCatChem* **2012**, *4*, 2026–2030.

CHAPTER 18

HEAT TRANSFER PERFORMANCE OF FUNCTIONALIZED GRAPHENE NANOPLATELET AQUEOUS NANOFLUIDS

Roberto Agromayor [1], David Cabaleiro [2], Angel A. Pardinas [1], Javier P. Vallejo [1,2], Jose Fernandez-Seara [1] and Luis Lugo[2]

[1]Área de Máquinas y Motores Térmicos, Escola de Enxeñería Industrial, Universidade de Vigo, Vigo E-36310, Spain

[2]Departamento de Física Aplicada, Facultade de Ciencias, Universidade de Vigo, Vigo E-36310, Spain

ABSTRACT

The low thermal conductivity of fluids used in many industrial applications is one of the primary limitations in the development of more efficient heat transfer systems. A promising solution to this problem is the suspension of nanoparticles with high thermal conductivities in a base fluid. These suspensions, known as nanofluids, have great potential for enhancing heat transfer. The heat transfer enhancement of sulfonic acid-functionalized graphene nanoplatelet water-based nanofluids is addressed in this work. A new experimental setup was designed for this purpose. Convection coefficients, pressure drops, and thermophysical properties of various nanofluids at different concentrations were measured for several operational conditions and the results are compared with those of pure water. Enhancements in thermal conductivity and in convection heat transfer coefficient reach 12% (1 wt %) and 32% (0.5 wt %), respectively. New correlations capable of predicting the Nusselt number and the friction factor of this kind of nanofluid as a function of other dimensionless quantities are developed. In addition, thermal performance factors are obtained from the experimental convection coefficient and pressure drop data in order to assess the convenience of replacing the base fluid with designed nanofluids.

INTRODUCTION

Energy is one of the most valuable resources in our society and, therefore, during the last years, many efforts have been made to increase energy efficiency and encourage the use of renewable energies. A wide variety of applications, such as those involved in the field of solar and geothermal energy, entails thermal energy transference from one fluid to another. Thus, heat management has emerged as one of the fields with the highest potential to improve thermal performance. In order to develop more efficient and compact heat transfer equipment, many heat transfer enhancement techniques have arisen since the mid-twentieth century. However, the low thermal conductivity of the fluids commonly used in industrial applications, such as water or glycols, has hindered any heat transfer enhancement. To overcome this problem, several authors have suggested the use of fluids with improved thermal properties.

Based on the fact that solids exhibit thermal conductivities various orders of magnitude larger than conventional heat

Citation: Roberto Agromayor, David Cabaleiro, Angel A. Pardinas, Javier P. Vallejo, Jose Fernandez-Seara, and Luis Lugo, Heat Transfer Performance of Functionalized Graphene Nanoplatelet Aqueous Nanofluids, Materials 2016, 9(6), 455; doi:10.3390/ma9060455

transfer industrial fluids, in the last decades of the 20th century some researchers proposed enhancing the thermal properties of fluids by dispersing millimetric and micrometric high-conductivity solid particles. Despite the thermal conductivity of these mixtures improving significantly with the addition of solid particles, these dispersions present multiple drawbacks that limit their use in industrial applications. Some of these problems are the erosion of heat transfer equipment, the sedimentation of the solid particles, or the huge increase of pressure drop associated with the higher viscosity in relation to the base fluid. Thus, the addition of solid particles was rejected as a heat enhancement alternative until the last years of the 20th century, when advances in materials science allowed the large-scale production of much smaller solid particles. In 1995, Choi and Eastman [1] resumed previous investigations about the increase of thermal conductivity through the addition of nanometric solid particles to base fluids. These samples, which they named nanofluids, have higher thermal conductivities than the initial base fluids and also lack the problems of the dispersions with millimetric and micrometric particles. Due to the small size of nanoparticles, heat transfer systems are not worn away and the sedimentation is much slower. Moreover, for small concentrations of nanoparticles, pressure drop increases are marginal compared with those of the base fluid [2,3]. The number of experimental studies and publications related to the heat transfer characteristics of nanofluids has increased exponentially in recent years [4,5,6,7]. Among the commonly used base fluids, water is one of the most widespread in heat exchange processes. For this reason, many water-based nanofluid studies have been carried out lately [8,9,10].

Despite the fact that nanofluids are two-phase systems (solid and liquid phases), many authors such as Pak and Cho [11], Xuan and Li [7], and Fard et al. [12] agree that their behavior is similar to that of a pure substance with increased thermal properties. In this work, the nanofluids studied are considered as a single-phase system. This approach assumes that the fluid phase and the nanoparticles are in thermal equilibrium and there is no slip in velocity between the base fluid and the solid particles.

Nanoparticles of different materials such as metals, metallic oxides, and carbon nanostructures have been employed to prepare nanofluids with enhanced thermal properties for many different applications in areas like thermal engineering, renewable energies, or medicine [13,14,15]. The two-dimensional structure of graphene, as a single-atom-thick nanosheet of sp2-bonded carbon atoms packed into a honeycomb lattice, exhibits a higher potential than other nanostructured carbon allotropes, such as one-dimensional nanotubes or zero-dimensional fullerenes [16]. Thus, graphene combines the advantages of highly ordered graphitic carbonhigh stability, abundance in source, and low cost—with the single-layer benefits of in-plane thermal conductivity as high as 3000 to 5000 $W \cdot m^{-1} \cdot K^{-1}$ [16,17,18]. Although graphene is hydrophobic and, consequently, cannot be dispersed in water for a long time without agglomerating [19], stable dispersions in aqueous/organic solvents can be prepared once the material has been functionalized (using an acid treatment or amino function) by means of proper sonication [20,21,22]. Yarmand et al. [22] stated that water nanofluids based on functionalized graphene nanoplatelets and silver (up to 0.1 wt %) are stable without the assistance of surfactants or ultrasonication.

Most of the studies about nanofluids focus on the analysis of their physical properties, both from an experimental and a theoretical point of view. The properties that play a major role in convection heat transfer are density, specific heat capacity, thermal conductivity, and viscosity. Although density and specific heat capacity can be easily predicted from a theoretical approach, thermal conductivity and viscosity must be carefully determined because there are not yet models that can accurately predict the transport properties of nanofluids [23,24]. Over the last years, several studies about the thermophysical and heat transfer performance of graphene nanoplatelet aqueous nanofluids have been reported in the literature [22,25,26,27,28,29,30]. However, most of them only analyze the effect of nanoplatelet addition at low mass concentrations. Thus, regarding thermophysical characterization, only Gupta et al. [25] and Hajjar et al. [26] reached 0.25 wt %, while Yu et al. [27] studied from 1 to 5 vol %.

Forced convection heat transfer is a complex phenomenon that combines velocity and temperature profiles, as well as the thermal properties of the fluid. Significant heat transfer enhancements are expected with the addition of high-conductivity nanometric particles. The increase of the effective thermal conductivity of the fluid is not the only factor that accelerates the heat transfer process [31]. The presence of these solid particles also increases the fluctuation and turbulence of nanofluids, further enhancing the heat transfer process. The increase of turbulence causes temperature gradients within the fluid to be reduced and the boundary layers to become thinner where transport phenomena are slower. Similarly, thanks to the small size of the solid particles, the increase of the pressure drop with respect to the base fluid is minimal. These characteristics of the fluid flow and heat transfer of nanofluids were also discussed by Xuan and Li [9]. Consequently, greater energy exchanges between the fluid and the solid walls of heat transfer equipment can be achieved for the same pumping power input. This is the reason why the use of nanofluids is one alternative with high potential for improving the energy efficiency of heat transfer processes. One of the current trends consists of the study of the effects and potential benefits resulting from the application of magnetic fields in heat transfer processes with nanofluids [10,32,33,34,35,36].

A difficulty faced when studying nanofluids is that the conventional heat transfer coefficient correlations of pure substances fail to predict their thermal behavior, even if the measured transport properties of the nanofluids are used in the calculations. This problem was previously reported by Xuan and Li [9], among others. It is caused by the presence of moving solid particles that modify the velocity and temperature profiles of the flow with respect to those of pure substances. Thus, new correlations that account for the presence of solid particles must be explored in order to quantitatively describe the thermal behavior of nanofluids.

The aim of this paper is to study the single-phase flow and thermal behavior of different sulfonic acid-functionalized graphene nanoplatelet aqueous nanofluids from an experimental point of view. Nanoplatelets and nanofluids are thermophysically characterized, and the heat transfer by forced convection is studied for different nanoparticle concentrations using a new device. The main results are discussed and compared with those obtained for water at the same operating conditions to investigate the heat transfer enhancement. Finally, new correlations to calculate the Nusselt number and the Darcy friction factor of this kind of nanofluids are proposed.

NANOFLUID CHARACTERIZATION

Material and Sample Preparation

Graphene nanoplatelets (GnPs) were provided by NanoInnova Technologies S.L. (Madrid, Spain) while Milli Q-Grade water was produced with a resistivity of 18.2 MΩ·cm by means of a Millipore system Milli-Q 185 Plus (Millipore Ltd., Watford, UK). Nanofluids were designed through a two-step method as dispersions of sulfonic acid-functionalized graphene nanoplatelets at nanoadditive mass concentrations of 0.25, 0.50, 0.75, and 1.00 wt % in water, which correspond to volume fractions of 0.19, 0.39, 0.59, and 0.79 vol %, respectively. The desired concentrations were prepared by weighing the powder in a Mettler AE-200 (Mettler Toledo, Greifensee, Switzerland), with an uncertainty of 10^{-4} g, and then stirring it into a predetermined volume of base fluid for 120 min. Afterwards, the suspensions were sonicated for 240 min by using an ultrasonic bath (Ultrasounds, JP Selecta S.A., Barcelona, Spain) operating at a sonication frequency of 20 kHz, with a maximum power output of 200 W.

Nanopowder Characterization

The morphology of the dry sulfonic acid-functionalized graphene nanopowder was analyzed through scanning electron microscopy by using a JEOL JSM-6700F field emission gun-SEM (JEOL, Tokyo, Japan) working at an accelerator voltage of 20 kV in backscattering electron image (Yttrium Aluminum Garnet type detector). This device is coupled to an energy dispersive X-ray (EDS) spectrometer Oxford Inca Energy 300 SEM (Oxford Instruments, Oxford, UK), which also allows for carrying out chemical microanalyses of the sample. SEM samples were prepared by depositing a drop of the nanopowder dispersed in analytical grade methanol (Sigma-Aldrich, Madrid, Spain) on the top of a silica support and drying under atmospheric conditions. Figure 1 and Figure 2 show the SEM image and EDS microanalysis of the studied GnP sample, respectively. Sulfonic acid-functionalized graphene exhibits a nanoplatelet-shape of up to some micrometers with wrinkled surfaces folding at the edges (Figure 1). The EDS spectrum shows the presence of carbon (C), oxygen (O), sulfur (S), and silica (Si), this last being due to impurities of physical support (Figure 2).

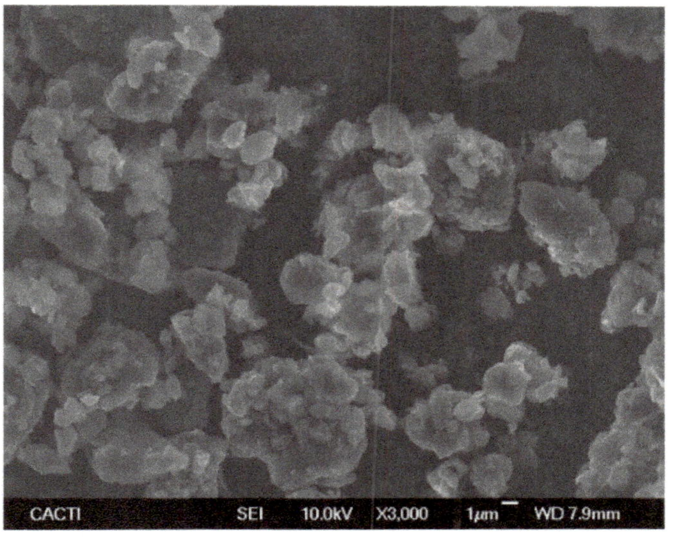

Figure 1. SEM image of sulfonic acid-functionalized graphene nanoplatelets at 3000× magnification.

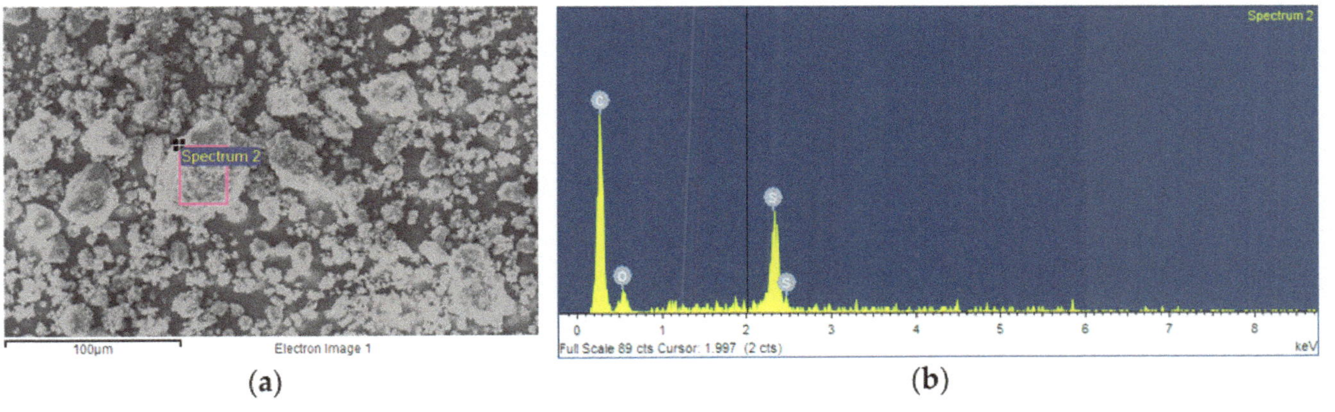

Figure 2: EDS microanalysis of sulfonic acid-functionalized graphene nanoplatelets: (**a**) studied area; (**b**) EDS spectrum.

Thermophysical Characterization

Regarding the thermophysical properties of the base fluid, those values from REFPROP [37] were considered. Concerning the dry GnP sample, the specific heat capacity was experimentally determined in this work by a quasi-isothermal temperature-modulated differential scanning calorimetry method (TDMSC) using a heat-flux differential scanning calorimeter, DSC, Q2000 (TA Instruments, New Castel, DE, USA) [38]. Otherwise, a density value of 1.27 $g \cdot cm^{-3}$ for GnP nanopowder was used. In this work, the densities, ρ, and specific heat capacities, c_p, of GnP nanofluids were obtained using the following weighted average equations [38]:

$$\rho_{nf} = \phi_v \times \rho_{GnP} + (1 - \phi_v) \times \rho_{bf} \quad (1)$$

$$c_{pnf} = \phi_m \times c_{pGnP} + (1 - \phi_m) \times c_{pbf} \quad (2)$$

where φ_v is the nanoadditive volume fraction, φ_m is the nanoadditive mass fraction, and the nf, GnP, and bf subscripts stand for nanofluid, graphene nanoplatelets, and base fluid, respectively. Obtained values for the density and specific heat capacity of the studied fluids in the temperature range from 20 to 40 °C are plotted together with the base fluid in Figure 3and Figure 4, respectively. Density increases with increasing graphene concentration, whereas the contrary occurs for the specific heat capacity. Modifications in the mentioned properties as regards the base fluid overall analyzed conditions are lower than 0.2% for ρ and 0.8% for c_p.

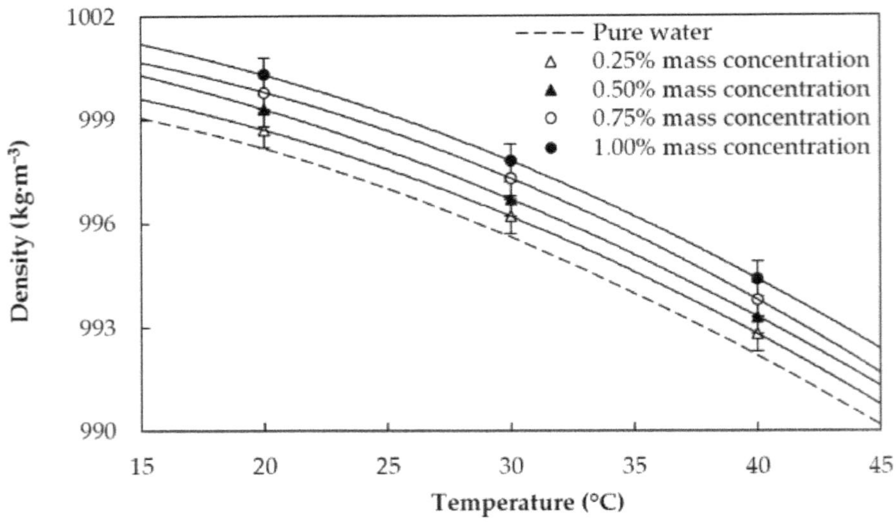

Figure 3: Density of the sulfonic acid-functionalized graphene nanoplatelet water-based nanofluids and water [37].

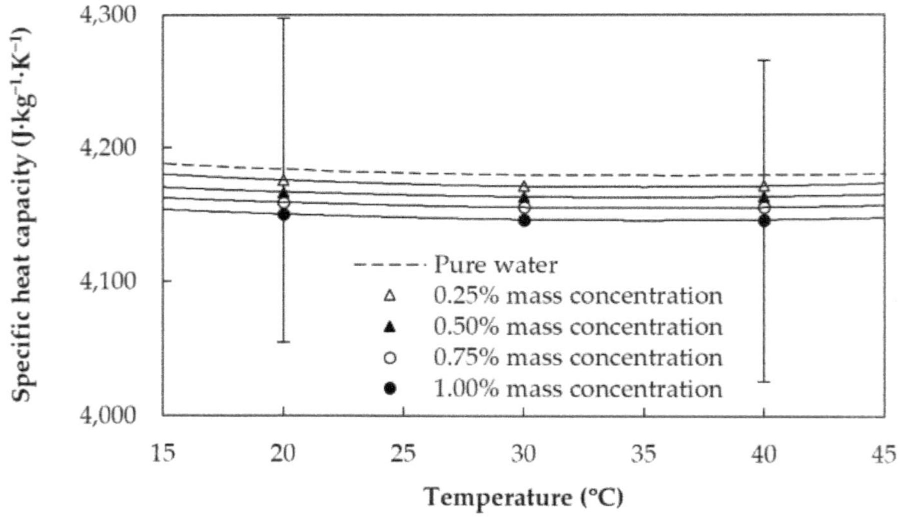

Figure 4: Specific heat capacity of the sulfonic acid-functionalized graphene nanoplatelet water-based nanofluids and water [37].

Effective thermal conductivities, k, were experimentally measured in this work from 20 to 40 °C by means of a KD2 Pro thermal analyzer (Decagon Devices, Inc., Pullman, WA, USA) together with a KS-1 probe of 1.3 mm diameter and 60 mm long. In order to precisely control the temperature, samples were fully immersed in a Grant GP200 (Grant Instruments, Cambridge, UK) oil bath. More details about the experimental device and measurement procedure can be found in previous works [39,40,41]. The temperature dependence of the thermal conductivity of the base fluid and the different GnP nanofluids is shown in Figure 5. This property rises as the nanoplatelet concentration increases, being 12% greater for the 1.00 wt % nanofluid than for the base fluid.

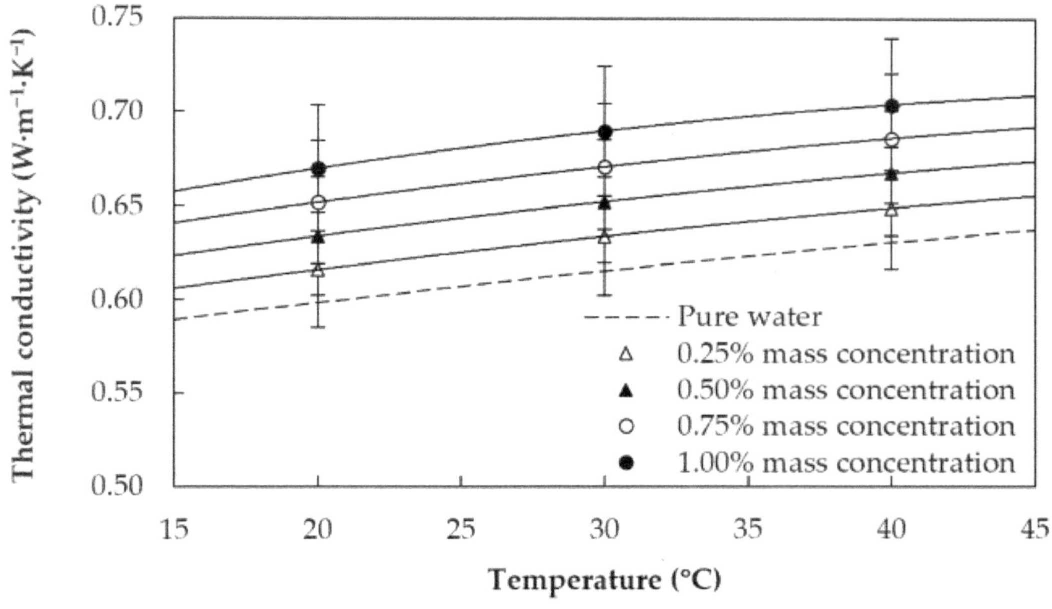

Figure 5. Thermal conductivity of the sulfonic acid-functionalized graphene nanoplatelet water-based nanofluids and water [37].

Finally, the dynamic viscosities of the fluid samples were determined in the temperature range between 20 and 40 °C by using a rotational Physica MCR 101 rheometer (Anton Paar, Graz, Austria) equipped with a cone-plate geometry with a cone angle of 1° and a diameter of 25 mm [2,41]. Both nanofluids and base fluid exhibit Newtonian behavior in the studied concentration range. Obtained values are presented in Figure 6 as a function of the temperature, where a remarkable increase in viscosity with the mass concentration can be seen.

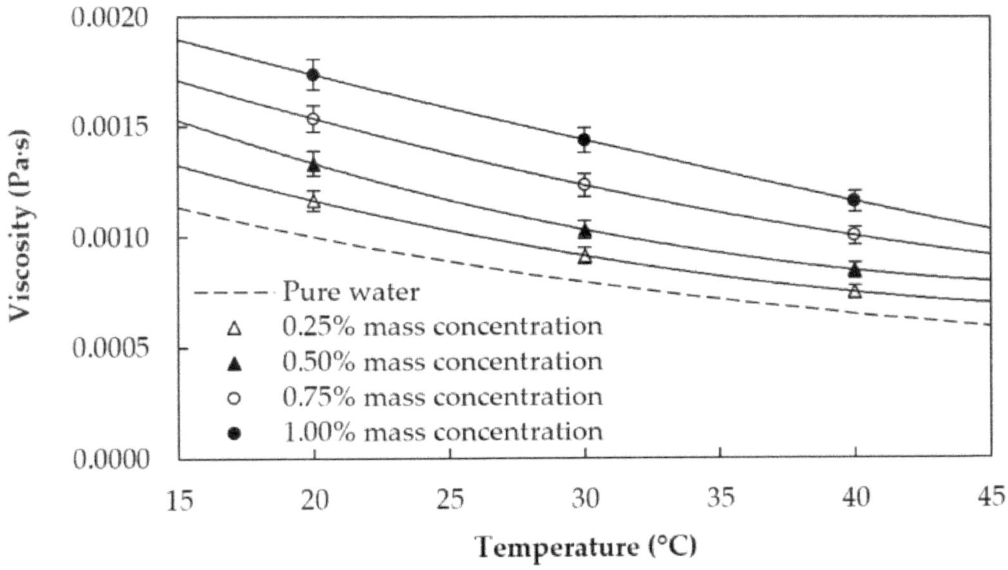

Figure 6. Viscosity of the sulfonic acid-functionalized graphene nanoplatelet water-based nanofluids and water [37].

HEAT TRANSFER COEFFICIENT DETERMINATION

Experimental Equipment

The new experimental facility is composed of two closed loops, using the tested fluid and hot water as working fluids, and an open refrigeration loop. The layout of the experimental equipment and the location of the devices used are shown inFigure 7. The main section of the experimental facility, where heat exchange and pressure drop are studied, consists of a stainless steel tube-in-tube heat exchanger with lengths of 930 mm and 1180 mm for the test sections of effective heating and pressure drop, respectively. The tested fluid is pumped through the inner tube and is heated by hot water from the other closed loop, which flows within the annular section. The internal tube has an inner diameter of 8 mm and an outer diameter of 10 mm, while the outer tube has an inner diameter of 15 mm. The test section was insulated to minimize the heat losses to the environment, which were estimated to be lower than 0.5% of the total heat transfer even for the most unfavorable operating conditions. Thus, the heat exchanger can be considered adiabatic, which greatly simplifies the analysis.

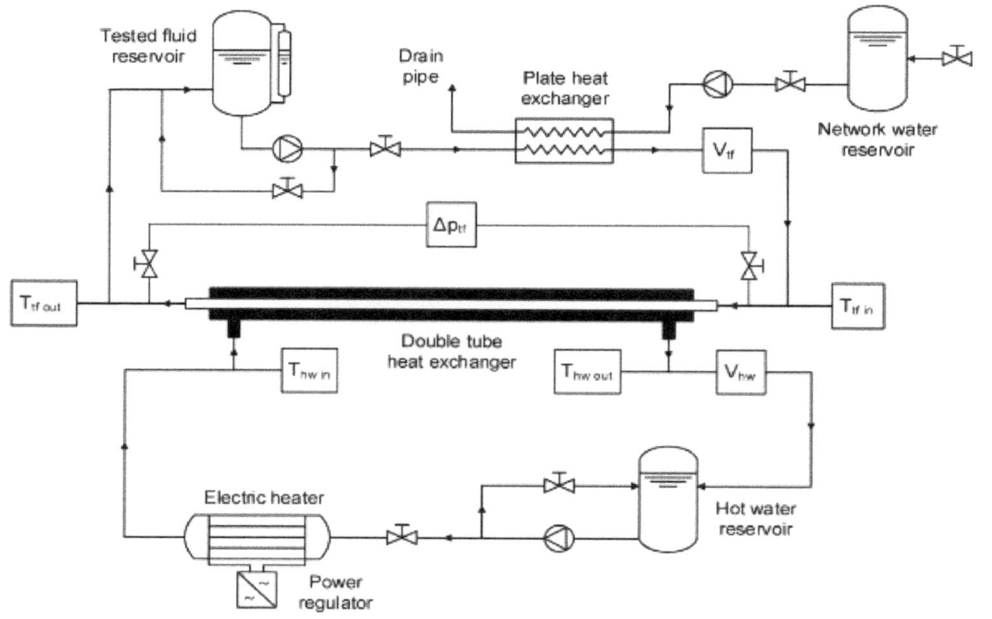

Figure 7: Layout of the experimental setup for the determination of heat transfer coefficients and pressure drop.

The hot water is stored in a 25 L reservoir that gives thermal inertia to the loop and is pumped through a 4.5 kW electric heater before it enters the annular section. Similarly, the tested fluid is pumped from a 3 L reservoir tank to a plate heat exchanger, where it transfers energy to the water of refrigeration. This tested fluid then flows into the tube-in-tube heat exchanger, where it receives energy from the hot water. The refrigeration loop consists of a reservoir tank where tap water is stored before being pumped through the plate heat exchanger to cool the tested fluid.

The flow rates of the hot water and tested fluid are adjusted both with valves and controlling the rotation speed of the centrifugal pumps with proportional-integral-derivative (PID) regulators. The temperature of the hot water is also controlled automatically with a PID that adjusts the electric power delivered by a power regulator to the electric heater. The flow rate of refrigeration water, and therefore the temperature of the tested fluid, is controlled manually with a needle valve.

The temperatures of hot water and tested fluid at the inlet and outlet of the test section are measured with four Pt100 A Class resistance temperature sensors. The flow rate of both fluids is measured with two different electromagnetic flowmeters. The pressure drop of the tested fluid through the tube-in-tube heat exchanger is measured with a differential pressure sensor. Lastly, the experimental setup is equipped with a data acquisition system based on a 16-bit acquisition card and a PC where the measured variables are displayed in real time and stored in a spreadsheet.

Experimental Procedure

With the aim of evaluating the enhancement of the convection heat transfer coefficients and the pressure drop of several sulfonic acid-functionalized graphene nanoplatelet aqueous nanofluids, a comparison between these results and those obtained for pure water was experimentally carried out. Thus, several sets of tests were conducted for nanofluids at 0.25%, 0.50%, 0.75%, and 1.00% mass concentrations of graphene nanoplatelets, together with those for pure water. Each test is defined through the values of the volumetric flow rate and average temperature in the heat exchanger of each fluid.

Heat transfer tests were conducted by pumping the fluid under study at different flow rates while the average temperature of both hot water and the tested fluid as well as the flow rate of hot water were kept constant. The pairs of studied average temperatures in relation to tested fluid/hot water were, in Celsius, 20/40, 30/40, 30/45, 30/50, and 40/60, and the studied fluid flow rate ranged from 200 to 700 L/h in 100 L/h steps. The hot water flow rate was set to 800 L/h for all the tests. The water flow rate was selected to be as small as possible to achieve a high temperature difference across the heat exchange, but high enough to ensure turbulent regime ($Re > 10^4$). This was done in order to calculate the heat transfer coefficient on the water side using widely accepted correlations for forced convection through annular sections. In order to assess the validity of the experimental measurements, the convection heat transfer coefficients for pure water were compared with the values predicted by the well-known Gnielinski correlations [42]. Deviations were less than 10% for all the tested conditions.

The pressure drop tests were also carried out keeping the temperature of the tested fluid constant and pumping at different flow rates. Two tests were conducted for average temperatures of 25 °C and 35 °C and the flow rate was varied from 250 to 700 $L \cdot h^{-1}$ in 50 $L \cdot h^{-1}$ steps. The deviations of the pressure drops measured and those calculated from Gnielinski correlations [42] were also less than 10%, finding slightly higher deviations for the lowest flow rates.

DATA ANALYSIS

The heat transfer capability of the graphene water-based nanofluids can be quantitatively described through the convection heat transfer coefficient. The convection coefficients of the different nanofluids are compared with those of pure water to analyze the heat transfer enhancement. Similarly, the pressure drops of the nanofluids and pure water are also compared.

The convection coefficient cannot be measured directly, but can be calculated from the measured data as follows. The temperatures of the tested fluid and the heating water at the inlet and the outlet of the test section ($T_{tf\ in}$, $T_{tf\ out}$, $T_{hw\ in}$, and $T_{hw\ out}$) were measured experimentally. The logarithmic mean temperature difference (LMTD), normally used in heat exchangers to describe the temperature driving force, is defined as:

$$LMTD = \frac{(T_{hwin} - T_{tfout}) - (T_{hwout} - T_{tfin})}{\ln\left(\frac{T_{hwin} - T_{tfout}}{T_{hwout} - T_{tfin}}\right)} \quad (3)$$

The overall thermal resistance in the test section, R_{ov}, was obtained as the quotient of the (LMTD) and the heat flow rate on the hot water side ($\dot{Q}$) according to the following equation:

$$R_{ov} = \frac{LMTD}{\dot{Q}} \quad (4)$$

where $\dot{Q}$ is given by:

$$\dot{Q} = \rho_{hw} \times c_{phw} \times \dot{V}_{hw} \times (T_{hwin} - T_{hwout}) \quad (5)$$

where $\dot{V}_{hw}$ is the hot water volume flow rate while ρ_{hw} and $c_{p\ hw}$ stand for its density and specific heat capacity, respectively. The calculations were carried out using the heat flow in the hot water side, Equation (5), because the uncertainties of the water properties are lower than those of the nanofluids.

The overall thermal resistance is the sum of the thermal resistance corresponding to the external forced convection (R_{hw}), the conductive thermal resistance of the steel tube (R_s), and the internal forced convection resistance (R_{tf}). Fouling resistances were assumed to be negligible in this work because the tube used was cleaned before the experiments. Therefore, the thermal resistance on the tested fluid side can be calculated using the following equation:

$$R_{tf} = R_{ov} - R_s - R_{hw} \quad (6)$$

The convection thermal resistance of the hot water side is given by the equation:

$$R_{hw} = (\pi \times d_2 \times L_h \times h_{hw})^{-1} \quad (7)$$

where d_2 corresponds to the outer diameter of the inner tube and L_h is the effective heating length of the test section. The heat transfer coefficient, h_{hw}, was calculated using the correlations proposed by Gnielinski [42] for fully developed turbulent flows in annular ducts with adiabatic outer surfaces. These relations are given by:

$$h_{hw} = k_{hw} \times Nu_{hw}/d_h \quad (8)$$

$$Nu_{hw} = \psi \times Nu^*_{hw} \quad (9)$$

$$\psi = 0.75 \times a^{-0.17} \quad (10)$$

$$a = d_2/d_3 \quad (11)$$

$$Nu^*_{w} = \frac{(f_{hw}/8) \times Re_{hw} \times Pr_{hw} \times (1 + (d_h/L_h)^{2/3})}{\omega + 12.7 \times (f_{hw}/8)^{1/2} \times (Pr_{hw}{}^{2/3} - 1)} \quad (12)$$

$$\omega = 1.07 + 900 \times Re_{hw}^{-1} - 0.63 \times (1 + 10 \times Pr_{hw})^{-1} \quad (13)$$

$$f_{hw} = (0.782 \times \ln(Re^*_{hw}) - 1.50)^{-2} \quad (14)$$

$$Re^*_{hw} = \frac{(1 + a^2) \times \ln(a) + (1 - a^2)}{(1 - a)^2 \times \ln(a)} \times Re_{hw} \quad (15)$$

where d_h, the hydraulic diameter for the annular section, is the inner diameter of the outer tube, d_3, minus the outer diameter of the inner tube, d_2; and k_{hw} is the thermal conductivity of hot water. The Nusselt number (Nu_{hw}), the Reynolds number (Re_{hw}), and the Prandtl number (Pr_{hw}) of the hot water were calculated according to their usual definitions. The thermophysical and transport properties of hot water were obtained from REFPROP [37], considering water to be a saturated liquid at an average temperature between the inlet and the outlet.

On the other hand, the conductive thermal resistance, R_s, was calculated assuming unidimensional heat transfer in the radial direction and according to the equation:

$$R_s = \frac{\ln(d_2/d_1)}{2\pi \times k_s \times L_h} \quad (16)$$

where d_1 is the inner diameter of the inner tube of the heat exchanger, made of AISI 316L stainless steel. Its thermal conductivity (k_s) was calculated using the equation proposed by Choong [43]:

$$k_s = 9.248 + 0.01571 \times (T + 273.15) \quad (17)$$

where the temperature of the tube, T, should be expressed in Celsius and the conductivity in $W \cdot m^{-1} \cdot K^{-1}$. The temperature of the tube for each test was considered as a constant given by:

$$T = \frac{1}{4} \times (T_{tfin} + T_{tfout} + T_{hwin} + T_{hwout}) \quad (18)$$

Consequently, once the inner thermal resistance, R_{tf}, due to convection was calculated using Equation (6), the convection coefficient was determined in a straightforward way using the following equation:

$$h_{tf} = \pi \times d_1 \times L_h \times R_{tf}{}^{-1} \quad (19)$$

On the other hand, the pressure drop of the fluid under study was measured directly and no further analysis is required to compare the results of the different nanofluids and water.

Heat transfer coefficient and pressure drop are dimensional variables and, although they can be convenient for comparing the thermal and hydrodynamic behavior of the nanofluids, they are not suitable to correlate the experimental data. Instead, dimensionless groups can be used to correlate the heat transfer and pressure drop with the parameters of the flow and the properties and composition of the nanofluid.

The tested fluid Nusselt number (Nu_{tf}) is the dimensionless form for the convection coefficient of the tested fluid (h_{tf}) and is given by the following equation:

$$Nu_{tf} = \frac{h_{tf} \times d_1}{k_{tf}} \tag{20}$$

Similarly, the Darcy friction factor of the tested fluid (f_{tf}), which represents the dimensionless pressure drop across the heat exchanger, is given by:

$$f_{tf} = \frac{d_1}{L_p} \times \frac{\Delta p_{tf}}{\rho_{tf} \times v_{tf}{}^2/2} \tag{21}$$

where L_p stands for the effective pressure drop length of the test section while Δp_{tf}, ρ_{tf}, and v_{tf} are the pressure drop, density, and average velocity of the tested fluid, respectively.

In this work, the Nusselt number of the tested fluid (Nu_{tf}) was related to its Reynolds number (Re_{tf}), Prandtl number evaluated at the bulk temperature (Pr_{tf}), Prandtl number evaluated at the wall temperature (Pr_{wall}), and volumetric concentration of nanoparticles (φ_v). The following equation is proposed in this work to correlate the heat transfer experimental data for the tested fluids:

$$Nu_{tf} = c_1 \times (1 + c_2 \times \phi_v)^{c_3} \times \mathrm{Re}_{tf}^{c_4} \times \mathrm{Pr}_{tf}^{c_5} \times (\mathrm{Pr}_{tf}/\mathrm{Pr}_{wall})^{c_6} \tag{22}$$

The Prandtl number, Pr_{wall}, at the wall is included in the correlation to account for the heat transfer effect on the temperature profile and so on the thermal properties. The temperature at the wall was considered as a constant given by:

$$T_{wall} = \frac{1}{2} \times (T_{tfin} + T_{tfout}) + R_{tf} \times \dot{Q} \tag{23}$$

It should be noted that the volumetric concentration of nanoparticles was used to correlate the experimental data. This is because the rheological behavior of suspensions, such as nanofluids, highly depends on hydrodynamic forces, which act on the surfaces of the particles; therefore, a geometric way to describe the concentration of the fluid is preferred. This is a common practice when studying nanofluids, as is also discussed by Pak and Cho [11].

Likewise, the Darcy friction factor of the tested fluid (f_{tf}) was related to the Reynolds number and the volumetric concentration of nanoadditives. The functional relation proposed for the friction factor follows the expression:

$$f_{tf} = c_7 \times (1 + c_8 \times \phi_v)^{c_9} \times \mathrm{Re}_{tf}^{c_{10}} \tag{24}$$

In order to further evaluate the thermal performance, the comparison of the Nusselt numbers of two fluids must be balanced by the friction factors, which are influenced by the nanoparticles' dispersion. A possible means of evaluating the thermal effectiveness of the different tested nanofluids is to use the thermal performance factor given by the following relation [44,45]:

$$\eta_{tf} = (Nu_{tf}/Nu_{bf}) \times (f_{tf}/f_{bf})^{1/3} \tag{25}$$

where f is the Darcy friction factor, Nu the Nusselt number, and tf and bf stand for tested and base fluids, respectively. The experimental results were analyzed using this parameter to compare the thermal performance of the different nanofluids.

The required thermophysical properties of the tested fluid, evaluated at the average temperature in the heat exchanger, were obtained from the aforementioned experimental values. A thorough uncertainty analysis of the measured and calculated magnitudes was performed following the recommendations of the "Guide to the Expression of Uncertainty in Measurements (GUM)" [46]. The results of this analysis are summarized later.

RESULTS

Firstly, the heat transfer coefficients and the pressure losses of the different nanofluids are compared and discussed in relation to water results. After that, the obtained numerical values for the Nusselt number and Darcy friction factor correlations are presented together with their validity ranges. In addition, the analysis of the thermal performance factor as a function of mass concentration is discussed.

Heat Transfer Performance

The experimental results for the analyzed sulfonic acid-functionalized graphene nanoplatelet water-based nanofluids are shown as an example of some of the numerous tests. The displayed behaviors are representative of the rest of the tests performed, the tendency of the data being similar along all the other tests.

Heat Transfer Coefficients

The experimental heat transfer coefficients of the different analyzed nanofluids and pure water in the 20 °C/40 °C tests at different flow rates are plotted in Figure 8 as a function of the mass concentration of graphene nanoplatelets at different flow rates. As could be expected, the heat transfer coefficients increase with the flow rate due to the increase of mixing and turbulence. The heat transfer coefficients for the 300 L·h^{-1} tests at different temperatures are plotted in Figure 9. Each curve represents the values measured at different tested fluid average temperatures, with 20 °C as heat exchanger LMTD. As can be observed, the heat transfer coefficient is higher as the temperature of the nanofluid rises, which agrees with the decrease of the tested fluid viscosity and the increase of its thermal conductivity.

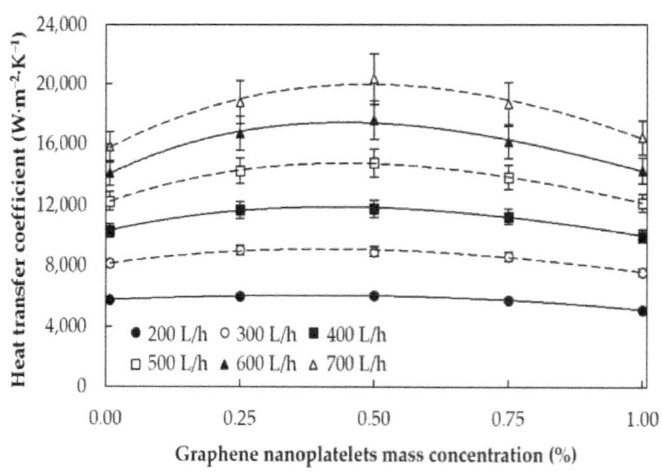

Figure 8: Heat transfer coefficient against graphene nanoplatelet concentration at different flow rates. The average temperatures of tested fluid/hot water are 20 °C/40 °C.

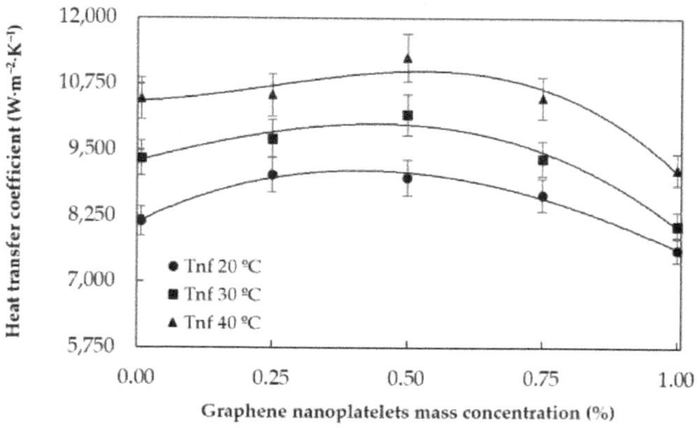

Figure 9: Heat transfer coefficient against graphene nanoplatelet concentration at different temperatures with a flow rate of 300 L·h^{-1}. The LMTD at the heat exchanger is 20 °C.

Upon further inspection of Figure 8 and Figure 9, it can be seen that the heat transfer coefficients have a smooth dependence with the concentration of graphene nanoplatelets. The heat transfer achieved with the 0.25% nanofluid is greater than that of pure water and the results were even better when the concentration was 0.50%. However, the heat transfer enhancement obtained with the 0.75% nanofluid was smaller and the heat transfer coefficients for the 1.00% mass concentration were even worse than those of pure water under many conditions. This may be because, at low nanoparticle concentrations, the heat transfer enhancement due to the thermal conductivity and the increased degree of mixing are more important than the penalty caused by the increase of the effective viscosity of the nanofluid. Nevertheless, at higher concentrations, the increase of viscosity due to the presence of solid particles predominates over the other two effects and, as a result, the heat transfer is reduced with respect to that achieved with the base fluid. This suggests the existence of an optimum nanoparticle concentration for a given set of operating conditions. The obtained relative heat transfer enhancements with respect to water for each nanofluid (minimum, average, and maximum) are summarized in Table 1.

Table 1: Heat transfer enhancement with respect to water for functionalized graphene nanoplatelet aqueous nanofluids

Nanofluid	Minimum	Average	Maximum
$\varphi_m = 0.25\%$	−1.7%	6.5%	18.6%
$\varphi_m = 0.50\%$	0.94%	15.0%	32.4%
$\varphi_m = 0.75\%$	−2.3%	7.0%	22.8%
$\varphi_m = 1.00\%$	−19.6%	-10.1%	4.0%

In relation to the heat transfer enhancements, we notice that Sadeghinezhad et al. [28] obtained increments in the Nusselt number up to 83% for graphene nanoplatelets nanofluids at 0.1 wt % under turbulent forced convection through a plain stainless steel tube. Yarmand et al. [29] quantified their maximum enhancement of the Nusselt number at 26.5% using a square stainless steel tube for samples at 0.1 wt %.

Pressure Losses

To determine if this kind of nanofluid is suitable for heat transfer applications, the pressure drop caused by the presence of the solid particles should also be analyzed. The measurements of the pressure drop of the 0.25%, 0.50%, and 0.75% nanofluids, as well as those of pure water, for different flow rates and at a test fluid temperature of 35 °C, are plotted inFigure 10, as an example. As expected, the pressure losses increase with the flow rate. Moreover, as can be seen in this figure, the pressure drop slightly increases with the mass concentration. This increase can be due to the higher viscosities of the nanofluid and, as the concentration of graphene increases, this effect becomes more important and the pressure losses rise.

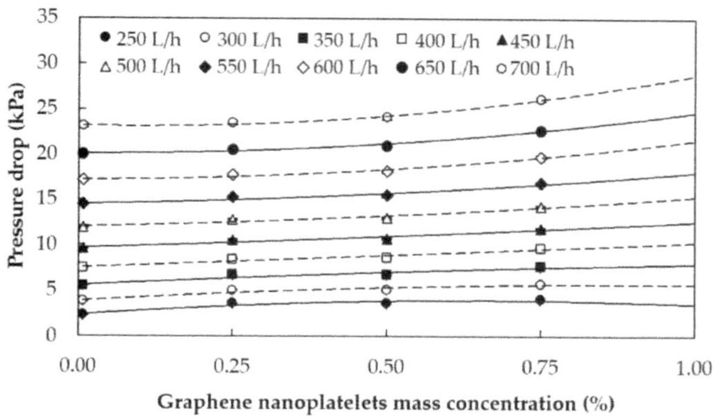

Figure 10: Pressure drops versus graphene nanoplatelet concentration at 35 °C test fluid temperature and for different flow rates.

Taking into account all the tests performed, it was also found that at a lower nanofluid temperature, e.g., 25 °C, the pressure losses were slightly higher, as a consequence of the higher viscosities. The relative experimental pressure increases with respect to pure water for each nanofluid (minimum, average, and maximum) are summarized in Table 2.

Table 2: Relative pressure drop in relation to pure water for graphene nanoplatelets water-based nanofluids

Nanofluid	Minimum	Average	Maximum
$\varphi_m = 0.25\%$	0.92%	14.5%	55.4%
$\varphi_m = 0.50\%$	4.11%	17.1%	56.9%
$\varphi_m = 0.75\%$	13.0%	28.0%	73.4%

These results are in agreement with those of Sadeghinezhad et al. [28], who found maximum increases in pressure drops of 14.6% for 0.1 wt %.

Dimensionless Analysis

Many correlations to calculate the Nusselt number and the Darcy friction factor for internal forced convection in circular ducts are available in the literature. However, those correlations were developed using measurements of pure substances and are not capable of predicting the behavior of nanofluids. Two new correlations suitable for calculating the Nusselt number and Darcy friction factor of sulfonic acid-functionalized graphene nanofluids were developed in this work.

Nusselt Number Correlation

The c_1–c_6 fitting coefficients in Equation (22) were determined using a least squares regression analysis. The proposed correlation for the Nusselt number and its validity ranges are given by the following equations:

$$Nu_{nf} = 0.011 \times (1 + 100 \times \phi_v)^{-0.095} \times \mathrm{Re}_{nf}^{0.886} \times \mathrm{Pr}_{nf}^{0.545} \times (\mathrm{Pr}_{nf}/\mathrm{Pr}_{wall})^{-0.495} \quad (26)$$

$$0.19\% \leqslant \phi_v \leqslant 0.79\% \quad (27)$$

$$5 \times 10^3 \leqslant Re_{nf} \leqslant 4 \times 10^4 \quad (28)$$

$$4.8 \leqslant Pr_{nf} \leqslant 10.8 \quad (29)$$

$$1.06 \leqslant Pr_{nf}/Pr_{wall} \leqslant 1.36 \quad (30)$$

It should be noted that the factor Pr_{nf}/Pr_{wall} that is used to account for the effect that the heat transfer has on the temperature profile is different if the fluid is either heated or cooled. During all the tests the heat was transferred from the hot water to the nanofluid and, therefore, the correlation given by Equation (26) should only be used when the nanofluid is being heated. The uncertainty of the experimental Nusselt number was evaluated and it ranged from 6% to 12% for all the measured values. The deviation between the Nusselt numbers correlated and those measured experimentally is 13.2% for the worst case (0.25% mass concentration nanofluid, 40 °C/60 °C, and 700 L·h^{-1}) and is less than 10% and 8% for 90% and 75% of the measured values, respectively.

Darcy Friction Factor Correlation

The procedure for the friction factor correlation was similar to that of the Nusselt number. The obtained fitting parameters from c_7 to c_{10} in Equation (24) are:

$$f_{nf} = 0.109 \times (1 + 100 \times \phi_v)^{0.215} \times Re_{nf}^{-0.159} \quad (31)$$

$$0.19\% \leqslant \phi_v \leqslant 0.59\% \quad (32)$$

$$8 \times 10^3 \leqslant Re_{nf} \leqslant 3.7 \times 10^4 \quad (33)$$

The uncertainty analysis of the experimental friction factor leads to values below 2% for all the measured values. The deviation between the Darcy friction factors correlated and those measured experimentally is 4.7% for the worst case (0.75% mass concentration nanofluid, 35 °C, and 300 L·h^{-1}) and is less than 2.5% and 2% for 90% and 75% of the measured values, respectively.

Thermal Performance Factor

An analysis of the thermal performance factor obtained for each graphene nanoplatelet mass concentration was carried out by combining the results of the different tests as gathered in Figure 11 for different flow rates.

A reference temperature of 30 °C for both experiments was selected in order to develop this comparison. Thus, Nusselt numbers were obtained from the 30 °C/40 °C heat transfer test (average temperature of the tested fluid at 30 °C) while friction factors were obtained through an interpolation of both 25 °C and 35 °C pressure drop tests. As can be observed inFigure 11, the thermal performance factor rises when the flow rate increases and all nanofluids present beneficial ratios except for the 1 wt %. The concentration dependence of this factor exhibits maximum values at 0.5% mass concentration, reaching a value of 1.27 for the highest flow rate. In agreement with previous results, the 1% sample shows the worst results, with thermal performance factors below the unit for all the analyzed flow rates.

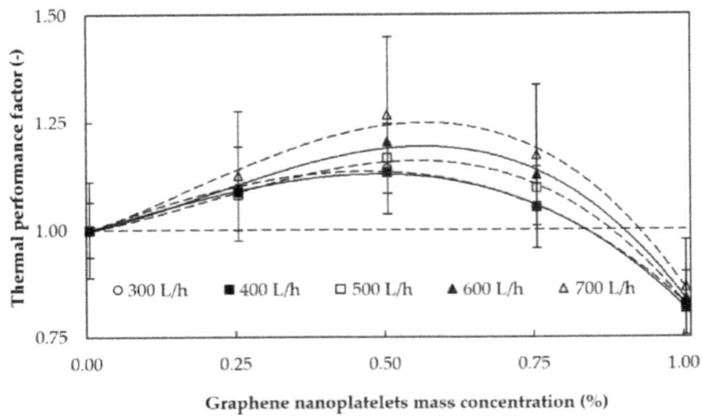

Figure 11: Thermal performance factor against graphene nanoplatelet concentration at 30 °C, with LMTD of 10 °C between fluids, for different flow rates.

CONCLUSIONS

In this work, a new experimental device was implemented to determine the heat transfer coefficients and the pressure drops of several sulfonic acid-functionalized graphene nanofluids in water up to 1 wt % (0.79 vol %). A rheometer, a transient hot wire technique, and a differential scanning calorimeter were used to measure viscosities, thermal conductivities, and heat capacities of the analyzed samples, respectively. Enhancements in thermal conductivity reach up to 12% for the maximum analyzed concentration.

Noticeable enhancements in the convection coefficients, h_{tf}, are achieved for all nanofluids except for 1 wt %. These coefficients increase with the flow rate and the temperature of the fluid. The mass concentration dependence shows an optimum at 0.5%, for which h_{tf} are up to 32% higher than those for water. The pressure losses monotonously increase with the nanoparticle concentration as a consequence of the viscosity of the tested fluid.

Two new correlations able to describe the Nusselt number and the friction factor of this kind of dispersion as a function of dimensionless numbers are presented. Deviations lower than 13.2% in the case of Nusselt and 4.7% for pressure losses are obtained. Moreover, heat transfer coefficient and pressure drop results were combined through the thermal performance factor to assess whether replacing the base fluid with the different designed nanofluids would be beneficial. The 0.5% nanofluid was found to exhibit the best results, reaching increases up to 27% in this factor. Hence, the use of nanofluids is an alternative with a high potential for heat transfer enhancement considering that, for an appropriate graphene nanoplatelet concentration, significant improvements in the heat transfer coefficients can be achieved, with low pressure drop increases.

The use of nanofluids would allow for developing more efficient and compacting heat transfer equipment. It can be especially significant in renewable and clean energy technologies, with a potentially interesting application in the field of solar and geothermal energy. This type of new material can lead to reductions in the temperature difference between fluids, increasing the efficiency of thermal machinery or decreasing the required flow rates, which would reduce the pumping power consumption. However, the field of nanofluids is still in its early stages and new works, both theoretical and experimental, should be conducted in order to understand and be able to predict their behavior.

ACKNOWLEDGMENTS

The authors acknowledge the functionalized graphene nanoplatelets powder provided by Nanoinnova Technologies S.L. (http://www.nanoinnova.com/Product). This work was supported by the Ministerio de Economía y Competitividad (Spain) and the FEDER program through ENE2012-32908 and ENE2014-55489-C2-2-R Projects. David Cabaleiro acknowledges the financial support under the FPU Program by the "Ministerio de Educación, Cultura y Deporte" (Spain) and Javier P. Vallejo acknowledges the FPI Program of the "Ministerio de Economía y Competitividad".

AUTHOR CONTRIBUTIONS

Ángel Á. Pardiñas participated in the implementation of the experimental setup to develop the heat transfer tests, Roberto Agromayor determined the experimental heat transfer coefficients and pressure drops of the nanofluids, and David Cabaleiro carried out the design of the nanofluids and their thermophysical characterization. Roberto Agromayor, Ángel Á. Pardiñas, David Cabaleiro, and Javier P. Vallejo analyzed the data and wrote the paper. José Fernández-Seara and Luis Lugo conceived the study, developed its design, and coordinated the redaction of the manuscript. All authors read and approved the final manuscript.

CONFLICTS OF INTEREST

The authors declare no conflict of interest. The founding sponsors had no role in the design of the study; in the collection, analyses, or interpretation of data; in the writing of the manuscript, and in the decision to publish the results.

REFERENCES

1. Choi, S.U.S.; Eastman, J.A. Enhancing thermal conductivity of fluids with nanoparticles. FED **1995**, 231, 99–105.
2. Cabaleiro, D.; Pastoriza-Gallego, M.J.; Gracia-Fernández, C.; Piñeiro, M.M.; Lugo, L. Rheological and volumetric properties of TiO_2-ethylene glycol nanofluids. Nanoscale Res. Lett. **2013**, 8, 1–13.
3. Sharma, A.K.; Tiwari, A.K.; Dixit, A.R. Rheological behaviour of nanofluids: A review. Renew. Sustain. Energy Rev.**2016**, 53, 779–791.
4. Huminic, G.; Huminic, A. Application of nanofluids in heat exchangers: A review. Renew. Sustain. Energy Rev. **2012**,16, 5625–5638.
5. Trisaksri, V.; Wongwises, S. Critical review of heat transfer characteristics of nanofluids. Renew. Sustain. Energy Rev.**2007**, 11, 512–523.
6. Lomascolo, M.; Colangelo, G.; Milanese, M.; de Risi, A. Review of heat transfer in nanofluids: Conductive, convective and radiative experimental results. Renew. Sustain. Energy Rev. **2015**, 43, 1182–1198.
7. Xuan, Y.; Li, Q. Heat transfer enhancement of nanofluids. Int. J. Heat Fluid Flow **2000**, 21, 58–64.
8. Gu, B.; Hou, B.; Lu, Z.; Wang, Z.; Chen, S. Thermal conductivity of nanofluids containing high aspect ratio fillers. Int. J. Heat Mass Transf. **2013**, 64, 108–114.
9. Mintsa, H.A.; Roy, G.; Nguyen, C.T.; Doucet, D. New temperature dependent thermal conductivity data for water-based nanofluids. Int. J. Therm. Sci. **2009**, 48, 363–371.
10. Ellahi, R.; Hassan, M.; Zeeshan, A. Shape effects of nanosize particles in Cu–H_2O nanofluid on entropy generation.Int. J. Heat Mass Transf. **2015**, 81, 449–456.
11. Pak, B.C.; Cho, Y.I. Hydrodynamic and heat transfer study of dispersed fluids with submicron metallic oxide particles. Exp. Heat Transf. Int. J. **1998**, 11, 151–170.
12. Fard, M.H.; Esfahany, M.N.; Talaie, M. Numerical study of convective heat transfer of nanofluids in a circular tube two-phase model versus single-phase model. Int. Commun. Heat Mass Transf. **2010**, 37, 91–97.
13. Rahman, S.U.; Ellahi, R.; Nadeem, S.; Zia, Q.M.Z. Simultaneous effects of nanoparticles and slip on jeffrey fluid through tapered artery with mild stenosis. J. Mol. Liq. **2016**, 218, 484–493.
14. Colangelo, G.; Favale, E.; Miglietta, P.; Milanese, M.; de Risi, A. Thermal conductivity, viscosity and stability of Al_2O_3-diathermic oil nanofluids for solar energy systems. Energy **2016**, 95, 124–136.
15. Colangelo, G.; Favale, E.; de Risi, A.; Laforgia, D. Results of experimental investigations on the heat conductivity of nanofluids based on diathermic oil for high temperature applications. Appl. Energy **2012**, 97, 828–833.
16. Lee, G.J.; Rhee, C.K. Enhanced thermal conductivity of nanofluids containing graphene nanoplatelets prepared by ultrasound irradiation. J. Mater. Sci. **2014**, 49, 1506–1511.
17. Balandin, A.A.; Ghosh, S.; Bao, W.; Calizo, I.; Teweldebrhan, D.; Miao, F.; Lau, C.N. Superior thermal conductivity of single-layer graphene. Nano Lett. **2008**, 8, 902–907.
18. Ahmad, I.; McCarthy, J.E.; Baranov, A.; Gun'ko, Y.K. Development of graphene nano-platelet based counter electrodes for solar cells. Materials **2015**, 8, 5953–5973.
19. Li, X.; Chen, Y.; Mo, S.; Jia, L.; Shao, X. Effect of surface modification on the stability and thermal conductivity of water-based SiO_2-coated graphene nanofluid. Thermochim. Acta **2014**, 595, 6–10.
20. Amiri, A.; Sadri, R.; Shanbedi, M.; Ahmadi, G.; Chew, B.T.; Kazi, S.N.; Dahari, M. Performance dependence of thermosyphon on the functionalization approaches: An experimental study on thermo-physical properties of graphene nanoplatelet-based water nanofluids. Energy Convers. Manag. **2015**, 92, 322–330.
21. Ijam, A.; Moradi Golsheikh, A.; Saidur, R.; Ganesan, P. A glycerol-water-based nanofluid containing graphene oxide nanosheets. J. Mater. Sci. **2014**, 49, 5934–5944.
22. Yarmand, H.; Gharehkhani, S.; Ahmadi, G.; Shirazi,

S.F.S.; Baradaran, S.; Montazer, E.; Zubir, M.N.M.; Alehashem, M.S.; Kazi, S.; Dahari, M. Graphene nanoplatelets-silver hybrid nanofluids for enhanced heat transfer. Energy Convers. Manag. **2015**, 100, 419–428.

23. Wen, D.; Lin, G.; Vafaei, S.; Zhang, K. Review of nanofluids for heat transfer applications. Particuology **2009**, 7, 141–150.

24. Rudyak, V.Y. Viscosity of nanofluids—Why it is not described by the classical theories. Adv. Nanopart. **2013**, 2, 266–279.

25. Gupta, S.S.; Siva, M.V.; Krishnan, S.; Sreeprasad, T.S.; Singh, P.K.; Pradeep, T.; Das, S.K. Thermal conductivity enhancement of nanofluids containing graphene nanosheets. J. Appl. Phys. **2011**, 110.

26. Hajjar, Z.; Rashidi, A.M.; Ghozatloo, A. Enhanced thermal conductivities of graphene oxide nanofluids. Int. Commun. Heat Mass Transf. **2014**, 57, 128–131.

27. Yu, W.; Xie, H.; Chen, W. Experimental investigation on thermal conductivity of nanofluids containing graphene oxide nanosheets. J. Appl. Phys. **2010**, 107.

28. Sadeghinezhad, E.; Togun, H.; Mehrali, M.; Nejad, P.S.; Latibari, S.T.; Abdulrazzaq, T.; Kazi, S.; Metselaar, H.S.C. An experimental and numerical investigation of heat transfer enhancement for graphene nanoplatelets nanofluids in turbulent flow conditions. Int. J. Heat Mass Transf. **2015**, 81, 41–51.

29. Yarmand, H.; Gharehkhani, S.; Shirazi, S.F.S.; Amiri, A.; Alehashem, M.S.; Dahari, M.; Kazi, S. Experimental investigation of thermo-physical properties, convective heat transfer and pressure drop of functionalized graphene nanoplatelets aqueous nanofluid in a square heated pipe. Energy Convers. Manag. **2016**, 114, 38–49.

30. Solangi, K.; Amiri, A.; Luhur, M.; Ghavimi, S.A.A.; Zubir, M.N.M.; Kazi, S.; Badarudin, A. Experimental investigation of the propylene glycol-treated graphene nanoplatelets for the enhancement of closed conduit turbulent convective heat transfer. Int. Commun. Heat Mass Transf. **2016**, 73, 43–53.

31. Ting, H.H.; Hou, S.S. Investigation of laminar convective heat transfer for Al_2O_3-water nanofluids flowing through a square cross-section duct with a constant heat flux. Materials **2015**, 8, 5321–5335.

32. Sheikholeslami, M.; Ellahi, R. Electrohydrodynamic nanofluid hydrothermal treatment in an enclosure with sinusoidal upper wall. Appl. Sci. **2015**, 5, 294–306.

33. Rashidi, S.; Dehghan, M.; Ellahi, R.; Riaz, M.; Jamal-Abad, M.T. Study of stream wise transverse magnetic fluid flow with heat transfer around an obstacle embedded in a porous medium. J. Magn. Magn. Mater. **2015**, 378, 128–137.

34. Sheikholeslami, M.; Ellahi, R. Three dimensional mesoscopic simulation of magnetic field effect on natural convection of nanofluid. Int. J. Heat Mass Transf. **2015**, 89, 799–808.

35. Sher Akbar, N.; Raza, M.; Ellahi, R. Influence of induced magnetic field and heat flux with the suspension of carbon nanotubes for the peristaltic flow in a permeable channel. J. Magn. Magn. Mater. **2015**, 381, 405–415.

36. Calvo-Bravo, J.; Cabaleiro, D.; Piñeiro, M.M.; Lugo, L. Magnetorheological behaviour of propylene glycol-based hematite nanofluids. Rheol. Acta **2015**, 54, 757–769.

37. Lemmon, E.W.; Huber, M.L.; McLinden, M.O. Reference Fluid Thermodynamic and Transport Properties (REFPROP); NIST Standard Reference Database; National Institute of Standards and Technology: Gaithersburg, MD, USA, 2010.

38. Cabaleiro, D.; Gracia-Fernández, C.; Legido, J.L.; Lugo, L. Specific heat of metal oxide nanofluids at high concentrations for heat transfer. Int. J. Heat Mass Transf. **2015**, 88, 872–879.

39. Cabaleiro, D.; Nimo, J.; Pastoriza-Gallego, M.J.; Piñeiro, M.M.; Legido, J.L.; Lugo, L. Thermal conductivity of dry anatase and rutile nano-powders and ethylene and propylene glycol-based TiO_2 nanofluids. J. Chem. Thermodyn. **2015**,83, 67–76.

40. Cabaleiro, D.; Pastoriza-Gallego, M.J.; Piñeiro, M.M.; Legido, J.L.; Lugo, L. Thermophysical properties of (diphenyl ether + biphenyl) mixtures for their use as heat transfer fluids. J. Chem. Thermodyn. **2012**, 50, 80–88.

41. Cabaleiro, D.; Pastoriza-Gallego, M.J.; Piñeiro, M.M.; Lugo, L. Characterization and measurements of thermal conductivity, density and rheological properties of zinc oxide nanoparticles dispersed in (ethane-1,2-diol + water) mixture. J. Chem. Thermodyn. **2013**, 58, 405–415.

42. Gnielinski, V. Heat Transfer in Concentric Annular and Parallel Plate Ducts. VDI Heat Atlas; Springer: Heidelberg, Germany, 2010; pp. 701–708.

43. Kim, C.S. Thermophysical Properties of Stainless Steels; Technical Report; Argonne National Lab.: Argonne, IL, USA, 1975.

44. Bianco, V.; Manca, O.; Nardini, S.; Vafai, K. Heat Transfer Enhancement with Nanofluids; CRC Press: Boca Raton, FL, USA, 2015.

45. Mehrali, M.; Sadeghinezhad, E.; Rosen, M.A.; Akhiani, A.R.; Tahan Latibari, S.; Mehrali, M.; Metselaar, H.S.C. Heat transfer and entropy generation for laminar forced convection flow of graphene nanoplatelets nanofluids in a horizontal tube. Int. Commun. Heat Mass Transf. **2015**, 66, 23–31.

46. Joint Committee for Guides in Metrology. Evaluation of Measurement Data—Guide to the Expression of Uncertainty in Measurements; JCGM 100: Sèvres, France, 2008.

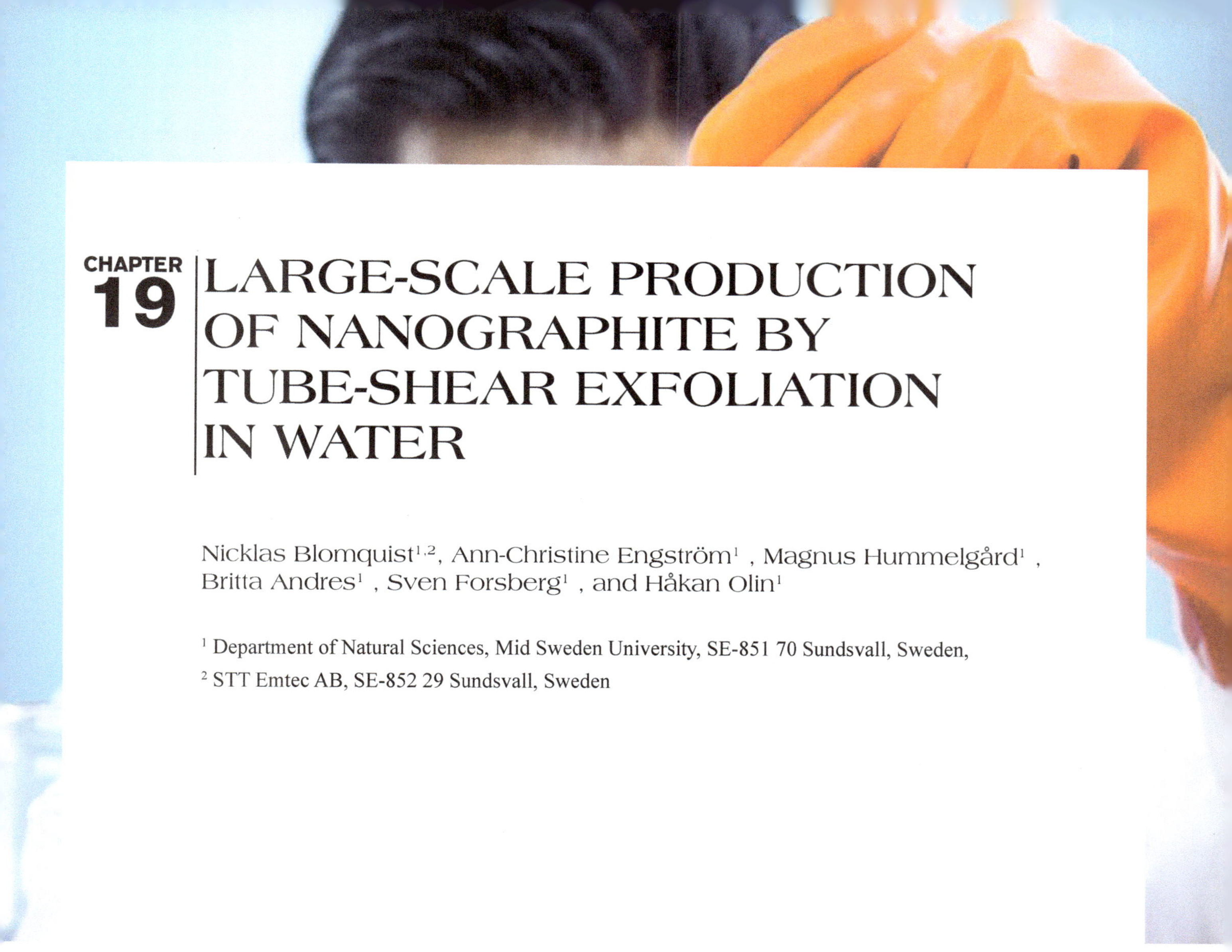

CHAPTER 19

LARGE-SCALE PRODUCTION OF NANOGRAPHITE BY TUBE-SHEAR EXFOLIATION IN WATER

Nicklas Blomquist[1,2], Ann-Christine Engström[1], Magnus Hummelgård[1], Britta Andres[1], Sven Forsberg[1], and Håkan Olin[1]

[1] Department of Natural Sciences, Mid Sweden University, SE-851 70 Sundsvall, Sweden,

[2] STT Emtec AB, SE-852 29 Sundsvall, Sweden

ABSTRACT

The number of applications based on graphene, few-layer graphene, and nanographite is rapidly increasing. A large-scale process for production of these materials is critically needed to achieve cost-effective commercial products. Here, we present a novel process to mechanically exfoliate industrial quantities of nanographite from graphite in an aqueous environment with low energy consumption and at controlled shear conditions. This process, based on hydrodynamic tube shearing, produced nanometer-thick and micrometer-wide flakes of nanographite with a production rate exceeding 500 gh^{-1} with an energy consumption about 10 Whg^{-1}. In addition, to facilitate large-area coating, we show that the nanographite can be mixed with nanofibrillated cellulose in the process to form highly conductive, robust and environmentally friendly composites. This composite has a sheet resistance below 1.75 Ω/sq and an electrical resistivity of 1.39×10^{-4} Ωm and may find use in several applications, from supercapacitors and batteries to printed electronics and solar cells. A batch of 100 liter was processed in less than 4 hours. The design of the process allow scaling to even larger volumes and the low energy consumption indicates a low-cost process.

INTRODUCTION

A cost-efficient and large-scale process of highly conductive carbon nanoparticles, such as graphene and nanographite, is essential to take the step from laboratory experiments to useful commercial products [1, 2]. Carbon nanoparticles and composites will be required for various applications, such as supercapacitors [3, 4], batteries [4, 5], printed electronics [6], or solar cells [4, 5]. These applications will need a cost-effective exfoliation process and preferably environmentally compatible solvents where the ideal is aqueous processing.

Exfoliated graphite comes in different qualities [7] from single-layer, bi-layer, few-layer (2–5 layers), multilayer graphene, and graphite nanosheets. Here, we define nanographite as a mix of all these qualities, with a certain size distribution. The requirement for electrodes, for supercapacitor and other applications, is demanding and rather different from applications that require only few-layer graphene. In these electrode applications the combined large-area and high

Citation: Blomquist N, Engström A-C, Hummelgård M, Andres B, Forsberg S, Olin H (2016) Large-Scale Production of Nanographite by Tube-Shear Exfoliation in Water. PLoS ONE 11(4): e0154686. doi:10.1371/journal.pone.0154686

conductivity as well as ability to coat are the most important ones. Most reports about graphite exfoliation processes aim to produce only graphene or few-layer graphene, and the partially exfoliated material are not used. This makes it hard to compare parameters like size distribution and production rate of various exfoliation processes designed for different uses and applications.

There is still no good method to precisely determine the yield, and thereby the production rate, in large-scale of a certain quality of exfoliated graphite, even if progress is emerging [8–12]. To avoid the time-consuming work of identify, specify and count every particle in the processed batch, UV-VIS spectroscopy is widely used to measure concentration together with TEM, AFM, Raman spectroscopy etc. to determine the particle quality [9–12]. It is worth noting that measurements on graphene and graphene-like particles are difficult, due to their propensity to agglomerate together and form few-layer, multi-layer, and finally graphite [12]. Therefor, to be able to compare the production rate of similar material, we need to estimate the ratio of few-layer produced in our process. This despite the fact that we, in our application, use the produced nanographite as is.

To exfoliate graphene from graphite in solution, sonication is the standard laboratory procedure. However, this method is difficult to scale due to that the concentration scales roughly inversely with liquid volume and the process is not energy efficient [9, 13, 14]. The extended treatment time results in a low throughput and the graphene sheets might be cut to smaller flakes during the exfoliation process. Other approaches to wet exfoliation are jet cavitation [15], ball milling [16], rotational dispersers [10, 17], wet grinding [18], and homogenizer processing [11]. These methods are potential candidates for large-scale production, however, as Paton et al. point out [10], the production rates of most of these methods is less than 0.4 gh^{-1}. To demonstrate large-scale exfoliation, Paton et al. [10] show that large quantities of defect free graphene can be made in N-methyl-2-pyrrolidone (NMP) suspension by high-shear rotational mixing of graphite. They show a few-layer graphene production rate of 5.3 gh^{-1} and estimate that it could be scaled up to a production rate exceeding 100 gh^{-1} for a 10 m^3 process scale. Nacken et al. [11] used a commercial high-pressure homogenizer to prove large-scale exfoliation of graphite in NMP and water-surfactant mixtures, with a production rate of 0.5 gh^{-1} of few-layer graphene.

In addition to a process with high throughput of conducting carbon nanoparticles, when making thin films and electrodes, the coating formulation will be of importance as well as adhesion properties. Therefore, the ability to form composites of the processed material will be essential. Graphene or nanographite composites have many applications for example in energy storage and harvesting [4] to phase change materials [19]. Malho et al. [2] have shown, by sonication, that graphene can be exfoliated directly from graphite in aqueous environment with nanofibrillated cellulose (NFC) as the only dispersant, forming a stiff, tough and strong nanocomposite. NFC can be used as binder in nanographite electrodes for supercapacitors to improve electrical and mechanical properties [3].

The combination of these problems point to the need to develop a process that is preferably water based, is suitable for large-scale production, and permit incorporation of binders to enable further processing. The process should also be energy-efficient, have high yield and meet the requirements of the specific application. For example, an electrode application such as supercapacitors require a certain size distribution and ability to add other carbon qualities, such as activated carbon, leading to high electrical conductivity combined with high active surface area.

In this paper we describe a novel type of process for large-scale exfoliation in an aqueous environment. We study the particle size distribution and the degree of delamination during the process. We also analyzed the sheet resistance and surface area of the nanographite-NFC composite to compare the material with other conducting carbon alternatives. The majority of the particles in the nanographite was nanometer-thin and micrometer-wide flakes with a graphene-like structure. The production rate was 500 gh^{-1} of nanographite.

MATERIALS AND METHODS

To prepare the suspension for exfoliation, 20 gL^{-1} thermally expanded graphite (SO#5-44-04) from Superior Graphite was mixed in water with an addition of 2 wt% polyacrylic acid, in relation to the graphite amount, as dispersant. The suspension volume was 100 L. The pH was adjusted to 3.5 by sulfuric acid in order to get a fixed and reproducible pH-value of the suspension during exfoliation. The expanded graphite has initially low pH when mixed with water, due to residual acid from the manufacturing process.

To exfoliate, the suspension was forced by a high-pressure pump through a 1 m long tube with an inner diameter of 2 mm. The pressure was held constant at 50 bar with an initial flow rate of 4.95 L min^{-1}, giving rise to a shear rate of 1×10^5 s^{-1}. This procedure was repeated 10 times (10 passes). The dynamic viscosity increased slightly for each passing leading to a decreased flow rate at constant pressure. The flow rate decreased from 4.95 L min^{-1} to 4.27 L min^{-1} during the 10 passes, which corresponds to a change in dynamic viscosity from 23.7 mPas to 27.6 mPas. The calculated shear rate went from 1.05×10^5 s^{-1} down to 0.91×10^5 s^{-1}. The calculated Reynolds number (Re) decreased from initially 2217 to 1642. To calculate the shear rate, $\dot{\gamma}$, for laminar flow in a straight tube we used [20]

$$\dot{\gamma} = \frac{4Q}{\pi r^3}, \quad (1)$$

where Q is the volumetric flow rate, and r is the hydraulic pipe radius. To calculate the Reynolds number we used [20]

$$Re = \frac{2Q\rho}{\mu\pi r}, \quad (2)$$

where ρ is the fluid density and μ is the dynamic viscosity of the fluid. The production rate, P_R, can be expressed by [14]

$$P_R = \frac{CV}{t}, \quad (3)$$

where C is the concentration or solid content of processed material, V is the liquid volume (batch volume), and t is the production time.

Samples of the suspension were taken during the process at 0, 5 and 10 passes to examine the change in particle structure. After 10 passes the pH was adjusted to 7 with sodium hydroxide and the suspension was divided in two batches, A and B. Fig 1 shows the exfoliation equipment.

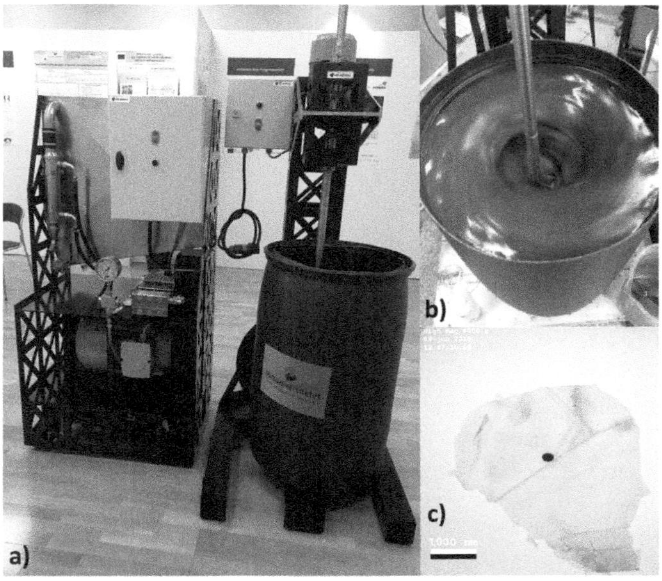

Figure 1: The exfoliation equipment. a) The hydrodynamic tube shearing device with dispersion barrel and stirrer, b) the dispersion barrel during stirring (to avoid graphite flotation) and, c) a TEM-image of the processed material.

To get robust freestanding films for measurements from the suspension, NFC was added as binder. The NFC used was TEMPO-oxidized kraft-pulp NFC made using the method described in [21]. In order to examine the addition of NFC for improved composite properties two different methods were used.

In batch A, 10 wt% NFC in relation to the graphite amount were added to the suspension during gentle stirring for 40 min. The suspension was then dispersed in two further passes in the tube shearing process with the parameters stated above in order to exfoliate the NFC and get a well-dispersed composite. In batch B, NFC was added in the same way but processed by high speed shearing with an IKA T25 digital Ultra Turrax disperser instead. 1 liter suspension was exfoliated with dispersing element S25N-10G in the Ultra Turrax at 12 000 rpm for 10 min.

Sample preparation and analysis

To analyze the structural change in the material, sample grids for a transmission electron microscope (TEM) were prepared. Suspensions from 0, 5 and 10 passes were diluted to 0.0047% solids content and one droplet (30 mg) was applied to the center of a TEM-grid. Five grids per dispersion were prepared. The microscope used was a JEOL-2000FX.

To analyze the particle size distribution in the suspensions from 5 and 10 passes, one droplet of the diluted suspensions (0.0047%) was applied on an aluminum scanning electron microscope sample stub. The particles were characterized in size by image analysis in a ZEISS EVO-50 SEM. In order to determine the particle size, predefined squares in the image were used where each particle could fit in. A total of 2645 particles were characterized from the two suspensions. To estimate the concentration of the smallest particles, UV-VIS spectroscopy was performed on the top phase of a sedimented sample from 10 passes with a Shimadzu 1800 spectrophotometer (absorbance at 660 nm with an extinction coefficient of 1060 mL $mg^{-1}m^{-1}$).

To analyze the particle thickness in the processed material, samples for Atomic Force microscopy (AFM) were prepared. Suspension from 10 passes were diluted to 0.0047% solids content and one droplet (30 mg) was applied to the center of 1×1 cm silicon wafer substrate. Samples from the top phase of a sedimented 10 pass suspension was also prepared. Four samples per dispersion were prepared. The microscope used was a Dimension AFM with a Nanoscope IIIa controller (Digital Instruments).

Composite films were made by filtering batch A and B respectively on Millapore Durapore Membrane Filter (filter type: 0.22 μm GV) in four different quantities, 0.1 g, 0.25 g, 0.5 g and 1 g, dry weight. Films of the unexfoliated dispersion (expanded graphite, water and polyacrylic acid) were also prepared in the same way, but due to its powdery unstable structure, no measurements could be done on these samples.

The sheet resistance of the films was measured using a four-point probe Keithley 2611A system after 24 h in room temperature. To analyze the surface area of the composite we used the Brunauer–Emmett–Teller (BET) method. The BET-samples were made from suspensions with 0 passes, batch A and B by instantly freezing the suspensions with liquid nitrogen followed by freeze-drying to form a dry powder. The BET measurements were made by Vesta Lab Sweden AB.

RESULTS

Particle structure analysis

Fig 2 shows the structure of the initial material. The thermally expanded graphite consists of large granulates, several hundred micrometers in size. These granulates have a partly cracked surface that looks like tightly packed graphite flakes.

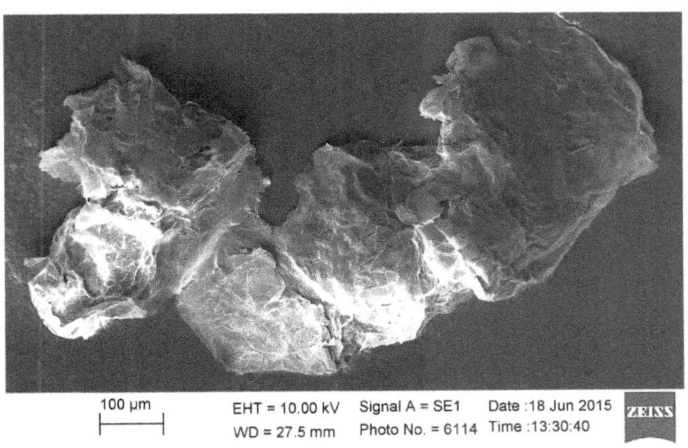

Figure 2: Initial material. SEM-image of the initial graphite, before exfoliation (0 passes).

Fig 3a shows a typical particle found in the suspension after 5 passes and Fig 3b is a typical particle found in the suspension after 10 passes. It can be seen in 3a that the material is partly exfoliated and has some cracks. After further exfoliation (10 passes), we can find a larger amount of well-exfoliated flakes that look like graphene, see Fig 3b. Note that the scale is different in the two images. Thin flakes can be found in both suspensions but we can clearly see that the amount of well-exfoliated flakes increases with increased amount of passes. The flakes are significantly large, and their transparency suggests that we obtained one or few layer graphene flakes.

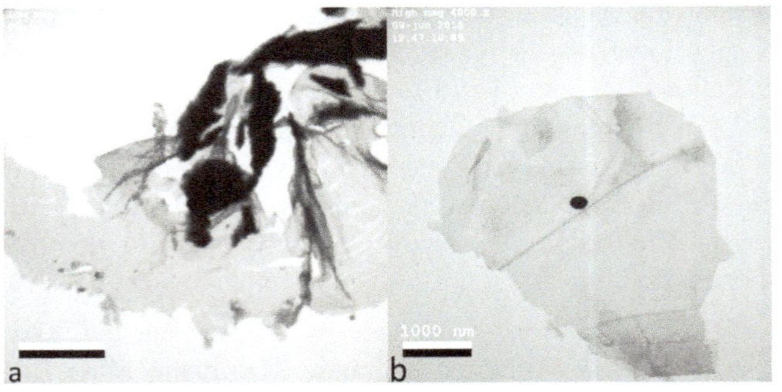

Figure 3: TEM-imaging. TEM-images of particles from the process. a) after 5 passes, a typical particle found in the suspension that fits the frame size 20×20 μm. The scale is 5 μm. b) after 10 passes, a typical particle that fits the frame size 2.5×2.5 μm. The scale is 1000 nm.

Fig 4 shows the particle size distribution from the SEM image analysis. It can clearly be seen on the SEM images, that the suspension from 10 passes has much fewer large and thick particles than the 5 pass sample and a larger amount of small thinner flakes. The frame size represents the surface size of the particles and does not state the particle thickness. The particles transparency in the SEM analysis was used to determine if they were thick or thin. The particles appear to become thinner with increasing number of passes, but retain a relatively high surface area. A selection of SEM images are found under section B. In TEM even smaller thin flakes, in the order of a few hundred nanometers, can be found. These were difficult to observe in the SEM due to the instrumental resolution, so the amount of flakes with surface size below 1 micrometer was probably much higher than the stated value in Fig 4. UV-VIS spectroscopy indicates that these fractions, frame size 2.5×2.5 μm^2 and smaller, corresponds to roughly 4% of the processed batch.

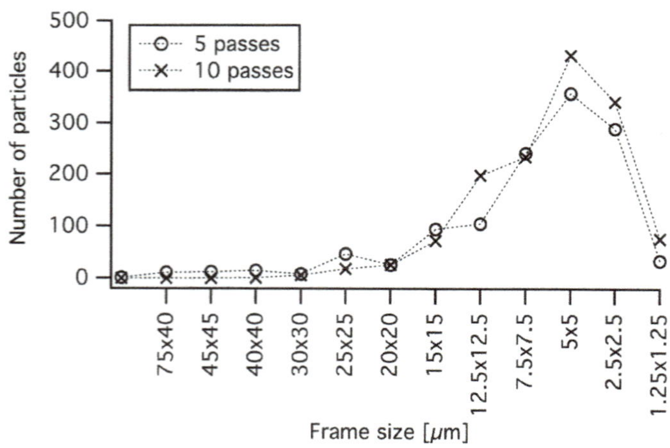

Figure 4: Particle size distribution. Particle size distribution from SEM image analysis.

Fig 5 shows a AFM-image of a typical particle found in the suspension after 10 passes, with the corresponding height profile. It can be seen from the AFM measurements that most of the nanographite flakes are partly folded or wrinkled and not flat against the silicon wafer substrate. The measured average flake thickness after 10 passes was in the range of 10 nm to 20 nm.

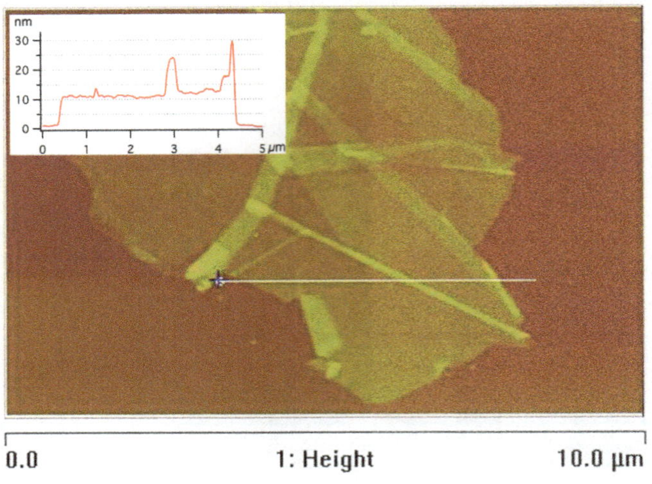

Figure 5: Thickness measurement. AFM image of a partly folded flake found in the suspension after 10 passes with corresponding height profile along the indication bar. The measured flake thickness was 9.12 nm and the flake fits frame size 7.5×7.5 μm.

Electrical properties

Sheet resistance was measured on each film from batch A and B. As shown in Fig 6, the sheet resistance decreased with increased grammage and batch A had an overall lower sheet resistance than batch B. The standard deviation was between 0.01 to 0.03 Ω/sq for batch A and 0.01 to 0.12 Ω/sq for batch B. The biggest deviation was measured on the films with the lowest grammage.

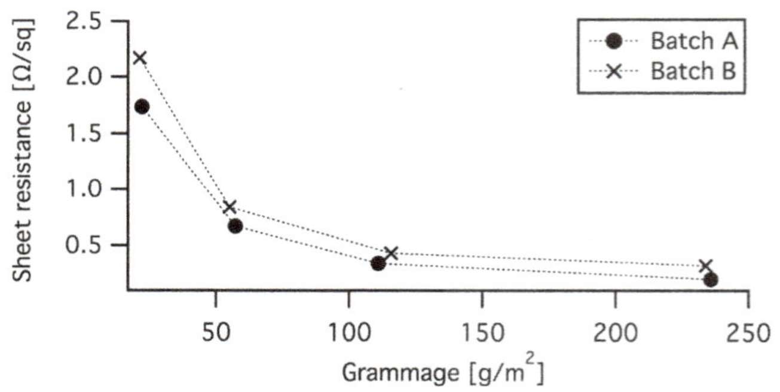

Figure 6: Sheet resistance. Sheet resistance Rs of composite films from batch A and B. The grammage was calculated from the weight and area of the each composite film. The standard deviation was too small to be visualized in the figure.

To compare the composite to commercial highly conductive carbon alternatives, the electrical resistivity was calculated by multiplying the sheet resistance with the film thickness. The electrical resistivity of Batch A was 1.39×10^{-4} Ωm, with a standard deviation of 0.25×10^{-4} Ωm. Batch A had an electrical resistivity of 1.21×10^{-4} Ωm, with a standard deviation of 0.39×10^{-4}Ωm. The highly conductive carbon alternatives used for comparison was battery graphite, carbon black and carbon ink. This types of carbon are often used to enhance the electric conductivity in electrodes and conducting layers. Comparative data was retrieved from the manufacturer and from published papers. The battery graphite was also mixed with NFC, in the same way as batch B, to see how the binder affects the electric properties. The comparison is presented in Table 1. The BET data is only added to be able to compare the resistivity to other carbon alternatives, since a very porous structure is expected to have a higher resistivity than a compact one.

Table 1: Comparison of electrical resistivity and BET surface area for alternative conducting carbons. ρ is the electrical resistivity, SSA is the specific surface area of the material, CB is carbon black and BG is battery graphite.

Sample	Binder	ρ [Ωm]	SSA [m^2g^{-1}]
BatchA	10% NFC	$1.39 \cdot 10^{-4}$	21.0
BatchB	10% NFC	$1.21 \cdot 10^{-4}$	22.4
BG, ABG2025[a]	10% NFC	$8.09 \cdot 10^{-4}$	not specified
BG, ABG2025[a,b]	compressed	$6.61 \cdot 10^{-4}$	19.8
CB, VulcanXC—72 [22]	compressed	$1.35 \cdot 10^{-3}$	254.0
CB, GPN991 [22]	compressed	$1.12 \cdot 10^{-2}$	9.8
LoctiteScreenprint—carbonink [b]	not specified	$3.56 \cdot 10^{-4}$	not specified

Table notes

[a] Ref. [3]

[b] According to the manufacturer.

doi:10.1371/journal.pone.0154686.t001

DISCUSSION

Production rate and scalability

To process a batch of 100 L nanographite with a solids concentration of 20 gL^{-1} under the described shear conditions, with a mean flow (Q_{mean}) of 4.6 $Lmin^{-1}$, took roughly 4 h and consumed about 20 kWh of electricity. This resulted in a production rate (P_R) of 500 gh^{-1}. To scale up the production we have to increase the amount of material passing trough the process without changing the shear conditions. Parameters that can be adjusted in the process are flow, graphite concentration, tube diameter, tube length and number of tubes. The concentration affects the dynamic viscosity and thus both pressure and Reynolds number. The tube diameter directly affects the shear rate and the tube length affects the pressure and shear time. To scale the process without any affect of the shear conditions we can adjust the flow according to the number of shear zones (tubes), N_{tube}. This can be expressed by

$$P_R = 500gh^{-1} \cdot N_{tube} \quad (4)$$

where the flow need to be adjusted as $Q_{mean} = N_{tube} \cdot 4.6\ Lmin^{-1}$. If the number of tubes, in parallel, doubles together with the flow rate, the shear conditions remains the same but the production rate is twice as high. Thus, scaling is achieved by simply increasing the number of tubes, while keeping the shear condition constant.

Comparison with other exfoliation methods

The exfoliation process, described above, enables a fully controlled production where all of the material in the suspension passes the shear zone one time per pass, and all parameters can be monitored and controlled. The parameters, especially pressure, flow rate and number of passes, used in the process are important for the quality of the output. During the optimization of the process, we saw that high pressures and turbulent flow (Reynolds numbers above 4000) together with numerous passes caused over-shearing and cracked the flakes into smaller fragments. We observed that the geometry of the shear zone was important for the quality of the nanographite; a long straight tube and laminar flow seemed to give rise to sufficient shear forces to exfoliate graphite to large sheets and minimize pinch force, which is likely the cause of cracking the sheets. Strong turbulent flow and poorly defined shear zones result in forces acting on the particles which are hard to calculate.

In contrast, to our well defined shear zone in the tube, a commercial homogenizer, which is designed to process food, is not optimal for exfoliation since both shear and pinch forces are present to tear and smash particles into smaller pieces resulting in too small graphene and nanographite flakes. This is not the case in tube-shearing. The same problem, as with homogenizers, may occur in processes with rotary dispersers (high-shear mixing). In these processes it is difficult to know the forces acting on the particles in the inlet and outlet of the direct shear zone. Another uncertainty with rotary dispersers is to determine how many times the particles have passed the shear zone, when exfoliated and unexfoliated graphite are located in the same container.

Two disadvantage with tube-shearing concerns limitation in the maximun concentration and the design restriction on the tube geometry. The limit of concentration is due to the need for pumping the suspension into the tube that have a maximum allowed viscosity. This problem is not necessarily the case for rotary dispenser but a commercial homogenizer faces the same problem. Design restriction on the tube geometry is due to the requirements, firstly, to be in the region of laminar flow to avoid cracking due to pinch forces, secondly, to maximize the shear rate to increase the exfoliation rate. However, as discussed above, scaling is simply achieved by increasing the number of tubes in the process equipment.

Composites for electrode applications

One significant difference between the composite films was the film density. In batch A, the NFC was added during tubes-shearing while in batch B the NFC was mixed with nanographite after tube-exfoliation, using a high-shear mixer. The density in batch A was the same for all films which indicates a well-dispersed and stable composite. In batch B we could see large variations in density for the four different films. This may indicate that the hydrodynamic tube shearing process is more suitable to get a well-dispersed composite than the high-shear mixing. This leads to superior coating ability of electrodes on large-area substrates with good mechanical and adhesion properties as compared to pure nanographite which leads to cracked films that do not adhere to the substrate, as seen in [3] for a slightly different system. The specific surface area of the composite was low, most likely because the flakes are stacked on each other forming a compact non-porous material.

There are various applications for exfoliated graphite composites. The nanographite composite produced in the described process had a low sheet resistance and low electrical resistivity, which suit applications as a highly conducting matrix around activated carbons in supercapacitor electrodes. In this application a wide size distribution can be beneficial as it can give rise to more pores and increase the surface area. In comparison with carbon black and battery graphite, the nanographite composite has lower electrical resistivity even though it is uncompressed and contains binder. Another suitable application for the composite is as electrical conductor in printed electronics due to its higher conductivity compared with carbon ink, together with a robust and durable structure.

CONCLUSION

We have demonstrated a hydrodynamic tube-shearing process suitable for producing large quantities of nanographite. This process produces micrometer-wide and nanometer-thick flakes of nanographite. The exfoliation occurs in an aqueous environment without any toxic chemicals or organic solvents, making it environmentally friendly. The energy consumption in the process is about 10 kWh per kilogram processed graphite, making it low cost. The process is easy to scale up even further, to increase the production rate, by using multiple tubes in parallel. The process is thus environmentally friendly, cost-efficient, and suitable for industrial implementation. We also demonstrated production of highly conductive and robust carbon composites by adding NFC during the process suitable for large-area coating of electrodes.

ACKNOWLEDGMENTS

We thank STT Emtec AB for construction and service of the equipment used in the exfoliation process. We also thank Vesta Lab Sweden AB for helping us with BET measurements.

AUTHOR CONTRIBUTIONS

Conceived and designed the experiments: NB. Performed the experiments: NB ACE. Analyzed the data: NB ACE MH BA. Contributed reagents/materials/analysis tools: NB ACE MH HO SF. Wrote the paper: NB HO. Responsible for funding application: SF. Proofreading and revising: BA ACE SF HO.

REFERENCES

1. Li D, Kaner RB. Graphene-based materials. Nat Nanotechnol. 2008;3:101. doi: 10.1126/science.1158180
2. Malho JM, Laaksonen P, Walther A, Ikkala O, Linder MB. Facile method for stiff, tough, and strong nanocomposites by direct exfoliation of multilayered graphene into native nanocellulose matrix. Biomacromolecules. 2012;13(4):1093–1099. doi: 10.1021/bm2018189. pmid:22372697
3. Andres B, Forsberg S, Dahlström C, Blomquist N, Olin H. Enhanced electrical and mechanical properties of nanographite electrodes for supercapacitors by addition of nanofibrillated cellulose. physica status solidi (b). 2014;251(12):2581–2586. doi: 10.1002/pssb.201451168.
4. Huang X, Qi X, Boey F, Zhang H. Graphene-based composites. Chemical Society Reviews. 2012;41(2):666–686. doi: 10.1039/C1CS15078B. pmid:21796314
5. Chang H, Wu H. Graphene-based nanocomposites: preparation, functionalization, and energy and environmental applications. Energy & Environmental Science. 2013;6(12):3483–3507. doi: 10.1039/c3ee42518e.
6. Tölle FJ, Fabritius M, Mülhaupt R. Emulsifier-Free Graphene Dispersions with High Graphene Content for Printed Electronics and Freestanding Graphene Films. Advanced Functional Materials. 2012;22(6):1136–1144. doi: 10.1002/adfm.201102888.
7. Bianco A, Cheng HM, Enoki T, Gogotsi Y, Hurt RH, Koratkar N, et al. All in the graphene family–a recommended nomenclature for two-dimensional carbon materials. Carbon. 2013;65:1–6. doi: 10.1016/j.carbon.2013.08.038.
8. Backes C, Smith RJ, McEvoy N, Berner NC, McCloskey D, Nerl HC, et al. Edge and confinement effects allow in situ measurement of size and thickness of liquid-exfoliated nanosheets. Nature communications. 2014;5. doi: 10.1038/ncomms5576. pmid:25099520
9. Ciesielski A, Samorì P. Graphene via sonication assisted liquid-phase exfoliation. Chemical Society Reviews. 2014;43(1):381–398. doi: 10.1039/C3CS60217F. pmid:24002478
10. Paton KR, Varrla E, Backes C, Smith RJ, Khan U, O'Neill A, et al. Scalable production of large quantities of defect-free few-layer graphene by shear exfoliation in liquids. Nature materials. 2014;13(6):624–630. doi: 10.1038/nmat3944. pmid:24747780
11. Nacken TJ, Damm C, Walter J, Rüger A, Peukert W. Delamination of graphite in a high pressure homogenizer. RSC Advances. 2015;5(71):57328–57338. doi: 10.1039/C5RA08643D.
12. Texter J. Graphene dispersions. Current Opinion in Colloid & Interface Science. 2014;19(2):163–174. doi: 10.1016/j.cocis.2014.04.004.
13. Cravotto G, Cintas P. Sonication-Assisted Fabrication and Post-Synthetic Modifications of Graphene-Like Materials. Chemistry–A European Journal. 2010;16(18):5246–5259. doi: 10.1002/chem.200903259.
14. Niu L, Coleman JN, Zhang H, Shin H, Chhowalla M, Zheng Z. Production of Two-Dimensional Nanomaterials via Liquid-Based Direct Exfoliation. Small. 2016;12(3):272–293. doi: 10.1002/smll.201502207. pmid:26663877
15. Shen Z, Li J, Yi M, Zhang X, Ma S. Preparation of graphene

by jet cavitation. Nanotechnology. 2011;22(36):365306. doi: 10.1088/0957-4484/22/36/365306. pmid:21844642

16. Zhao W, Wu F, Wu H, Chen G. Preparation of colloidal dispersions of graphene sheets in organic solvents by using ball milling. Journal of Nanomaterials. 2010;2010:6. doi: 10.1155/2010/528235.
17. Jeong SY, Kim SH, Han JT, Jeong HJ, Yang S, Lee GW. High-performance transparent conductive films using rheologically derived reduced graphene oxide. Acs Nano. 2011;5(2):870–878. doi: 10.1021/nn102017f. pmid:21261292
18. Knieke C, Berger A, Voigt M, Taylor RNK, Röhrl J, Peukert W. Scalable production of graphene sheets by mechanical delamination. Carbon. 2010;48(11):3196–3204. doi: 10.1016/j.carbon.2010.05.003.
19. Shi JN, Ger MD, Liu YM, Fan YC, Wen NT, Lin CK, et al. Improving the thermal conductivity and shape-stabilization of phase change materials using nanographite additives. Carbon. 2013;51:365–372. doi: 10.1016/j.carbon.2012.08.068.
20. McCabe WL, Smith JC, Harriott P. Unit operations of chemical engineering. vol. 5. McGraw-Hill New York; 1993.
21. Saito T, Hirota M, Tamura N, Kimura S, Fukuzumi H, Heux L, et al. Individualization of nano-sized plant cellulose fibrils by direct surface carboxylation using TEMPO catalyst under neutral conditions. Biomacromolecules. 2009;10(7):1992–1996. doi: 10.1021/bm900414t. pmid:19445519
22. Pantea D, Darmstadt H, Kaliaguine S, Sümmchen L, Roy C. Electrical conductivity of thermal carbon blacks: Influence of surface chemistry. Carbon. 2001;39(8):1147–1158. doi: 10.1016/S0008-6223(00)00239-6.

CHAPTER 20

RECENT PROGRESS IN THE GROWTH AND APPLICATIONS OF GRAPHENE AS A SMART MATERIAL: A REVIEW

Brahim Aïssa[1,2], Nasir K. Memon[1], Adnan Ali[1] and Marwan K. Khraisheh[1]

[1]Qatar Environment and Energy Research Institute (QEERI), Qatar Foundation, Doha, Qatar

[2]Department of Smart Materials and Sensors for Space Missions, MPB Technologies Inc., Montreal, QC, Canada

Innovative breakthroughs in fundamental research and industrial applications of graphene material have made its mass and low-cost production as a necessary step toward its real world applications. This one-atom thick crystal of carbon, gathers a set of unique physico-chemical properties, ranging from its extreme mechanical behavior to its exceptional electrical and thermal conductivities, which are making graphene as a serious alternative to replace many conventional materials for various applications. In this review paper, we highlight the most important experimental results on the synthesis of graphene material, its emerging properties with reference to its smart applications. We discuss the possibility to successfully integrating graphene directly into device, enabling thereby the realization of a wide range of applications, including actuation, photovoltaic, thermoelectricity, shape memory, self-healing, electrorheology, and space missions. The future outlook of graphene is also considered and discussed.

INTRODUCTION

Graphene material is considered as the first lab-made 2D atomic crystal. Because of their unique physical and chemical properties – such as mechanical stiffness, strength and elasticity, and extremely high electrical and thermal conductivity (Geim and Novoselov, 2007a; Geim, 2009) – graphene is described to be a serious alternative to replace many conventional materials in various applications, and could enable many disruptive innovation and potentially existing markets. For example, the combination of optical transparency, electrical and thermal conductivities, and mechanical elasticity will find application either in flexible electronics and/or transparent coatings, and the list of such combinations is continuously growing (Figure 1). Basically, graphene is a single 2D layer of carbon atoms, with a typical thickness of 0.34 nm. It is sp^2 hybridized, where carbon atoms are covalently bonded to three other atoms in a hexagonal lattice structure (Geim, 2009;Layek and Nandi, 2013). Recently, graphene has been extensively investigated, both in terms of fundamental research and R&D applications. Graphene was isolated for the first time by Novoselov et al. (2004) what was worth to them the 2010 Nobel Prize in Physics for their groundbreaking work. Their unprecedented structural and physico-chemical properties (especially its mechanical and electrical behaviors) in addition to its carrier mobility – the highest know to date, at room temperature – makes the research on graphene one of the most important topics in all

Citation: Aïssa B, Memon NK, Ali A and Khraisheh MK (2015) Recent progress in the growth and applications of graphene as a smart material: a review. Front. Mater. 2:58. doi: 10.3389/fmats.2015.00058

materials science fields (Basu and Bhattacharyya, 2012). On the other hand, graphene' structure serves as the basic shape of almost all other carbonaceous materials, including fullerene (Muge and Chabal, 2011), single and multi-walled carbon nanotubes (Hassan, 2012), and even graphite, which is simply a multiple layers graphene (Basu and Bhattacharyya, 2012). Literature survey shows that graphene material has numbers of potential applications, including nanoelectronic like-devices, gas sensors, hydrogen storage, and polymer-based nanocomposites (Boukhvalov et al., 2008;Ponnamma et al., 2010; Schwierz, 2010; Casolo et al., 2011), and could serve as an ideal prototype to investigate the properties of many other 2D nanosystems, such as 2D silicon and silicon carbide (2D-*SiC*), zinc oxide, boron nitride, and germanium (Elias et al., 2009; Bekaroglu et al., 2010;Houssa et al., 2010; Tang and Cao, 2010; Voon et al., 2010; Zhang et al., 2010a).

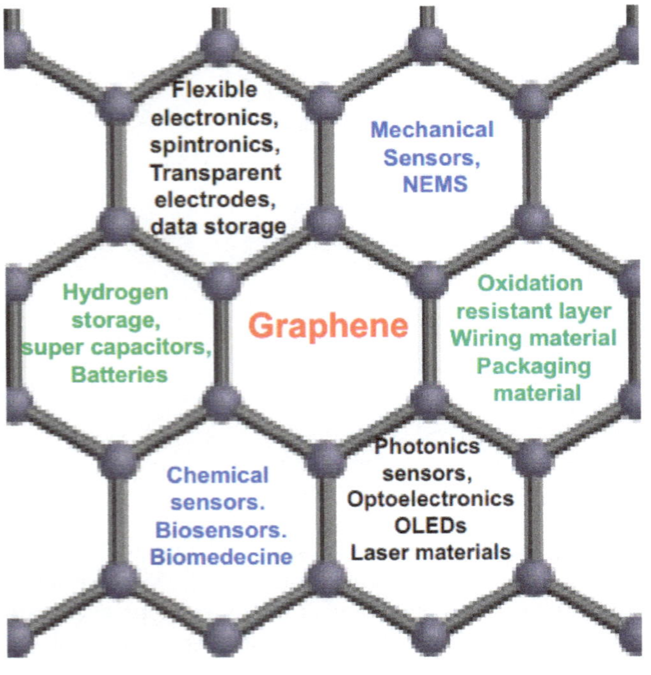

Figure 1: Various potential applications of graphene material.

Experimentally measured properties of graphene have not only exceeded those obtained in any other material but also reached very often its theoretically predicted limits. A typical example is its room-temperature carrier mobility, of 2.5×10^5 cm²/Vs, which is found to be very close to the theoretical limit of 2×10^5 cm²/Vs (Mayorov et al., 2011). Many other representative examples could be found in the relevant literature, just to cite a few: a Young's modulus of 1 TPa very close to that predicted by theory (Liu et al., 2007; Lee et al., 2008; Morozov et al., 2008); a thermal conductivity of 3000 W mK^{-1} (Balandin, 2011); an optical absorption of 2.3% in the infrared (Nair et al., 2008); its property to be completely impermeable to gases (Bunch et al., 2008); its ability to carry one million time higher densities of electrical current than copper (Moser et al., 2007), and its potential to be chemically functionalized (Elias et al., 2009; Loh et al., 2010; Nair et al., 2010). It is worth noting that the majority of these properties have been experimentally measured for a high-quality graphene samples, deposited on specific substrates, such as hexagonal boron nitride (Dean et al., 2010; Mayorov et al., 2011). However, similar properties have not been observed so far on graphene material prepared using classical techniques, although these conventional processes are continuously improving (Neto et al., 2009; Sarma et al., 2011). We conclude then that the challenge related to find markets of graphene applications is mainly related to the real progress realized in its mass production with appropriate characteristics. The number of publications related to the graphene material is continually growing has increased dramatically especially in last years (from about 4000 in 2010 to more than 14,000 in 2014) (Choi et al., 2010). Figure 2 shows how the number of refereed articles dealing with graphene material has steadily increased since 2004, based on data collected from the Engineering Village web-based information service.

The same tendency is also recorded with patents applications, which have downright doubled within 2 years only [i.e., from 2010 to 2012 with a total of 8416 patents worldwide by February 2013 (Zhang et al., 2013)].

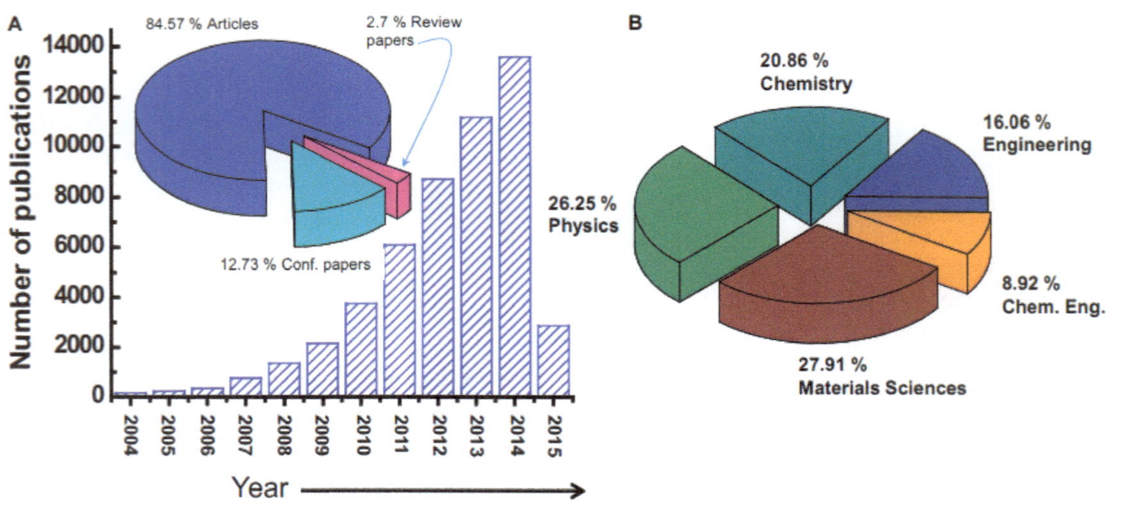

Figure 2: (A) Number of the publications/year on the graphene materials. The inset is the distribution of the document type, where only 2.7% of the publications are related to a review work. **(B)** Distribution of the publications by subject area.

In the recent years, there have been many review works, related either to theoretical and/or experimental studies, discussing the topics of synthesis and application of graphene material. To cite just few recent works, Neto et al., (2009) reviewed the electrical properties of graphene, and then focused on its electronic transport properties (Neto et al., 2009). Other experimental reviews included detailed discussions of synthesis (Zhang et al., 2010a) and Raman spectroscopy processes of transport mechanisms (Ni et al., 2008; Avouris, 2010; Giannazzo et al., 2011), related to electronic applications graphene, including transistors like-devices, bandgap engineering (Loh et al., 2010), and optoelectronic technologies (Bonaccorso et al., 2010; Schwierz, 2010). However, among all the published articles on the matter, only 18 review works have been conducted on the smart applications of the graphene materials (Figure 3), and to the best of our knowledge, only one review-article has been published on 2015. In sum, along with the increase in the number of publications in this area comes a need for a comprehensive review article, and the objective of this paper is to address this need. The literature is indeed lacking a comprehensive review of the recent experimental advancements on graphene material and its smart applications. This is the aim of our article. However, due to the huge number of various works that are involved, and often the unavailability of access to many conference proceedings, the emphasis of this paper was on the most accessible refereed journal articles. Obviously, it was not possible and practical to cover all of these articles, especially since a lot of them had already been covered by previous paper review; an attempt was made to select representative articles in each of the relevant categories. This review should be particularly well suited to graduate students who desire an introduction to the study of graphene that will provide them with many references for further reading.

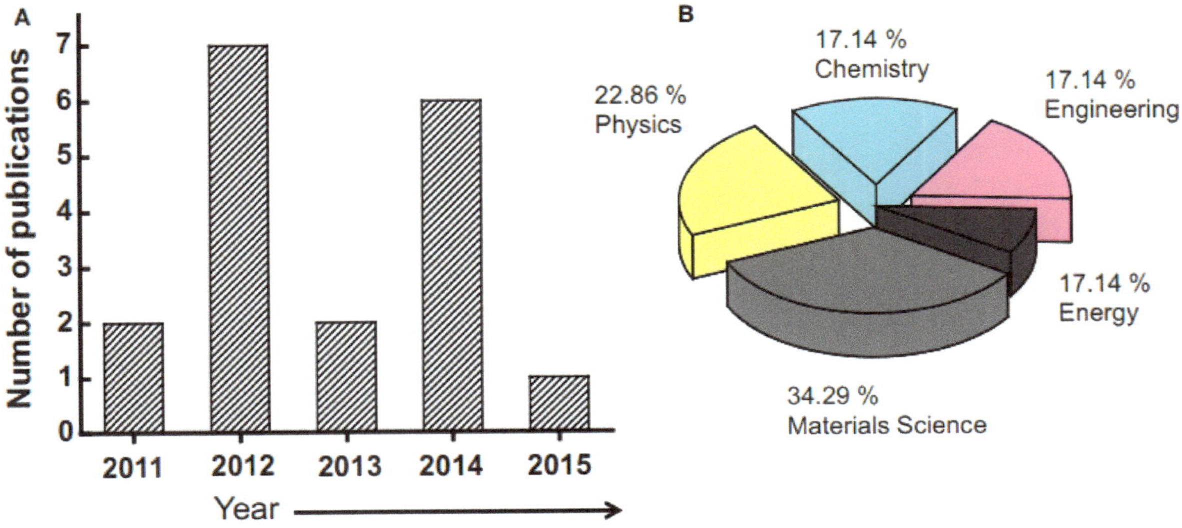

Figure 3: (A) Number of the review publications/year on the graphene materials. **(B)**Distribution of the review publications per subject area.

GRAPHENE SCALABLE SYNTHESIS PERSPECTIVES

Initially discovered by micromechanical exfoliation of graphite (Geim and Novoselov, 2007b), graphene has generated widespread interest as a smart material. However, for graphene to make a significant impact within industry, it is important to develop methods for scalable synthesis of high-quality graphene. The current common production methods for graphene include liquid exfoliation, ultrahigh vacuum processes, annealing of silicon carbide (*SiC*), and obviously, chemical vapor deposition (CVD). Other methods, which could be used for scalable graphene synthesis, include plasma enhanced CVD, flame synthesis, and pulsed laser deposition (PLD). These methods will be discussed in this review with a focus on identifying processes that can be translated for commercial production of graphene. This review does not cover methods for the production of graphene oxide.

Micromechanical Exfoliation

Micromechanical exfoliation involves peeling highly ordered pyrolytic graphite (HOPG) using adhesive tape (Novoselov et al., 2005). Since each layer of graphene is bonded to the other layer by van der Waals bonding, it is feasible to cleave HOPG. Normally, the peeling is performed several times. This process can also be used to produce few layers graphene (FLG). While this is the simplest method for the production of graphene and is commonly used in laboratory experiments, the production method is not scalable for large-scale graphene growth.

Liquid-Phase Exfoliation

Liquid-phase exfoliation (LPE) involves using a solvent to exfoliate graphite by ultrasonication (Hernandez et al., 2008; Lotya et al., 2009). Commonly used solvents include acetic acid, sulfuric acid, and hydrogen peroxide (Singh et al., 2011). The ultrasonication time is typically 60 min with a power of 250–500 W. Green and Hersam (2009) reported the use of sodium cholate as a surfactant for the exfoliation of graphene. Moreover, they were able to separate the sheets by density gradient ultracentrifugation, which enabled the isolation of graphene from FLG. LPE can be used for the production of graphene nanoribbons (GNRs) (Li et al., 2008), where the width is <10 nm. While LPE represents a scalable method for the production of graphene, large scale film growth remains really challenging.

Chemical Vapor Deposition Based Synthesis

Chemical vapor deposition of graphene involves the use of transition metals, where nickel (Ni) (Reina et al., 2008; Chae et al., 2009; Kim et al., 2009; Losurdo et al., 2011) and copper (Cu) (Li et al., 2009; Bae et al., 2010; Guermoune et al., 2011; Suk et al., 2011; Wang et al., 2011) are suitable for large scale production of graphene.

Graphene growth based on CVD has shown exceptional device properties (Figure 4) (Bae et al., 2010), with electron mobility of 7350 $cm^2\ V^{-1}s^{-1}$ (Novoselov et al., 2005). In addition, large scale production of 30″ graphene films was demonstrated using roll-to-roll CVD (Mattevi et al., 2011). The graphene obtained from this process was of high quality, with a sheet resistance of ~125 Ω/square and 97.4% optical transmittance.

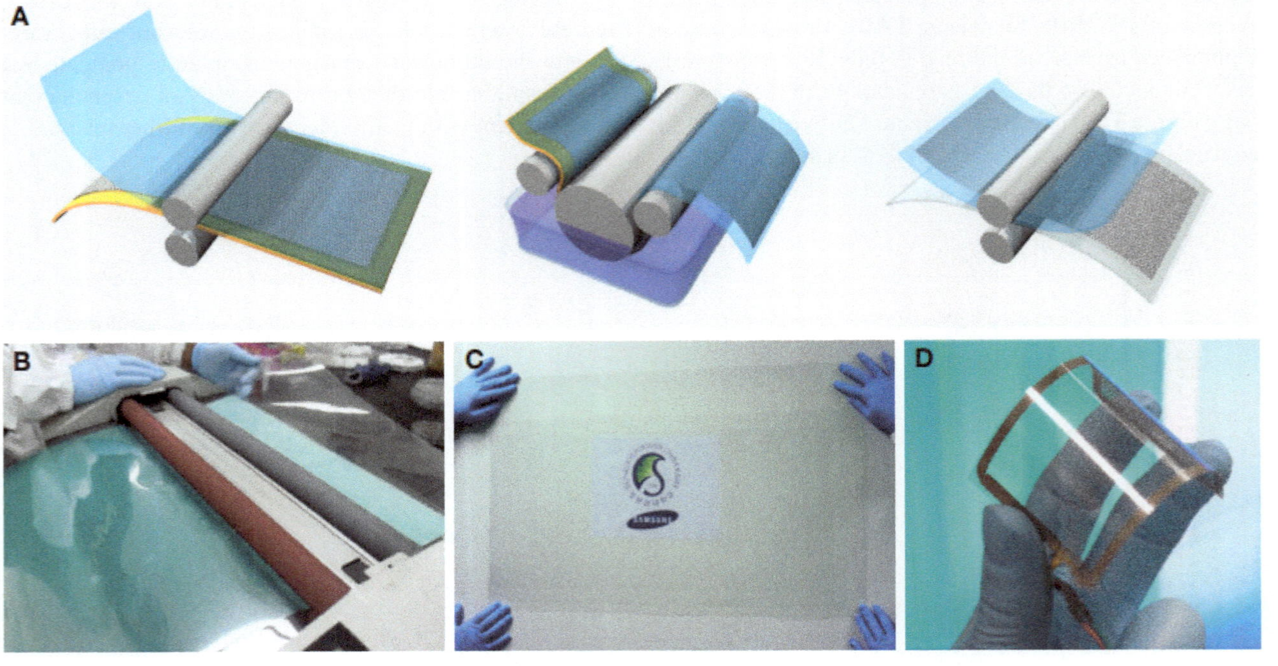

Figure 4: (A) Schematic of the roll-to-roll production of graphene films grown on a copper foil. **(B)** Roll-to-roll transfer of graphene films from a thermal release tape to a PET film at 120°C. **(C)** A transparent large-area graphene film transferred on a 35″ PET sheet. **(D)** An assembled graphene/PET touch panel showing mechanical flexibility. Reproduced with permission of Bae et al. (2010). Copyright 2010, Nature Nanotechnology.

Graphene growth using CVD is fairly straightforward, where a copper or nickel substrate is placed in a reactor at temperatures normally around 1000°C. The initial step in the process is to introduce hydrogen in the reactor. This step is critical to eliminate any oxide layer present in the metal, for the case of Cu this will reduce any native layers of CuO and Cu_2O. The hydrogen atmosphere also enables the growth of grain boundaries (Mattevi et al., 2011), which is necessary for the growth of high-quality graphene. Afterwards, a hydrocarbon gas (typically methane) is added to the reactor. The hydrocarbon gas provides the necessary carbon species used in the growth of graphene. The hydrocarbon gas to hydrogen ratio plays an important role in the growth of graphene. If insufficient hydrogen is present, this could result in oxidized metal layers being present, which can lead to a disordered graphene structure. By contrast, excess hydrogen has shown to etch away graphene. On polycrystalline substrates, the graphene flakes tend to have different lattice orientations.

Using CVD, graphene is grown onto transition metals, which enables a low-energy pathway by forming intermediate compounds for the growth of graphene. The first row of transition metals Fe, Co, Ni, and Cu is of great interest due to their low cost and high availability. The difference in the carbon solubility between these metals impacts the growth quality, where Fe has the highest and Cu has the lowest carbon solubility. For this reason, Cu is an ideal metal for growing single layer graphene. When using Ni and Co it is common to get up to 10 layers of graphene. Similarly, on Fe it is common to have FLG.

Most practical applications of graphene require that the underlying surface be insulating. For this reason, graphene must be transferred to an insulating surface, such as SiO_2 (Bhaviripudi et al., 2010). Additionally, this transfer is required to measure the optoelectronic properties of the synthesized graphene. The commonly used process to transfer graphene is to first deposit and cure poly (methylmethacrylate) (PMMA) on the metal sheet. Afterwards, etch the Cu metal sheet using iron chloride. This gives a floating sheet of PMMA and graphene, which is rinsed in deionized water. Subsequently, transfer this layer to an insulating surface and use acetone to remove the PMMA layer.

Plasma Enhanced Chemical Vapor Deposition Synthesis

Plasma enhanced CVD (PECVD) is another method used for the synthesis of graphene that is comparable to the thermal CVD process (Zhu et al., 2007; Yuan et al., 2009; Kim et al., 2011; Bo et al., 2013). PECVD is based on a number of plasma sources, such as microwave (MW) (Malesevic et al., 2008), radio frequency (RF) (Wang et al., 2004), and direct current (dc) arc discharge (Krivchenko et al., 2012) have been utilized in the growth of graphene. Copper and nickel are typically used as the substrate for PECVD graphene growth; however, a number of additional substrates have also been used (Dato et al., 2008; Bo et al., 2013). A particularly exciting technique is a substrate free method based on the decomposition of ethanol in MW-based PECVD reactor (Guermoune et al., 2011). Such methods provide can be used for scalable production of graphene powder. Typical growth conditions of PECVD graphene on a substrate are 5–100% CH_4 in H_2 with a substrate temperature of 500–800°C (Singh et al., 2011; Bo et al., 2013). The power of the plasma is 900 W. Such processes can enable the growth of graphene at lower temperatures and shorter duration (<5 min). However, the quality of the graphene film is typically lower when compared to thermal CVD.

The growth of graphene films at low temperature is important for a number of applications. For instance, fabrication of graphene for high-performance display glass has to be at a temperature lower than 660°C (Ren et al., 1998; Bo et al., 2013). Furthermore, production at lower temperature can provide new manufacturing opportunities in the area of flexible electronics based on plastics (Tung et al., 2009). Critical challenges still need to be overcome, as at lower temperatures the graphene film tends to have a higher disorder.

Flame Synthesis

Flame synthesis is extensively used to produce commercial quantities of nanoparticles. Of the most widely used nanoparticles, carbon black, fumed silica, and titania, flame synthesis is the dominant method in the production of these materials. Volumetric production of the flame synthesis industry is on the order of 100 metric tons per day (Kammler et al., 2001). A key advantage of flames is that it readily provides the high temperature necessary for gas-phase synthesis along with a carbonizing or oxidizing environment.

With respect to graphene, flame synthesis is not as commonly studied when compared to CVD, but it offers several important advantages, such as scalability and cost effectiveness. The most commonly used flame types include premixed, normal diffusion, inverse diffusion, and co-flow (Inoue et al., 2010; Memon et al., 2013). Since early 2000s, a number of researchers have focused on the use of flames for CNT synthesis (Inoue et al., 2010). However, the development of flame synthesis for graphene is still in its early stage. In addition to flame type, other parameters including temperature, species concentration, and velocity impact the growth process. Graphene being a two-dimensional material requires large-scale production across a substrate. Due to the temperature and species gradients that occur in most flames it is difficult to scale the growth of graphene across an entire substrate. Moreover, a reduced environment with carbon rich species, which is necessary for graphene growth, is difficult to achieve in most flames. Nevertheless, flame synthesis has the potential to economically enable the mass production of graphene.

Similar to earlier CNT flame synthesis papers, where the growth of CNTs was observed near the soot region of a premixed flame, carbon particles containing graphene were observed in Bunsen (propane) flame (Ossler et al., 2010). These particles were collected by placing a transmission electron microscopy grid 2 cm above the tip of the burner. The grid was held within the flame for 10–50 ms. The graphene films were several hundreds of nanometers in size.

In an attempt to grow graphene on copper, Li et al. (2011b) investigated the growth of graphene using an ethanol burner. The substrate was placed within the flame at a temperature of 550–700°C and the flame was extinguished using a cap to prevent the oxidation of the copper foil. The growth of an amorphous carbon film was observed on the substrate and XPS confirmed the formation of sp^2, sp^3, and C–O bonded atoms. Graphene was not observed due to the low temperature and the presence of oxygen within the flame. In a different experiment, Li et al. (2011a) were able to synthesis graphene successfully on nickel. The process utilized two different burners (burner 1 and burner 2), with the substrate situated within the interior region of the flame structure itself (Figure5). Burner 1 (alcohol burner) surrounded the substrate for the entire time, where it prevented air oxidation and served as the carbon source. Burner 2 (butane-fueled Bunsen burner) provided the additional heating of the substrate and served as the carbon source for graphene growth. The flame was extinguished using a cap. There are still numerous challenges in using flame synthesis for the growth of graphene, specifically in developing methods that result in higher quality graphene.

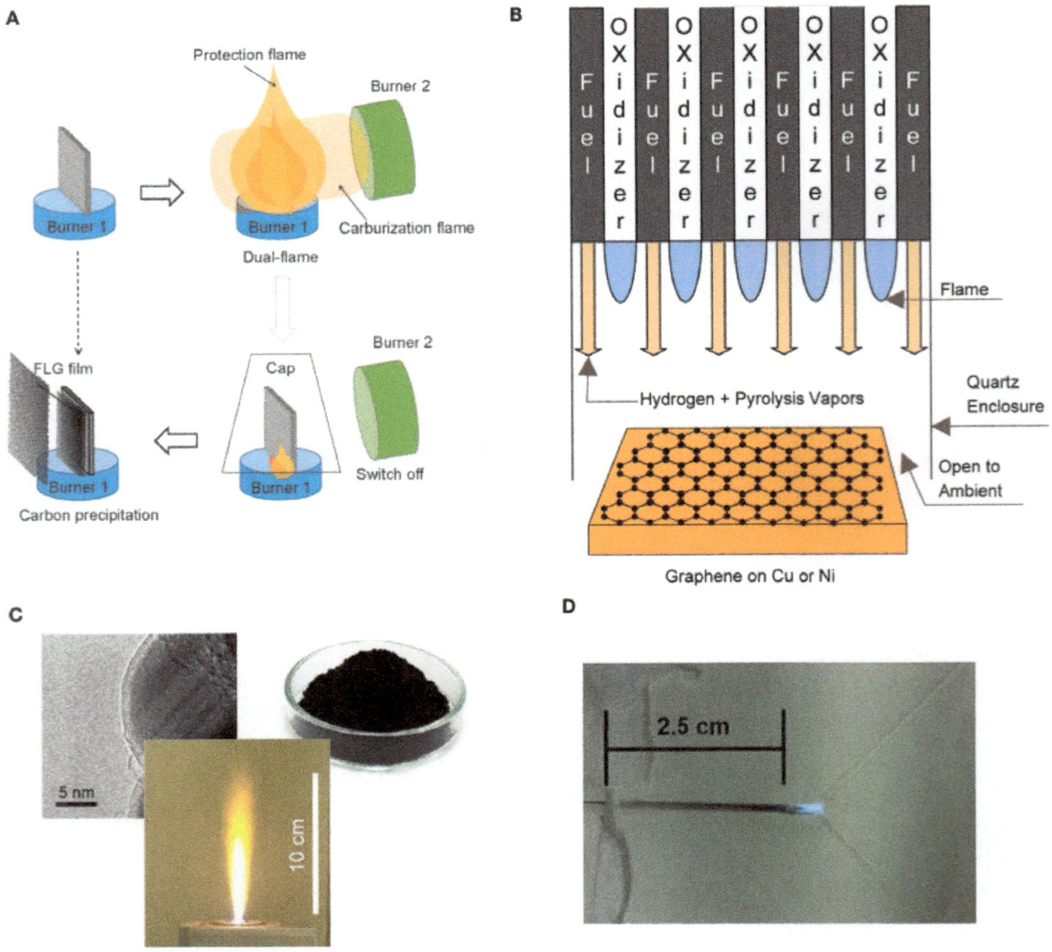

Figure 5: Different configurations used for the flame synthesis of graphene, (A) dual flame configuration, (B) multiple inverse diffusion flames, (C) flame spray pyrolysis and (D) microcombustor. Reproduced with permissions fromLuechinger et al. (2008), Li et al. (2011b), Memon et al. (2011), and Kellie et al. (2013).

Epitaxial Growth on Silicon Carbide Substrate

Yannopoulos et al. (2012) have investigated the thermal decomposition of *SiC* surface, which was providing an epitaxial growth of graphene material (Figure 6). They reported a new process using a CO_2 laser as the heating step for a fast and one-step growth process of large uniform graphene film on *SiC*. This method can control the stacking order of epitaxial graphene and is cost-effective since it does not involve any pretreatment step or high-vacuum process. The decomposition operated at low temperature and proceeded in the second time scale, thus providing a means to engineering graphene patterns on *SiC* by focused laser beams.

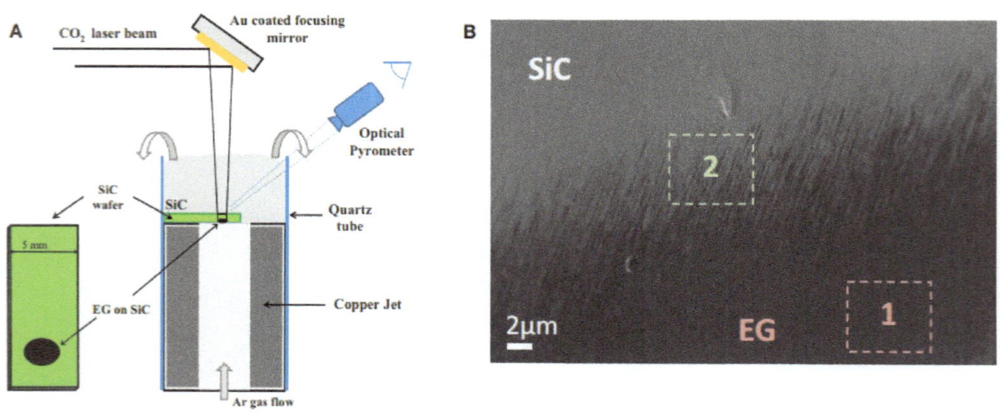

Figure 6: (A) Schematic diagram of the CO_2 laser induced epitaxial growth of graphene on*SiC* wafers. **(B)** SEM micrograph showing the formation of epitaxial growth graphene (Zone 1) on 6H-*SiC* (0001). Reproduced with the permission from Yannopoulos et al. (2012). Copyright 2012, Wiley-VCH.

Pulsed Laser Deposition

The PLD process is definitely considered as one of the most versatile growth approaches. Since the laser energy source is located outside the deposition chamber; the use of either ultrahigh vacuum or ambient gas becomes possible (Krebs et al., 2003). Combined with a stoichiometry transfer between ablated target and substrate where the material is deposited, this flexibility allows depositing theoretically all possible kinds of materials, including polymers or fullerenes (Eason et al., 2006). This technique was first employed by Smith et al. (1965) (Krebs et al., 2003) in 1965 to elaborate semiconductors and dielectric thin films. In 1987, it was then fully developed byDijkkamp et al. (1987) for the deposition of high-temperature superconductors. Their work allowed to define the main characteristics of PLD, namely, the stoichiometry transfer between target and deposited film (Smith et al., 1965; Dijkkamp et al., 1987; Chrisey and Hubler, 1994; Eason et al., 2006). Since the work of Dijkkamp et al., the deposition technique has been extensively used for all kinds of oxides, nitrides, carbides and also for preparing metallic systems and even polymers or fullerenes (Krebs et al., 2003). During PLD, almost all experimental parameters can be adjusted to control the film growth, and ranging from the laser parameters (Fluence, wavelength, pulse-duration, and repetition rate), to the deposition conditions (target-to-substrate distance, temperature, nature of the gas, pressure, etc.).

A representative schematic diagram for PLD (Krebs et al., 2003) is shown in Figure 7. Inside the vacuum chamber (ultrahigh vacuum, UHV), targets of elementary or alloy elements are struck at an angle of 45° by a high energy focused pulsed laser beam. The atoms and ions ablated from the target(s) are then deposited directly on the substrate (Krebs et al., 2003). In the majority of the cases, the substrates are attached with their surfaces parallel to the target(s) surfaces at a distance of 2–10 cm.

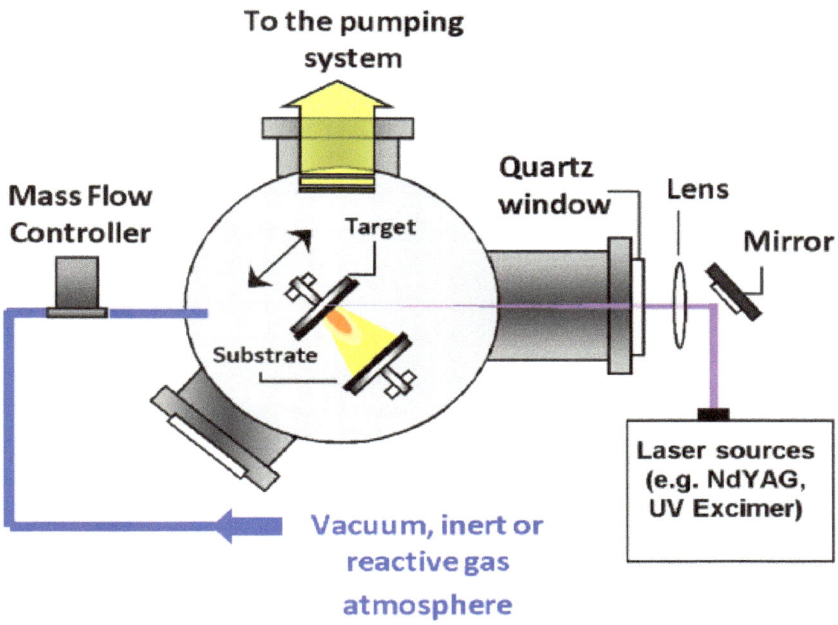

Figure 7: Schematic of a representative laser deposition tool.

To the benefit of the reader, Figure 8 summarizes the main first-deposited materials since the introduction of PLD in 1987, with respect to the year for deposition and corresponding reference.

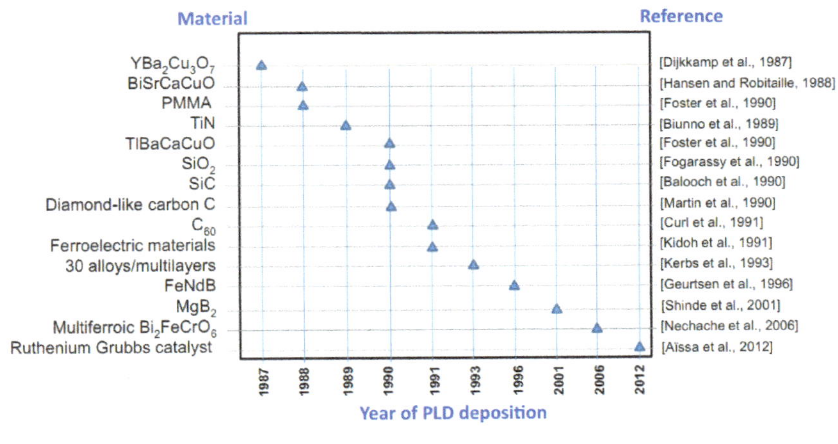

Figure 8: List of the materials deposited for the first time by PLD after 1987.

As mentioned earlier, in the PLD process, one of the main advantages is the fact that during deposition, the stoichiometry of the deposited material is very close to the target (Krebs et al., 2003). Consequently, it is possible to prepare stoichiometric thin films from a single alloy bulk target (Krebs et al., 2003).

In the context of the graphene growth, and in parallel to the CVD deposition method, physical vapor deposition has also been reported for the growth of graphene (Koh et al., 2010; Zhang and Feng, 2010b). In UHV, PLD chambers graphite is normally used as the target with a transition metal as the substrate (Figure 9). A substrate temperature of 1300°C was reported for 1–2 layers of high-quality graphene (Figure 9) (Zhang and Feng, 2010b). No carbide formation occurs at the interface of graphene and the metal (Zhang and Feng, 2010b). While numerous metals can be used as a catalyst, nickel appears to be the most promising for low temperature growth. Numerous parameters, such as the cooling rate and laser power, impact the quality of graphene films (Koh et al., 2010).

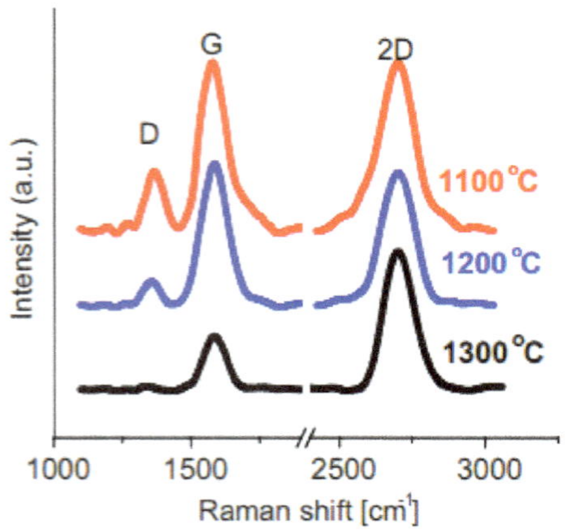

Figure 9: Graphene deposition by means of PLD. Measured Raman spectra with respect to the growth temperature. Reproduced with permission from Zhang and Feng (2010b). Copyright 2010, Elsevier.

Laser-Based Chemical Vapor Deposition

A continuous wave (CW) laser is utilized for laser-based CVD in an enclosed chamber (Figure 10A) (Park et al., 2011). The precursor gases used include methane and hydrogen, with Ni foil as the substrate. The synthesis mechanism is based on a vapor–liquid–solid that only takes nanoseconds to picoseconds. The spectra of the Raman spectroscopy showing the different graphene layers are illustrated in Figure 10B. A key advantage of this process is that it can be used for graphene lithography, where the laser can be scanned on specific areas of a metal catalyst enabling direct growth.

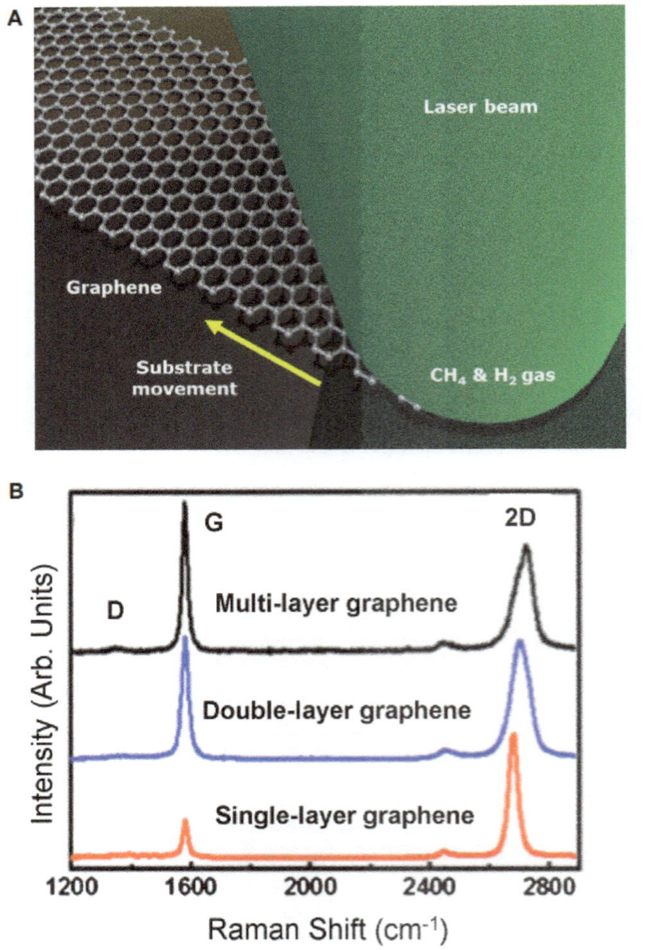

Figure 10: **(A)** Schematic of the laser induced CVD process, and **(B)** mea-sured Raman as a function of the number of graphene layers. Reprinted with permission from Park et al. (2011). Copyright 2011, American Institute of Physics.

Laser Growth Directly on Silicon and Quartz Substrates

Sun et al. (2010) produced graphene on Cu and Ni film using PMMA. Due to the existence of the metal films, the graphene films need be transferred to other substrate using polydimethylsiloxane (PDMS) or PMMA (Kim et al., 2009; Reina et al., 2009; Sun et al., 2010). Silicon wafer is the most important single-crystal substrate used for semiconductor devices and integrated circuits (ICs).Suemitsu et al. (2010) produced epitaxial graphene on Si substrate. In their approach, a *SiC* film of about 100 nm-thick was deposited on the Si wafer before growth, so graphene was grown on the*SiC* surface. Direct growth of graphene on bare Si substrate without any other material is very attractive. Graphene films can form a Schottky junction with Si, which can produce a built-in electric field and realize electron-hole separation, and has been used to fabricate solar cells (Li et al., 2010a; Gunst et al., 2011; Karamitaheri et al., 2011).

Wei and Xu (2012) have demonstrated the growth of FLG directly on a silicon substrate using a laser irradiation. Silicon substrates were coated with PMMA, which was then evaporated using a CW laser beam. The laser beam also melts the silicon surface and carbon atoms from PMMA separates from the silicon upon cooling to form FLG. A substrate of 1 cm × 2 cm p-type (111)-oriented Si wafer was used to grow graphene. The silicon wafer was first cleaned and the native oxide layer was removed in buffer hydrofluoric acid (HF) solution to form H-terminated silicon surface. A PMMA layer was coated on the Si surface by spin coating, then covered by a quartz wafer of the same size, and then fixed on a sample stage using two spring clamps. The purpose of using the quartz wafer is to maintain a high enough concentration of carbon after PMMA is evaporated and dissociated by laser irradiation. The growth was conducted in a vacuum chamber using a CW that was directed on the Si surface to melt the surface of the Si wafer.

For the synthesis of graphene on metal, two main growth mechanisms were proposed. On Ni, graphene was produced via carbon dissolution and precipitation (Yu et al., 2008). On Cu, the growth can be explained by surface-catalyzed process, which involves carbon nucleation on the Cu surface, and the growth of graphene with the addition of carbon to the edges (Wei and Xu 2012). However, both of these growth mechanisms cannot explain the graphene growth on Si. Cu or Ni maintains solid in the graphene growth process. They found that if the laser power was below the melting point of silicon, there was no graphene grown on the silicon surface.

Similar work has been achieved by Wei et al. (2013), who grow FLG (2–3 layers) on quartz substrate by using a continuous-wave laser by suing a photoresist S-1805 coated on the quartz wafer (thickness 30 nm).

APPLICATIONS OF GRAPHENE MATERIALS IN FUNCTIONAL DEVICES

Graphene research has skyrocketed since the Nobel Prize winners Andre Geim and Konstantin Novoselov published a number of papers on the discovery of graphene 11 years ago. Since then, there has been a sharp increase in our scientific knowledge of graphene, as shown by the number of publications and patent applications on graphene (Figure 2). However, only a handful of graphene-based devices have entered the market, thus we are in an early stage in the commercialization of this material. Further research and development of graphene are critical to help achieve the full potential of this material and the sections below relate to research work of using graphene in smart applications. Figure 11 shows attempt of classification of some potential smart applications of graphene material as a function to the corresponding technology readiness level.

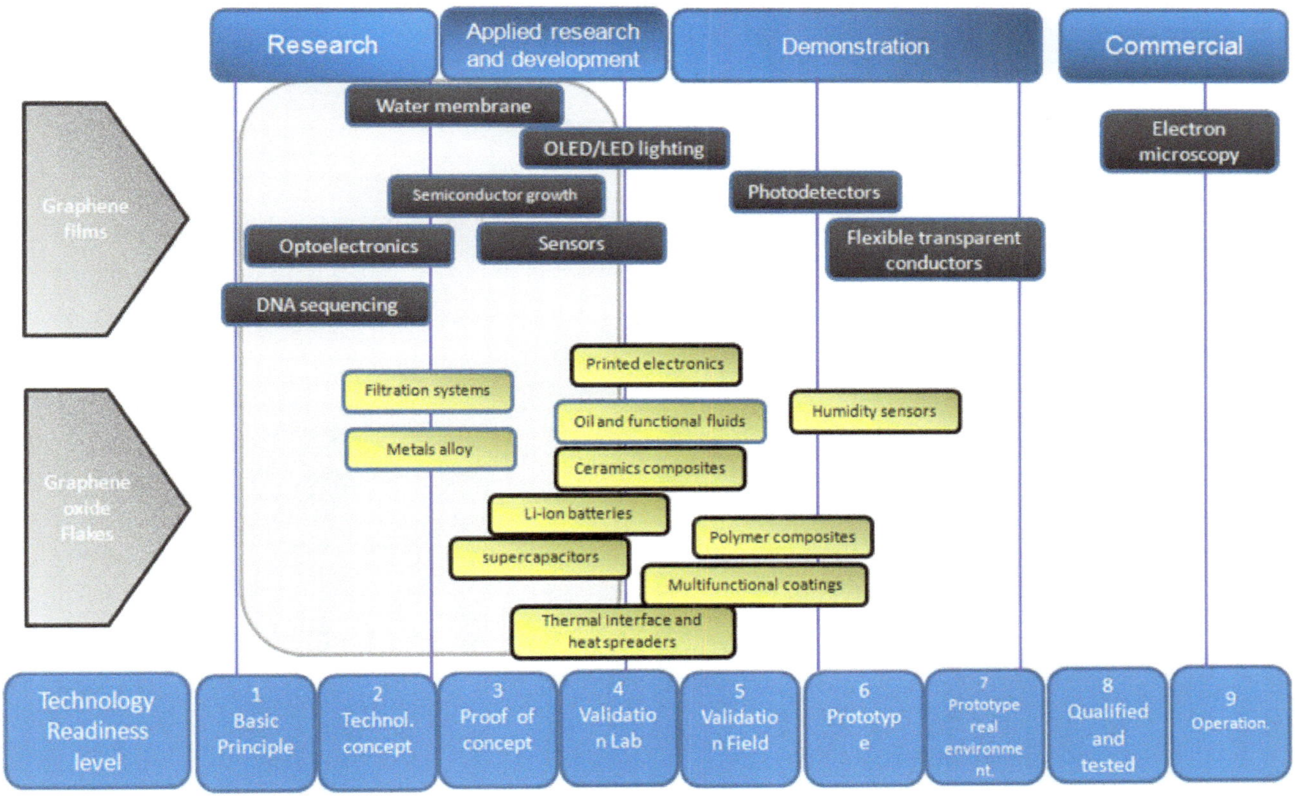

Figure 11: Potential smart applications of graphene material as a function to the technology readiness level.

Photovoltaic Cell

Photovoltaic (PV) cell is the device which converts light to electricity (Chapin et al., 1954; Wang et al., 2008). So far, graphene-based solar cells have been demonstrated in dye-sensitized PV cells (Choe et al., 2010; Jo et al., 2012), organic bulk-heterojunction PV cells (Li et al., 2010b; Yin et al., 2010), hybrid ZnO/poly(3-hexylthiophene) (P3HT) PV cells (Li et al.,

2010c), Si based PV cells (Shim et al., 2011), and InGaN p-i-n PV cells (Gomez De Arco, 2010).

The functionalization of graphene material either during synthesis process (*in situ*) or post-treatment has demonstrated not only the possibility to control the properties of the surfaces and interfaces but also tailoring its work function (Jo et al., 2010; Guo et al., 2011; He et al., 2011; Wan et al., 2011). In an organic PV cell, the difference in work function between the two conductors creates an electrical field in the organic layer (Jo et al., 2012). To date, the power conversion efficiency (PCE) of graphene-electrode organic solar cells (OSCs) has been reported to be in the range of 0.08–2.60% (Wan et al., 2011; Jo et al., 2012), which is indeed much lower than those of conventional OSCs made with ITO electrodes (8.37%) (Stankovich et al., 2006; Chen et al., 2010a;He et al., 2011; Wan et al., 2011; Yu et al., 2012; Saravanakumar et al., 2013; Yokomizo et al., 2013;Sharma et al., 2014). However, the PCE of graphene-electrode OSCs needs to be improved to make graphene film a serious candidate for OSCs (Jo et al., 2012).

Transparent and Flexible Electronics

Currently, most of the research groups in electronics devices fabrication are investigating different routes to fabricate flexible and transparent electronic devices for various types of applications, including smart windows, IC cards, displays, LEDs, solar cells, etc. (Wager et al., 2003; Yu et al., 2011; Jo et al., 2012). In materials, graphene is one of the viable aspirants, which have all the required properties, at the same time, such as optical transparency, mechanical flexibility, and high conductivity. In recent years, different research groups have reported the integration of graphene-based composite electrodes, including graphene/pentacene and graphene/SWCNT, into transparent and flexible electronic like-devices, such as transistors, memory devices, and ICs (Wager et al., 2003; Ji et al., 2011; Lee et al., 2011; Yu et al., 2011; Jo et al., 2012).

Pentacene-based organic FETs-devices were elaborated onto a flexible substrate by means of patterning and transfer or transfer and patterning processes (P-T) and (T-P) techniques that are defining the graphene electrodes (Figure 12A) (Lee et al., 2011; Jo et al., 2012). As a gate electrodes, a plastic substrate based on PEDOT/PSS was used, and poly-4-vinylphenol (PVP) as cross-linker. The typical transfer response of these organic FET is shown in the Figure 12A, where a carrier mobility of 0.01 and 0.12 cm^2/Vs were systematically estimated for the (T-P) and (P-T) processes, respectively (Jo et al., 2012). In addition to the field effect carrier mobility, the overall electronic characteristics of the organic transistors dealing with the (P-T) technique of graphene are superior (Jo et al., 2012). Figure 12B is showing the variation of the electrical resistance as a function of the number of graphene layers. However, there is large room for improving these electronic properties, especially the carrier mobility and to make it more feasible to use in large scale applications (Jo et al., 2012).

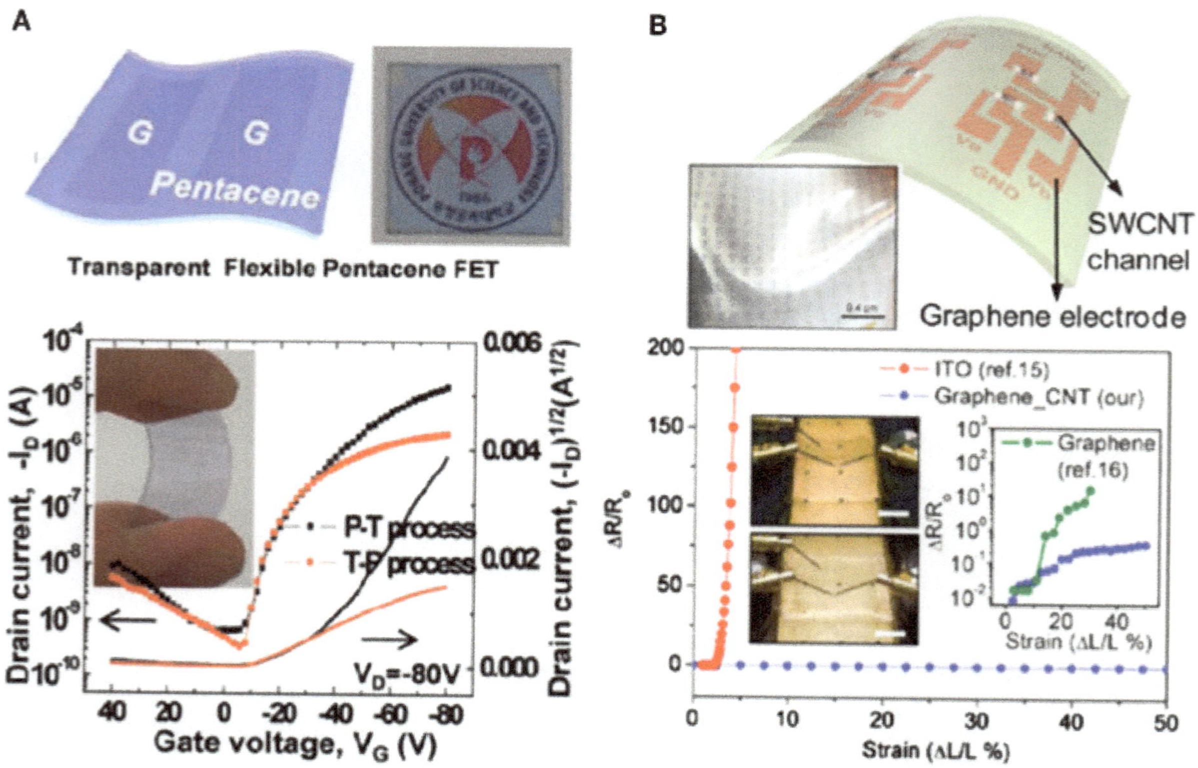

Figure 12: **(A)** Graphical representation and optical photo of an optically transparent and mechanically flexible FETs using graphene as source and drain electrodes. **(B)** Graphical representation composite graphene/SWCNT composite electrode network in FETs and its photograph. Inset in **(A)** shows an image of a mechanically stretched graphene/SWCNT composite based transistor, at the 0 and 50% strain, respectively; the right inset **(B)** is showing the variation of the electrical resistance w/r to the graphene number of layers. Reproduced with permission of Lee et al. (2011). Copyright 2013, Wiley-VCH.

SMART APPLICATIONS OF GRAPHENE

Thermoelectric Application of Graphene

Thermoelectric materials (TEM) achieve the conversion between thermal and electrical energy and vice versa. This field has regained renewed attention because of the huge potential of TEM to be applied in Peltier coolers and thermoelectric power generators. It is well established that the performance of TEM are determined mainly by its dimensionless figure of merit, namely ZT. To date, the performance research of TEM have mainly focused on inorganic semiconductors, such as PbTe, Bi_2Te_3, $CoSb_3$, SnSe, and theirs alloys or composites (Li et al., 2010d). The challenge to develop TEM for a crystalline system with high performance is know how to tailor the interconnected thermoelectric physical parameters, including Seebeck coefficient, the electrical conductivity, and thermal conductivity.

The efficiency of a TEM-based device is usually characterized by the following figure of merit:

$$ZT = S^2\sigma T/\kappa,$$

where S is the Seebeck coefficient (or thermopower, μV/K), σ is the electronic conductance (S/m), T is the thermal conductivity including contributions from both phonons and electrons (W/mK), and κ is the absolute temperature (K).

Figure 13 shows a schematic illustration of a thermoelectric module for (a) power generation (Seebeck effect) and (b) active generation (Peltier effect). Figure 13A shows an applied temperature difference, which causes charge carriers in the material (electron or holes), to diffuse from the hot side to the cold one, resulting in a current flow through the circuit. Figure 13B is schematic of the heat that evolves at the upper junction and is adsorbed at the lower junction when a current is made to flow through the circuit.

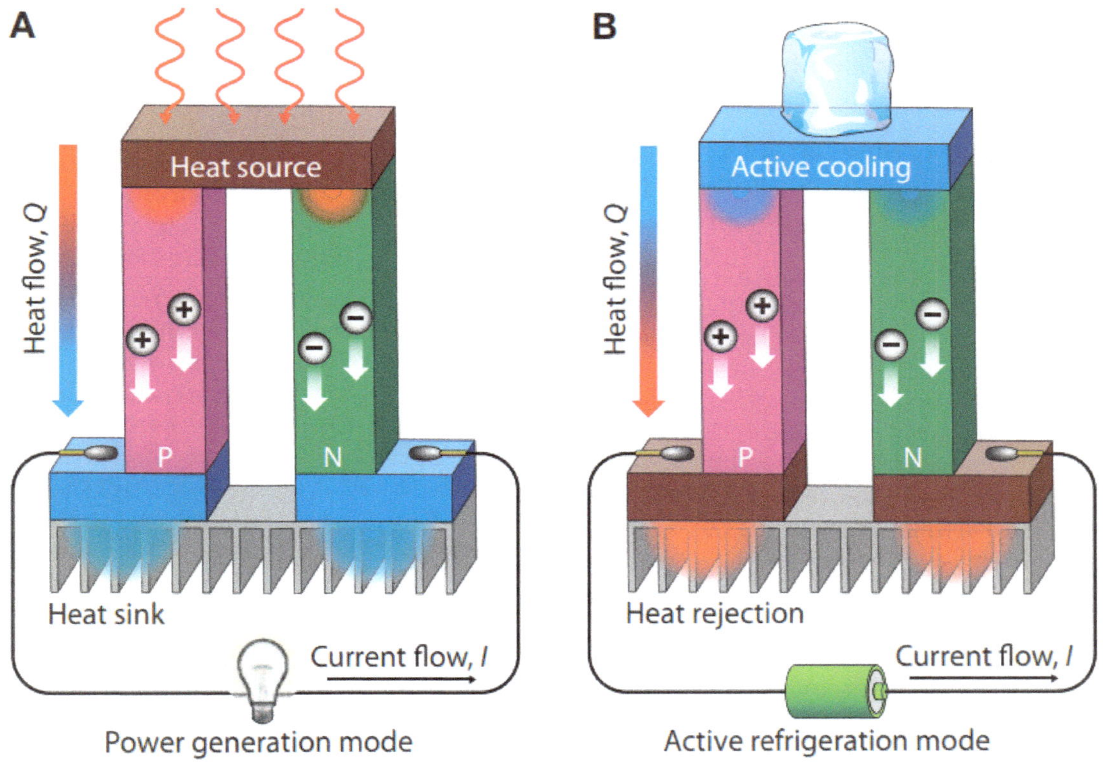

Figure 13: Schematic illustration of a thermoelectric module for (A) power generation (seebeck effect) and (B) active generation (Peltier effect). Reproduced with the Permission from Li et al. (2010d). Copyright 2010, NPG Asia Mater.

Recently, great effort has been made in improving the TEM dimensionless figure of merit (ZT). The difficulty in to simultaneously optimizing them, which causes thermoelectric research to stagnate for a while, until great reductions in thermal conductivity were theoretically and experimentally proven in nanostrtuctured materials.

Ghosh et al. (2010) examined the evolution of the thermal properties of FLG with respect to increasing thickness (i.e., the number of graphene layer, n). The results obtained by Raman spectroscopy have shown that, overall, the thermal conductivity decreases with increasing the FLG thickness, approaching that of the bulk graphite limit (Figure 14) (Ghosh et al., 2010). The experimental data points in Figure 14 were all normalized to the same width of graphene sheet of 5 μm (Ghosh et al., 2010). The detailed procedure is described in Ghosh et al. (2010).

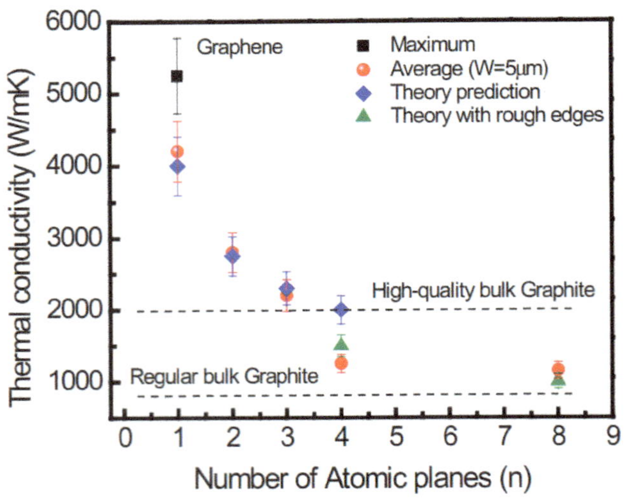

Figure 14: Thermal conductivity with respect to the number for atomic planes of quasi-2D carbon materials. Experimental and simulated thermal conductivity of suspended FLG as a function of number of atomic layer, *n*, taken at the fixed width, *W* of the graphene flake. Reproduced with the permission from Ghosh et al. (2010). Copyright 2011, Nature Materials.

The evolution observed from 2D graphene to bulk graphite is explained by the cross-plane coupling of the low-energy phonons and changes in the phonon scattering, since more states are available for scattering owing to the increased number of phonon branches. The thermoelectric power (TEP) is the voltage developed across a sample when a constant temperature gradient is applied. TEP of 80 μV/K was recently measured in graphene at room temperature (300 K) (Zuev et al., 2009). Similar to the quantum Hall effect in electronic transport, quantized TEP has also been observed in graphene at high-magnetic fields (Zuev et al., 2009). The TEP can be tuned in graphene, even to negative values, under the application of a gate bias or chemical potential (Wei et al., 2009). Very large TEP values have been predicted for GNRs, for instance, 4 mV/K for a 1.6 nm wide ribbon (Haskins et al., 2009). In comparison, the highest value experimentally reported so far is 850 μV/K for two-dimensional electron gases in $SrTi_2O_3$ heterostructures (Ohta et al., 2007), while only a few μV/K has been reported for bulk graphite (Tyler et al., 1953). The TEP power of single walled carbon nanotubes (SWNTs) has been theoretically and experimentally shown to be 60 μV/K (Hone et al., 1998), inferior to that of graphene. A giant thermoelectric coefficient of 30 mV/K was reported in metallic electrodes periodically patterned over graphene, deposited on SiO_2 substrate (Chang et al., 2007).

On the other hand, thermoelectric properties of graphene have attracted increased interest as well, since it can convert heat to electricity and vice versa. A high thermopower value of 80 μV/K was reported for graphene (Zhu et al., 2009; Zuev et al., 2009). Various structures of graphene have been examined, including nanoribbons (GNRs) (Haskins et al., 2009; Sevinçli et al., 2009; Ouyang and Hu, 2010; Huang et al., 2011; Mazzamuto et al., 2011), quantum dots (Yan et al., 2012), graphene junctions, and chevron-type structures (Chen et al., 2010b). It is worth noted here that ZT of zigzag GNRs can exceed 3 (Ouyang and Hu, 2010). Recently, many works have been conducted on the creation of graphene band gap by means of making an array of holes (antidots) into the graphene layer (Pedersen et al., 2008; Ouyang et al., 2011; Petersen et al., 2011; Chang and Nikolic, 2012). This is an inevitable property for the integration of graphene material directly into transistor architecture (Ouyang et al., 2011; Petersen et al., 2011). The gap can be engineered by controlling the lattice geometry (i.e., the antidot size and the hole-to-hole separation) (Shen et al., 2008; Eroms and Weiss, 2009; Bai et al., 2010; Kim et al., 2010; Gunst et al., 2011; Karamitaheri et al., 2011). Further, thermoelectric properties of these structures (2D graphene antidot lattices) have also been investigated where ZT up to 0.3 was found (Li et al., 2010d).

Graphene in Shape Memory Materials

Shape memory polymers (SMPs) are defined as smart materials, with the faculty to respond to an external stimulus, typically heat, and have a wide-range of applications from biomedical devices to space materials (Han and Chun., 2014). Polyurethane (PU) is the most attractive SMP material, commonly referred to as SMP polyurethane (SMPU) (Lee et al., 2014). Based on the specific molecular design, SMPU can form either a crystalline or amorphous structure, with a broad actuation temperature range from −20 to 150°C (Kim et al., 2014). SMP composites based on nanofillers have shown to increase the overall shape memory properties. An exciting research area is the use of carbon-based nanomaterials, particularly graphene, as nanofillers for SMPUs.

Han and Chun (2014) prepared a graphene/PU composite material by functionalizing graphene with diazonium salts carrying phenethyl alcohol. The resulting material showed enhanced shape memory, thermal, and mechanical properties. The resulting material showed shape fixity up to 98 with 94% shape recovery ratio after four cycles. Hysteresis loss can be low as 2%. Rana et al. (2013)prepared a flexible and conductive shape memory composite based on PU and functionalized graphene sheets. The graphene sheets were functionalized using phenyl isocyanate and poly diol. The composite resulted in 97% shape recovery and 95% shape fixity (see Figure 15). Modifications to PU can also improve the shape memory performance with graphene. Thakur et al. (2013) have demonstrated that castor oil-modified hyperbranched PU and graphene-oxide (without functionalization) can result in a shape recovery of ~99.5% with shape fixity of ~90%. Park et al. (2014) have demonstrated that PU and allyl isocyanate modified graphene oxide can also be actuated by infrared absorption, with a resulting shape recovery of 90%.

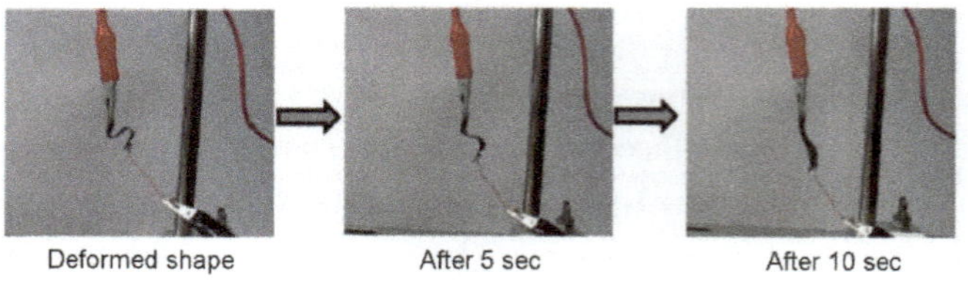

Figure 15: Shape recovery behavior of graphene-cross linked PU composites, for a temporary helix shape (left) to a permanent shape when applying a constant voltage of 50 V. Reproduced with permission from Rana et al. (2013). Copyright 2013, Royal Society of Chemistry.

Graphene in Self-Healing Materials

Long-term stability and durability of materials are one of the main challenges faced today for structural and coating applications, especially when using polymeric composites materials. Indeed, there are many parameters affecting the degradation of materials, including environmental conditions of large types (temperature gradient, UV irradiation, oxygen erosion, corrosion, etc.). As indicated by their name, self-healing materials are first conceived and then designed and elaborated to have the potential to heal mainly the mechanical properties of the materials when damaged. Representative applications of self-healing materials are found in composite polymers, various metals, ceramics, and their related composites, which are subjected to a wide variety of healing principles.

Recently, a few studies have demonstrated the potential of graphene as an additive for self-healing materials. Dong et al. (2013) synthesized a composite material of poly(acrylamide) (PAM), poly(acrylic acid) (PAA), and graphene that exhibits shape memory effect and self-healing ability. Graphene content was in the range of 10–30% and the material can be recovered after 20 cycles of cut and self-healing. The self-healing capabilities are illustrated in Figure 16, where 10 wt.% graphene sample is cut in the middle and the two pieces were healed together at 37°C for 20 min. The results suggest that the self-healing ability and shape memory effect occur due to a "zipper effect" of PAM–PAA that forms or dissociate the hydrogen-bond network, where such effects are limited without the addition of graphene. Huang et al. (2013) have demonstrated the use of FLG with thermoplastic polyurethane (TPU) as self-healing material initiated using an electric stimulus. When the FLG loading rate is 5 wt.%, the material can be healed to an efficiency higher than 98% in 3 min. Furthermore, upon increasing the FLG loading rate to 8 wt.%, the same healing efficiency of 98% can be achieved in 15 s only. It is expected that graphene efficiently converts electrical energy to thermal energy, which improves the self-healing ability of TPU diffusion and re-entanglement of the TPU chains. More recently, Wang et al. (2013) and Sullivan et al. (1977) have recently reported the fabrication of a composite based on a cross-linked (CL) hydrogen bonding polymer with graphene oxide. This composite material was found to enable a fast self-healing capability, with high efficiency, occurring at room-temperature. More importantly, the healing reaction was produced without the need of any external stimuli, such as electrical bias, light, and heat (Zhang et al., 2014). The introduction of the graphene material in this CL-based polymer was found to be crucial to reduce the needed amount of CL sites that are necessary for the healing reaction. It is worth noting here that a huge amount of CL sites are not only negatively impacting the mechanical properties of the polymer (including flexibility) but also its dynamic characteristics (Zhang et al., 2014). Such nanocomposite materials involving CL-hydrogen bonding polymer with graphene-oxide can be useful in many applications, e.g., protecting barrier for electronic devices, sealing layer for gas systems, and stretchable self-healing conductive wires (Zhang et al., 2014).

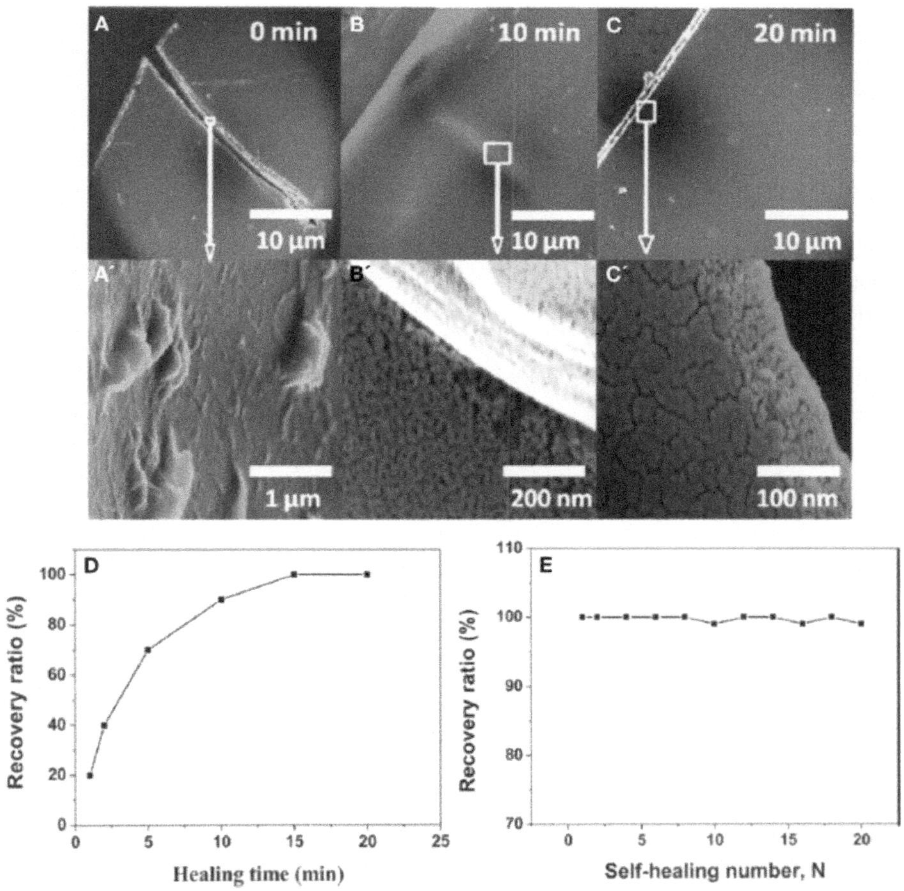

Figure 16: SEM image of G-PAM-PAA strip heating at 37°C in different times, (A) 0 min, (B) 10 min, and (C) 20 min. (A′–C′) are close-up view of the factures occurred in **(A–C)**, respectively. Percentage of the recovery ratio as a function of **(D)**healing-time and **(E)** self-healing number (N). Reproduced with permission from Dong et al. (2013). Copyright 2013, Wiley-VCH.

Graphene in Photomechanical Actuators

Actuators are materials, which change their shape or dimensions under the application of external stimulus. To date, the well-recognized materials for actuation are piezoelectrics (Nakamura et al., 1989), ferro-electrics (Kuribayashi et al., 1989), shape memory alloys (Damjanovic and Newnham, 1992), electrostrictive materials (Damjanovic and Newnham, 1992), and conducting polymers (Smela et al., 1999). Recently, large stresses and strains from low-voltage electromehcanical actuation have exhibited by carbon nanotubes (CNTs) (Baughman et al., 1999) and porous metallic nanoparticles (Ahir and Terentjev, 2005). Beside this, both single walled CNT (SWCNT) and multiwalled CNT (MWCNT)-based composites have been reported to undergo photomechanical actuation (Dreyer et al., 2010).

Several articles have reported about graphene-based composites, mostly in sheets form and have derived from graphite oxide or graphite intercalation compounds (GICs). It has been observed that intrusion of graphite oxide or graphite intercalation compounds derived fillers can enhance electrical conductivities of polymeric matrices (Gómez-Navarro et al., 2008), the Young's moduli (Potts et al., 2011), and could easily be functionalized to tailor to host polymer properties (Sun et al., 2009). Now, it is well-known that by tailoring the number of layers of graphene nanoplateletes (GNP) and GNR, the overall properties of composite material change [such as saturable absorption (Casiraghi et al., 2007) and electric field assisted band gaps (Zhang et al., 2009)].

Loomis et al. (2012) have reported that a simple polymer composite system with photomechanical responses is realized solely by incorporation of a homogeneous dispersion of GNPs within a PDMS elastomer matrix. It has been observed that the actuation responses of GNPs/PDMS composites depend on the initial applied pre-strain as in CNT/PDMS composites. Photomechanical stress change 2.4–3.6 times is greater for GNP/PDMS composites, compared to any other tested form of nanocarbon. These stress changes reported are usable and recoverable work achieved by the actuators. Energy conversion factor (ηM) of 7–9 MPa W^{-1} for optical-to-mechanical is obtained during testing. Until now, this is a largest energy conversion factor of an extraordinary photomechanical effect exhibited by any material so far.

Graphene in Piezoelectric Materials

Piezoelectric material has the property of converting mechanical movement into electrical movement and vice versa. Traditionally, piezoelectricity is considered to be an intrinsic property of a particular material phase. Piezoelectric materials have been frequently used in a wide variety of applications from pressure sensors (Pereira and Castro Neto, 2009), to acoustic transducers (Guinea et al., 2010), to high voltage generators (Bunch et al., 2007) for dynamical control of material deformation by application of external electric force.

Luk'yanchuk et al. (2015) have found that piezoelectricity can be concocted into intrinsically non-piezoelectric materials such as graphene. This is a nanoscale new phenomenon and lacking a direct bulk analog (Luk'yanchuk et al., 2015). This new phenomenon has provided room for practical approach toward manipulation and dynamic control different concerns in nanodevices, such as optical, chemical, and electronic. Luk'yanchuk et al. (2015) have reported an extraordinary two dimensional piezoelectric effect, both on a strained and unstrained graphene junction. Interestingly, it has been formally attested that this 2D piezo effect is a direct consequence of the difference in the two work functions (of the two type of graphene) and hence to the charge transfer occurring from the biaxial strain when putting the two graphene types together (i.e., bend band structure). The effect has termed as the band-piezoelectric effect, which exhibits a massive magnitude due to the ultrathin structure of graphene (Luk'yanchuk et al., 2015).

Using this new type of piezo effect, a piezoelectric nanogenerator and a piezoresistive pressure sensor within a graphene nano-electro-mechanical-system (NEMS) platform have been demonstrated. In this novel device, the deformation caused by an AFM tip resulted in charge separation, due to the modified band structure of the bent membrane (Luk'yanchuk et al., 2015). Consequently, by using appropriate metal electrodes with suited work function, we can even collect the cumulated charge and produce an electrical voltage (Luk'yanchuk et al., 2015).

Graphene in Electrorheology Materials

Electrorheology (ER) material is a type of smart material where the rheological properties of the material can be reversibly transformed with the application of an external electric field (Zhang et al., 2010c, 2012; Yin et al., 2012; Zhang and Choi, 2014). An ER material is composed of polarizable particles suspended in an insulating medium and after applying an electric field the particles are polarized and form a column like structure. This modifies the rheological properties of the material, such as viscosity, shear stress, and dynamic modulus. Applications of such materials include damper systems, ER polishing, tactile displays, medical devices, and robotic actuators. Recently, graphene, r-GO, and GO, due to their unique properties, have gathered the interest of the scientific community as an additive for ER materials (Zhang and Choi, 2014). Zhang et al. (2010c) prepared colloidal r-GO using a modified Hummers method, which was used to prepare a nanocomposite material comprising of GO and polyaniline (PANI). The resulting material showed adjustable electrical conductivity, which has potential for use as an ER material. Additionally, Zhang et al. (2012) also prepared GO particles suspended in silicone oil, where the ER fluid is polarized and exhibits viscoelastic properties at various strains under an electric field.

Yin et al. (2012) prepared a novel ER suspension comprising of two dimensional PANI decorated GO sheets. With the application of an electric field, the suspension containing PANI and GO, shows higher ER effect when compared to pure PANI. The performance of the material is dependent on the thickness of the PANI coating applied to the GO sheets. Furthermore, Yin et al. (2013) prepared mesoporous silica-coated r-GO nanosheets as a dispersal for ER fluids. Silica coating limits the conductivity of graphene, which enables high polarization and ER response, with the application of an electric field particularly high-frequency AC electric fields. The silica shells also limit the restacking of graphene. Li et al. (2015) studied the ER effects of non-conducting GO and conducting r-GO comprise an insulating SiO_2 shell. The results show that the GO/SiO_2 has a higher ER response to DC and low-frequency AC electric fields, while the r-GO/SiO_2 has a higher ER response to high-frequency AC electric fields. The different behavior can be explained by the impact of the polarization rate with regards to the inter-particle interaction.

Multifunctional Graphene Nanocomposite Foams for Space Applications

Space is the new frontier and new materials, devices, and technologies for aerospace applications represent now an emerging sector with significant employment prospects and opportunities for profit. Materials and devices used in space applications (e.g., satellites) should combine functionality with low weight and reduced volume, to optimize cost effectiveness. The main cost of a satellite is its launch into orbit (estimated at about 7 M$), therefore reducing weight and volume can significantly decrease the overall cost. Weight savings can be achieved by replacing heavy copper wiring, which accounts for example up to 4000 lbs of weight on a Boeing 747 and about one-third of the weight of large satellites (Meador et al., 2010), with low density carbon-based wiring cables. In addition, Joule heating from metallic parts requires additional components for cooling (radiators), thus adding to the overall weight and cost. The combination of superior electronic and thermal properties of graphene materials could potentially revolutionize the design and fabrication of light weight electrical and electronic devices to be used in space applications. In this context, graphene is a promising candidate because its ballistic electron transport limits Joule heating and should allow reductions in weight, volume, and subsequently total cost.

Another potential space application of graphene is its use as nanofiller to fabricate light weight and robust nanocomposite, to be used, e.g., as thermal barrier. Recently, graphene nanoplatelets having FLG with 1–5 layers and typical diameters ranging from 1 to over 100 µm have been successfully grown by a team from Michigan State University[1]. These nanoplatelets show, overall, similar properties to a single graphene layer while being mechanically much robust. In addition, their cost production can be lowered to be competitive to other carbonaceous additives and fillers. These nanoplatelets could be used as a nanoadditive into various polymer foams, increasing thereby their thermal, mechanical, and electrical properties, while the foam still maintains its unique structure and low density. This combination of unique properties has direct potential application as flame resistant in space technology.

CONCLUSION AND OUTLOOKS

Research on graphene has considerably enriched our understandings and applications of two-dimensional atomic crystal. The combination of unprecedented physical and chemical properties, including its extremely high strength, thermal and electrical conductivity, and mechanical flexibility, have harried scientists to investigate in-depth its real potential in delivering improvements to different devices in many technological and scientific fields, ranging from materials science to physics, chemical engineering, and even biology. In this review, we have presented an overview of the main gas-phase synthesis of graphene and its applications as smart material systems. The possibility of successfully integrating graphene directly into device, could not only improve the electrical and/or mechanical properties but also enable the realization of a wide range of applications, such as actuation, thermoelectricity, shape memory, and self-healing. However, this success is conditioned by the prior success to address the following issues: (i) developing a cost-effective growth process to synthesize functional graphene in a reasonable scale with acceptable degree of reproducibility, for the realization of practical applications; (ii) considering the unique electrical, mechanical, and optical properties of graphene, incorporation of these characteristics into the existing smart systems from an interdisciplinary point of view should be highly valued; (iii) finally, from practical view-point, its highly desirable to fabricate graphene multifunctional systems, that are responsive to multiple stimuli. Considering the current and ongoing achievements, it is believed that the smart applications of graphene systems with more functionality are expected to emerge as new complements – and/or even replacements – to existing conventional systems, especially in fields of electronics, energy, and space.

CONFLICT OF INTEREST STATEMENT

The authors declare that the research was conducted in the absence of any commercial or financial relationships that could be construed as a potential conflict of interest.

ACKNOWLEDGMENTS

Authors thank the financial support of the Qatar Environment and Energy Research Institute and Qatar Foundation.

REFERENCES

1. Ahir, S. V., and Terentjev, E. M. (2005). Photomechanical actuation in polymer-nanotube composites. *Nat. Mater.* 4, 491–495. doi: 10.1038/nmat1391
2. Aïssa, B., Nechache, R., Haddad, E., Jamroz, W., Merle, P. G., and Rosei, F. (2012). Ruthenium Grubbs' catalyst nanostructures grown by UV-excimer-laser ablation for self-healing applications. *Appl. Surf. Sci.* 258, 9800–9804. doi:10.1510/icvts.2010.255588
3. Avouris, P. (2010). Graphene: electronic and photonic properties and devices. *Nano Lett.* 10, 4285–4294. doi:10.1021/nl102824h
4. Bae, S., Kim, H., Lee, Y., Xu, X., Park, J. S., Zheng, Y., et al. (2010). Roll-to-roll production of 30-inch graphene films for transparent electrodes. *Nat. Nanotechnol.* 5, 574–578. doi:10.1038/nnano.2010.132
5. Bai, J., Zhong, X., Jiang, S., Huang, Y., and Duan, X. (2010). Graphene nanomesh. *Nat. Nanotechnol.* 5, 190–194. doi:10.1038/nnano.2010.8
6. Balandin, A. A. (2011). Thermal properties of graphene and nanostructured carbon materials. *Nat. Mater.* 10, 569–581. doi:10.1038/nmat3064
7. Balooch, M. T., Tench, R. J., Siekhaus, W. J., Allen, M. J., Connor, A. L., and Olander, D. R. (1990). Deposition of SiC films by pulsed excimer laser ablation. *Appl. Phys. Lett.* 57, 1540–1542. doi:10.1063/1.103346

8. Basu, S., and Bhattacharyya, P. (2012). Recent developments on graphene and graphene oxide based solid state gas sensors.*Sens. Actuators B Chem.* 173, 1–21. doi:10.1016/j.snb.2012.07.092

9. Baughman, R. H., Cui, C., Zakhidov, A. A., Iqbal, Z., Barisci, J. N., Spinks, G. M., et al. (1999). Carbon nanotube actuators.*Science* 284, 1340. doi:10.1126/science.284.5418.1340

10. Bekaroglu, E., Topsakal, M., Cahangirov, S., and Ciraci, S. (2010). First-principles study of defects and adatoms in silicon carbide honeycomb structures. *Phys. Rev. B* 81, 075433. doi:10.1103/PhysRevB.81.075433

11. Bhaviripudi, S., Jia, X., Dresselhaus, M. S., and Kong, J. (2010). Role of kinetic factors in chemical vapor deposition synthesis of uniform large area graphene using copper catalyst. *Nano Lett.* 10, 4128–4133. doi:10.1021/nl102355e

12. Biunno, N., Narayan, J., Hofmeister, S. K., Srivatsa, A. R., and Singh, R. K. (1989). Low-temperature processing of titanium nitride films by laser physical vapor deposition. *Appl. Phys. Lett.* 54, 1519. doi:10.1063/1.101338

13. Bo, Z., Yang, Y., Chen, J., Yu, K., Yan, J., and Cen, K. (2013). Plasma-enhanced chemical vapor deposition synthesis of vertically oriented graphene nanosheets. *Nanoscale* 5, 5180–5204. doi:10.1039/c3nr33449j

14. Bonaccorso, F., Sun, Z., Hasan, T., and Ferrari, A. C. (2010). Graphene photonics and optoelectronics. *Nat. Photon* 4, 611–622. doi:10.1038/nphoton.2010.186

15. Boukhvalov, D. W., Katsnelson, M. I., and Lichtenstein, A. I. (2008). Hydrogen on graphene: electronic structure, total energy, structural distortions and magnetism from first-principles calculations. *Phys. Rev. B* 77, 035427. doi:10.1103/PhysRevB.77.035427

16. Bunch, J. S., van der Zande, A. M., Verbridge, S. S., Frank, I. W., Tanenbaum, D. M., Parpia, J. M., et al. (2007). Electromechanical resonators from graphene sheets. *Science* 315, 490–493. doi:10.1126/science.1136836

17. Bunch, J. S., Verbridge, S. S., Alden, J. S., van der Zande, A. M., Parpia, J. M., Craighead, H. G., et al. (2008). Impermeable atomic membranes from graphene sheets. *Nano Lett.* 8, 2458–2462. doi:10.1021/nl801457b

18. Casiraghi, C., Hartschuh, A., Lidorikis, E., Qian, H., Harutyunyan, H., Gokus, T., et al. (2007). Rayleigh imaging of graphene and graphene layers. *Nano Lett.* 7, 2711–2717. doi:10.1021/nl071168m

19. Casolo, S., Martinazzo, R., and Tantardini, G. F. (2011). Band engineering in graphene with superlattices of substitutional defects. *J. Phys. Chem. C* 115, 3250–3256. doi:10.1021/jp109741s

20. Chae, S. J., GüneŞ, F., Kim, K. K., Kim, E. S., Han, G. H., Kim, S. M., et al. (2009). Synthesis of large-area graphene layers on poly-nickel substrate by chemical vapor deposition: wrinkle formation. *Adv. Mater. Weinheim* 21, 2328–2333. doi:10.1002/adma.200803016

21. Chang, P.H. and Nikolic, B.K. (2007). Giant thermoelectric effect in graphene. *Appl. Phys. Lett.* 91, 203116. doi:10.1021/nl500755m

22. Chang, P. H., and Nikolic, B. K. (2012). Edge currents and nanopore arrays in zigzag and chiral graphene nanoribbons as a route toward high-ZT thermoelectrics. *Phys. Rev. B* 86, 041406. doi:10.1103/PhysRevB.86.041406

23. Chapin, D. M., Fuller, C. S., and Pearson, G. L. (1954). A new silicon p-n junction photocell for converting solar radiation into electrical power. *J. Appl. Phys.* 25, 676–677. doi:10.1063/1.1721711

24. Chen, Z., Berciaud, S., Nuckolls, C., Heinz, T. F., and Brus, L. E. (2010a). Energy transfer from individual semiconductor nanocrystals to graphene. *ACS Nano* 4, 2964–2968. doi:10.1021/nn1005107

25. Chen, Y., Jayasekera, T., Calzolari, A., Kim, K. W., and Nardelli, M. B. (2010b). Thermoelectric properties of graphene nanoribbons, junctions and superlattices. *J. Phys. Condens. Matter.* 22, 372202. doi:10.1088/0953-8984/22/37/372202

26. Choe, M., Lee, B. H., Jo, G., Park, J., Park, W., Lee, S., et al. (2010). Efficient bulk-heterojunction photovoltaic cells with transparent multi-layer graphene electrodes. *Org. Electron.* 11, 1864–1869. doi:10.1016/j.orgel.2010.08.018

27. Choi, W., Lahiria, I., and Seelaboyinaa, R. (2010). Synthesis of graphene and its applications: a review. *Crit. Rev. Solid State Mater. Sci.* 35, 52–71. doi:10.1080/10408430903505036

28. Chrisey, D. B., and Hubler, G. K. (1994). *Pulsed Laser Deposition of Thin Films*. New York, NY: John Wiley.

29. Curl, R. F., and Smalley, R. E. (1991). *Scientific American*, Vol. 33. 32.

30. Dato, A., Radmilovic, V., Lee, Z., Phillips, J., and Frenklach, M. (2008). Substrate-free gas-phase synthesis of graphene sheets.*Nano Lett.* 8, 2012–2016. doi:10.1021/nl8011566

31. Damjanovic, D. and Newnham, R. E. (1992). Electrostrictive and piezoelectric materials for actuator applications. *J. Int. Mat. Sys. Struct.* 3, 190–208. doi:10.1177/1045389X9200300201

32. Dean, C. R., Young, A. F., Meric, I., Lee, C., Wang, L., Sorgenfrei, S., et al. (2010). Boron nitride substrates for high-quality graphene electronics. *Nat. Nanotechnol.* 5, 722–726. doi:10.1038/nnano.2010.172

33. Dijkkamp, D., Venkatesan, T., Wu, X. D., Shaheen, S. A., Jisrawi, N., Min-Lee, Y. H., et al. (1987). Preparation of Y-Ba-Cu oxide superconductor thin films using pulsed laser evaporation from high Tc bulk materials. *Appl. Phys. Lett.* 51, 619. doi:10.1063/1.98366

34. Dong, J., Ding, J., Weng, J., and Dai, L. (2013). Graphene enhances the shape memory of poly (acrylamide-co-acrylic acid) grafted on graphene. *Macromol. Rapid Commun.* 34, 659–664. doi:10.1002/marc.201200814

35. Dreyer, D. R., Park, S., Bielawski, C. W., and Ruoff, R. S. (2010). The chemistry of graphene oxide. *Chem. Soc. Rev.* 39, 228–240. doi:10.1039/b917103g

36. Eason, R. (2006). *Pulsed Laser Deposition of Thin Films:*

Applications-Led Growth of Functional materials. Hoboken, NJ: John Wiley & Sons, Inc.

37. Elias, D. C., Nair, R. R., Mohiuddin, T. M., Morozov, S. V., Blake, P., Halsall, M. P., et al. (2009). Control of graphene's properties by reversible hydrogenation: evidence for graphane. *Science* 323, 610–613. doi:10.1126/science.1167130

38. Eroms, J., and Weiss, D. (2009). Weak localization and transport gap in graphene antidot lattices. *New J. Phys.* 11, 095021. doi:10.1088/1367-2630/11/9/095021

39. Fogarassy, E., Fuchs, C., Slaoui, A., and Stoquert, J. P. (1990). SiO2 thin-film deposition by excimer laser ablation from SiO target in oxygen atmosphere. *Appl. Phys. Lett.* 57, 664–666. doi:10.1063/1.104253

40. Foster, C. M., Voss, K. F., Hagler, T. W., Mihailovi, D., Heeger, A. J., Eddy, M. M., et al. (1990). Infrared reflection of epitaxial Tl2Ba2CaCu2O8 thin films in the normal and superconducting states. *Solid State Commun.* 76, 651. doi:10.1016/0038-1098(90)90108-N

41. Geim, A. K. (2009). Graphene: status and prospects. *Science* 324, 1530–1534. doi:10.1126/science.1158877

42. Geim, A. K., and Novoselov, K. S. (2007a). The rise of graphene. *Nat. Mater.* 6, 183–191. doi:10.1038/nmat1849

43. Geim, A. K., and Novoselov, K. S. (2007b). The rise of graphene. *Nat. Mater.* 6, 183–191. doi:10.1038/nmat1849

44. Geurtsen, A. J. M., Kools, J.C.S., Wit, L., and Lodde, J.C. (1996). Pulsed laser deposition of permanent magnetic Nd2Fe14B thin films. *Appl. Surf. Sci.* 9698, 887–890. doi:10.1016/0169-4332(95)00541-2

45. Ghosh, S., Bao, W., Nika, D. L., Subrina, S., Pokatilov, E. P., Lau, C. N., et al. (2010). Dimensional crossover of thermal transport in few-layer graphene. *Nat. Mater.* 9, 555–558. doi:10.1038/nmat2753

46. Giannazzo, F., Raineri, V., and Rimini, E. (2011). "Transport properties of graphene with nanoscale lateral resolution," in*Scanning Probe Microscopy in Nanoscience and Nanotechnology 2*, ed. Bhushan B. (Berlin: Springer), 247–285.

47. Gomez De Arco, L. (2010). Continuous, highly flexible, and transparent graphene films by chemical vapor deposition for organic photovoltaics. *ACS Nano* 4, 2865–2873. doi:10.1021/nn901587x

48. Gómez-Navarro, C., Burghard, M., and Kern, K. (2008). Elastic properties of chemically derived single graphene sheets. *Nano Lett.* 8, 2045–2049. doi:10.1021/nl801384y

49. Green, A. A., and Hersam, M. C. (2009). Solution phase production of graphene with controlled thickness via density differentiation. *Nano Lett.* 9, 4031–4036. doi:10.1021/nl902200b

50. Guermoune, A., Chari, T., Popescu, F., Sabri, S. S., Guillemette, J., Skulason, H. S., et al. (2011). Chemical vapor deposition synthesis of graphene on copper with methanol, ethanol, and propanol precursors. *Carbon* 49, 4204–4210. doi:10.1016/j.carbon.2011.05.054

51. Guinea, F., Katsnelson, M. I., and Geim, A. K. (2010). Energy gaps and a zero-field quantum Hall effect in graphene by strain engineering. *Nat. Phys.* 6, 30–33. doi:10.1038/nphys1420

52. Gunst, T., Markussen, T., Jauho, A.-P., and Brandbyge, M. (2011). Thermoelectric properties of finite graphene antidot lattices.*Phys. Rev. B* 84, 155449. doi:10.1103/PhysRevB.84.155449

53. Guo, C. X., Guai, G. H., and Ming, C. (2011). Graphene based materials: enhancing solar energy harvesting. *Adv. Energy Mater.*1, 448–452. doi:10.1002/aenm.201100119

54. Han, S., and Chun, B. C. (2014). Preparation of polyurethane nanocomposites via covalent incorporation of functionalized graphene and its shape memory effect. *Compos. Part A. Appl. Sci. Manuf.* 58, 65–72. d

55. Hansen, S. G., and Robitaille, T. E. (1988). Formation of polymer films by pulsed laser evaporation. *Appl. Phys. Lett.* 52, 81–83. doi:10.1063/1.99332

56. Haskins, J., Kınacı, A., Sevik, C., Sevinçli, H., Cuniberti, G., and Cağın, T. (2009). A theoretical study on thermoelectric properties of graphene nanoribbons. *Appl. Phys. Lett.* 94, 263107. doi:10.1021/nn200114p

57. Hassan, R. (ed.) (2012). "Graphene nanoelectronics," in *Metrology, Synthesis Properties and Applications* (Berlin: Springer-Verlag), 598.

58. He, Z., Zhong, C., Huang, X., Wong, W.-Y., Wu, H., Chen, L., et al., (2011). Simultaneous enhancement of open-circuit voltage, short-circuit current density, and fill factor in polymer solar cells. *Adv. Mater. Weinheim* 23, 4636–4643. doi:10.1002/adma.201103006

59. Hernandez, Y., Nicolosi, V., Lotya, M., Blighe, F. M., Sun, Z., De, S., et al. (2008). High-yield production of graphene by liquid-phase exfoliation of graphite. *Nat. Nanotechnol.* 3, 563–568. doi:10.1038/nnano.2008.215

60. Hone, J., Ellwood, I., Muno, M., Mizel, A., Cohen, M. L., Zettl, A., et al. (1998). Thermoelectric power of single-walled carbon nanotubes. *Phys. Rev. Lett.* 80, 1042–1045. doi:10.1103/PhysRevLett.80.1042

61. Houssa, M., Pourtois, G., Afanas'ev, V. V., and Stesmans, A. (2010). Electronic properties of two-dimensional hexagonal germanium. *Appl. Phys. Lett.* 96, 082111. doi:10.1063/1.3332588

62. Huang, L., Yi, N., Wu, Y., Zhang, Y., Zhang, Q., Huang, Y., et al. (2013). Multichannel and repeatable self-healing of mechanical enhanced grapheme-thermoplastic polyurethane composites. *Adv. Mater. Weinheim* 25, 2224–2228. doi:10.1002/adma.201204768

63. Huang, W., Wang, J.-S., and Liang, G. (2011). Theoretical study on thermoelectric properties of kinked graphene nanoribbons.*Phys. Rev. B* 84, 045410. doi:10.1103/PhysRevB.84.045410

64. Inoue, K., Yanagisawa, R., Koike, E., Nishikawa, M., and Takano, H. (2010). Combustion synthesis of carbon nanotubes and related nanostructures. *Prog. Energy Combust. Sci.* 36, 696–727. doi:10.1016/j.freeradbiomed.2010.01.013

65. Ji, Y., Lee, S., Cho, B., Song, S., and Lee, T. (2011). Flexible organic memory devices with multilayer graphene electrodes. *ACS Nano* 5, 5995–6000.

doi:10.1021/nn201770s

66. Jo, G., Na, S.-I., Oh, S.-H., Lee, S., Kim, T.-S., Wang, G., et al. (2012). The application of graphene as electrodes in electrical and optical devices. *Nanotechnology* 23, 112001. doi:10.1088/0957-4484/23/11/112001

67. Jo, G., et al. (2010). Tuning of a graphene-electrode work function to enhance the efficiency of organic bulk heterojunction photovoltaic cells with an inverted structure. *Appl. Phys. Lett.* 97, 213301. doi:10.1063/1.3514551

68. Kammler, H. K., Mäer, L. and Pratsinis, S. E. (2001). Flame synthesis of nanoparticles. *Chem. Eng. Technol.* 24, 583–596. doi:10.1002/1521-4125(200106)24:6<583::AID-CEAT583>3.0.CO;2-H

69. Karamitaheri, H., Pourfath, M., Faez, R., and Kosina, H. (2011). Geometrical effects on the thermoelectric properties of ballistic graphene antidot lattices. *J. Appl. Phys.* 110, 054506. doi:10.1063/1.3629990

70. Kellie, B. M., Silleck, A. C., Bellman, K., Snodgrass, R. and Prakash, S. (2013). Deposition of few-layered graphene in a microcombustor on copper and nickel substrates. *RSC Adv.* 3, 7100–7105. doi:10.1039/c3ra40632f

71. Kerbs, H. U., and Bremert, O. (1993). Pulsed-laser deposition of thin metallic alloys. *Appl. Phys. Lett.* 62, 2341. doi:10.1021/am3022976

72. Kidoh, H., Ogawa, T., Morimoto, A. and Shimizu, T. (1991). Ferroelectric properties of lead-zirconate-titanate films prepared by laser ablation. *Appl. Phys. Lett.* 58, 2910. doi:10.1063/1.104719

73. Kim, J., et al. (2011). Low-temperature synthesis of large-area graphene-based transparent conductive films using surface wave plasma chemical vapor deposition. *Appl. Phys. Lett.* 98, 091502. doi:10.1063/1.3561747

74. Kim, J.T., Kim, B.K., Kim, E.Y., Park, H.C., and Jeong, H.M. (2014). Synthesis and shape memory performance of polyurethane/graphene nanocomposites. *React. Funct. Polym.* 74, 16–21. doi:10.1016/j.reactfunctpolym.2013.10.004

75. Kim, K. S., Zhao, Y., Jang, H., Lee, S. Y., Kim, J. M., Kim, K. S., et al. (2009). Large-scale pattern growth of graphene films for stretchable transparent electrodes. *Nature* 457, 706–710. doi:10.1038/nature07719

76. Kim, M., Safron, N. S., Han, E., Arnold, M. S., and Gopalan, P. (2010). Fabrication and characterization of large-area, semiconducting nanoperforated graphene materials. *Nano Lett.* 10, 1125–1131. doi:10.1021/nl9032318

77. Koh, A. T. T., Foong, Y. M. and Chua, D. H. C. (2010). Cooling rate and energy dependence of pulsed laser fabricated graphene on nickel at reduced temperature. *Appl. Phys. Lett.* 97, 114102. doi:10.1063/1.3489993

78. Krebs, H. U., Weisheit, M., Faupel, J., Süske, E., Scharf, T., Fuhse, C., et al. (2003). Pulsed laser deposition (PLD), a versatile thin film technique. *Adv. Solid State Phys.* 505–518. doi:10.1007/978-3-540-44838-9_36

79. Krivchenko, V.A., Dvorkin, V.V., Dzbanovsky, N.N., Timofeyev, M.A., Stepanov, A.S., Rakhimov, A.T., et al., (2012). Evolution of carbon film structure during its catalyst-free growth in the plasma of direct current glow discharge. *Carbon N. Y.* 50, 1477–1487. doi:10.1016/j.carbon.2011.11.018

80. Kuribayashi, K., et al. (1989). Millimeter-sized joint actuator using a shape memory alloy. *Sens. Actuators* 20, 57–64. doi:10.1016/0250-6874(89)87102-1

81. Layek, R. K., and Nandi, A. K. (2013). A review on synthesis and properties of polymer functionalized graphene. *Polymer* 54, 5087–5103. doi:10.1016/j.polymer.2013.06.027

82. Lee, C., Wei, X., Kysar, J. W., and Hone, J. (2008). Measurement of the elastic properties and intrinsic strength of monolayer graphene. *Science* 321, 385–388. doi:10.1126/science.1157996

83. Lee, S.K. and Kim, B.K. (2014). Synthesis and properties of shape memory graphene oxide/polyurethane chemical hybrids.*Polym. Int.* 63, 1197–1202. doi:10.1002/pi.4617

84. Lee, W. H., Park, J., Sim, S. H., Jo, S. B., Kim, K. S., Hong, B. H., et al. (2011). Transparent flexible organic transistors based on monolayer graphene electrodes on plastic. *Adv. Mater. Weinheim* 23, 1752–1756. doi:10.1002/adma.201004099

85. Li, J. F., et al. (2010d). High-performance nanostructured thermoelectric materials. *NPG Asia Mater.* 2, 152–158. doi:10.1038/asiamat.2010.138

86. Li, L., et al. (2015). Graphene oxide vs. reduced graphene oxide as core substrate for core/shell-structured dielectric nanoplates with different electro-responsive characteristics. *J. Mater. Chem. C* 3, 5098–5108. doi:10.1039/C5TC00474H

87. Li, S. S., Tu, K. H., Lin, C. C., Chen, C. W., and Chhowalla, M. (2010b). Solution-processable graphene oxide as an efficient hole transport layer in polymer solar cells. *ACS Nano* 4, 3169–3174. doi:10.1021/nn100551j

88. Li, X., Zhu, H., Wang, K., Cao, A., Wei, J., Li, C., et al. (2009). Large-area synthesis of high-quality and uniform graphene films on copper foils. *Science* 324, 1312–1314. doi:10.1126/science.1171245

89. Li, X., et al. (2010c). Graphene-on-silicon Schottky junction solar cells. *Adv. Mater. Weinheim* 22, 2743–2748. doi:10.1002/adma.200904383

90. Li, X., Zhu, H., Wang, K., Cao, A., Wei, J., Li, C., et al (2008). Chemically derived, ultrasmooth graphene nanoribbon semiconductors. *Science* 319, 1229–1232. doi:10.1126/science.1150878

91. Li, X., Zhu, H., Wang, K., Cao, A., Wei, J., Li, C., et al (2010a). Graphene-on-silicon Schottky junction solar cells. *Adv. Mater.* 22, 2743. doi:10.1002/adma.200904383

92. Li, J.-F., Liu, W.-S., Zhao, L.-D., and Zhou, M., (2011a). Ethanol flame synthesis of highly transparent carbon thin films. *Carbon N. Y.* 49, 237–241. doi:10.1016/j.carbon.2010.09.009

93. Li, Z., Zhu, H., Xie, D., Wang, K., Cao, A., Wei, J., et al. (2011b). Flame synthesis of few-layered graphene/graphite films. *Chem. Commun.* 47, 3520–3522. doi:10.1039/c0cc05139j

94. Liu, F., Ming, P., and Li, J. (2007). Ab initio calculation of ideal strength and phonon instability of graphene under tension.*Phys. Rev. B* 76, 064120. doi:10.1103/PhysRevB.76.064120

95. Loh, K. P., Bao, Q., Anga, P. K., and Yang, J. (2010). The chemistry of graphene. *J. Mater. Chem.* 20, 2277–2289. doi:10.1039/b920539j

96. Loomis, J., King, B., Burkhead, T., Xu, P., Bessler, N., Terentjev, E., et al. (2012). Graphene-nanoplatelet-based photomechanical actuators. *Nanotechnology* 23, 045501. doi:10.1088/0957-4484/23/4/045501

97. Losurdo, M., Giangregorio, M. M., Capezzuto, P., and Bruno, G. (2011). Graphene CVD growth on copper and nickel: role of hydrogen in kinetics and structure. *Phys. Chem. Chem. Phys.* 13, 20836–20843. doi:10.1039/c1cp22347j

98. Lotya, M., Hernandez, Y., King, P. J., Smith, R. J., Nicolosi, V., Karlsson, L. S., et al. (2009). Liquid phase production of graphene by exfoliation of graphite in surfactant/water solutions. *J. Am. Chem. Soc.* 131, 3611–3620. doi:10.1021/ja807449u

99. Luechinger, N. A., Athanassiou, E. K., and Stark, W. J. (2008). Graphene-stabilized copper nanoparticles as an air-stable substitute for silver and gold in low-cost ink-jet printable electronics. *Nanotechnology* 19, 445201. doi:10.1088/0957-4484/19/44/445201

100. Luk'yanchuk, I. A., Varlamov, A. A., and Kavokin, A. V. (2015). Observation of a giant two-dimensional band-piezoelectric effect on biaxial-strained graphene. *NPG Asia Mater* 7, e154. doi:10.1038/am.2014.124

101. Malesevic, A., Vitchev, R., Schouteden, K., Volodin, A., Zhang, L., Tendeloo, G. V., et al. (2008). Synthesis of few-layer graphene via microwave plasma-enhanced chemical vapour deposition. *Nanotechnology* 19, 305604. doi:10.1088/0957-4484/19/30/305604

102. Martin, J. A., Vazquez, L., Bernard, P., Comin, F. and Ferrer, S. (1990). Epitaxial growth of crystalline, diamond-like films on Si(100) by laser ablation of graphite. *Appl. Phys. Lett.* 57, 1742–1744. doi:10.1063/1.104053

103. Mattevi, C., Kima, H., and Chhowalla, M. (2011). A review of chemical vapour deposition of graphene on copper. *J. Mater. Chem.* 21, 3324–3334. doi:10.1039/C0JM02126A

104. Mayorov, A. S., Gorbachev, R. V., Morozov, S. V., Britnell, L., Jalil, R., Ponomarenko, L. A., et al. (2011). Micrometer-scale ballistic transport in encapsulated graphene at room temperature. *Nano Lett.* 11, 2396–2399. doi:10.1021/nl200758b

105. Mazzamuto, F., Hung Nguyen, V., Apertet, Y., Ca, C., Chassat, C., Saint-Martin, J., et al., (2011). Enhanced thermoelectric properties in graphene nanoribbons by resonant tunneling of electrons. *Phys. Rev. B* 83, 235426. doi:10.1103/PhysRevB.83.235426

106. Meador, M. A., et al. (2010). *DRAFT Nanotechnology Roadmap Technology Area 10. National Aeronautics and Space Administration (NASA)*. 5. Available at: http://www.nasa.gov/pdf/501325main_TA10-Nanotech-DRAFT-Nov2010-A.pdf

107. Memon, N. K., Tse, S. D., Al-Sharab, J. F., Yamaguchi, H., Goncalves, A.-M. B., Kear, B. H., et al., (2011). Flame synthesis of graphene films in open environments. *Carbon N. Y.* 49, 5064–5070. doi:10.1016/j.carbon.2011.07.024

108. Memon, N. K., Anjum, D. H., and Chung, S. H. (2013). Multiple-diffusion flame synthesis of pure anatase and carbon-coated titanium dioxide nanoparticles. *Combust. Flame* 160, 1848–1856. doi:10.1016/j.combustflame.2013.03.022

109. Morozov, S. V., Novoselov, K. S., Katsnelson, M. I., Schedin, F., Elias, D. C., Jaszczak, J. A., et al. (2008). Giant intrinsic carrier mobilities in graphene and its bilayer. *Phys. Rev. Lett.* 100, 016602. doi:10.1103/PhysRevLett.100.016602

110. Moser, J., Barreiro, A., and Bachtold, A. (2007). Current-induced cleaning of graphene. *Appl. Phys. Lett.* 91, 163513. doi:10.1063/1.2789673

111. Muge, A., and Chabal, Y. J. (2011). Nature of graphene edges: a review. *Jpn. J. Appl. Phys.* 50, 070101. doi:10.1021/ar500306w

112. Nair, R. R., Blake, P., Grigorenko, A. N., Novoselov, K. S., Booth, T. J., Stauber, T., et al. (2008). Fine structure constant defines visual transparency of graphene. *Science* 320, 1308. doi:10.1126/science.1156965

113. Nair, R. R., Ren, W., Jalil, R., Riaz, I., Kravets, V. G., Britnell, L., et al. (2010). Fluorographene: a two-dimensional counterpart of teflon. *Small* 6, 2877–2884. doi:10.1002/smll.201001555

114. Nakamura, K., Fukazawa, K., Yamada, K., and Saito, S. (1989). Hysteresis-free piezoelectric actuators using linbo3 plates with a ferroelectric inversion layer. *Ferroelectrics* 93, 211–216. doi:10.1080/00150198908017348

115. Nechache, R., Harnagea, C., Pignolet, A., Normandin, F., Veres, T., Carignan, L.-P. et al. (2006). Growth, structure, and properties of epitaxial thin films of first-principles predicted multiferroic Bi2FeCrO6. *Appl. Phys. Lett.* 89, 102902–102902–3. doi:10.1063/1.2346258

116. Neto, A. H. C., Guinea, F., Peres, N. M. R., Novoselov, K. S., and Geim, A. K. (2009). The electronic properties of graphene.*Rev. Mod. Phys.* 81, 109–162. doi:10.1103/RevModPhys.81.109

117. Ni, Z., Wang, Y., Yu, T., and Shen, Z. (2008). Raman spectroscopy and imaging of graphene. *Nano Res.* 1, 273–291. doi:10.1007/s12274-008-8036-1

118. Novoselov, K. S., Geim, A. K., Morozov, S. V., Jiang, D., Katsnelson, M. I., Grigorieva, I. V., et al. (2005). Two-dimensional gas of massless dirac fermions in graphene. *Nature* 438, 197–200. doi:10.1038/nature04233

119. Novoselov, K. S., Geim, A. K., Morozov, S. V., Jiang, D., Zhang, Y., Dubonos, S. V., et al. (2004). Electric field effect in atomically thin carbon films. *Science* 306, 666–669. doi:10.1126/science.1102896

120. Ohta, H., Kim, S., Mune, Y., Mizoguchi, T., Nomura, K., Ohta, S., et al. (2007). Giant thermoelectric Seebeck coefficient of a two-dimensional electron gas in

SrTiO3. *Nat. Mater.* 6, 129–134. doi:10.1038/nmat1821

121. Ossler, F., Wagner, J., Canton, S., and Wallenberg, R. (2010). Sheet-like carbon particles with graphene structures obtained from a Bunsen flame. *Carbon N. Y.* 48, 4203–4206. doi:10.1016/j.carbon.2010.07.013
122. Ouyang, F., Peng, S., Liu, Z., and Liu, Z. (2011). Bandgap opening in graphene antidot lattices: the missing half. *ACS Nano* 5, 4023–4030. doi:10.1021/nn200580w
123. Ouyang, T., and Hu, M. (2010). Enhanced thermoelectric figure of merit in edge-disordered zigzag graphene nanoribbons. *Phys. Rev. B* 81, 113401. doi:10.1088/0957-4484/25/24/245401
124. Park, J. and Kim, B.K. (2014). Infrared light actuated shape memory effects in crystalline polyurethane/graphene chemical hybrids. *Smart Mater. Struct.* 23, 025038. doi:10.1088/0964-1726/23/2/025038
125. Park, J. B., Xiong, W., Gao, Y., Qian, M., Xie, Z. Q., Mitchell, M., et al., (2011). Fast growth of graphene patterns by laser direct writing. *Appl. Phys. Lett.* 98, 123109. doi:10.1063/1.3569720
126. Pedersen, T. G., Flindt, C., Pedersen, J., Mortensen, N. A., Jauho, A. P., and Pedersen, K. (2008). Graphene antidot lattices: designed defects and spin qubits. *Phys. Rev. Lett.* 100, 136804. doi:10.1103/PhysRevLett.100.136804
127. Pereira, V. M., and Castro Neto, A. H. (2009). Strain engineering of graphene's electronic structure. *Phys. Rev. Lett.* 103, 046801. doi:10.1103/PhysRevLett.103.046801
128. Petersen, R., Pedersen, T. G., and Jauho, A. P. (2011). Clar sextet analysis of triangular, rectangular, and honeycomb graphene antidot lattices. *ACS Nano* 5, 523–529. doi:10.1021/nn102442h
129. Ponnamma, D., Guo, Q., Krupa, I., Al-Maadeed, M. A., K, T. V., Thomas, S., et al. (2010). Recent advances in graphene based polymer composites. *Prog. Polym. Sci.* 35, 1350–1375. doi:10.1039/c4cp04418e
130. Potts, J. R., Dreyer, D. R., Bielawski, C. W., and Ruoff, R. S. (2011). Graphene-based polymer nanocomposites. *Polymer* 52, 5–25. doi:10.1016/j.polymer.2010.11.042
131. Rana, S., Cho, J. W. and Tan, L. P., (2013). Graphene-crosslinked polyurethane block copolymer nanocomposites with enhanced mechanical, electrical, and shape memory properties. *RSC Adv.* 3, 13796–13803. doi:10.1039/c3ra40711j
132. Reina, A., Jia, X., Ho, J., Nezich, D., Son, H., Bulovic, V., et al. (2008). Large area, few-layer graphene films on arbitrary substrates by chemical vapor deposition. *Nano Lett.* 9, 30–35. doi:10.1021/nl801827v
133. Reina, A., Jia, X., Ho, J., Nezich, D., Son, H., Bulovic, V., et al. (2009). Large area, few-layer graphene films on arbitrary substrates by chemical vapor deposition. *Nano Lett.* 9, 30–35. doi:10.1021/nl801827v
134. Ren, Z. F., Huang, Z. P., Xu, J. W., Wang, J. H., Bush, P., Siegal, M. P., et al. (1998). Synthesis of large arrays of well-aligned carbon nanotubes on glass. *Science* 282, 1105–1107. doi:10.1126/science.282.5391.1105
135. Saravanakumar, B., Mohan, R., Kim, S.-J. (2013). Facile synthesis of graphene/ZnO nanocomposites by low temperature hydrothermal method. *Mater. Res. Bull.* 48, 878–883. doi:10.1016/j.materresbull.2012.11.048
136. Sarma, S. D., Adam, S., Hwang, E. H., and Rossi, E. (2011). Electronic transport in two-dimensional graphene. *Rev. Mod. Phys.*83, 407–470. doi:10.1103/RevModPhys.83.407
137. Schwierz, F. (2010). Graphene transistors. *Nat. Nanotechnol.* 5, 487–496. doi:10.1038/nnano.2010.89
138. Sevinçli, H., Sevik, C., Caın, T., and Cuniberti, G. (2009). Disorder enhances thermoelectric figure of merit in armchair graphane nanoribbons. *Appl. Phys. Lett.* 95, 192114. doi:10.1038/srep01228
139. Sharma, R., Alam, F., Sharma, A. K., Dutta, V. and Dhawan, S. K. (2014). ZnO anchored graphene hydrophobic nanocomposite-based bulk heterojunction solar cells showing enhanced short-circuit current. *J. Mater. Chem. C*2, 8142–8151. doi:10.1039/C4TC01056F
140. Shen, T., Wu, Y. Q., Capano, M. A., Rokhinson, L. P., Engel, L. W. and Ye, P. D. (2008). Magneto-conductance oscillations in graphene antidot arrays. *Appl. Phys. Lett.* 93, 122102. doi:10.1063/1.2988725
141. Shinde, S. R., Ogale, S. B., Greene, R. L., Venkatesan, T., Canfield, P. C., Bud'ko, S. L., et al. (2001). Superconducting MgB2 thin films by pulsed laser deposition. *Appl. Phys. Lett.* 79, 227–229. doi:10.1063/1.1385186
142. Shim, J.-P., Choe, M., Jeon, S.-R., Seo, D., Lee, T. and Lee, D.-S., (2011). InGaN-based p–i–n solar cells with graphene electrodes. *Appl. Phys. Express* 4, 052302. doi:10.1143/APEX.4.052302
143. Singh, V., Joung, D., Zhai, L., Das, S., Khondaker, S. I., and Seal, S. (2011). Graphene based materials: past, present and future.*Prog. Mater. Sci.* 56, 1178–1271. doi:10.1016/j.pmatsci.2011.03.003
144. Smela, E. and Gadegaard, N. (1999). Surprising volume change in PPy(DBS): an atomic force microscopy study. *Adv. Mater.* 11, 953. doi:10.1002/(SICI)1521-4095(199908)11:11<953::AID-ADMA953>3.3.CO;2-8
145. Smith, H.M. and Turner, A.F. (1965). Vacuum deposited thin films using a ruby laser. *Appl. Opt.* 4, 147–148. doi:10.1364/AO.4.000147
146. Stankovich, S., Dikin, D. A., Dommett, G. H., Kohlhaas, K. M., Zimney, E. J., Stach, E. A., et al. (2006). Graphene-based composite materials. *Nature* 442, 282–286. doi:10.1038/nature04969
147. Suemitsu, M. and Fukidome, H. (2010). Epitaxial graphene on silicon substrates. *J. Phys. D Appl. Phys.* 43, 374012. doi:10.1088/0022-3727/43/37/374012
148. Suk, J. W., Kitt, A., Magnuson, C. W., Hao, Y., Ahmed, S., An, J., et al. (2011). Transfer of CVD-grown monolayer graphene onto arbitrary substrates. *ACS Nano* 5, 6916–6924. doi:10.1021/nn201207c
149. Sullivan, T. D., and Powers, J. M. (1977). Flexural disc piezoelectric polymer hydrophones. *J. Acoust. Soc. Am.* 60, S47–S47.
150. Sun, Z., Hasan, T., Torrisi, F., Popa, D., Privitera, G., Wang, F. et al., (2009). Graphene mode-locked ultrafast

laser. *ACS Nano*4, 803. doi:10.1021/nn901703e

151. Sun, Z., Yan, Z., Yao, J., Beitler, E., Zhu, Y., and Tour, J. M. (2010). Growth of graphene from solid carbon sources. *Nature* 468, 549–552. doi:10.1038/nature09579
152. Tang, S., and Cao, Z. (2010). Structural and electronic properties of the fully hydrogenated boron nitride sheets and nanoribbons: insight from first-principles calculations. *Chem. Phys. Lett.* 488, 67–72. doi:10.1016/j.cplett.2010.01.073
153. Thakur, S. and Karak, N. (2013). Bio-based tough hyperbranched polyurethane–graphene oxide nanocomposites as advanced shape memory materials. *RSC Adv.* 3, 9476–9482. doi:10.1039/c3ra40801a
154. Tung, V. C., Chen, L. M., Allen, M. J., Wassei, J. K., Nelson, K., Kaner, R. B., et al. (2009). Low-temperature solution processing of graphene– carbon nanotube hybrid materials for high-performance transparent conductors. *Nano Lett.* 9, 1949–1955. doi:10.1021/nl9001525
155. Tyler, W.W. and Wilson, A.C. (1953). Thermal conductivity, electrical resistivity, and thermoelectric power of graphite. *Phys. Rev.* 89, 870–875. doi:10.1103/PhysRev.89.870
156. Voon, L. C., Sandberg, E., Aga, R. S., and Farajian, A. A. (2010). Hydrogen compounds of group-IV nanosheets. *Appl. Phys. Lett.*97, 163114. doi:10.1063/1.3495786
157. Wager, J.F. (2003). Transparent electronics. *Science* 300, 1245–1246. doi:10.1126/science.1085276
158. Wan, X., Long, G., Huang, L., and Chen, Y. (2011). Graphene – a promising material for organic photovoltaic cells. *Adv. Mater. Weinheim* 23, 5342–5358. doi:10.1002/adma.201102735
159. Wang, C., Liu, N., Allen, R., Tok, J. B., Wu, Y., Zhang, F., et al. (2013). A rapid and efficient self-healing thermo-reversible elastomer crosslinked with graphene oxide. *Adv. Mater. Weinheim* 25, 5785–5790. doi:10.1002/adma.201302962
160. Wang, J., Zhu, M., Outlaw, R. A., Zhao, X., Manos, D. M., and Holloway, B. C. (2004). Synthesis of carbon nanosheets by inductively coupled radio-frequency plasma enhanced chemical vapor deposition. *Carbon N. Y.* 42, 2867–2872. doi:10.1016/j.carbon.2004.06.035
161. Wang, X., Zhi, L., and Müllen, K. (2008). Transparent, conductive graphene electrodes for dye-sensitized solar cells. *Nano Lett.*8, 323–327. doi:10.1021/nl072838r
162. Wang, Y., Zheng, Y., Xu, X., Dubuisson, E., Bao, Q., Lu, J., et al. (2011). Electrochemical delamination of CVD-grown graphene film: toward the recyclable use of copper catalyst. *ACS Nano* 5, 9927–9933. doi:10.1021/nn203700w
163. Wei, D. and Xu, X. (2012). Laser direct growth of graphene on silicon substrate. *Appl. Phys. Lett.* 100, 023110. doi:10.1063/1.3675636
164. Wei, D., Mitchell, J. I., Tansarawiput, C., Nam, W., Qi, M., Ye, P. D., et al. (2013). Laser direct synthesis of graphene on quartz.*Carbon N. Y.* 53, 374–379. doi:10.1016/j.carbon.2012.11.026
165. Wei, P., Bao, W., Pu, Y., Lau, C. N., and Shi, J. (2009). Anomalous thermoelectric transport of dirac particles in graphene. *Phys. Rev. Lett.* 102, 166808. doi:10.1103/PhysRevLett.102.166808
166. Yan, Y., Liang, Q.-F., Zhao, H., and Wu, C.-Q. (2012). Thermoelectric properties of hexagonal graphene quantum dots. *Phys. Lett. A* 376, 1154–1158. doi:10.1016/j.physleta.2012.02.013
167. Yannopoulos, S. N., Siokou, A., Nasikas, N. K., Dracopoulos, V., Ravani, F. and Papatheodorou, G. N. (2012). CO2-Laser-induced growth of epitaxial graphene on 6H-SiC(0001). *Adv. Funct. Mater.* 22, 113–120. doi:10.1002/adfm.201101413
168. Yin, J., Wang, X., Chang, R. and Zhao, X. (2012). Polyaniline decorated graphene sheet suspension with enhanced electrorheology. *Soft Matter* 8, 294–297. doi:10.1039/C1SM06728A
169. Yin, J., Chang, R., Kai, Y. and Zhao, X. (2013). Highly stable and AC electric field-activated electrorheological fluid based on mesoporous silica-coated graphene nanosheets. *Soft Matter* 9, 3910–3914. doi:10.1039/c3sm27835b
170. Yin, Z., Wu, S., Zhou, X., Huang, X., Zhang, Q., Boey, F., et al. (2010). Electrochemical deposition of ZnO nanorods on transparent reduced graphene oxide electrodes for hybrid solar cells. *Small* 6, 307–312. doi:10.1002/smll.200901968
171. Yokomizo, Y., Krishnamurthy, S. and Kamat, P. V. (2013). Photoinduced electron charge and discharge of graphene–ZnO nanoparticle assembly. *Catal. Today* 199, 36–41. doi:10.1016/j.cattod.2012.04.045
172. Yu, H., Wang, T., Wen, B., Lu, M., Xu, Z., Zhu, C., et al., (2012). Graphene/polyaniline nanorod arrays: synthesis and excellent electromagnetic absorption properties. *J. Mater. Chem.* 22, 21679–21685. doi:10.1039/c2jm34273a
173. Yu, Q., Lian, J., Siriponglert, S., Li, H., Chen, Y. P. and Pei, S.-S., (2008). Graphene segregated on Ni surfaces and transferred to insulators. *Appl. Phys. Lett.* 93, 113103. doi:10.1063/1.2982585
174. Yu, W. J., Lee, S. Y., Chae, S. H., Perello, D., Han, G. H., Yun, M., et al. (2011). Small hysteresis nanocarbon-based integrated circuits on flexible and transparent plastic substrate. *Nano Lett.* 11, 1344–1350. doi:10.1021/nl104488z
175. Yuan, G.D., Zhang, W.J., Yang, Y., Tang, Y.B., Li, Y.Q., Wang, J.X., et al., (2009). Graphene sheets via microwave chemical vapor deposition. *Chem. Phys. Lett.* 467, 361–364. doi:10.1016/j.cplett.2008.11.059
176. Zhang, H. and Feng, P. X. (2014). Environmentally responsive graphene systems. *Small* 10, 2151–2164. doi:10.1002/smll.201303080
177. Zhang, W. L. and Choi, H. J. (2014). Graphene oxide based smart fluids. *Soft Matter* 10, 6601–6608. doi:10.1039/c4sm01151a
178. Zhang, W. L., Liu, Y. D., Choi, H. J. and Kim, S. G. (2012). Electrorheology of graphene oxide. *ACS Appl.*

Mater. Interfaces 4, 2267–2272. doi:10.1021/am300267f

179. Zhang, Y., Tang, T. T., Girit, C., Hao, Z., Martin, M. C., Zettl, A., et al. (2009). Direct observation of a widely tunable bandgap in bilayer graphene. *Nature* 459, 820–823. doi:10.1038/nature08105
180. Zhang, Y., Tiwarya, P., Scott Parenta, J. and Kontopoulou, M. (2013). Crystallization and foaming of coagent-modified polypropylene: nucleation effects of cross-linked nanoparticles. *Polymer* 54, 4814–4819. doi:10.1016/j.polymer.2013.07.020
181. Zhang, Y., Wu, S., Wen, Y.-H. and Zhu, Z. (2010a). Surface-passivation-induced metallic and magnetic properties of ZnO graphitic sheet. *Phys. Lett.* 96, 223113. doi:10.1063/1.3442507
182. Zhang, H. and Feng, P. X. (2010b). Fabrication and characterization of few-layer graphene. *Carbon N. Y.* 48, 359–364. doi:10.1016/j.carbon.2009.09.037
183. Zhang, W. L., Park, B. J. and Choi, H. J. (2010c). Colloidal graphene oxide/polyaniline nanocomposite and its electrorheology.*Chem. Commun.* 46, 5596–5598. doi:10.1039/c0cc00557f
184. Zhu, L., Ma, R., Sheng, L., Liu, M. and Sheng, D. N. (2009). Thermopower and Nernst effect in graphene in a magnetic field.*Phys. Rev. B* 80, 081413. doi:10.1103/PhysRevB.80.081413
185. Zhu, M., Wang, J., Holloway, B. C., Outlaw, R. A., Zhao, X., Hou, K., et al. (2007). A mechanism for carbon nanosheet formation.*Carbon N. Y.* 45, 2229–2234. doi:10.1016/j.carbon.2007.06.017
186. Zuev, Y. M., Chang, W. and Kim, P. (2009). Thermoelectric and magnetothermoelectric transport measurements of graphene.*Phys. Rev. Lett.* 102, 096807. doi:10.1103/PhysRevLett.102.096807

CHAPTER 21

RECENT ADVANCES ON CARBON NANOTUBES AND GRAPHENE REINFORCED CERAMICS NANOCOMPOSITES

Iftikhar Ahmad [1], Bahareh Yazdani [2], and Yanqiu Zhu [2]

[1]Center of Excellence for Research in Engineering Materials, Advanced Manufacturing Institute, King Saud University, Riyadh 11421, Saudi Arabia

[2]College of Engineering, Mathematics and Physical Sciences, University of Exeter, Exeter EX4 4QF, UK

ABSTRACT

Ceramics suffer the curse of extreme brittleness and demand new design philosophies and novel concepts of manufacturing to overcome such intrinsic drawbacks, in order to take advantage of most of their excellent properties. This has been one of the foremost challenges for ceramic material experts. Tailoring the ceramics structures at nanometre level has been a leading research frontier; whilst upgrading via reinforcing ceramic matrices with nanomaterials including the latest carbon nanotubes (CNTs) and graphene has now become an eminent practice for advanced applications. Most recently, several new strategies have indeed improved the properties of the ceramics/CNT nanocomposites, such as by tuning with dopants, new dispersions routes and modified sintering methods. The utilisation of graphene in ceramic nanocomposites, either as a solo reinforcement or as a hybrid with CNTs, is the newest development. This article will summarise the recent advances, key difficulties and potential applications of the ceramics nanocomposites reinforced with CNTs and graphene.

INTRODUCTION

Ceramics are potential contestants for diverse sophisticated engineering applications, and plenty of attentions have been focused to further improve their properties by adopting emerging technologies. As a result, much deeper understandings and significant amounts of improvements in their structures and properties have been achieved after decades of efforts. However, many challenge issues limit their wide applications, such as the degradation of high temperature mechanical properties of non-oxide silicon nitride (Si_3N_4) and silicon carbide (SiC), and low fracture toughness, poor creep, deprived thermal shock resistance of oxide ceramics like alumina (Al_2O_3) and zirconium oxide (ZrO_2) [1]. To date, ceramics have found some niche applications, from high speed cutting tools, dental implants, chemical and electrical insulators, to wear resistance parts and various coatings, due to their high hardness, chemical inertness and high electrical and thermal insulating properties [2]. Low fracture toughness restricts ceramics for applications in aircraft engine parts and in extreme environments for space engineering [3]. Presence of impurities, pores and cracks cause pure ceramics extremely brittle, and complex/expensive processing technology is needed to reduce such fatal drawbacks. For decades, the addition of a second reinforcing

Citation: Iftikhar Ahmad, Bahareh Yazdani, and Yanqiu Zhu, Recent Advances on Carbon Nanotubes and Graphene Reinforced Ceramics Nanocomposites, Nanomaterials 2015, 5(1), 90-114; doi:10.3390/nano5010090

phase in ceramics has been an effective practice to improve their toughness, converting brittle ceramics to practical engineering materials. Recent advances in nanomaterials have offered the opportunity to tailor the ceramic structures at nanometre scale, for the development of new classes of stronger, tougher engineering ceramics with added functionalities. Chosen nanomaterials with distinct morphologies and properties have been used to reinforce monolithic ceramics [4,5,6]. In particular, the exceptional mechanical behaviour and outstanding multifunctional features of carbon nanotubes (CNTs) and graphene have made them the wonder materials, standing out from many other nanomaterials, among different research communities. There has been much documented research attempting to incorporate both types of CNT in brittle ceramics to convert them into tough, strong, electric and thermal conductive materials [7,8,9,10,11,12,13,14,15]. Graphene, known as a monolayer of carbon atoms arranged in a honeycomb lattice, has shown similar properties to carbon nanotubes with impressive thermal, mechanical, and electrical properties, and is a promising alternative of CNTs in various applications [16,17,18]. Compared with CNTs, graphene also have large specific surface areas and they do not form agglomerates in a matrix when handled appropriately, thus an ideal nano-filler for composite materials. In this regard, the low-cost, high quality and commercially more viable a-few-layer-thick graphene nanosheets, designated as graphene nanoplatelets (GNPs) are more promising for practical engineering applications, thus attracted considerable research interests for advanced ceramic matrices. Indeed, various crucial ceramics such as Al_2O_3, Si_3N_4 and ZrO_2 have been reinforced by the GNP fillers and obvious improvements in fracture toughness, thermal and electrical properties have been obtained [19,20,21,22,23,24,25,26]. However, research of ceramic-GNP nanocomposites is in its infancy, and more thorough and systematic studies are required.

Nevertheless, ceramics reinforced with CNTs, graphene and GNT (CNTs/graphene hybrid) have indeed showed significant improvements in the fracture toughness and other mechanical properties by following complex toughening mechanisms, although wide variations in the results still remain problematic. In fact, processing ceramic nanocomposites is complicated due to the introduction of a second reinforcement phase of nanometric scale. Conventional rules and benefits associated with microscopic reinforcement phases need to be modified carefully and validated fully before being applied directly. In this context, recent advances in the fabrication technology, mechanical properties and potential applications of typical ceramic nanocomposites reinforced with CNTs and graphene are presented in this paper. The main purpose of this review is to provide a comprehensive picture of the current state of research progresses and challenges concerning the graphene and CNTs-reinforced ceramic composites, to assist the ceramic community for further developments.

CNTS-REINFORCED CERAMICS NANOCOMPOSITES

Pre-Processing for Good Dispersion

A statistical summary of the varieties of processes opted to fabricate CNTs containing ceramic nanocomposites is graphically presented in Figure 1, and Table 1 gives further details of these processes. It is evident that 88% of the reported cases used the readily available and economically feasible multi-walled carbon nanotubes (MWCNTs) as the reinforcement, against single-walled carbon nanotubes (SWCNTs). Nearly 40% adopted a wet oxidation process (treatment with concentrated H_2SO_4 and HNO_3 in 3:1 ratio) to purify the CNTs in an effort to remove unwanted impurities including amorphous carbon nanofibres, carbon nanoparticles, amorphous carbon coating layers, and metallic catalyst residues; and about 33% of the reports attempted pristine CNTs; whilst the others tried oxidation through annealing [27,28,29,30]. The wet oxidation purification method for CNTs offers two folded advantages, realizing purification and simultaneous attachments of carboxyl functional groups onto the CNT surfaces which facilitates their mixing with and dispersions into matrices. It is a fact that CNT clusters prevent each individual CNT forming the ideal interconnection desired with the matrix, leading to ill-constructed interface and microstructures, detrimental to the final mechanical properties. Thus homogenous CNT dispersion within the matrix is extremely imperative. In addition, the actual quality of CNT dispersion is the foremost factor in tumbling the densities of CNT-reinforced ceramics, because homogenous dispersion is attainable only at low CNT concentrations (<2 wt%), and higher than that normally ended up with severe agglomerations [31].

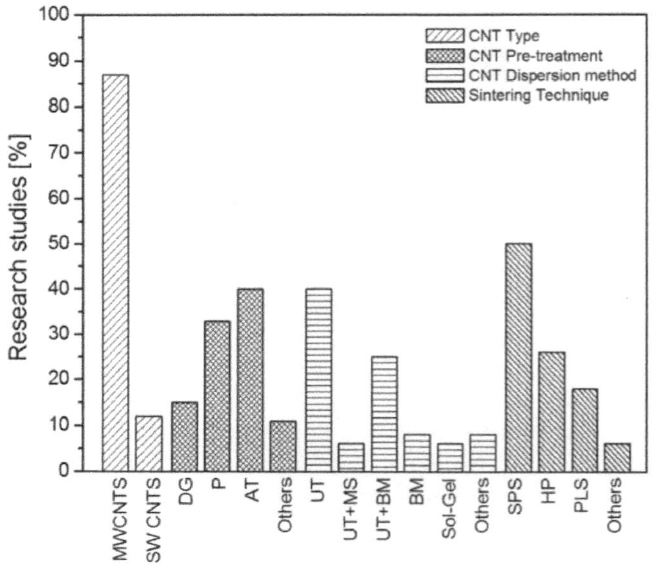

Figure 1: Statistical analysis of the carbon nanotubes (CNTs)-reinforced ceramic nanocomposites.

To combat this dispersion issue, as shown in Figure 1, the most dominant (40%) technique involving colloidal technology (ultra-sonication of CNTs for different durations into different solvents with or without surfactants). Until recently, attempts increasingly focused on a combined process (colloidal technique and ball milling) which have produced better results and was more reproducible than other techniques (ball milling, sol-gel, planetary centrifuge mixing, magnetic stirring, tape casting, *etc.*), as shown in Table 1. Moreover cationic, anionic and neutral surfactants have greatly contributed to the detangling of CNT clusters, of which SDS (sodium dodcyle sulphate) seems to be the most used one [11]. In addition, several reports described the growth of CNTs directly onto the surface of ceramics nanoparticles using a standard chemical vapour deposition (CVD) technique; however, this method failed to create high quality coverage, on top of the low yield issues [13].

Table 1: Processing details of CNTs-reinforced ceramics nanocomposites.

Reference	Matrix	CNT types	Purification methods	Dispersion procedures	Sintering techniques
[10]	Si_3N_4	SW	P	UT of CNTs with surfactant (C16TAB) and Si_3N_4	SPS under vacuum
[12]	Al_2O_3	MW	Oxidation at 500 °C for 90 min	UT of CNTs in ethanol	SPS at 1500 °C for 10 min under 50 MPa
[32]	Al_2O_3	MW	AT (H_2SO_4 + HNO_3)	UT of CNTs into water and SDS then incubation for 2 weeks	HP at 1600 °C, 60 min, 40 MPa
[33]	Al_2O_3	MW	AT (H_2SO_4 + HNO_3) for 3 h	24 h BM of ball Al_2O_3 powder and 30 min UT of CNTs in water and then BM of CNTs/Al_2O_3 mixture	PLS at 1500–1600 °C, 120–240 min, Ar
[34]	Al_2O_3	MW	Pristine	UT of CNTs for 1 h in alcohol	CIP at150MPa and PLS at 1500 °C, and 1700 °C with 2 h
[35]	Al_2O_3	MW	AT (heating in 65% HNO_3 at 80 °C for 8 h)	BM and Surfactant (Darvan C–N)	PLS at 1500 °C for 2 h using Ar
[36]	Mulite	MW	P	CNTs dispersion into ethanol by MS and UT	HP at 1600 °C for 60 min under Ar atmosphere at 30 MPa
[37]	Si_3N_4	MW	P	24 h ball milling the CNTs and Si_3N_4 slurry	HP at 1750 °C for 60 min under 30 MPa
[38]	ZrB_2–SiC	MW	P	20 min UT of CNTs and matrix with subsequent 24 h ball milling	HP at 1900 °C for 60 min under 30 MPa
[39]	$BaTiO_3$	MW	P	-	HP, 1200 °C, 60 min
[40]	Al_2O_3	MW	-	DG (CVD at 750 °C for 15 min for direct CNTs growth on Al_2O_3 nano-particles)	SPS at 1150 °C for 10 min under 100 MPa
[41]	Al_2O_3	SW	Pristine	UT of CNTs in ethanol	SPS at 1520 °C under 80 MPa
[42]	Al_2O_3	MW	P	35 h UT in water	SPS at 1300 °C, 20 min, 90 MPa
[43]	Al_2O_3	MW	AT	UT of CNTs and Al_2O_3 in water followed by 2 h and BM of CNTs/Al_2O_3	PLS at 1600 °C, 15 min, Ar
[44]	Al_2O_3	MW	AT (HNO_3 for 30 min)	5 h BM of CNTs and 1 h UT of CNTs. 5 h BM of CNT/Al_2O_3 in ethanol	PLS at 1550 °C, Ar
[45]	Al_2O_3	MW	AT (H_2SO_4 + HNO_3 in 3:1 for 7 h)	surfactant (SDS) using combination of UT and 24 h BM	HP at 1550 °C for 1 h under 30 MPa using Ar gas
[46]	Al_2O_3 + ZrO	MW	AT (heating in 65% HNO3 at 80 °C for 8 h))	2 min UT of CNTs with surfactant (SDS)and 24 BM then freezing with Nitrogen	HP at 1500 °C for 2 h under 30 MPa in Ar atmosphere
[47]	Al_2O_3	SW	AT (H_2SO_4 + HNO_3)	UT for 24 h	SPS at 1300 °C for 5 min under 75 MPa

Notes: SW: Single-walled CNTs; MW: Multi-walled CNTs; UT: Ultrasonication; BM: Ball milling; HP: Hot-pressing; SPS: Spark plasma sintering; PLS: Pressureless sintering; SDS: Sodium dodecyle sulphate; CIP: cold isostatic pressing; P: Pristine; MS: Magnetic stirring.

Cultivation of CNTs in porous ceramics is intriguing process, and numerous efforts have been documented for highly ordered CNT growth within the pores of thin SiO_2 and Al_2O_3 membranes, which led to novel CNTs-reinforced porous ceramics with potential applications as field emitters, nanocapacitors, and scanning microscope probes [48,49,50,51,52]. To prepare CNTs carrying porous ceramic composites, Fan *et al.* [27] first embedded catalyst inside the SiO_2 pores of the ceramics pores, then allowed for the carbon source to diffuse into and deposit inside the pores; whilst Kyotani *et al.* [48] grew CNTs on an anodic aluminium oxide (AAO) porous membrane with and without catalyst. The CNT growth mechanism in a catalyst free CNTs-porous ceramic is still not completely understood. The AAO membrane template may itself catalysed the CNT growth by deposition of carbon atoms on the internal pore surface of the complex three dimensional structure, as proposed by Sui *et al.* [49]. Since catalyst facilities carbon source decomposition, thus further deposition of atomic carbon tends to result in more ordered or well-crystallised structures, leading to better quality nanocomposites than the non-catalysed process. Patterning and lithography technology enabled Bae *et al.* to deposit a thin Si layer on an AAO substrate for enhanced CNT growth [50]. Parham *et al.* have recently prepared a 3 wt% CNT-containing composite using Al_2O_3 and SiO_2 porous ceramics, and resulting composites exhibited a high efficiency for yeast cell filtration (98%), a 100% heavy metal ions removal from water and excellent particulate filtration performance from air [51,52].

Despite these achievements for CNT dispersion in various ceramic matrices, some known issues still remain. For example, SWCNTs are always produced in the form of bundles of tens of nanotubes, and their separation into individual tubes is still extremely difficult, concerning the nanocomposite fabrication. This area thus needs further investigations, because SWCNTs have promising applications in biomedical engineering, composite technology and nanodevices [30]. Dispersion of CNTs within ceramic matrices is generally assessed by the microscopic images of the fractured surfaces of nanocomposite samples taken from desired areas of interest, and the representativeness of this method sometimes is a concern, as it may not be a true reflection of the CNT dispersions for other locations [53,54,55]. Therefore, the standardization of CNT dispersion assessments in composites beyond ceramics is vital for quality control in manufacturing and industrial applications.

Densification Processes

Achieving near full density, without damaging the CNT structure and morphology, is a fundamental requirement and another important challenge in ceramic matrix nanocomposite technology, as most of the mechanical properties are strongly affected by the density. CNTs hinder the ceramic grains coalescence by existing at the grain boundaries, which tends to lead to poorly densified microstructures [56]. For this reason, pressure-assisted consolidification processes are generally be used to counter this problem. Figure 1 shows that about 76% nanocomposites were consolidated by pressure-assisted sintering processes, in which spark plasma sintering (SPS) and hot-pressing (HP) have a share of 50% and 26%, respectively. Hot-pressing provides simultaneous high pressures and high temperatures to powder systems, which in turn gives high densities, thus good mechanical properties to either pure ceramics and their composites. Coble *et al.* explained that the enhanced densities were associated with accelerate densification due to higher stresses caused by external pressure, and this phenomenon consolidated the grains to a desirable density; unfortunately, damage to the CNTs during the matrix grain growth could occur due to prolong sintering at extremely high temperatures which was a potential big shortcoming of HP [32,57]. The structural damage problem of CNTs during HP can be avoided by using SPS technique, in which near full densification is achievable at lower sintering temperatures with substantially short holding time. The microscopic images (Figure 2) showed that the CNTs were mainly located at the ceramic grain boundaries, well adhered with the matrix without apparent damage to the structure and morphology [58]. Pressureless sintering (PLS) offers convenient and cheaper consolidation alternations, but wide variation in earlier results have made this technique unattractive and debatable. For example, Zhan *et al.* [15] and Ahmad *et al.* [59] claimed widely different densities for similar samples, as high as 99% and as low as <90% for 1 wt% CNTs-reinforced Al_2O_3, respectively. However in recent reports, Sarkar *et al.* [34]

densified Al_2O_3containing 0.3 vol% of MWCNTs to >99% at 1700 °C using PLS sintering; and similarly Michalek *et al.* [35] and Ghobadi *et al.*[60] obtained 99.9% and >98% densities for Al_2O_3 reinforced with 0.1 wt% and 1 vol% MWCNTs, respectively. Regarding other ceramics, Tatami *et al.* [61] achieved Si_3N_4-MWCNTs composites by pressureless sintering during which the constituents including sintering additives were initially pressed uniaxially and subsequently sintered in furnace at 1700 °C under N_2 atmosphere. They recorded a drop in relative densities from 100% to 90% for 0 to 5 wt% CNT additions; whilst they obtained much higher densities >96% for 5 wt% MWCNT additions by using HP. Microwave sintering is another inspiring and "green" sintering technique, with the advantage of lower densification temperatures and shortened processing time. This technique has been successfully used to consolidate most mainstream industrial ceramics (e.g., Al_2O_3, ZrO_2, Si_3N_4), both pure and composite forms, and resulted in high densities, due to rapid microwaves heating characteristics [62,63]. Nevertheless, the mixed large and fine grained microstructure of the final ceramic consolidated by microwave sintering has made it bit divisive, but this technique exhibits great potentials for CNTs-reinforced ceramics' densification. The advantageous features such as short sintering time and low densification temperatures are not deleterious for CNT structures; furthermore the localize heating at the grain boundaries may be helpful in constructing strong interfaces between CNTs and the ceramic matrices; finally, the grain coarsening seems not a big issue in microwave sintered CNTs-reinforced ceramics, possibly owing to the grain refining tendency of CNTs [64,65]. Table 1 covers the HP, PLS and SPS these main methods.

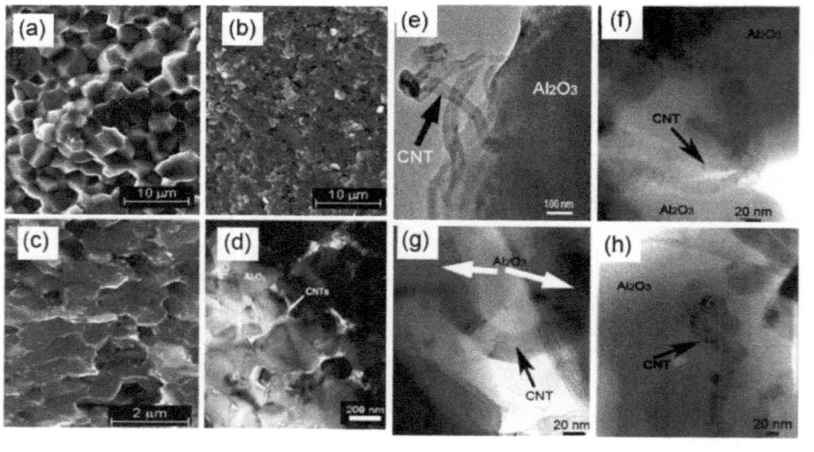

Figure 2: Structural features of (**a**) Monolithic Al_2O_3 showing large grains with inter-granular fracture; (**b**) CNTs/Al_2O_3 nanocomposites with fine grains; (**c**) Trans-granular fracture mode in CNTs/Al_2O_3 nanocomposites; and (**d**) Single-walled (SW)CNTs at grain boundary of Al_2O_3 matrix. TEM images exhibiting the CNT–ceramic interactions (**e**) Multi-walled (MW)CNTs (black arrow) showing their morphology in nanocomposite; (**f**) A single MWCNT existing at grain boundary; (**g**) in porosity and (**h**) Embedded within a single ceramic grain. Adapted from References [32] and [66] with permissions. Copyright 2010, Elsevier Ltd.

In seeking of highly dense composite structure, new techniques are always attempted, and we will summarise a few diverse and interesting methods here. For example, in order to protect the CNT structures by preventing reactions with SiC during high temperature integration, Thostenson *et al.* first prepared a preform of SiC nanoparticles and CNTs (duly dispersed in polymer matrix) then carbonized the perform followed by infiltrating molten Si into the preform under vacuum at 1400 °C to claim dense nanocomposites; whilst Wang *et al.* packed SiC nanoparticles and CNTs in a cylinder and arranged graphite heater inside the cylinder followed by heating up to 1700 °C to consolidate the CNTs-SiC mixture [36,67,68,69].

Microstructural Analysis

Sharp reduction from coarser grains in monolithic ceramics (Figure 2a) to finer grains in CNTs-reinforced ceramics (Figure 2b) is a principle feature of structural change, occurred due to the pinning of matrix grains by CNTs which restricted the grain growth during sintering [66]. Fracture mode alteration from inter-granular in monolithic ceramics (Figure 2a) to trans-granular in the CNTs-reinforced ceramics (Figure 2c) is another interesting change being revealed. The morphological analyses of fractured surfaces are helpful in depicting the mechanisms behind such changes. In the case of monolithic Al_2O_3, it shows clearly the edge and corner fractural features (Figure 2a), representing the typical inter-granular fracture mode; and conversely a blurry and glaze-like surface appears for CNTs-reinforced Al_2O_3 (Figure 2c), indicating the trans-granular mode of fracture [66]. These observations mean that CNTs, as the second phase, must be responsible for altering the fracture modes. Indeed, when CNTs were homogenously dispersed within the ceramic matrix, they arranged themselves at various locations such as along grain boundaries (Figure 2f), across grains boundaries (Figure 2g), inside single grains (Figure 2h), contributing to strengthening the composites at nanometre level by making bridges across grains and sharing the grains, as discussed in prior studies [66,70]. Presumably, all these interesting arrangements of CNTs in ceramic matrices promoted the trans-granular fracture, rather than inter-granular fracture as did in the pure ceramic. In a very recent report, Ahmad *et al.*obtained 5-fold finer grain size in MWCNTs/Al_2O_3 nanocomposites by 300 ppm Y_2O_3 doping than its undoped Al_2O_3counterpart, and mixed inter/intra fracture mode in Y_2O_3 doped nanocomposites was observed [31,37]. However this fracture mode change phenomenon is another grey area that is not fully understood for CNTs-reinforced ceramics, which offers opportunities for prospective thinking and further research work.

Mechanical and Functional Properties

In view of the vast applications of the economically viable Al_2O_3 ceramics in industry, lots of studies have been done to improve their fracture toughness by CNT additions. However, inconsistent results (Table 2) put question marks on these triumphs and core issues in such discrepancies were found in the CNT dispersion methods, choice of sintering process and techniques adopted for characterisation [34,40,41,50,59,61,71,72]. For example, Table 2 shows that the higher fracture toughness values of MWCNTs-reinforced Al_2O_3 were obtained at lower CNT additions (<2 wt%), and declining trend can be seen at higher CNT levels in all cases, except from the values reported by Zhan *et al.* [15]. Furthermore, the composite fracture toughness reported by Zhan *et al.* [15] was the highest in Table 2. However, this value may be due to various factors: (1) the use of SPS techniques (positive); (2) reinforcement phase being SWCNTs (positive); and (3) the assessment of the fracture toughness by an unreliable direct crack method, DCM, (negative). Yamamoto *et al.* [12] used the SPS to sinter similar composite reinforced with MWCNTs, and used the single-edged notched beam (SENB) method to assess the fractures toughness, however the results were now as good as the results reported by Zhan *et al.* [15] and Ahmad *et al.* [31]. In case of the high values reported by Ahmad *et al.* [66], it is probably due to the better dispersion of CNTs within the matrix, as they adopted a unique method. Further, Huang *et al.* [39] showed tremendous improvements in fracture toughness (57%, 114% and 328%) values for $BaTiO_3$ ceramic after reinforced with (0.5, 1 and 3 wt%) MWCNTs; whereas a 15% improvement was recorded by Tian *et al.* [38] for 2 wt% MWCNTs-reinforced ZrB_2-SiC ceramics.

Table 2: Properties of CNTs-reinforced ceramics.

Reference	Matrix	CNT contents	Relative density (%)	Hardness (GPa)	Flexural strength (MPa)	Fracture toughness (MPa. $m^{1/2}$)
[10]	Si_3N_4	0	99.2	15.7	1046	4.8
		1 wt%MWCNTs	98.7	15.0	996	6.6
[12]	Al_2O_3	0	95.6	17.3	500	4.4
		0.5 wt% MWCNTs	99.2	16.8	685	5.9
		1 wt% MWCNTs	98.9	15.9	650	5.7
[15]	Al_2O_3	0	-	-	-	3.3
		3 wt% SWCNTs	-	-	-	7.9
[27]	Al_2O_3	0	97.7	-	326	3.08
		6 wt% MWCNTs	95.4	-	314	5.55
[32]	Al_2O_3	0	99.8	16	356	3.5
		2 wt% MWCNTs	99.5	18	402	6.8
		5 wt% MWCNTs	99.1	-	423	5.7
[34]	Al_2O_3	0	99.5	17.5	222	3.92
		0.15 vol% MWCNTs	98.4	21.4	242	5.27
[35]	Al_2O_3	0	-	16.9	-	5.5
		1 vol% MWCNTs	-	13.5	-	6.0
[36]	Mulite ($3Al_2O_3$ + $2SiO_2$)	0	-	-	466	2.0
		2 wt% MWCNTs	-	-	512	3.3
[69]	SiC	0	939	-	303	3.3
		10 wt% MWCNTs	94.7	-	321	3.8
[38]	ZrB_2-SiC	0	-	15.8	582	4
		2 wt% MWCNTs	-	15.5	616	4.6
[39]	$BaTiO_3$	0	98.5	-	-	0.7
		98.50	98.5			0.7
		0.5 wt% MWCNTs	97.3			1.1
		1 wt% MWCNTs	99.2			1.5
		3 wt% MWCNTs	98.6			3.0
[73]	Al_2O_3	0	-	-	395	4.41
		20 vol% MWCNTs	-	-	403	4.62
[74]	Al_2O_3	0	-	-	-	3
		1 wt% MWCNTs	-	-	-	5
[75]	Al_2O_3	0	-	15.71	-	3.24
		5 wt% MWCNTs	-	0.72	-	4.14
[76]	Al_2O_3	0	-	18.2	-	4.5
		2.5 wt% MWCNTs	-	15.75	-	11.4
[77]	Al_2O_3	0	99.9	22.9	-	3.54
		10 vol% MWCNTs	97.4	11	-	2.76

Recently, Sarkar *et al.* [34] calculated fracture toughness values of the Al_2O_3–MWCNT nanocomposites by employing DCM method using Niihara and Liang models, and reported better fracture toughness values than those obtained using SENB technique; whereas Ahmad *et al.* [66], reported higher fracture toughness values attained from SENB method than those obtained from DCM method using Chantikul model. These conflicting reports suggest that engineering components cannot be validated for structural load-bearing applications using DCM method; however, this convenient method is widely employed for fracture toughness comparisons [31]. Similar inconclusive and controversial fracture toughness values regarding CNTs-reinforced Si_3N_4 were also reported by Corral *et al.* [10] and Pasupuleti *et al.* [37] Both consolidate $Si_3N_4$with 1 wt% CNTs and obtained 30% reduction (by SENB method) and 40% increment (by ISB method) in fracture toughness, respectively.

Regarding other mechanical properties such as hardness and elastic modulus, Yamamoto *et al.* [12] investigated a range of MWCNT additions in Al_2O_3 ceramics and concluded a drop in hardness and rise in flexural strengths at low MWCNT additions and further reduction in both properties by adding more MWCNTs, and consistent results were reported by many others for the same material system, as shown in Table 2. Pasupuleti *et al.* [37] showed a small decrease in hardness (4%) and flexural strength (5%) for 1 wt% MWCNTs-reinforced Si_3N_4, however, Corral *et al.* [10] reported a much severer 45% reduction in hardness for same reinforcement contents in Si_3N_4.

The dual role of CNTs, indirectly enhancing the mechanical properties and directly acting as lubricant, converts ceramic composites into an attractive wear resistance material, and various reports demonstrated the steady reduction of friction coefficient with CNT additions [78]. High thermal and electrical properties of the CNTs have been predicted and several attempted to incorporate CNTs into insulated ceramics in order convert them into highly electrical and thermally conductive materials [73]. Ceramics exhibited higher electric conductivity (EC) when reinforced with SWCNTs (10^6 S/m) than with MWCNTs (10^3–10^5 S/m) [40,79]. Sarkar *et al.* reported that the EC of MWCNTs-reinforced composites was dependent on the formation of electrically conductive networks by dispersing the CNTs homogenously in the matrix, and on the grain sizes of the final nanocomposites, as larger grain size with less grain boundaries showed better results [80]. So far Estili *et al.* has obtained the highest EC of 4816 S/m for Al_2O_3-20 vol% MWCNTs, which is 43% higher than that reported by Zhan *et al.*[73,79]. In addition, Kumari *et al.* obtained an exceptional value of 3336 S/m by reinforcing Al_2O_3 with 19 wt% MWCNTs nanocomposites, however, at the cost of poor mechanical properties [41]. In contrast to MWCNTs, the SWCNTs reinforcement into the Al_2O_3 matrix offered better conductivity of 3345 S/m without compromising mechanical properties [71]. Zaman *et al.* studied the effects of surface functionalization of the SWCNTs on EC and reported that the hydroxyl group functionalized SWCNT offered ~10 times higher EC in 1 wt% SWCNTs-reinforced Al_2O_3 nanocomposites than those functionalized by carboxylic acid group [72]. Moreover, Bi *et al.* reported a drop in the electrical percolation by increasing the aspect ratios of MWCNTs [81]. Although the thermal conductivity (TC) of SWCNTs and MWCNTs are ranges from 3000 to 6000 W/m·K, however, their nanocomposite with ceramics barely demonstrated good thermal performance. Compared to unreinforced Al_2O_3, Zhan *et al.* reported lower (7.3 W/m·K) TC in nanocomposites reinforced with 15 vol% SWCNTs than their monolithic counterpart (27.3 W/m·K) [71]. Both Kumari *et al.* and Bakshi *et al.* reported higher TC values (63.52 and 6 W/m·K) in nanocomposites with (8 and 4 wt%) MWCNTs than those of pure Al_2O_3 (19.96 and 5.37 W/m·K) samples, respectively [82]. This area of research is therefore more complicated and interesting.

CNTs/Ceramic Interface and the Toughening Mechanism

Reinforcement (fibres or whiskers) pullout is the main toughening mechanism in conventional ceramics, which is further associated with weak interfacial connection of reinforcement with matrix. Same classical approach was proposed for toughness mechanism in CNTs-reinforced ceramics in several initial reports [83,84,85]. However, in later research Padture *et al.* and many others observed that micro-structural features of CNTs-ceramics were immensely dissimilar from conventional composites, and these observations strongly suggested that existing microscale mechanism may not be fully applicable to CNTs–ceramic systems [18,23,61]. Microstructure of conventional ceramic composites consists of inflexible and straight reinforcement, and the interface is optimally designed in such a way that it debonds on applied load [86]. Imagine, when a reinforcement encounters a crack then it bridges the crack in its wake, pullout does frictional work and these together effectively make crack propagation more difficult, in addition to this large reinforcement dimensions lead to longer crack-wake bridging zones and consequently resulted in higher toughness [85,87]. Meanwhile, these large reinforcements prompt larger flaws and turn strength to lower values. In contrast, CNTs are highly flexible, hollow nanometre sized fibres, therefore the toughening mechanism may be entirely different from conventional ones. Frictional pullout of fibres occurred in classical composites may not be the only toughening mechanism in CNTs-reinforced ceramics. Accordingly, new concepts and philosophies of uncoiling and elastic stretching of CNTs during the crack propagation were proposed as main toughness mechanisms by Padture *et al.* [7]. During crack propagation, initial uncoiling of CNTs occurs in the crack wake, and when the crack further propagates the uncoiled CNT stretches elastically serving as stretched CNT bridges instead of conventional frictional pull-out bridges, thus impedes the crack propagation, as shown in Figure 2g [66]. These concepts are convincingly identified the role of CNTs as an individual entity, also applied to the cluster form. Surface damages to SWCNTs during purification and subsequent sintering process are well-known, and in this picture the role of CNT's elastic stretching in toughness is slightly litigious. In this regard, mathematical modelling will be a helpful tool in explaining the role of CNTs in strengthening ceramics and predicting their behaviour in services.

Recent developments in the electron microscope technology are changing the research approaches, and attentions are now tending to focus on tailoring the interface structure at atomic level, to construct defect-free structures with interesting functionalities. FIB-SEM (focused ion beam scanning electron microscopy) has made the scientists' life not as hard as ages ago. A TEM (transmission electron microscope) sample of hard materials, e.g., ceramics can be prepared in hours, which was a laborious task to arrange for days and even weeks earlier. For CNTs-Al_2O_3 nanocomposites, due to the interesting reaction of Al_2O_3 with alkaline, a simple powder etching process can always be used to collect CNTs with a thin layer of Al_2O_3 residue, for interface study under TEM. Similar results are obtained when compared with FIB-SEM [66].

Back to CNTs-reinforced ceramics, where the interface controls the CNT debonding, pullout and crack-bridging at micron and nanometre level, these different mechanisms act as energy dissipative processes during mechanical loading. Physically, an interface is a complex transitional region layered between the reinforcement and parent matrix thus, the control of interface chemistry and tailoring smart microstructures are essential steps for producing exceptionally tough and strong nanocomposites. Dedicated efforts have been done to explore the CNTs-ceramic transition region and each addressed in interesting way [12,66]. Indeed, rough surface of CNTs produces the required frictional forces which resist in detaching CNTs from the ceramic matrix. Yamamoto *et al.* proposed that acid treatment does not significantly damage the overall structures of MWCNTs however, localized etches of the cylindrical body at different locations create nanoscale defects (nano-pits) along the tube axis, as shown in Figure 3b [12]. These nano-pits having depths of ~15 nm are anchored by the matrix grains (Figure 3c), forming locks and resistance in MWCNTs' sliding over the matrix, thus leading to good connection of composite constituents at the interface [12]. Further, a close cross-sectional look of the MWCNT shown inFigure 3d of the high resolution TEM image reveals its uneven surface, hollow core and graphitic layers. These layers are not concentrical on a long distance and many compartments exist, which is a typical feature of MWCNTs synthesized by CVD. Ahmad *et al.* [78] postulated that high surface roughness of the CNTs could result in two potential advantages like chemically highly reactive and physically difficult to slide out of the matrix, compared with a smooth surface. The former could help to improve the interfacial bonding with the matrix and the latter can pose much larger friction forces to stop the CNT pullout [78].

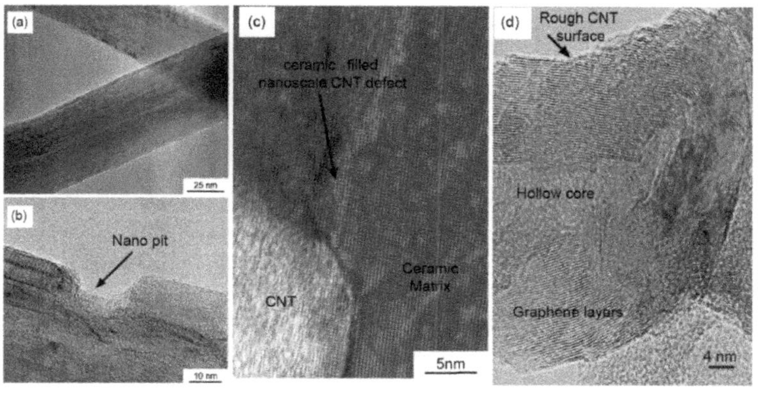

Figure 3: (**a**) TEM image of the pristine MWCNTs; (**b**) High-magnification TEM image of the acid-treated MWCNT surface, arrow indicates nano-pit; (**c**) Nano-pit on the acid-treated MWCNTs is filled up with Al_2O_3 crystal; and (**d**) Rough surface of MWCNT produced by chemical vapour deposition (CVD) method. Adapted from References [12] and [32] with permissions. Copyright 2009, Elsevier Ltd. and 2008 IOP Publishing Ltd.

The CNT's surface unevenness and its anchoring with the ceramics matrix are a good physical explanation of enhanced frictional forces at the interface. However, the chemical interactions of CNTs with the ceramics remained unattended for several years. Estili *et al.* [88] rigorously studied the interfacial areas of CNTs/Al_2O_3 nanocomposites using high resolution-TEM, but unable to identify any interfacial phases or intermediate compounds at the CNTs/Al_2O_3 interface. A recent attempt addressed this topic and explained the chemical activity took place at the CNTs/Al_2O_3 interface during HP process, and reported the formation of an extremely thin (1–2 nm) intermediate phase of Al_2OC, which is possibly produced due to the carbothermal reduction of Al_2O_3 by CNTs [78]. Figure 4a–b show clear evidence of a CNT sticking with Al_2O_3 at the interface.

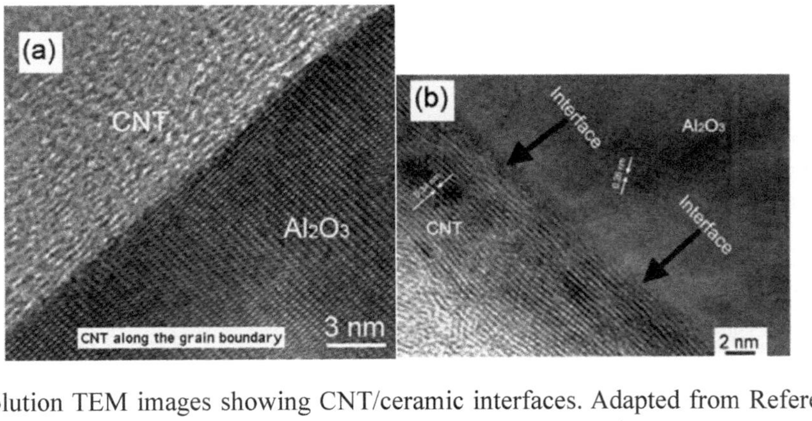

Figure 4: (**a**,**b**) High-resolution TEM images showing CNT/ceramic interfaces. Adapted from References [15] and [32] with permissions. Copyright 2005 Advanced Study Center Co. Ltd. and 2010 Elsevier Ltd.

Owning to the multi-layer graphene structure of MWCNTs, the possibility of such chemical and physical reactions with the accommodation of nano-pits and eating of few outer layers for Al_2OC or Al_4C_3 formation can be justified. However, this may not be true for SWCNTs which contain only a single graphene layer while forming the tubular structure, even plenty of studies claimed tremendous improvements in ceramics properties [22,26]. This raises one big question as to being only one layer how it reacts with the matrix to form a good interface following the toughening mechanisms proposed above. Therefore, this mystery remains unresolved. The understanding of the nanostructure characteristics and the interfacial relationship between SWCNTs and the ceramic matrices is far from satisfactory, which opens new windows of potential research in this advanced area of nanotechnology [18,26,40,61].

GRAPHENE REINFORCED CERAMIC NANOCOMPOSITES

Raw Materials

As a cousin of CNTs, the 2D graphene bears many similarities to CNTs in terms of nanocomposites application. For bulk engineering nanocomposite applications which require large volume amount, a few layered graphene platelets or flakes, including the reduced graphene oxide (GO), are far more viable and economical than single layered graphene. Therefore, the term graphene in this context refers to graphene nanoplatelets (GNPs).

In composite applications, both the mechanical exfoliation and reduction from GO have been used successfully [89,90,91,92,93,94]. In the mechanical cleavage method, commercial graphite powder (Aldrich) has been milled intensively in high efficient attritor mill in the presence of ethanol for 10 h, then the produced GNPs were mixed with ceramic powder [89]. The second method uses the Hummers' process to produce GO, then using this water soluble GO to mix with ceramic powders [90]. In general, both mechanical milling and hummers method suffers from various sizes and thicknesses for the former due to lack of control on the milling energy and from surface structural damage for the latter originated from the oxidation [91], which will have negative effects on the final properties of the ceramic composites. Therefore, better quality control of the GNPs is of fundamentally importance for high quality nanocomposites development.

GNS Dispersions Processes

As discussed above for CNTs, mixing step is an equally challenging step in preparing graphene-reinforced ceramics composites. To avoid any damage and reduce agglomeration of GNPs will help to achieve high mechanical and physical properties. In essence, the dispersion of GNPs in fact is easier than CNTs, as the difficulties accompanied in CNT's dispersion such as high aspect ratio and van der Waals interactions which cause CNT bundling are absence for GNPs. In addition, high specific area and 2D geometry of GNPs offer better disperseability in ceramic matrices. As a younger cousin to CNTs, the gained knowledge for CNTs can generally be used as a reference for GNPs-reinforced ceramic composites. Thus in this context, focus will be mainly on the different features with comparison. Wet powder mixing are successful to disperse CNTs in ceramic matrixes [92,93,94,95], whilst for GNPs the choice of solvents are much wider than processing CNTs. Isopropyl alcohol NMP, DMF have all be used to mix with various ceramic matrices such as Al_2O_3, Si_3N_4, and ZrO_2 powders. This drawback of this technique is energy consuming, and might cause damage to the GNP reinforcements. Colloid processing is a modified wet mixing process, and the key is to produce stabilized suspensions from GNTs and ceramic particles by changing their surface chemistry which facilitates homogeneous dispersion of GNPs. Anionic or cationic surfactants are generally used to alter the surface charge of GNPs, to positive or negative respectively, followed by adding them to a ceramic suspension with the same/opposite charges to form an homogenous ceramic-GNP dispersion. This hetero-coagulation process is a very effective route for well-dispersed ceramic composites [12,91]. Starting with GO, Wang *et al.* [20] used such electrostatic attractions between GO and Al_2O_3 particles to obtain homogenous dispersions of GO in Al_2O_3 powder first, followed by subsequent reduction of GO, who achieved a 53% and 13 orders of magnitude improvement in fracture toughness and conductivity. Walker *et al.* [25] used CTAB in both the GNPs and Si_3N_4 suspensions for mixing, and resulted in a 235% improvement in fracture toughness with only 1.5 vol% GNP addition.

Sintering Techniques

The densification of GNPs-reinforced ceramic nanocomposites also includes pressureless sintering, HP, SPS and HIP (hot-isostic pressing). The low temperature requirement and fast sintering rate advantages of SPS made it widely used for ceramic nanocomposites filled with carbon nano-fillers [20,24,43,95,96,97,98]. However, there are a few groups reported very good GNPs-reinforced ceramic nanocomposites based on HP densification [97,98,99,100]. For example, the GNPs-Si_3N_4nanocomposites reported by Rutkowski *et al.* [99] showed improved thermal properties. After comparing the HP and SPS processes for GNPs-Al_2O_3 nanocomposites, Inam *et al.* [98] found out that the structural integrity of graphene from HP process is better than SPSed samples, with higher crystallinity, thermal stability and electrical conductivity, and was attributed to the thermally induced graphitization caused by longer sintering condition in a HP.

Structural Features, Mechanical Properties and Toughening Mechanisms

Toughening ceramic is one of the main research objectives for GNP nanocomposites, whist other benefits such as flexural strength and hardness can also be obtained. Using only 1.5 vol% the flexible 2D GNPs as a reinforcement for Si_3N_4, Walker*et al.* reported a 235% improvement in the toughness, and found GNPs anchoring with or wrapping around Si_3N_4 grains [25], thus blocking the crack propagation through the GNPs. This is the first time that such toughening mechanism was observed, and is a major different from the 1D CNTs. The same effective anchoring toughening was also confirmed by Liu *et al.* [24] in their GNPs/Al_2O_3 system, documented a 30.75% increase in flexural strength and a 27.20% increase in fracture toughness. These securely anchored GNPs around Al_2O_3 grains can form large area of interfaces with the matrix, increasing the interfacial friction, therefore the energy required for pulling out GNPs from the matrix will be greater than pulling out CNTs. They also successfully extended their process to a GNPs-reinforcing the ZrO_2-Al_2O_3 system using SPS [24], in comparison with CNTs. The authors believed that due to similar mechanical properties to CNTs, and better dispersability GNPs are an effective alternative for CNTs in ceramic composites. In Si_3N_4 matrix, Tapaszto *et al.* [100] showed that GNPs indeed outperformed CNTs. However, it should be noted that, due to the larger contact area between GNPs and the matrix grains, the interface quality plays a more important role in toughening the ceramics than CNTs and other reinforcement phases.

The different roles of CNTs and GNPs separately in ceramic matrices were well-documented, however within the same matrix could be more complex. A combination of the various advantages of different reinforcement phases, the very nature of the composite concept, could lead to superior properties, however this has rarely been investigated so far in ceramics. Very recently, Yazdani *et al.* [26] reported both 63% and 17% improvements in fracture toughness and flexural strength by using such a hybrid (MWCNTs + GNPs = GNTs) reinforcement phase in Al_2O_3 matrix. In their report, the role of GNPs and CNTs has been investigated separately. As evident in Figure 5, a large GNP rolled around Al_2O_3 grain due to its flexibility and produced large area of interface with Al_2O_3 matrix; therefore, it increased the required pull out energy during fracture and strengthened the grain boundaries so the fracture occurred through the Al_2O_3 grains rather than along the grain boundaries, whist MWCNTs contributed more to the bridging effect due to their higher aspect ratio. It is believed that MWCNTs can be stretched much longer than GNPs before collapsing during crack propagation. These roles are complementary with each other at appropriate concentrations, allowing for absorbing more energy during crack propagations.

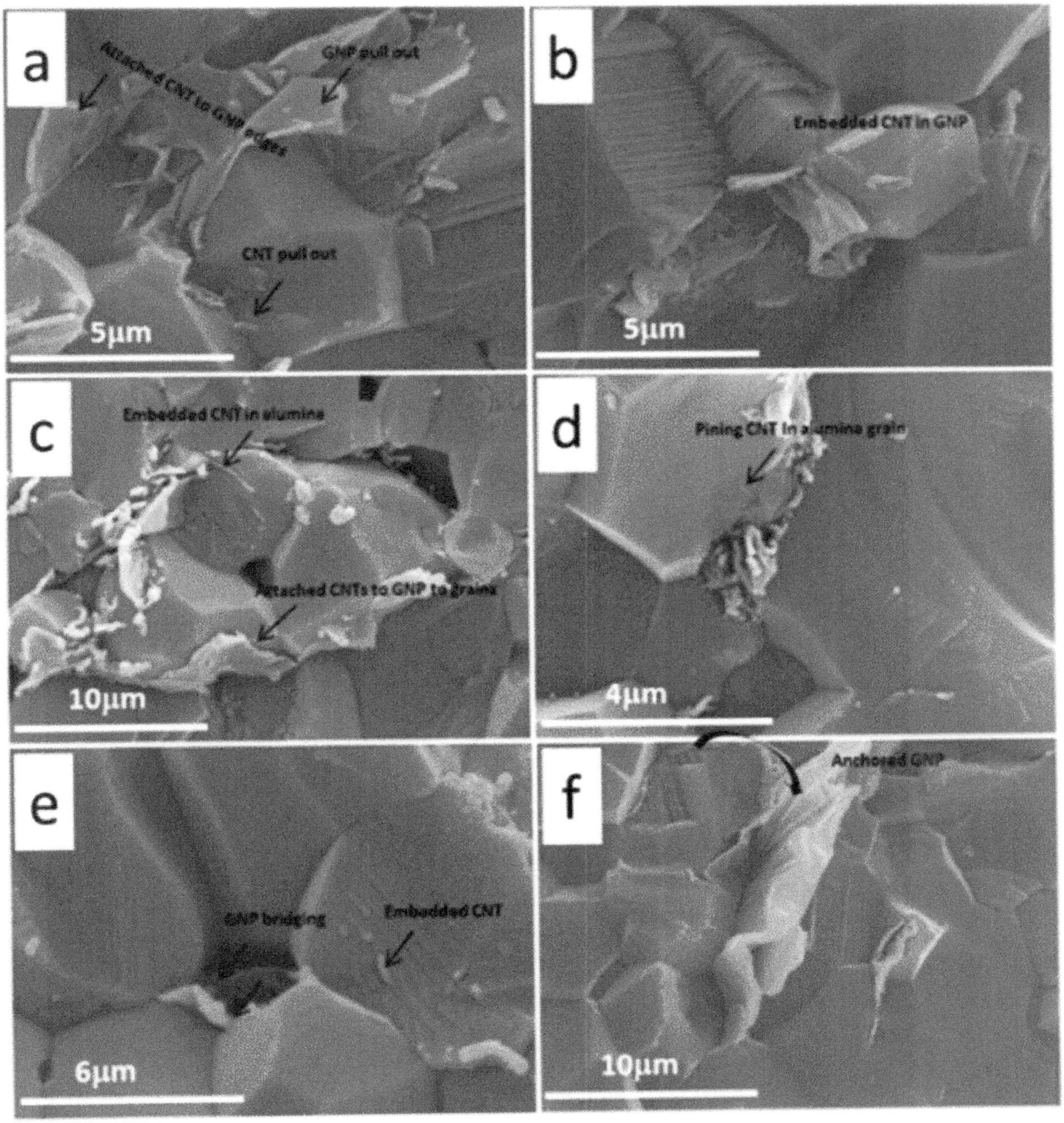

Figure 5: SEM images from fractured surface of GNT-Al_2O_3 nanocomposites with various GNP/CNT ratio, (**a** and **b**); Al_2O_3-(0.5 wt% GNP + 1 wt% CNT), (**c–e**); Al_2O_3-(0.5 wt% GNP + 0.5 wt% CNT), (**f**); Al_2O_3-0.5 wt% GNP. Adapted from Reference [26] with permission. Copyright 2014 Elsevier Ltd.

POTENTIAL APPLICATIONS

Owing to the improved fracture toughness and ancillary benefits of electrical and thermal properties, ceramics reinforced with CNTs and GNPs are promising for numerous prospective applications in the field of photonics, biomedical, automotive and aerospace engineering. Firstly, associated with the enhanced mechanical performance of Al_2O_3, the significantly improved wear resistance property of these composites could be suitable for a number of wear and sliding applications in automobile industry like cylinder lines, valve seat and piston rings [101]. Secondly, the SiC, Si_3N_4 and $BaTiO_3$ systems filled with CNTs made them suitable for structural applications, such as bearings, seals, armour, liners, nozzles and cutting tools. Thirdly, the thermally and chemically stable ceramic composites could revise their high thermal conductivity and be suitable for high temperature components such as in jet engine and brake disks for aircrafts [102]. Further, CNTs/GNPs can also convert ceramics into functional materials for aerospace and automobile industries, such as knock sensors, seat pressure sensors, temperature sensors, oil sensors, impact sensors and road surface sensors, whilst the outstanding electrical properties of CNTs/GNPs make Al_2O_3 ceramic attractive for specific applications like heating elements, electrical igniters, electromagnetic/antistatic shielding of electronic components, electrode for fuel cells, crucibles for vacuum induction furnaces and electrical feed through [44,74,86,103,104,105]. Table 3 summarises the potential industries may have benefits from ceramic nanocomposites reinforced with CNTs and graphene. As the research is progressing in this important area, novel CNTs-reinforced ceramics with stunning properties are expected and may substituted several automobile and aerospace components in future furthermore, owning functionalities these have potential for third generation nanodevices.

Table 3: Potential application of key ceramics nanocomposites reinforced with CNTs and graphene

References	Ceramic matrix	Reinforcing agent	Key properties	Parts/Components	Potential industries
[101]	Al_2O_3	CNTs/graphene	Wear resistance, high toughness, electrical properties, thermal properties	Cutting tools, corrosion/erosion resistance pipes, electrical contacts, armour plates	Automobile, petrochemical industry, electric component manufacturing, defence industry
[106]	Si_3N_4	CNTs/graphene	Excellent mechanical, chemical, and thermal properties	Gas turbines, aircraft engine components and bearings	Power generation, aerospace, automobile sector
[107]	$BaTiO_3$	CNTs/graphene	Ferroelectrics, piezoelectric and colossal magnetoresistor properties	Electric generator, computer hard disks, sensors	Renewable energy, power generation, electronic, computer manufacturing, data storage, aerospace industry
[108–110]	ZrO_2	CNTs/graphene	High mechanical properties, excellent fracture toughness, elevated temperature stability, high breakdown electrical field and large energy bandgap	Solid oxide fuel cells, oxygen sensors and ceramic membranes	Renewable energy, chemical industry, water desalination sectors
[111–113]	TiN and FeN	CNTs/graphene	Excellent electrical properties	Capacitors, electronic conductor in electronic devices	Electrochemical industry, power and electronic sector, aerospace and automobile industries
[114]	Mulite	CNTs/graphene	High in electric and optical properties	Sensor	Electronic industry, aerospace sector and automobile industry

CONCLUSIONS

Advanced in the ceramics reinforced with carbon nanostructures (CNTs and graphene) have been thoroughly reviewed. Successes in the purification and dispersions of MWCNTs are somehow satisfactory, however SWCNTs need further research and standards for CNT dispersion are vital for addressing the quality and reliability with confidence. Microwave sintering has potential for producing dense nanocomposites and may eliminate the CNT damage problem associated with the hot-pressing, and by adopting to standard testing methods fracture toughness discrepancies could be reduced. CNTs-reinforced ceramics follow the combined advanced toughening mechanisms of CNT's stretching/uncoiling and the classical fibre pullout theory, as an energy dissipating process. Rough surface and nanopits of MWCNTs explain the strong interface connections with ceramic matrix and the confirmation of the formation of intermediate Al_2OC or Al_4C_3 phases at the interface further strengthens these explanations. Conclusively, problems of reinforcing MWCNTs into ceramics have been solved to some extend; however, the addition of SWCNTs still carries questions. Despite challenges and controversial issues, CNTs have successfully enhanced the toughness and other properties of brittle ceramics and converted them into useful materials for next generation applications.

It is clear that graphene can play an important role as filler in ceramics according to publications. In addition to the exceptional mechanical properties of GNPs which are similar to CNTs, researches have shown that GNPs can be more easily dispersed in ceramic matrix than CNTs which is the key challenge in preparing ceramic composites. Additionally its 2D and flexible microstructure introduced a new toughening mechanism in the ceramic matrix (anchoring around the grain) that could significantly absorb energy against crack propagation and delay the fracture. However, work on graphene ceramic composites is in its early stages and there are still considerable works that need to be done in order to optimise their processing, microstructure and interfacial properties to obtain better multifunctional properties from graphene-ceramic composites.

ACKNOWLEDGMENTS

Iftikhar Ahmad gratefully acknowledges the technical and financial support of Research Center of College of engineer, Deanship of Scientific Research, King Saud University, Riyadh, Kingdom of Saudi Arabia.

AUTHOR CONTRIBUTIONS

Iftikhar Ahmad contributed to the carbon nanotube part, Bahareh Yazdani contributed to the graphene part, and Yanqiu Zhu involved in all stages of the article preparation.

REFERENCES

1. Niihara, K. New design concept of structural ceramics-ceramics nanocomposites. *J. Ceram. Soc. Jpn.* **1991**, *99*, 974–982.
2. Osayande, L.; Okoli, O.I. Fracture toughness enhancement for Al_2O_3 system: A review. *Int. J. Appl. Ceram. Technol.***2008**, *5*, 313–323.
3. Ohnabe, H.; Masaki, S.; Sasa, T. Potential application of ceramics matrix composites to aero-engine components. *Compos. A* **1999**, *30*, 489–496.
4. Llorca, J.; Elices, M.J.; Celemin, A. Toughness and microstructural degradation at high temperature in SiC fiber-reinforced ceramics. *Acta Mater.* **1998**, *46*, 2441–2453.
5. Yongqing, F.; Gu, Y.W.; Hejun, D. SiC whisker toughened Al_2O_3-(Ti,W)C ceramic matrix composites. *Scr. Mater.* **2001**,*44*, 111.D–116.D.
6. Garcıa, E.; Schicker, S.; Bruhn, J.; Janssen, R.; Claussen, N. Processing and mechanical properties of pressureless-sintered niobium—Al_2O_3-matrix composites. *J. Am. Ceram. Soc.* **1998**, *81*, 429–432.
7. Padture, N.P. Multifunctional composites of ceramics and single-walled carbon nanotubes. *Adv. Mater.* **2009**, *21*, 1767–1770.
8. Sheldon, B.W.; Curtin, W.A. Nanoceramics composites tough to test. *Nat. Mater.* **2004**, *3*, 505–506.
9. Peigney, A. Tougher ceramics with nanotubes. *Nat. Mater.* **2003**, *2*, 15–16.
10. Corral, E.; Bell, N.; Stuecker, J.; Perry, N.; Garay, J.; Barrera, E.V. Engineered nanostructures for multifunctional single-walled carbon nanotube reinforced silicon nitride nanocomposites. *J. Am. Ceram. Soc.* **2008**, *91*, 3129–3137.
11. Fan, J.; Zhao, D.; Song, J. Preparation and microstructure of multi-walled carbon nanotubes toughened Al_2O_3composite. *J. Am. Ceram. Soc.* **2006**, *89*, 750–753.
12. Yamamoto, G.; Omori, M.; Hashida, T.; Kimura, H. A novel structure for carbon nanotube reinforced Al_2O_3composites with improved mechanical properties. *Nanotechnology* **2008**, *19*, 315708.
13. Xia, Z.H.; Lou, J.; Curtin, W.A. A multiscale experiment on the tribological behavior of aligned carbon nanotube/ceramic composites. *Scr. Mater.* **2008**, *58*, 223.
14. Wie, T.; Fan, Z.; Wie, F. A new structure for multi-walled carbon nanotubes reinforced Al_2O_3 nanocomposite with high strength and toughness. *Mater. Lett.* **2008**, *62*, 641–644.
15. Zhan, G.D.; Mukherjee, A. Processing and characterization of nanoceramic composites with interesting structural and functional properties. *Rev. Adv. Mater. Sci.* **2005**, *10*, 185–196.
16. Costa, J.; Flacker, A.; Nakashima, M.; Fruett, F.; Zampieri, M.; Longo, E. Integration of microfabricated capacitive bridge and thermistor on the alumina substrates. *ECS Trans.* **2012**, *49*, 451–458.
17. Martin, C.A.; Lee, G.F.; Fedderly, J.J. Composite Armor System Including a Ceramic-Embedded Heterogeneously Layered Polymeric Matrix. U.S. Patent 8387510 B1, 2013.
18. Baron, B.; Kumar, C.; le Gonidec, G.; Hampshire, S. Comparison of different alumina powders for the aqueous processing and pressureless sintering of Al_2O_3-SiC nanocomposites. *J. Eur. Ceram. Soc.* **2002**, *22*, 1543–1552.
19. Ipek, M.; Zeytin, S.; Bindal, C. An evaluation of Al_2O_3-

ZrO_2 composites produced by coprecipitation method. *J. Alloys Compd.* **2011**, *509*, 486–489.

20. Wang, K.; Wang, Y.F.; Fan, Z.J.; Yan, J.; Wei, T. Preparation of composites by spark plasma sintering. *Mater. Res. Bull.***2011**, *46*, 315–318.
21. Ramirez, C.; Garzón, L.; Miranzo, P.; Osendi, M.; Ocal, C. Electrical conductivity maps in graphene nanoplatelet/silicon nitride composites using conducting scanning force microscopy. *Carbon* **2011**, *49*, 3873–3880.
22. Balazsi, C. Silicon nitride composites with different nanocarbon additives. *J. Korean Ceram. Soc.* **2012**, *49*, 352–362.
23. Kvetkova, L.; Duszova, A.; Hvizdos, P.; Dusza, J.; Kun, P.; Balazsi, C. Fracture toughness and toughening mechanisms in graphene platelet reinforced Si_3N_4 composites. *Scr. Mater.* **2012**, *66*, 793–796.
24. Liu, J.; Yan, H.; Reece, M.J.; Jiang, K. Toughening of zirconia/alumina composites by the addition of graphene platelets. *J. Eur. Ceram. Soc.* **2012**, *32*, 4185–4193.
25. Walker, L.S.; Marotto, V.R.; Rafiee, M.A.; Koratkar, N.; Corral, E.L. Toughening in graphene ceramic composites. *ACS Nano* **2011**, *5*, 3182–3190.
26. Yazdani, B.; Xia, Y.; Ahmad, I.; Zhu, Y. Graphene and carbon nanotube (GNT)-reinforced alumina nanocomposites. *J. Eur. Ceram. Soc.* **2015**, *35*, 179–186.
27. Fan, Y.; Wang, L.; Li, J.; Sun, S.; Chen, F.; Chen, L.; Jiang, W. Preparation and electrical properties of graphene nanosheet/Al_2O_3 composites. *Carbon* **2010**, *48*, 743–1749.
28. Ebbesen, T.W.; Ajyan, P.M. Large scale synthesis of carbon nanotubes. Nature, 1992, 358, 220–222.
29. Hiura, H.; Ebbesen, T.W.; Tanigaki, K. Opening and purification of carbon nanotubes in high yields. *Adv. Mater.* **1995**,*7*, 275–276.
30. Tohji, K.; Takashaki, H.; Nishina, Y. Purification procedure for single-walled nanotubes. *J. Phys. Chem.* **1997**, *B101*, 1974–1978.
31. Ahmad, I.; Islam, M.; Almajid, A.; Yazdani, B.; Zhu, Y.Q. Investigation of yttria-doped Al_2O_3 nanocomposites reinforced by multi-walled carbon nanotubes. *Ceram. Int.* **2014**, *40*, 9327–9335.
32. Ahmad, I.; Kennedy, A.; Zhu, Y.Q. Carbon nanotubes reinforced Al_2O_3 nanocomposites: Mechanical properties and interfacial investigations. *J. Comput. Sci. Technol.* **2010**, *70*, 1199–1206.
33. Zhang, S.C.; William, G.; Hilmas, G.E.; Edward, J.Y. Pressureless sintering of carbon nanotube-Al_2O_3 composites. *J. Eur. Ceram. Soc.* **2010**, *30*, 1373–1380.
34. Sarkar, S.; Das, P.K. Microstructure and physic-mechanical properties of pressure-less sintered multi-walled carbon nanotubes/Al_2O_3 nanocomposites. *Ceram. Int.* **2012**, *38*, 423–432.
35. Michalek, M.; Lkova, M.; Sedla, J.; Galusek, D. Al_2O_3/MWCNTs composites by aqueous slip casting and pressureless sintering. *Ceram. Int.* **2013**, *139*, 6543–6550.
36. Wang, J.; Kou, H.; Liu, X.; Pan, Y.; Guo, J. Reinforcement of mullite matrix with multi-walled carbon nanotubes. *Ceram. Int.* **2007**, *33*, 719–722.
37. Pasupuleti, S.; Peddetti, R.; Halloran, J.P. Toughening behavior in a carbon nanotube reinforced silicon nitride composite. *Mater. Sci. Eng.* **2008**, *A491*, 224–229.
38. Tian, W.; Kan, Y.; Wang, P. Effect of carbon nanotubes on the properties of ZrB_2-SiC ceramics. *Mater. Sci. Eng.* **2008**,*487*, 568–573.
39. Huang, Q.; Gao, L.; Sun, J. Effect of adding carbon nanotubes on microstructure, phase transformation and mechanical properties of $BaTiO_3$ ceramics. *J. Am. Ceram. Soc.* **2005**, *88*, 3515–3518.
40. Kumari, L.; Zhang, T.; Du, G.H.; Li, W.Z.; Wang, Q.W.; Datye, A.; Wu, K.H. Synthesis, microstructure and electrical conductivity of carbon nanotube–alumina nanocomposites. *Ceram. Int.* **2009**, *35*, 1775–1781.
41. Echeberria, J.; Rodríguez, N.; Bocanegra-Bernal, M.H. Hard and tough carbon nanotube-reinforced zirconia-toughened Al_2O_3 composites prepared by spark plasma sintering. *Carbon* **2012**, *50*, 706–717.
42. Kim, S.W.; Chung, W.S.; Sohn, K.S.; Son, C.Y.; Lee, S. Improvement of flexure strength and fracture toughness in alumina matrix composites reinforced with carbon nanotubes. *Mater. Sci. Eng.* **2009**, *A517*, 293–299.
43. Bakhsh, N.; Khalid, F.A.; Hakeem, A.S. Effect of sintering temperature on densification and mechanical properties of pressureless sintered CNT–Al_2O_3 nanocomposites. *Mater. Sci. Eng.* **2014**, *60*, 012059.
44. Li, T. Improving the antistatic ability of polypropylene fibers by inner antistatic agent filled with carbon nanotubes.*Comput. Sci. Tech.* **2004**, *64*, 2089–2096.
45. Hanzel, O.; Sedlácek, J.; Sajgalík, P. New approach for distribution of carbon nanotubes in Al_2O_3 matrix. *J. Eur. Ceram. Soc.* **2014**, *34*, 1845–1851.
46. Michalek, M.; lkova, M.; Sedla, J.; Galusek, D. Mechanical properties and electrical conductivity of Al_2O_3/MWCNT and Al_2O_3/zirconia/MWCNT composites. *Ceram. Int.* **2014**, *40*, 1289–1295.
47. Poyato, R.; Gallardo-López, A.; Gutiérrez-Mora, F.; Morales-Rodríguez, A.; Muñoz, A.; Domínguez-Rodríguezb, A. Effect of high SWNT content on the room temperature mechanicalproperties of fully dense 3YTZP/SWNT composites.*J. Eur. Ceram. Soc.* **2014**, *34*, 1571–1579.
48. Kyotani, T.; Tsai, L.F.; Tomita, A. Preparation of ultrafine carbon tubes in nanochannels of an anodic aluminum oxide film. *Chem. Mater.* **1996**, *8*, 2109–2113.
49. Sui, Y.C.; Acosta, D.R.; Cui, B.Z. Structure, thermal stability, and deformation of multibranched carbon nanotubes synthesized by CVD in the AAO template. *J. Phys. Chem.* **2001**, *B105*, 1523–1527.
50. Bae, E.J.; Choi, W.B.; Park, G.S.; Song, S. Selective growth of carbon nanotubes on pre-patterned porous anodic aluminum oxide. *Adv. Mater.* **2002**, *14*, 277–279.
51. Parhama, H.; Bates, S.; Xia, Y.; Zhu, Y. A highly efficient and versatile carbon nanotube/ceramic composite filter.

Carbon **2013**, *54*, 215–223.

52. Parhama, H.; Kennedy, A.; Zhu, Y. Preparation of porous Al_2O_3—Carbon nanotube composites via direct growth of carbon nanotubes. *J. Comput. Sci. Technol.* **2011**, *71*, 1739–1745.

53. Sun, J.; Gao, L. Development of a dispersion process for carbon nanotubes in ceramic matrix by hetero-coagulation.*Carbon* **2003**, *41*, 1063–1068.

54. Chan, B.; Seung, I. Fabrication of CNT-reinforced Al_2O_3 matrix nanocomposites by sol-gel. *Mater. Sci. Eng.* **2005**, *395*, 124–128.

55. Sun, J.; Gao, L. Reinforcement of Al_2O_3 matrix with multi-walled CNT. *Ceram. Int.* **2004**, 893–896.

56. Gao, L.; Jiang, L.; Sun, J. Carbon nanotube-ceramic composites. *J. Electroceram.* **2006**, *17*, 51–55.

57. Coble, R.L. Diffusion Models for hot pressing with surface energy and pressure effects as driving forces. *J. Appl. Phys.***1970**, *41*, 4798–4808.

58. Legorreta, G.; Estournes, C.; Peigney, A.; Weibel, A.; Flahaut, E.; Laurent, C. Spark-plasma-sintering of double-walled carbon nanotube–magnesia nanocomposites. *Scr. Mater.* **2009**, *60*, 741–744.

59. Ahmad, I.; Dar, M.A. Structure and properties of Y_2O_3-doped Al_2O_3-MWCNT nanocomposites prepared by PL-sintering and hot-pressing. *J. Mater. Eng. Perform.* **2014**, *23*, 2110–2119.

60. Ghobadi, H.; Ali, N.; Ebadzade, T.; Sadeghian, Z.; Barzegar-Bafrooei, H. Improving CNT distribution and mechanical properties of MWCNT reinforced Al_2O_3 matrix. *Mater. Sci. Eng.* **2014**, *A617*, 110–114.

61. Tatami, J.; Katashima, T.; Komeya, K.; Meguro, T.; Wakihara, T. Electrically conductive CNT-dispersed silicon nitride ceramics. *J. Am. Ceram. Soc.* **2005**, *88*, 2889–2893.

62. Katz, J.D.; Blake, R.D. Microwave sintering of multiple alumina and composite components. *J. Am. Ceram. Soc.* **1991**,*70*, 1304.

63. Sheppard, L.M. Firing technology heats up for the 90s. *J. Am. Ceram. Soc.* **1988**, *67*, 1656.

64. De, A.; Ahmad, I.; Whitney, E.D.; Clark, D.E. Microwaves theory and applications. *Mater. Process.* **1991**, *21*, 329–339.

65. Fujitsu, S.; Ikegami, M.; Hyashi, T. Sintering of partially stabilized zirconia by microwave heating using ZnO–MnO_2–Al_2O_3 plates in a domestic microwave oven. *J. Am. Ceram. Soc.* **2000**, *83*, 2085–2087.

66. Ahmad, I.; Cao, H.; Chen, H.; Zhao, H.; Kennedy, A.; Zhu, Y. Carbon nanotube toughened aluminium oxide nanocomposites. *J. Eur. Ceram. Soc.* **2009**, *30*, 865–873.

67. Valecillos, M.C.; Hirota, M.; Brite, M.E.; Hirao, K. Sintering of alumina by 2.45 GHz microwave heating. *J. Eur. Ceram. Soc.* **2004**, *24*, 387–391.

68. Thostenson, P.G.; Karandikar, T.W. Fabrication and characterization of reaction bonded silicon carbide/carbon nanotube composites. *J. Appl. Phys.* **2005**, *38*, 3962–3965.

69. Wang, Y.; Voronin, G.A.; Zerda, T.W.; Winiarski, A. SiC–CNT nanocomposites: High pressure reaction synthesis and characterization. *J. Phys. Condens. Matter.* **2006**, *18*, 275–282.

70. Kristen, H.B.; Gary, L.M.; Dinesh, K.A. Microwave sintering of alumina at 2.45 GHz. *J. Am. Ceram. Soc.* **2003**, *86*, 1307–1312.

71. Lopez, A. Hardness and flexural strength of single-walled carbon nanotubes/Al_2O_3 composites. *J. Mater. Sci.* **2014**, *49*, 7116–7123.

72. Zaman, A.C.; Kaya, F.; Kaya, C. OH and COOH functionalized single walled carbon nanotubes-reinforced alumina ceramic nanocomposites. *Ceram. Int.* **2012**, *38*, 1287–1293.

73. Estili, M.; Kawasaki, A.; Sakka, Y. Highly concentrated 3D macrostructure of individual carbon nanotubes in a ceramic environment. *Adv. Mater.* **2012**, *24*, 4322–4326.

74. Martinlli, J.R.; Sene, F.F. Electrical resistivity of ceramic-metal composite materials: Application in crucibles for induction furnaces. *Ceram. Int.* **2000**, *26*, 325–335.

75. Puchy, V.; Hvizdos, P.; Dusza, J.; Kovac, F.; Inam, F.; Reece, M.J. Wear resistance of Al_2O_3–CNT ceramic nanocomposites at room and high temperature. *Ceram. Int.* **2013**, *39*, 5821–5826.

76. Lee, K.; Mo, C.B.; Park, S.B.; Hong, S.H. Mechanical and electrical properties of multiwalled CNT–alumina nanocomposites prepared by a sequential two-step processing of ultrasonic spray pyrolysis and spark plasma sintering. *J. Am. Cream. Soc.* **2011**, *94*, 3774–3779.

77. Thomson, K.E.; Jiang, D.; Yao, W.; Ritchie, R.O.; Mukherjee, A.K. Characterization and mechanical testing of alumina-based nanocomposites reinforced with niobium and/or carbon nanotubes fabricated by spark plasma sintering. *Acta Mater.* **2012**, *60*, 622–632.

78. Ahmad, I.; Kennedy, A.; Zhu, Y. Wear resistance properties of multi-walled carbon nanotubes reinforced Al_2O_3nanocomposite. *Wear* **2010**, *269*, 71–78.

79. Zhan, G.D.; Mukherjee, A.K. Carbon nanotube reinforced alumina-based ceramics with novel mechanical, electrical, and thermal properties. *Int. J. Appl. Ceram. Technol.* **2004**, *1*, 161–171.

80. Sarkar, S.; Das, P.K. Effect of sintering temperature and nanotube concentration on microstructure and properties of carbon nanotube/alumina nanocomposites. *Ceram. Int.* **2014**, *40*, 7449–7458.

81. Bi, S.; Su, X.; Hou, G.; Liu, C.; Song, W.L.; Cao, M.S. Electrical conductivity and microwave absorption of shortened multi-walled carbon nanotube/alumina ceramic composites. *Ceram. Int.* **2013**, *39*, 5979–5983.

82. Bakshi, S.R.; Balani, K.; Agarwal, A. Thermal conductivity of plasma-sprayed aluminum oxide—Multiwalled carbon nanotube composites. *J. Am. Cream. Soc.* **2008**, *91*, 942–947.

83. Laurent, C.; Peigney, A.; Rousset, A. Carbon nanotubes-Fe-Al_2O_3 nanocomposites. Part II: Microstructure and mechanical properties of the hot-pressed composites. *J. Eur. Ceram. Soc.* **1998**, *18*, 2005–2013.

84. Siegel, R.W.; Chang, S.K.; Ajayan, P.M. Mechanical behaviour of polymer and ceramic matrix nanocomposite. *Scr. Mater.* **2001**, *44*, 2061–2064.

85. Hoagland, R.G. A treatment of inelastic deformation around a crack tip due to micro cracking. *J. Am. Ceram. Soc.* **1980**,*63*, 404–410.

86. Kawamura, H.; Yamamoto, S. *Improvement of Diesel Engine Startability by Ceramic Glow Plug Start System*; Society of Automotive Engineers: New York, NY, USA, 1983.

87. Zheng, G.; Sano, H.; Cheng, H.M. A TEM study of the microstructure of carbon fiber/polycarbosilane-derived SiC composites. *Carbon* **1999**, *37*, 2057–2062.

88. Estili, M. The homogeneous dispersion of surfactantless, slightly disordered, crystalline, multiwalled carbon nanotubes in α-alumina ceramics for structural reinforcement. *Acta Mater.* **2008**, *56*, 4070–4079.

89. Knieke, C.; Berger, A.; Voigt, M.; Taylor, R.N.K.; Röhrl, J.; Peukert, W. Scalable production of graphene sheets by mechanical delamination. *Carbon* **2010**, *48*, 3196–3204.

90. Hummers, J.; William, S.; Offeman, R.E. Preparation of graphitic oxide. *J. Am. Chem. Soc.* **1958**, *80*, 1339–1339.

91. Inam, F.; Yan, H.; Reece, M.J.; Peijs, T. Dimethylformamide: an effective dispersant for making ceramic-carbon nanotube composites. *Nanotechnology* **2008**, *19*, 195710.

92. Balázsi, C.; Wéber, F.; Arató, P.; Fényi, B.; Hegman, N.; Kónya, Z.; Kiricsi, I.; Vértesy, Z.; Biró, L.P. Development of CNT-silicon nitrides with improved mechanical and electrical properties. *Adv. Sci. Technol.* **2006**, *45*, 1723–1728.

93. Milsom, B.; Viola, G.; Gao, Z.; Inam, F.; Peijs, T.; Reece, M.J. The effect of carbon nanotubes on the sintering behaviour of zirconia. *J. Eur. Ceram. Soc.* **2012**, *32*, 4149–4156.

94. Guo, S.; Sivakumar, R.; Kitazawa, H.; Kagawa, Y. Electrical properties of silica-based nanocomposites with multiwall carbon nanotubes. *J. Am. Ceram. Soc.* **2007**, *90*, 1667–1670.

95. Liu, J.; Yan, H.; Jiang, K. Mechanical properties of graphene platelet-reinforced alumina ceramic composites. *Ceram. Int.* **2013**, *39*, 6215–6221.

96. Ramirez, C.; Osendi, M.I. Characterization of graphene nanoplatelets-Si_3N_4 composites by Raman spectroscopy. *J. Eur. Ceram. Soc.* **2013**, *33*, 471–477.

97. Kvetková, L.; Duszová, A.; Kašiarová, M.; Orčáková, F.; Dusza, J.; Balázsi, C. Influence of processing on fracture toughness of Si_3N_4 + graphene platelet composites. *J. Eur. Ceram. Soc.* **2013**, *33*, 2299–2304.

98. Inam, F.; Vo, T.; Bhat, B.R. Structural stability studies of graphene in sintered ceramic nanocomposites. *Ceram. Int.***2014**, *40*, 16227–16233.

99. Rutkowski, P.; Stobierski, L.; Górny, G. Thermal stability and conductivity of hotpressed Si_3N_4–graphene composites*J. Therm. Anal. Calorim.* **2014**, *116*, 321–328.

100. Tapaszto, O.; Kun, P.; Weber, F.; Gergely, G.; Balazsi, K.; Pfeifer, J.; Arato, P.; Kidari, A.; Hampshire, S.; Balazsi, C. Silicon nitride based nanocomposites produced by two different sintering methods. *Ceram. Int.* **2011**, *37*, 3457–3461.

101. Evans, A.G. Perspective on the development of high-toughness ceramics. *J. Am. Ceram. Soc.* **1990**, *73*, 187–206.

102. Ritchie, R.O. The quest for stronger tougher materials. *Science* **2008**, *320*, 448–452.

103. Curtin, W.A.; Sheldon, B.W. CNT-reinforced ceramics and metals. *Mater. Today* **2004**, *7*, 44–49.

104. Kramer, P.; White, K. Effect of sintering parameters and composition on the resistivity of a cermet used as an electrical feed through. *Ceram. Eng. Sci. Proc.* **1982**, *3*, 512–518.

105. Tajima, Y. Development of high performance silicon nitride ceramics and their applications. *Mater. Res. Soc. Symp. Proc.* **1993**, *287*, 98–201.

106. Komeya, H.K. Development of ceramic antifriction bearing. *JSAE Rev.* **1986**, *7*, 72–79.

107. Huang, Q.; Gao, L. Manufacture and electrical properties of multiwalled carbon nanotube/$BaTiO_3$ nanocomposite ceramics. *J. Mater. Chem.* **2004**, *14*, 2536–2541.

108. Lee, S.Y.; Kim, H.; McIntyre, P.C.; Saraswat, K.C.; Byun, J.S. Atomic layer deposition of ZrO_2 on W for metal-insulator-metal capacitor application. *Appl. Phys. Lett.* **2003**, *823*, 2874–2876.

109. Dusza, J.; Tomasek, K.; Blugan, G.; Kuebler, J. Microstructure and properties of carbon nanotube/zirconia composite*J. Eur. Ceram. Soc.* **2008**, *28*, 1023–1027.

110. Patsalas, P.; Logothetidis, S. Optical, electronic, and transport properties of nanocrystalline titanium nitride thin films*J. Appl. Phys.* **2001**, *90*, 4725–4734.

111. Carmalt, C.J.; Whaley, S.R.; Lall, P.S.; Cowley, A.H.; Jones, R.A. Titanium (IV) azido and imido complexes as potential precursors to titanium nitride. *J. Chem. Soc. Dalton Trans.* **1998**, *1998*, 553–558.

112. Janes, R.A.; Aldissi, M.; Kaner, R.B. Controlling surface area of titanium nitride using metathesis reactions. *Chem. Mater.* **2003**, *15*, 4431–4435.

113. Kim, S.; Kumta, P.N. Hydrazide sol–gel synthesis of nanostructured titanium nitride: Precursor chemistry and phase evolution. *J. Mater. Chem.* **2003**, *13*, 2028–2035.

114. Cao, M.S.; Wang, R.G.; Fang, X.Y.; Cui, Z.X.; Chang, T.J.; Yang, H.J. Preparing γ′-Fe4N ultrafine powder by twice-nitriding method. *Powder Technol.* **2001**, *115*, 96–98.

CHAPTER 22

APPLICATIONS AND NANOTOXICITY OF CARBON NANOTUBES AND GRAPHENE IN BIOMEDICINE

Caitlin Fisher,[1,2] Amanda E. Rider,[1,2] Zhao Jun Han,[1,2] Shailesh Kumar,[1,2] Igor Levchenko,[1,2] and Kostya (Ken) Ostrikov[1,2]

[1]Plasma Nanoscience Centre Australia (PNCA), CSIRO Materials Science and Engineering, Lindfield, NSW 2070, Australia

[2]Plasma Nanoscience at Complex Systems, School of Physics, The University of Sydney, Sydney, NSW 2006, Australia

ABSTRACT

Owing to their unique mechanical, electrical, optical, and thermal properties, carbon nanostructures including carbon nanotubes and graphenes show great promise for advancing the fields of biology and medicine. Many reports have demonstrated the promise of these carbon nanostructures and their hybrid structures (composites with polymers, ceramics, and metal nanoparticles, etc.) for a variety of biomedical areas ranging from biosensing, drug delivery, and diagnostics, to cancer treatment, tissue engineering, and bioterrorism prevention. However, the issue of the safety and toxicity of these carbon nanostructures, which is vital to their use as diagnostic and therapeutic tools in biomedical fields, has not been completely resolved. This paper aims to provide a summary of the features of carbon nanotube and graphene-based materials and current research progress in biomedical applications. We also highlight the current opinions within the scientific community on the toxicity and safety of these carbon structures.

INTRODUCTION

Carbon nanotubes (CNTs) and graphene are very promising candidates to form the basis of new biological and medical devices. Carbon nanotubes can be thought of as rolled-up graphene sheets with no overlapping edges [1]. Their diameters typically vary from 1 to 100 nm and their lengths can be several orders of magnitude larger, up to millimeters, even centimeters long [2]. Various orientations of CNTs are shown in Figures 1(a)–1(c): randomly oriented, vertically aligned, and in a "dandelion-like" structure, respectively. The well-documented beneficial mechanical, electrical and chemical characteristics of CNTs and graphene [1,3–7] as well as their ability to be hybridized with a wide range of organic and inorganic materials make them ideal candidates for many biomedical applications such as biosensing [8–12], tissue engineering [13–15], and drug delivery [16, 17].

Citation: Caitlin Fisher, Amanda E. Rider, Zhao Jun Han, Shailesh Kumar, Igor Levchenko, and Kostya (Ken) Ostrikov, "Applications and Nanotoxicity of Carbon Nanotubes and Graphene in Biomedicine," Journal of Nanomaterials, vol. 2012, Article ID 315185, 19 pages, 2012. doi:10.1155/2012/315185

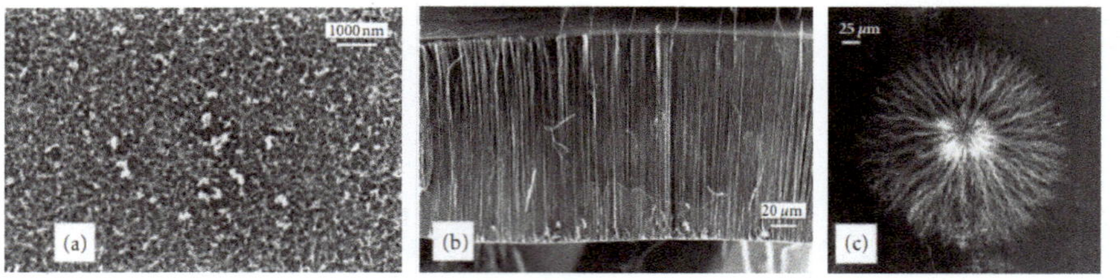

Figure 1: Various orientations of CNTs grown using chemical vapor deposition on Si substrates: (a) entangled, randomly orientated CNTs; (b) vertically aligned CNTs; (c) dense "dandelion-like" CNT structure grown using plasma-enhanced chemical vapor deposition on an etched, catalyst-free Si substrate. Details on the growth of similar structures can be found in [40].

In the past two decades, intense efforts have been directed at providing specificity, selectivity, reproducibility, and robustness to these carbon nanostructures in biologically relevant environments [18–22]. However, the issue of toxicity of CNTs and graphene in living biological systems, which is vital for the successful incorporation of these materials into functional biomedical devices, remains unsolved at macroscopic, cellular, and intracellular levels [23–25].

In this paper we will, in Section 2, discuss the role of CNTs in biosensing, tissue engineering, and drug delivery. Aspects of the toxicity of CNTs in living biological systems are then discussed in Section 3 and the emerging graphene-related biomedical applications and associated safety issues are briefly presented in Section 4. Finally, a summary of this work and an outlook for future research is provided.

APPLICATIONS OF CARBON NANOTUBES IN BIOMEDICINE

Due to the chemical inertness of graphitic walls, functionalization of CNTs and graphene is often the key step required in any application of these materials. Let us first briefly consider the possible ways of performing such functionalization.

Functionalization of Carbon Nanotubes

Successful functionalization should maintain the integrity of CNT's mechanical, electrical, and chemical properties as well as the activity of the biological species being attached. In general, there are two CNT functionalization methods: covalent bonding and noncovalent wrapping [26–28].

Covalent bonding involves chemical attachment of the desirable species to the CNT [26], often at the tube ends or at defect sites [1, 29]. Oxidation processes are often used as a preparation step to create chemically active sites for covalent bonding [30]. These oxidation processes can be performed through wet chemical or dry plasma routes [30], amongst others. A variety of biological species have been covalently bonded to CNTs [26, 31, 32]. However, this often alters the intrinsic structure and properties of CNTs as well as the properties of attached biomolecules [33]. An example of a covalently functionalized CNT is shown in Figure 2(a), which shows a scanning electron microscope (SEM) image of multiwalled CNTs (MWCNTs) covalently functionalized with ferritin [34].

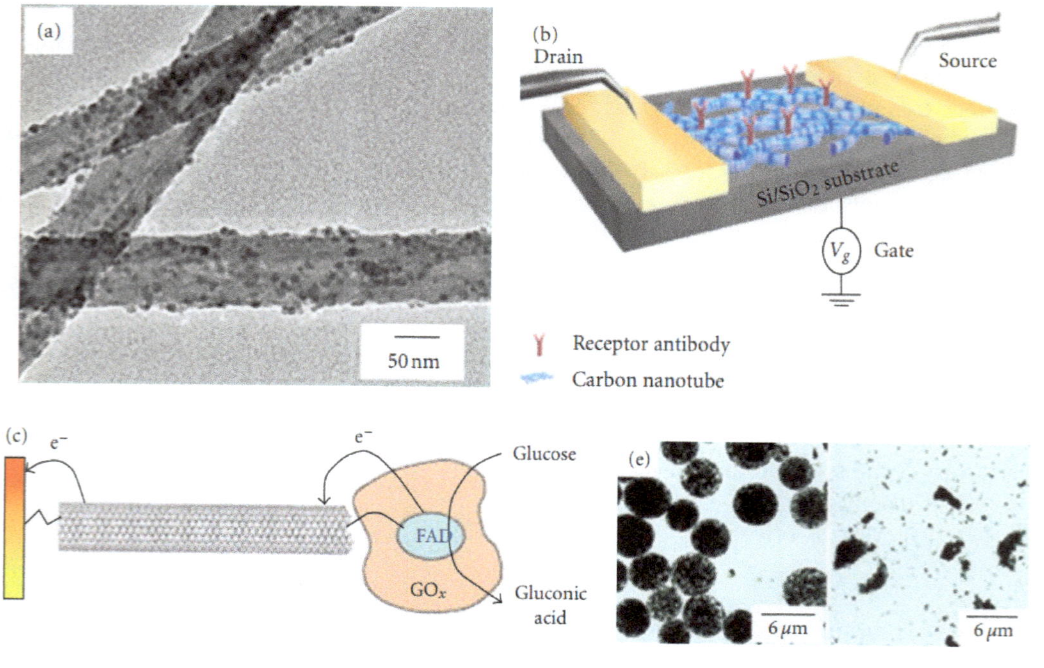

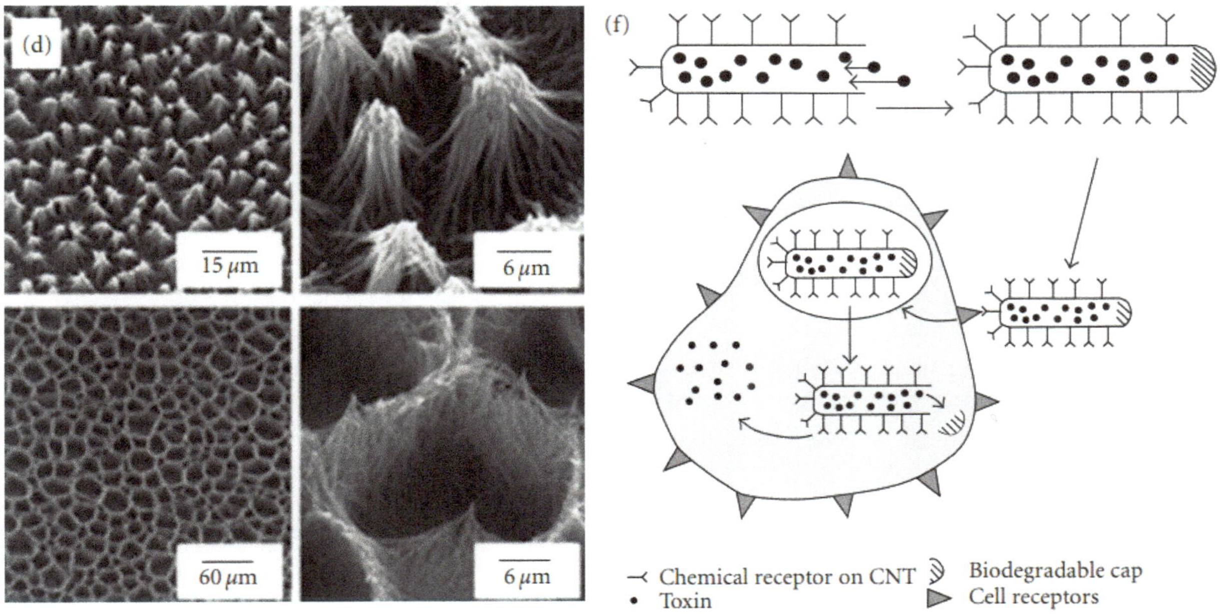

Figure 2: (a) Transmission electron microscopy image of ferritin-functionalized MWNT (reproduced with permission from [34]); (b) CNT-FET device for biosensing using antibody receptors (reproduced with permission from [22]); (c) Electrochemical sensor for glucose consisting of an Au electrode bonded to a CNT-G *Ox* which detects redox reactions between FAD and glucose (reproduced with permission from [51]); (d) "basket-like" periodic MWNT lattices into which mice fibroblast is to be implanted (reproduced with permission from [52]); (e) alginate-polylysinealginate microcapsule destruction by the photoacoustic effect by folate acid—functionalized SWNTs inside the cells (reproduced with permission from [53]); (f) loading and unloading of molecules inside CNTs for drug delivery; drugs are released when a CNT is uncapped within the cytoplasm of the cell (reproduced with permission from [54]).

Noncovalent bonding, or physical adsorption, on the other hand, is the process of wrapping a species around the CNT walls [27]. This method is preferable to covalent bonding in many cases as it causes less structural damage to the CNT and the wrapped species, and the chemical environment required during fabrication is less harsh [28]. Over the years, the noncovalent wrapping of CNTs by polysaccharides [33], DNA [35, 36], proteins, polypeptides [37], and synthetic polymers has been widely reported.

The degree of CNT functionalization is commonly characterized by atomic force microscopy, Raman spectroscopy, SEM, transmission electron microscopy (TEM), ultraviolet-visible light spectroscopy, Fourier-transform infrared spectroscopy, thermal gravimetric analysis, and gel electrophoresis [38, 39]. The bioactivity of the attached biomolecules can be characterized by the immunochemical methods such as enzyme-linked immunosorbent assay (ELISA) [38]. It should be noted that whilst the bioactivity is strongly dependent on the bonding between the biomolecule and the CNT, the stoichiometry and the loading ratio are also important factors that must be considered [32].

Carbon Nanotubes for Biosensing

The incorporation of CNTs in biosensing devices has made a significant progress in the last decade [39]. By definition, a biosensor is an analytic device which consists of a receptor that interacts with the targeted analyte to be measured and a transducer (or detector) that transforms the signal from the interaction into a form that can be easily measured. The one-dimensional (1D) structure of CNTs allows signals to be transported in a confined space, making them extremely sensitive to electrical and chemical changes in their immediate environment [8]. There are generally two configurations of CNT-based biosensors: CNT field-effect transistors (CNT-FETs) [41, 42] and CNT electrochemical sensors [3, 43, 44]. Here, we concentrate on these two types of biosensors and provide details on the role of CNTs in these devices and the common approaches employed to improve their sensitivity.

Carbon nanotube FET biosensors have a current carrying channel connected to a source and a drain, which can be regulated by a gate voltage [3]. A typical CNT-FET setup is shown in Figure 2(b). The current carriers (electrons or holes) running through the channel are highly sensitive to changes in external electric fields and as such can be used to detect electrical signals produced by biological activity or biochemical interactions [45]. The conductive channel in CNT-FET devices can be either an individual semiconducting single-walled CNT (SWCNT) [46] or a randomly distributed bundle of CNTs [8]. The former was first introduced by Martel et al. [47] and has shown superior performance compared to traditional metal-oxide-semiconductor FETs (MOSFETs) [48]. Because of their high sensitivity, CNT-FET biosensors are well suited for the detection of very low analyte concentrations [8, 49]. For example, ultrasensitive detection of DNA at concentrations of 100 fM has been detected with SWCNT-FETs [35, 50].

Semiconducting SWCNTs are often used in FET biosensors as opposed to metallic SWCNTs since their conductivity can be gate-modulated by electrical changes in the external environment [49]. Various purification processes have been reported to separate metallic and semiconducting SWCNT mixtures, such as electrophoresis, centrifugation, chromatography, and solubilisation [55–57]. However, purification of semiconducting SWNTs can add considerable time and resources to the production process [56], and as such should be taken into account when considering the economic viability of CNT device fabrication.

Several approaches have been attempted to further amplify the signals and improve the specificity and sensitivity of CNT-FETs. One of these approaches is to bind specific receptor species, which can undergo selective interactions with the analyte, to the CNT channels. Many types of receptors have been successfully bound to functionalized CNT-FETs, such as aptamers [58], antibodies [59], sugars [60], DNA [50], and proteins [49, 61]. Furthermore, decorating the CNTs with conductive metal nanoparticles (typically Pt and Au) can also amplify the detection signals [62]. Rajesh et al. [63] demonstrated that the sensitivity of DNA detection was 2.5 times greater after the addition of ZnS nanoparticles into CNT-FETs.

Controlling the Schottky barrier effect is another common way to amplify detection in CNT-FETs. Changes in the conductivity of CNT-FETs are mostly due to gate coupling effects and the Schottky barrier effect. The Schottky barrier effect arises from differences between the work functions of the CNT and the metal contacts [64] and is often hard to predict, and is therefore detrimental to the sensor's performance [8]. Methods such as coating the metal contacts with polymers have been used to minimize this effect and stabilize the response [64]. However, it should also be noted that one can actually amplify the CNT-FET signals for sensing applications by controlling the Schottky barrier effect [3, 65].

Real-time measurement is vital for the monitoring of dynamic systems and is a promising approach for applications in fields such as bioterrorism, food safety, drug testing, and on-the-spot medical diagnosis [65–67]. Carbon nanotube FET biosensors have also been used to monitor biological interactions in real time [8], including interactions involving DNA [35], proteins [64], immunoglobulin [58], morphine [45], and biological processes such as phagocytosis [62]. Again, decoration by metal nanoparticles can be used to improve the performance of CNT-FETs in real-time measurements, as has been demonstrated in the cases of glucose [35] and heroin [45].

On the other hand, CNT-based electrochemical biosensors have been used to detect chemical redox interactions [39, 68]. In this sensing mode the electrical properties and small size of each CNT is preferred, as it allows them to act as tiny electrodes with direct contact to biological systems. For example, a SWCNT with a diameter of ~1 nm is comparable to the size of DNA and the active sites of proteins [69]. The 1D structure of CNTs also allows them to interact with one species at a time, ideal for single-species biodetection and biosensing [62, 70]. Carbon nanotube electrochemical biosensors have been used to detect DNA [68], glucose [71, 72], proteins [49], enzymes [72], RNA [73–75], H_2O_2 [76], and many other biomolecules. Analogous to the case of CNT-FET biosensors, functionalization of CNTs is important to increase the specificity of the electrochemical biosensors.

Figure 2(c) shows an example of an electrochemical biosensor, a gold electrode-CNT-glucose oxidase (G *Ox*) biosensor for detecting redox reactions between glucose (i.e., the analyte) and the redox active center of the G *Ox*, flavin adenine dinucleotide (FAD) [51]. The orientation of the CNTs on the electrode has been found to be important for sensing performance, with several studies showing that using aligned CNTs increased the sensitivity of the device [18, 69]. Other biosensors rely on an array of CNTs perpendicular to the electrode where the properties of the biosensor sensitivity are dependent on the spatial distribution and relative diameters of the CNTs in the array [50, 77]. Recently, spun CNT fibers have also been attached to electrodes for biosensing [78] and CNT fiber sensors have been reported to detect glucose with greater sensitivity than traditional Pt-Ir sensors [44].

Many different methods have been used to improve the sensitivity of CNT electrochemical biosensors. The most common way has been to combine metal nanoparticles with CNTs to increase the conductivity and redox activity of the sensor [64, 71, 72, 78]. Other methods include gas treatment [79], oxidation, and plasma etching [80, 81], which can add extra functional groups (e.g., –COOH, $-NH_2$) to the CNTs. These functional groups act as active reaction sites for analyte binding [82]. Lastly, optimizing the ratio of semiconducting and metallic CNTs can also improve the sensitivity [50] and even the response speed [78] of CNT electrochemical sensors. For example, it has been shown that biosensing devices using a mixture of metallic and semiconducting CNTs performed better than those using pure semiconducting CNTs alone [50].

Since many of these electrochemical biosensors are used in real biological environments, it is often necessary to prevent nonspecific binding of proteins, which can interfere with the measurements. Polymer coatings such as poly(ethylene glycol) (PEG) and poly(ethylene oxide) (PEO) are often used to reduce nonspecific binding [8, 49, 83, 84]. However, care must be taken to choose the correct polymer which does not distort the CNT and the biological receptor [36].

It should be noted that there are many nanostructures already used in biosensors, for example, Au nanoparticles [85]. Au nanoparticles are gold clusters ranging from 1 to 100 nm in diameter [86] and are easily functionalized with biomolecules such as DNA, enzymes, proteins, peptides, oligonucleotides, glucose, and RNA [85, 86]. Biosensing with Au nanoparticles is quite well established, especially by enhanced surface plasmon resonance [86] and enhanced Raman spectroscopy [86]. The toxicity of Au nanoparticles is currently deemed minimal compared to CNTs [85, 87]. The main advantage of using CNTs in biosensors instead of Au nanoparticles is that CNTs have many more parameters that can be varied, which potentially allows a wider variety of biosensors to be produced.

Carbon Nanotubes in Tissue Engineering

Tissue engineering aims to repair, regenerate, and replace diseased or damaged tissue. In tissue engineering, scaffolds are often used to promote cell adhesion, growth, differentiation, and proliferation in a three-dimensional (3D) matrix. These scaffolds also provide mechanical strength and add some degrees of control over location and orientation of the attached cells. Using CNTs as scaffolds has attracted great interest due to their mechanical strength [13], chemical stability [14], and biological inertness [52, 88]. Typically, carbon nanotubes are grown into a 3D porous structure, or else coat an existing 3D porous structure, for example, collagen [89] and then seeded with cells [90]. The cells are allowed to grow over the scaffold until they become self-supporting [90]. Carbon nanotube scaffolds can be fabricated into many different structures [91] with dimensions comparable to biological cellular scaffolds [88, 92], which allows them to support a wide range of biological species.

To promote cell growth and adhesion, CNTs are often functionalized with, for example, carboxyl groups, polymers, and sugars [13–15]. It has been shown that surface charge, functional groups, and hydrophilicity are important in determining cell adhesion and growth [93, 94]. For example, in a study by Zhang et al., hydrophilic, neutrally charged amylose-CNT hybrid matrices supported cell growth and proliferation as compared to chitosan-CNT, sodium alginate-CNT, or chitosan-sodium-alginate-CNTs [33].

The periodicity, size, and structure of the CNT scaffolds can also affect the cellular interactions with the scaffold [52, 95]. Vertically aligned MWCNTs can be fabricated into 3D scaffold matrices of periodic "basket-like" cavities, where the cavity size and density depend on the dimensions of the original array as shown in Figure 2(d) [52]. Mice fibroblast cell lines L929 were successfully cultured on these periodic matrices [52]. Patterns of CNTs can be fabricated by lithography and used to direct the growth of cells such as human mesenchymal stem cells (hMSCs) and neurons [96, 97]. In particular, hMSCs have been found to grow and differentiate well on 3D matrices of fibronectin (FN) functionalized CNTs [98], but their growth was inhibited when seeded on carboxylated CNT scaffolding [99]. The polarization of neurons is also extremely sensitive to their environment; hence CNTs could be used to direct neural connections and interactions, potentially for in vivo applications [100].

Apart from using CNTs solely to construct scaffolds, they can also be added to other scaffolding materials to produce heterostructures. For example, porous polymer composites can be produced by dispersing CNTs into PEG or poly(propylene fumarate) (PPF) [101–103]. The addition of CNTs allows the polymer composites to be electrically conductive, which is useful for stimulating cell growth [104]. These composites have generally been used to promote osteoblastic cell growth in vitro for bone regeneration research [104–107] and also for neural regeneration [108, 109]

We also note that another carbon nanostructure that has recently emerged as a potential scaffold material for tissue engineering is nanodiamond. For example, a monolayer of nanodiamonds has been shown to support neuronal cell growth [110]. Nanodiamond crystals range from 2 to 10 nm in diameter [111], are mechanically stable [110, 112], have a large surface area [112], and are nontoxic [111]. However, like the Au nanoparticles mentioned in Section 2.2, nanodiamonds have significantly fewer tunable characteristics compared to CNTs, thus offer fewer possibilities for tissue scaffold designs.

Carbon Nanotubes in Drug Delivery

In delivering drugs, the aim is to use a carrier molecule functionalized with a receptor to carry a drug around the body until it attaches to the problematic site, only then releasing the drug [113]. Receptor-functionalized CNTs have been suggested as targeted vehicles for drug delivery, where they are perceived to have several advantages. Firstly, the nanometer size of CNTs allows them to permeate into cellular membranes, making them ideal for inserting drugs directly into cells [16, 114, 115]. Secondly, each CNT can be functionalized to detect and interact with a single cell, improving the delivery efficiency and reducing the drug dosage [17,116]. It has been demonstrated that 5 million species can be bonded to an 80 nm long CNT [54]. The combination of size and ease of functionalization allows CNT to deliver drugs to cells, such as neurons and cardiomyocytes, which are difficult to reach by traditional drug-delivery methods [117]. Lastly, drugs can be encapsulated into CNTs [54, 118–120], where release of the drug in the desired cell compartment requires the chemical disintegration of the CNT cap, as illustrated in Figure 2(f) [22]. Successful examples include anticancer drugs and IR-emitting molecules for direct heat treatment in vivo [121].

The CNT must have a drug-unloading mechanism for the drug delivery to function effectively [122]. Drug unloading from CNT carriers can by triggered by environmental changes, such as changes in temperature and pH [54]. For example, intracellular pH is lower than extracellular pH; and CNTs, which cross the membrane, can be activated to release their drug load and influence intracellular processes [123]. This allows gene delivery straight into the nuclei of cells, possible by "injection" of CNTs into cells [124]. Drugs may also be released by optical stimulation using near-infrared (700–1100 nm) wavelengths, which are not absorbed by most biological structures, in particular skin [125].

Apart from delivering specific drugs, CNTs can also be functionalized for therapeutic applications. For example, CNTs functionalized with folic acid can bind to cancer cells, which can be killed by using infrared radiation to induce vibration, that is, forming cellular "bombs" [53]. Similar research has been carried out in other studies [125–128]. Figure 2(e) includes two images showing folate-functionalized SWCNTs inside alginate-polylysine-alginate microcapsules [53]. The first image shows the microcapsules before IR irradiation and the second image shows the obliterated microcapsules after irradiation [53]. This specific targeting of cells reduces damage done to surrounding biological systems and is more effective at destroying malignant cells.

NANOTOXICITY OF CARBON NANOTUBES

Background and Motivation

The diagnostic and therapeutic applications of CNT-based materials mentioned above will only be trialed clinically after detailed information on their environmental and health and safety effects in host biological systems is obtained [129–131]. A few preliminary tests have showed that CNTs are biologically benign to certain cells, tissues, and organs under limited conditions [132–134], while further studies have indicated that CNTs are potential hazards that can cause both acute and chronic adverse effects to many living systems [4,24, 135]. Nevertheless, at this stage, it appears that the biological effects of CNTs are sample specific and must be assessed on a case-by-case basis. The nanotoxicity of CNTs, therefore, requires continuing and extensive investigations and, indeed, this will be required by regulatory bodies before CNTs can be used in clinical environments as functional biomaterials and biomedical devices.

Despite several years of research, definitive findings regarding the extent of toxicological risks arising from using nanotubes are far from complete. Continuing research is required to determine, for example, how CNTs enter cells, where CNTs are internalized, which the cytotoxic mechanisms are relevant, and how the nanotoxicity is affected by a variety of physicochemical characteristics, such as diameter, length, presence of impurities, surface functionalization, and surface wettability. In this section, we give a brief overview of the progress made to date on understanding the nanotoxicity of CNTs, including the exposure, cellular uptake, subcellular localization, and intracellular trafficking, as well as mechanisms that may result in the mitigation and inhibition of nanotoxicity.

The Production of and Exposure to CNTs

Despite their relatively recent discovery [136], production of CNTs had already reached 4000 tons by 2010 and could exceed 12000 tons by 2015 [137]. Such large-scale production has inevitably led to exposure risks for both animals and human beings. The most common ways for CNTs to enter the host include inhalation, ingestion from food and water, and absorption through skin wounds or scars [138]. In laboratory-related exposure experiments, intravenous injection [4, 139, 140], intratracheal administration [141], and abdominal implantation [142], are often employed to study the nanotoxicity of CNTs in different organs including the lungs.

The Cellular Uptake of CNTs

The uptake of CNTs into cells plays a critical role in determining their cytotoxicity and genotoxicity. The outermost layer of the cell, the cellular membrane, consists of a phospholipid bilayer [129], which serves to segregate the subcellular compartments from the external medium, and to regular the transport of foreign materials, including CNTs, into cells [143].

Experimental results indicate that CNTs can be internalized by a variety of cells. Although systematic knowledge is still lacking, it is in general considered that there are two possible pathways for CNTs to cross the cellular membrane and enter cells [129]. One pathway is passive transport, which includes diffusion, membrane fusion, and direct pore transport [24, 144]. Individually dispersed CNTs in aqueous solutions have been experimentally demonstrated to be able to enter the cytoplasm of cells by directly crossing the membrane [145, 146], despite recent modeling showing that the energy cost of entering the cellular membrane via rupture and diffusion was high compared to that of the energy of thermal motion of CNTs [147].

A more common pathway for the cellular uptake of CNTs is active transport via endocytosis, which includes phagocytosis and pinocytosis [129, 143, 148]. Endocytosis involves the enclosing of foreign objects in vesicles or vacuoles pinched off from the cellular membrane. In general, long CNTs (>1 μm in length) were taken up by phagocytosis, which was mainly conducted by macrophages, monocytes, and neutrophils [143]. Shorter CNTs of length from a few to several hundred nanometers, on the other hand, were mainly internalized bypinocytosis, such as macropinocytosis, clathrin-mediated endocytosis, and caveolin-driven endocytosis [129,143]. Endocytosis is an energy-dependent process, and the orientation of CNT entry can be controlled by the interplay between the tip recognition through receptor binding and the rotation driven by asymmetric elastic strain at the nanotube-phospholipid bilayer interface, as demonstrated recently by numerical modeling [149]. In the most common case, a near-perpendicular orientation resulted in a minimum energy barrier [147].

The exact cellular uptake pathway of CNTs is complex and depends on many experimental parameters, such as the size, length, hydrophobicity, surface chemistry, and the cell culture medium. For example, the uptake of functionalized SWCNTs in phagocytotic cells was found to occur via endocytosis if they were longer than 400 nm and via diffusion-controlled internalization if smaller than 400 nm [150]. Lee and Geckeler [129] have also shown that individual MWCNTs entered cells through direct penetration in vitro, whereas bundled MWCNTs entered via endocytosis.

Surface chemistry of SWCNTs also influences the cellular uptake pathway [151]. It was demonstrated that when SWCNTs were grafted with folate using PEGylation and linked by phospholipid bilayer, they could enter HeLa cells bound with the folate receptor (FR, a specific tumor marker) but not those without FR. Similarly, when SWCNTs were grafted with $\alpha_v\beta_3$, they could enter integrin- (a receptor of $\alpha v\beta 3$) positive U87-MG cells but not integrin-negative cells [152]. If non-cell-targeting molecules were grafted, the entry mode of CNTs will depend on the properties of the conjugated molecules, for example, large molecules such as bovine serum albumin (BSA) enter cells only through endocytosis [152].

Additionally, the hydrophobic surface of CNTs can interact with components in cell growth medium and affect the cellular uptake. Serum proteins in the cell growth medium can bind to CNT walls through π - πinteractions or electrostatic attractions, forming a protein coating [129]. The "screening effect" of such protein coatings, known as the "protein corona," allows functionalized CNTs to experience a similar cellular uptake pathway [129, 153].

Finally, the culture medium itself can affect the cellular uptake of CNTs. For example, single-walled CNTs grafted with fluorescein isothiocyamate (FITC) were found to be taken up by cells in a pH 5.8 medium, but the uptake was inhibited in a slightly alkaline medium (pH 7.2) [154]. It was suggested that under the alkaline condition, the anionic form of FITC could dominate the neutral form and hamper the cellular uptake of CNTs [154]. If properly exploited, this property may facilitate the removal of the internalized SWCNTs, though much more research is required.

Subcellular Localization and Intracellular Trafficking of CNTs

Once taken into the cell, CNTs are often localized in one of a number of different subcellular compartments, for example, cytoplasm, cytoskeleton, mitochondria, lysosomes, endoplasmic reticulum, vesicles, and nuclei and can be translocated between these compartments [143]. Because of their small size and the weak contrast between these cellular components and CNTs, it is often difficult to characterize the subcellular distribution of internalized CNTs. Over the years, techniques, such as confocal and fluorescent microscopy, TEM [4, 24], SEM with focused-ion beam (FIB) [155], Raman spectroscopy [4, 140], and laser and photobleaching [154], have been developed. For example, TEM is a very effective tool to see CNTs (and any other nanoparticles) inside frozen cells. It was utilized by Porter et al. [24] to show that SWCNTs were localized within lysosomes after 2 days, whereas after 4 days, bundles of SWCNTs were localized in endosomes, translocated across the nuclear membrane, and localized within the nucleus.

One of the major findings made in recent years in this field is that the subcellular localization of CNTs depends on how CNTs enter the cells. When functionalized CNTs directly crossed the cell membrane, they were localized exclusively inside mitochondria, whereas if being endocytosized, they were located inside the lysosomes and phagososome [144, 148, 150]. Small CNTs that entered the cellular membrane through diffusion were found mainly in the cell cytoplasm [129, 150]. These preliminary findings have shed light on the selective translocation and localization of SWCNTs to desired subcellular components.

Figure 3 illustrates the main uptake pathways, the subcellular localization, and the intracellular trafficking of differently functionalized CNTs [129]. It is noted that the internalized CNTs may be translocated between different subcellular components through carrier-mediated transport [154]. For example, endocytosized MWCNTs accumulated in the endosomes could be transported to the endoplasmic reticulum, from which they translocated into the cytosol [148]. Intracellular trafficking can also be effectively controlled by attaching a suitable functional tag on the CNTs. For instance, FITC-SWCNTs were taken up in vacuoles through carrier-mediated transport, but when functionalized with an inhibitor of the carrier-mediated transport, they mostly accumulated in the cytoplasm [154]. Further trafficking of functionalized SWCNTs into nucleoplasm and the nucleus were also observed [129].

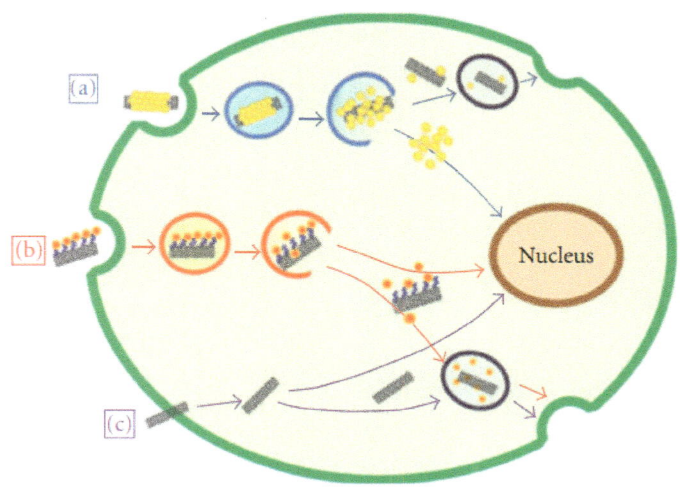

Figure 3: The cellular uptake pathways, subcellular localization and intracellular trafficking of differently functionalized CNTs. (a) Supermolecularly functionalized CNT via endocytosis, (b) covalently functionalized CNT bound with drugs via endocytosis, and (c) individual or specifically functionalized CNT via direct penetration (reproduced with permission from [130]).

Nanotoxicity Mechanisms in the Lungs

One of the earliest concerns over the toxicity of CNTs arose from the similarity of their structure (i.e., a fibrous shape and a high aspect ratio) and biopersistence to asbestos, an infamous carcinogenic material known to cause mesothelioma [141, 156, 157]. Asbestos has a fibrous structure of 20 to several hundred nanometers in diameter. The toxicity of asbestos in the lungs is thought to be mediated through the generation of reactive oxygen species (ROS), which can induce the activation of antioxidant defenses, causing the release of proinflammatory and profibrotic cytokines from inflammatory and epithelial cells, and activation of the apoptotic (cell death) pathway [141, 158].

Carbon nanotubes in the lungs are often considered less toxic than asbestos [141]. However, many experiments have confirmed that they indeed show many asbestos-like behaviors, although it is still unknown if they can cause mesothelioma. The characteristics of cell deaths were observed when lung cells were exposed to CNTs [24]. When CNTs were inhaled, they were able to reach the subpleura and cause subpleural fibrosis in mice [135]. Chronic exposure to SWCNTs was also shown to cause the malignant transformation of human lung epithelial cells, which is evident for CNT-induced carcinogenesis [156].

Besides the ability to generate ROS in lung cells [4, 159], other cytotoxic mechanisms of CNTs also include blocking ion channels, regulating the intracellular calcium level, binding to subcellular organelles and proteins to stop their functions, and attacking the nucleus and damaging DNA, which can induce apoptosis or necrosis cell deaths and/or mutational cellular events [24, 160, 161]. These effects are in turn strongly influenced by the type, size, shape, surface chemistry, and the route of administration of CNTs [162].

The cytotoxic effect of size and shape of CNTs is best represented by their high aspect ratio, which results in incomplete phagocytosis by the mononuclear cells because the CNTs are too large [129]. This induced incomplete or frustrated phagocytosis can result in macrophage activation and granulomatous inflammation. In fact, it has been hypothesized that the failure of resident macrophages to clear CNTs is the main reason for the activation of proinflammatory pathways that induce lung fibrosis, lung cancer, and malignant mesothelioma [149]. In addition, the aggregation of CNTs by van der Waals' interactions could also affect profibrogenic cellular responses and contribute to the pulmonary toxicity of CNTs in vivo [141, 163].

In vivo studies using a mouse model also showed that the pathological origin of CNTs was dependent on the route of administration [164]. For the case of intratracheal instillation, the agglomerates of CNTs with different size and morphology were observed in bronchi of mice, which led to inflammation in 24 days. On the other hand, when CNTs were inhaled, aggregation of CNTs was observed on the lining wall of the bronchi but no inflammation was induced [164].

Surface wettability is another factor that affects the cytotoxicity of CNTs. When macrophages are in contact with hydrophilic CNTs, less inflammation was observed compared to those in contact with hydrophobic CNTs or titanium (a biocompatible metal) [165]. It was found that less proinflammatory cytokines (e.g., tumor necrosis factor-alpha, or TNF-alpha) and interleukin-6 (IL-6) were secreted from macrophages containing hydrophilic CNTs. As a result, hydrophobic CNTs are more toxic than their hydrophilic counterparts.

A final remark on the pulmonary toxicity of CNTs is that many nanoparticle impurities (e.g., Ni, Fe, and Co) carried by CNTs are highly toxic. Removing these nanoparticle impurities is necessary to show the intrinsic toxicity of CNTs [129, 135, 166]. However, conventional purification routes for removing these impurities may generate a high density of functional groups or defects on CNTs, which in turn influences the cytotoxicity. As demonstrated in a recent study, purified CNTs have been shown to have the strongest adverse effects among pristine CNTs, carbon graphite, active carbon, and carbon black.

Biodistribution and Nanotoxicity of CNTs in Other Organs

Many in vivo studies have shown that CNTs delivered to a specific area in the body are not confined to that area [139]. For example, intravenously injected CNTs were shown to be taken up both by the liver and the spleen and then excreted rapidly through the kidney [139, 140]. In contrast, SWCNTs injected into the bloodstream of mice persisted within liver and spleen macrophages (Kupffer cells) [4]. Singh et al. [167] also reported that intravenously injected ammonia-functionalized SWCNTs were excreted mainly via the renal route without uptake in the liver and spleen in mice [167].

Because of the migration of CNTs in biological systems, their toxicity to a variety of other organs should also be tested. In many cases, macrophages, which form the first line of defense against foreign materials, will interact with the administrated CNTs [24]. This is why macrophages are one of the mostly studied cells in in vitro investigations of CNT toxicity. After CNTs have been ingested by macrophages, they can enter into the blood and lymph circulation at a later stage [24]. Carbon nanotubes can also be dispersed by mucins (glycosylated proteins produced by epithelial tissues) in certain cases and cleaned away in a physiological solution, before they can interact with cells [168].

The long-term accumulation of SWCNTs in organs was studied by Yang et al., who showed that no apoptosis was induced in the main organs [4]. On the other hand, a low percentage of early miscarriages and fetal malformations was observed in female mice exposed to pristine SWCNTs and ROS were detected in the placentas of malformed fetuses [169]. Carbon nanotubes could also induce actin (fibrous proteins that can form filaments and higher-order network structures of the cytoskeleton and that perform essential functions such as force generation, motility, and division) bundling and reduced cell proliferation, which may cause chronic changes to cellular functions [170]. Inhaled MWCNTs were also shown to be able to suppress the immune function of the spleen through the signals coming from the lungs of the exposed mice [171, 172].

The Mitigation and Inhibition of CNT Nanotoxicity

In attempting to fully utilize the excellent properties of CNTs without being hampered by their adverse effects, several strategies have been proposed to mitigate or inhibit the toxicity of CNTs. The most common method is through surface functionalization [129, 143]. Noncovalently PEGylated CNTs have been shown to be less toxic than oxidized and pristine CNTs [4, 140, 169]. Repeated administrations of carboxylated (CNT–COOH) and amine-CNTs (CNT–NH_2) in male mice caused only reversible testis damage with no effect on their fertility [173]. Other examples include cell-adhesion peptides bound MWCNTs, which did not interfere with neuronal functionality [174], and amine-functionalized SWCNTs, which even protected the neurons [96].

Another way to reduce the cytotoxicity is through the dispersion of CNTs in a biocompatible block copolymer [141, 175, 176]. In vivo experiments showed that SWCNTs functionalized with Pluronic polymers can be gradually cleared from the body by alveolar macrophages through mucociliary clearance, reducing risk of lung fibrosis [141]. Lastly, CNTs have been shown to be biodegradable to certain enzymes, such as plant peroxidases, where the degraded CNT fragments can then be effectively phagocytosized by surrounding cells [177, 178]. It has further been shown that the biodegraded CNT fragments, when aspirated into the lungs of mice, did not generate an inflammatory response [178].

GRAPHENE: BIOMEDICAL APPLICATIONS AND NANOSAFETY

Introductory Remarks

Graphene, a 2D carbon nanomaterial with a honeycomb-like structure, has been the subject of a considerable interest after being the subject of the 2010 Nobel Prize for Physics. Its unique properties, including ballistic electron transport [179, 180] at room temperature, tunable band-gap (for few-layer graphene), high chemical and mechanical stability, low electrical noise, high thermal conductivity, and biocompatibility, have led it to be used in many advanced devices ranging from ultra-capacitors to spintronic devices [181–186]. In line with the purpose of this paper, however, we will concentrate on the emerging biomedical-related applications of graphene and its derivatives (i.e., pristine graphene, graphene oxide, metal nanoparticle decorated graphenes, vertical graphene nanosheets, and many other hybrid structures) in biosensors, biocompatible scaffolds, tumor treatment, and drug delivery. We will first present an overview of how and why graphene is used in these cutting edge applications, followed by a brief discussion of how to make graphene for these applications using plasma-based fabrication and other methods.

Graphene Biosensing and Biomedical Applications

As noted in Section 2, biosensors consist of two fundamental parts—a biomolecular recognition element (receptor) and a transducer. The former interacts with the analyte, whereas the latter processes the "sensed" information and translates it into a useable signal. Graphene is a particularly versatile material as it can play a role in both of these components, in plasmonic/optical- and electrical-based sensors. For example, the affinity of graphene for aromatic ring containing biomolecules has been utilized in the biomolecular recognition element in surface plasmon resonance (SPR) sensors. Wu et al. [189] demonstrated that a graphene-Au SPR biosensor, shown in Figure 4(d), was more sensitive than a conventional Au-only SPR biosensor due to (1) the greater adsorption efficiency of ring-based biomolecules on graphene and (2) increased sensitivity to refractive index (RI) changes (a 25% increase in sensitivity to RI change for 10 graphene layers compared to the Au only case). This affinity can also be used as the basis for graphene surface-enhanced Raman scattering (SERS) sensors [190]. Ling et al. [187] reported that using a graphene substrate resulted in a SERS enhancement (see Figures 4(a)–4(c)) of common SERS probes including Rhodamine 6G and crystal violet. This enhancement is chemical, rather than electromagnetic, in nature. Specifically, it is due to the π - πstacking that occurs when the ring-containing molecule aligns itself parallel to the graphene basal plane—this stacking means that charge transfer between graphene and the biomolecule occurs easily, resulting in a chemical SERS enhancements of around 2–17 times [187]. This chemical enhancement, however, is markedly less than the electromagnetic SERS enhancement achieved when Ag or Au nanostructures are used as SERS substrates, despite graphene's higher bioaffinity.

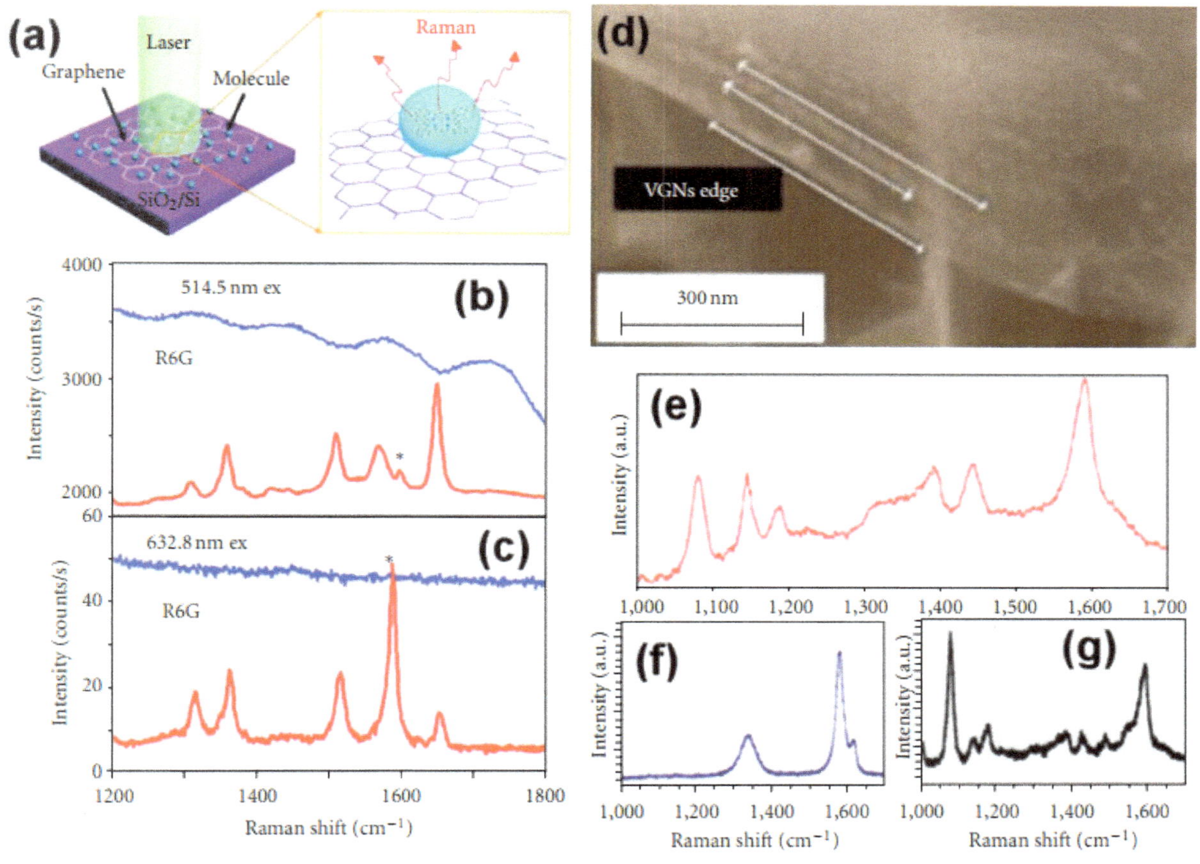

Figure 4: (a) Schematic of graphene as a SERS substrate. Graphene as a SERS substrate for Rhodamine 6G detection using (b) 514.4 nm laser and (c) 632.8 nm laser. The blue line is a SiO_2/Si substrate and the red line is graphene. Note that (a–c) reproduced with permission from [187]; (d) a quasi-linear arrangement of Au nanoparticles on vertical graphene nanosheets (VGNs); SERS spectra of 10^{-2}m 4-aminothiophenol on (e) Au-VGNs array, (f) VGNs only, and (g) Au nanoparticles only. Note that (d–g) reproduced with permission from [188].

Vertical graphenes (such as those in Figures 5(a) and 5(b)) can, however, be used in combination with metal nanoparticles to make 3D metal-graphene nanohybrid SERS platforms. Rider et al. [188] presented a novel SERS substrate consisting of vertical graphenes decorated with Au nanoparticles (see Figures 5(d–g)). Using vertical, rather than horizontal, graphenes provides a markedly higher effective "bookshelf" like area to which Au nanoparticles can attach—a conservative estimate puts the bookshelf nanoparticle density as 224% greater than that for the "flat sensor area" [188]. This means that there is a much greater area to which analyte species can attach, compared to typical horizontal sensor architectures. Gold-decorated graphene composites have also been used in electrodes for electrochemical sensors, due to their high electrocatalytic activity and electrochemical stability [195]. In addition, making use of the high fluorescence quenching efficiency of graphene, Chang et al. [196] constructed a graphene fluorescence resonance energy transfer (FRET) aptasensor, with a reported thrombin detection limit two orders of magnitude lower than CNT-based FRET sensors.

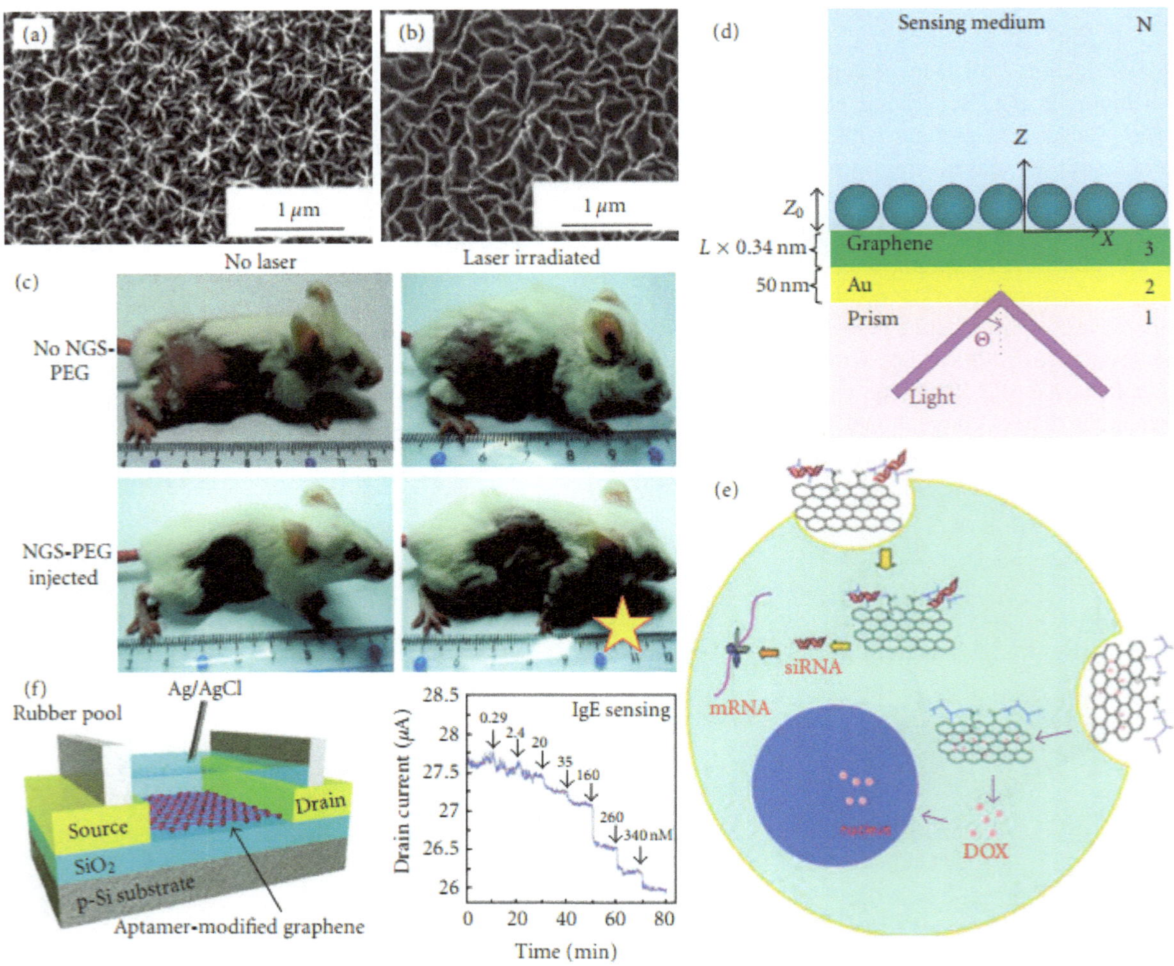

Figure 5: (a, b) Turnstile-like VGNs and maze-like VGNs, respectively. Reproduced with permission from [191]; (c) efficient tumor ablation in mice using PEG-graphene nanosheets where the best tumor ablation occurred in the mice injected with PEG-graphene nanosheets and laser treated (indicated by the yellow star). Reproduced with permission from [192]; (d) graphene as a biomolecular recognition element in SPR sensor. Reproduced with permission from [189]; (e) use of graphene oxide as a drug delivery vehicle for anticancer drug DOX. Reproduced with permission from [193]; (f) use of graphene in FRET-based sensor, specifically for detecting IgE; reproduced with permission from [194].

Electrical sensors that have incorporated graphene include electrochemical-impedance based sensors [197], electrochemical sensors (in which graphene-related materials form the electrode) [198], and FET sensors [194, 199, 200]. Wan et al. [197] constructed immunosensors based on electrochemical impedance with reduced graphene sheets as electron conductors [197]. Ohno et al. [194, 201] showed that graphene FETs (such as those shown in Figure 5(f)) could be used for label-free biosensing—specifically, they demonstrated electrical detection (via a change in the drain current) of solution pH, protein adsorption and specific biomolecules (such as IgE). Zhang et al. noted that the hydrophobic interaction between certain proteins and chemically reduced graphene oxide is promising for protein immobilization and as a result could be used in a biosensor [202].

Aside from sensors, the biocompatibility and chemical inertness (the basal plane, not the edges, which are quite reactive) of graphene surfaces have led to their use as biocompatible scaffolds for the growth of human osteoblasts [203] as well as components in drug delivery and tumor treatment routes [193, 204–206]. Use of graphene oxide as a drug delivery vehicle for anticancer drug doxorubicin [207] is shown in Figure 5(e). Due to the strong optical absorbance of graphene nanosheets in the near infrared, they have been used as a crucial component in photothermal treatment [192, 208, 209] with reported efficient tumor ablation; this can be clearly seen in Figure 5(c), where the best tumor ablation was observed in mice treated with PEG-graphene nanosheets, indicated by the star [192]. The passage of single molecules through nanochannels or nanopores is important for many biological diagnosis processes [210] including DNA sequencing [211, 212]—nanopores in graphene have recently been used for single DNA molecule translocation [213].

Diagnostics and therapies do not exist in a vacuum, they are informed by each other—leading to mutual benefits—this is reflected in the emerging term "theranostics," focusing on individualized medical treatments [214]. This is an area in which graphene-related materials, as already indicated by Yang et al. [192, 215], can play a very important role as their properties lend them both to sensing and intervention/treatment methods.

Nanosafety and Nanotoxicity of Graphene

As mentioned in Section 3, for any biological-related

application, particularly in vivo applications, great care must be taken to ensure that the toxicity of the nanomaterial is well characterized and understood. Whilst this has been extensively done for carbon nanotubes, markedly fewer studies [158, 216–222] are available for graphenes (e.g., an ISI Web of Knowledge topic search on 03/01/2012 gave 59 hits for "graphene" and "toxicity" compared to 1668 hits for "nanotube" and "toxicity"). There is even less consensus as to the sagacity of using this next-generation material as an integral component in in vivo applications. Zhang et al. [158] compared the cytotoxicity level of graphene to that of carbon nanotubes in the case of neuronal PC12 cells. They found that toxicity was shape and composition dependent, with graphene overall having a lower toxicity than CNTs; however the toxicity of graphene was curiously found to be inverse to concentration [223], with graphene exhibiting a higher toxicity than CNTs at low concentrations [223]. Studies on the uptake of PEG-coated graphene nanosheets in mice and subsequent photothermal treatment of cancerous tumors did not show any adverse toxic effects [192, 215]. In other studies, however, sharp graphene nanosheet edges [216] have been shown to cause considerable damage to the cell membrane of bacteria, although this antibacterial property has the potential to be useful. Moreover, hydrophilic carboxyl-functionalized graphenes have been shown to be able to be internalized in cells without any toxic effects, in contrast to hydrophobic pristine graphene [224]. The biocompatibility of graphene oxide has also been studied, with toxicity shown to be dose-dependent in both humans and animals [225], with little to no effect for low and medium doses in mice [225]. Graphene oxide nanosheets were demonstrated to be biocompatible with yeast cells [226]. With the wide range of morphologies, coatings, and hybrid structures available for graphenes, more detailed and longer-term studies are required before serious in vivo biomedical graphene applications are implemented.

Fabrication of Vertically Aligned Graphene Structures

Many different fabrication methods have been used to make graphene-related materials, from the first successful isolation via micromechanical cleavage [227], to chemical reduction of graphene oxide [228, 229], chemical vapour deposition (CVD) [230, 231], thermal decomposition of SiC [232], plasma nanofabrication techniques including plasma-enhanced CVD [233, 234] and arc discharges [235], as well as unzipping of CNTs via a variety of methods (e.g., chemical treatment, plasma etching) [236, 237]. For a detailed description of the relative merits of these fabrication methods, we refer the interested reader to two comprehensive reviews [238, 239]. As noted in Section 4.2, vertically aligned graphene structures have beneficial properties for sensing and other biomedical devices. However, so far there have been limited successes in fabricating these structures. In particular, plasma-based self-organization [240] is a promising way to grow catalyst-free, high-quality vertical graphenes [191, 241]. Papers by Seo and Kumar et al. [20, 191,241] have shown that it is possible to grow graphene nanosheets with a variety of morphologies (e.g., unidirectional, or see Figure 5(a) for turnstile and Figure 5(b) for maze-like) with different optoelectronic properties by modifying the plasma parameters such as gas composition and the degree of ionization. Plasma-based fabrication routes reduce human exposure to any hazardous byproducts since the growth is conducted under vacuum. Moreover, the nanostructures are typically surface-bound, which means they are less likely to be ingested or inhaled, as discussed for CNTs in Section 3 [131].

SUMMARY AND OUTLOOK

In this paper, we have discussed the promising future of incorporating CNTs into the field of biomedicine, specifically their current roles in biosensing, drug delivery, and tissue engineering. The benefits of CNTs were presented, together with potential nanotoxicity and harmful effects of CNTs, on biological systems. Also discussed were the potential uses for graphene for similar biomedical applications as well as the problems associated with graphene's toxicity and safety. There are many challenges ahead that must be addressed before CNTs and graphene can be successfully integrated into biomedical devices and technology. The main advances required, in our opinions, include the following.(i)Advanced techniques and facile methods are needed to increase the sensitivity of CNT biosensors towards single-molecule detection.(ii)More efficient loading and unloading methods for drug delivery would refine overall performance of CNTs as carriers. (iii)Further research is required into various CNT hybrid scaffolds to promote cell adhesion, growth, differentiation, and proliferation.(iv)More specialized coatings and CNT-functionalization to minimize nonspecific bonding are needed. (v)Protocols and further experiments should be conducted to determine the exact nature of the nanotoxicity of CNT-based and graphene-based materials.(vi)Innovative ideas and further experiments are needed to further develop the use of graphene in advanced biomedical applications.(vii)Innovative solutions are required to reduce fabrication and running costs of CNT and graphene biomedical devices to make them economically viable.

The long-term goals associated with incorporating CNTs and graphene into biomedical technology suggest that further research is required before these carbon nanostructured devices reach sufficient performance standards.

ACKNOWLEDGMENTS

The authors would like to thank Q. Cheng, S. Yick, T. van der Laan, D. H. Seo, and W. Yan for insightful discussion. C. Fisher acknowledges the CSIRO OCE Honors Scholarship for Electroactive Biomaterials, A. Rider and Z. Han acknowledge the CSIRO OCE Postdoctoral Fellowship and support from the CSIRO Sensors and Sensor Network TCP. This work is supported by the Australian Research Council (ARC) and CSIRO's OCE Science Leadership Program.

REFERENCES

1. M. S. Dresselhaus, G. Dresselhaus, J. C. Charlier, and E. Hernández, "Electronic, thermal and mechanical properties of carbon nanotubes," Philosophical Transactions of the Royal Society A, vol. 362, no. 1823, pp. 2065–2098, 2004.
2. S. Chakrabarti, H. Kume, L. Pan, T. Nagasaka, and Y. Nakayama, "Number of walls controlled synthesis of

millimeter-long vertically aligned brushlike carbon nanotubes," Journal of Physical Chemistry C, vol. 111, no. 5, pp. 1929–1934, 2007.

3. S. N. Kim, J. F. Rusling, and F. Papadimitrakopoulos, "Carbon nanotubes for electronic and electrochemical detection of biomolecules," Advanced Materials, vol. 19, no. 20, pp. 3214–3228, 2007. ·
4. S. T. Yang, X. Wang, G. Jia et al., "Long-term accumulation and low toxicity of single-walled carbon nanotubes in intravenously exposed mice," Toxicology Letters, vol. 181, no. 3, pp. 182–189, 2008.
5. C. Fisher, Z. J. Han, I. Levchenko, and K. Ostrikov, "Control of dense carbon nanotube arrays via hierarchical multilayer catalyst," Applied Physics Letters, vol. 99, no. 14, Article ID 143104, 2011.
6. Z. J. Han, I. Levchenko, S. Yick, and K. Ostrikov, "3-Orders-of-magnitude density control of single-walled carbon nanotube networks by maximizing catalyst activation and dosing carbon supply,"Nanoscale, vol. 3, no. 11, pp. 4848–4853, 2011.
7. Z. J. Han, S. Yick, I. Levchenko et al., "Controlled synthesis of a large fraction of metallic single-walled carbon nanotube and semiconducting carbon nanowire networks," Nanoscale, vol. 3, no. 8, pp. 3214–3220, 2011.
8. G. Gruner, "Carbon nanotube transistors for biosensing applications," Analytical and Bioanalytical Chemistry, vol. 384, no. 2, pp. 322–335, 2006.
9. M. O'Connor, N. K. Sang, A. J. Killard et al., "Mediated amperometric immunosensing using single walled carbon nanotube forests," Analyst, vol. 129, no. 12, pp. 1176–1180, 2004.
10. M. Pumera, "The electrochemistry of carbon nanotubes: fundamentals and applications," Chemistry, vol. 15, no. 20, pp. 4970–4978, 2009.
11. V. Vamvakaki, M. Fouskaki, and N. Chaniotakis, "Electrochemical biosensing systems based on carbon nanotubes and carbon nanofibers," Analytical Letters, vol. 40, no. 12, pp. 2271–2287, 2007.
12. A. Merkoçi, "Carbon nanotubes in analytical sciences," Microchimica Acta, vol. 152, no. 3-4, pp. 157–174, 2006.
13. B. S. Harrison and A. Atala, "Carbon nanotube applications for tissue engineering," Biomaterials, vol. 28, no. 2, pp. 344–353, 2007.
14. L. Zhang and T. J. Webster, "Nanotechnology and nanomaterials: promises for improved tissue regeneration," Nano Today, vol. 4, no. 1, pp. 66–80, 2009. ·
15. T. Dvir, B. P. Timko, D. S. Kohane, and R. Langer, "Nanotechnological strategies for engineering complex tissues," Nature Nanotechnology, vol. 6, no. 1, pp. 13–22, 2011.
16. O. C. Farokhzad and R. Langer, "Impact of nanotechnology on drug delivery," ACS Nano, vol. 3, no. 1, pp. 16–20, 2009.
17. A. Bianco, K. Kostarelos, and M. Prato, "Applications of carbon nanotubes in drug delivery," Current Opinion in Chemical Biology, vol. 9, no. 6, pp. 674–679, 2005.
18. Y. Lin, F. Lu, Y. Tu, and Z. Ren, "Glucose biosensors based on carbon nanotube nanoelectrode ensembles," Nano Letters, vol. 4, no. 2, pp. 191–195, 2004. ·
19. L. Lacerda, A. Bianco, M. Prato, and K. Kostarelos, "Carbon nanotubes as nanomedicines: from toxicology to pharmacology," Advanced Drug Delivery Reviews, vol. 58, no. 14, pp. 1460–1470, 2006. ·
20. S. Kumar and K. Ostrikov, "Unidirectional arrays of vertically standing graphenes in reactive plasmas,"Nanoscale, vol. 3, no. 10, pp. 4296–4300, 2011.
21. M. Bottini, N. Rosato, and N. Bottini, "PEG-modified carbon nanotubes in biomedicine: current status and challenges ahead," Biomacromolecules, vol. 12, no. 10, pp. 3381–3393, 2011.
22. D. W. H. Fam, A. Palaniappan, A. I. Y. Tok, B. Liedberg, and S. M. Moochhala, "A review on technological aspects influencing commercialization of carbon nanotube sensors," Sensors and Actuators B, vol. 157, no. 1, pp. 1–7, 2011.
23. C. W. Lam, J. T. James, R. McCluskey, S. Arepalli, and R. L. Hunter, "A review of carbon nanotube toxicity and assessment of potential occupational and environmental health risks," Critical Reviews in Toxicology, vol. 36, no. 3, pp. 189–217, 2006.
24. A. E. Porter, M. Gass, K. Muller, J. N. Skepper, P. A. Midgley, and M. Welland, "Direct imaging of single-walled carbon nanotubes in cells," Nature Nanotechnology, vol. 2, no. 11, pp. 713–717, 2007. ·
25. C. P. Firme and P. R. Bandaru, "Toxicity issues in the application of carbon nanotubes to biological systems," Nanomedicine: Nanotechnology, Biology, and Medicine, vol. 6, no. 2, pp. 245–256, 2010.
26. X. Peng and S. S. Wong, "Functional covalent chemistry of carbon nanotube surfaces," Advanced Materials, vol. 21, no. 6, pp. 625–642, 2009.
27. M. Shim, N. W. S. Kam, R. J. Chen, Y. Li, and H. Dai, "Functionalization of carbon nanotubes for biocompatibility and biomolecular recognition," Nano Letters, vol. 2, no. 4, pp. 285–288, 2002.
28. C. Y. Hu, Y. J. Xu, S. W. Duo, R. F. Zhang, and M. S. Li, "Non-covalent functionalization of carbon nanotubes with surfactants and polymers," Journal of the Chinese Chemical Society, vol. 56, no. 2, pp. 234–239, 2009.
29. J. N. Coleman, U. Khan, W. J. Blau, and Y. K. Gun'ko, "Small but strong: a review of the mechanical properties of carbon nanotube-polymer composites," Carbon, vol. 44, no. 9, pp. 1624–1652, 2006.
30. N. Karousis, N. Tagmatarchis, and D. Tasis, "Current progress on the chemical modification of carbon nanotubes," Chemical Reviews, vol. 110, no. 9, pp. 5366–5397, 2010.
31. W. Huang, S. Taylor, K. Fu et al., "Attaching proteins to carbon nanotubes via diimide-activated amidation," Nano Letters, vol. 2, no. 4, pp. 311–314, 2002. ·
32. S. S. Karajanagi, A. A. Vertegel, R. S. Kane, and J. S.

Dordick, "Structure and function of enzymes adsorbed onto single-walled carbon nanotubes," Langmuir, vol. 20, no. 26, pp. 11594–11599, 2004. ·

33. X. Zhang, L. Meng, and Q. Lu, "Cell behaviors on polysaccharide-wrapped single-wall carbon nanotubes: a quantitative study of the surface properties of biomimetic nanofibrous scaffolds," ACS Nano, vol. 3, no. 10, pp. 3200–3206, 2009.
34. K. Jiang, L. S. Schadler, R. W. Siegel, X. Zhang, H. Zhang, and M. Terrones, "Protein immobilization on carbon nanotubes via a two-step process of diimide-activated amidation," Journal of Materials Chemistry, vol. 14, no. 1, pp. 37–39, 2004.
35. J. W. Ko, J. M. Woo, A. Jinhong et al., "Multi-order dynamic range DNA sensor using a gold decorated SWCNT random network," ACS Nano, vol. 5, no. 6, pp. 4365–4372, 2011.
36. Y. Weizmann, D. M. Chenoweth, and T. M. Swager, "Addressable terminally linked DNA-CNT nanowires," Journal of the American Chemical Society, vol. 132, no. 40, pp. 14009–14011, 2010.
37. K. A. Williams, P. T. M. Veenhuizen, B. G. De la Torre, R. Eritja, and C. Dekker, "Nanotechnology: carbon nanotubes with DNA recognition," Nature, vol. 420, no. 6917, p. 761, 2002.
38. W. Huang, S. Taylor, K. Fu et al., "Attaching proteins to carbon nanotubes via diimide-activated amidation," Nano Letters, vol. 2, no. 4, pp. 311–314, 2002. ·
39. W. Yang, P. Thordarson, J. J. Gooding, S. P. Ringer, and F. Braet, "Carbon nanotubes for biological and biomedical applications," Nanotechnology, vol. 18, no. 41, Article ID 412001, 2007.
40. S. Kumar, I. Levchenko, K. Ostrikov, and J. A. McLaughlin, "Plasma-enabled, catalyst-free growth of carbon nanotubes on mechanically-written Si features with arbitrary shape," Carbon, vol. 50, no. 1, pp. 325–329, 2012.
41. X. Guo, L. Huang, S. O'Brien, P. Kim, and C. Nuckolls, "Directing and sensing changes in molecular conformation on individual carbon nanotube field effect transistors," Journal of the American Chemical Society, vol. 127, no. 43, pp. 15045–15047, 2005.
42. Z. Kuang, S. N. Kim, W. J. Crookes-Goodson, B. L. Farmer, and R. R. Naik, "Biomimetic chemosensor: designing peptide recognition elements for surface functionalization of carbon nanotube field effect transistors," ACS Nano, vol. 4, no. 1, pp. 452–458, 2010. ·
43. X. Yu, D. Chattopadhyay, I. Galeska, F. Papadimitrakopoulos, and J. F. Rusling, "Peroxidase activity of enzymes bound to the ends of single-wall carbon nanotube forest electrodes," Electrochemistry Communications, vol. 5, no. 5, pp. 408–411, 2003.
44. Z. Zhu, W. Song, K. Burugapalli, F. Moussy, Y. L. Li, and X. H. Zhong, "Nano-yarn carbon nanotube fiber based enzymatic glucose biosensor," Nanotechnology, vol. 21, no. 16, Article ID 165501, 2010. ·
45. S. G. Mhaisalkar, J. N. Tey, S. Gandhi et al., "Direct detection of heroin metabolites using a competitive immunoassay based on a carbon-nanotube liquid-gated field-effect transistor," Small, vol. 6, no. 9, pp. 993–998, 2010.
46. D. W. H. Fam and A. I. Y. Tok, "Mono-distributed single-walled carbon nanotube channel in field effect transistors (FETs) using electrostatic atomization deposition," Journal of Colloid and Interface Science, vol. 338, no. 1, pp. 266–269, 2009.
47. R. Martel, T. Schmidt, H. R. Shea, T. Hertel, and P. Avouris, "Single- and multi-wall carbon nanotube field-effect transistors," Applied Physics Letters, vol. 73, no. 17, pp. 2447–2449, 1998.
48. S. Roy and Z. Gao, "Nanostructure-based electrical biosensors," Nano Today, vol. 4, no. 4, pp. 318–334, 2009.
49. R. J. Chen, S. Bangsaruntip, K. A. Drouvalakis et al., "Noncovalent functionalization of carbon nanotubes for highly specific electronic biosensors," Proceedings of the National Academy of Sciences of the United States of America, vol. 100, no. 9, pp. 4984–4989, 2003.
50. D. Fu, H. Okimoto, C. W. Lee et al., "Ultrasensitive detection of DNA molecules with high on/off single-walled carbon nanotube network," Advanced Materials, vol. 22, no. 43, pp. 4867–4871, 2010. ·
51. A. Merkoçi, M. Pumera, X. Llopis, B. Pérez, M. Del Valle, and S. Alegret, "New materials for electrochemical sensing VI: carbon nanotubes," TrAC—Trends in Analytical Chemistry, vol. 24, no. 9, pp. 826–838, 2005.
52. M. A. Correa-Duarte, N. Wagner, J. Rojas-Chapana, C. Morsczeck, M. Thie, and M. Giersig, "Fabrication and biocompatibility of carbon nanotube-based 3D networks as scaffolds for cell seeding and growth," Nano Letters, vol. 4, no. 11, pp. 2233–2236, 2004.
53. B. Kang, D. Yu, Y. Dai, S. Chang, D. Chen, and Y. Ding, "Cancer-cell targeting and photoacoustic therapy using carbon nanotubes as "bomb" agents," Small, vol. 5, no. 11, pp. 1292–1301, 2009.
54. T. A. Hilder and J. M. Hill, "Encapsulation of the anticancer drug cisplatin into nanotubes," inInternational Conference on Nanoscience and Nanotechnology (ICONN '08), pp. 109–112, February 2008.
55. N. Komatsu and F. Wang, "A comprehensive review on separation methods and techniques for single-walled carbon nanotubes," Materials, vol. 3, no. 7, pp. 3818–3844, 2010.
56. R. Krupke, F. Hennrich, H. V. Löhneysen, and M. M. Kappes, "Separation of metallic from semiconducting single-walled carbon nanotubes," Science, vol. 301, no. 5631, pp. 344–347, 2003.
57. P. Łukaszczuk, E. Borowiak-Paleń, M. H. Rümmeli, and R. J. Kaleńczuk, "Gel-based separation of single-walled carbon nanotubes for metallic and semiconducting fractions," Materials Research Bulletin, vol. 46, no. 10, pp. 1535–1539, 2011.
58. K. Maehashi, T. Katsura, K. Kerman, Y. Takamura, K. Matsumoto, and E. Tamiya, "Label-free protein biosensor based on aptamer-modified carbon nanotube

field-effect transistors," Analytical Chemistry, vol. 79, no. 2, pp. 782–787, 2007.

59. K. Bradley, M. Briman, A. Star, and G. Gruner, "Charge transfer from adsorbed proteins," Nano Letters, vol. 4, no. 2, pp. 253–256, 2004.

60. L. N. Cella, W. Chen, N. V. Myung, and A. Mulchandani, "Single-walled carbon nanotube-based chemiresistive affinity biosensors for small molecules: ultrasensitive glucose detection," Journal of the American Chemical Society, vol. 132, no. 14, pp. 5024–5026, 2010.

61. R. J. Chen, H. C. Choi, S. Bangsaruntip et al., "An investigation of the mechanisms of electronic sensing of protein adsorption on carbon nanotube devices," Journal of the American Chemical Society, vol. 126, no. 5, pp. 1563–1568, 2004.

62. I. Heller, W. T. T. Small, S. G. Lemay, and C. Dekker, "Probing macrophage activity with carbon-nanotube sensors," Small, vol. 5, no. 22, pp. 2528–2532, 2009.

63. Rajesh, B. K. Das, S. Srinives, and A. Mulchandani, "ZnS nanocrystals decorated single-walled carbon nanotube based chemiresistive label-free DNA sensor," Applied Physics Letters, vol. 98, no. 1, Article ID 013701, 2011.

64. I. Heller, A. M. Janssens, J. Männik, E. D. Minot, S. G. Lemay, and C. Dekker, "Identifying the mechanism of biosensing with carbon nanotube transistors," Nano Letters, vol. 8, no. 2, pp. 591–595, 2008.

65. F. N. Ishikawa, B. Stauffer, D. A. Caron, and C. Zhou, "Rapid and label-free cell detection by metal-cluster-decorated carbon nanotube biosensors," Biosensors and Bioelectronics, vol. 24, no. 10, pp. 2967–2972, 2009.

66. T. H. Kim, S. H. Lee, J. Lee et al., "Single-carbon-atomic-resolution detection of odorant molecules using a human olfactory receptor-based bioelectronic nose," Advanced Materials, vol. 21, no. 1, pp. 91–94, 2009.

67. G. Peng, E. Track, and H. Haick, "Detecting simulated patterns of lung cancer biomarkers by random network of single-walled carbon nanotubes coated with nonpolymeric organic materials," Nano Letters, vol. 8, no. 11, pp. 3631–3635, 2008.

68. H. Wang, R. Yang, L. Yang, and W. Tan, "Nucleic acid conjugated nanomaterials for enhanced molecular recognition," ACS Nano, vol. 3, no. 9, pp. 2451–2460, 2009.

69. J. J. Gooding, R. Wibowo, J. Liu et al., "Protein electrochemistry using aligned carbon nanotube arrays,"Journal of the American Chemical Society, vol. 125, no. 30, pp. 9006–9007, 2003.

70. H. Jin, D. A. Heller, M. Kalbacova et al., "Detection of single-molecule H_2O_2 signalling from epidermal growth factor receptor using fluorescent single-walled carbon nanotubes," Nature Nanotechnology, vol. 5, no. 4, pp. 302–309, 2010.

71. X. B. Yan, X. J. Chen, B. K. Tay, and K. A. Khor, "Transparent and flexible glucose biosensor via layer-by-layer assembly of multi-wall carbon nanotubes and glucose oxidase," Electrochemistry Communications, vol. 9, no. 6, pp. 1269–1275, 2007.

72. R. Cui, H. Huang, Z. Yin, D. Gao, and J. J. Zhu, "Horseradish peroxidase-functionalized gold nanoparticle label for amplified immunoanalysis based on gold nanoparticles/carbon nanotubes hybrids modified biosensor," Biosensors and Bioelectronics, vol. 23, no. 11, pp. 1666–1673, 2008.

73. T. Dastagir, E. S. Forzani, R. Zhang et al., "Electrical detection of hepatitis C virus RNA on single wall carbon nanotube-field effect transistors," Analyst, vol. 132, no. 8, pp. 738–740, 2007.

74. X. Tang, S. Bansaruntip, N. Nakayama, E. Yenilmez, Y. I. Chang, and Q. Wang, "Carbon nanotube DNA sensor and sensing mechanism," Nano Letters, vol. 6, no. 8, pp. 1632–1636, 2006.

75. H. M. So, D. W. Park, E. K. Jeon et al., "Detection and titer estimation of Escherichia coli using aptamer-functionalized single-walled carbon-nanotube field-effect transistors," Small, vol. 4, no. 2, pp. 197–201, 2008.

76. E. Nossol and A. J. G. Zarbin, "A simple and innovative route to prepare a novel carbon nanotube/prussian blue electrode and its utilization as a highly sensitive H_2O_2 amperometric sensor,"Advanced Functional Materials, vol. 19, no. 24, pp. 3980–3986, 2009.

77. F. N. Ishikawa, M. Curreli, C. A. Olson et al., "Importance of controlling nanotube density for highly sensitive and reliable biosensors functional in physiological conditions," ACS Nano, vol. 4, no. 11, pp. 6914–6922, 2010.

78. J. Wang, "Electrochemical glucose biosensors," Chemical Reviews, vol. 108, no. 2, pp. 814–825, 2008. ·

79. X. Xu, S. Jiang, Z. Hu, and S. Liu, "Nitrogen-doped carbon nanotubes: high electrocatalytic activity toward the oxidation of hydrogen peroxide and its application for biosensing," ACS Nano, vol. 4, no. 7, pp. 4292–4298, 2010.

80. C. Fernández-Sánchez, E. Pellicer, J. Orozco, C. Jiménez-Jorquera, L. M. Lechuga, and E. Mendoza, "Plasma-activated multi-walled carbon nanotube-polystyrene composite substrates for biosensing,"Nanotechnology, vol. 20, no. 33, Article ID 335501, 2009. ·

81. Z. Wu, Y. Xu, X. Zhang, G. Shen, and R. Yu, "Microwave plasma treated carbon nanotubes and their electrochemical biosensing application," Talanta, vol. 72, no. 4, pp. 1336–1341, 2007. ·

82. K. Jiang, A. Eitan, L. S. Schadler et al., "Selective attachment of gold nanoparticles to nitrogen-doped carbon nanotubes," Nano Letters, vol. 3, no. 3, pp. 275–277, 2003.

83. S. Srivastava and J. LaBaer, "Nanotubes light up protein arrays," Nature Biotechnology, vol. 26, no. 11, pp. 1244–1246, 2008.

84. A. Star, J. C. P. Gabriel, K. Bradley, and G. Grüner, "Electronic detection of specific protein binding using nanotube FET devices," Nano Letters, vol. 3, no. 4, pp. 459–463, 2003.

85. P. Ghosh, G. Han, M. De, C. K. Kim, and V. M. Rotello, "Gold nanoparticles in delivery applications,"Advanced Drug Delivery Reviews, vol. 60, no. 11, pp. 1307–1315,

2008.

86. S. Zeng, K. T. Yong, I. Roy, X. Q. Dinh, X. Yu, and F. Luan, "A review on functionalized gold nanoparticles for biosensing applications," Plasmonics, vol. 6, no. 3, pp. 491–506, 2011.

87. E. Boisselier and D. Astruc, "Gold nanoparticles in nanomedicine: preparations, imaging, diagnostics, therapies and toxicity," Chemical Society Reviews, vol. 38, no. 6, pp. 1759–1782, 2009. ·

88. H. L. Hsu, L. J. Teng, Y. C. Chen et al., "Flexible UV-ozone-modified carbon nanotube electrodes for neuronal recording," Advanced Materials, vol. 22, no. 19, pp. 2177–2181, 2010.

89. E. Hirata, M. Uo, H. Takita, T. Akasaka, F. Watari, and A. Yokoyama, "Multiwalled carbon nanotube-coating of 3D collagen scaffolds for bone tissue engineering," Carbon, vol. 49, no. 10, pp. 3284–3291, 2011.

90. J. V. Veetil and K. Ye, "Tailored carbon nanotubes for tissue engineering applications," Biotechnology Progress, vol. 25, no. 3, pp. 709–721, 2009.

91. S. Y. Park, D. S. Choi, H. J. Jin et al., "Polarization-controlled differentiation of human neural stem cells using synergistic cues from the patterns of carbon nanotube monolayer coating," ACS Nano, vol. 5, no. 6, pp. 4704–4711, 2011.

92. M. Endo, S. Koyama, Y. Matsuda, T. Hayashi, and Y. A. Kim, "Thrombogenicity and blood coagulation of a microcatheter prepared from carbon nanotube-Nylon-based composite," Nano Letters, vol. 5, no. 1, pp. 101–105, 2005.

93. H. Fan, P. Chen, R. Qi et al., "Greatly improved blood compatibility by microscopic multiscale design of surface architectures," Small, vol. 5, no. 19, pp. 2144–2148, 2009.

94. W. Tutak, M. Chhowalla, and F. Sesti, "The chemical and physical characteristics of single-walled carbon nanotube film impact on osteoblastic cell response," Nanotechnology, vol. 21, no. 31, Article ID 315102, 2010.

95. N. A. Kotov, B. S. Shim, and J. Zhu, "Carbon nanotubes composites made by the layer-by-layer assembly: from ultrastrong materials to solar cells and devices for neural interface," Abstracts of Papers of the American Chemical Society, vol. 237, 2009.

96. H. J. Lee, J. Park, O. J. Yoon et al., "Amine-modified single-walled carbon nanotubes protect neurons from injury in a rat stroke model," Nature Nanotechnology, vol. 6, no. 2, pp. 121–125, 2011.

97. M. P. Mattson, R. C. Haddon, and A. M. Rao, "Molecular functionalization of carbon nanotubes and use as substrates for neuronal growth," Journal of Molecular Neuroscience, vol. 14, no. 3, pp. 175–182, 2000.

98. S. Namgung, T. Kim, K. Y. Baik, M. Lee, J. M. Nam, and S. Hong, "Fibronectin-carbon-nanotube hybrid nanostructures for controlled cell growth," Small, vol. 7, no. 1, pp. 56–61, 2011.

99. D. Liu, C. Yi, D. Zhang, J. Zhang, and M. Yang, "Inhibition of proliferation and differentiation of mesenchymal stem cells by carboxylated carbon nanotubes," ACS Nano, vol. 4, no. 4, pp. 2185–2195, 2010.

100. H. Hu, Y. Ni, V. Montana, R. C. Haddon, and V. Parpura, "Chemically functionalized carbon nanotubes as substrates for neuronal growth," Nano Letters, vol. 4, no. 3, pp. 507–511, 2004.

101. X. Shi, J. L. Hudson, P. P. Spicer, J. M. Tour, R. Krishnamoorti, and A. G. Mikos, "Injectable nanocomposites of single-walled carbon nanotubes and biodegradable polymers for bone tissue engineering," Biomacromolecules, vol. 7, no. 7, pp. 2237–2242, 2006.

102. T. R. Nayak, L. Jian, L. C. Phua, H. K. Ho, Y. Ren, and G. Pastorin, "Thin films of functionalized multiwalled carbon nanotubes as suitable scaffold materials for stem cells proliferation and bone formation," ACS Nano, vol. 4, no. 12, pp. 7717–7725, 2010. ·

103. A. Eitan, K. Jiang, D. Dukes, R. Andrews, and L. S. Schadler, "Surface modification of multiwalled carbon nanotubes: toward the tailoring of the interface in polymer composites," Chemistry of Materials, vol. 15, no. 16, pp. 3198–3201, 2003.

104. M. Vila, J. L. Hueso, M. Manzano et al., "Carbon nanotubes—mesoporous silica composites as controllable biomaterials," Journal of Materials Chemistry, vol. 19, no. 41, pp. 7745–7752, 2009.

105. B. Sitharaman, X. Shi, X. F. Walboomers et al., "In vivo biocompatibility of ultra-short single-walled carbon nanotube/biodegradable polymer nanocomposites for bone tissue engineering," Bone, vol. 43, no. 2, pp. 362–370, 2008.

106. S. Giannona, I. Firkowska, J. Rojas-Chapana, and M. Giersig, "Vertically aligned carbon nanotubes as cytocompatible material for enhanced adhesion and proliferation of osteoblast-like cells," Journal of Nanoscience and Nanotechnology, vol. 7, no. 4-5, pp. 1679–1683, 2007.

107. Z. J. Han, K. Ostrikov, C. M. Tan, B. K. Tay, and S. A. F. Peel, "Effect of hydrophilicity of carbon nanotube arrays on the release rate and activity of recombinant human bone morphogenetic protein-2," Nanotechnology, vol. 22, no. 29, Article ID 295712, 2011. ·

108. M. A. Shokrgozar, F. Mottaghitalab, V. Mottaghitalab, and M. Farokhi, "Fabrication of porous chitosan/poly(vinyl alcohol) reinforced single-walled carbon nanotube nanocomposites for neural tissue engineering.," Journal of biomedical nanotechnology, vol. 7, no. 2, pp. 276–284, 2011.

109. H. J. Lee, O. J. Yoon, D. H. Kim et al., "Neurite outgrowth on nanocomposite scaffolds synthesized from PLGA and carboxylated carbon nanotubes," Advanced Engineering Materials, vol. 11, no. 12, pp. B261–B266, 2009.

110. V. N. Mochalin, O. Shenderova, D. Ho, and Y. Gogotsi, "The properties and applications of nanodiamonds," Nature Nanotechnology, vol. 7, no. 1, pp. 11–23, 2012.

111. A. M. Schrand, H. Huang, C. Carlson et al., "Are diamond nanoparticles cytotoxic?" Journal of Physical Chemistry B, vol. 111, no. 1, pp. 2–7, 2007.

112. V. N. Khabashesku, J. L. Margrave, and E. V. Barrera, "Functionalized carbon nanotubes and nanodiamonds for engineering and biomedical applications," Diamond and Related Materials, vol. 14, no. 3–7, pp. 859–866, 2005.

113. W. Wu, R. Li, X. Bian et al., "Covalently combining carbon nanotubes with anticancer agent: preparation and antitumor activity," ACS Nano, vol. 3, no. 9, pp. 2740–2750, 2009.

114. J. Chen, S. Chen, X. Zhao, L. V. Kuznetsova, S. S. Wong, and I. Ojima, "Functionalized single-walled carbon nanotubes as rationally designed vehicles for tumor-targeted drug delivery," Journal of the American Chemical Society, vol. 130, no. 49, pp. 16778–16785, 2008.

115. J. Rojas-Chapana, J. Troszczynska, I. Firkowska, C. Morsczeck, and M. Giersig, "Multi-walled carbon nanotubes for plasmid delivery into Escherichia coli cells," Lab on a Chip, vol. 5, no. 5, pp. 536–539, 2005.

116. C. R. Martin and P. Kohli, "The emerging field of nanotube biotechnology," Nature Reviews Drug Discovery, vol. 2, no. 1, pp. 29–37, 2003.

117. M. S. Ladeira, V. A. Andrade, E. R. M. Gomes et al., "Highly efficient siRNA delivery system into human and murine cells using single-wall carbon nanotubes," Nanotechnology, vol. 21, no. 38, Article ID 385101, 2010.

118. S. J. Son, X. Bai, and S. B. Lee, "Inorganic hollow nanoparticles and nanotubes in nanomedicine. Part 1. Drug/gene delivery applications," Drug Discovery Today, vol. 12, no. 15-16, pp. 650–656, 2007.

119. R. Prakash, S. Washburn, R. Superfine, R. E. Cheney, and M. R. Falvo, "Visualization of individual carbon nanotubes with fluorescence microscopy using conventional fluorophores," Applied Physics Letters, vol. 83, no. 6, pp. 1219–1221, 2003.

120. M. Ritschel, A. Leonhardt, D. Elefant, S. Oswald, and B. Büchner, "Rhenium-catalyzed growth carbon nanotubes," Journal of Physical Chemistry C, vol. 111, no. 24, pp. 8414–8417, 2007.

121. R. Klingeler, S. Hampel, and B. Büchner, "Carbon nanotube based biomedical agents for heating, temperature sensoring and drug delivery," International Journal of Hyperthermia, vol. 24, no. 6, pp. 496–505, 2008.

122. T. A. Hilder and J. M. Hill, "Modeling the loading and unloading of drugs into nanotubes," Small, vol. 5, no. 3, pp. 300–308, 2009.

123. Q. Mu, D. L. Broughton, and B. Yan, "Endosomal leakage and nuclear translocation of multiwalled carbon nanotubes: developing a model for cell uptake," Nano Letters, vol. 9, no. 12, pp. 4370–4375, 2009.

124. A. Nunes, N. Amsharov, C. Guo et al., "Hybrid polymer-grafted multiwalled carbon nanotubes for in vitro gene delivery," Small, vol. 6, no. 20, pp. 2281–2291, 2010.

125. H. K. Moon, S. H. Lee, and H. C. Choi, "In vivo near-infrared mediated tumor destruction by photothermal effect of carbon nanotubes," ACS Nano, vol. 3, no. 11, pp. 3707–3713, 2009.

126. A. Joshi, S. Punyani, S. S. Bale, H. Yang, T. Borca-Tasciuc, and R. S. Kane, "Nanotube-assisted protein deactivation," Nature Nanotechnology, vol. 3, no. 1, pp. 41–45, 2008.

127. S. Ghosh, S. Dutta, E. Gomes et al., "Increased heating efficiency and selective thermal ablation of malignant tissue with DNA-encased multiwalled carbon nanotubes," ACS Nano, vol. 3, no. 9, pp. 2667–2673, 2009.

128. B. Kang, Y. Dai, S. Chang, and D. Chen, "Explosion of single-walled carbon nanotubes in suspension induced by a large photoacoustic effect," Carbon, vol. 46, no. 6, pp. 978–981, 2008.

129. Y. Lee and K. E. Geckeler, "Carbon nanotubes in the biological interphase: the relevance of noncovalence," Advanced Materials, vol. 22, no. 36, pp. 4076–4083, 2010.

130. P. Ghafari, C. H. St-Denis, M. E. Power et al., "Impact of carbon nanotubes on the ingestion and digestion of bacteria by ciliated protozoa," Nature Nanotechnology, vol. 3, no. 6, pp. 347–351, 2008. ·

131. Z. J. Han, I. Levchenko, S. Kumar et al., "Plasma nanofabrication and nanomaterials safety," Journal of Physics D, vol. 44, no. 17, Article ID 174019, 2011.

132. A. Huczko, "Synthesis of aligned carbon nanotubes," Applied Physics A, vol. 74, no. 5, pp. 617–638, 2002.

133. D. Elgrabli, M. Floriani, S. Abella-Gallart et al., "Biodistribution and clearance of instilled carbon nanotubes in rat lung," Particle and Fibre Toxicology, vol. 5, article no. 20, 2008.

134. Y. Usui, K. Aoki, N. Narita et al., "Carbon nanotubes with high bone-tissue compatibility and bone-formation acceleration effects," Small, vol. 4, no. 2, pp. 240–246, 2008.

135. J. P. Ryman-Rasmussen, M. F. Cesta, A. R. Brody et al., "Inhaled carbon nanotubes reach the subpleural tissue in mice," Nature Nanotechnology, vol. 4, no. 11, pp. 747–751, 2009.

136. S. Iijima, "Helical microtubules of graphitic carbon," Nature, vol. 354, no. 6348, pp. 56–58, 1991.

137. Innovative Research and Products Incorporated, "Production and applications of carbon nanotubes, carbon nanofibres, fullerenes, graphene and nanodiamonds: a global technology survey and market analysis," 2011, http://www.innoresearch.net/report_summary.aspx?id=77&pg531&rcd=et-113&pd=2/1/2011.

138. C. Buzea, I. I. Pacheco, and K. Robbie, "Nanomaterials and nanoparticles: sources and toxicity,"Biointerphases, vol. 2, no. 4, pp. MR17–MR71, 2007.

139. X. Deng, G. Jia, H. Wang et al., "Translocation and fate of multi-walled carbon nanotubes in vivo,"Carbon, vol. 45, no. 7, pp. 1419–1424, 2007.

140. M. L. Schipper, N. Nakayama-Ratchford, C. R. Davis et al., "A pilot toxicology study of single-walled carbon nanotubes in a small sample of mice," Nature Nanotechnology, vol. 3, no. 4, pp. 216–221, 2008. ·

141. G. M. Mutlu, G. R. S. Budinger, A. A. Green et al., "Biocompatible nanoscale dispersion of single-walled carbon nanotubes minimizes in vivo pulmonary toxicity," Nano Letters, vol. 10, no. 5, pp. 1664–1670, 2010.

142. C. A. Poland, R. Duffin, I. Kinloch et al., "Carbon nanotubes introduced into the abdominal cavity of mice show asbestos-like pathogenicity in a pilot study," Nature Nanotechnology, vol. 3, no. 7, pp. 423–428, 2008.

143. F. Zhao, Y. Zhao, Y. Liu, X. Chang, C. Chen, and Y. Zhao, "Cellular uptake, intracellular trafficking, and cytotoxicity of nanomaterials," Small, vol. 7, no. 10, pp. 1322–1337, 2011.

144. F. Zhou, D. Xing, B. Wu, S. Wu, Z. Ou, and W. R. Chen, "New insights of transmembranal mechanism and subcellular localization of noncovalently modified single-walled carbon nanotubes," Nano Letters, vol. 10, no. 5, pp. 1677–1681, 2010.

145. D. Pantarotto, J. P. Briand, M. Prato, and A. Bianco, "Translocation of bioactive peptides across cell membranes by carbon nanotubes," Chemical Communications, vol. 10, no. 1, pp. 16–17, 2004.

146. S. Liu, L. Wei, L. Hao et al., "Sharper and faster "Nano darts" kill more bacteria: a study of antibacterial activity of individually dispersed pristine single-walled carbon nanotube," ACS Nano, vol. 3, no. 12, pp. 3891–3902, 2009.

147. S. Pogodin and V. A. Baulin, "Can a carbon nanotube pierce through a phospholipid bilayer?" ACS Nano, vol. 4, no. 9, pp. 5293–5300, 2010.

148. M. Wang, S. Yu, C. Wang, and J. Kong, "Tracking the endocytic pathway of recombinant protein toxin delivered by multiwalled carbon nanotubes," ACS Nano, vol. 4, no. 11, pp. 6483–6490, 2010.

149. X. Shi, A. Von Dem Bussche, R. H. Hurt, A. B. Kane, and H. Gao, "Cell entry of one-dimensional nanomaterials occurs by tip recognition and rotation," Nature Nanotechnology, vol. 6, no. 11, pp. 714–719, 2011.

150. A. Antonelli, S. Serafini, M. Menotta et al., "Improved cellular uptake of functionalized single-walled carbon nanotubes," Nanotechnology, vol. 21, no. 42, Article ID 425101, 2010.

151. Q. Mu, D. L. Broughton, and B. Yan, "Endosomal leakage and nuclear translocation of multiwalled carbon nanotubes: developing a model for cell uptake," Nano Letters, vol. 9, no. 12, pp. 4370–4375, 2009.

152. F. Zhou, D. Xing, B. Wu, S. Wu, Z. Ou, and W. R. Chen, "New insights of transmembranal mechanism and subcellular localization of noncovalently modified single-walled carbon nanotubes," Nano Letters, vol. 10, no. 5, pp. 1677–1681, 2010.

153. K. Kostarelos, L. Lacerda, G. Pastorin et al., "Cellular uptake of functionalized carbon nanotubes is independent of functional group and cell type," Nature Nanotechnology, vol. 2, no. 2, pp. 108–113, 2007.

154. M. F. Serag, N. Kaji, E. Venturelli et al., "Functional platform for controlled subcellular distribution of carbon nanotubes," ACS Nano, vol. 5, no. 11, pp. 9264–9270, 2011.

155. Y. Yao, L. K. L. Falk, R. E. Morjan, O. A. Nerushev, and E. E. B. Campbell, "Cross-sectional TEM investigation of nickel-catalysed carbon nanotube films grown by plasma-enhanced CVD," Journal of Microscopy, vol. 219, no. 2, pp. 69–75, 2005.

156. L. Wang, S. Luanpitpong, V. Castranova et al., "Carbon nanotubes induce malignant transformation and tumorigenesis of human lung epithelial cells," Nano Letters, vol. 11, no. 7, pp. 2796–2803, 2011. ·

157. M. J. Osmond-McLeod, C. A. Poland, F. Murphy et al., "Durability and inflammogenic impact of carbon nanotubes compared with asbestos fibres," Particle and Fibre Toxicology, vol. 8, article no. 15, 2011.

158. Y. Zhang, S. F. Ali, E. Dervishi et al., "Cytotoxicity effects of graphene and single-wall carbon nanotubes in neural phaeochromocytoma-derived pc12 cells," ACS Nano, vol. 4, no. 6, pp. 3181–3186, 2010.

159. J. Palomäki, E. Välimäki, J. Sund et al., "Long, needle-like carbon nanotubes and asbestos activate the NLRP3 inflammasome through a similar mechanism," ACS Nano, vol. 5, no. 9, pp. 6861–6870, 2011.

160. N. M. Schaeublin, L. K. Braydich-Stolle, A. M. Schrand et al., "Surface charge of gold nanoparticles mediates mechanism of toxicity," Nanoscale, vol. 3, no. 2, pp. 410–420, 2011.

161. J. Muller, I. Decordier, P. H. Hoet et al., "Clastogenic and aneugenic effects of multi-wall carbon nanotubes in epithelial cells," Carcinogenesis, vol. 29, no. 2, pp. 427–433, 2008.

162. K. Kostarelos, A. Bianco, and M. Prato, "Promises, facts and challenges for carbon nanotubes in imaging and therapeutics," Nature Nanotechnology, vol. 4, no. 10, pp. 627–633, 2009. ·

163. X. Wang, T. Xia, S. Addo Ntim et al., "Dispersal state of multiwalled carbon nanotubes elicits profibrogenic cellular responses that correlate with fibrogenesis biomarkers and fibrosis in the murine lung," ACS Nano, vol. 5, no. 12, pp. 9772–9787, 2011.

164. J. G. Li, W. X. Li, J. Y. Xu et al., "Comparative study of pathological lesions induced by multiwalled carbon nanotubes in lungs of mice by intratracheal instillation and inhalation," Environmental Toxicology, vol. 22, no. 4, pp. 415–421, 2007.

165. Y. W. Chun, W. Wang, J. Choi et al., "Control of macrophage responses on hydrophobic and hydrophilic carbon nanostructures," Carbon, vol. 49, no. 6, pp. 2092–2103, 2011.

166. G. Cellot, L. Ballerini, M. Prato, and A. Bianco, "Neurons are able to internalize soluble carbon nanotubes: new opportunities or old risks?" Small, vol. 6, no. 23, pp. 2630–2633, 2010.

167. R. Singh, D. Pantarotto, L. Lacerda et al., "Tissue biodistribution and blood clearance rates of intravenously administered carbon nanotube radiotracers," Proceedings of the National Academy of Sciences of the United States of America, vol. 103, no. 9, pp. 3357–3362, 2006.

168. B. Belgorodsky, E. Drug, L. Fadeev, N. Hendler, E.

Mentovich, and M. Gozin, "Mucin complexes of nanomaterials: first biochemical encounter," Small, vol. 6, no. 2, pp. 262–269, 2010.

169. A. Pietroiusti, M. Massimiani, I. Fenoglio et al., "Low doses of pristine and oxidized single-wall carbon nanotubes affect mammalian embryonic development," ACS Nano, vol. 5, no. 6, pp. 4624–4633, 2011. ·

170. B. D. Holt, P. A. Short, A. D. Rape, Y. L. Wang, M. F. Islam, and K. N. Dahl, "Carbon nanotubes reorganize actin structures in cells and ex Vivo," ACS Nano, vol. 4, no. 8, pp. 4872–4878, 2010.

171. L. A. Mitchell, J. Gao, R. V. Wal, A. Gigliotti, S. W. Burchiel, and J. D. McDonald, "Pulmonary and systemic immune response to inhaled multiwalled carbon nanotubes," Toxicological Sciences, vol. 100, no. 1, pp. 203–214, 2007.

172. L. A. Mitchell, F. T. Lauer, S. W. Burchiel, and J. D. McDonald, "Mechanisms for how inhaled multiwalled carbon nanotubes suppress systemic immune function in mice," Nature Nanotechnology, vol. 4, no. 7, pp. 451–456, 2009.

173. Y. Bai, Y. Zhang, J. Zhang et al., "Repeated administrations of carbon nanotubes in male mice cause reversible testis damage without affecting fertility," Nature Nanotechnology, vol. 5, no. 9, pp. 683–689, 2010.

174. C. Gaillard, G. Celiot, S. Li et al., "Carbon nanotubes carrying cell-adhesion peptides do not interfere with neuronal functionality," Advanced Materials, vol. 21, no. 28, pp. 2903–2908, 2009.

175. E. B. Malarkey, R. C. Reyes, B. Zhao, R. C. Haddon, and V. Parpura, "Water soluble single-walled carbon nanotubes inhibit stimulated endocytosis in neurons," Nano Letters, vol. 8, no. 10, pp. 3538–3542, 2008.

176. X. Shi, B. Sitharaman, Q. P. Pham et al., "In vitro cytotoxicity of single-walled carbon nanotube/ biodegradable polymer nanocomposites," Journal of Biomedical Materials Research A, vol. 86, no. 3, pp. 813–823, 2008.

177. B. L. Allen, P. D. Kichambare, P. Gou et al., "Biodegradation of single-walled carbon nanotubes through enzymatic catalysis," Nano Letters, vol. 8, no. 11, pp. 3899–3903, 2008.

178. V. E. Kagan, N. V. Konduru, W. Feng et al., "Carbon nanotubes degraded by neutrophil myeloperoxidase induce less pulmonary inflammation," Nature Nanotechnology, vol. 5, no. 5, pp. 354–359, 2010.

179. S. V. Morozov, K. S. Novoselov, and A. K. Geim, "Electron transport in graphene," Physics-Uspekhi, vol. 51, no. 7, pp. 727–748, 2008.

180. B. Rosenstein, M. Lewkowicz, H. C. Kao, and Y. Korniyenko, "Ballistic transport in graphene beyond linear response," Physical Review B, vol. 81, no. 4, Article ID 041416, 2010.

181. C. Liu, Z. Yu, D. Neff, A. Zhamu, and B. Z. Jang, "Graphene-based supercapacitor with an ultrahigh energy density," Nano Letters, vol. 10, no. 12, pp. 4863–4868, 2010.

182. B. Chitara, L. S. Panchakarla, S. B. Krupanidhi, and C. N. R. Rao, "Infrared photodetectors based on reduced graphene oxide and graphene nanoribbons," Advanced Materials, vol. 23, no. 45, pp. 5419–5424, 2011.

183. Q. Zeng, J. Cheng, L. Tang et al., "Self-assembled graphene-enzyme hierarchical nanostructures for electrochemical biosensing," Advanced Functional Materials, vol. 20, no. 19, pp. 3366–3372, 2010.

184. S. K. Saha, M. Baskey, and D. Majumdar, "Graphene quantum sheets: a new material for spintronic applications," Advanced Materials, vol. 22, no. 48, pp. 5531–5536, 2010.

185. J. R. Miller, R. A. Outlaw, and B. C. Holloway, "Graphene double-layer capacitor with ac line-filtering performance," Science, vol. 329, no. 5999, pp. 1637–1639, 2010.

186. M. S. Lundstrom, "Graphene: the long and winding road," Nature Materials, vol. 10, no. 8, pp. 566–567, 2011.

187. X. Ling, L. Xie, Y. Fang et al., "Can graphene be used as a substrate for Raman enhancement?" Nano Letters, vol. 10, no. 2, pp. 553–561, 2010.

188. A. E. Rider, S. Kumar, S. A. Furman, and K. K. Ostrikov, "Self-organized Au nanoarrays on vertical graphenes: an advanced three-dimensional sensing platform," Chemical Communications, vol. 48, no. 21, pp. 2659–2661, 2012.

189. L. Wu, H. S. Chu, W. S. Koh, and E. P. Li, "Highly sensitive graphene biosensors based on surface plasmon resonance.," Optics express, vol. 18, no. 14, pp. 14395–14400, 2010.

190. X. Ling and J. Zhang, "Interference phenomenon in graphene-enhanced Raman scattering," Journal of Physical Chemistry C, vol. 115, no. 6, pp. 2835–2840, 2011. ·

191. D. H. Seo, S. Kumar, and K. Ostrikov, "Control of morphology and electrical properties of self-organized graphenes in a plasma," Carbon, vol. 49, no. 13, pp. 4331–4339, 2011.

192. K. Yang, S. Zhang, G. Zhang, X. Sun, S. T. Lee, and Z. Liu, "Graphene in mice: ultrahigh in vivo tumor uptake and efficient photothermal therapy," Nano Letters, vol. 10, no. 9, pp. 3318–3323, 2010.

193. L. Zhang, Z. Lu, Q. Zhao, J. Huang, H. Shen, and Z. Zhang, "Enhanced chemotherapy efficacy by sequential delivery of siRNA and anticancer drugs using PEI-grafted graphene oxide," Small, vol. 7, no. 4, pp. 460–464, 2011.

194. Y. Ohno, K. Maehashi, and K. Matsumoto, "Label-free biosensors based on aptamer-modified graphene field-effect transistors," Journal of the American Chemical Society, vol. 132, no. 51, pp. 18012–18013, 2010.

195. W. Hong, H. Bai, Y. Xu, Z. Yao, Z. Gu, and G. Shi, "Preparation of gold nanoparticle/graphene composites with controlled weight contents and their application in biosensors," Journal of Physical Chemistry C, vol. 114, no. 4, pp. 1822–1826, 2010.

196. H. Chang, L. Tang, Y. Wang, J. Jiang, and J. Li, "Graphene

fluorescence resonance energy transfer aptasensor for the thrombin detection," Analytical Chemistry, vol. 82, no. 6, pp. 2341–2346, 2010.

197. Y. Wan, Z. Lin, D. Zhang, Y. Wang, and B. Hou, "Impedimetric immunosensor doped with reduced graphene sheets fabricated by controllable electrodeposition for the non-labelled detection of bacteria,"Biosensors and Bioelectronics, vol. 26, no. 5, pp. 1959–1964, 2011.

198. T. Kuila, S. Bose, P. Khanra, A. K. Mishra, N. H. Kim, and J. H. Lee, "Recent advances in graphene-based biosensors," Biosensors and Bioelectronics, vol. 26, no. 12, pp. 4637–4648, 2011.

199. Y. A. Akimov, K. Ostrikov, and E. P. Li, "Surface plasmon enhancement of optical absorption in thin-film silicon solar cells," Plasmonics, vol. 4, no. 2, pp. 107–113, 2009.

200. I. Heller, S. Chatoor, J. Männik, M. A. G. Zevenbergen, C. Dekker, and S. G. Lemay, "Influence of electrolyte composition on liquid-gated carbon nanotube and graphene transistors," Journal of the American Chemical Society, vol. 132, no. 48, pp. 17149–17156, 2010.

201. H. Ohno, D. Takagi, K. Yamada, S. Chiashi, A. Tokura, and Y. Homma, "Growth of vertically aligned single-walled carbon nanotubes on alumina and sapphire substrates," Japanese Journal of Applied Physics, vol. 47, no. 4, pp. 1956–1960, 2008.

202. Y. Zhang, J. Zhang, X. Huang, X. Zhou, H. Wu, and S. Guo, "Assembly of graphene oxide-enzyme conjugates through hydrophobic interaction," Small, vol. 8, no. 1, pp. 154–159, 2012.

203. J. S. Czarnecki, K. Lafdi, and P. A. Tsonis, "A novel approach to control growth, orientation, and shape of human osteoblasts," Tissue Engineering A, vol. 14, no. 2, pp. 255–265, 2008.

204. H. Pandey, V. Parashar, R. Parashar, R. Prakash, P. W. Ramteke, and A. C. Pandey, "Controlled drug release characteristics and enhanced antibacterial effect of graphene nanosheets containing gentamicin sulfate," Nanoscale, vol. 3, no. 10, pp. 4104–4108, 2011.

205. V. K. Rana, M. C. Choi, J. Y. Kong et al., "Synthesis and drug-delivery behavior of chitosan-functionalized graphene oxide hybrid nanosheets," Macromolecular Materials and Engineering, vol. 296, no. 2, pp. 131–140, 2011.

206. X. M. Sun, Z. Liu, K. Welsher et al., "Nano-graphene oxide for cellular imaging and drug delivery,"Nano Research, vol. 1, no. 3, pp. 203–212, 2008.

207. L. Zhang, J. Xia, Q. Zhao, L. Liu, and Z. Zhang, "Functional graphene oxide as a nanocarrier for controlled loading and targeted delivery of mixed anticancer drugs," Small, vol. 6, no. 4, pp. 537–544, 2010.

208. W. Zhang, Z. Guo, D. Huang, Z. Liu, X. Guo, and H. Zhong, "Synergistic effect of chemo-photothermal therapy using PEGylated graphene oxide," Biomaterials, vol. 32, no. 33, pp. 8555–8561, 2011.

209. B. Tian, C. Wang, S. Zhang, L. Feng, and Z. Liu, "Photothermally enhanced photodynamic therapy delivered by nano-graphene oxide," ACS Nano, vol. 5, no. 9, pp. 7000–7009, 2011.

210. C. Dekker, "Solid-state nanopores," Nature Nanotechnology, vol. 2, no. 4, pp. 209–215, 2007.

211. S. K. Min, W. Y. Kim, Y. Cho, and K. S. Kim, "Fast DNA sequencing with a graphene-based nanochannel device," Nature Nanotechnology, vol. 6, no. 3, pp. 162–165, 2011.

212. Y. Cho, S. K. Min, W. Y. Kim, and K. S. Kim, "The origin of dips for the graphene-based DNA sequencing device," Physical Chemistry Chemical Physics, vol. 13, no. 32, pp. 14293–14296, 2011.

213. C. A. Merchant, K. Healy, M. Wanunu et al., "DNA translocation through graphene nanopores," Nano Letters, vol. 10, no. 8, pp. 2915–2921, 2010.

214. S. S. Kelkar and T. M. Reineke, "Theranostics: combining imaging and therapy," Bioconjugate Chemistry, vol. 22, no. 10, pp. 1879–1903, 2011.

215. K. Yang, J. Wan, S. Zhang, Y. Zhang, S. T. Lee, and Z. Liu, "In vivo pharmacokinetics, long-term biodistribution, and toxicology of pegylated graphene in mice," ACS Nano, vol. 5, no. 1, pp. 516–522, 2011.

216. O. Akhavan and E. Ghaderi, "Toxicity of graphene and graphene oxide nanowalls against bacteria,"ACS Nano, vol. 4, no. 10, pp. 5731–5736, 2010.

217. Y. Chang, S. T. Yang, J. H. Liu et al., "In vitro toxicity evaluation of graphene oxide on A549 cells,"Toxicology Letters, vol. 200, no. 3, pp. 201–210, 2011.

218. K. M. Garza, K. F. Soto, and L. E. Murr, "Cytotoxicity and reactive oxygen species generation from aggregated carbon and carbonaceous nanoparticulate materials," International Journal of Nanomedicine, vol. 3, no. 1, pp. 83–94, 2008.

219. K.-H. Liao, Y. S. Lin, C. W. MacOsko, and C. L. Haynes, "Cytotoxicity of graphene oxide and graphene in human erythrocytes and skin fibroblasts," ACS Applied Materials and Interfaces, vol. 3, no. 7, pp. 2607–2615, 2011.

220. M. Pumera, "Graphene in biosensing," Materials Today, vol. 14, no. 7-8, pp. 308–315, 2011.

221. S. K. Singh, M. K. Singh, M. K. Nayak, S. Kumari, J. J. Grácio, and D. Dash, "Characterization of graphene oxide by flow cytometry and assessment of its cellular toxicity.," Journal of biomedical nanotechnology, vol. 7, no. 1, pp. 30–31, 2011.

222. X. Zhang, J. Yin, C. Peng et al., "Distribution and biocompatibility studies of graphene oxide in mice after intravenous administration," Carbon, vol. 49, no. 3, pp. 986–995, 2011.

223. L. Yan, F. Zhao, S. Li, Z. Hu, and Y. Zhao, "Low-toxic and safe nanomaterials by surface-chemical design, carbon nanotubes, fullerenes, metallofullerenes, and graphenes," Nanoscale, vol. 3, no. 2, pp. 362–382, 2011.

224. A. Sasidharan, L. S. Panchakarla, P. Chandran, et al., "Differential nano-bio interactions and toxicity effects of pristine versus functionalized graphene.," Nanoscale, vol. 3, no. 6, pp. 2461–2464, 2011.

225. K. Wang, J. Ruan, H. Song et al., "Biocompatibility of graphene oxide," Nanoscale Research Letters, vol. 6, no. 1, pp. 1–8, 2011.
226. S. H. Yang, T. Lee, E. Seo, E. H. Ko, I. S. Choi, and B.-S. Kim, "Interfacing living yeast cells with graphene oxide nanosheaths," Macromolecular Bioscience, vol. 12, no. 1, pp. 61–66, 2012.
227. K. S. Novoselov, A. K. Geim, S. V. Morozov et al., "Electric field in atomically thin carbon films,"Science, vol. 306, no. 5696, pp. 666–669, 2004.
228. K. Liu, J. J. Zhang, F. F. Cheng, T. T. Zheng, C. Wang, and J. J. Zhu, "Green and facile synthesis of highly biocompatible graphene nanosheets and its application for cellular imaging and drug delivery," Journal of Materials Chemistry, vol. 21, no. 32, pp. 12034–12040, 2011.
229. S. Alwarappan, A. Erdem, C. Liu, and C. Z. Li, "Probing the electrochemical properties of graphene nanosheets for biosensing applications," Journal of Physical Chemistry C, vol. 113, no. 20, pp. 8853–8857, 2009.
230. A. N. Obraztsov, E. A. Obraztsova, A. V. Tyurnina, and A. A. Zolotukhin, "Chemical vapor deposition of thin graphite films of nanometer thickness," Carbon, vol. 45, no. 10, pp. 2017–2021, 2007.
231. Q. Yu, J. Lian, S. Siriponglert, H. Li, Y. P. Chen, and S. S. Pei, "Graphene segregated on Ni surfaces and transferred to insulators," Applied Physics Letters, vol. 93, no. 11, Article ID 113103, 2008.
232. Y. Q. Wu, P. D. Ye, M. A. Capano et al., "Top-gated graphene field-effect-transistors formed by decomposition of SiC," Applied Physics Letters, vol. 92, no. 9, Article ID 092102, 2008.
233. A. N. Obraztsov, A. A. Zolotukhin, A. O. Ustinov, A. P. Volkov, Y. Svirko, and K. Jefimovs, "DC discharge plasma studies for nanostructured carbon CVD," Diamond and Related Materials, vol. 12, no. 3–7, pp. 917–920, 2003.
234. M. Zhu, J. Wang, B. C. Holloway et al., "A mechanism for carbon nanosheet formation," Carbon, vol. 45, no. 11, pp. 2229–2234, 2007.
235. K. S. Subrahmanyam, L. S. Panchakarla, A. Govindaraj, and C. N. R. Rao, "Simple method of preparing graphene flakes by an arc-discharge method," Journal of Physical Chemistry C, vol. 113, no. 11, pp. 4257–4259, 2009.
236. L. Jiao, L. Zhang, X. Wang, G. Diankov, and H. Dai, "Narrow graphene nanoribbons from carbon nanotubes," Nature, vol. 458, no. 7240, pp. 877–880, 2009. ·
237. D. V. Kosynkin, A. L. Higginbotham, A. Sinitskii et al., "Longitudinal unzipping of carbon nanotubes to form graphene nanoribbons," Nature, vol. 458, no. 7240, pp. 872–876, 2009.
238. W. Choi, I. Lahiri, R. Seelaboyina, and Y. S. Kang, "Synthesis of graphene and its applications: a review,"Critical Reviews in Solid State and Materials Sciences, vol. 35, no. 1, pp. 52–71, 2010.
239. H. Jiang, "Chemical preparation of graphene-based nanomaterials and their applications in chemical and biological sensors," Small, vol. 7, no. 17, pp. 2413–2427, 2011.
240. A. E. Rider and K. Ostrikov, "Assembly and self-organization of nanomaterials," in Plasma Processing of Nanomaterials, R. M. Sankaran, Ed., pp. 371–392, CRC Press, Boca Raton, Fla, USA, 2011.
241. D. H. Seo, S. Kumar, and K. Ostrikov, "Thinning vertical graphenes, tuning electrical response: from semiconducting to metallic," Journal of Materials Chemistry, vol. 21, no. 41, pp. 16339–16343, 2011.

INDEX

N

O

P

R

S

T

W

X